Lecture Notes in Computer Science　9247

Commenced Publication in 1973
Founding and Former Series Editors:
Gerhard Goos, Juris Hartmanis, and Jan van Leeuwen

Editorial Board

Sergey Balandin · Sergey Andreev
Yevgeni Koucheryavy (Eds.)

Internet of Things, Smart Spaces, and Next Generation Networks and Systems

15th International Conference, NEW2AN 2015
and 8th Conference, ruSMART 2015
St. Petersburg, Russia, August 26–28, 2015
Proceedings

 Springer

Editors
Sergey Balandin
FRUCT Oy
Helsinki
Finland

Sergey Andreev
Tampere University of Technology
Tampere
Finland

Yevgeni Koucheryavy
Tampere University of Technology
Tampere
Finland

ISSN 0302-9743   ISSN 1611-3349 (electronic)
Lecture Notes in Computer Science
ISBN 978-3-319-23125-9   ISBN 978-3-319-23126-6 (eBook)
DOI 10.1007/978-3-319-23126-6

Library of Congress Control Number: 2015946749

LNCS Sublibrary: SL5 – Computer Communication Networks and Telecommunications

Springer International Publishing AG Switzerland is part of Springer Science+Business Media
(www.springer.com)

Preface

We welcome you to the joint proceedings of the 15th NEW2AN (Next-Generation Wired/Wireless Advanced Networks and Systems) and 8th conference on Internet of Things and Smart Spaces ruSMART (Are You Smart) held in St. Petersburg, Russia, during August 26–28, 2015.

Originally, the NEW2AN conference was launched by ITC (International Teletraffic Congress) in St. Petersburg in June 1993 as an ITC-Sponsored Regional International Teletraffic Seminar. The first edition was entitled "Trafffic Management and Routing in SDH Networks" and held by R&D LONIIS. In 2002, the event received its current name, NEW2AN. In 2008, NEW2AN acquired a new companion in Smart Spaces, ruSMART, hence boosting interaction between researchers, practitioners, and engineers across different areas of ICT. From 2012, the scope of ruSMART conference has been extended to cover the Internet of Things and related aspects.

Presently, NEW2AN and ruSMART are well-established conferences with a unique cross-disciplinary mixture of telecommunications-related research and science. NEW2AN/ruSMART is accompanied by outstanding keynotes from universities and companies across Europe, USA, and Russia.

The 15th NEW2AN technical program addresses various aspects of advanced wireless networks. This year, particular attention was paid to mobile ad hoc, sensor, and cloud networks, emerging cellular systems and their components, as well as contemporary signal and circuit design. In particular, the contributors have proposed novel and innovative systems and techniques for streaming, video, and other higher-layer applications, as well as for optical and satellite systems. It is also worth mentioning the rich coverage of business and services aspects of next-generation communication networks, advanced materials for communication systems and their properties, and information security.

The 8th conference on Internet of Things and Smart Spaces, ruSMART 2015, provided a forum for academic and industrial researchers to discuss new ideas and trends in the emerging areas of Internet of Things and smart spaces that create new opportunities for fully customized applications and services. The conference brought together leading experts from top affiliations around the world. This year, we saw good participation from representatives of various players in the field, including academic teams and industrial world-leader companies, particularly representatives of Russian R&D centers, which have a good reputation for high-quality research and business in innovative service creation and applications development.

This year, the first day of the NEW2AN/ruSMART technical program started with the keynote talk on "Securing the Internet of Things: Opportunities and Challenges with Scaling IoT Solutions" given by Rob van den Dam, who is Global Telecommunications Industry Leader, at IBM Institute for Business Value in The Netherlands.

We would like to thank the Technical Program Committee members, as well as the associated reviewers, for their hard work and important contribution to the conference.

This year, the conference program met the highest-quality criteria with an acceptance ratio of around 35%.

The current edition of NEW2AN/ruSMART was organized in cooperation with Open Innovations Association FRUCT, IEEE, St. Petersburg State Polytechnical University, Tampere University of Technology, Technopark and ISST lab of NRU ITMO, St. Petersburg State University of Telecommunications, and Popov Society. The support of these organizations is gratefully acknowledged.

We also wish to thank all those who contributed to the organization of the conference. In particular, we are grateful to Roman Florea for his substantial work in supporting the conference website and his excellent job on the compilation of camera-ready papers and interaction with Springer.

We believe that the 15th NEW2AN and 8th ruSMART conferences delivered an informative, high-quality, and up-to-date scientific program. We also hope that participants enjoyed both the technical and social conference components, the Russian ways of hospitality, and the beautiful city of St. Petersburg.

August 2015   Sergey Balandin
                                                                     Sergey Andreev
                                                             Yevgeni Koucheryavy

Organization

NEW2AN International Advisory Committee

Igor Faynberg	Alcatel Lucent, USA
Jarmo Harju	Tampere University of Technology, Finland
Villy B. Iversen	Technical University of Denmark, Denmark
Andrey Koucheryavy	St. Petersburg State University of Telecommunications, Russia
Kyu Ouk Lee	ETRI, Republic of Korea
Sergey Makarov	St. Petersburg State Polytechnical University, Russia
Mohammad S. Obaidat	Monmouth University, USA
Andrey I. Rudskoy	St. Petersburg State Polytechnical University, Russia
Manfred Sneps-Sneppe	Ventspils University College, Latvia
Michael Smirnov	Fraunhofer FOKUS, Germany
Sergey Stepanov	Sistema Telecom, Russia

NEW2AN Technical Program Committee

Alexander F. Kriachko	St. Petersburg State Polytechnical University, Russia
Alexander Sayenko	NOKIA, Finland
Andreas Kassler	Karlstad University, Sweden
Andrey Turlikov	State University of Aerospace Instrumentation, Russia
Antonino Orsino	University Mediterranea of Reggio Calabria, Italy
Arvind Swaminathan	Qualcomm Inc, USA
Burkhard Stiller	University of Zürich, Switzerland
Christian Tschudin	University of Basel, Switzerland
Chrysostomos Chrysostomou	Frederick University, Cyprus
Dieter Fiems	Ghent University, Belgium
Dirk Staehle	University of Würzburg, Germany
Dmitri Moltchanov	Tampere University of Technology, Finland
Dmitry Tkachenko	IEEE St. Petersburg BT/CE/COM Chapter, Russia
Dr Nitin	Jaypee Institute of Information Technology, India
Edmundo Monteiro	University of Coimbra, Portugal
Evgeni Osipov	Lulea University of Technology, Sweden
Eylem Ekici	Ohio State University, USA
Francisco Barcelo-Arroyo	Universitat Politecnica de Catalunya (UPC), Spain
George Pavlou	University of Surrey, UK
Giovanni Giambene	University of Siena, Italy

Vitaly Gutin	Popov Society, Russia
Vitaly Li	Kangwon National University, Republic of Korea
Vladimir S. Zaborovsky	St. Petersburg State Polytechnical University, Russia
Wei Koong Chai	University College London, UK
Weilian Su	Naval Postgraduate School, USA
Yevgeni Koucheryavy	Tampere University of Technology, Finland
Zhefu Shi	University of Missouri - Kansas City, USA

ruSMART Executive Technical Program Committee

Sergey Boldyrev	Nordea, Helsinki, Finland
Nikolai Nefedov	ETH Zurich, Switzerland
Ian Oliver	Nokia Networks Helsinki, Finland
Alexander Smirnov	SPIIRAS, Russia
Vladimir Gorodetsky	SPIIRAS, Russia
Michael Lawo	TZI Center for Computing Technologies, University of Bremen, Germany
Michael Smirnov	Fraunhofer FOKUS, Germany
Dieter Uckelmann	University of Applied Sciences in Stuttgart, Germany
Cornel Klein	Siemens Corporate Technology, Germany

ruSMART Technical Program Committee

Sergey Balandin	FRUCT, Finland
Michel Banatre	IRISA, France
Mohamed Baqer	University of Bahrain, Bahrain
Sergei Bogomolov	LGERP R&D Lab, Russia
Mu-Song Chen	University of Texas, USA
Gianpaolo Cugola	Politecnico di Milano, Italy
Alexey Dudkov	NRPL Group, Finland
Harry Fulgencio	Leiden University, The Netherlands
Kim Geun-Hyung	Dong Eui University, Republic of Korea
Didem Gozupek	Bogazici University, Turkey
Victor Govindaswamy	Texas A&M University, USA
Andrei Gurtov	Aalto University, Finland
Prem Jayaraman	Monash University, Australia
Jukka Honkola	Innorange Oy, Finland
Dimitri Konstantas	University of Geneva, Switzerland
Alexey Kashevnik	SPIIRAS, Russia
Kim Geunhyung	Dong Eui University, Korea
Cornel Klein	Siemens, Germany
Dmitry Korzun	Petrozavodsk State University, Russia
Kirill Krinkin	Academic University of Russian Academy of Science, Russia
Juha Laurila	University of Turku, Finland
Johan Lilius	Abo Academia, Finland

Pedro Merino   University of Malaga, Spain
Ilya Paramonov   Yaroslavl State University, Russia
Luca Roffia   University of Bologna, Italy
Bilhanan Silverajan   Tampere University of Technology, Finland
Nikolay Shilov   SPIIRAS, Russia
Markus Taumberger   VTT, Finland
Dieter Uckelmann   Hochschule für Technik Stuttgart, Germany

Sponsoring Institutions

Technopark
of ITMO University

СПб|ГУТ)))

◆IEEE

IEEE
COMMUNICATIONS
SOCIETY
Russia Northwest Chapter

Contents

NEW2AN

ruSMART

The Monitoring of Information Security of Remote Devices of Wireless Networks

Ilya Lebedev and Viktoria Korzhuk[(✉)]

Saint Petersburg National Research University of Information Technologies,
Mechanics and Optics (ITMO University), Saint-Petersburg, Russia
{lebedev,vika}@cit.ifmo.ru

Abstract. The issues of information security monitoring of remote of devices self-organizing wireless networks are discussed. The model of interaction of the remote devices is shown, the analyzed characteristics of identification of node abnormal behavior for different types of topologies are identified. The approach to the assessment with the use of the selected features of the system is suggested. The experiment, providing the obtainment of statistical information about the remote devices in different modes of ad-hoc networks is revealed. The results of the system activity for a broadcast packet of network scanning with given characteristics are presented. The dependences of the responses from sleeping and waking devices for variable system structure based on ad-hoc wireless networks are given. The feature of this approach is that it allows using the built-in set of commands that gives the opportunity to avoid additional costs when constructing the control system for wireless sensor networks.

Keywords: Information security · Wireless networks · Multi-agent systems · Abnormal behavior · The vulnerability · The model of information security

1 Introduction

The widespread advent of wireless networks and the possibility of its detection outside the controlled area make a very attractive target for attempts to realise the variety of attacks. The implementation of a large number of projects based on the technology of Bluetooth, ZigBee, Wi-Fi and its use in the intelligent transportation systems, local networks and sensor networks raises the need to ensure the required security level of circulating data.

The introduction of ad-hoc wireless networks is accompanied by the necessity to solve additional problems of ensuring information security [1,2]. Among the main vulnerabilities we can highlight the possibility of listening to channels, sending "external" packages, the realization of the physical access to the node by attacker, lack of standardization of intelligent routing algorithms that take into account the state of the network. A large number of devices providing intelligent transfer, collection, processing of information packages, its relative remoteness, autonomy of operation, dynamically changing topology, weak development of models, methods and algorithms of rapid detection of

S. Balandin et al. (Eds.): NEW2AN/ruSMART 2015, LNCS 9247, pp. 3–10, 2015.
DOI: 10.1007/978-3-319-23126-6_1

incorrect information from compromised nodes determine the complexity of creating classic protection system [3,4,5]. Insufficient capacity of the computing resources of the components of ad-hoc wireless networks, the distribution over a large area and often in inaccessible places lead to the fact that the keys are distributed using broadcast packets, so the application of encryption protocols such as WEP and WPA only does not allow to provide the required level of protection fully.

Thus, there is a necessity to develop some models, methods of information security monitoring of remote devices of wireless networks.

2   A Probabilistic Model of Device Interaction

When considering the number of projects involving the use of wireless ad-hoc networks it is necessary to analyze its technical characteristics to identify the weak spots and to evaluate opportunities of device interaction [6].

For example, studies of various topologies [7,8] of ad-hoc wireless networks show the need to review the information receipt, processing and transfer. In practice there is often a "bottleneck" where a significant amount of transmitted information "flock" to a single node and forms a connection between two elements, the availability of which may be affected by adjacent elements that communicate with each other (Fig. 1).

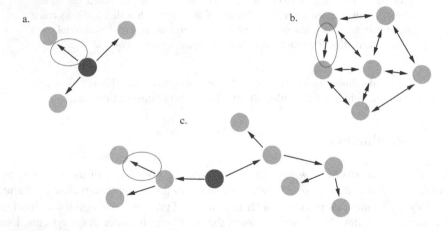

Fig. 1. Analyzed interaction of devices of ad-hoc network topology (a) star; (b) multi-cell network; (c) cluster tree

Different models are being developing for the evaluation of characteristics and levels of system functioning indicators [9,10,11].

If the network consists of two nodes then analyzed interaction between the two devices of ad-hoc networks can be represented in the form of a simple Markov chain with two states:

S_0 - the interaction between devices is possible.

S_1 - the interaction between devices is not possible.

λ_{01} – the intensity of events changing the state of interaction. The intensity depends on the characteristics of the system when event occurs, such as loading the buffer of the receiver node, processing the received command, the priority need of the packet transmission, changing different operating modes.

μ_{10} – the intensity of events that transforms the channel into a state of possible interactions caused by clearing the resources, which are important for current operations, or by changing the device state to "ready".

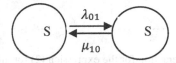

Fig. 2. The channel state scheme

The event, which causes the interaction between the two devices of ad-hoc network to the impossible state, can occur not only as a result of the exchange of information between the two devices [12].

It can be associated with the necessity of information exchange between adjacent devices in current frequency range, with algorithms of communications protocols, with noises related to the work of other networks and devices. Thus, the intensity of events that takes the interaction between the two devices self-organizing network to the impossible state we can describe as:

$$\lambda_{01} = \lambda_{node} + \lambda_{out} \tag{1}$$

where λ_{node} – where is the intensity of the events connected with the direct device interaction, and λ_{out} – is the intensity of the external events such as noise from other systems.

Then the probability of interaction can be determined by:

$$P_0(t) = \frac{\mu}{\lambda_n + \lambda_{out} + \mu} + \frac{(\lambda_n + \lambda_{out})e^{-(\lambda_n + \lambda_{out} + \mu)t}}{\lambda_n + \lambda_{out} + \mu} \tag{2}$$

If the number of devices and systems that may influence the processes of receiving, processing and transmission between nodes is quite large, the intensity of events, which takes the interaction between the two devices of ad-hoc network to the impossible state, can be considered as the sum of the intensities of the node and the intensities of external influences.

For example, if n is the number of devices competing for interaction with each other when sending information, the expression (1) can be converted to the form:

$$P_0(t) = \frac{\mu}{n\lambda_n + \lambda_{out} + \mu} + \frac{(n\lambda_n + \lambda_{out})e^{-(n\lambda_n + \lambda_{out} + \mu)t}}{n\lambda_n + \lambda_{out} + \mu} \tag{3}$$

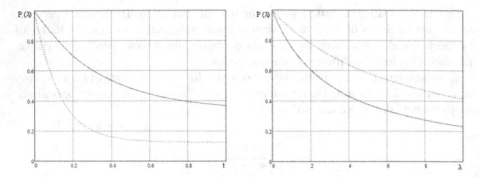

Fig. 3. The probability of interaction for the expression (2) for different t and ratios μ and λ

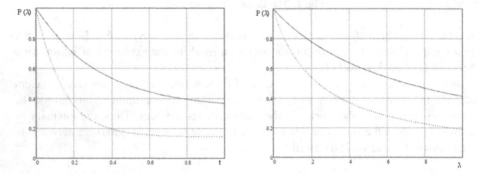

Fig. 4. The probability of interaction for the expression (3) for different t and ratios μ and λ

Such models allow us to estimate the probabilistic state of the system and levels of indicators, that are specific to its functioning in different modes and environments.

3   Anomaly Detection in the Ad-Hoc Wireless Networks

One of the problematic issues in the organization of the monitoring process of information security is the choice of the investigated characteristics that correlated with probabilistic model of the interaction and the functioning of the remote autonomous devices controlled of the system [13].

The first type of characteristics used for the system control is different identifier of internal devices (information about addresses, serial numbers, preset roles of devices in the network, information in the registers). The collection, processing and storage of this information allow us to control the appearance of "new" devices and detect attempts to inject the extraneous element into the network.

However, there is the possibility of making changes to the functioning algorithm by reprogramming the device by an attacker or to exert external impact on the sensors [14].

To detect abnormal behavior we must use the characteristics reflecting the system state that can be used in statistical analysis [15]. For ad-hoc wireless networks (Zig-Bee) based on standard sets of commands, for example, it can be:

- the response time of a broadcast packet;
- the response time of an address packet;
- the duration of the session;
- the number of outgoing packets;
- the frequency characteristics of the information exchange initiation.

Thus, the controlled system O consists of many elements o_i, $i = 1 \dots n$. Each element o_i corresponds to a tuple of studied characteristics $x =< x_1, \dots, x_m >$. When monitoring it is expected that each team, each packet processed in ad-hoc network, fills statistical database of the analyzed characteristics.

For wireless networks developers provide specific tools that make the transmission and reception of various types of broadcast packets enable for all stages of network construction and functioning. While sending commands during the functioning, the characteristics of the received messages are changing (the response time, the possibility of packet waste).

4   The Experiment

The forming of statistical tables on the destination node of the data collection as one of the possible solutions involves broadcast from the router to remote devices. First, it allows detecting asymmetric communication between devices and, second, provides data for statistical analysis. Figure 5 shows the scheme of the experiment.

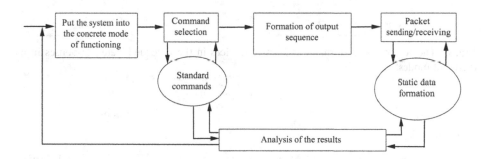

Fig. 5. Scheme of the experiment

The purpose of the experiment is to obtain the dependence of quantitative indicators of the responses of the devices in the system for different modes and different commands. In this aim we carry out a series of actions.

1.   Put the system into the desired mode of functioning. In the simplest case, the frequency of message exchange increases or decreases at the specified speed for the statistics.

2.   Option of the command sent by the coordinator to other devices affects the size of the sent packet and has a different reaction time, packet processing time on the remote network device that allows us to create statistical data.

3.   The formation of sequence, the determination of the number of commands in the sequence and distribution provide an opportunity to evaluate the response of remote devices on the ratio of the number of processed commands to the total command number.

4.   Receiving packets from devices ensures the accumulation of statistics, also is used to generate a database of the studied parameters for sending commands and responses from the devices.

5.   The statistical analysis of accumulated data, the formation of the next sequence of commands, the changeover of the system and its remote devices in various modes of functioning are used to generate further action in the experiment and the accumulation of statistics.

Figures 6-7 presents the results of the system functioning for a broadcast network scan packet at a speed of sending 19200 bit/s per port router, the transfer rate of the wireless channel is 250 Kbit/s, the exit time of the terminal device from a sleep mode 128 MS, where

M - the number of consecutively sent messages;

n - number of received responses;

m - the number of expected responses.

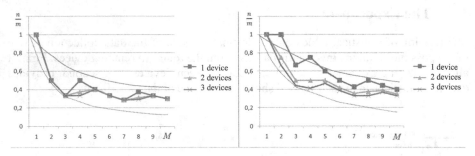

Fig. 6. The responses of dormant and active devices in the unloaded state. Responses in the different modes.

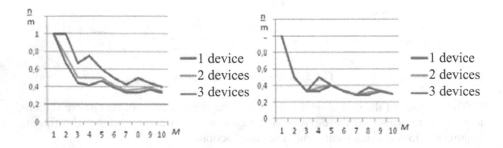

Fig. 7. Responses of the active ending devices. Responses of dormant devices.

Considering that m is the number of possible outcomes and n is the number of occurrences of the event "getting the answer", then

$$P(O) = \frac{n}{m} \qquad (4)$$

Assuming that M allows to determine the intensity of events that translates into the impossible state of interaction in the concrete mode, we select the values of μ and λ and overlay analytical dependencies on the experimental results. The combination of graphs and simulation results for the expression (4) can be used to find the limiting performance of the system.

The evaluation can determine the device functioning mode. According to the graphic (Fig. 6), it is possible to monitor if the mode of operation is changed: whether it is in "dormant" or "active" mode. Unauthorized transfer of device mode from the attacker can affect both the speed of information processing and energy-saving. The graphs in figure 7 show the differences in the functioning of the system in a star topology consisting of the "active" and "dormant" elements.

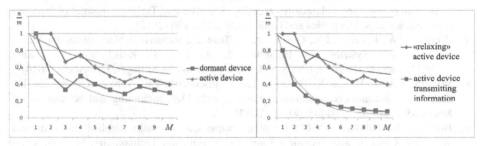

Fig. 8. The comparison of modeling and statistical results

To evaluate qualitative characteristics (like in the Figure 8) it is essential to make the selection of different indicators and their groups. However, even the obtained statistics for broadcast packets on the basis of the conducted experiment shows a different results of the responses of the analyzed system.

5   Conclusion

The approach to the detection of abnormally functioning devices in self-organizing wireless networks based on statistical analysis of patterns of responses from remote devices. The peculiarity of this approach is that it allows you to use the built-in set of commands that gives you the opportunity to avoid additional costs when building control systems for wireless sensor networks.

As a result of the experiment was obtained statistics for broadcast packets, which shows that building external monitoring systems is possible based on the built-in tools.

The results can be used in the design and analysis of the characteristics of external systems, information security monitoring.

References

1. Kumar, P., Ylianttila, M., Gurtov, A., Lee, S.-G., Lee, H.-J.: An Efficient and Adaptive Mutual Authentication Framework for Heterogeneous Wireless Sensor Networks-based Applications. MDPI Sensors **14**(2), 2732–2755 (2014)
2. Sridhar, P., Sheikh-Bahaei, S., Xia, S., Jamshidi, Mo.: Multi agent simulation using discrete event and soft-computing methodologies. In: Proceedings of the IEEE International Conference on Systems, Man and Cybernetics, vol. 2, pp. 1711–1716 (2003)
3. Page, J., Zaslavsky, A., Indrawan, M.: Countering security vulnerabilities using a shared security buddy model schema in mobile agent communities. In: Proc. of the First International Workshop on Safety and Security in Multi-Agent Systems (SASEMAS 2004), pp. 85–101 (2004)
4. Peters, J.F.: Approximation spaces for hierarchical intelligent behavioral system models. Advances in Soft Computing (28), 13–30 (2005)
5. Zikratov, I.A., Kozlova, E.V., Stratowa, T.V.: The analysis of vulnerabilities of robotic systems with swarm intelligence. Scientific and Technical Journal of Information Technologies, Mechanics and Optics **5**(87), 149–154 (2013)
6. Gvozdev, V.A., Zikratov, I.A., Lebedev, I.S., Lapshin, S.V., Solov'ev, I.N.: Prognostic evaluation of software architectures security. Scientific and Technical Journal of Information Technologies, Mechanics and Optics. 4(80), 126–130 (2012)
7. Zikratov, I.A., Lebedev, I.S., Gurtov, A.V.: Trust and Reputation Mechanisms for Multi-agent Robotic Systems. In: Balandin, S., Andreev, S., Koucheryavy, Y. (eds.) NEW2AN/ruSMART 2014. LNCS, vol. 8638, pp. 106–120. Springer, Heidelberg (2014)
8. Wyglinski, A.M., Huang, X., Padir, T., Lai, L., Eisenbarth, T.R., Venkatasubramanian, K.: Security of autonomous systems employing embedded computing and sensors. IEEE Micro **33**(1), art. no. 6504448, 80–86 (2013)
9. Bobtsov, A., Nikiforov, V.O.: Adaptive output control: tasks, applied problems and solutions. Scientific and Technical Journal of Information Technologies, Mechanics and Optics. **1**(83). S.1–S.14 (2013)
10. Maltsev, G.N., Dzhumkov, V.V.: A generalized model of a discrete communication channel with grouping errors. Information and Control Systems **1**, 27–33 (2013)
11. Prabhakar, M., Singh, J.N., Mahadevan, G.: Nash Equilibrium and Marcov Chains to Enhance Game Theoretic Approach for Vanet Security. In: Aswatha Kumar, M., Selvarani, R., Kumar, T.V.S. (eds.) Proceedings of ICAdC. AISC, vol. 174, pp. 191–199. Springer, Heidelberg (2013)
12. Komov, S. A. and others: Terms and definitions in the field of information security. – M., AC-trust, 2009. -304 S.
13. Koval, E.N., Lebedev, I.S.: Total security model of the robotic systems. Scientific and Technical Journal of Information Technologies, Mechanics and Optics **4**(86) S.153–S.154 (2013)
14. Zikratov, I.S., Stratowa, T.V., Lebedev, I.S., Gurtov, V.A.: Building a Model of trust and reputation to the objects of multi-agent robotic systems with decentralized control. Scientific and Technical Journal of Information Technologies, Mechanics and Optics **3**(91), 30–39 (2014)
15. Shago, F.N., Zikratov, I.A.: The optimization system of planning of audit of information security management. Scientific and Technical Journal of Information Technologies, Mechanics and Optics **2**(90), 111–118 (2014)

Synthesis of the Wireless Sensor Network Structure in the Presence of Physical Attacks

Vladimir A. Mochalov$^{(\boxtimes)}$

Institute of Cosmophysical Research and Radio Wave Propagation, Moscow, Russia
sensorlife@mail.ru
http://www.ikir.ru/en/

Abstract. Under a physical attack we will understand bringing all transit nodes and information collection centers out of operation within a certain area exposed to the attack. A finite set of different intercompatible types of functional nodes, transit nodes and information collection centers is known. We consider a task of allocating different types of transit nodes of the sensor network in a heterogeneous space with a known allocation of functional nodes and information collection centers, when at most N physical attacks can be performed simultaneously. To solve this task, in the paper we propose an algorithm of a wireless sensor network synthesis, utilizing natural computations and taking into account the possible simultaneous execution of at most N physical attacks.

Keywords: Sensor networks · Network synthesis · Physical attacks

1 Introduction

Ubiquitous wireless sensor networks (WSNs) [1] are one of the promising technologies of the 21 century. The process of constructing sensor networks requires the solution of many complex problems related to different fields of research. One of the main problems is to ensure information security (IS). The importance of this problem especially for sensor networks is determined, on the one hand, by the responsibility of their applications, and on the other hand, by a large number of network vulnerabilities due to the wireless data transmission environment, the autonomy of the functioning of a large number of network nodes, software and hardware limitations, as well as power consumption limitations of network nodes. Due to the limited capabilities of the nodes, the classical complex mechanisms of the IS cannot be fully implemented in the WSN, so an energy-efficient implementation of the IS system is an important requirement for a WSN.

At the present day, an urgent task is to develop efficient algorithms for evaluation of the information security of the WSN and construction of the WSN basing on the developed evaluations. The design process of the WSN of considered class is quite complex and requires the implementation of the strategies ensuring information security at different levels of the WSN functioning. To date, a number of methods and algorithms for information security protection of the WSN have been developed; however, there is no overall view of the entire

© Springer International Publishing Switzerland 2015
S. Balandin et al. (Eds.): NEW2AN/ruSMART 2015, LNCS 9247, pp. 11–22, 2015.
DOI: 10.1007/978-3-319-23126-6_2

construction process of a WSN protected from unauthorized access that would define the place of various existing and newly developed methods in this process. The considered direction of research is very young.

The main challenges to meet when constructing a WSN, in view of the requirements to the IS, are the development of: the algorithms for evaluation of the information security of the WSN; the algorithms for synthesis of the topological structure of the WSN with regard to the IS; the algorithms for constructing an enduring structure of the WSN; the algorithms for implementing the IS strategies at different levels of its structure; the algorithms for implementing various software and hardware security systems to the WSN structure; the protected energy-efficient operation protocols for the WSN; the algorithms for simulation modeling of the WSN operation.

2   The Model of the Wireless Sensor Network

In the work we use a model of the WSN structure, where on the functional level the following types of WSN nodes can be defined: (1) functional nodes (F-nodes) that collect information in some neighborhood of their location; (2) transit nodes (T-nodes) that manage routing and retransmit the information collected by F-nodes to the information collection centers (ICC) to be utilized further; (3) ICCs that manage the WSN and process information collected by the WSN.

A wireless sensor network is allocated on an object distributed in 2-dimensional heterogeneous space. The points of the object that must have F-nodes allocated at them are given. The heterogeneous space defines spatial restrictions on allocating the WSN nodes and the function of electromagnetic signal attenuation in this space. The use of two-dimensional space does not influence the generality of the reasoning for three-dimensional space, but simplifies theoretical consideration of the proposed method and software implementation of the algorithm for designing fault-tolerant structure of WSN. In general case there can be multiple ICCs in the WSN, and the information that has arrived into each of them is available to one or multiple users for making decisions and performing certain actions. It means that information received by F-nodes should be retransmitted, with a required degree of reliability, to several ICCs by means of transit nodes allocated within the given object in a certain way.

We will consider that this information is used by a generalized "end user", for instance, is being sent from each ICC to the Internet.

We know the description of the WSN allocation object, spatial restrictions for allocating the WSN nodes, a finite set of different, intercompatible types of functional nodes N_F, transit nodes N_T and information collection centers N_C. Also, we know the allocation of different types of F-nodes and information collection centers (Fig. 1). It is necessary to allocate T-nodes in such way (Fig. 2), that the designed WSN structure would have the "desired properties" assigned by a designer. In this work we consider the endowment of the WSN with properties of fault-tolerance and vitality as minimizing the cost.

Vitality is the property of the system to execute the specified amount of functions in conditions of the environmental impact and system components failures within the prescribed limits. In this work, by environmental impact we shall mean physical attacks. By a physical attack we shall mean bringing all transit nodes and ICC out of operation within a certain area exposed to the attack.

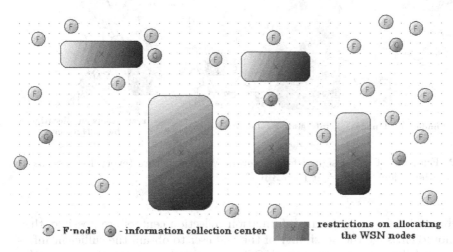

Fig. 1. Example of a known allocation of F-nodes and information collection centers

Fault-tolerance, or the property of the system to be tolerant to the failures of its elements, is the ability of the system to remain functional (including the recoverability in an appropriate time interval) upon failures of any non-empty subset of its elements.

In the process of the WSN operation there may arise failures of both the nodes and the communication channels. By the node failure we mean an event when the node does not function due to either the failures of its components or the battery discharge. By the channel failure we mean an event the consequence of which is impossibility to use the channel for any data transmission. As a confidence factor of a node, we take the probability of fault-free operation during time interval (the given operation time) providing that the nodes are irrecoverable and their failures are independent. As a confidence factor of a wireless communication channel between two WSN nodes in a heterogeneous space, we take the probability of existence of the channel with required properties in time T.

By the WSN failure we mean an event that bring the WSN into a state where reliable information from a certain number of F-nodes cannot be delivered to the ICC. WSN failure may be caused by failures of F-nodes, T-nodes and ICC as well as of the channels connecting these nodes.

Let us consider the situation when, along with failures of F-nodes and T-nodes, ICC failures are also possible. Let us consider also the task of ensuring the required reliability of delivering to the users the information from the object controlled by the WSN in this case.

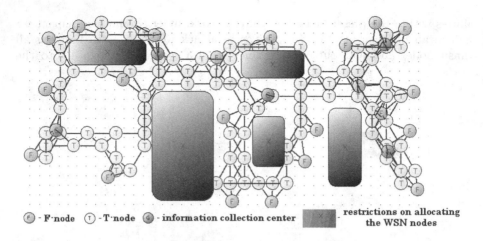

Ⓕ - **F-node**   Ⓣ - **T-node**   Ⓖ - **information collection center**   ▨ - **restrictions on allocating the WSN nodes**

Fig. 2. Example of the designed fault-tolerant WSN structure (every F-node has at least 3 independent paths to at least 3 ICC)

Obviously, if failures of the information collection centers are possible, then one should envisage the possibility of the end user to obtain the sufficient information at failures of any number of ICCs not exceeding a certain maximum (as specified in the WSN design). This number determines the required degree of fault tolerance of the WSN in relation to the ICC failures.

We believe that the delivery of information from an ICC to the end user is carried out without losses and distortions, and that the information obtained from any F-node by the end-user is sufficient for him, if it arrives from at least one of the K ICCs the F-node is associated with. Let $q = q(T)$ be the known probability of failure of one ICC in time T. Then the probability of at least one of K ICCs to be operative is calculated in the following manner: $P(Q) = 1 - q^K$.

In case the requirement $P(Q) \geq P^*$ is set, where P^* is a defined value, one can define the necessary ICCs number K solving the inequality $1 - q^K \geq P^*$ for K. In order to allow the information to be delivered from all F-nodes of the network to the end user upon failures of any $(K - 1)$ ICCs, one needs to construct (with help of suitable T-nodes allocation) such WSN structure that would assure the defined minimal probability of each F-node to be connected with some K ICCs from the total number of L after simultaneous realization no more than N physical attacks.

In order to evaluate the probability of the F-node connectivity to the ICC, one can use the algorithm based on selection of independent paths (IP) from the F-node to the ICC and utilizing in the next step the lower bound of Litvak-Ushakov [2]. In such a way, the optimization parameters in the assigned task are the WSN structural reliability (increasing the probability of the F-nodes connectivity to K ICCs) and cost (minimizing the number of allocated T-nodes).

For the *search of independent paths* from the F-node to the ICC, one can apply the Ford-Falkerson algorithm [3] which implements calculation of the maximal flow between the F-node and the ICC. Thus, in work [4] the authors provide a procedure of finding independent paths basing on transformation of a nondirected graph representing the WSN into an oriented graph, after which the Ford-Falkerson algorithm is applied.

Transformation of a nondirected graph into an oriented one is performed by changing all pairs of T-nodes connected by nondirected edges into four T-nodes cyclically connected with oriented edges. In such way, one T-node is transformed into two T-nodes, and all ingoing connections from other nodes come onto the first T-node, while the outgoing connections come from the second T-node (Fig. 3).

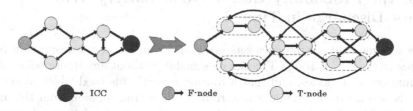

ICC   F-node   T-node

Fig. 3. An example of transformation of a nondirected graph into a directed graph

The lower bound of the probability of connectivity of an F-node F_c to the ICC is calculated as [5]: $P'_{fs} = (1 - Q_{fs})$, where Q_{fs} is the probability that all IPs from the given F-node F_c to the ICC are inoperable:

$$Q_{fs} = \prod_{\mu_{fs}^k \in M_{fs}} q_{fs}^k$$

where q_{fs}^k is the probability of the given IP μ_{fs}^k between the F-node F_c and the ICC being inoperable, $q_{fs}^k = 1 - p_{fs}^k$, where p_{fs}^k, is the probability of the k-th independent path μ_{fs}^k between the F-node F_c and the ICC being operable. Here p_{fs}^k, is defined as the probability of the operable state of all channels and all nodes composing this path, that is

$$p_{fs}^k = \prod_{c_{ij} \in C_{fs}^k} p_{ij} \prod_{b_m \in B_{fs}^k} p_m$$

where $p_m p_{ij}$ are the probabilities of the operable state (that is, the absense of failure) of, correspondingly, the node b_m and the channel c_{ij} (between some nodes b_i and b_j); C_{fs}^k is the set of all channels included in the path μ_{fss}^k; B_{fs}^r is the set of all nodes included in the path μ_{fss}^k. The probability of the operable state of the node b_m is defined as $p_m = p_A \cdot p_E$, where p_A is the probability of no hardware failure of the node components in time T, and p_E is the probability that the battery does not discharge in time T. We will assume that for a hardware failure of the WSN node the exponential low of reliability holds (the intensity of hardware failures of the components $\lambda_0 = const$), that is $p_A = e^{-\lambda_0 T}$.

In order to calculate p_E one can use the procedure of simulation modeling of the WSN operation by the network operation models selected by the designer and the algorithms of information collection, routing, self-organization etc.

In order to calculate p_E one can also use the approximate estimates that do not require to perform simulation modeling of the WSN operation. One of the possible examples of such estimate is given below.

3   An Example of Approximate Estimation of the Probability that T-Node Battery Will not Discharge in Time T

Let us assume that all data packets are sent by means of retransmission from F-nodes to the ICC through T-nodes on schedule without any intermediate data combination on T-nodes, and that a routing algorithm is used which at certain intervals sets for each F-node a new route of data transmission from this node to the BS. The route here is being selected from a set of existing independent paths.

By retransmission when the WSN operates on schedule we mean a time-distributed operation on receiving and transmitting a data packet. The number of retransmissions performed by a T-node during one period of information collection from the WSN depends on the number of data transmission routes from F-nodes to the ICC, passing through the given T-node.

We know the required time T of the WSN operation until failure (*hour*) and the period T_p of information collection by the end nodes (*hour*). We assume that power voltage of T-nodes is constant and that autonomous power sources with the low self-discharge current are utilized (for example, the Lithium Thionyl Chloride ones) and with the known E_{ALL} and E_{DIS}, where E_{ALL} is the battery capacity of the T-node ($mA \cdot h$) and E_{DIS} is the losses in time T caused by discharge of the autonomous power source ($mA \cdot h$).

We assume that T-node can be in two states: 1) in the low energy consumption mode (further, the sleep mode); 2) in the mode of performing retransmission operation. We know the useful current I_{SLEEP} in the sleep mode (mA) and the average useful current I_{RETR} in the retransmission mode (mA). As $I_{SLEEP} \ll I_{RETR}$, one can calculate the "relative" available for retransmission operations in time T battery capacity of the T-node E_{MU} in the following way: $E_{MU} = E_{ALL} - E_{DIS} - T \cdot I_{SLEEP}$. Herewith, the consequent calculations on current consumption in the retransmission mode should be held in respect to I_{SLEEP}, that is, $I'_{RETR} = I_{RETR} - I_{SLEEP}$.

We assume that the mathematical expectation μ and the dispersion δ^2 of the random variable of "relative" consumable charge E_R ($mA \cdot h$) are known (or calculated considering I'_{RETR} with known $\mu_{\triangle T}$ and $\delta_{\triangle T^2}$ that define, correspondingly, the mathematical expectation and the dispersion of one retransmission duration) for one retransmission of the WSN T-node with the distribution law selected by the designer.

In order to evaluate k, the number of retransmissions performed by the T-node during one period of collecting information from all F-nodes for sending it to the ICC, we execute the OE-2 algorithm [6]. In such way, the evaluation of the total number of retransmissions K performed by the T-node in time T will be equal to $K = k \cdot (T/T_p)$.

After adding up K independent equally distributed variables E_R (in accordance with the central limit theorem) we obtain the normal distribution $N(K \cdot \mu, K \cdot \delta^2)$ of the random variable of the total consumable charge E_K in K retransmissions. Expressing the integral of the density of the random variable E_K by the error function erf, we calculate the probability p_E that the WSN node will not discharge during K retransmissions (that is, we calculate the probability $P(E_K \leq E_{MU})$): $p_E = \frac{1}{2}(1 + erf(\frac{E_{MU} - K \cdot \mu}{\sqrt{2 \cdot K \cdot \delta^2}}))$.

4   Algorithm of WSN Synthesis Taking Into Account the Possible Simultaneous Execution of at Most N Physical Attacks

In work [7] we propose generalized bio-inspired algorithms for structure synthesis of a wireless sensor network. Further we consider applying one of the algorithms [7] for solving the task set in the work.

In recent years, the research area of $NaturalComputing$ is rapidly developing. It unites mathematical methods in which the principles of natural mechanisms of decision making are embedded [8]. Scientists have developed bio-inspired algorithms (BA) modeling animals behavior for solving various optimization problems that either do not have exact solution, or the solutions search space is vary large and complex constraints of the objective function are present, as well as NP-complete.

The described recommendations on applying BA and the proof in [9] that even the constrained variant of the problem of minimal coverage on plane is NP-complete allow to conclude about the possibility to apply self-organizing bio-inspired algorithms for a self-organizing WSN structure design.

In works [7],[10] the author propose a generalized functional flow chart of a multi-agent bio-inspired design of the WSN structure, which is based on the use by bio-inspired agents (B-agents) the shared global memory of stored pheromone (SGMSP). That is, SGMSP acts as a repository of knowledge (experience) for all B-agents. To use the following flow chart it is needed to define rules to move through the T-nodes for each type of B-agent and to take into account the use of SGMSP in these rules.

Step 1-3: Determine the model and the evaluation function for probability of presence of wireless channel with the desired properties in a heterogeneous space between two arbitrary points of the WSN allocation object in time T.

Step 2: Define a set M_{ALL} of all optimization parameters; the functions for calculating the parameters of M_{ALL}; a subset of optimization parameters $M_1 \subset M_{ALL}$.

Step 3: Determine the membership functions of fuzzy sets that characterize the optimization parameters of the M_{ALL}; fuzzy expert systems to derive the confidence factor to meet the functional requirements of the designer.

Create an empty set of the best solutions Ω_{BEST}. Determine the maximum number of solutions b_K that will be stored in Ω_{BEST}.

Create a variable k for varying the density of T-nodes of the mesh which is initialized with the value 1 ($k = 1$); compute the smallest distance D_{LOW} of the confident radio transmission among all nodes (F-nodes, T-nodes and ICC).

Step 4: In RAM, form using triangular cells a mesh grid of T-nodes covering the object of the WSN (for example Fig. 4). When forming the covering mesh grid, do not set T-nodes at the coordinates of barriers. Set the length of the cell side to the value $D = \frac{D_{LOW}}{k}$. In each grid cell, place one T-node of each type. Create an empty set Ω_T and add all of the T-nodes of the mesh to it.

Form AN, the array of sets containing all possible sets of the WSN nodes (T-nodes and ICCs) which can be set out of operation due to simultaneous execution of at most N physical attacks.

Create the array of sets GN with number of elements equal to $|AN|$.

Create the boolean variable $b = true$.

In a loop, change the value of variable i in the interval $[0; |AN|)$ and on each iteration perform the following steps:

1. Form the set $GN[i]$ by adding into it all the F-nodes and the ICCs;
2. Remove from the set $GN[i]$ all ICCs belonging to $AN[i]$;
3. Add T-nodes of the difference of the sets Ω_T and $AN[i]$ into $GN[i]$, that is, $GN[i] = GN[i] \cup (\Omega_T \setminus AN[i])$;
4. Form the network structure S_S from $GN[i]$ nodes;
5. With use of a fuzzy expert system, calculate the reliability coefficient K_{D1} of meeting the requirements of the designer for parameters of a set M_1 of structure S_S;
6. If $K_{D1} < p_1$ (where p_1 is a threshold) then set $b = false$ and exit the loop.

If b equals false, then either increase k ($k = k + 1$) and go to **Step 4** or exit with notification of failure.

Step 5: Execute bio-inspired multi-agent algorithms. We suppose that there is a fitness function $C(n)$ of multi-criteria evaluation in the interval $[0, 1]$, defining quality of the T-node of any type. We suppose that there is a choice function $F_L(k)$ using to select the set Ω_{Lk} of favorite types of T-nodes for the next k-th agent.

5.1. Create a two-dimensional array $feromoneDif$ to store changes in the pheromone. All values of $feromoneDif$ must be initialized as zeros.

Define the number of different ant agents m, the strategy for choosing the initial location of the agent and other parameters needed for the agent to perform the work.

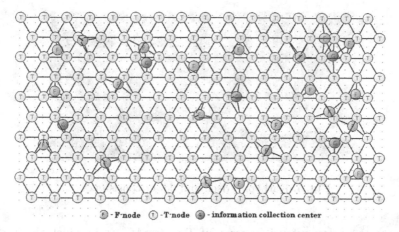

⊙ - F-node   Ⓣ - T-node   ⊚ - information collection center

Fig. 4. Example of covering the object of the WSN using triangular cells and only one type of T-node

5.2. Karl von Frisch in his work [11] noted the critical importance of flower scent to be found by bees. So, for example, Karl von Frisch writes that "if you look closely at the bees collecting honey on a meadow (Fig. 5), you may notice the striking fact: a bee flies in a hurry from a clover on a clover, paying no attention to the rest of the flowers; the other one at the same time flies from a thyme to a thyme, and the third one acts as if it is interested only in forget-me-nots. Biologists call such behavior "flower consistency". This applies of course only to individuals and not to the entire family; when one group of bees collects nectar from clovers, other worker bees from the same hive can choose the aims of their flights to the forget-me-nots, thymes or other flowers. Flower consistency is beneficial both for bees and plants. For bees this is because they are, standing by the certain flowers, meet the same working conditions which they have become accustomed to. One must see how long does a bee that had flown on a specific flower for the first time touch it with its proboscis, until it finds the hidden droplets of nectar, and how smartly does she reach the goal afterwards: only then one can understand what a saving of time does flower consistency provide. This is because the more often does one repeat specific action, the better it performs. But even more important is such behavior of bees for flowers, because they are dependent on rapid and successful pollination; it is clear that the pollen of clover, for example, would be totally unsuitable for thyme" [11].

For each ant agent, perform the following steps including the "flower consistency" heuristics (the code can be parallelized, i.e. to run in a separate thread for each agent):

A. Define the set Ω_{Lk} for the ant agent, using $F_L(k)$;

B. Create the set $J_{i,k}$ of T-nodes and add all T-nodes from Ω_T to it;

C. Create an empty set L_k of T-nodes and add all located in a mesh network T-nodes having types of Ω_{Lk} to it. Applying the "flower consistency" heuristics for ants movement consists in that initially the ant visits T-nodes of the set L_k;

Fig. 5. Karl von Frisch and flower consistency

D. Create an empty extensible array of T-nodes M_{STR}, in which the T-nodes of the designed structure will be placed;

E. Select the initial location of ant (starting T-node T_C) and add T_C to M_{STR}. Remove T-node T_C from $J_{i,k}$;

F. If the ant has walked on all T-nodes having type of Ω_{Lk} (i.e., $|J_{i,k} \cap L_k| == 0$), then $L_k = J_{i,k}$ (i.e. the ant will further walk on T-nodes of "non-favorite" types). If $|J_{i,k}| == 0$, then go to step **J**;

G. Using the following modification of a known probabilistic-proportional rule [12] of ant movement, define the next T-node T_N of the ant's route.

$$P_{ij,k}(t) = \begin{cases} \dfrac{[\tau_{ij}(t)]^{\alpha} \cdot [\eta_{ij}]^{\beta} \cdot c_j^{\gamma}}{\sum\limits_{l \in J_{i,k} \cap L_k} [\tau_{il}(t)]^{\alpha} \cdot [\eta_{il}]^{\beta} \cdot c_l^{\gamma}}, & \text{if } j \in J_{i,k} \cap L_k \\ 0, & \text{if } j \notin J_{i,k} \cap L_k \end{cases}$$

where $P_{ij,k}(t)$ is probability of the k-th ant moving from node i to node j on the t-th iteration; $\tau_{ij(t)}$ is the pheromone trace on the edge (i,j) in the global memory of the stored pheromone ($feromoneNetwork$ array); η_{ij} is the ant's visibility; c_j is the estimate of T-node j quality; α, β and γ are parameters defining weights of the pheromone trace, of visibility when selecting the route and T-node quality; $J_{i,k}$ is the set of nodes that the ant k being in node i still has to visit.

In addition to the strategies of calculating ant's visibility η_{ij} proposed in [10] ("keep away from ICC", "keep away from the node", "keep close to ICC", "keep close to the node", "absent") we add one more possible strategy "neighbour nodes", i.e.

$$\eta_{ij} = \begin{cases} 1, & \text{if node } j \text{ is adjacent to at least one node from } M_{STR} \\ 0, & \text{otherwise} \end{cases}$$

H. Add T_N to M_{STR}. Remove T-node T_N from $J_{i,k}$.

I. If $|J_{i,k}| > 0$, then go to step **F**;

J. Steps of eliminating optimization and confidence factor calculation:

J.1. Select the strategy of eliminating optimization: (a) step-by-step optimization with consideration of optimization parameters M_1. The fuzzy expert estimation of the structural parameters M_1 is used; (b) step-by-step optimization with consideration of all optimization options MALL. The unit of simulation modeling and complex assessment of the network is used;

J.2. Revert the M_{STR} array;

J.3. Create the array GN_{COPY} by means of copying the array GN;

J.4. In a loop, temporarily exclude each T-node $T_O \in M_{STR}$ from M_{STR} and from each set of GN_{COPY}. Then compute for each set of GN_{COPY} the confidence factor K_D of meeting the requirements of the parameters of eliminating optimization strategy. If in absence of T-node T_O evaluation of at least one of network structures from GN_{COPY} stops meeting the designer requirements, then put T_O back to all sets of GN_{COPY} and to M_{STR} into its place.

J.5. Create and init variable $K_{DMINALL} = 1$. For each set of GN_{COPY} do: form the network structure S_S; perform simulation modeling of S_S. The results of the modeling and structural-parametric estimates of the various parameters are the input to the complex expert system for evaluation of network structure. Calculate with the latter the confidence factor K_{DALL} of meeting all the requirements of the designer. If $K_{DALL} < K_{DMINALL}$ then $K_{DMINALL} = K_{DALL}$.

J.6. Revert the M_{STR} array.

J.7. If $K_{DMINALL} > 0$ then, in accordance with one of the following strategies, increase the pheromone amount in the array $feromoneDif$:
(a) consequent update - increase the amount of pheromone on the edges of the agent sequential traveling on T-nodes of M_{STR} by the value equal to $\Delta\tau_{ij,k}(t) = Q_{agent} \cdot K_{DMINALL}$, where Q_{agent} is the amount of pheromone secreted by the agent on one edge;
(b) full-mesh update - increase the amount of pheromone on all edges of the fully connected graph constructed on the basis of T-nodes of M_{STR} by the value equal to $\Delta\tau_{ij,k}(t) = Q_{agent} \cdot K_{DMINALL}$.

J.8. If $K_{DMINALL}$ is greater that the estimate of the worst solution from Ω_{BEST}, or ($|\Omega_{BEST}| < b_k$ and $K_{DMINALL} > 0$), then add into Ω_{BEST} the current solution. By the solution we mean the couple (M_{STR}, $K_{DMINALL}$). If $|\Omega_{BEST}| \geq b_k$ then leave in Ω_{BEST} only b_K best solutions.

5.3. After all agents have performed step **5.2.**, update $SGMSP$ ($feromoneNetwork$ array) in accordance with the following well-known rule [12]: $\tau_{ij}(t+1) = (1-p)\cdot\tau_{ij}(t) + \Delta\tau_{ij}(t)$, where $\Delta\tau_{ij}(t)$ is the amount of pheromone on edge (i,j) in the array of pheromone changes $feromoneDif$, and $p \in [0,1]$ is the coefficient of pheromone evaporation. To enhance the intermediate best solutions, the amount of pheromone on the edges of the routes of the best solutions Ω_{BEST} should be increased (an example is using "elite" ants).

5.4. If the stopping criterion is not met, go to **Step 4**.

Step 6: If it is necessary to continue the search, then increase k $(k = k + 1)$ and go to **Step 4**.

Step 7: Return the best solution from Ω_{BEST}.

5   Conclusions

The bioinspired algorithm proposed in this work can be used to find approximate solutions to the problem of synthesis of the wireless sensor network structure in the presence of physical attacks. The algorithm was implemented programmatically in *Java* and became the part of the system of wireless sensor networks design. Since the algorithm during synthesis of the WSN structure performs a full search of all possible consequences of the simultaneous execution of N physical attacks the use of the proposed algorithm is only possible for small number of attacks (N) to design not very big WSN structures.

Acknowledgments. The reported study was supported by RFBR, research project No15 07-09431A "Development of the principles of construction and methods of self-organization for Flying Ubiquitous Sensor Networks".

References

1. Faludi, R.: Building Wireless Sensor Networks, p. 320. O'Reilly Media (2010)
2. Rainshke, K., Ushakov, I.A.: A reliability estimation of systems with use of the graphs. M.: Radio and communications, p. 208 (1988)
3. Cormen, T.H., Leiserson, C.E., Rivest, R.L., Stein, C.: Introduction to Algorithms. MIT Press and McGraw-Hill, pp. 651–664 (2001)
4. Ahlberg, M., Vlassov, V., Terumasa, Y.: Router placement in wireless sensor networks. IEEE International Conference on Mobile Adhoc and Sensor Systems (MASS), pp. 538–541 (2006)
5. Mochalov, V.A.: The method of fault-tolerant sensor network structure synthesis with restrictions on placement of nodes in heterogeneous space. T-Comm. **10**, 71–75 (2012)
6. Mochalov, V.A.: Algorithms for increase sensor network total working time to the moment of its failure. Devices and systems. Management, control, diagnostics, vol. 7, pp. 12–19 (2010)
7. Mochalov, V.A.: Multi-agent Bio-inspired Algorithms for Wireless Sensor Network Design. In: 17th International Conference on Advanced Communication Technology (ICACT 2015)
8. Rozenberg, G., Bäck, T., Kok, J.N.: Handbook of Natural Computing. Springer (2012). ISBN 978-3-540-92909-3
9. Fowler, R.J.: Optimal packing and covering in the plane are NP-complete. Inf. Process. Lett. **12**(3), 133–137 (1981)
10. Mochalov, V.A.: Hybrid bionic algorithm for wireless sensor network structure synthesis. T-Comm. **10**, 72–77 (2013)
11. Frisch, K.V.: Bees, their vision, chemical senses, and language, p. 152. Cornell University Press, Ithaca (1971)
12. Bonavear, F., Dorigo, M., Theraulaz, G.: Swarm Intelligence: from Natural to Artificial Systems, p. 320. Oxford University Press, New York (1999)

Parent-Aware Routing for IoT Networks

Necip Gozuacik$^{(\boxtimes)}$ and Sema Oktug

Department of Computer Engineering, Istanbul Technical University,
Istanbul, Turkey
{gozuacikn,oktug}@itu.edu.tr

Abstract. The deployment of wireless sensor networks (WSNs) accessible through the Internet has caused a growing trend for IoT (Internet of Things). RPL (IPv6 Routing Protocol for Low-Power and Lossy Networks) is proposed by IETF (Internet Engineering Task Force) for IPv6 (Internet Protocol Version 6) constrained IoT networks as the routing protocol. Here, Objective Function (OF) determines how RPL nodes translate metrics into ranks and select routes in a network. This paper introduces a solution to have a load balanced network based on Parent-Aware Objective Function (PAOF). PAOF uses both ETX (Expected Transmission Count) and parent count metrics to compute the best path for routing. This paper evaluates the proposed solution by implementing in Contiki OS (Operating System) with Cooja simulation. MRHOF (Minimum Rank with Hysteresis Objective Function) is used for comparison. Simulation results verify that PAOF gives better parent load density, delay and parent diversity.

Keywords: IoT · Wireless sensor network · RPL · Objective function · Load balancing

1 Introduction

A Wireless Sensor Network is a distributed, self- organized network of small, energy-constrained nodes that collect and generate data [1]. With the rising of IoT platforms [2], wireless sensors are employed for various fields like transport, manufacturing, building, agriculture, biomedical etc. [3]

6LoWPAN (IPv6 over Low power Wireless Personal Area Networks) [4] is one of the most popular technology standard used in WSNs employing IoT platforms. In 6LoWPAN, the routing protocol is very important to have a connection especially with outside networks, Internet cloud. Here, RPL [5] is the mostly preferred routing protocol. RPL determines the routes using objective functions (called the OF methodology) [6]. The mostly used objective function is MRHOF [7] which decides to select paths based on the ETX [8] value. More detail about RPL will be given in the next section.

Increasing in the number of sensor nodes in a wireless sensor network makes it difficult to utilize all the nodes effectively [9]. In this work we introduce a

© Springer International Publishing Switzerland 2015
S. Balandin et al. (Eds.): NEW2AN/ruSMART 2015, LNCS 9247, pp. 23–33, 2015.
DOI: 10.1007/978-3-319-23126-6_3

load balancing technique to be used with RPL in order to enhance the load distribution of the network along the network lifetime [10].

In order to achieve load balancing, we propose a new method called PAOF to be used in RPL. The performance of PAOF is studied under various network topologies and compared with that of MRHOF. It is observed that PAOF generates a better load balanced network, diversity of parent selection and reduced end-to-end delay.

The remainder of the paper is organized as follows: Section 2 describes the routing in IoT networks by giving details of RPL. In Section 3, we present the proposed approach for optimization in RPL. The simulation environment, the scenarios employed in simulations and the results obtained for PAOF and MRHOF are presented in Section 4. Finally, Section 5 concludes the paper by giving future directions.

2   Routing in IoT Networks

Routing is considered as one of the critical items in 6LoWPAN networks. In the past, there have been several routing protocols for 6LoWPAN-compliant LLNs (Low Power and Lossy Networks), such as Hydro [11], Hilow [12], and Dymolow [13]. Unfortunately, these solutions are not able to fulfill all the requirements expected from IoT networks.

In order to address most of the requirements and open issues [14], the IETF ROLL (Routing Over Low Power and Lossy Networks) working group [15] has proposed a routing protocol called RPL. RPL is designed for networks with lossy links, which are those exposed to high Packet Error Rate (PER) [16] and link outages.

RPL is a distance vector and a source routing protocol. operates on top of several link layer mechanisms including IEEE 802.15.4 [17] PHY (Physical) and MAC (Media Access Control) layers.

RPL is based on the topological concept of Directed Acyclic Graphs (DAGs). The DAG defines a tree based structure that identifies the routes between nodes in the network. However, a DAG structure is more than a classical tree in the sense that a node might associate with multiple parent nodes in the DAG, in contrast to basic trees where only one parent is allowed. More specifically, RPL organizes nodes as Destination-Oriented DAGs (DODAGs), where most popular destination nodes (i.e. sinks) or those providing a default route to the Internet (i.e. gateways) act as the roots of the DAGs. In Figure 1, a DODAG structure generated using RPL can be seen.

When forming paths to route packets, each node identifies a stable set of parents on a path toward the DODAG root/sink node, and associates itself with a preferred parent, which is selected based on OF. OF defines how RPL nodes map metric(s) into ranks, and how to select and optimize routes in a DODAG. It is responsible for rank computation based on specific routing metrics (e.g. delay, link quality, connectivity, etc.) and also specify routing constraints and optimization objectives.

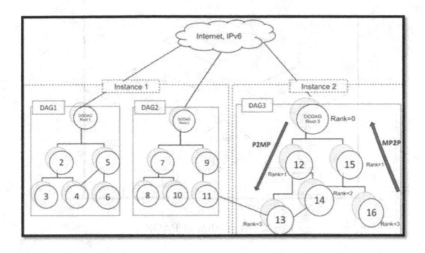

Fig. 1. RPL with DODAG [6]

A couple of designed OF implementation are the Objective Function Zero (OF0) which uses a hop count-based metric [18]; the MRHOF also known as OF1 which selects the path with the smallest ETX value; the Energy-Aware OF (EAOF) which uses energy level of the nodes and ETX value [19].

3   Proposed Routing Technique

RPL identifies the best paths to route packets through the network according to the OF and a set of metrics as described in the previous section. These metrics can be either node attributes, such as hop-count, remaining node energy; or link attributes, such as link quality, latency, and ETX.

Among these metrics, ETX is widely used to design reliable routing protocols for WSNs since it reflects the quality of the paths employed. In addition to this, hop count, energy level are also used metrics/constraints. However, none of the existing OFs do not consider parent count as a metric.

The proposed technique PAOF employs both the ETX value and parent count (the number of candidate parents) in order to compute the most efficient path to the sink. Here, the ETX values are still the key items being considered. We use the MinHopRankIncrease parameter defined in RPL Control message DIO (DODAG Information Object) [20] as a reference point. PAOF algorithm interests in the parent count metric only if ETX delta between two candidate parents is smaller than MinHopRankIncrease value. If so, the algorithm compares the number of parent counts and selects the minimum one as preferred parent. Within this, more nodes will have chance to be selected although their ETX values are not best. Hence, we are able to utilize more nodes as preferred parent. The details about the algorithm are given in Figure 2.

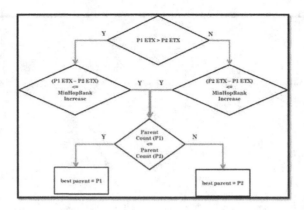

Fig. 2. Algorithm of PAOF

4   Performance Results

Many network simulators are used to measure and compare the performance of routing protocols for WSNs [21]. In this work, we decided to use Cooja [22]. Cooja is a flexible WSNs simulator designed for simulating networks running Contiki OS [23].

In order to show the performance of PAOF, this section provides several simulation scenarios and the corresponding results. We implemented PAOF on top of ContikiRPL in Contiki OS. We compare the results obtained using PAOF with those of MRHOF using the ETX metric which is also known default OF in ContikiRPL.

In the figures comparing the results, OF1 represents MRHOF, and OF2 represents PAOF.

4.1   Performance Metrics

When comparing the results obtained using MRHOF and PAOF, the following metrics were employed:

Average Parent Load Density: Here, the aim is to compute the average load density [24] on all selected preferred parents. This value is computed as

$$(Number of Delivered Successful Packets)/(Number of Preferred Parents) \tag{1}$$

Average Packet Delay: To measure the end-to-end delay between time the packet generated and reached to the sink node.

$$((Packet Arrival Time)-(Packet Generation Time))/(Number of Total Packets) \tag{2}$$

Number of DIO Messages: DIO message is used by the nodes to send their rank information to siblings during DODAG construction. As PAOF interacts on directly DIO message generation, this metric will be useful to have an idea from the point of introducing overhead.

Number of DAO (Destination Advertisement Object) Messages: DAO message is used by the nodes to send routing tables to their preferred parent nodes during DODAG construction. As PAOF aims to increase the number of preferred parents, this metric will show us if there is an overhead in the total messaging.

Parent Diversity: This metric shows us how many different nodes [25] can be selected as preferred parent in a network topology. This value is computed as

$$(Number of Preferred Parents)/(Number of Total Nodes) \qquad (3)$$

4.2 Simulation Setup

Figure 3 gives the details about the parameter values employed in the Cooja and Contiki based simulation environment.

In this work, we employed three network topologies, which are given in Figures 4, 5 and 6, in order to study the performance of the method proposed. In all of these three topologies, there exist one sink node and 24 sensor nodes. Sink node, which is colored green, is placed at the center in the first topology, while it is located in the middle top position in the other two topologies as shown in the figures.

Parameter Name	Parameter Value
Number of nodes	1 sink and 24 sensors
Radio range	50 m
Network Layer	IPv6 with 6LoWPAN
Transport Layer	UDP
Routing Protocol	RPL
Channel Check Rate	8
RPL Mode	Storing Mode
Network Setup Time	60 s
Simulation Time	960 s

Fig. 3. Simulation network parameters in Cooja

In the simulations, each node generated a payload data of length 30 bytes at the time intervals determined by the negative exponential [26] function with lambda value 0.2.

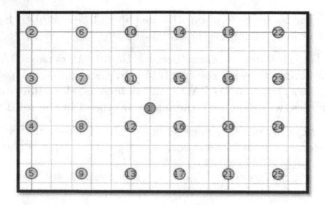

Fig. 4. Topology 1

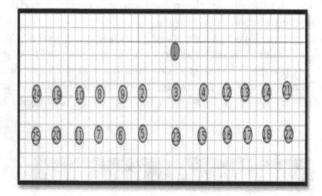

Fig. 5. Topology 2

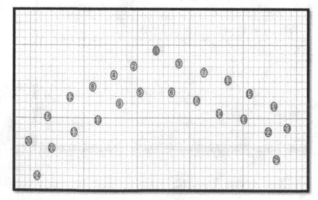

Fig. 6. Topology 3

The traffic started out after 60 seconds. The first minute is left for the RPL control messages DIO, DAO and DIS (DODAG Information Solicitation) traffic in order to setup a stable DODAG graph. After than each node generated 20 packets for each traffic scenario.

Simulation were run three times for each topology and the results show the average values with 95% Confidence Interval.

4.3 Simulation Results

The results by employing PAOF show significant improvements considering the average parent load density and parent diversity. We achieved lower parent load density and higher parent diversity because we increased the possibility of some intermediate nodes to be selected as the preferred parent with the proposed PAOF approach. Hence, more nodes are tagged as the preferred parent in the network where all of them are used to transmit packets toward the sink node.

We have some promising results for average packet delay as the number of preferred parents is increased in the network leading to less collisions.

However, in the proposed PAOF approach the number of DIO and DAO messages increases because we give chance to more nodes to be considered as the preferred parents when the nodes have similar ETX values.

The charts summarizing the simulation results can be found in Figure 7-11.

In Fig. 7, we can see that average parent load density is lower in OF 2 than OF 1 for all of the topologies.

In Fig. 8 we can also observe that average packet delay is better in OF 2 for Topology 1 and 2.

In Fig. 11, OF 2 ensures that number of selected preferred parents is higher than OF 1.

The main drawback of the proposed PAOF routing technique is that it leads more DIO and DAO message generation as shown in Fig. 9 and 10. However, as DIO transmission is governed by Trickle Timer [27], we can evaluate that

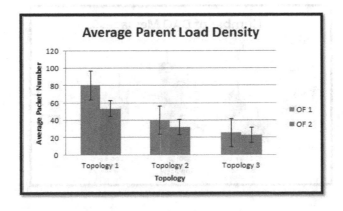

Fig. 7. Average Parent Load Density

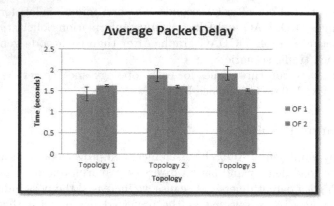

Fig. 8. Average Packet Delay

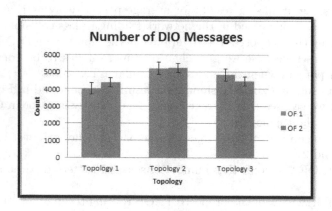

Fig. 9. Number of DIO Messages

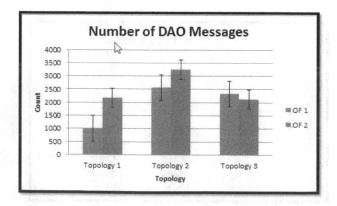

Fig. 10. Number of DAO Messages

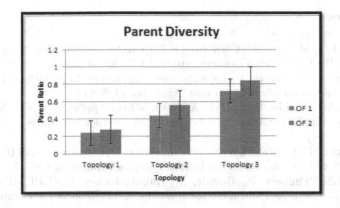

Fig. 11. Parent Density

DIO/DAO message generation will be still under control and stable considering the network lifetime.

5   Conclusion

In this paper, we introduced a new OF approach for RPL to be used in IoT networks, called PAOF. We compared the performance of this new technique with MRHOF employing the Cooja simulation tool with Contiki OS.

From the simulation results, we could conclude that the proposed routing technique PAOF performs better where the topology is mid-level sparse and the sink node is not located in the center.

The obtained results show that PAOF makes significant improvements in parent load density and diversity as compared to MRHOF employing ETX. PAOF ensures that the network will become load balanced, hence, have longer network lifetime and behave tolerantly in case of congestion.

The future work on the proposed technique will be to investigate the effectiveness of the proposed technique employing various IoT applications generating periodic, self-similar, hybrid traffic. Moreover, parent count in this work is considered as a metric with deriving from Layer 3. Similarly, child count can be considered as a new candidate node metric with deriving from Layer 2. This may also pioneer to a cross layer algorithm utilizing both Layer 2 and 3.

In future, PAOF algorithm can also be combined with MRHOF so an Adaptive OF can be introduced into literature where OF is selected dynamically based on node metrics and constraints.

Acknowledgments. We would like to thank Netas for the constructive support given to Necip Gozuacik during the period of this research work.

References

1. Akyildiz, I.F., Melodia, T., Chowdury, K.R.: Wireless multimedia sensor networks: A survey. IEEE Wireless Communications **14**(6), 32–39 (2007)
2. Milinkovic, A., Milinkovic, S., Lazic, L.: Some experiences in building IoT platform. In: 22nd Telecommunications Forum Telfor, pp. 1138–1141 (2014)
3. Chen, S., Xu, H., Liu, D., Hu, B.: A Vision of IoT: Applications, Challenges, and Opportunities With China Perspective. IEEE Internet of Things Journal **1**(4), 349–359 (2014)
4. Montenegro, G., Kushalnagar, N., Hui, J., Culler, D.: Transmission of IPv6 packets over IEEE 802.15.4 networks. In: Internet Proposed Standard RFC 4944 (2007)
5. Winter, I.T., Thubert, P., Brandt, A., Hui, J., Kelsey, R.: RPL: IPv6 routing protocol for low power and lossy networks. In: IETF Request for Comments 6550 (2012)
6. Gaddour, O., Koubaa, A.: RPL in a nutshell: A survey. Elsevier Compueter Networks **56**(14) (2012)
7. The Minimum Rank with Hysteresis Objective Function. https://tools.ietf.org/html/rfc6719
8. The ETX Objective Function for RPL. http://tools.ietf.org/html/draft-gnawali-roll-etxof-00
9. Colistra, G., Pilloni, V., Atzori, L.: Objects that agree on task frequency in the IoT: A lifetime-oriented consensus based approach. In: IEEE World Forum on Internet of Things (WF-IoT), pp. 383–387 (2014)
10. Kafi, M.A., Djenouri, D., Ben-Othman, J., Badache, N.: Congestion Control Protocols in Wireless Sensor Networks: A Survey. IEEE Communications Surveys and Tutorials **16**(3), 1369–1390 (2014)
11. Tavakoli, M.: HYDRO: A hybrid routing protocol for lossy and low power networks. In: IETF Internet Draft: draft-tavakoli-hydro-01 (2009)
12. Kim, K., Yoo, S., Park, J., Park, S.D., Lee, J.: Hierarchical routing over 6LoWPAN (HiLow). In: IETF: Internet Draft: draft-deniel-6lowpan-hilow-hierarchical-routing-00.txt, vol. 38 (2005)
13. Kim, K., Park, S., Chakeres, I., Perkins, C.: Dynamic MANET on-demand for 6LoWPAN (DYMO-low) routing. In: Internet Draft: draft- montenegro-6lowpan-dymo-low-routing-03 (2007)
14. IoT Workshop RPL Tutorial. https://www.iab.org/wp-content/IAB-uploads/2011/04/Vasseur.pdf
15. Routing Over Low Power and Lossy Networks (ROLL). https://datatracker.ietf.org/wg/roll/charter
16. Han, B., Lee, S.: Efficient packet error rate estimation in wireless networks. In: Testbeds and Research Infrastructure for the Development of Networks and Communities (TridentCom) (2007)
17. Wireless medium access control (MAC) and physical layer (PHY) specications for low-rat wireless personal area networks (LR-WPANs). In: IEEE 802.15.4 Standard, Part 15.4 (2003)
18. Objective Function Zero for the Routing Protocol for Low-Power and Lossy Networks (RPL). https://tools.ietf.org/html/rfc6552
19. Abreu, C., Ricardo, M., Mendes, P.M.: Energy-aware routing for biomedical wireless sensor networks. Journal of Network and Computer Applications **40**, 270–278 (2014)

20. RPL: IPv6 Routing Protocol for Low-Power and Lossy Networks. https://tools. ietf.org/html/rfc6550
21. Wu, D.: QoS provisioning in wireless networks. In: Wireless Communications and Mobile Computing (2005)
22. Osterlind, F., Dunkel, A., Eriksson, J., Finne, N.: Cross-Level sensor network simulation with COOJA. In: 31st IEEE Conference on Local Compueter Networks, pp. 641–648 (2006)
23. Dunkels, A., Gronvall, B., Voigt, T.: Contiki - a lightweight and flexible operating system for tiny networked sensors. In: 29th Annual IEEE International Conference on Local Computer Networks, pp. 455–462(2004)
24. Aljawawdeh, H., Almomani, I.: Dynamic load balancing protocol (DLBP) for wireless sensor networks. In: IEEE Jordan Conference on Applied Electrical Engineering and Computing Technologies (AEECT), pp. 1–6 (2013)
25. Del-Valle-Soto, C., Mex-Perera, C., Orozco-Lugo, A., Galvan-Tejada, G.M., Olmedo, O., Lara, M.: An efficient multi-parent hierarchical routing protocol for WSNs. In: Wireless Telecommunications Symposium (WTS), pp. 1–8 (2014)
26. Rahmani, A.M., Kamali, I., Lotfi-Kamran, P., Afzali-Kusha, A.: Negative exponential distribution traffic pattern for power/performance analysis of network on chips. In: 22nd International Conference on VLSI Design, pp. 157–162 (2009)
27. Clausen, T., Verdiere, A.C., Jiazi, Y.: Performance analysis of Trickle as a flooding mechanism. In: 15th IEEE International Conference on Communication Technology (ICCT), pp. 565–572 (2013)

A Multi-Broker Platform for the Internet of Things

Alfredo D'Elia[1]([✉]), Fabio Viola[1], Luca Roffia[2], and Tullio Salmon Cinotti[1,2]

[1] ARCES - University of Bologna, Bologna, Italy
alfredo.delia4@unibo.it
[2] DISI - University of Bologna, Bologna, Italy

Abstract. The emerging paradigm of the Internet of Things with millions of devices dynamically interconnected to share data imposes new requirements to applications and infrastructures. New challenges raise in terms of connectivity, resource discovery and support for multidomains distributed applications. Furthermore, the connection parameters of involved devices can dynamically change over the time and must be properly discovered. This paper proposes a multi broker platform (MBP) built on top of an existing interoperability platform, Smart-M3. MBP enables the original platform to manage multi-domain scenarios, it also provides a semantic mechanism for context-broker discovery and a virtualization interface to reach remote nodes.

Keywords: Internet of things · Smart-M3 · Multi-broker · Context-awareness

1 Introduction

The Internet of Things vision [AI1] imposes many requirements to middleware solutions for information sharing. Millions of devices should be not only connected to the network, but also discoverable, controllable and, as much as possible, autonomous and behaving in a "smart" way to simplify user lives. In the IoT, users, devices and software programs are interacting parts of an active ecosystem: they understand each other and seamlessly interoperate to concretize personalized, prompt and even proactive services. Many middleware solutions proposed in literature, by corporate research labs or in the context of largely funded European research projects, aim to smart context aware service provision through context broker (CB) based software infrastructures [Fi1,So1,Fa1]. In these infrastructures, conceived and improved since the times of the context toolkit [SD1], the main concept is that the CB gives access to all the relevant information characterizing the target application scenarios. Devices and software agents can behave smartly if they are able to understand their relevant context subset and be promptly informed on its changes. Among the existing solutions [TD1,HS1], our study is directed on a specific software architecture supporting information semantics, SPARQL query language and publish-subscribe mechanism, i. e. Smart-M3 [HL1]. Typical Smart-M3 scenarios cover a wide range

© Springer International Publishing Switzerland 2015
S. Balandin et al. (Eds.): NEW2AN/ruSMART 2015, LNCS 9247, pp. 34–46, 2015.
DOI: 10.1007/978-3-319-23126-6_4

of heterogeneous domains (e.g., HealthCare [Ve1], ElectricMobility [Be1], Smart Conferences [KG1], e-Tourism [SK1]) thanks to an active community continuously improving the original platform proposed by Nokia [HL1]. Interoperability at information level is granted by the use of RDF grounded on ontological domain descriptions. The platform is versatile, but limitations occur when the CB has not public address or when the address of the CB is not known by the software agents and should be discovered. In particular there are three main assumptions typical of Smart-M3 scenarios that should be addressed to improve the platform usability and generality. First, the address of the CBs is supposed to be known by every agent and device. Second, all the entities in a typical Smart-M3 scenario are supposed to be in the same subnetwork of the CB (this is not true if the CB is installed on a mobile device or if it is on a local network behind a proxy or firewall). Third, all the interactions are between an agent and a single CB preventing the content of more physically distinct CB to be considered as a single logical entity (e.g. if more CBs host sensor data of a person in different situations and the end user is a doctor who needs to monitor all the data of that person). To address these three main points, we propose a Multi-Broker Platform (MBP) consisting of a set of software components and mechanisms to virtualize a CB allowing:

- CB discovery based on content or CB profile properties to solve the issue of hardcoding CB addresses into software agents.
- CB virtualization to allow public access to CB that may be behind a firewall, a network address translation (NAT) or on a mobile network.
- CB logical aggregation to create the abstraction of a logical CB corresponding to a set of single physical CBs.
- Virtual CB distribution on multiple servers to distribute the computational load of virtual entities and routing network traffic according to available resources and effective usage.

To the best of our knowledge, even if these contributions can be identified in existing systems, they are never present all together in a SPARQL publish-subscribe platform. Discussion is organized as follows: after analyzing related works, the software architecture of the proposed middleware is introduced. Then details on the main architectural components and on the most important mechanisms implemented are given. In the evaluation, performance tests related to data volume, to the effects of subscriptions and to scalability of aggregated virtual CB with respect to the number of components are discussed. Eventually, conclusions are drawn.

2   Related Work

IoT paradigm requires supporting an high number of heterogeneous software agents so that it is often impossible to connect them to a single context broker. The need to handle large scenarios, involving also mobile nodes and local networks, mainly motivates and inspires the emerging research interest on multi-broker architectures and on federation of context brokers. The research work presented in this paper refers to three different research topics: connection to remote hosts (e.g. behind a NAT or on mobile networks), resource discovery mechanisms, and Multi-Broker architectures. It is very common for devices to be hidden from the public network because of the presence of a NAT or a firewall. Mobile devices, nodes of wireless sensor networks and devices in enterprise local networks are examples of nodes that could be relevant but difficult to reach from a networking perspective. To solve this problem, NAT traversal techniques can be used like TCP reversal, TCP relaying or TCP hole punching [SF1]. Wacker et al. in [WS1] propose an hybrid approach to overcome the four possible types of NAT. The current implementation of MBP uses the TCP relaying approach and can be extended in future versions to apply Wacker's method. Resource discovery mechanisms are commonly implemented to finddevices and their supported features [WP1]. The discovery operation in the field of ubiquitous or pervasive computing significantly differs from the one used in web services due to the attitude of focusing on interaction only between applications, ignoring people [ZM1]. Multiple solutions have been designed and implemented in this application field, both in academic research or in enterprise environments [Ni1, AJ1]. An MBP specific service discovery has been designed to allow smart applications to discover the proper context broker through its profile properties. As in [TC1], the MBP discovery process relies on semantic profiles based on a machine interpretable description of the resources to be discovered. MBP discovery can be also compared with the one proposed in [SR1] where a BPEL based approach is presented. Both MBP and the BPEL-based solutions imply a graph matching but in the MBP there is no translation of BPEL service descriptions and of user requests into a graph and vice versa. Multi broker and broker federations have been deeply studied in literature over the last years. Kiani et al. [KA1, KA2] introduce a Context Broker Federation to enhance the scalability of the applications, reduce communication overheads and simultaneously split applications into multiple domains, allowing a seamless interaction of context consumers with the whole data set. In these works, context consumers and context producers interact with a single context broker, typically the nearest. We propose a different approach where the chosen broker is not always the nearest and the interaction of a context consumer or producer with remote context brokers or with multiple brokers is allowed through a virtual interface. The most related research work at platform level is the one proposed by Kiljander et al. in [Ki1] from which the main idea of an architecture enabling IoT applications through multiple interconnected SIBs has been taken. Despite the discovery mechanism is similar (i.e., a semantic description of SIB profiles), there are however substantial differences both in the motivations and in the implementation. The work described in [Ki1] is mainly aimed at defining a wise software architecture, it does not deal in detail with practical problems related to communication among public and local nodes and it makes large use of ucodes. Instead, in this paper is presented an implementation proposal and description, taking into account specific communication problems among heterogeneous nodes in different sub-networks. MBP also proposes mechanisms for logical union of distinct knowledge bases, for abstract platform status description and for dynamic load balancing through resource distribution on different servers. In the MBP current implementation, ucode usage is not planned while the role of the publisher is more complex and functional for the platform with respect to the advertiser described in [Ki1].

3   Multi-broker Platform Architecture

The Multi-Broker Platform (MBP) here presented is based on Smart-M3 [HL1, MR1, Ma1], a semantic publish-subscribe platform where clients (named Knowledge Processors, KPs) can update and subscribe to changes in a RDF store through SPARQL 1.1 Update and Queries. Communication between the clients and the platform (named Semantic Information Broker, SIB) is granted by a Smart Space Access Protocol (SSAP) over TCP/IP sockets. MBP extends Smart-M3 with:

- SIB discovery based on static or dynamic profile properties or on SIB content;
- connection to SIBs without a public IP address (we name these remote SIBs);
- querying multiple SIBs at a time;
- platform dynamic distribution over multiple servers.

Figure 1 shows the main components designed and their interactions. Continuous lines represent SSAP messages i.e. the messages commonly used by legacy KPs to interact with the SIB. Dashed lines represent instead platform-specific or intra MBP communication that is implemented through a JSON based protocol opportunely designed and called Multi Broker Platform Protocol (MBPP). The following subsections will describe with more detail each MBP component.

Manager. The manager is the coordination unit of MBP. It is a multithreaded server whose purpose is to handle two kinds of requests: to discovery SIBs and to administrate the platform. Platform administration requests are used, for example, to register new SIBs and administrate the platform components (i.e., new virtualizers, virtual SIBs, virtual multi SIBs that will be detailed later on in this section). All the requests for creation, modification and deletion of SIBs are MBPP messages validated by the manager. The MBPP interface exposed by the manager provides the following classes of commands:

- creation commands (C): to register virtualizers, public SIBs and to create and register virtual SIBs (VSIBs) and virtual multi SIBs (VMSIBs). Resource allocation is made following load balancing policies. The load of a virtualizer is defined on the basis of resource usage (e.g. CPU usage, RAM, number of threads); in the current implementation it is defined as the number of managed VSIBs and VMSIBs. New processes are allocated to the virtualizer with the lowest load;
- discovery commands (R): to discover CBs with simple or refined search. A KP performs a discovery request with a proper MBPP primitive sent to the manager. The content of the request is transformed into a query that

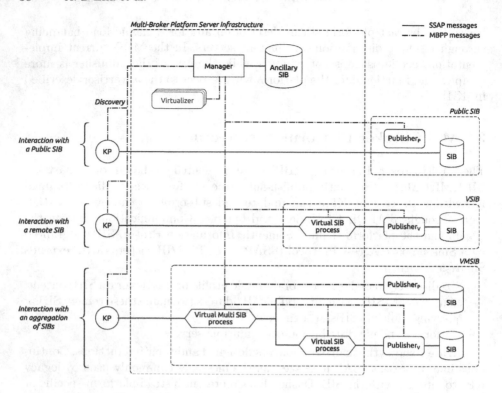

Fig. 1. Software architecture of the Multi Broker Platform

the manager will execute on the ancillary SIB to retrieve all the available SIBs matching the search criteria;
- editing commands (U): to update the information of a registered entity or to update the composition of a multi SIB;
- deletion commands (D): to undo entities registration. In case of de-registration of a virtualizer, the processes hosted by it will be transferred to the other available virtualizers always following load balancing policies.

Ancillary SIB. The ancillary SIB holds the status of the platform with information about the available virtualizers and all the registered SIBs. The KPs using the platform interact with the ancillary SIB always through the manager mediation. The information in the ancillary SIB is specified according to the MBP Ontology which has been designed to represent all the MBP entities and their status. Three types of SIB have been defined. The public SIB is publicly reachable and it is registered to the MBP in order to be discovered. A virtual SIB (VSIB) is a process running on a virtualizer and exploited by a remote SIB to become reachable. It is a process running on a public server; it exposes the same interface of a SIB and allows to reach a SIB even if it is behind a NAT. A virtual multiple SIB (VMSIB) is the aggregation of more SIBs through

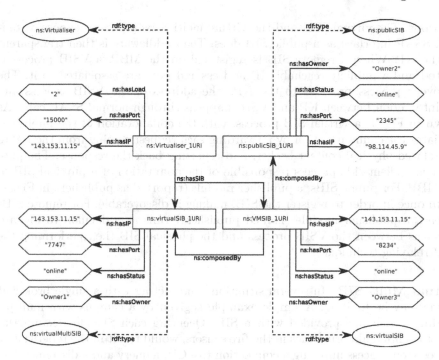

Fig. 2. Example of the ancillary SIB content

a proper virtualization process running on a virtualizer (to which each client connects to). Figure 2 shows an example of the content of the Ancillary SIB highlighting classes, instances, data and object properties in the case of one virtualizer and three registered SIBs of the three different kinds.

Virtualizer. The virtualizer is the software module delegated to run virtual SIB and virtual multi SIB processes. At least one virtualizer should be instantiated and it must be run on a public server. Other virtualizers can be registered in order to improve the scalability of the platform. When a new virtual SIB is requested, the manager selects the best virtualizer according to load balancing policies. The virtualizer only receives and sends intra MBP messages so it is provided only with a JSON MBPP interface. The support of MBP for multiple virtualizers makes MBP a distributed and scalable platform: when more computational power is needed a new virtualizer can be registered to include also the resources of a new server. Virtualizers are the basis on which a federation of SIBs can be built and managed.

Publisher and Virtual SIB. For the backward compatibility, the way a KP interacts with a SIB, no matter if it is a virtualized one or a real one, should be unchanged. The development of the middleware has then been guided by this consideration. New primitives are needed only to activate MBP specific functions. The virtual SIBs and the virtual multi SIBs processes expose a standard SSAP interface as a public SIB does. The middleware is then transparent for the KP. When a remote SIB is registered on the MBP, a VSIB process is started and a publicly reachable IP address and port are associated to it. The discovery process provides to the KPs the address of the VSIB process and the interaction between KP and VSIB happens through normal SSAP calls. As shown in Fig. 1, a virtual SIB process, with the collaboration of the publisher, mediates the communication by forwarding each request coming from the KP to the related physical context broker, and reporting back the response. The publisher is a client-side process responsible of the adaptation of a physical SIB to the MBP. For public SIBs, a publisher module (reported as publisher$_p$ in Fig. 1) is run once in order to register the SIB, making it discoverable. For remote SIBs instead, a publisher module (named publisher$_v$) runs as a daemon to forward messages between the VSIB process and the physical SIB (i.e., performing the TCP relaying technique).

Virtual Multi SIB. In several situations, interacting with a single broker at a time may be limiting. A simple example is given by a campus with multiple buildings, each one provided with a SIB. Querying each SIB, for example to detect eventual abnormalities in the fire sensors, would result in the repetition of a three step process implying a connection to a CB, a query and a disconnection. MBP provides a virtualization mechanism in order to aggregate multiple SIB on a single logical entity which will multicast the queries and subscriptions to its components. A virtual multi SIB process is spawned by a virtualizer and acts as a packet forwarder towards its CBs. No constraints are given about the type of each component context brokers that can be a public SIB, a virtual SIB or either a virtual multi SIB. No differences will be noticed by any KP interacting with the virtual multi SIB, since it exposes an SSAP interface as a real semantic information broker would do. A VMSIB composition may change dynamically depending on specific requests to the MBP manager or on network events like the abrupt disconnection of one of the SIBs. In every case the actual status of the VMSIB components is tracked in the ancillary SIB and can be queried at any time. A VMSIB is used to query multiple SIBs at a time, but, as the SIB contents are independently managed by KPs, a duplicate removal algorithm eliminates possible identical results.

4   Evaluation

The evaluation is carried out using two personal computers. The first one is a Samsung RC530 notebook with an Intel i7-2670QM eight core 2.2 GHz processors and 8 GB RAM. The operating system running on this test machine is Linux Mint 17 Qiana. The second test computer is an IBM Thinkpad T400 with an Intel Core 2 Duo P8400 2 × 2.26 GHz processor, 2 GB RAM, running Linux Mint Nadia. Both two computers are connected to a dedicated wireless 802.11n LAN. The infrastructure is executed on the first host and composed by a single manager, a single virtualizer, from 2 to n instances of the semantic information broker RedSib[1]: one playing the role of the ancillary SIB, the other $n-1$ acting as real SIBs. On the second host the performance tests are run.

4.1   Evaluation of a Virtual SIB

Before evaluating the performances of the virtual SIB, we present a comparison between the number of messages exchanged in two scenarios: direct interaction with the SIB and interaction mediated by a virtual SIB process. Figure 3a shows the classic case of interaction of a KP with a SIB. A request is sent from the KP to the SIB which processes it and sends a response. In figure 3b shows message flow in case of interaction with a virtual SIB. The time required to complete a request can be formalized, in the two cases, as:

$$T_{SIB} = T_{kp-sib} + T_{elab\_SIB} + T_{sib-kp} \tag{1}$$

$$T_{virtualSIB} = T_{kp-vsib} + T_{elab\_vsib} + T_{vsib-pub} + T_{elab\_pub} + T_{pub-SIB} + \\ T_{elab\_SIB} + T_{SIB-pub} + T_{elab\_pub} + T_{pub-vsib} + T_{elab\_vsib} + T_{vsib-kp} \tag{2}$$

where T_{kp-sib}, T_{sib-kp}, $T_{kp-vsib}$, $T_{vsib-kp}$, $T_{vsib-pub}$, $T_{pub-SIB}$, $T_{SIB-pub}$ and $T_{pub-vsib}$ are transmission times. With reference to the test scenario, the first four transmission times can be referred to as external transmission times since they involve the communication between the two hosts, while the others can be referred to as internal transmission times, since they are related to intra-machine communication. The two interation times can be rewritten as:

$$T_{SIB} = 2 \cdot T_{ext} + T_{elab\_SIB} \tag{3}$$

$$T_{virtualSIB} = 2 \cdot T_{elab\_vsib} + 2 \cdot T_{elab\_pub} + T_{elab\_SIB} + 2 \cdot T_{ext} + 4 \cdot T_{int} \tag{4}$$

From 3 and 4, $T_{virtualSIB}$ becomes:

$$T_{virtualSIB} = T_{SIB} + 4 \cdot T_{int} + 2 \cdot T_{elab\_vsib} + 2 \cdot T_{elab\_pub} \tag{5}$$

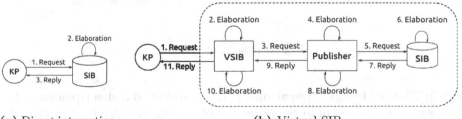

(a) Direct interaction   (b) Virtual SIB

Fig. 3. Message flow of the interaction with a real and a virtual SIB

---

[1] http://sourceforge.net/projects/smart-m3/files/Smart-M3-RedSIB_0.9/

Test 1: Time of Insert Operation vs. the Number of Inserted Triples.
The first test evaluates the time required for the insertion of a block of n triples.
The test is run both with a direct connection to the desired SIB, and with a
connection to the relative virtual SIB. The reported graph shows no significative
difference between the direct interaction and the one mediated by a virtual SIB.
The insertion through a virtual SIB is about 25% slower than the direct one
(Fig. 4 a). Considering the most frequent use case, where no more than a hundred
triples are inserted at once, difference is even lower, making the interaction
through a Virtual SIB process not so heavier than the classic one. When the
number of the inserted triples grows, the trasmission times become longer, the
time needed by the virtual SIB and the publisher to reassemble the TCP packets
and parse the SSAP message grows too. The confirm messages are not affected
thanks to their shortness. The same test has been repeated increasing the load
of the virtualizer with up to 15 VSIB processes running (Fig. 4 b). As the
VSIB processes are independent, the results are, as expected from the previous
analysis, almost the same.

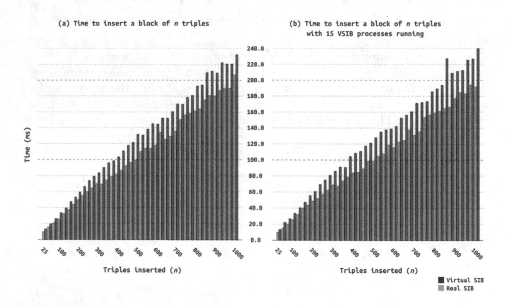

Fig. 4. Results of Test 1

Test 2: Time of Insert Operation vs. the Number of Subscriptions. The
behaviour of the platform in presence of subscriptions has also been evaluated.
In this test has been calculated the time elapsed to insert 5 triples the number of subscription (from 50 to 500). The subscriptions taken into account are
SPARQL subscription to any triple (SELECT ?s ?p ?o WHERE { ?s ?p ?o }).
The results (Fig. 6 a) shows a 25% platform overhead which remains unchanged
as it depends from the amount of data and not from the number of subscriptions.

This result shows that the platform scales well with the number of subscriptions as these do not introduce additional delay to the message forwarding.

4.2   Evaluation of a Virtual Multi SIB

From the message flow shown in Fig. 5 and following the same analysis done for a single virtual SIB:

$$T_{vmsib} = T_{ext} + T_{int} + 2 \cdot T_{elab\_vmsib} + 2 \cdot T_{elab\_vsib}$$
$$+2 \cdot (T_{elab\_pub} + T_{elab\_SIB}) \tag{6}$$

$$T_{vmsib} = T_{SIB} + 2 \cdot T_{elab\_vmsib} + 2 \cdot T_{elab\_vsib_i} + 2 \cdot T_{elab\_pub_i} + T_{internal_i} \tag{7}$$

It's worthwhile that among the elaboration times the one carried out by the VMSIB process cannot be a priori excluded from the analysis, due to the duplicate removal algorithm.

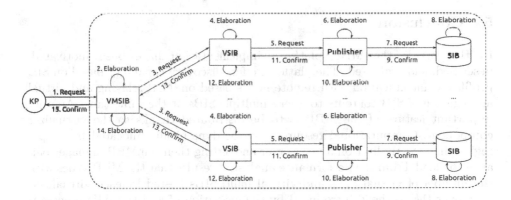

Fig. 5. Message flow of the interaction through virtual multi SIB

Test 3: Evaluation of the Query Mechanism. When a query is performed on a VMSIB, the process waits for all the replies coming from its components in order to produce a single response message that aggregates query results removing duplicates, if any. In Test 3 (Fig. 6 b) we evaluate the duplicate removal algorithm implemented by the VMSIB process. The test considers an increasing number of component SIBs all loaded with the same data set. The test has been executed with a number of component SIBs varying from one to fifteen. Each component SIB contains one thousand triples, four hundreds of which are returned by the query. This worst-case analysis highlights the good performance of the duplicate removal algorithm since the time needed to provide a response grows slowly if the number of SIBs is less or equal than six (up to 2400 duplicates), then the delay starts to grow slightly faster.

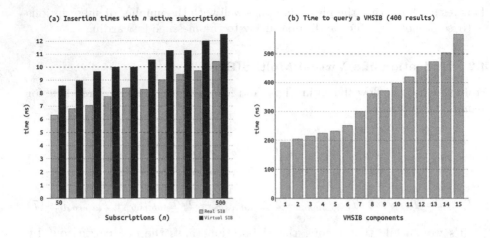

Fig. 6. Test 2 on a virtual SIB (a) and Test 3 on a VMSIB (b)

5   Conclusion

In this article the MBP platform implementation have been motivated, described, and evaluated. The platform offers discovery capability based on SIB profiles, communication with remote nodes based on TCP relaying and virtual aggregation of SIB contents to query multiple SIBs at the same time. Another important feature of the MBP is to be distributed thanks to the virtualizer component. Computational resources can be dynamically allocated by simply registering a virtualizer on a new server and letting then the MBP manager balance the load. From the performance analysis it results that the MBP scales with the number of virtual SIBs and virtual multi SIBs hosted by each virtualizer. Moreover the overhead introduced by the mediation of a virtual SIB process is negligible in the majority of the use cases letting the MBP being transparent from the KP perspective.

References

[AI1] Atzori, L., Iera, A., Morabito, G.: The Internet of Things: A survey. Comput. Networks **54**, 2787–2805 (2010)

[Fi1] FI-WARE project. http://www.fi-ware.eu/

[So1] SOFIA 2 home. http://sofia2.com/home_en.html

[Fa1] Falcarin, P., et al.: Context data management: An architectural framework for context-aware services. Serv. Oriented Comput. Appl. **7**, 151–168 (2013)

[SD1] Salber, D., Dey, A.K., Abowd, G.D.: The context toolkit: aiding the development of context-enabled. In: Proc. SIGCHI Conf. Hum. Factors Comput. Syst., CHI 1999, pp. 434–441 (1999). doi:10.1145/302979.303126

[TD1]   Truong, H.L., Dustdar, S.: A survey on context-aware web service systems. Int. J. Web Inf. Syst. **5**, 5–31 (2009)

[HS1]   Hong, J., Suh, E., Kim, S.J.: Context-aware systems: A literature review and classification. Expert Syst. Appl. **36**, 8509–8522 (2009)

[HL1]   Honkola, J., Laine, H., Brown, R., Tyrkkö, O.: Smart-M3 information sharing platform. In: Proceedings - IEEE Symposium on Computers and Communications, pp. 1041–1046 (2010). doi:10.1109/ISCC.2010.5546642

[Ve1]   Vergari, F., et al.: An integrated framework to achieve interoperability in person-centric health management. Int. J. Telemed. Appl. (2011)

[Be1]   Bedogni, L., et al.: An interoperable architecture for mobile smart services over the internet of energy. In: 2013 IEEE 14th Int. Symp. a World Wireless, Mob. Multimed. Networks, WoWMoM 2013, pp. 1–14 (2013). doi:10.1109/WoWMoM.2013.6583495

[KG1]   Korzun, D.G., Galov, I.V., Kashevnik, A.M., Shilov, N.G., Krinkin, K., Korolev, Y.: Integration of smart-M3 applications: blogging in smart conference. In: Balandin, S., Koucheryavy, Y., Hu, H. (eds.) NEW2AN 2011 and ruSMART 2011. LNCS, vol. 6869, pp. 51–62. Springer, Heidelberg (2011)

[SK1]   Smirnov, A., Kashevnik, A., Balandin, S.I., Laizane, S.: Intelligent mobile tourist guide. In: Balandin, S., Andreev, S., Koucheryavy, Y. (eds.) NEW2AN 2013 and ruSMART 2013. LNCS, vol. 8121, pp. 94–106. Springer, Heidelberg (2013)

[SF1]   Srisuresh, P., Ford, B., Kegel, D.: State of peer-to-peer (P2P) communication across network address translators (NATs). Internet Engineering Task ForceRequest for Comments **5128**, 1–32 (2008)

[WS1]   Wacker, A., Schiele, G., Holzapfel, S., Weis T.: A NAT traversal mechanism for peer-to-peer networks. In: Peer-to-Peer Computing, pp. 81–83 (2008)

[WP1]   Want, R., Pering, T.: System challenges for ubiquitous & pervasive computing. In: Proceedings 27th Int. Conf. Softw. Eng. ICSE 2005, pp. 9–14 (2005). doi:10.1109/ICSE.2005.1553532

[ZM1]   Zhu, F., Mutka, M.W., Ni, L.M.: Service Discovery in Pervasive Computing Environments. Pervasive Comput., 81–90 (2005). doi:10.1109/MPRV.2005.87

[Ni1]   Nidd, M.: Service discovery in DEAPspace. IEEE Pers. Commun. **8**, 39–45 (2001)

[AJ1]   Avancha, S., Joshi, A., Finin, T.: Enhanced service discovery in Bluetooth. Computer **35**, 96–99 (2002)

[TC1]   Toninelli, A., Corradi, A., Montanari, R.: Semantic-based discovery to support mobile context-aware service access. Comput. Commun. **31**, 935–949 (2008)

[SR1]   Suarez, L.J., Rojas, L.A., Corrales, J.C., Steller, L.A.: Service discovery in ubiquitous computing environments. In: The Sixth International Conference on Internet and Web Applications and Services, ICIW 2011, pp. 1–9 (2011)

[KA1]   Kiani, S.L., Anjum, A., Bessis, N., Hill, R.: Large-scale context provisioning: a use-case for homogenous cloud federation. In: Proc. 2012 6th Int. Conf. Complex, Intelligent, Softw. Intensive Syst., CISIS 2012, pp. 241–248 (2012). doi:10.1109/CISIS.2012.161

[KA2]   Kiani, S.L., Anjum, A., Knappmeyer, M., Bessis, N., Antonopoulos, N.: Federated broker system for pervasive context provisioning. J. Syst. Softw. **86**, 1107–1123 (2013)

[Ki1]   Kiljander, J., et al.: Semantic Interoperability Architecture for Pervasive Computing and Internet of Things. IEEE Access **2**, 856–873 (2014)

[MR1] Morandi, F., Roffia, L., D'Elia, A., Vergari, F., Salmon Cinotti, T.: Redsib: a smart-M3 semantic information broker implementation. In: Balandin, S., Ovchinnikov, A. (eds.) Proc. 12th Conf. of Open Innovations Association FRUCT and Seminar on e-Tourism, SUAI, pp. 86–98 (2012)

[Ma1] Manzaroli, D., et al.: Smart-M3 and OSGi: the interoperability platform. In: Proc. - IEEE Symp. Comput. Commun., pp. 1053–1058 (2010). doi:10.1109/ISCC.2010.5546622

Cloud IoT Platforms: A Solid Foundation for the Future Web or a Temporary Workaround?

Sergey Efremov[✉], Nikolay Pilipenko, Leonid Voskov, and Mikhail Komarov

National Research University Higher School of Economics,
20 Myasnitskaya str., 101000 Moscow, Russia
{sefremov,npilipenko,lvoskov,mkomarov}@hse.ru

Abstract. Middleware platforms have become an essential part of the modern Internet of Things (IoT) and more recently Web of Things (WoT). These platforms provide the necessary tools for data storage, processing and visualization as well as for discovery, management, and interconnection of devices, which are used by manufacturers, system integrators and end users.

In this paper we first present our view on the evolution of the Web and propose a new definition of Web 3.0 based on the number of content originators, which to the best of our knowledge hasn't been formalized in the same way before.

We then give an overview of the existing IoT platforms, including our own solution, outlining technological trends on the one side and important concerns on the other. In the end, we discuss the possible direction towards the Web of Services as a potential way to solve one of the problems.

Keywords: Internet of Things · Web of Things · Web 3.0 · Cloud IoT platforms · Internet of Services · Web of Services

1 Introduction

The World-Wide-Web has evolved significantly since its introduction in 1989. The evolution was mainly driven by a gradual progress in technology, both hardware and software. However the Web as a whole had several distinctive transitions. The terms Web 1.0 and Web 2.0, marking one such transition, have become widely accepted. In particular, the second generation of the Web, according to the definition popularized by O'Reilly [1] is connected with the social user generated content becoming significant in amount and value. Web 2.0 was enabled by a numbers of technologies, including AJAX, content syndication, stylesheets, DOM, REST and others.

The next generation has been introduced in a number of ways, however it still lacks a globally accepted definition. Even before Web numbering started, Tim Berners-Lee et al. [2] proposed the idea of a Semantic Web, which would integrate meaningful and related information from multiple sources and store it in machine-readable form. Some researchers connect Web 3.0 to the changing end-user experience and technologies behind this change [3]. Other opinions exist as well [4].

S. Balandin et al. (Eds.): NEW2AN/ruSMART 2015, LNCS 9247, pp. 47–55, 2015.
DOI: 10.1007/978-3-319-23126-6_5

We propose our own view on the evolution of Web by considering the number of content originators (COs) as the key indicator as opposed to specific functionality and technologies. Thus any transition to the next Web generation can be determined by a qualitative leap of the indicator. In Web 1.0 the main content was created by professional users and enterprises. Web 2.0 started as ordinary users became primary suppliers of new information. We suggest that Web 3.0 can be defined as the new generation of Web, in which various internet-connected devices form the largest group of content originators in terms of the number of individual sources.

Estimating the exact number of devices publishing content is not a straightforward task. According to statistical data from CISCO research groups [5] in 2014 the M2M sector accounted over 7 billion of connected devices, each a potential content source. However some additional clarification is required. Often an internet-connected device, such as a smartphone, is utilized to transfer user-generated information to the Web, for instance, to publish a blog post, status update, etc. In this case we don't consider it a CO. At the same time a smartphone contains a number of sensors, e.g. GPS, accelerometer, which can generate data without participation from an end user. In the latter sense a smartphone can also be accounted for as a content source.

Raw device data can be considered as content when it is published and may be accessed by many consumers, which can be both human users and devices in the new Web. However with the current state-of-the-art of the Internet of Things, most of devices don't have enough capabilities to store, process and provide constant accessibility to all of their generated content, as well as to form dynamic links between each other. Therefore special cloud services are required [6]. We believe that middleware data platforms are becoming the key part of modern Web 3.0, similar to social networks, blogs and other popular services determining the image of Web 2.0. In the next section we provide an overview of what we think are the most significant solutions in the field, and present our own cloud-based WoT platform.

The important question however is whether these platforms are a temporary solution caused by constraints of many internet-connected devices or a solid foundation that will work for many decades ahead. In section 3 we give our vision of the problem, outlining the main issues that still exist and possible solutions to them.

2   Overview of Modern IoT Platforms

The primary goal of cloud IoT/WoT platforms is to simplify storage, control, visualization and data exchange between different smart devices. The core of any such platform is a centralized data storage, which provides different services on top of itself. Using a centralized platform gives a number of advantages.

Manufacturers get a cloud storage for their devices' data and hence don't need to manage a server infrastructure. They also get additional functionality through the platform, such as integration with social networks and possibility of interaction with "foreign" devices.

End users are provided with a central point to manage multiple devices, often of different brands. Raw data differs from device to device, however inside the platform it is visualized in a flexible form and can be accessed from different types of user interfaces such as web, smartphone or tablet.

Machine-to-machine communication often requires solving several additional tasks, such as discovery and implementing different connectivity patterns [6]. Centralized platforms greatly simplify these tasks.

Over the last 5 years several dozen IoT platforms have emerged (see table 1). Xively (former COSM) [7] is the most popular middleware IoT solution on the market today. It uses the PaaS model, offering developers a platform with a broad range of supported tools and technologies, e.g. HTTP, MQTT, classic TCP Socket, Web Socket, SSL support and a number of data formats - JSON, XML, CSV. To simplify development Xively provides a number of ready-to-use libraries to access its API from different programming languages. The end functionality is still somewhat limited. The main function of data acquisition and storage is implemented in the form of data channels (datastreams), each logging changes of a particular property from a physical device. Hence any device comprises a set of such channels. The end user is given a possibility to develop new devices within the platform (defining channel sets, monitoring and controlling properties) and to manage products (sets of devices for distribution). The service can be extended by using triggers - HTTP requests executed towards a given URL in case of certain events. Basic visualization tools are provided as well.

The EVRYTHING platform [8] is functionally similar to Xively apart from a smaller number of supported protocols and data formats. EVRYTHING allows developing external applications, which can communicate with the devices using the OAuth protocol.

Paraimpu [9] is a platform primarily for developers of do-it-yourself gadgets with the aim to establish meaningful connections between them. Unlike Xively, there is no need to describe properties and channels of a device. Instead each device contains a single text field that can incorporate any type of data, including unstructured one. The platform imposes a strict distinction between sensors and actuators, so that connections can be made easily. As any device's data is represented as a single string, preprocessing can be done with the help of regular expressions. As of today, the platform does not provide any toolkits for developing and distributing products or applications and can be considered as a testing environment for technological hobbyists.

For the last 2 years we have been developing our own IoT platform - Thinger [10]. The key idea behind this project was to simplify development of customized IoT solutions. The main features related to data acquisition and processing are similar to most of the other systems. Data Providers enable to connect off-the-shelf devices from different manufacturers with OAuth-based APIs. Data Handlers were added to build customizable scenarios of data processing and device interconnection. These scenarios can be written using C# (core language of the platform) or Python programming languages or by means of a special visual constructor.

More significant improvements include the following:

- an integrated script engine, which helps to implement various scenarios of data processing and device interconnection
- dynamically installable applications enabling high level of customization for each device or group of devices
- graphical tools for designing device widgets and managing device interconnection scenarios targeted at users without programming skills
- flexible access and user control system for manageable M2M communication
- specialized algorithm for dynamic protocol selection aimed at optimizing constrained resources of remote devices

The Thinger platform, currently being on the verge of a public launch, already provides its services for several commercial customers. One particular example is a network of self service car washes, which relies on Thinger for a wide range of tasks – from remote monitoring of appliances to managing customer loyalty cards.

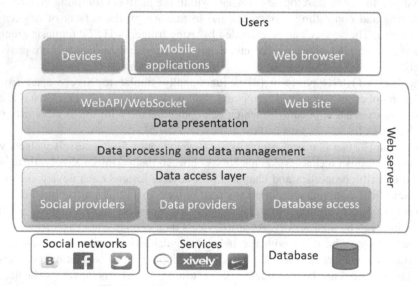

Fig. 1. Architecture of the "Thinger" platform

Although most of the current platforms are IPv4 oriented, it is clear that the future of IoT and Internet in general will rely on IPv6. Another important emerging feature is support of novel HTML5 and WebSocket technologies, which enable asynchronous data transfer for the event driven communucation model and which are backward compatible with existing web applications. JSON became the primary textual data format in the IoT due to its minimalism, however binary formats are also partly supported.

As the current paper is not intended to provide a detailed survey of all available IoT platforms, we omit further description. Instead the reader is referred to [11, 12]. In the next section we give an overview of the main issues of IoT platforms still pending a solution.

Table 1. Comparison of modern IoT platforms

Platform	Protocols	Network	Data formats	Encryption	Authentication
Xively	HTTP, TCP Socket, WebSocket, MQTT	IPv4, IPv6	JSON, CSV, XML	SSL	API Key
ThingWorx	HTTP, WebSocket, MQTT, OData	IPv4	NOT AVAILABLE	NOT AVAILABLE	NOT AVAILABLE
ThingSpeak	HTTP	IPv4	JSON, CSV, XML	SSL	API Key
Paraimpu	HTTP, WebSocket	IPv4	JSON, CSV	SSL	Token
Open Sen.se	HTTP, WebSocket	IPv4	JSON	Not supported	API Key
WoTKit	HTTP	IPv4	JSON	Not supported	API Key, HTTP Basic Authentication, OAuth
EVRYTHNG	HTTP	IPv4	JSON	SSL	API Key
SmartThings	HTTP	IPv4	JSON	SSL	API Key, OAuth
ioBridge	HTTP, WebSocket	IPv4	XML, JSON, JSONP	SSL	NOT AVAILABLE
nimbits	HTTP, WebSocket	IPv4	JSON	SSL	API Key
Thinger	HTTP, WebSocket, SSE	IPv4	JSON	SSL	API Key, HMAC-SHA1 signature

3   Discussion

Security remains one of the major concerns for the Web of Things and middleware platforms in particular. Table 1 shows that most of the IoT platforms use an API key for device authentication. Such key is integrated into a device and is transmitted with each transaction to prove device's identity. Although SSL is supported on the cloud side, unsecure HTTP connections are also allowed, as end devices may not have SSL implemented, for example, due to resource limitations.

It is clear that an IoT security solution has to be complex, covering not only channel protection, which today is well developed and standardized, but also data protection on end devices and servers. Special attention should be paid to minimizing human factor, as an accidental error in a system that manages numerous automated devices can be much more dramatic in its consequences than losing access to an e-mail or social network account.

Another important issue to be dealt with is the difference of data representation for various devices. Even for a joint hardware platform devices might be described differently in terms of their data access. Moving towards the Internet of Services (IoS) could be a possible solution. The concept of the Internet of Services addresses the question of how we communicate and engage with the Internet of Things [13-16]. It suggests that along with every connected system having its own unique API, the metadata collected from a group of systems should also provide APIs. It will simplify consumption of key information and events, decision making and management of all services. The Internet of Services will replace the platforms with APIs so that the big amount of data produced by things could be shared with data consumers: apps, people or other things.

The term "Web of Services" makes an accent on the technical implementation of the IoS. Such web makes services accessible to and processable for machines, having a semantic architecture in common and following the main web principles, such as decentralization, modularity, simplicity and addressability via URIs.

The main problem of the WoS today is the absence of a clear definition of what constitutes a service at a conceptual level. Hence there is no unified way to describe a service semantically, though different approaches are developed (OWL-S, WSMO, WDSL-S). So as long as there are lots of services available on the web, there is no opportunity to find them automatically, without human interaction.

If a unified way to describe a service is applied, a wide range of possible applications will be accessible, such as:

* Service discovery - a machine will be able to find a service, which could solve a problem automatically, without consulting a person;
* Contracting and execution - a machine could choose the best option among the available services in terms of execution details and contracting, for example, the price of service or the frequency of necessity;
* Billing or revenue sharing - a machine could be able to make a best deal with the service provider on things such as billing or revenue sharing for service usage;

- Replacement on failure, based on experience - if the chosen service falls short of user's expectations, a machine would replace it with a better one. There may also be an opportunity for it to rate the service, making other machines know the quality of used services;
- Service detalization - a machine could be able to split a task into subtasks and find a service for each of them. These services may be done simultaneously if it cuts the costs or is logically possible.

As of today, the Web of Services, which would provide means to find and use available services automatically, has not formed. Yet, there might be an opportunity to share services or the experience of using them between people via a social network. This could be a transition period between the IoS and the WoS, while researchers are working on the semantic approach to service description or any other solution to the problem.

In the fully developed Web of Services libraries are used to accommodate information about different services [17-18]. These service libraries are utilized by different groups of users.

Solution architects query a service library to find existing services, which they can use as building blocks for service-based applications. They may also use descriptions of services, which are available in the service library, to complete a set of building blocks. If they need an element that is not available in the service library, they can form service metadata that describe the service interface. The service can then be implemented by developers.

The possible ultimate outcome of the above-described facts could be gradual transformation of current data platforms to service libraries or at least integration between them.

4   Conclusion

In this paper we presented our vision of the state-of-the-art of the Internet of Things with respect to cloud based data platforms. We argue that the huge increase in the number of machine content originators can define the new web generation. In this sense middleware data platforms play a key role, transforming raw data from numerous devices to content, usable by people or machines. These platforms provide a number of flexible tools for device manufacturers, system integrators and end users.

Despite the large number of available solutions, most of them have a similar set of core capabilities: data storage, open API for external access, user management, authorization etc. We started developing our own Thinger platform with the idea of providing a better end-user experience and stronger customization possibilities. To achieve this, dynamically installable device applications as well as several graphical tools were implemented.

We envision the future of IoT/WoT platforms as follows. Unification of device representation inside the middleware will soon become a major task. We believe that implementing service-based approach may be one of the possible solutions.

Furthermore in many cases interaction between smart devices occurs only when they are in proximity to each other. Communication through the cloud in such scenario is highly inefficient. As D2D communication technologies, e.g. described in [19], continue to evolve, some functionality will be brought downward from the cloud to the device level. As a result, special gateways may appear, which will take part of the cloud platform functionality, mainly related to inter-device communication, temporary data storage and providing external access to data.

However the middleware will be still required to provide the interface for the end user as well as a centralized storage of device metadata for discovery purposes. The Internet of Services will replace platforms with APIs so that the huge amount of data produced by smart things could be shared with various data consumers: human users, apps or other things, which can then become 3rd party service providers.

References

1. O'Reilly, T., Battelle, J.: Opening Welcome: State of the Internet Industry, San Francisco, CA (2004)
2. Berners-Lee, T., Hendler, J., Lassila, O.: The semantic web. Scientific American **284**(5), 28–37 (2001)
3. Cabage, N., Zhang, S.: Web 3.0 has begun. Interactions **20**(5), 26–31 (2013)
4. Silva, J.M., Mahfujur Rahman, A.S.M., El Saddik, A.: Web 3.0: a vision for bridging the gap between real and virtual. In: Proceedings of the 1st ACM International Workshop on Communicability Design and Evaluation in Cultural and Ecological Multimedia System (CommunicabilityMS 2008), pp. 9–14. ACM, New York (2008)
5. Cisco Visual Networking Index: Global Mobile Data Traffic Forecast Update (2014–2019)
6. Efremov, S., Pilipenko, N., Voskov, L.: An Integrated Approach to Common Problems in the Internet of Things. Procedia Engineering **100**, 1215–1223 (2015)
7. Xively. http://xively.com (accessed April 15, 2015)
8. Evrything. http://evrything.com (accessed April 15, 2015)
9. Pintus, A., Carboni, D., Piras, A.: PARAIMPU: a platform for a social web of things. In: Proc. 21st Int. Conf. Companion World Wide Web, pp. 401–404 (2012)
10. Pilipenko, N., Voskov, L.: THINGER: web-oriented platform for interaction between smart things. In: Distributed Computer and Communication Networks: Control, Computation, Communications (DCCN 2013), pp. 289–293 (2013)
11. Balamuralidhara, P., Prateep, M., Arpan, P.: Software Platforms for Internet Of Things And M2M. Journal of the Indian Institute of Science **93**(3), 487–497 (2013)
12. Emeakaroha, V.C., Fatema, K., Healy, P., Morrison, J.P.: Contemporary Analysis and Architecture for a Generic Cloud-based Sensor Data Management Platform. Sensors & Transducers **185**(2), 100–112 (2015)
13. Evans, D.: The Internet of Things. How the Next Evolution of the Internet Is Changing Everything. Cisco Internet Business Solutions Group. https://www.cisco.com/web/about/ac79/docs/innov/IoT_IBSG_0411FINAL.pdf (accessed April 20, 2015)
14. Casagras IOT Definition. Casagras. http://cordis.europa.eu/news/rcn/30283_en.html (accessed April 20, 2015)
15. SAP IOT Definition. SAP Research. http://services.future-internet.eu/images/1/16/A4_Things_Haller.pdf (accessed April 20, 2015)

16. ETP EPOSS IOT Definition. ETP EPOSS. http://www.smart-systems-integration.org/public/internet-of-things (accessed April 20, 2015)
17. Sulistyo, S., Prinz, A.: PMG-Pro: a model-driven development method of service-based applications. In: Ober, I., Ober, I. (eds.) SDL 2011. LNCS, vol. 7083, pp. 138–153. Springer, Heidelberg (2011)
18. Komarov, M.M., Nemova, M.D.: Emerging of new service-oriented approach based on the internet of services and internet of things. In: IEEE 10th International Conference on e-Business Engineering (ICEBE), Ch. 66, pp. 429–434. IEEE Computer Society Conference Publishing Services (CPS), Los Alamitos (2013)
19. Corson, M.S., et al.: Toward proximity-aware internetworking. IEEE Wireless Communications 17(6), 26–33 (2010)

The Smart-M3 Platform: Experience of Smart Space Application Development for Internet of Things

Dmitry G. Korzun[1], Alexey M. Kashevnik[2,3(✉)], Sergey I. Balandin[4], and Alexander V. Smirnov[2,3]

[1] Petrozavodsk State University (PetrSU), Petrozavodsk, Russia
dkorzun@cs.karelia.ru
[2] SPIIRAS, St. Petersburg, Russia
[3] ITMO University, St. Petersburg, Russia
{alexey,smir}@iias.spb.su
[4] FRUCT Oy, Helsinki, Finland
Sergey.Balandin@fruct.org

Abstract. Efficient resource utilization in the Internet and in appearing Internet of Things (IoT) environments needs "smart applications". They operate over shared resources of the computing environment to construct services sensitive to the users and their needs. Smart spaces support services that actively involve surrounding digital devices and Internet services. In this paper, we consider the Smart-M3 platform—an open source solution for creating smart spaces with ontology-driven information sharing. This study makes a next step for evolving smart space application development. We systemize the key properties for application development using Smart-M3 in IoT settings. The properties are analyzed on selected use cases, covering such emerging IoT application domains as collaborative work and e-Tourism. Our experimental evaluation confirms the applicability of analyzed solutions for today's computing environments.

Keywords: Smart space · Smart-M3 · Ontologies · Applications development

1   Introduction

The amount of information and services is growing in the Internet so fast that users cannot efficiently utilize the available multitude of resources. A lot of work exists on making smart applications, see [1–3] and references therein. Low communication between exiting services results in high fragmentation, i.e., information produced by one service is rarely accessible in another. The challenge is for application to operate intelligently over all resources available in the computing environment. On one hand, service construction necessitates involvement of surrounding devices as active information producers and processors. On the other hand, the users become more interested in personalized and context-aware services, including their proactive delivery.

© Springer International Publishing Switzerland 2015
S. Balandin et al. (Eds.): NEW2AN/ruSMART 2015, LNCS 9247, pp. 56–67, 2015.
DOI: 10.1007/978-3-319-23126-6_6

Smart spaces can be thought as advanced computing environments that acquire and apply knowledge to adapt services in order to enhance user experience [4–5]. Smart space participant is an autonomous information processing unit, able to run on surrounding devices [6]. Services are constructed as a result of interaction over shared information [7]. The Internet of Things (IoT) supports ubiquitous connectivity property for smart spaces [6, 8]. The most common view of IoT refers to the connection of physical (smart) objects [9], while the core of technology is in information interconnection and convergence [10]. Each physical object is made a "smart" digital device. They operate in their environment realizing continued processing of many data flows, originated from various sources and consumed by multiple applications.

Smart space applications for IoT environments require a platform to provide infrastructural support of service operation, delivery, and use. First of all, ensuring large-scale interoperability is unavoidable step in further evolution of the information world [11, 12]. In this paper, we consider the Smart-M3 platform—an open source solution for creating smart spaces with ontology-driven information sharing. M3 stands for Multidevice, Multidomain, and Multivendor. The focus is on the semantic interoperability: heterogeneous participants create a common understanding of the collected corpus of information and derived knowledge. One of the key Smart-M3 advantages is its ground on the Semantic Web. The runtime information and majority of the underlying mechanisms are visible and manageable via RDF triple stores and SPARQL endpoints, which provide interoperability and knowledge reasoning. The use of ontological knowledge representation and pub/sub coordination models makes the Smart-M3 platform well suitable for the IoT settings. The scalability is achieved by localization of all needed instantly information in a common knowledge base, where ontological relations between data heterogeneous sources are primarily used instead of data duplication in a centralized store.

This paper summarizes our experience of smart spaces application development using Smart-M3 and oriented to deployment in IoT environments. We show that open source solutions are an effective choice for the considered class of smart spaces application. The presented study takes into account many of our pilot applications [13–18] in order to systemize the key properties that provide architectural patterns and interaction models. The applicability and feasibility are demonstrated on two use cases, selected from our wide pool of pilot applications. They cover such emerging IoT application domains as collaborative work and e-Tourism. The first case is for mobile users collaborating in a spatially-localized digitally-equipped environment. In the second case, the smart space expands to a large geographical area with travelling mobile users; each individually consumes touristic information from multiple Internet services. For these two cases we introduce reference architectures and ontologies, which can be transferred to development of various smart space applications. Our study performs scalability experiments for estimating the Smart-M3 response time in dependence on the size of smart space. This experimental evaluation confirms the applicability of Smart-M3 platform for today's computing environments.

The rest of the paper is organized as follows. Section 2 introduces our approach for smart spaces application development using Smart-M3. Section 3 demonstrates the applicability and feasibility based on two selected use cases. In Section 4, we evaluate the pilot implementations. Section 5 summarizes our key findings.

2   Smart Spaces and Enabler Technologies

The IoT paradigm [9] grew from ubiquitous computing [19] and now evolves to the architectural model of loosely coupled decentralized system of objects; each is augmented with sensing/actuation, processing, and network capabilities [6, 10]. Communication can utilize various means of wired and wireless network connectivity. Physical entities (e.g., everyday embedded digital equipment and consumer electronics) are transformed into smart objects. This kind of "smart" makes each object acting autonomously, making own decisions, sensing the environment, communicating with other objects, accessing resources in the Internet, and interacting with end-users.

A new type of ubiquitous computing environment is formed (IoT-environment), consisting of many objects on the variety of networked devices: sensors, data processors, actuators, consumer electronics, personal mobile devices, multimodal systems, etc. In addition to relatively common IoT objects such as embedded devices and consumer electronics, an essential class of smart objects emerges because of personal mobile devices such as tablets, smartphones, and various wearable gadgets. Heterogeneous equipment of the environment forms a hardware basement for development of the next generation of service-oriented applications called "smart services". They are naturally constructed within a multi-agent system of interacting objects, making the underlying IoT environment "smart" as we explained below.

According to [4] a smart environment is able to acquire and apply knowledge about the environment and its users in order to improve their experience of service consumption in that environment. This kind of "smartness" is due to the service construction using a rich corpus of potentially available information. In an IoT-environment, its smart object reasons about, controls, and adapts to physical surroundings and user context. Object's operation follows a cycle of (i) perceiving the state of the environment, (ii) reasoning about the state together with application goals and outcomes of possible actions, and (iii) acting upon the environment to change the state. A smart object is a provider of one or more services; each service is constructed solely based on direct observations (e.g., by sensors the device has) and direct interaction with other objects. For instance, an indoor climate control system senses recent temperature in the room and changes the air flow appropriately.

More advanced interaction model assumes that service construction is a cooperative activity: several objects collect and share common knowledge about the environment and its ongoing processes. In particular, space-based model [20] declares this style for service-oriented applications, which is closely related to multi-agent systems. Each agent can publish and retrieve information into/from a common (shared) information space. An application is a system of distributed software agents interacting indirectly by sharing information. Smart objects can be hosts for such agents, along with agents hosted on traditional desktop and server computers. If a single smart object is associated with a service as its provider then it has on its own to acquire and apply information to construct the service (interacting with other objects if needed). In contrast, information sharing allows a group of participants to interact implementing a collective service provider. Each object can deposit its own piece to the information space, even without knowledge how this information is then used and for which services. As a result, "smartness" of services can be implemented as follows. Knowledge about the environment and its users is acquired collectively by participants themselves (self-generation is possible as well) and this knowledge can be applied so flexibly that the variants of provided services are limited only by the imagination.

Presence of multiple devices leads to the interoperability problem. Its solution is in concretizing a way of information representation. Triple space computing [20] applies the Semantic Web technologies, which aim at making information on the Web readable by both humans and machines [21]. RDF triples are basic information representation units in the form of subject-predicate-object [22]. They in turn are organized into RDF graphs with subjects and objects as nodes and predicates as links, i.e., forming semantic web databases or RDF triplestores. The content is described in a semantic-aware structured manner, with ontology as the primary building tool. Semantic queries to the informational content become possible, utilizing RDF reasoning capabilities and semantic query languages like SPARQL. As a result, a semantic infrastructure appears with a localized knowledge base targeted to the needs of applications. The semantic technology provides a clear way to achieve interoperability and to enable integration, communication, and coordination of many autonomous, distributed, and heterogeneous service providers and their clients. This semantic approach is then extended to other space-based applications apart from web services only [23].

For interaction of multiple objects pure query/response mechanisms are not enough. New development opportunities appear due to event-based programming [24]. Changes in the shared informational content lead to reaction of appropriate participants. The subscription operation provides means to detect changes in the shared content and then to react accordingly.

We consider smart spaces deployed in localized IoT environments; each is typically associated with a physical spatial-restricted place (office, room, home, city square, etc.). The environment is equipped with variety of devices: sensors, data processors, actuators, consumer electronics, personal mobile devices, multimodal systems, etc. Each device participates in the smart space via its software agent running on the device. The informational content virtualizes the environment representing entities of physical and informational worlds, including the participants themselves [25].

Information stored in the same space can be further processed, providing deduced knowledge that otherwise cannot be available from a single source. This property allows methods of multi-agent systems: any agent can infer new facts as a reaction to knowledge that has been published by other agents. This iterative knowledge self-generation model leads to the opportunity to create increasingly complex and flexibly deployed applications with emphasis on "smartness", including such properties as service adaptability and personalization to users, context-aware recommendation, and proactive service delivery.

This notion of smart space was adopted by the M3 architecture and implemented in Smart-M3 platform [26]. The key architectural component of Smart-M3 is Semantic Information Broker (SIB) that implements an information hub for agents of a given environment. Agents are also called knowledge processors (KPs); They run on devices of the environment. Some of them act on behalf of external data sources, resources, and services. Network communication between a KP and its SIB uses Smart Space Access Protocol (SSAP) for exchanging messages. The protocol is session-based: every session starts with a join operation and a leave operation ends the session.

3   Application Development and Use Cases

This section presents Smart-M3-based application development for such emerging IoT application domains as collaborative work and e-Tourism. For each domain we describe our development of a research prototype system. The presented results provide evaluation of the approach, reference architectures, and ontologies for smart space based application development and application services interaction to enrich their semantic interoperability for increasing the quality of these applications.

Collaborative Work

The new generation of personal devices, such as smartphones and tablet computers, supports people to effectively communicate for working together, to provide own resources to the collective solving process, and access assisting services in the IoT-aware environment (workspace). For this emerging IoT application domain of collaborative work let us consider the SmartRoom application [16, 27], which illustrates the needs of collaborative activity and motivates employing smart spaces.

In such activity as conferencing, meetings or seminars, a virtual shared workspace (conference room) can be constructed to support activity operation for local and remote participants [4]. Computing equipment is localized in a room and WLAN provides the network connectivity. Examples of devices are interfaces for media information (e.g., projector–computer pairs, TV panels, interactive boards), sensing devices (e.g., physical sensors and actuators, network activity detectors, microphones, cameras), user access and control devices (e.g., laptops, netbooks, smartphones), WLAN infrastructure (e.g., Wi-Fi access points).

The smart space manages the semantics of the activity, provides for the services and surrounding equipment possibility of information sharing, and relates all information that the activity needs and collaboratively produces. The services are accessible and the activity is controllable from personal devices and room equipment. The most important information, such as current speaker's presentation and activity agenda, is displayed on big public screens. The initial idea was evaluated in SmartConference system [16]. Now it evolves to the SmartRoom application [27], which supports several types of activity in an IoT-aware room-localized environment.

The architectural scheme is shown in Fig. 1. The core services—Agenda and Presentation—maintain the activity program and digital presentational content of each speaker, respectively. Two public screens are user interfaces for the services.

If a service is intended for all attendees then its information is visualized on a public screen, possibly composed with related information from some other services. SmartRoom services can be accessed personally using clients (e.g., Android and Windows Phone). Although all services are potentially applicable for any participant, a specific service subset is offered for each user based on her/his preferences and current context. It aims at personalization and enables proactive delivery of services.

The SmartRoom WLAN is attached to the Internet, allowing external systems to participate in the smart space. This property supports the following important extensions. (1) Resource-consuming processing can be delegated, e.g., to nearby servers of the corporate computing system or to cloud systems. (2) A rich set of existing Internet services can be used, extending the functionality of the SmartRoom application.

The SmartRoom space acts as an informational hub to relate all data sources and participants. The ontology defines how the information are related to different services and participants are represented (Fig. 2). It defines the main concepts and relationships for the Smartroom application and its services interaction. It contains classification of the SmartRoom services (subclasses of the class *"Service"*), different kinds of notifications used in the system (subclasses of class *"Notifications"*), information about participants (subclasses of the class *"ParticipantInfo"*), different kinds of materials that SmartRoom users can present (subclasses of the class *"Materials"*), structure of activities in SmartRoom (subclasses of the class *"ActivityStructure"*), and different types of activities supported by SmartRoom application (subclasses of the class *"Activity"*). Other relationships between presented classes are depicted in Fig. 2.

The following key properties have been determined for the ontology concepts. Personal information about the participant is represented with the class *"Person"*: properties *"age"*, *"img"*, *"language"*, *"mbox"*, *"name"*, *"organization"*, *"phone"*, *"status"*. Properties *"username"* and *"password"* of the class *"UserProfile"* keep system information of the SmartRoom user.

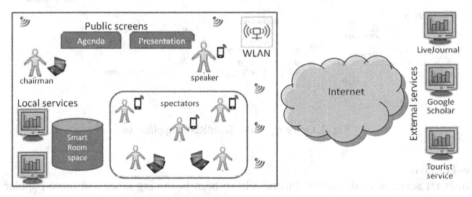

Fig. 1. Architectural view on the SmartRoom application

Properties *"currentPresentationSlideCount"*, *"currentSlideImg"*, *"currentSlide-Num"* of the class *"PresentationService"* characterize the current presentation. Properties *"PresentationTags"*, *"PresentationURL"*, and *"PresentationTitle"* describe information about a presentation. Properties *"SectionDate"*, *"SectionStartTime"*, *"SectionTitle"* of the class *"Section"* describe characteristics of the activity section. User profile keeps username and password for participant identification and authentication. Context represents mutable characteristics of participant (activity level, presence in the room, etc.) while person information describes permanent or long-term user properties. Person information is based on the FOAF ontology, which provides a widespread vocabulary for describing people and relations between them. Note that the use of well-known standardized ontologies aims at simpler integration with other systems. Each profile forms a personal subspace in the smart space. Similarly each service defines its own service subspace. Although such subspaces distinguish information in the global smart space and make it fragmented, subspaces are linked to support interactions. For instance, information on the current presentation in the Presentation service space is borrowed from the personal space of the speaker. Files with presentations are not duplicated in the smart space; it keeps only metadata and URLs to resources located in the external file sharing services.

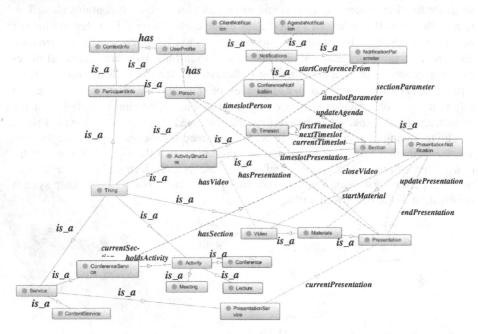

Fig. 2. Ontology of the SmartRoom application

e-Tourism

Internet services in the tourist business have been becoming more and more popular in recent years. Tourists use such services to book hotels, buy flights, and search for attractions, instead of traditional booking of complete tours from a travel agency. Augmentation of such services with personalized and related situational information is a base for new non-trivial scenarios for use by tourists online and in mobile mode.

Developed mobile tourist guide application is a recommendation system based on Smart-M3 platform. Mobile clients are developed for Android-based devices. The system determines the current tourist location and provides context-aware recommendations about attractions around (e.g., museums, monuments, parks, squares) and their textual and photo description. The user browses attractions and makes decisions on attendance. The personal preferences and the current situation in the region are taken into account. The application exploits external Internet services for the source of actual information about attractions based on tourists' ratings.

The application consists of the following services (see Fig. 3):

- client application installed to the user mobile device that shares tourist context with the smart space and provides the tourist results of application operation;

- attraction information service that implements retrieving and caching the information about attractions;
- recommendation service that evaluates attraction/image/ description scores based on ratings that have been saved to internal database earlier (detailed description of developed recommendation method is given in [17]);
- region context service that acquires and provides information about current situation in the considered region (e.g., weather, traffic jams, closed attractions);
- ridesharing service that finds matching of the tourist movement with accessible drivers; it provides the tourist possibilities of comfort transportation to the preferred attraction (detailed description of developed algorithm is presented in [18]);
- public transport service that finds information about public transport to reach the preferred attraction.

The ontology of the mobile tourist guide is presented in Fig. 4. The ontology defines the main concepts and relationships for the mobile tourist guide services interaction. It contains classification of attractions (subclasses of the class "Attraction"), different kinds of attraction description (subclasses of class "AttractionDescription"), classification of different transportation means (subclasses of the class "Transportation"), user description (class "User") that can be a tourist (class "Tourist"). Location description (class "Location"), and route description (class "Route") that consists of points (class "Point"), which is related to the class "Location". Other relationships between presented classes are depicted in Fig. 4.

The following properties have been determined for the ontology concepts. Property *"user_id"* for the class *"User"* specifies identifier for the user in the intelligent mobile tourist guide.

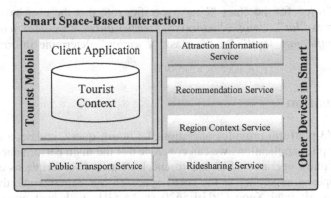

Fig. 3. General architecture of intelligent mobile tourist guide

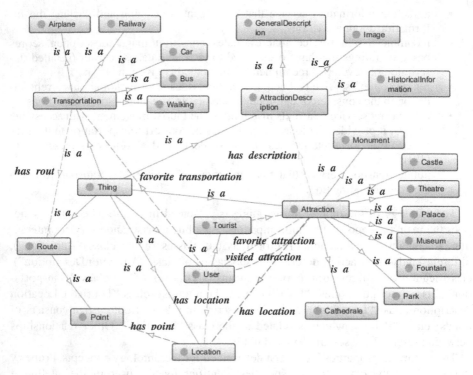

Fig. 4. Ontology for intelligent mobile tourist guide services interaction

Property *"role"* for the class *"User"* specifies role that the user have at the moment. Properties *"lat"*, *"long"*, *"address"*, *"temperature"*, *"weather_icon"*, *"weather"*, and *"traffic_jam_level"* for the class *"Location"* determine information about location.

4   Evaluation

The Smart-M3 platform provides operations for insert, query, update, and subscribe. Based on our pilot applications we can conclude that the main performance bottleneck is the subscription operation: SIB has to check all active subscriptions every time a new RDF triple is appearing in the smart space. In our experiments, the following computer is used: Intel Xeon CPU E5620 @ 2.4 GHz with host operation system Windows Server 2008, virtual operation system is Debian 7.6 64 bit, hypervisor: Hyper-V, allocated RAM: 1,4 Gb, allocated CPU cores: 1, network: Ethernet: 1000 Mbit/s. The maximum number of subscriptions is on the order of ten thousands. In the e-Tourism case, every tourist is described by three subscriptions, so the application can support more than 3000 simultaneously connected tourists. In the developed pilot applications, the number of subscribe transactions depends on the number of online clients. In the SmartRoom application, the number of subscription depends on the number of participants (who are becoming SmartRoom users).

The Smart-M3 platform has been estimated by dependency of the response time on the number of triples in a smart space. The dependency of query transaction execution time based on the increasing of RDF triples in the smart space is shown in Fig. 5. For the experiments, a test e-Tourism application has been developed that generates triples that describe tourists in a smart space (one tourist is described by approximately 30 triples) and calculates query & insert transaction execution time. Insert transaction execution time does not depend on the number of triples in the smart space while a query transaction has a linear dependency (see Fig. 5). The solid line shows the measured average query execution time in dependence on the number of tourists. Every value is measured by 100 times, after that values corresponded to the normal distribution with 10% border are considered. For this values average, minimum and maximum values are calculated. Minimal and maximum measured deviation is used for the vertical error bar. The dashed line in Fig. 5 shows linear regression for these data. Calculated coefficient R2 is 0.95, showing a good correspondence of the calculated linear regression and measured data.

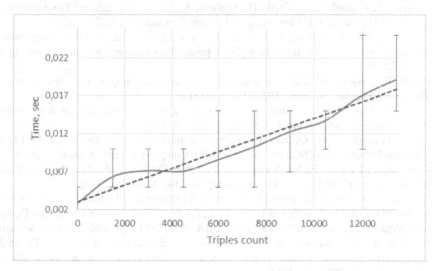

Fig. 5. Response time depends on the number of tourists in a Smart Space

5   Conclusion

The paper discussed our experience of smart spaces application development based on the open source Smart-M3 information sharing platform. We present our approach and demonstrate its applicability and feasibility based on two selected use cases that cover such emerging IoT application areas as collaborative work and e-Tourism. The first case is for mobile users collaborating in a spatially-localized digitally-equipped environment. The second case aims at provision of mobile users with touristic information from multiple Internet services. The presented architectural and ontological models can be used as reference solutions for many service-oriented systems in various IoT application domains. Our study is also supported with experimental evaluation of the pilot implementations. We can conclude that the Smart-M3 platform is applicable for application development even for the case of IoT settings.

Acknowledgements. The presented results are part of the research carried out within the project funded by grants # 14-07-00252, 13-07-00336, 13-07-12095 of the Russian Foundation for Basic Research and by Program P40 "Actual Problems of Robotics" of the Presidium of the Russian Academy of Sciences. This work is partially financially supported by the Ministry of Education and Science of Russia within project # 648-14 of the basic part of state research assignment for 2014-2016 and by the Government of Russian Federation within Grant 074-U01.

References

1. Augusto, J., Callaghan, V., Cook, D., Kameas, A., Satoh, I.: Intelligent Environments: a manifesto. Human-centric Computing and Information Sciences **3**(12) (2013)
2. Palviainen, M., Kuusijärvi, J., Ovaska, E.: A semi-automatic end-user programming approach for smart space application development. Pervasive and Mobile Computing **12**, 17–36 (2014)
3. Patela, P., Cassou, D.: Enabling high-level application development for the Internet of Things. Systems and Software **103**, 62–84 (2015)
4. Cook, D.J., Das, S.K.: How Smart are our Environments? An Updated Look at the State of the Art, Pervasive and Mobile Computing **3**(2), 53–73 (2007)
5. Oliver, I., Boldyrev, S.: Operations on spaces of information. In: Proc. IEEE Int'l Conf. Semantic Computing (ICSC), pp. 267–274. IEEE Computer Society, September 2009
6. Kortuem, G., Kawsar, F., Sundramoorthy, V., Fitton, D.: Smart objects as building blocks for the internet of things. IEEE Internet Computing **14**(1), 44–51 (2010)
7. Balandin, S., Oliver, I., Boldyrev, S., Smirnov, A., Kashevnik, A., Shilov, N.: Anonymous agents coordination in smart spaces. In: Proc. 4th Int'l Conf. Mobile Ubiquitous Computing, Systems, Services and Technologies (UBICOMM 2010), pp. 242–246, October 2010
8. Kiljander, J., Ylisaukko-oja, A., Takalo-Mattila, J., Eteläperä, M., Soininen, J.-P.: Enabling semantic technology empowered smart spaces. Journal of Computer Networks and Communications **2012**, 14 (2012)
9. Gershenfeld, N., Krikorian, R., Cohen, D.: The Internet of Things. Scientific American **291**(4), 76–81 (2004)
10. Atzori, L., Iera, A., Morabito, G.: The Internet of Things: A Survey. Computer Networks **54**(15), 2787–2805 (2010)
11. Kiljander, J., D'Elia, A., Morandi, F., Hyttinen, P., Takalo-Mattila, J., Ylisaukko-Oja, A., Soininen, A., Cinotti, T.: Semantic Interoperability Architecture for Pervasive Computing and Internet of Things. IEEE Access **2**, 856–873 (2014)
12. Whitmore, A., Agarwal, A., Xu, L.: The Internet of Things—A survey of topics and trends. Information Systems Frontiers **17**(2), 261–274 (2015)
13. Smirnov, A., Kashevnik, A., Shilov, N., Teslya, N., Shabaev, A.: Mobile application for guiding tourist activities: tourist assistant – TAIS. In: proc. of the 16th Conference of Open Innovations Association FRUCT, pp. 94–100. IEEE, Oulu (2014)

14. Varfolomeyev, A., Korzun, D., Ivanovs, A., Petrina, O.: Smart personal assistant for historical tourism. In: Proc. 2nd Int'l Conf. on Environment, Energy, Ecosystems and Development, Athens, Greece, pp. 9–15 (2014)
15. Smirnov, A., Shilov, N., Kashevnik, A., Teslya, N.: Smart logistic service for dynamic ridesharing. In: Andreev, S., Balandin, S., Koucheryavy, Y. (eds.) NEW2AN/ruSMART 2012. LNCS, vol. 7469, pp. 140–151. Springer, Heidelberg (2012)
16. Korzun, D.G., Galov, I.V., Kashevnik, A.M., Shilov, N.G., Krinkin, K., Korolev, Y.: Integration of smart-M3 applications: blogging in smart conference. In: Balandin, S., Koucheryavy, Y., Hu, H. (eds.) NEW2AN 2011 and ruSMART 2011. LNCS, vol. 6869, pp. 51–62. Springer, Heidelberg (2011)
17. Smirnov, A., Kashevnik, A., Ponomarev, A., Teslya, N., Shchekotov, M., Balandin, S.I.: Smart space-based tourist recommendation system. In: Balandin, S., Andreev, S., Koucheryavy, Y. (eds.) NEW2AN/ruSMART 2014. LNCS, vol. 8638, pp. 40–51. Springer, Heidelberg (2014)
18. Smirnov, A., Shilov, N., Kashevnik, A., Teslya, N., Laizane, S.: Smart space-based ridesharing service in e-tourism application for Karelia region accessibility. Ontology-based Approach and Implementation. In: proc. 8th Int. Joint Conference on Software Technologies, Reykjavik, Iceland, pp. 591–598, July 29–31, 2013
19. Weiser, M.: The Computer for the Twenty-First Century. Scientific American **265**(3), 94–104 (1991)
20. Nixon, L.J.B., Simperl, E., Krummenacher, R., Martin-recuerda, F.: Tuplespace-Based Computing for the Semantic Web: A Survey of the State-of-the-Art. Knowl. Eng. Rev. **23**, 181–212 (2008)
21. Horrocks, I.: Ontologies and the Semantic Web. Commun. ACM **51**(12), 58–67 (2008)
22. Gutierrez, C., Hurtado, C.A., Mendelzon, A.O., Pérez, J.: Foundations of Semantic Web Databases. J. Comput. Syst. Sci. **77**(3), 520–541 (2011)
23. Martín-Recuerda, F.: Towards cspaces: a new perspective for the semantic web. In: Bramer, M., Terziyan, V. (eds.) IASW. IFIP, vol. 188, pp. 113–139. Springer, Heidelberg (2005)
24. Korzun, D., Lomov, A., Vanag, P., Honkola, J., Balandin, S.: Generating modest high-level ontology libraries for Smart-M3. In: Proc. 4th Int'l Conf. Mobile Ubiquitous Computing, Systems, Services and Technologies (UBICOMM 2010), pp. 103–109, October 2010
25. Korzun, D., Balandin, S.: A peer-to-peer model for virtualization and knowledge sharing in smart spaces. In: Proc. 8th Int'l Conf. on Mobile Ubiquitous Computing, Systems, Services and Technologies (UBICOMM 2014), pp. 87–92, August 2014
26. Honkola, J., Laine, H., Brown, R., Tyrkkö, O.: Smart-M3 information sharing platform. In: Proc. IEEE Symp. Computers and Communications (ISCC 2010), pp. 1041–1046. IEEE Computer Society, June 2010
27. Korzun, D.G., Balandin, S.I., Gurtov, A.V.: Deployment of smart spaces in internet of things: overview of the design challenges. In: Balandin, S., Andreev, S., Koucheryavy, Y. (eds.) NEW2AN 2013 and ruSMART 2013. LNCS, vol. 8121, pp. 48–59. Springer, Heidelberg (2013)

Multi-level Robots Self-organization
in Smart Space: Approach and Case Study

Alexander V. Smirnov[1,2], Alexey M. Kashevnik[1,2(✉)], Sergey Mikhailov[2],
Mikhail Mironov[2], and Olesya Baraniuc[2]

[1] SPIIRAS, St. Petersburg, Russia
{smir,alexey}@iias.spb.su
[2] ITMO University, St. Petersburg, Russia
{mikhaylovsergeyandreevich,mironoff.togo}@gmail.com,
ob@itc.vuztc.ru

Abstract. This paper presents an approach and case study for multi-level robots self-organization in smart space. The presented approach benefits from integration of such technologies as multi-level self-organization and knowledge fusion. A Smart-M3 information sharing platform is used for creating smart spaces with ontology-driven information sharing based on Semantic Web. To provide semantic interoperability, the RDF ontologies for the robots participating in the scenario, have been built and ontology matching technique is used. The scenario implementation is based on Lego® Mindstorms EV3 set for robot construction, which at the moment is one of the most popular sets for education.

Keywords: Robots · Self-organization · Smart space

1   Introduction

Smart space is a new research and development technology that becomes more and popular last years. It is a computational environment consisting of multiple heterogeneous participants (electronic and computational devices, Internet pages, data based, etc.) which has intelligent behavior and can proactively provide services taking into account current situation. This technology [1, 2] aims in the seamless integration of different devices by developing ubiquitous computing environments, where different services can share information with each other, make different computations and interact for joint tasks solving. Thereby, applications based on smart spaces has to integrated self-organization mechanisms to allow their participants interact with each other and jointly solve tasks. Self-organizing systems are characterized by their capacity to spontaneously (without external control) produce a new organization in case of environmental changes. These systems are particularly robust, since they adapt to changes, and are able to ensure their own survivability [3]. Self-organization of robots in smart space requires of their operation in physical part while their interaction has to be organized in cyber part. The papers [4, 5] considers approaches for organizing robots behavior for implementing joint tasks.

© Springer International Publishing Switzerland 2015
S. Balandin et al. (Eds.): NEW2AN/ruSMART 2015, LNCS 9247, pp. 68–79, 2015.
DOI: 10.1007/978-3-319-23126-6_7

The open source Smart-M3 platform has been used for organization of smart space infrastructure for robots self-organization. The use of this platform enables to significantly simplify further development of the system, include new information sources and services, and to make the system highly scalable. The key idea of this platform is that the formed smart space is device-, domain-, and vendor-independent. Smart-M3 assumes that devices and software entities can publish their embedded information for other devices and software entities through simple, shared information brokers. The Smart-M3 platform consists of two main parts: information agents and kernel [6]. The kernel consists of two elements: Semantic Information Broker (SIB) and information storage. Information agents are software entities, installed on mobile devices of the smart space users and other devices, which host smart space services. These agents interact with SIB through the Smart Space Access Protocol (SSAP). The SIB is the access point for receiving the information to be stored, or retrieving the stored information. All this information is stored in the information storage as a graph that conforms with the rules of the Resource Description Framework (RDF) [7, 8]. In accordance with these rules all information is described by triples "Subject - Predicate - Object".

Case study, developed in this paper, solves the task of pick-and-place an object from pipeline to warehouse. There are three kinds of robots and two types of smart space services are participated in the case study. The waterproof slower manipulating robot and non-waterproof quick one can solve the task of pick the object from pipeline robot move it and place at warehouse. The pipeline robot move object from production to the place where it can be picked by a manipulating robot. The pipeline robot scans object characteristics publish them to the smart space. The first smart space service is a user interface that can change / generate a policy for taking it into account by manipulating and pipeline robots. Another service is a weather service that change / generate the policy based of information about current weather (e.g., in case of rain the object has to be moved by the waterproof manipulating robot).

In the proposed case study it can be highlighted two levels of self-organization: level of robots (they have to decide who will move the object) and level of operation planning services (they have to generated feasible policy for robots). Presented case study enhances the research work [9] by the implementation of multi-level self-organization approach for generating behavioral policies. This policies are taking into account by robots in physical level during the self-organization process.

The rest of the paper is structured as follows. Developed multi-level self-organization approach is presented in Section 3. Section 3 describes developed reference model robots interaction in smart space. Section 4 presents robots self-organization case study implementation The results are summarized in Conclusion.

2   Multi-level Self-organization Approach

For operating efficiently smart space-based systems have to be provided with self-organization mechanisms for their participants. The analysis of literature related to organizational behavior & team management has showed that the most efficient teams are self-organizing teams working in the organizational context. For example, social self-organization has been researched by Hofkirchner [10], Fuchs [11], etc. However, in this case there is a significant risk for the group to choose a wrong strategy preventing from achieving desired goals. For this purpose, self-organizing groups / systems need to have a certain guiding control from an upper level. This consideration produces the idea of multi-level self-organization.

The process of self-organization of a network assumes creating and maintaining a logical network structure on top of a dynamically changing physical network topology. This logical network structure is used as a scalable infrastructure by various functional entities like address management, routing, service registry, media delivery, etc. The autonomous and dynamic structuring of components, context information and resources is the essential work of self-organization [12]. The network is self-organized in the sense that it autonomically monitors available context in the network, provides the required context and any other necessary network service support to the requested services, and self-adapts when context changes.

To guide self-organizing groups / systems, the guiding control via policy transfer from an upper level is used in the proposed approach. This control enables a more efficient self-organization based on the "top-to-bottom" configuration principle, which assumes conceptual configuration followed by parametric configuration. In this regard, each level can be considered as a scenario-based decision arena following certain complex knowledge patterns related to adaptable business models.

The key mechanisms supporting self-organizing networks are self-organization mechanisms and negotiation models. The following self-organization mechanisms are usually selected [13]: intelligent relaying, adaptive cell sizes, situational awareness, dynamic pricing, intelligent handover.

The following negotiation models can be mentioned [14]:

- Different forms of spontaneous *self-aggregation*, to enable both multiple distributed services / agents to collectively and adaptively provide a distributed service, e.g. a holonic (self-similar) aggregation.
- *Self-management* as a way to enforce control in the ecology of services / agents if needed (e.g. assignment of "manager rights" to an service / agent.
- *Situation awareness* – organization of situational information and their access by services / agent, promoting more informed adaptation choices by them and advanced forms of stigmergic (indirect) interactions.

To guide such self-organizing groups / systems a certain guiding control is needed (e.g. via policy transfer) from an upper level. The multi-level self-organization has not been addressed yet in research. This approach would enable a more efficient self-organization based on the "top-to-bottom" configuration principle, which assumes conceptual configuration followed by parametric configuration. In this regard, each level can be considered as a scenario-based decision arena following certain complex knowledge patterns related to adaptable business models.

The approach is based on the following principles: self-management and responsibility, decentralization, as well as integration of chain policy transfer (a formal chain of policies running from top to bottom) with network organization (without any social hierarchy of command and control within a level), initiative from an upper level and co-operation within one level. The idea can be interpreted as producing "guided order from noise". In accordance with [15] such system falls into the class of purposeful systems.

Intra-level self-organization is considered as a threefold process of (i) cognition (where subjective context-dependent knowledge is produced), (ii) communication (where system-specific objectification or subjectification of knowledge takes place), and (iii) synergetical co-operation (where objectified, emergent knowledge is produced). The Individually acquired context-dependent (subjective) knowledge is put to use efficiently by entering a social co-ordination and co-operation process. The objective knowledge is stored in structures and enables time-space distanciation of social relationships.

The resources that are parts of a system permanently change their joint environment what results in a synergetic collaboration and leads to achieving a certain level of collective intelligence. This is also supported by the fact that individual resource behavior is partially determined by the social environment the resources are contributing to (called "norms"). For this purpose a protocol has been developed based on the BarterCast approach that originates from the following ideas: (i) each service builds a network representing all interactions it knows about; (ii) the reputation of a service depends on the reputation of other services in the path between this service and the service connecting to it.

The overall scheme of the approach is shown in Fig. 1. Since the structures and self-organization models of all the levels are identical, the developed framework is fully scalable. This makes it possible to perform conceptual development of the smart space participants, i.e. to define kinds of participants needed, their characteristics, etc. Then, at the implementation stage, the particular behavior and functionality of the participants may vary in different application domains.

The interoperability between the agents at the technological level is provided via usage of common standards and protocols, the interoperability at the level of semantics is ensured via usage of ontologies.

3 Robots Interaction in Smart Space

Fig. 2 presents the reference model of robots interaction in smart space. The system is divided into two main paths: physical space and smart space. In physical part robots implement actions while in cyber part they interact with each other. The three kinds of robots and two types of smart space services are participated in the case study. The waterproof slower manipulating robot and non-waterproof quick one can solve the task of pick the object from pipeline robot move it and place at warehouse. The pipeline robot move object from production to the place where it can be picked by a manipulating robot. The pipeline robot scans object characteristics publish them to the smart space.

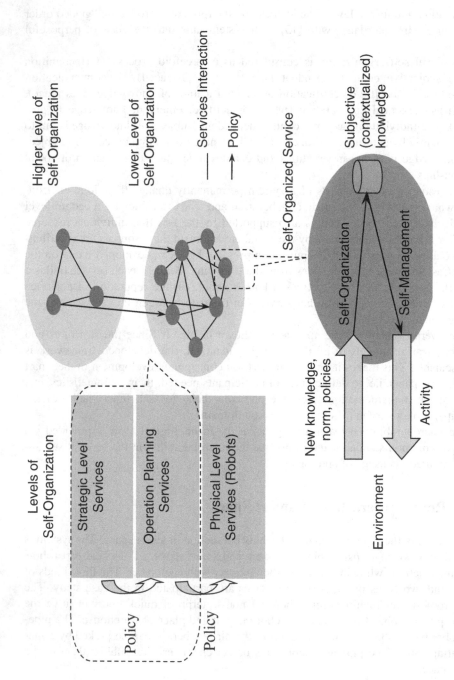

Fig. 1. Approach overview

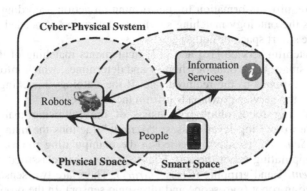

Fig. 2. Reference Model of Robots Interaction in Smart Space

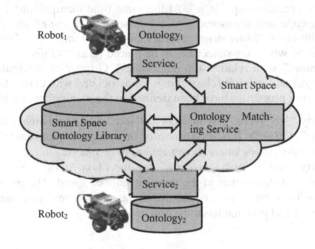

Fig. 3. Robots Interaction in Smart Space Based on Ontology Matching

The first smart space service is a user interface that can change / generate a policy for taking it into account by manipulating and pipeline robots. Another service is a weather service that change / generate the policy based of information about current weather (e.g., in case of rain the object has to be moved by the waterproof manipulating robot). To provide semantic interoperability between robots their interaction in smart space is based on ontologies. Each robot uploads its ontology to the smart space ontology library when it connects to the system. The ontology represents the robot. It contains information about robot requirements and possibilities. Robot requirements represent the information, which the robot needs for starting its scenario. Robot possibilities is the information that robots can provide in scope of the considered system. Proposed robots interaction scheme in the smart space is presented in Fig. 3. Robot$_1$ and robot$_2$ connect to the system and upload their ontologies to smart space ontology library. For this purpose special services are used that implement robots logic and interaction in the smart space. When a service has information that can be helpful for other services in the smart space, it uploads this information according to previously uploaded own ontology.

If a service requires information for performing an action according to the system scenario, it uses the ontology matching service to determine, if needed information is accessible in the smart space or not.

Ontology matching service [16] and [17] implements matching of the service ontology with the smart space ontology library and determines, which information in the smart space corresponds to the required one. If the ontology matching service finds this information, the service downloads it from the smart space.

The manipulating robot ontology consists of the following main classes (see Fig. 4). There are three top level classes: *"Actions"* (actions the manipulation robot implements), *"Sensors"* (sensors installed in the manipulating robot), and *"Object"* (object the manipulating robot moves). Class *"Actions"* is classified as movements (class *"Movement"*) and gripping (Class *"Grip"*). There are two sensors is installed to the manipulating robot (gyroscope and ultrasonic sensor). In the proposed ontology the corresponding classes (*"Gyroscope"* and *"Ultrasonic"*) is associated with class *"Sensors"* with relationship *"is_a"*. At the same time manipulating robot movement requires gyroscope and ultrasonic sensor (classes *"Gyroscope"* and *"Ultrasonic"* is associated with class *"Movement"* with relationship *"requires"*). Type of object determines position where the object has to be placed (class *"Object"* is associated with class *"Movement"* with relationship determines). Gripper of manipulating robot has to take the object and move it (class *"Grip"* is associated with class *"Object"*).

The following properties have been defined in the manipulating robot ontology:

- Property *"has_color"* for the class *"Object"*. List of possible values are (*"red"*, *"green"*, *"blue"*). Based on value of this property the manipulating robot determines the places where the object has to be placed.
- Property *"has_gripping_velocity"* for the class *"Grip"*. Value is a positive number that determines at the moment gripper speed. The property is used by the pipeline robot to understand that at the moment manipulating robot is taking the object and pipeline has to be remain stopped.

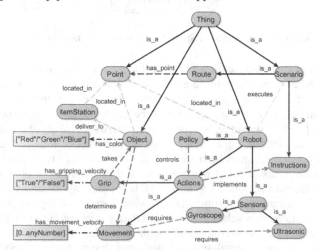

Fig. 4. Manipulating Robot Ontology

- Property "has_movement_velocity" for the class "Movement". Value is a positive number that determines at the moment manipulating robot movement speed. The property can be used by other system services to determine the time when the object can be accessible in destination position.

The pipeline robot ontology consists of the following main classes (see Fig. 5). There are three top level classes: *"Movement"* (different types of movement that pipeline robot can implement), *"Sensors"* (sensors installed in the manipulating robot), and *"Object"* (object the pipeline robot moves and determine color). Class *"Pipeline"* has taxonomical relationship *("is_a")* with the class *"Movement"* as this type of movement is supported by the pipeline robot. Class *"ColorSensor"* has taxonomical relationship *("is_a")* with the class *"Sensor"* as this sensor type is installed for the pipeline robot. Sensors are installed in the pipeline (class *"Sensors"* has *"installed_in"* relationship with the class *"Pipeline"*). Object is placed in pipeline that moves it (class *"Object"* has *"placed_in"* relationship with the class *"Pipeline"*). For determining color of the object when it moves in pipeline the color sensor is used (class *"ColorSensor"* has relationship *"determine_color"* with class *"Object"*).

The following properties have been defined in the pipeline robot ontology.

- Property *"has_color"* for the class *"Object"*. List of possible values are (*"red"*, *"green"*, *"blue"*). Based on the color sensor output the property is take on the appropriate value.
- Property *"has_velocity"* for the class *"Pipeline"*. Value is a positive number that determines at the moment the pipeline speed. The property can be used for smart space services for estimation of time when the object will be ready for manipulation.
- Property *"is_ready_for_manipulation"* for the class object. Value of this property is a Boolean variable that is *"true"* if pipeline robot is completing move the object and *"false"* if the object is moving.

4   Robots Self-organization Case Study

Two types of manipulating robots (Fig. 6) are participated in the scenario: waterproof manipulating robot with a roof and manipulating robot without a roof.

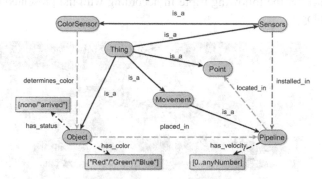

Fig. 5. Pipeline Robot Ontology

Fig. 6. Manipulating robots

The first one has protection of objects against precipitation, but has a low speed. The second robot on the contrary has a high speed, but has no object protection. Both of them subscribes to the information in smart space about current policy of manipulation and shares their IP-address in smart space:

("Robot_[robot number]", "has_ip", [robot ip]).

This policy is generating by two services: user service and weather service. The first one is Windows application that starts by User. After connection to smart space app shares triple about it:

("User", "is_here", 1).

After that the second service update current weather information using robots *ip address* and checks precipitation. Based on the parameters obtained from the user service (User choice) and current weather information services generating action policy for robots self-organization and share it in smart space.

The pipeline robot (Fig. 7) is stationary and has a pipeline that moves objects from the location to the destination. It has a color sensor that determines the color of the moved object. When the robot is moving the object, the pipeline velocity is shared with smart space by the following triple in according with the presented bellow pipeline robot ontology.

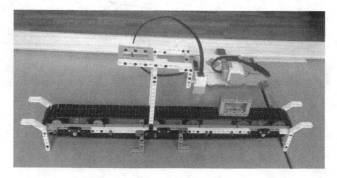

Fig. 7. The pipeline robot

("Pipeline", "has_velocity", [pipeline velocity])

When the color is determined, it is shared with smart space by the following way.

("Object", "has_color", [object color])

When the object has been moved to the destination point and is ready for manipulation by the manipulating robot, the related triple is shared with smart space by pipeline robot.

("Object", "is_ready_for_manipulation", 1)

("Pipeline", "has_velocity", 0)

The second robot has possibility to load an object from pipeline at the location, move to destination, unload the object, and return to the initial location. The manipulating robot subscribes to the information in smart space if an object is ready for manipulation.

("Object", "is_ready_for_manipulation", None)

When the pipeline robot moves the object to the destination point and shares with the smart space appropriate information the manipulating robot gets notification and moves to the object location. It shares with the smart space the movement velocity.

("Movement", "has_movement_velocity", velocity)

Then it takes the object using the on-board gripper. When the gripper is taking the object the robot shares the gripping velocity with the smart space.

("Grip", "has_gripping_velocity", velocity)

The manipulating robot queries information of object color from the smart space to determine the place, where the object has to be unloaded.

("Object", "has_color", None)

Then the robot moves the object to the place according to the object color obtained from smart space (Fig. 8). The overall sequence diagram of a manipulating robot and service interaction is presented in Fig. 9.

Fig. 8. Pick-and-Place System Scenario

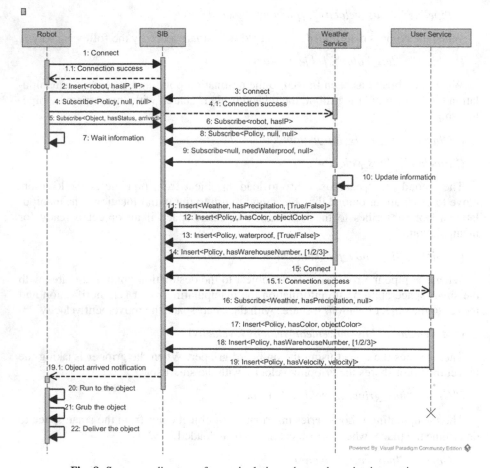

Fig. 9. Sequence diagram of a manipulating robot and service interaction

5   Conclusion

The paper presents an approach and case study implementation for multi-level robots self-organization in smart space. Developed approach allows to self-organize smart space participants in different levels, three levels have been proposed (physical level for the robots, operation planning level for the operational service and strategic level for the services which form the strategy of the case study) and two levels have been implemented. Implementation is based on Smart-M3 information sharing platform that provides possibilities of information sharing based on Semantic Web.

Acknowledgements. The presented results are part of the research carried out within the project funded by grants # 13-07-00336, 13-07-12095, 13-01-00286 of the Russian Foundation for Basic Research, program P40 "Actual Problems of Robotics" of the Presidium of the Russian Academy of Sciences. The work has been done at Technopark of ITMO University and partially financially supported by Government of Russian Federation, Grant 074-U01.

References

1. Cook, D.J., Das, S.K.: How smart are our environments? an updated look at the state of the art. Pervasive and Mobile Computing **3**(2), 53–73 (2007)
2. Balandin, S., Waris, H.: Key properties in the development of smart spaces. In: Stephanidis, C. (ed.) UAHCI 2009, Part II. LNCS, vol. 5615, pp. 3–12. Springer, Heidelberg (2009)
3. Serugendo, G., Gleizes, M., Karageorgos, A.: Self-Organisation and Emergence in MAS: An Overview. Informatica **30**, 45–54 (2006)
4. Baca, J., Pagala, P., Rossi, C., Ferre, M.: Modular robot systems towards the execution of cooperative tasks in large facilities. Robotics and Autonomous Systems **66**, 159–174 (2015)
5. Rodić, A., Jovanović, M., Stevanović, I., Karan, B., Potkonjak, V.: Building Technology Platform Aimed to Develop Service Robot with Embedded Personality and Enhanced Communication with Social Environment, Digital Communications and Networks (2015). doi:10.1016/j.dcan.2015.03.002
6. Honkola, J., Laine, H., Brown, R., Tyrkko, O.: Smart-M3 information sharing platform. In: Proc. ISCC 2010, pp. 1041–1046. IEEE Comp. Soc., June 2010
7. Berners-Lee, T., Fielding, R., Masinter, L.: RFC 3986 – Uniform Resource Identifier (URI): Generic Syntax. URL: http://tools.ietf.org/html/rfc3986
8. Resource Description Framework (RDF). W3C standard, Web: http://www.w3.org/RDF/
9. Smirnov, A., Kashevnik, A., Teslya, N., Mikhailov, S., Shabaev, A.: Smart-M3-based robots self-organization in pick-and-place system. In: Proceedings of the 17th Conference of the Open Innovations Association FRUCT, Yaroslavl, Russia, pp. 210–215, April 20–24, 2015
10. Hofkirchner, W.: Emergence and the Logic of Explanation, An Argument for the Unity of Science. Acta Polytechnica Scandinavica, Mathematics, Computing and Management in Engineering Series **91**, 23–30 (1998)
11. Fuchs, C.: Globalization and Self-Organization in the Knowledge-Based Society. TripleC **1**(2), 105–169 (2003). (http://triplec.uti.at)
12. Ambient Networks Phase 2, Integrated Design for Context, Network and Policy Management, Deliverable D10.-D1 (2006). http://www.ambient-networks.org/Files/deliverables/D10-D.1_PU.pdf
13. Telenor R&D, Report, Project No TFPFAN, Program Peer-to-peer computing (2003). http://www.telenor.com/rd/pub/rep03/R_17_2003.pdf
14. De Mola, F., Quitadamo, R.: Towards an agent model for future autonomic communications. In: Proceedings of the 7th WOA 2006 Workshop From Objects to Agents, Catania, Italy, September 26–27, 2006. http://sunsite.informatik.rwth-aachen.de/Publications/CEUR-WS/Vol-204/P07.pdf
15. Jantsch, E.: Design for Evolution. George Braziller, New York (1975)
16. Smirnov, A., Kashevnik, A., Shilov, N., Balandin, S., Oliver, I., Boldyrev, S.: On-the-fly ontology matching in smart spaces: a multi-model approach. In: Balandin, S., Dunaytsev, R., Koucheryavy, Y. (eds.) ruSMART 2010. LNCS, vol. 6294, pp. 72–83. Springer, Heidelberg (2010)
17. Balandin, S., Boldyrev, S., Oliver, I.J., Turenko, T., Smirnov, A.V., Shilov, N.G., Kashevnik, A.M.: Method and apparatus for ontology matching. US 2012/0078595 A1 (2012)

High Capacity Trucks Serving as Mobile Depots for Waste Collection in IoT-Enabled Smart Cities

Theodoros Anagnostopoulos[1,3](✉), Arkady Zaslavsky[1,2],
Stefanos Georgiou[1], and Sergey Khoruzhnikov[1]

[1] Department of Infocommunication Technologies, ITMO University,
49 Kronverksky Pr., St. Petersburg 197101, Russia
stefanos1316@gmail.com, xse@vuztc.ru
[2] CSIRO Computational Informatics, CSIRO, Box 312, Clayton South, VIC 3169, Australia
Arkady.Zaslavsky@csiro.au
[3] Community Imaging Group, University of Oulu, 90570 Oulu, Finland
tanagnos@ee.oulu.fi

Abstract. Internet of Things (IoT) enables Smart Cities with novel services. Waste collection in Smart Cities becomes a dynamic process with the proliferation of sensors and actuators embedded on real waste bins. Heterogeneous fleets of trucks are used for efficient waste collection exploiting the diverse road network. In this paper we propose a novel approach by incorporating Low Capacity Trucks (LCTs) and High Capacity Trucks (HCTs). However, HCTs are serving as Mobile Depots (MDs) which are cost efficient and decongest traffic in Smart Cities. A detailed system overview illustrates the architecture of the proposed approach. We also propose novel algorithms which support dynamic waste collection with MDs. Scheduling and routing are transformed to dynamic models. Specifically, we propose a novel scheduling algorithm while we customize an existing routing algorithm. The models where experimentally evaluated with real and synthetic data from the city of St. Petersburg, Russia. The results were promising and proved that the incorporation of MDs is efficient for waste collection in IoT-enabled Smart Cities. Finally, we perform an economic analysis in order to define the economic impact of the proposed solution to the municipality budget for an ownership cost for a period of five years; in which the proposed solution proved to be cost efficient.

Keywords: Smart cities · Internet of things · Waste collection · Dynamic models · Mobile depots

1 Introduction

Cities which incorporate IoT are transformed to Smart Cities. One such city is St. Petersburg, Russia. Smart Cities is the future of civil habitation, since by 2050 the vast amount of earth population (i.e., 70 percent) will move to urban areas thus forming vast cities [1]. These cities will incorporate smart infrastructure in order to

© Springer International Publishing Switzerland 2015
S. Balandin et al. (Eds.): NEW2AN/ruSMART 2015, LNCS 9247, pp. 80–94, 2015.
DOI: 10.1007/978-3-319-23126-6_8

manage their needs for fundamental and advanced services [2]. The use of Future Internet with the enhancement of IPv6 (i.e., 6LoWPN) as well as sensors and wireless sensor networks enable IoT to reform Smart City municipality activity in every aspect of daily life [3], [4]. One such a service that has great impact on citizen quality of life is the efficient waste collection [5]. In past years waste collection was treated in a rather static approach, nowadays with the proliferation of sensors and actuators; IoT enable dynamic solutions as well [6].

In order to understand in depth the concept of Smart Cities, a definition must be provided. A lot of definitions have been provided in the literature. Let us consider the more representative:

- *"We define a Smart City as one that uses information and communications technology to make both its critical infrastructure and its components and public services more interactive, efficient and better known to its residents. In the broadest sense, a city can be regarded as 'smart' when its investments in human and social capital and in its communications infrastructure foment sustainable economic development and high standard of living along with the wise management of natural resources by an engaged government."* [7],
- *"A Smart City combines ICT and Web 2.0 technology with other organizational, design and planning efforts to de-materialize and speed up bureaucratic processes and help to identify new, innovative solutions to city management complexity, in order to improve sustainability and livability."* [8],
- *"A Smart City is a city that uses Smart Computing technologies to make the critical infrastructure components and services of a city – which include city administration, education, healthcare, public safety, real estate, transportation, and utilities – more intelligent, interconnected, and efficient."* [9].

In this paper it is used the definition best suits to the IoT-enabled waste collection in smart cities, which is [10]: "A Smart City is a city well performing in a forward-looking way in the following fundamental components (i.e., Smart Economy, Smart Mobility, Smart Environment, Smart People, Smart Living, and Smart Governance), built on the 'smart' combination of endowments and activities of self-decisive, independent and aware citizens". This definition contains the fundamental component of Smart Environment which is relevant to environmental pollution [11]. A municipality service which acts as a countermeasure to environmental pollution within the Smart City is the dynamic waste collection.

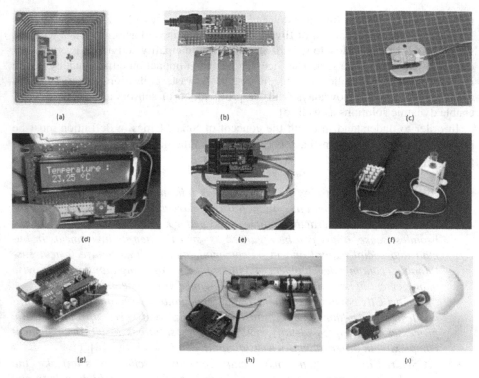

Fig. 1. IoT mechanisms: (a) RFIDs, (b) capacity sensors, (c) weight sensors, (d) temperature sensors, (e) humidity sensors, (f) chemical sensors, (g) pressure sensors, and (h),(i) actuators

We propose a novel approach for waste collection which incorporates Low Capacity Trucks (LCTs) and High Capacity Trucks (HCTs). Specifically, HCTs are serving as Mobile Depots (MDs) which are cost efficient and decongest traffic in Smart Cities. A detailed system overview illustrates the architecture of the proposed approach. IoT components such as sensors and actuators are the backbone of the proposed architecture, see Fig. 1. We also propose novel algorithms which support waste collection with MDs. Scheduling and routing are transformed to dynamic models. Specifically, we propose a novel scheduling algorithm while we customize an existing routing algorithm. The models where experimentally evaluated with real and synthetic data from the city of St. Petersburg. The results are encouraging and prove that the proposed models are applicable to real waste collection with MDs within the IoT-enabled Smart City of St. Petersburg. Finally, we perform an economic analysis in order to define the economic impact of the proposed solution to the municipality budget for an ownership cost for a period of five years. The results prove that the proposed solution is cost efficient.

The rest of the paper is structured as follows. In Section 2 it is reported the related work. Section 3 presents the IoT-enabled system overview. Section 4 describes the dynamic scheduling and routing algorithms. In Section 5 it is performed the experimental evaluation. Section 7 discusses the economic impact of the proposed solution, while Section 7 concludes the paper.

2   Related Work

We report on methods which adopt dynamic models for waste collection. Authors in [12] develop routing with a mobile measuring system on the trucks. They perform stochastic dynamic routing which makes corrections during or after the execution of the existing routes. Authors in [13] introduce a memetic algorithm to perform routing enforced with time windows and conflicts context. Model incorporates a combination of flow and set partitioning formulation to achieve multi-objective optimization. Authors in [14] consider dynamic scheduling over a set of previously defined collection trips. The main objective of the approach is to minimize the total operational and fixed truck costs. A mathematical formulation methodology is proposed in [15] developing a plan of service areas, defining routing, and designing scheduling taking into consideration possible new alternative solutions in managing the system as a whole. In [16] authors evaluate dynamic planning methods applied for waste collection of underground bins. Model reduces the amounts of carbon dioxide released in the environment from trucks by making dynamic routing more effective.

In [17] authors introduce a dynamic routing model based on fuzzy demands by assuming the demands of the customers as fuzzy variables. Model incorporates a heuristic approach based on fuzzy credibility theory. In [18], authors propose routing with time windows which analyze the logistics activity within a city. Model finds the cost optimal routes in order the trucks to empty the bins with an adaptive large neighborhood search algorithm. Authors in [19] introduce a rollon-rolloff routing, serving multiple disposal facilities, with huge amounts of waste at construction sites and shopping districts. It is applied large neighborhood search with iterative heuristics algorithms. In [20] authors incorporate discrete event simulation for waste collection from underground bins. Model applies dynamic planning to exploit information transmitted through motion sensors embedded in the underground bins.

Model in [21] is specialized in waste collection of plastic waste which is differentiated from the other municipal solid waste. It is achieved heuristic redesign of the collection routes using an eco-efficiency metric balancing the trade-off between the costs and environmental issues. A heuristic solution is proposed in [22] where authors state the waste collection as a periodic truck routing problem with intermediate waste depots. The model incorporates variable neighborhood search and dynamic programming in order to achieve optimal solution. Dynamic scheduling and routing model in [23] applies capacity sensors and wireless communication infrastructure thus to be aware of the bins state. It incorporates analytical modeling and discrete-event simulation in order to achieve real-time dynamic routing and scheduling. Authors in [24] use improved dynamic route planning. They enhance a guided variable neighborhood threshold meta heuristic adapted to the problem of waste collection.

In [25] authors propose a genetic algorithm to solve dynamic routing problem. Specifically, model assumes that the waste collection problem could be treated as a Traveling Salesman Problem (TSP). Then the genetic algorithm solves the TSP optimally. Authors in [26] propose a heuristic method for dynamic routing considering several tunable parameters. Sensors enable reverse inventory routing in more dense waste networks. Heuristics deal with uncertainty of daily and seasonal effects.

Authors in [27] propose a routing model which incorporates Ant Colony System (ACS) algorithm in order to achieve dynamic routing. They treat the location of the bins as a spatial network and apply k-means in order to cluster the bins distribution into a set of partial clusters. In [28] authors combine routing and scheduling optimization. Historical data applied to bins individually establish the daily circuits of collection points to be visited. Planning is applied to scheduling for better system management.

In [29] authors propose a novel IoT-enabled dynamic routing model for waste collection in a Smart City. The proposed model is robust in case of emergency (i.e., a road under construction, unexpected traffic congestion). Finally, authors in [30] propose a robust waste collection model exploiting cost efficiency of IoT potentiality in Smart Cities. The research extends [29] by introducing a dynamic routing algorithm which is robust and copes with cases of truck replacement due to overload or damage in the city of St. Petersburg in Russia. Related research in waste collection focuses on dynamic scheduling and routing models. However less research states the waste collection as a Smart City service. Specifically, only in [29] and [30] it is addressed the waste collection as a problem which can be solved with IoT infrastructure; incorporated in Smart Cities. In this paper we extend the research proposed in [29] and [30] by introducing a system overview for IoT-enabled waste collection in a Smart City which incorporates MDs. We also introduce a dynamic scheduling and routing algorithm which handles the heterogeneous fleet of LCTs and HCTs exploiting data collected from sensors and actuators embedded on bins. The model is evaluated on real and synthetic data proving the advantages of MDs as a part of real time waste collection in Smart Cities.

3   System Overview

The proposed system is composed of several parts see Fig. 2. IoT components such as sensors and actuators embedded in bins are aware of the bins' status. We incorporate the following IoT components: (a) RFIDs for bin tagging and identification, (b) capacity sensors for measuring the volume of waste, (c) weight sensors for measuring the weight of waste, (d) temperature sensors for measuring the temperature within the bin, (e) humidity sensors for measuring humidity within the bin, (f) chemical sensors for measuring the gas emissions produced from chemical reactions, (g) pressure sensors for measuring the pressure of waste to the bin walls, and (h) actuators which lock the bin lid when it becomes full of waste. Waste is collected from Low Capacity Trucks (LCTs) and transferred to High Capacity Trucks (HCTs) which serve as Mobile Depots (MDs). Specifically, MDs are HCTs which replace the use of common depots. Common depots have the weakness to be static and require certain space within the Smart City area. In addition common depots degrade the quality of life of the specific municipality areas within the Smart City. We use MDs because there are cost efficient since less trucks are required for transferring waste from local areas of the city to dumps outside the city. In European cities the local roads and backyards are small and cause heavy traffic during waste collection. The use of MDs as an economically efficient solution within the cities is not convenient due to their large size which would result to even worse traffic congestion. However, by incorporating MDs in highways out of the city and LCTs in local roads in the city it is achieved traffic de-congestion since fewer trucks are used within the Smart Cities. MDs transport waste to dumps and recycling/processing plants outside the city. Truck drivers are equipped with smart phones taking navigation instructions for dynamic routing.

Fig. 2. Reference Model System Overview: (a) Smart Bins, (b) Dynamic Scheduling and Routing, (c) Municipality Services, (d) Smart Phone Navigation, (e) Low Capacity Trucks, (f) High Capacity Mobile Depot Trucks, (g) Dumps, and (h) Recycling and Processing

Waste collection is considered as a Cloud Service where data produced from IoT components (i.e., sensors and actuators) are aggregated and integrated into a Decision Support System (DSS) which is managed from the municipality of St. Petersburg. Dynamic scheduling and routing is separated according to its use in two categories: (a) waste bins' dynamic scheduling and routing, and (b) trucks' dynamic scheduling and routing. In [29] and [30] we focused in the first category while in this paper we focus in the second category. Specifically, waste bins' dynamic scheduling and routing address the problem of collecting waste from bins and empty it to trucks. However, in trucks' dynamic scheduling and routing it is addressed the problem of collecting waste from LCTs and empty it to MDs. Note that LCTs are moving in the narrow roads and backyards within the city while MDs are moving through highways connecting the city with the dumps and the recycling/processing plants. There are certain collection crossroad match points within the city that LCTs are transferring waste to MDs but these are dynamically changing according to the current scheduling and routing during the period of time. In the next section it is presented the novel dynamic scheduling algorithm and the customized existing routing algorithm which are responsible for the transportation of waste with LCTs and MDs.

4   Algorithms

Let us consider a set of l LCTs and a set of h HCTs serving as MDs. The novel scheduling algorithm is presented in Table 1. Specifically, the algorithm checks the status of HCTs and LCTs. If the capacity of a certain HCT reaches a threshold θ then the HCT is emptied in the dump and/or a recycling/processing plant. Consequently if the capacity of a certain LCT reaches a threshold θ then the algorithm performs certain actions: (a) finds the nearest HCT to the specified LCT, (b) finds the collection crossroad match point p in the highway, (c) computes the shortest path route r_l from LCT to p, (d) computes the shortest path route r_h from the nearest HCT to p. The algorithm returns the r_l, r_h routes. Consequently, the nearest algorithm is presented in Table 2. Specifically, this algorithm has the following inputs: (a) the specific LCT to get empty, and (b) the set of h HCTs serving as MDs. It is assumed that the LCT moves in local roads and backyards while HCTs move in highways. The algorithm performs a sequential search in order to find the nearest HCT from the h set to the current LCT according to the minimum Euclidean distance. The algorithm returns the specific nearest HCT.

Table 1. Scheduling

Input: $\sum_{i=1}^{m} l_i$ //l_i's are the low capacity trucks moving in local roads
$\quad\quad\sum_{j=1}^{q} h_j$ //h_j's are the high capacity trucks moving in highways
Output: r_l //Shortest path route from l_i to matching point p
$\quad\quad r_h$ //Shortest path route from h_j to matching point p
While $(true)$ **do**
$\quad$ **If** $(c_h > \theta)$ **then** //Capacity of h_j reaches a threshold θ
$\quad\quad c_h \leftarrow 0$ //h_j get emptied in the dump
$\quad$ **End If**
$\quad$ **If** $(c_l > \theta)$ **then** //Capacity of l_i reaches threshold θ
$\quad\quad n_h \leftarrow nearest(l_i, \sum_{j=1}^{q} h_j)$ //Find the nearest h_j to l_i
$\quad\quad p \leftarrow match(l_i, n_h)$ //Find the matching point p in the highway
$\quad\quad r_l \leftarrow routng(l_i, p)$ //Compute the shortest path route from l_i to p
$\quad\quad r_h \leftarrow routng(n_h, p)$ //Compute the shortest path route from n_h to p
$\quad\quad c_h \leftarrow c_h + c_l$
$\quad\quad c_l \leftarrow 0$
$\quad$ **End If**
End While

Table 2. Nearest

Input: l_i //Low capacity truck moving in local roads
$\sum_{j=1}^{q} h_j$ //h_j's are the high capacity trucks moving in highways
Output: n_h //Nearest h_j to l_i
$n \leftarrow first(\sum_{j=1}^{q} h_j)$ //Set the first h_j
$min_h \leftarrow distance(l_i, n)$ //Minimum distance between l_i and n
For $(h_j \in \sum_{j=1}^{q} h_j)$ **do** //Find the n_h with minimum distance from l_i
   $n \leftarrow next(\sum_{j=1}^{q} h_j)$ //Get the next h_j
   If $(min_h > disance(l_i, n))$ **then**
      $min_h \leftarrow disance(l_i, n)$
      $n_h \leftarrow n$ //Get the nearest h_j to l_i
   End If
End For

Table 3. Collection Match

Input: l_i //Low capacity truck moving in local roads
n_h //High capacity truck h_j moving in highways nearest to l_i
Output: p //Crossroad match point in junction of local road and highway between l_i and n_h
$\sum_{k=1}^{s} p_s \leftarrow radius(l_i, n_h)$ // Crossroad points in junction of local roads and highways between l_i and n_h
$temp \leftarrow first(\sum_{k=1}^{s} p_s)$ //Set the first p_s
$min_l \leftarrow distance(temp, l_i)$ //Minimum distance between p_s and l_i
$min_h \leftarrow distance(temp, n_h)$ //Minimum distance between p_s and n_h
For $(p_s \in \sum_{k=1}^{s} p_s)$ **do** //Find the match point p with minimum distance between both l_i and n_h
   $temp \leftarrow next(\sum_{k=1}^{s} p_s)$ //Get the next p_s
   If $\big((min_l + min_h) > (distance(temp, l_i) + istance(temp, n_h))\big)$ **then**
      $min_l \leftarrow distance(temp, l_i)$
      $min_h \leftarrow distance(temp, n_h)$
      $p \leftarrow temp$ //Get the nearest p from both l_i and n_h
   End If
End For

Table 4. Routing

Input: v, p // $v \in \{l_i, n_h\}$ is either a l_i or n_h truck
Output: r_v //route from v to p. Implements the shortest path routing algorithm in [31]
$u \leftarrow init(\sum_{b=1}^{d} u_b)$ //Initialize the set of vertices to be equal to the set of crossroads between v and p
$a \leftarrow init(\sum_{e=1}^{f} a_e)$ //Initialize the set of links to be equal to the set of roads between v and p
$w \leftarrow init(\sum_{g=1}^{y} w_g)$ //Initialize the set of weights to be equal to the set of distances between v and p
$r_v \leftarrow shortest\_path(u, a, w)$ //Compute r_v route according to the customized algorithm in [31]

The collection match algorithm is presented in Table 3. Specifically, this algorithm has the following inputs: (a) the specific LCT to get empty, and (b) the nearest HCT serving as MD to the specific LCT. It is assumed that the collection crossroad match point p is a junction between a local road and a highway between LCT and HCT, respectively. In order to find the collection crossroad match point it is gathered all the candidate crossroad points within a certain radius between LCT and the nearest HCT. Then it is computed the Euclidean distance of every pair between the LCT to p and p to the nearest HCT. Consequently, it is performed sequential search to find the minimum pair of distances. The algorithm returns the collection crossroad match point p which has the minimum pair of distances between LCT and the nearest HCT. Finally, the customized existing routing algorithm is presented in Table 4. Specifically, this algorithm has the following inputs: (a) a vehicle v which is either a LCT or a nearest HCT, and (b) a collection crossroad match point p. It is assumed that the route to be computed is the minimum set of crossroads, roads and Euclidean distances between the vehicle v and p. However, the routing algorithm implements the Resource Constraint Shortest Path Algorithm in Path Planning for Fleet Management presented in [31], thus in the current context the specific algorithm is customized to our environment. Concretely, it is feed with all the sets of crossroads, roads and Euclidean distances between the vehicle v and p given a distance threshold radius. The algorithm returns the shortest path r_v between the vehicle v and p.

5   Experimental Evaluation

In this section we perform experimental evaluation of the proposed algorithms based on real and synthetic data from the city of St. Petersburg, Russia [32]. It is assumed that the algorithms were tested on a set of $l = 1000$ LCTS and a set of $h = 400$ HCTs serving as MDs. LCTs have capacity $c_l = 3000$ kilograms while the capacity threshold is $\theta = 2500$ kilograms. The distance needed to traverse a LCT through its route from the local roads and backyards to a collection crossroad match point p is between 2 to 7 kilometers on average. Time spend to collect waste during this trip is between 10 to 15 minutes. Fuel consumed during that period is between 1 to 2 litters. Table 5 presents the average statistics of capacity transferred, distance covered, time spent and fuel consumed during a LCT waste routing trip r_l for 10 repetitions of the algorithms.

HCTs have capacity $c_h = 12000$ kilograms while the capacity threshold is $\theta = 10000$ kilograms. The distance needed to traverse a HCT which serves as a MD through its route from the highways and collection crossroad match points p to dumps and recycling/processing plants outside the city is between 10 to 15 kilometers on average. Time spend to collect waste during this trip is between 20 to 30 minutes. Fuel consumed during that period is between 2 to 4 litters. Table 6 presents the average statistics of capacity transferred, distance covered, time spent and fuel consumed during a HCT waste routing trip r_h for 10 repetitions of the algorithms.

Table 5. Average LCTs capacity, distance, time and fuel required during r_l

Repetitions	Capacity (kilograms)	Distance (kilometers)	Time (minutes)	Fuel (litters)
1	2534	5.2	12.25	1.2
2	2641	7.0	14.05	2.0
3	2511	2.8	10.55	1.0
4	2570	3.1	11.40	1.1
5	2586	5.4	13.25	1.5
6	2623	6.3	15.00	1.8
7	2591	6.2	12.35	1.9
8	2620	7.0	15.00	2.0
9	2527	4.5	11.50	1.3
10	2530	4.7	12.55	1.6

Table 6. Average HCTs capacity, distance, time and fuel required during r_h

Repetitions	Capacity (kilograms)	Distance (kilometers)	Time (minutes)	Fuel (litters)
1	11121	14.4	27.45	3.7
2	10299	10.6	20.15	2.0
3	10390	12.2	25.30	2.7
4	11264	14.7	30.00	3.9
5	10303	11.8	21.05	2.2
6	10441	12.5	23.25	2.6
7	10502	15.0	30.00	4.0
8	11158	14.6	29.00	3.8
9	10407	14.3	28.05	3.5
10	11092	13.8	26.55	3.2

Table 7. Average number of LCTs served by a certain number of nearest HCTs

Repetitions	LCTs	Nearest HCTs
1	530	215
2	740	296
3	360	163
4	1000	400
5	590	244
6	820	312
7	270	104
8	910	371
9	452	175
10	680	284

Since the number of LCTs is higher than the number of nearest HCTs it is inferred that certain number of LCTs will be served by a smaller number of the nearest HCTs. In Table 7 it is presented the number of LCTs being served by the nearest HCTs on average for 10 repetitions of the algorithms. It can be observed that the number of LCTs is always greater that the number of the nearest HCTs. This is explained due to the high capacity of HCTs compared to the low capacity of LCTs.

Moreover, during a scheduling trigger and its handling from a LCT it is required a period of response time which varies between 5 to 8 minutes. In addition during a scheduling trigger and its handling from a HCT it is required a period of response time which varies between 10 to 15 minutes. In Fig. 3 it is presented the number of LCTs and HCTs response times on average for 10 repetitions of the algorithms. It is inferred that the response time of LCTs is more stable compared with the response time of the nearest HCTs. Specifically, response time in the case of LCTs is converged to time value of 6 minutes on average with standard deviation around 1 minute. This is explained due to the local movement of LCTS in local roads and backyards within the city which cover bounded areas of the Smart City. On the contrary in the case of nearest HCTs the distances to get covered are much longer which leads the routing trip to hardly converge to the time value of 12.5 minutes with standard deviation around 2.5 minutes. Specifically, this is explained since there is not

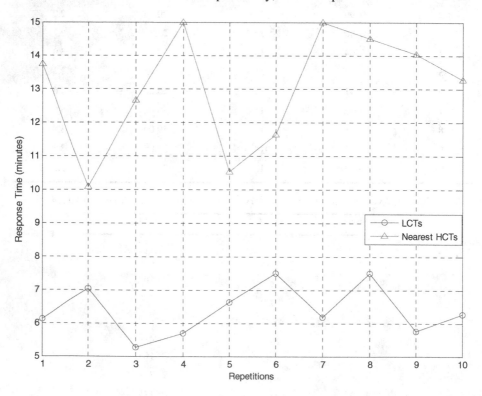

Fig. 3. Average response time between a trigger and its handling from LCTs and nearest HCTs

such a bounded locality in movement during driving in highways that could lead to a smooth time outcome. In addition, since standard deviation is only 2.5 minutes it is negligible compared with the big deviations in time of the traffic congestion in Smart Cities. The proposed solution is proved that can be applied in real time traffic behavior for efficient waste collection.

6   Economic Impact

The proposed solution is cost efficient since it reserves the ownership costs low for a period of five years. Specifically, in the case of St. Petersburg; which is a city of 5 million citizen population, the average solid waste produced within a year is 2.5 billion kilograms. Recently, in daily basis municipality of St. Petersburg occupies a homogenous fleet of 750 LCTs. The capacity of each LCT is 3000 kilograms and the ownership cost to purchase is $10,000 on average. In addition, the capacity of an HCT is 12000 kilograms and the ownership cost to purchase is $20,000 on average. According to the study in [30] a homogenous fleet of 20 LCTs can be replaced with a smaller heterogeneous fleet of 4 HCTs and 10 LCTs. In the case of St. Petersburg this indicates a heterogeneous fleet of 525 trucks; 150 HCTs and 375 LCTs. Currently the ownership cost for purchasing LCTs for a period of five years is $7,500,000. With the proposed solution the ownership cost for purchasing LCTs and HCTs for a period of five years is $3,500,000 and $3,000,000, respectively; thus $6,500,000 in total. This proves that the total municipality benefit of purchasing the proposed heterogeneous fleet of LCTs and HCTs for a period of five years is $1,000,000; thus the municipality benefit per year is $200,000.

7   Conclusion

Smart Cities is the future of civil habitation. Internet of Things enables Smart Cities with sensors and actuators which can be used for proposing new services or efficiently redesigning existing ones. Waste collection is one of the services that can be redesigned incorporating dynamic models, since in the past it was treated with static approaches. In this paper we propose a system overview and architecture which incorporates LCTs and HCTs which are serving as MDs. The strengths of this approach are the low operational cost during waste collection and the reduction of traffic within the Smart City. In addition the use of MDs reduces the problem of degradation to specific areas in the city thus guarantees better quality of life for the citizens. We also propose a dynamic scheduling algorithm and customize an existing routing algorithm for efficient waste collection between LCTs and MDs. The models were experimentally evaluated on real and synthetic data from the city of St. Petersburg, Russia. The results prove that MDs constitute an efficient innovation for IoT-enabled waste collection within Smart Cities. We also perform an economic analysis in order to define the economic impact of the proposed solution to the municipality budget for an ownership cost for a period of five years. The proposed solution proved to be cost efficient. Future work comprise the incorporation of Manhattan distance during the calculation of distances between LHTs and MDs regarding a certain collection crossroad match point p. This is explained since the route trips of the vehicles are mapped to city block distances that best describe the movement within the Smart City.

Acknowledgements. The research was carried out with the financial support of the Ministry of Education and Science of the Russian Federation under grant agreement #14.575.21.0058.

References

1. Fazio, M., Paone, M., Puliafito, A., Villari, M.: Heterogeneous sensors become homogeneous things in smart cities. In: 6th IEEE International Conference on Innovative Mobile and Internet Services in Ubiquitous Computing (IMIS), Palermo, Italy, pp. 775–780, July 2012
2. Balakrishna, C.: Enabling technologies for smart city services and applications. In: 6th IEEE International Conference on Next Generation Mobile Applications, Services and Technologies (NGMAST), Paris, France, pp. 223–227, September 2012
3. Sanchez Lopez, T., Ranasinghe, D.C., Harrison, M., Mcfarlane, D.: Adding sense to the Internet of Things. Personal and Ubiquitous Computing 16(3), 291–308 (2012)
4. Jara, A.J., Lopez, P., Fernandez, D., Castillo, J.F., Zamora, M.A., Skarmeta, A.F.: Mobile digcovery: discovering and interacting with the world through the Internet of Things. Personal and Ubiquitous Computing 18(2), 323–338 (2014)
5. Suakanto, S., Supangkat, S.H., Suhardi, Saragih, R.: Smart city dashboard for integrating various data of sensor networks. In: IEEE International Conference on ICT for Smart Society (ICISS), Jakarta, Indonesia, pp. 1–5, June 2013
6. Carli, R., Dotoli, M., Pellegrino, R., Ranieri, L.: Measuring and managing the smartness of cities: a framework for classifying performance indicators. In: IEEE International Conference on Systems, Man, and Cybernetics (SMC), Manchester, UK, pp. 1288–1293, October 2013
7. Priano, F.H., Guerra, C.F.: A framework for measuring smart cities. In: The Proceedings of the 15th Annual ACM International Conference on Digital Government Research, dg.o 2014, Aguascalientes, Mexico, pp. 44–54, June 2014
8. Nam, T., Pardo, T.A.: Smart city as urban innovation: focusing on management, policy, and context. In: The Proceedings of the 5th ACM International Conference on Theory and Practice of Electronic Governance, ICEGOV 2011, Tallinn, Estonia, pp. 185–194, September 2011
9. Nam, T., Pardo, T.A.: Conceptualizing smart city with dimensions of technology, people and institutions. In: the Proceedings of the 12th Annual ACM International Digital Government Research Conference: Digital Government Innovation in Challenging Times, dg.o 2011, College Park, MD, USA, pp. 282–291, June 2012
10. Giffinger, R., Fertner, C., Kramar, H., Kalasek, R., Pichler-Milanovic, N., Meijers, E.: Smart Cities: Ranking of European medium-sized cities, Centre of Regional Science (SRF). Vienna University of Technology, Vienna (2007). http://www.smart-cities.eu (Accessed on March 18, 2015)
11. Samaras, C., Vakali, A., Giatsoglou, M., Chatzakou, D., Angelis, L.: Requirements and architecture design principles for a smart city experiment with sensor and social networks integration. In: The Proceedings of the 17th Panhellenic Convference on Informatics, Thessaloniki, Greece, PCI 2013, pp. 327–334, September 2013

12. Milić, P., Jovanović, M.: The Advanced System for Dynamic Vehicle Routing in the Process of Waste Collection. Facta Universitatis, Series: Mechanical Engineering **9**(1), 127–136 (2011)
13. Minh, T.T., Van Hoai, T., Nguyet, T.T.N.: A memetic algorithm for waste collection vehicle routing problem with time windows and conflicts. In: Murgante, B., Misra, S., Carlini, M., Torre, C.M., Nguyen, H.-Q., Taniar, D., Apduhan, B.O., Gervasi, O. (eds.) ICCSA 2013, Part I. LNCS, vol. 7971, pp. 485–499. Springer, Heidelberg (2013)
14. Li, J.Q., Borenstein, D., Mirchandani, P.B.: Truck Scheduling for Solid Waste Collection in the City of Porto Alegre, Brazil. Omega **36**, 1133–1149 (2008)
15. Ramos, P.T.R., Gomes, M.I., Povoa, A.P.B.: Assessing and Improving Management Practices when Planning Packaging Waste Collection Systems. Resources Conservation and Recycling **85**, 116–129 (2014)
16. Stellingwerff, A.: Dynamic Waste Collection: Assessing the Usage of Dynamic Routing Methodologies. Master Thesis, Industrial Engineering & Management, University of Twente, Twente Milieu (2011)
17. Nadizadeha, A., Nasaba, H.H.: Solving the Dynamic Capacitated Location-Routing Problem with Fuzzy Demands by Hybrid Heuristic Algorithm. European Journal of Operational Research (2014) (in press available online, Elsevier)
18. Buhrkal, K., Larsen, A., Ropke, S.: The Waste Collection Vehicle Routing Problem with Time Windows in a City Logistics Context. Procedia Social and Behavioral Sciences **39**, 241–254 (2012)
19. Juyoung, W., Byung-In, K., Seongbae, K.: The rollon–rolloff waste collection vehicle routing problem with time windows. European Journal of Operational Research **224**(3), 466–476 (2013)
20. Mes, M.: Using simulation to assess the opportunities of dynamic waste collection. In: Bangsow, S. (ed.) Use Cases of Discrete Event Simulation. Non-series, vol. 109, pp. 277–307. Springer, Heidelberg (2012)
21. Bing, X.: Vehicle routing for the eco-efficient collection of household plastic waste. Waste Management **34**(4), 719–729 (2014)
22. Hemmelmayr, V., Doerner, K.F., Hartl, R.F., Rath, S.: A heuristic solution method for node routing based solid waste collection problems. Journal of Heuristics **19**(2), 129–156 (2013)
23. Johansson, O.M.: The effect of dynamic scheduling and routing in a solid waste management system. Waste Management **26**, 875–885 (2006)
24. Nuortio, T., Kytojoki, J., Niska, H., Braysy, O.: Improved route planning and scheduling of waste collection and transport. Expert Systems with Applications **30**, 223–232 (2006)
25. Von Poser, I., Awad, A.R.: Optimal Routing for Solid Waste Collection in Cities by using Real Genetic Algorithm. Information and Communication Technologies, ICTTA **1**, 221–226 (2006)
26. Mes, M., Schutten, M., Rivera, A.P.: Inventory routing for dynamic waste collection. Beta conference, WP No. 431, Eindhoven, Netherlands (2013)
27. Reed, M., Yiannakou, A., Evering, R.: An ant colony algorithm for the multi-compartment vehicle routing problem. Applied Soft Computing **15**, 169–176 (2014)
28. Zsigraiova, Z., Semiao, V., Beijoco, F.: Operation Costs and Pollutant Emissions Reduction by Definition of new Collection Scheduling and Optimization of MSW Collection Routes using GIS. The Case Study of Barreiro, Portugal. Waste Management **33**, 793–806 (2013)

29. Anagnostopoulos, T.V., Zaslavsky, A.: Effective waste collection with shortest path semi-static and dynamic routing. In: Balandin, S., Andreev, S., Koucheryavy, Y. (eds.) NEW2AN/ruSMART 2014. LNCS, vol. 8638, pp. 95–105. Springer, Heidelberg (2014)
30. Anagnostopoulos, T., Zaslavsky, A., Medvedev, A.: Robust waste collection exploiting cost efficiency of IoT potentiality in smart cities. In: IEEE 1st International Conference on Recent Advances in Internet of Things (RIoT) (Accepted on February 4, 2015)
31. Avella, P., Boccia, M., Sforza, A.: Resource Constraint Shortest Path Problems in Path Planning for Fleet Management. Journal of Mathematical Modeling and Algorithms **3**, 1–17 (2004)
32. Real allocation distribution of bins within the city of Saint Petersburg. http://wikimapia. org/ (Accessed on: March 27, 2015)

Big Data Governance for Smart Logistics: A Value-Added Perspective

Jae Un Jung[1]([⊠]) and Hyun Soo Kim[2]

[1] BK21Plus Groups, Dong-A University, Busan, Korea
imhere@dau.ac.kr
[2] Department of MIS, Dong-A University, Busan, Korea
hskim@dau.ac.kr

Abstract. For the last five years, worldwide curiosity concerning big data has sharply increased, along with cloud computing and the Internet of Things (IoT). However, without reasonable governance of the role of and value-sharing among stakeholders, it is difficult to recognize big data's potential value, beyond the raw data. This research suggests refinements to this governance, in terms of cross-sector collaboration for the benefit of all. This is particularly true for those in the logistics sector, which is one of the most promising, although it is composed of complicated and exclusive industries. For the development/evaluation of such governance, we designed a new business model that governs stakeholder roles and responsibilities, and presented a case study on shipping reefer (refrigerated) containers. We found that our model is applicable to both the private and public sectors for smart logistics, and feasible for resolving conventional logistical challenges such as security, smuggling, and indemnification disputes.

Keywords: Big data · Cross-Sector collaboration · Governance · Smart logistics

1   Introduction

Advances in information and communication technologies enable many things that were once considered impossible, using smart devices (products) and services, and which cross both offline and online boundaries. The smart world brings convenience to the individual, but, in the business sector, heterogeneous and vast data, generated by freewheeling smart activities, increases the maintenance costs [1]. Recently, advanced research studies [2] have suggested ways to overcome these challenges (reduce cost and simplify technical maintenance) and, by extension, create value added from big data digital sources. Many challengers or opinion leaders define big data; however, from the business perspective, it is difficult to discuss it without considering its value [3,4].

According to [5], the logistics and distribution industry is destined to be one of the most profitable when big data investments are realized.

If logistics companies can attach smart sensors or devices to their freight and equipment (facility), intelligent and smart logistics systems can be implemented.

© Springer International Publishing Switzerland 2015
S. Balandin et al. (Eds.): NEW2AN/ruSMART 2015, LNCS 9247, pp. 95–103, 2015.
DOI: 10.1007/978-3-319-23126-6_9

Recently, the IoT and machine-to-machine (M2M) technologies have been heavily researched [6,7]. Such approaches inspire logistics companies to open new businesses, but also cause problems in the management and governance of data that are constantly being generated. Additionally, in the logistics industry, developing or adopting a new concept that crosses sector boundaries requires the agreement of diverse stakeholders, from consignors (shippers) to governments.

To create a new business model for the use of big data in the logistics field, governance that is acceptable to diverse stakeholder perspectives requires definition, and the model must incorporate reasonable benefits and incentives to induce stakeholder interest from multiple sectors.

This research defines the governance that is acceptable for the logistics industry. We design a new business model, which requires changing the existing roles and responsibilities of stakeholders, as well as new rules for data governance, and present a case concerning shipping reefer containers.

2   IT and Business Challenges for Smart Logistics

2.1   IT Trends and Big Data

Gartner, an IT opinion leader, presented the top 10 technologies expected to exert a significant impact in the next three years, for the following three categories: merging the real and virtual worlds (computing everywhere, the IoT, 3D printing), intelligence everywhere (advanced, pervasive, and invisible analytics, as well as context-rich systems and smart machines), and the emergence of the new IT reality (cloud/client computing, software-defined applications and infrastructure, web-scale IT, risk-based security, and self-protection) [8,9]. The top 10 technologies for the smart world, with the exception of 3D printing, which makes physical artifacts, are those that are closely connected; these are summarized in Fig. 1.

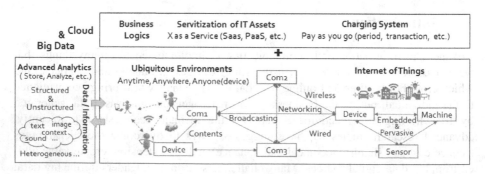

Fig. 1. Overview of IT Trends

In time, autonomous networking using sensors embedded in items, and even living things, will pervade and enable increasingly agile services with advanced analytics (ubiquitous environment and IoT). Rich intelligent assets will be servitized (defined as services) by IT business strategies, and on-demand intelligent services will be accessible anytime and anywhere on pervasive networks with diverse charging systems (cloud or cloud computing) [10]. Heterogeneous information activities will generate vast data, thereby prompting the development of management solutions (advanced analytics) for big data. Big data resources are targeted to find new intelligence and value; therefore, big data solutions will represent innovative service models in their own right (big data and cloud).

We analyzed worldwide interest in big data through query analysis, using Google Trends, as presented in Fig. 2. The results show that big data queries, now ubiquitous because of public interest, have increased rapidly since 2011. The most queries were retrieved in India, which is one of the countries well-known for offshore IT outsourcing. The second and third highest numbers of queries were conducted in Singapore and South Korea, respectively, both of which are famous for having global shipping hubs or ports. Moreover, Singapore is positioning itself as an international financial and tourism center, along with Hong Kong, which ranked fourth in the results [11,12].

Therefore, we infer that interest in big data is diffused among the IT industry, as well as other industries linked with big data such as logistics and finance.

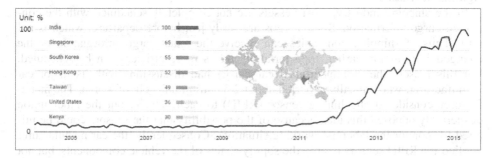

Fig. 2. Global Interest in Big Data

2.2 Conflict of Interests and Shareable Data

For Internal Business Intelligence. One of the major worldwide car carriers (shipping companies), headquartered in Korea, operates approximately 100 vessels for worldwide schedules. However, just five years ago, exact profits and losses could not be calculated by the end of each business or fiscal year. The calculation was typically delayed until the following year because the company could not integrate all of the business data from the complex processes that were globally scattered. Large and heterogeneous data are the key to optimizing service quality and operational resources; however, this was a secondary concern. In short, because the financial problem was the priority, the operational data required for revenue and expense accounts, such as oil consumption and anchorage, were integrated and, now, the profits and losses can be analyzed annually, but not yet quarterly or bi-annually.

Accounting data represent the operational results for the improvement of profitability. In addition, the optimization of intelligence planning is required for routing and stowage. Therefore, collaborative research to improve previous routing and stowage planning was conducted, but the artifacts did not meet the industrial requirements.

Separately, the IT team developed another integrated solution, but it was also confronted with the same issues and planners' opposition, and the project could not continue. However, the IT team persuaded the planning team to accept the integrated planning system, on the condition that the IT project team removes all of the automatic planning functions from the new planning system. The ostensible reason was the functional insufficiency of the planning algorithms, but, internally, there was a struggle for leadership between the planners and automatic planning algorithms. Because planning knowledge was central to the shipping company and successful planning, the company relented. Ultimately, the business intelligence improved because the integrated data management functions supported the planners' complicated decisions to a greater extent than before. This case illustrates that an algorithmic approach to big data, to improve internal business intelligence, is not necessarily the optimal solution.

For Cross-Sector Stakeholders. If rich sensors and devices are attached to all of the logistics processes, logistics data or information can be synchronized with the physical distribution processes; however, except for on the technical side, there are some significant issues.

In the shipping industry, most vessels are not completely scheduled with the confirmed cargo information. Some cargos are ready just before departure, which is not rare. Thus, if shipping companies can receive the exact cargo information or the changed information earlier, transportation that is more efficient can be scheduled. For this reason, the aforementioned car carrier had discussions with its major car manufacturer, who annually exported about 2.6 million units out of Korea. The manufacturer considered attaching a sensor (RFID) to each vehicle, but the labor union desperately opposed this idea because of the possibility that the sensors could be utilized to monitor the laborers' working situations. Consequently, the car manufacturer applied the RFID sensors to only the supply chains of the vehicle components, but not to the transportation process of the completed vehicles.

Though the car manufacturer can utilize sensors, it may be asked whether the vehicle information can be delivered by sensor without any problems, such as ownership/responsibility for operating intelligent infrastructures and managing/governing data. Most logistics data are interrelated with the business strategies on customizing, pricing/bidding, routing, etc. Therefore, if a new innovative solution can be adopted, governing strategies are required to adjust complicated interests of stakeholders, in terms of business. Without the solution, cross-sector stakeholders in the logistics industry will not allow their logistics data to flow over into other sectors.

Fig. 3 describes the levels and steps of big data analytics for smart logistics. To implement context-aware intelligence with big data, as shown in Fig. 3(a), data gathering and delivery are rudimentary at each section in Fig. 3(b). However, both of the two episodes mean that, to generate/share data, it is rudimentary to realize the IoT, but it also difficult, with the exception of finding sharable value in big data in the logistics industry.

As such, this research aims to approach the utilization of big data, in terms of governance and value added that is acceptable to the real world. Thus, we design a big data business model, which crosses the diverse stakeholders in both the private and public sectors with the cold chain of reefer containers [13].

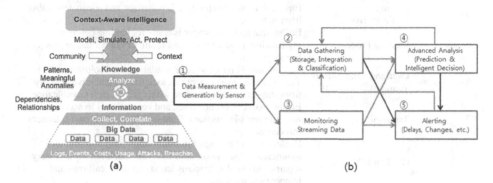

Fig. 3. Big Data Analytics Levels and Steps (Sources: [9,14], edited)

3   Scenario-Based Value-Added Services and Governance

3.1   Basic Model

Table 1 describes the overall process for shipping reefer containers from a consignor, or the origin of the produce, to a consignee, or final destination. According to [15], South Korea imports the most shrimp from Vietnam; thus, we can design a basic model showing that frozen shrimp are refrigerated and imported to Korean fishery markets from Vietnam. This model only focuses on how to transport a reefer container from a consignor to a consignee, as shown in Fig. 4.

In this model, exact and stable transportation is crucial. The requirements are that the temperature in a reefer container has to be kept at -18°C or colder, and the approximate shelf life in ambient air should not exceed 10 to 12 months, etc. However, a consignor or consignee does not usually have an interest in such constraints prior to his or her cargo encountering a big problem, like distortion or spoilage.

Shrimp are frozen and packed as an ice block, which is white or transparent in color. However, if, upon opening a reefer container, a consignee sees green ice blocks, the question becomes what process is required to fix the problem.

This case was discovered about five years ago in Korea. There was a problem with the power supply of the reefer container, so the ice blocks of shrimp were repeatedly frozen and melted. This repetition caused the contraction and expansion of the green brains of the shrimp, which burst and permeated the ice blocks.

This problem can be solved by real-time monitoring with smart sensors or devices. Therefore, in this basic model, the management of real-time data and handling of streaming data are required to be developed as a service.

Table 1. Process for the Case of Shipping Reefer Containers [13]

Sequence	Process	Responsibility
1	Farm (Land/Sea)	Shipper should maintain the (marine) product at the desired temperature to maintain freshness.
2	Truck	Trucks deliver within the shortest possible time.
3	Pre-cooling Room	Temperature is set correctly and condensation minimized.
4	Packing Facilities/Houses	Ensure that (marine) produce is palletized and adequately stabilized in the cool room.
5	Loading Process	Ensure that reefer container is ready for loading before removing (marine) produce from cool room. Container has to pass pre-trip inspection.
6	Overland Transport to Terminal	Vehicle should have effective insulation and cooling system. Trailers must have proper twist lock fittings and travel directly from shipper place to terminal within the shortest possible time.
7	Terminal	Ensure that temperature is set, and ventilation is in accordance with shipper's instructions. Container has to pass post-treatment inspection.
8	On Board (Vessel)	Reefers are to be plugged in immediately after loading, and monitored at four- to six-hour intervals. Look out for necessary repairs. Terminal technicians are to record all call-outs and temperature readings.
9	Direct Destination/ Transshipment Port	Reefer containers are to be unplugged only when the loading process is ready. Reefer containers must be powered on immediately upon delivery to reefer yard.
10	Transshipment Terminal	Cargo must be monitored while waiting for on-carriage.
11	On Carriage (Vessel)	Reefer shipment is always given priority to load on first available carriage because of limited holding shelf life.
12	Discharge Port Terminal	Reefer temperature must be regularly monitored before consignee collects cargo.
13	Land Transport to Consignee	Trucking time to consignee location must be minimized.
14	Consignee Premise	Cargo must be unstuffed and loaded into cold room immediately to minimize heat exposure to the (marine) produce. If unstuffing only takes place at a later stage, the laden reefer container must be plugged in to ensure power supply is on to maintain temperature.

Fig. 4. Basic Model for Smart Logistics (Sources: [16], edited)

3.2   Extended Model

Even if there is something wrong with the cargo, a complaint is not considered acceptable to transport company, forwarder, and insurer until the cause is discovered using objective data. The importer of the frozen shrimp can claim compensation, but it may not receive 100% credit for the damage. In this situation, only 10% of the real cargo price was compensated because the reefer container was declared to be worth only 10% of its actual price, following the bad practice in the industry. Such a practice causes individual loss or insurance fraud in the private sector and, in the public sector, it can be considered smuggling.

For this reason, we designed a service center, as shown in Fig. 5, which plays a role in adding value, by composing diverse companies for logistics functions and IT technologies (platform, application, data management system, etc.), and to coordinate their conflict of interest on each side. In this service center, logistics information can be shared through data, across discrete sectors, by charging system for each data transaction. Moreover, customs agencies and the government also require this function to detect and protect against smuggling (by price or weight deception). If a customs agency participates in this model as a client, the service center secures stable revenue from this process. In addition, the service center can provide, on demand, any type of service such as analytics, information analysis, classified data, etc. Therefore, this model merits consideration from insurers when they inspect indemnification claims for cargo damages.

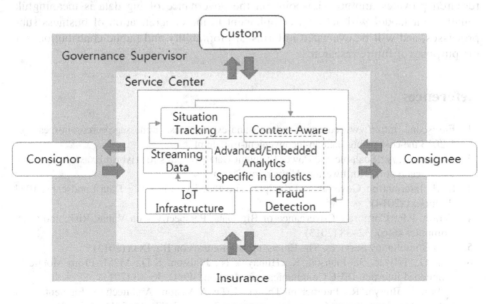

Fig. 5. Extended Model for Smart Logistics

To keep the service center going, there are two strategies. First, profits from its operation should be utilized to coordinate stakeholders' interest; otherwise, private stakeholders will not participate in the service model. For this reason, it is difficult to solicit investments in IoT infrastructures and big data because almost all of the logistics players are in the private sectors. Finding profits from sharable data is the first governance to coordinate each concern naturally. Second, the governance supervisor plays an essential and administrative role in running the service center.

Logistics data contain critical and sensitive security information; thus, the data should be strictly utilized, and an authorized supervising organization is required to supervise whether the service center performs as a proper governing system, using both government and non-government experts.

4   Conclusion

This research introduced a business model for big data, to coordinate logistics of the stakeholders, based on the insight into and necessity for governing conflict, with regard to stakeholders' interests, using the situations involving the aforementioned car manufacturer and shipping company. From our discussion of the IT and business challenges for smart logistics, in Section 2, we see that big data seem to have a lot of potential; however, in the realization of such potential value, practical conflicts of interest acted as obstacles. Thus, we focused on the governance of big data in terms of the value added, and reinterpreted the governance issue as a collaborative IT business model, not limited only to the resources of big data. Consequently, the fact that our research provides another viewpoint on the governance of big data is meaningful. Finally, our model will act as a complement to the concretization of business sub-processes, and will be evaluated in terms of profitability and public contribution, for the purposes of future research.

References

1. Eurescom. http://www.eurescom.eu/news-and-events/eurescommessage/eurescom-message-1-2013/the-cost-of-big-data-an-inconvenient-truth.html
2. Choudhary, A.: Scalable and Power Efficient Data Analytics for Hybrid Exascale Systems. Technical report, Northwestern University (2015)
3. IBM, Information Governance Principles and Practices for a Big Data Landscape. IBM Redbooks (2014)
4. Tallon, P.P.: Corporate Governance of Big Data: Perspectives on Value, Risk, and Cost. Computer 46(6), 32–38 (2013)
5. Tata Consultancy Services. The Emerging Big Returns on Big Data (2013)
6. Wu, G., Talwar, S., Johnson, K., Himayat, N., Johnson, K.D.: M2M: From Mobile to Embedded Internet. IEEE Communication Magazine 49(4), 36–43 (2011)
7. Gubbi, J., Buyya, R.: Internet of Things (IoT): A Vision, Architectural Elements, and Future Directions. Future Generation Computer Systems 29(7), 1645–1660 (2013)
8. Gartner. http://www.gartner.com/smarterwithgartner/gartners-top-10-strategic-technology-trends-for-2015/
9. Cearly, D.: Top 10 Strategic Technology Trends for 2015, Gartner Symposium ITXPO (2014)

10. Sultan, N.: Servitization of the IT Industry: the Cloud Phenomenon. Strategic Change **23**(5–6), 375–388 (2014)
11. CNBC. http://www.cnbc.com/id/101963981
12. Euromonitor International. http://blog.euromonitor.com/2015/01/top-100-city-destinations-ranking.html
13. Pacific International Lines. https://www.pilship.com/cn-guidelines-on-refrigerated-cargoes/166.html
14. Jung, J.U., Kim, H.S.: Deployment of Cloud Computing in Logistics Industry. Journal of Digital Convergence **12**(2), 163–171 (2014)
15. Zarrouki, K.: Inside South Korea: the Fish and Seafood Trade. Global Analysis Report, Agriculture and Agri-Food Canada (2015)
16. Kelvin Cold Chain Logistics. http://www.kelvincoldchain.com/faqs.htm

Waste Management as an IoT-Enabled Service in Smart Cities

Alexey Medvedev[1(✉)], Petr Fedchenkov[1], Arkady Zaslavsky[1,2],
Theodoros Anagnostopoulos[1,3], and Sergey Khoruzhnikov[1]

[1] ITMO University, Kronverkskiy Pr., 49, St.-Petersburg, Russia
{alexey.medvedev,petr_fedchenkov}@niuitmo.ru, xse@vuztc.ru
[2] CSIRO, Melbourne, Australia
arkady.zaslavsky@csiro.au
[3] Community Imaging Group, University of Oulu, 90570 Oulu, Finland
tanagnos@ee.oulu.fi

Abstract. Intelligent Transportation Systems (ITS) enable new services within Smart Cities. Efficient Waste Collection is considered a fundamental service for Smart Cities. Internet of Things (IoT) can be applied both in ITS and Smart cities forming an advanced platform for novel applications. Surveillance systems can be used as an assistive technology for high Quality of Service (QoS) in waste collection. Specifically, IoT components: (i) RFIDs, (ii) sensors, (iii) cameras, and (iv) actuators are incorporated into ITS and surveillance systems for efficient waste collection. In this paper we propose an advanced Decision Support System (DSS) for efficient waste collection in Smart Cities. The system incorporates a model for data sharing between truck drivers on real time in order to perform waste collection and dynamic route optimization. The system handles the case of ineffective waste collection in inaccessible areas within the Smart City. Surveillance cameras are incorporated for capturing the problematic areas and provide evidence to the authorities. The waste collection system aims to provide high quality of service to the citizens of a Smart City.

Keywords: Waste collection · Smart city · Internet of Things (IoT) · Intelligent transportation systems · Surveillance systems

1 Introduction

Recent advances in production of mobile computers and smartphones, smart sensors and sensor networks in connection with next generation mobile networks opened vast opportunities for researchers and developers of various systems and application in the field of Smart Cities and ITS. Thought some areas like application for monitoring public transport are already well researched, other areas are still working with outdated technologies and models. One of such areas is the management of solid

A. Zaslavsky is an International Adjunct Professor at ITMO University since 2012.

S. Balandin et al. (Eds.): NEW2AN/ruSMART 2015, LNCS 9247, pp. 104–115, 2015.
DOI: 10.1007/978-3-319-23126-6_10

waste collection process. In a Smart City collection of waste is a crucial point for environment and its quality should be considered seriously. In order to understand the concept of Smart Cities in depth, a suitable definition is provided. In this research we use the most suitable definition for the IoT-enabled waste collection in Smart Cities, which is [1]: "A Smart City is a city well performing in a forward-looking way in the following fundamental components (i.e., Smart Economy, Smart Mobility, Smart Environment, Smart People, Smart Living, and Smart Governance), built on the 'smart' combination of endowments and activities of self-decisive, independent and aware citizens". In this definition we can see important component - Smart Environment - which is tightly connected to environmental pollution. The main countermeasure to environmental pollution in terms of a Smart City is the IoT-enabled waste collection. The following definition of IoT is used in this paper [2]: "The Internet of Things allows people and things to be connected Anytime, Anyplace, with Anything and Anyone, ideally using Any path/network and Any service". IoT technologies enable new services and reshape the existing ones in Smart Cities [3]. For instance static waste collection is redesigned to Waste Collection as a Service. As the result this enables online dynamic scheduling and routing of the trucks [4]. Issues connected to dynamic waste collection could be divided into 2 main problems: (i) when to collect waste form bins (i.e., scheduling), and (ii) what route the trucks will follow (i.e., routing).

In this paper we propose waste collection system enhanced with IoT services which enable dynamic scheduling and routing in a Smart City. We also present the design of a cloud system for organization of waste collection process and applications for waste truck drivers and managers. The proposed system also features an on-board surveillance system which raises the process of problem reporting and evidence collection to a higher level.

The rest of the paper is structured as follows. Section 2 presents related work in the area of IoT-enabled waste collection in Smart Cities. Section 3 describes the main features of the system and some scenarios of usage. Section 4 presents the system model architecture and two applications. One is a mobile application for the waste truck driver and another is a web application for waste management company. Section 5 contains the evaluation on one possible scenario and section 6 concludes the paper and discusses future work.

2 Related Work

The area of route planning and optimizing for logistic purposes is well-researched and hundreds of ITS already exist. There are also a number of projects aiming to provide an effective system specializing on waste collection needs. A Geographical Information System (GIS) transportation model for solid waste collection that elaborates plans for waste storage, collection and disposal has been proposed in [5] for the city of Asansol in India. In [6] an enhanced routing and scheduling waste collection model is proposed for the Eastern Finland, featuring the usage of a guided variable neighbourhood thresholding metaheuristic. In the city of Porto Alegre in Brazil authors propose [7] a truck-scheduling model for solid waste collection. The aim of the research was to develop an optimal schedule for trucks on defined collection routes. Examples of other systems are described in [8],[9],[10] and [11].

A survey presented in [12] reviews the researches done on waste collection in developing countries from 2005 to 2011 and considers challenges for developing countries in waste collection sphere. The research focuses on determination the stakeholders' actions/behaviour and evaluation of influential factors defining their role in waste collection process. The models in the survey were tested on real data. In [13] a survey considering system approaches for solid waste collection in developing countries is presented. The research compares the history and the current practices, presented from 1960s to 2013. Information about the challenges and complexities is also given there. The output of the survey is drawing a conclusion that developing and implementing solid waste collection approaches in developing countries are of a great importance. The main issue is that waste collection does not include innovation that IoT can provide. Models do not use real time information of the waste collection, although some approaches use advanced scheduling and routing via exploiting modern ICT algorithms. Information about bins status was not considered as part of waste collection. All the reviewed surveys do not propose a model that will use IoT technology for Smart Cities, though they consider different approaches for waste collection.

Moreover, enabling combined participation of stakeholders like road police and city administrations in one system is not covered. Finally, the concept of implementing an on-board surveillance system for fast problem reporting and evidence collecting is not implemented in mentioned projects, but is described in [14] and [15] separately from waste management topic. All this allows stating the need for development a system facilitating the usage of IoT data, dynamic routing models and participation of diverse types of stakeholders.

3   Main Features and Scenarios of Usage

System architecture aims to suit two main targets. First target is providing software-as-a-service (SaaS) products for customers. Mainly, these customers are private companies that are involved in waste collection, owning waste trucks, organize work of drivers, get contracts from municipalities and pass wastes to recycling organizations or city dumps. Second main target is developing a system, which makes possible mutually beneficial communication between all the stakeholders involved in the chain of supplying goods and utilizing solid waste in smart city.

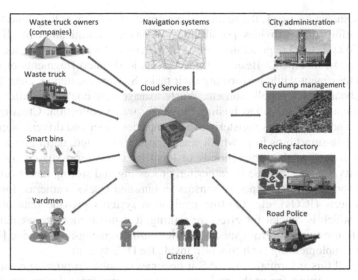

Fig. 1. The big picture of a waste collection management system

A list of possible stakeholders of the system and brief description of their needs, business rules, possibilities and connections with others is presented below:

- City administration needs understanding of the big picture, generating reports, control over pricing etc.
- District administrations are interested in controlling the process of waste collection, checking quality of service (all waste collected, all in time, waste collected cleanly, waste transported to special places), quick and legal ways for solving disputes and problems. Municipalities can also deploy and maintain smart city infrastructure like capacity sensors in waste bins and wireless networks for data transferring.
- Waste trucks owning companies need platform for organizing and optimization of their business process in general without serious investments in developing, deploying and supporting their own system. Such system must include effective dynamic routing based on IoT data for the truck fleet. Besides, controlling drivers and tracking the fleet is also an important issue.
- Waste truck drivers need navigation system for fulfilling their tasks. Another issue is reporting problems and passing them to the operators in the office instead of thinking how to solve the problem, this can sufficiently save time of a driver and vehicle. Drivers also need evidence that their work was done correctly and cleanly.
- Managers of dumps and recycling factories can publish their possibilities or needs in acquiring certain amount of waste for storing or recycling
- Staff that is responsible for trash bins in the current yards needs communications with waste management companies and truck drivers.
- Road police can get reports about inaccurate car parking that leads to impossibility of waste collection.
- Citizens want to have better service, lower cost and having easy accessible reports on what has been done and how much it cost.

As it is shown on fig. 1, the main component of the system is the cloud based DSS. It is a platform that provides possibilities for intercommunication of all the stake-holders. Waste trucks generate sensor data about their capacity, location, fuel available and consumed. Besides, truck drivers load video fragments or pictures of problems they meet while performing their tasks. Sensors located in smart bins generate data about capacity, pollution etc. Waste management companies after registering in the system create rules and business logic for waste collection. Creating the business logic and rules means registering the companies' fleet and drivers, registration of smart and non-smart bins from which waste must be collected, defining time windows for waste collection that corresponds to local laws and terms of contract with the municipality. Important issue is gathering, processing and storing data from heterogeneous sensors, including capacity sensors in bins and trucks, cameras, Internet Connected Objects (ICOs), etc. On-line navigation systems provide data about traffic situation, which is crucial for effective routing. It is much more convenient and cost effective to use this data from special services using sensing-as-a-service [16] model, rather than implementing such function inside the DSS system.

Having all this information in DSS it becomes possible to provide customers with best possible routing for each truck. Moreover, reports from drivers when they encounter a problem on the road are processed semi-automatically that results in a much faster problem solving. It is possible to count plenty of system usage scenarios but due to the lack of space we present and evaluate only one, which is presented in the "Scenario – inaccessible waste bin" section of this chapter.

3.1   Scenario – Inaccessible Waste Bin

Waste truck drivers report about his inability to drive inside the yard or approach the waste bin with the truck and get wastes. Usual reason for it is inaccurate car parking, which is shown in fig. 2.

Fig. 2. Inaccessible waste bin scenario

This report includes video or picture of a problem made with drivers' smartphone or tablet on which the client android app is running. This data is annotated by a voice message, GPS coordinates and other available data. Then this report is processed in the DSS and if it is correct it is sent to organizers of waste collection in this particular place and to the road police. The truck driver doesn't waste time for waiting, he/she goes to the next point and the route is dynamically recounted. When the problem is solved the system recounts the route for one of the available trucks and the waste from unlocked bin is collected. It is combined with dynamic routing algorithms [17] to maximize the efficiency of waste collection. As it is stated in [17] static models do not fit the idea and IoT-enabled potentiality of a smart city. It is often faster and cheaper to make a longer route saving time from traffic congestions or waiting for a road problem to be solved; thereafter the need for IoT-enabled dynamic routing engine for the fleet becomes one of the main features of the designed DSS. Schematically old static and new dynamic approaches are shown on fig. 3.

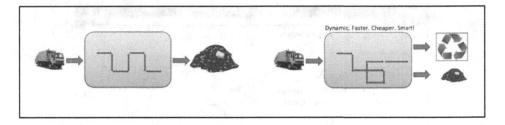

Fig. 3. Static and dynamic routing approaches

4   System Model Architecture and Applications

As most Intelligent Transportation Systems, the designed system also implements the engine for storing, rendering, updating and displaying maps   as one of the main components. Some of the criteria for choosing the engine were licence independence, possibility for making changes in existing maps and possibility to build a separate instance in a private own cloud. As a result OpenStreetMap [18] has been chosen as the main technology for acquiring maps data and for displaying maps and routes both for the drivers' android application and web application for managers and other clients. Nominatim [19] is a part of OpenStreetMap project; it is used for geocoding − finding latitude and longitude by in OSM data by name and address.

As it was already mentioned above, a typical client of the described system is a waste management company that owns a heterogeneous fleet of vehicles and needs to service a number of points in a city. This is a well-known problem in logistics and transportation - the vehicle routing problem (VRP) [20] and its objective is to minimise the total route cost. There are several variations and specializations of the VRP but their description is omitted in this paper due to space limitations. A number of open source and commercial projects exist enabling fast solution of VRP. Examples of such projects are JSPRIT [21], Open-VRP, OptaPlanner, SYMPHONY, VRP Spreadsheet Solver etc. JSPRIT [21]–java based, open source toolkit for solving rich traveling salesman (TSP)[26] and vehicle routing problems (VRP) has been chosen the main library used for solving VRP and building initial routes due its lightness, flexibility and ease of use. Another advantage of JSPRIT library is its easy extensibility that will be significantly useful while adding special features and algorithms specific for waste collection. GraphHopper [27] is a fast and memory efficient Java road routing engine. It is used for calculating optimized routes for waste trucks based on OpenStreetMap data.

A web-based application for waste management companies is presented in fig. 4. It provides managers and operators with facilities like registering the infrastructure and vehicles, tracking the fleet, mark waste bins as blocked and unblocked etc. A mobile Android-based application for a waste truck driver is shown in fig. 7. As the main feature it delivers smart navigation options to the driver. Secondly, the application provides an option of reporting a problem. In fig. 7. (right) a process of making a report about a blocked by car waste bin is shown. Noticeably, we implement a feature of annotating a report with voice that allows not distracting the driver from his work.

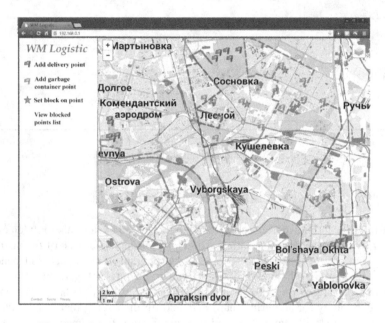

Fig. 4. Web-based application for waste management companies

4.1   Surveillance System

As it was already mentioned in Introduction section, one of the main features of the proposed system is a waste truck based surveillance system that serves several purposes. These purposes are:

- Evidence collection system for easy accident analysis [14]
- Reporting road and other problems [15]
- Proof of correctly and cleanly done work

First two scenarios are based on CityWatcher application and are described in [14] and [15]. CityWatcher is an android based application for smartphones, which acts as an IoT car black box. It records video of the situation on the road and annotates it with time and coordinates. The difference with other black boxes is the ability of authorized personnel to make requests to local storage of all participating in the system devices to search for an evidence of road accidents.

The distinction from CityWatcher application is in the number of cameras used. CityWatcher was designed to use with the camera built into smartphone. This was a suitable solution for a civil purpose, but it may be not enough for professional service. In case of waste truck surveillance system several wired or wireless cameras can be used simultaneously.

```xml
<vehicles>
   <vehicle>
      <id>1</id>
      <typeId>1</typeId>
      <startLocation>
         <id>[x=60.0269214918535][y=30.2791301660061933]</id>
         <coord x="60.0269214918535" y="30.2791301660061933"/>
      </startLocation>
      <endLocation>
         <id>[x=60.0269214918535][y=30.2791301660061933]</id>
         <coord x="60.0269214918535" y="30.2791301660061933"/>
      </endLocation>
      <timeSchedule>
         <start>0.0</start>
         <end>1.7976931348623157E308</end>
      </timeSchedule>
      <returnToDepot>false</returnToDepot>
   </vehicle>
</vehicles>
<services>
   <service id="1" type="pickup">
      <location>
         <id>[x=60.00118322593268][y=30.253984456547315]</id>
         <coord x="60.00118322593268" y="30.253984456547315"/>
      </location>
      <capacity-dimensions>
         <dimension index="0">23</dimension>
      </capacity-dimensions>
      <duration>10000.0</duration>
      <timewindows>
         <timewindow>
            <start>1800000.0</start>
            <end>1.7976931348623157E308</end>
         </timewindow>
      </timewindows>
   </service>
</services>
```

Fig. 5. XML file with vehicle and task description

5   Experimental Evaluation

We use real and synthetic data to evaluate the proposed system. Real data is the road graph of St.-Petersburg and waste bins location. In the first experiment we use 6 trucks for collecting waste from 24 bins. The task for the JSPRIT library algorithms is described as an XML file, a part of which is presented on fig. 5. The XML file contains the description of one vehicle and one pickup point, which represents a smart waste bin.

The result of the VRP solvation is graphically presented in fig. 6. As a result after the first experiment we have routes for several trucks, distances, time and fuel consumed. This is the best-case scenario, as all the bins in this experiment are treated as accessible (not blocked).

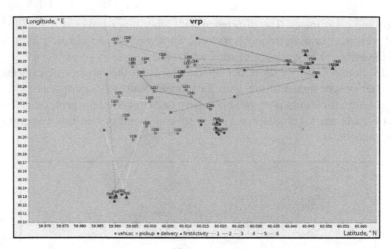

Fig. 6. The result of the VRP solvation

Routes for this experiment are presented in fig. 4 (all trucks for manages) and fig. 7 (one route for a truck driver).

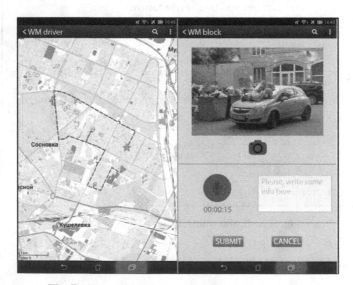

Fig. 7. Navigation (left) and problem reporting (right)

In the second experiment several random bins are blocked. We use this approach several times on different percentage of blocked bins.

When the truck driver finds a blocked bin or other problem that makes it impossible to collect the waste he/she loses several minutes for reporting the problem via telephone and leaves the place. When all accessible points are collected the driver makes one more round for collecting waste from bins that are assumed to be unblocked now. This is the worst case scenario, as the diver loses time for driving into the yard, recognizing the problem, reporting it and returning to the same place later.

In the third experiment we assume that when a bin is blocked the truck reports the problem with a mobile application and continues the trip. For example, the waste collection point in the yard contains four bins – for plastic, glass, paper and organics. While the truck that reports the problem (e.g. collecting plastic) does not get significant resource economy, other three trucks are informed about the problem and automatically exclude current point from their route. When the problem is marked as solved by police or municipality staff the route is dynamically rebuilt and one of the available trucks gets a task for collecting waste from that point. Line graph for this experiment are presented on fig. 8. The lowest line represents the ideal scenario without blocked bins. It is easy to see, that total time (and accordingly cost) used by informed waste trucks (red line) in comparison with the scenario without informing drivers about blocked bins (green line) is significantly lower. This experimental evaluation showed that our approach for coping with blocked bins scenario is cost-effective.

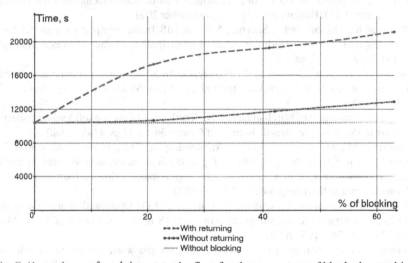

Fig. 8. Dependence of total time spent by fleet for the percentage of blocked waste bins

6   Conclusion and Future Work

In this paper we have presented a novel cloud-based system for waste collection in smart cities. The system aims to provide services for different kind of stakeholders involved in this area - from city administrations to citizens. Still, the design focuses mostly on providing SaaS services to commercial waste management companies. Development of applications for city administrations, municipal staff, recycling factories and other stakeholders is planned to be done in future. We have evaluated the proposed system and shown that implementing on-board surveillance cameras for problem reporting in conjunction with a cloud DSS system and dynamic routing models can give a significant increase of cost-effectiveness, which is one of the most indicating criteria for the Smart City.

Acknowledgement. The research has been carried out with the financial support of the Ministry of Education and Science of the Russian Federation under grant agreement #14.575.21.0058.

References

1. Centre of Regional Science. Vienna University of Technology. Smart Cities. Ranking of European Medium-Sized Cities (accessed on: December 23, 2014). http://www.smart-cities.eu
2. Guillemin, P., Friess, P.: Internet of things strategic research roadmap. The Cluster of European Research Projects, Tech. Rep., September 2009 (accessed on: December 23, 2014). http://www.internet-of-things-research.eu
3. Delicato, F.C., Pires, P.F., Batista, T., Cavalcante, E., Costa, B., Barros, T.: Towards an IoT ecosystem. In: The Proceedings of the 1st ACM International Workshop on Software Engineering for Systems-of-Systems, SESoS 2013, Montpellier, France, pp. 25–28, July 2013
4. Lingling, H., Haifeng, L., Xu, X., Jian, L.: An intelligent vehicle monitoring system, based on internet of things. In: IEEE 7th International Conference on Computational Intelligence and Security (CIS), Hainan, pp. 231–233, December 2011
5. Ghose, M.K., Dikshit, A.K., Sharma, S.K.: A GIS based transportation model for solid waste disposal – A case study on Asansol municipality. Journal of Waste Management 26(11), 1287–1293 (2006)
6. Nuortio, T., Kytojoki, J., Niska, H., Braysy, O.: Improved route planning and scheduling of waste collection and transport. Journal of Expert Systems with Applications 30(2), 223–232 (2006)
7. Li, J.Q., Borenstein, D., Mirchandani, P.B.: Truck scheduling for solid waste collection in the City of Porto Alegre, Brazil. Journal of Omega 36(6), 1133–1149 (2008)
8. Zamorano, M., Molero, E., Grindlay, A., Rondriquez, M.L., Hurtado, A., Calvo, F.J.: A planning scenario for the application of geographical information systems in municipal waste collection: A case of Churriana de la Vega (Granada, Spain). Journal of Resources, Conservation and Recycling 54(2), 123–133 (2009)
9. Tavares, G., Zsigraiova, Z., Semiao, V., Carvalho, M.G.: Optimisation of MSW collection routes for minimum fuel consumption using 3D GIS modeling. Journal of Waste Management 29(3), 1176–1185 (2009)
10. Benjamin, A.M., Beasley, J.E.: Metaheuristics for the waste collection vehicle routing problem with time windows, driver rest period and multiple disposal facilities. Journal of Computers & Operations Research 37(12), 2270–2280 (2010)
11. Son, L.H.: Optimizing Municipal Solid Waste collection using Chaotic Particle Swarm Optimization in GIS based environments: A case study at Danang city, Vietnam. Journal of Expert Systems with Applications 41(18), 8062–8074 (2014)
12. Guerrero, L.A., Maas, G., Hogland, W.: Solid waste management challenges for cities in developing countries. Journal of Waste Management 33(1), 220–232 (2013)
13. Marshall, R.E., Farahbakhsh, K.: Systems approaches to integrated solid waste management in developing countries. Journal of Waste Management 33(4), 988–1003 (2013)
14. Medvedev, A., Zaslavsky, A., Grudinin, V., Khoruzhnikov, S.: Citywatcher: annotating and searching video data streams for smart cities applications. In: Balandin, S., Andreev, S., Koucheryavy, Y. (eds.) NEW2AN/ruSMART 2014. LNCS, vol. 8638, pp. 144–155. Springer, Heidelberg (2014)
15. Medvedev, A., Zaslavsky, A., Khoruzhnikov, S., Grudinin, V.: Reporting road problems in smart cities using OpenIoT framework. In: Podnar Žarko, I., Pripužić, K., Serrano, M. (eds.) FP7 OpenIoT Project Workshop 2014. LNCS, vol. 9001, pp. 169–182. Springer, Heidelberg (2015)

16. Perera, C., Zaslavsky, A., Christen, P., Georgakopoulos, D.: Sensing as a service model for smart cities supported by Internet of Things. Transactions on Emerging Telecommunications Technologies **25**(1), pp. 81–93
17. Anagnostopoulos, T.V., Zaslavsky, A.: Effective waste collection with shortest path semi-static and dynamic routing. In: Balandin, S., Andreev, S., Koucheryavy, Y. (eds.) NEW2AN/ruSMART 2014. LNCS, vol. 8638, pp. 95–105. Springer, Heidelberg (2014)
18. OpenStreetMap Webpage (accessed on 03 March 2015). https://www.openstreetmap.org
19. Nominatim Wiki Webpage (accessed on 03 March 2015). http://wiki.openstreetmap.org/wiki/Nominatim
20. Dantzig, G.B., Ramser, J.H.: The Truck Dispathing problem. Management Science **6**(1), 80–91 (1959)
21. JSPRIT Webpage (accessed on 15 April 2015). http://jsprit.github.io/
22. Open-VRP Webpage (accessed on 15 April 2015). https://github.com/mck-/Open-VRP/
23. OptaPlanner Webpage (accessed on 15 April 2015). http://www.optaplanner.org/
24. SYMPHONY Webpage (accessed on 15 April 2015). https://projects.coin-or.org/SYMPHONY/
25. VRP Spreadsheet Solver Webpage (accessed on 27 April 2015). http://verolog.deis.unibo.it/vrp-spreadsheet-solver/
26. TSP Webpage (accessed on 27 April 2015). http://en.wikipedia.org/wiki/Travelling_salesman_problem/
27. GraphHopper Webpage (accessed on 27 April 2015). https://graphhopper.com/

Service Intelligence Support for Medical Sensor Networks in Personalized Mobile Health Systems

Dmitry G. Korzun[1]([⊠]), Ilya Nikolaevskiy[2], and Andrei Gurtov[3,4]

[1] Department of Computer Science, Petrozavodsk State University (PetrSU),
33, Lenin Ave., Petrozavodsk 185910, Russia
dkorzun@cs.karelia.ru
[2] Department of Computer Science, Aalto University,
PO Box 19800, 00076 Aalto, Finland
ilya.nikolaevskiy@aalto.fi
[3] Helsinki Institute for Information Technology HIIT, Espoo, Finland
gurtov@hiit.fi
[4] ITMO University, Saint Petersburg, Russia

Abstract. Mobile health (m-Health) scenarios form an important direction for enhancing "traditional" healthcare systems. The latter implement backend services for use primarily by medical personnel and typically at hospitals. Current development meets with the challenge of personal data inclusion to the whole healthcare system with subsequent "smart" service construction and delivery. This paper makes a step towards the concept development of intelligence support in personalized m-Health systems. We study a reference architectural model that aims at intelligent utilization of personal mobile data in generic health services. Each personalized m-Health system contains the patient's medical sensor network (MSN). To support the service intelligence we employ the smart spaces paradigm with its prominent technologies adopted from the Internet of Things (IoT) and Semantic Web.

Keywords: Medical sensor network · Mobile healthcare · m-Health · Service intelligence · Smart spaces · Internet of Things · Semantic Web

1 Introduction

The recent advances in bioengineering and the proliferation of wireless sensor platforms have allowed the realization of pervasive and mobile health (m-Health) systems [1,2]. Sensors, wearable by a patient or implantable within the body, form a medical sensor network (MSN) accessible to medical personnel and the patient herself. The mobile terminal centric view states that end-user device (such as smartphone) becomes a personalized access point and service hub from the patient's MSN to enhanced healthcare system.

Being dynamically utilized, rich of multi-source data, and dependent on changeable surrounding environment, healthcare services have to be made smart [3,4]. It requires a possibility to understand the recent situation and latest

S. Balandin et al. (Eds.): NEW2AN/ruSMART 2015, LNCS 9247, pp. 116–127, 2015.
DOI: 10.1007/978-3-319-23126-6_11

sensed data, and then to react in a best-effort manner. Intelligent access to personal patient's data offers many opportunities to enhance the delivery of services, to improve the patient experience, and to advance integrated health [5,6].

This paper opens a discussion on the challenge of intelligence support in personalized m-Health systems. Our approach is based on emerging technologies of Smart Spaces, Internet of Things (IoT), and Semantic Web. We present a reference architectural model that supports intelligence in attached personalized MSN-based m-Health systems. The discussion involves into the theoretical consideration the objectives and properties that reflect the key demands of enhanced healthcare systems [1,3,7].

Our solutions provide smart space based support for service adaptation, personalization, and proactive delivery. Dynamic relation of multi-source data forms a smart space [8,9] as a specific case of an information hub. It supports semantics-based analysis of collected data and derived knowledge directly in this space. The smart space allows feeding health services with non-medical data. Enhanced health applications are enabled, which are not based purely on electronic health records [5,10,11]. Our approach is based on principles developed in the open research pilot for smart spaces—Smart-M3 platform [12]. The platform is oriented to a wide range of IoT-aware multi-domain applications.

This study makes a step towards the concept development of intelligence support in personalized m-Health systems. We consider the following results the main contribution of this development.

- Reference architectural model for healthcare systems enhanced with personalized MSN-based m-Health systems. The focus is on intelligence support that can be realized based on this architecture. Other aspects (e.g., security) are considered in detail in other publications.
- Properties of the service intelligence support achieved due to the use of smart spaces for enhancing healthcare systems. The focus is on semantics a personalized m-Health system requires to be represented in the smart space.

The rest of the paper is organized as follows. Section 3 provides background information on smart spaces, motivating their role in development of healthcare systems. Section 4 describes our reference architectural model with intelligence support. Section 5 provides the system design based on the Smart-M3 platform. Section 6 summarizes the paper.

2   Related Work

Surveys [1,2] make introduction to MSN and their use in m-Health systems and in IoT settings. Jara *et al.* [13] provide an interconnection IoT-aware framework for personalized m-Health systems. The proposed solution is based on a novel interconnection protocol YOAPY (it means *connect* and *link* in Guarani language). The YOAPY preprocessing and data aggregation module uses domain specific methods to make real-time data transmission feasible (e.g. special compression methods for ECG). The solution made continuous and remote vital sign monitoring feasible and introduced technological innovations for empowering health monitors and patient devices with Internet capabilities. The work, however, does not concern service intelligence in generic m-Health scenarios.

Architectural solutions for personalized m-Health systems based on personal mobile gateways are studied in [1,14]. In our study, we adapt the generic IoT-aware system architecture from [15], which enables security of personal patient's data and their transfer to backend healthcare services.

Intelligence for generic IoT-aware scenarios was discussed in [16]. The role of smart spaces for such scenarios was shown in [8]. The need of intelligence for advancing services and improving their security is recognized important, e.g., in [17,18]. Intelligence for healthcare systems was analyzed in [3,4]. In particular, work [3] proposes a smart healthcare systems framework for conceptualizing data-driven and mobile- and cloud-enabled smart healthcare systems. Survey [4] examines the infrastructure and technology that ambient intelligence techniques require in healthcare scenarios.

In the healthcare domain, the general discussion on IoT application development can be found in [4,6,19]. As a particular example, the intelligence of healthcare monitoring was considered in [11], where two pilot applications were developed: one is for bedside monitoring of cardiac patients at hospitals, the other is for homecare monitoring of patients after a revascularisation therapy. Mileo *et al.* [20] presented an intelligent home environment to context-aware monitoring and to control the evolution of patient's health and home environment. Castillejo *et al.* [21] presented an everyday life application involving a WSN as the base of a novel context-awareness sports scenario that includes a smartwatch, a physiological monitoring device, and a smartphone.

Application of smart spaces for healthcare system was studied in [9], solving interoperability issues in IT-based support of the entire healthcare cycle. The role of ontologies for structured representation of multi-source data for effective processing in medical systems is presented in [22,23]. This paper focuses on solutions for enhancing existing healthcare system with personalized m-Health subsystems. The enhancement is based on advanced properties that the smart spaces approach provides in the form of service intelligence support.

3   Smart Spaces

The smart spaces approach aims at constructing advanced service-oriented systems for ubiquitous computing and IoT-aware environments [8]. In contrast to the giant global graph approach of Semantic Web, smart spaces apply the localization principle [24]: do not duplicate data from the whole world but relate locally the problem-aware fragments from multiple sources. Massive data are kept in appropriate external databases and information systems on backend servers. A service is constructed by interaction of software agents. They access the smart space as a shared semantic information hub and use the publish/subscribe coordination model for interaction. We expect that the approach is reasonable for constructing personalized m-Health systems, similarly as happening now in such application domains as collaborative work [25] and e-Tourism [26].

Commercial examples of IoT platforms aimed for managing and storing sensor data in the cloud include Xively and ThingWorx. Medical data processing is covered by strict privacy laws such as Health Insurance Portability and Accountability Act (HIPPA) in the USA and Directive on Data Protection in EU. The legislation puts special conditions on smart processing of patient data, for example by requiring that data never leaves national borders.

The semantics describe relations between information fragments kept in the smart space. From the point of view of ontology modeling, information fragments are objects representing concepts of the problem domain and their instances. In the healthcare case, objects represent patients, medical personnel, diseases, medical supplies, recommendations, etc. Semantic relations represent such knowledge as a patient is ill with pneumonia or as correspondence of diseases to appropriate medicines. Ontology is an effective description tool for this kind of knowledge. For instance, study [22] introduced an ontology for the personalized management of chronically ill patients at home that integrates general medical knowledge with knowledge specific for this kind of patients. The advantage of ontological representation for integrating medical, social, and familiar resources was shown in [23] using senior homecare assistance as a case study.

The ontological representation of available knowledge within a smart space addresses the clear goal: enabling the integral care of the patient in her/his own environment. The relevance of this approach was demonstrated in [9] based on linking the following information layers of person-centric healthcare systems.

- User plane: local patient monitoring and feedback (interactions by and with the patients).
- Medical plane: assessment of clinical data, diagnosis, treatment planning and execution, and feedback to the patient (interactions by and with the doctors).
- Statistical plane: external knowledge management (interactions with medical researchers).

Based on ontological representation, agents derive knowledge by understanding the semantics from the smart space. Then agents apply this knowledge for service construction and delivery. In particular, they are able to recognize and react on situations such as

- service clients appear,
- data processing over suitable databases is needed,
- service outcome is ready for delivering.

The above situations are important when a healthcare system aims at provision of so-called "smart" services [3]. Criteria of "smart healthcare service" is still open for discussion; the details is out of the scope of this paper.

A possible open source platform to construct and deploy smart spaces is Smart-M3 [12]. The central architectural element is a semantic information broker (SIB). It runs on a dedicated Internet host. SIB maintains an RDF-based knowledge base (an RDF triplestore with information search extensions) for interoperable information sharing. The content is accessible by agents directly communicating with their SIB. Read/write operations with the content are provided, where RDF triples are parameters. Subscription operation allows an agent to detect changes in a specified part of the shared content. Smart-M3 suggests the term "knowledge processor" (KP) to distinguish this class of agents from general multi-agent systems. The emphasis is on asynchronous collective knowledge generation and utilization via information sharing.

The basic coordination model is reacting on observations in the shared and collectively-generated content.

1. Reading data: by instant query or subscription.
2. Making autonomous decisions: based on shared data and local knowledge.
3. Reaction: actions to the environment or publishing new content.

This type of interaction evolves the generic IoT vision on smart objects as augmented with sensing/actuation, processing, and network capabilities [16]. Each acts autonomously making own decisions, senses the environment, communicates with other objects, accesses the Internet and Web, and interacts with users. This IoT potential for m-Health systems is indicated in [11,13]. Note that autonomous decisions do not prevent rule-based system decisions, which are important in healthcare domain (e.g., human may activate a critical decision)

4   Architectural Model

The architectural model is derived from [15], see Fig. 1. We address a wide spectrum of healthcare systems that can be created using existing health services. A patient's personalized m-Health system is formed on top of a MSN. Then this system is attached to the entire healthcare system via the patient's gateway. A smart space is created where private MSNs and healthcare services can cooperate based on information collecting and sharing. The idea is to collect information from the participants and to support its shared intelligent use.

The architectural model supports the user mobility. The digital service environment on the side of medical facilities (e.g., in hospital) is enhanced with remote participants (i.e., patients and medical personnel). Based on the smart spaces approach, services become accessible in a semantic rendezvous style. Recent system's state and semantics of medical data are dynamically represented in the smart space such that users, their data and context are interrelated with services. For instance, if a user and a service are represented as informational objects in the smart space then a link connecting the objects can be also stored in the smart space to represent that the user needs the service. Activities of users become synchronized with multiple services that the healthcare system provides. That is, the smart space provides the semantics for deciding which services are appropriate at the moment for a particular user.

All MSN devices form a private information space of the patient. In general, it covers Body Area Network (BAN) and Personal Area Network (PAN). The

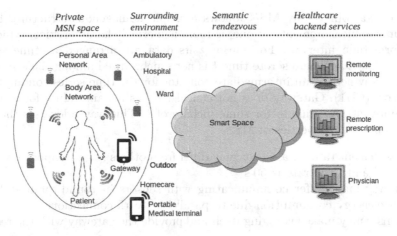

Fig. 1. A healthcare system is enhanced with personalized m-Health systems.

BAN subsystem consists of wearable and implantable devices. ECG or glucose sensors, insulin pumps and accelerometers are examples of such components. The PAN subsystem is composed of environmental sensors deployed around as well as of portable and mobile devices that belong to the patient. Temperature and humidity sensors, tablet PCs, and smartphones are rich sources of personal and environmental information.

MSN includes a gateway, which becomes responsible for communication between devices inside the MSN and the outside world [15]. The gateway aggregates, processes, and shares information in the smart space. Which data from the private MSN space to share in the smart space is also decided by the gateway. Gateway also receives information from smart space. This enables two-way communications between patient and medical personnel. For use in emergency situations, where Internet access is absent for gateway, each medical team should be equipped with a Portable Medical Terminal (PMT). A PMT is very similar to the gateway (e.g., smartphone or tablet PC) and is used by medical personnel to directly access the patient's MSN.

Healthcare services run on backend servers in medical facility or in clouds. Services acquire input data from the MSN, surrounding environment, and status/outcome of other services.

All m-Health systems together constitute the personalized part of the entire healthcare system. On behalf of its patient, the MSN space acts as a health data source for the smart space. Health services become exploiting private MSN data. The smart space collects data from many patients, semantically relates the data with other informational sources, supports knowledge reasoning over this multidimensional and multi-domain information content, and provides deduced knowledge for use in healthcare services. In turn, each patient receives services using a gateway—user interface (UI) functions. Similar functions are performed using an appropriate PMT by medical personnel on the patient side.

As a data source, any MSN follows a simple homogenous structure. Each sensor makes measurements of one or more numerical characteristics, typically at uniform time intervals. For sensor i its data are arranged as time series $v_{i1}, v_{i2}, \ldots, v_{it}, \ldots$. The discrete time t is not synchronized among the sensors.

A gateway acts as an intermediate collector for the time series coming from the sensors (CH1). Gateway does not consider v_{it} as instant values for given t. They represent an estimate for some vicinity of t. This way addresses the following important properties of MSN.

1. Measurement rate for a sensor may differ from others (e.g., temperature rate is 1 s^{-1} and ECG rate is 50 s^{-1}).
2. Gateway may prefer communicating with sensors in round-robin style to reduce energy consumption due to parallel network connections.
3. A sensor may make averaging itself and provide the gateway with the result.

Therefore, a kind of measurements synchronization is performed: time-close data are aggregated on a interval (vicinity of t). If no measurement is available for t then the latest measured value is used. The intervals provide common time for all samples. This way supports computationally simple decision-making based on analysis of vectors

$$\mathbf{v}(t) = (v_1, v_2, \ldots, v_n),$$

where t is the time of gateway, a synchronized time point for measurements from the n sensors.

The analysis extracts local semantics relating the n measured parameters between themselves and in short-range time horizon. The processing can be implemented directly at the gateway (in parallel to transferring the data to the smart space). Methods of pattern recognition can be applied or simple rule-based reasoning. An example is alert appearing on the gateway UI when a certain characteristic exceeds a bound. Another example is emergency response triggered (e.g., SMS is sent on behalf of the patient) when several vital signals become changing in correlation (e.g., simultaneous high peaks). Methods for extracting such local semantics of sensor-measured data are well-elaborated in healthcare monitoring systems [1].

This way of sensor data accumulation and local semantics extraction supports limited utilization of the gateway capacity. The gateway always keeps a limited-size fragment of the time series. From the security point of view, even being lost the gateway does not cause much leakage of private information. As for the capacity objective, no massive data processing is performed on the gateway, all resources-expensive analysis is delegated to the backend service infrastructure.

Smart space represents data from multiple sources. Long time series as well as other massive data are stored in dedicated information systems at the backend servers. The smart space keeps representation of objects of the problem domain: patient, her or his health parameters, context, etc. For accessing additional data the object provides references to appropriate information systems.

Consequently, traditional analysis of individual time series of a patient can be performed at the backend servers. In turn, the result is published in the smart space as derived knowledge. For example, personalized recommendations on motional activity are associated with the patient in an adaptive style.

In addition, combined multi-person analysis is possible. For instance, searching similar time series among different patients can be performed at the backend servers. Then the smart space keeps the result to relate semantically the patients with similar symptoms. That is, the smart space provides a way to form person-to-person relationships in the data based on semantics. Similarly, the measurements can also be related with other available data, e.g., patient profiles and medical notes from physicians [5].

This heterogeneous informational content in the smart space can be further semantically interlinked to support knowledge reasoning. The known effective technology for representing such semantics is RDF from Semantic Web. In this case, establishing a relation becomes a simple act of publishing few RDF triples. We apply this technology in our Smart-M3 based solution for the above smart space construction.

5   Smart-M3 based Multi-agent Design

We employ Smart-M3 platform [12] as a testbed infrastructure. Several KP types are introduced to implement a personalized m-Health system, see Table 1. The healthcare system SIB runs on a dedicated backend server.

On the patient side, the semantic-oriented MSN data representation is simple. Values v_{it} and derived conclusions can be encoded with a small set of RDF triples. The triples contain numerical values and relate them with other attributes (e.g., time, sensor ID). In the smart space, keeping time series data in RDF is inefficient. The raw numerical data are stored in traditional databases, as it happens in many healthcare information systems. The smart space keeps links to relate the patient's representation (profile) with all personal data and derived knowledge collected from the databases. This solution follows the semantic hub property of smart spaces, allowing easy linking of knowledge for any given person. The emerging potential of medical information integration in a smart space was explained in [9]. Note that many ontologies are already developed for representing patients and medical data.

Table 1. Smart-M3 KPs for smart space based healthcare system.

Type	Device	Role
MSN data collector	Gateway, PMT	KP collects health data from the patient and forwards them to the healthcare system and the smart space.
Service	Backend server	KP activates appropriate service and mediator KPs to construct the service when there are clients. Its outcome is represented in the smart space to deliver to client KPs.
Mediator	Backend server	KP runs certain data processing over its database or information system and makes the outcome represented in the smart space.
UI agent	Gateway, PMT	KP shows results from the healthcare services to the user based on current situation in the smart space and at the patient side.

Now we can define the following generic iterations. The life cycle is started and ended at patient's gateway.

1. MSN data collector iteratively feeds the system with health data. The status of this regular process is published in the smart space.
2. Personal data are collected in healthcare system databases. Service KPs recognize their semantics (relations) and publish the knowledge in the smart space.
3. Service KP recognizes a situation in the smart spaces when the service is needed. The notification is represented in the smart space, and all relevant services and mediator KPs start to cooperate in the service construction.
4. Mediator KP initiates task-specific processing in the database using known methods (e.g., time series analysis, pattern recognition). The derived knowledge is published in the smart space (relating with already available content).
5. UI agent KP responds to the service outcome and visualizes it appropriately on the patient side.

Clearly, for each particular health service and its parent healthcare system the iterations stated above depend on the concrete problem and require problem-specific design and implementation. In this paper we address a generic solution framework based on the proposed architectural model. Consider some advantages of the generic solution.

Overhead on the patient side is preserved low in terms of memory and processing consumption. Complex processing (e.g., data mining, construction of semantics, knowledge derivation) is delegated to the backend infrastructure. Gateway constructs an aggregated picture of measurements from sensors and keeps a short-term snapshot due to the limited-size window. Processing of such short-term snapshots is made cheap due to simple structure of the numerical data. The whole time series are transferred to the backend infrastructure. The Semantic Web technologies provide methods and software of practically reasonable performance within a single RDF triplestore. Smart-M3 preserves the efficiency in operation with dynamic mobile data coming in parallel from multiple users. When collected in databases of the backend infrastructure, the medical data can be efficiently processed using already developed methods, as it happens now in many medical information systems, see reviews in [4,6,10]. Moreover, our architectural model allows certain data processing to be delegated to cloud infrastructure if very high performance is needed.

The mobility support of a patient is also intelligent. When the network connection is temporarily absent, the data publication process is postponed. If the absence is short then the gap in data is filled after reconnection. For long absence of connection the data from this period are not stored in the healthcare system. This data collection irregularity is reflected on the system side (process status in the smart space), and healthcare system can react properly, e.g., notifying the responsible medical personnel.

Our Smart-M3 based solution intentionally supports the following properties for the service intelligence. All they are mostly due to the semantic-aware representation of smart space content and the proposed KP roles.

Adaptation: The smart space is regularly fed up with recent knowledge on the involved participants and environment. This knowledge is accessible to a service, which can adapt its operation properly. For example, a health monitoring service provides motional activity recommendation online.

Context-Awareness: Data coming from patients and medical personnel include contextual data such as geolocation and status. For example, in emergency response the target patient can be associated with all medical personnel that are (i) nearby (fast reachability from the current location) and (ii) qualified (required competences and equipment for the case).

Personalization: Every patient and medical personnel has personal representation in the smart space (both factual data and semantic relations). A service applies this knowledge to personalize operation and delivery. For example, diet and medicine recommendations take into account observed cases of allergy. Moreover, a personal gateway can make own context-aware interpretation of the service outcome and find its best visualization for the user.

Proactive Delivery: Mediator KPs explicitly represent in the smart space such situations when a service is needed to a client. This representation is detected by service KPs to start appropriate services. For example, a patient is notified about nearby hospitals when she or he is walking far from the home.

In summary, this intelligence support makes so many opportunities for advancing health services that the number is limited by our imagination only.

On one hand, the smart space preserves the composition of content as coming from particular patients. Each person-related part of information, indeed, can be enhanced with semantic links to other informational pieces. On the other hand, the modular structure of KPs supports construction of a service oriented to a given person. It includes proactive service delivery when a service KP detects a need of the service for a given user.

6   Conclusion

This paper analyzed a reference architectural model for inclusion of personalized MSN-based m-Health systems, as mobile components, to an entire healthcare system. The model enhances health services for out-of-hospital settings with best-effort and personalized delivery to patients and their medical personnel. We present properties of the service intelligence support, which are achieved due to the use of smart spaces. This result provides the conceptual means of making MSN-measured patient's data be shared in the smart space. The content is enhanced with derived knowledge that relates information fragments from multiple sources, including conclusions from patient's data analysis. The presented properties of the service intelligence support allow integrated use of personal data and other relevant medical information and open new opportunities for making smart healthcare services. We plan to apply the considered solutions for implementing particular m-Health service scenarios, see [27] for scenario details.

Acknowledgments. This work is financially supported by the Ministry of Education and Science of the Russian Federation within project # 14.574.21.0060 (RFMEFI57414X0060) of Federal Target Program "Research and development on priority directions of scientific-technological complex of Russia for 2014–2020".

References

1. Alemdar, H., Ersoy, C.: Wireless sensor networks for healthcare: A survey. Computer Networks **54**(15), 2688–2710 (2010)
2. Kumar, P., Lee, H.J.: Security issues in healthcare applications using wireless medical sensor networks: A survey. Sensors **12**(1), 55–91 (2012)
3. Demirkan, H.: A smart healthcare systems framework. IT Professional **15**(5), 38–45 (2013)
4. Acampora, G., Cook, D.J., Rashidi, P., Vasilakos, A.V.: A survey on ambient intelligence in healthcare. Proceedings of the IEEE **101**, 2470–2494 (2013)
5. Mandl, K.D., Mandel, J.C., Murphy, S.N., Bernstam, E.V., Ramoni, R.L., Kreda, D.A., McCoy, J.M., Adida, B., Kohane, I.S.: The SMART platform: early experience enabling substitutable applications for electronic health records. Journal of the American Medical Informatics Association **19**(4), 597–603 (2012)
6. Yang, C.C., Leroy, G., Ananiadou, S.: Smart health and wellbeing. ACM Trans. Manage. Inf. Syst. **4**(4), 15:1–15:8 (2013)
7. Mingyu, W., Qiang, Z., Weimo, Z., Jijiang, Y., Qing, W., Weiyi, Q., Dingcheng, X., Minwei, Z., Yan, T., Hao, C., Jian, L., Xiaoqian, L., Hongdi, W., Geng, L., Qiang, G.: Remote rehabilitation model based on BAN and cloud computing technology. In: IEEE 14th Int'l. Conf. on e-Health Networking, Applications and Services (Healthcom), pp. 119–123. IEEE Computer Society, October 2012
8. Korzun, D.G., Balandin, S.I., Gurtov, A.V.: Deployment of smart spaces in internet of things: overview of the design challenges. In: Balandin, S., Andreev, S., Koucheryavy, Y. (eds.) NEW2AN 2013 and ruSMART 2013. LNCS, vol. 8121, pp. 48–59. Springer, Heidelberg (2013)
9. Vergari, F., Cinotti, T.S., D'Elia, A., Roffia, L., Zamagni, G., Lamberti, C.: An integrated framework to achieve interoperability in person-centric health management. Int. J. Telemedicine Appl. **2011**, 5:1–5:10 (2011)
10. Hovenga, E.J.S., Kidd, M.R., Garde, S., Cossio, C.H.L. (eds.): Health Informatics: An Overview. Studies in Health Technology and Informatics, vol. 151. IOS Press, Amsterdam (2010)
11. Nee, O., Hein, A., Gorath, T., Hulsmann, N., Laleci, G.B., Yuksel, M., Olduz, M., Tasyurt, I., Orhan, U., Dogac, A., Fruntelata, A., Ghiorghe, S., Ludwig, R.: SAPHIRE: intelligent healthcare monitoring based on semantic interoperability platform: pilot applications. IET Communications **2**(2), 192–201 (2008)
12. Honkola, J., Laine, H., Brown, R., Tyrkkö, O.: Smart-M3 information sharing platform. In: Proc. IEEE Symp. on Computers and Communications (ISCC 2010), pp. 1041–1046. IEEE Computer Society, June 2010
13. Jara, A.J., Zamora-Izquierdo, M.A., Skarmeta, A.F.: Interconnection framework for mHealth and remote monitoring based on the Internet of Things. IEEE Journal on Selected Areas in Communications **31**(9), 47–65 (2013)
14. Kuptsov, D., Nechaev, B., Gurtov, A.: Securing medical sensor network with HIP. In: Nikita, K.S., Lin, J.C., Fotiadis, D.I., Arredondo Waldmeyer, M.-T. (eds.) MobiHealth 2011. LNICST, vol. 83, pp. 150–157. Springer, Heidelberg (2012)

15. Nikolaevskiy, I., Korzun, D.G., Gurtov, A.: Security for medical sensor networks in mobile health systems. In: Proc. IEEE Int'l. Symp. on a World of Wireless, Mobile and Multimedia Networks (WoWMoM 2014), pp. 1–6. IEEE Computer Society (2014)
16. Kortuem, G., Kawsar, F., Sundramoorthy, V., Fitton, D.: Smart objects as building blocks for the internet of things. IEEE Internet Computing 14(1), 44–51 (2010)
17. Gurtov, A., Nikolaevskiy, I., Lukyanenko, A.: Using HIP DEX for key management and access control in smart objects. In: Proc. of Workshop on Smart Object Security, March 2012 (Position paper)
18. Evesti, A., Suomalainen, J., Ovaska, E.: Architecture and knowledge-driven self-adaptive security in smart space. Computers 2(1), 34–66 (2013)
19. Whitmore, A., Agarwal, A., Xu, L.: The Internet of Things–A Survey of Topics and Trends. Information Systems Frontiers 17(2), 261–274 (2015)
20. Mileo, A., Merico, D., Bisiani, R.: Support for context-aware monitoring in home healthcare. J. Ambient Intell. Smart Environ. 2(1), 49–66 (2010)
21. Castillejo, P., Martínez, J., López, L., Rubio, G.: An Internet of Things approach for managing smart services provided by wearable devices. International Journal of Distributed Sensor Networks 2013 (2013)
22. Riaño, D., Real, F., López-Vallverdú, J.A., Campana, F., Ercolani, S., Mecocci, P., Annicchiarico, R., Caltagirone, C.: An ontology-based personalization of healthcare knowledge to support clinical decisions for chronically ill patients. J. of Biomedical Informatics 45(3), 429–446 (2012)
23. Valls, A., Gibert, K., Sánchez, D., Batet, M.: Using ontologies for structuring organizational knowledge in home care assistance. International Journal of Medical Informatics 79(5), 370–387 (2010)
24. Korzun, D.: Service formalism and architectural abstractions for smart space applications. In: Proc. 10th Central & Eastern European Software Engineering Conference in Russia (CEE-SECR 2014). ACM, October 2014
25. Korzun, D., Galov, I., Kashevnik, A., Balandin, S.: Virtual shared workspace for smart spaces and M3-based case study. In: Balandin, S., Trifonova, U. (eds.) Proc. 15th Conf. of Open Innovations Association FRUCT, pp. 60–68. ITMO Univeristy, April 2014
26. Smirnov, A., Kashevnik, A., Balandin, S.I., Laizane, S.: Intelligent mobile tourist guide. In: Balandin, S., Andreev, S., Koucheryavy, Y. (eds.) NEW2AN 2013 and ruSMART 2013. LNCS, vol. 8121, pp. 94–106. Springer, Heidelberg (2013)
27. Borodin, A., Zavyalova, Y., Zaharov, A., Yamushev, I.: Architectural approach to the multisource health monitoring application design. In: Proc. 17th Conf. of Open Innovations Association FRUCT, pp. 36–43. ITMO Univeristy, April 2015

Synthesis of Multi-service Infocommunication Systems with Multimodal Interfaces

O.O. Basov[1], D.A. Struev[1], and A.L. Ronzhin[2,3(✉)]

[1] Academy of FAP of Russia, 35, Priborostroitelnaya, Orel 302034, Russia
{oobasov,dima-orel86}@mail.ru
[2] SPIIRAS, 39, 14th line, St. Petersburg 199178, Russia
ronzhin@iias.spb.su
[3] SUAI, 67, Bolshaya Morskaya, St. Petersburg 190000, Russia

Abstract. In the paper, the necessity of creating infocommunication systems of state government providing all the subscribers' communication interaction aspects (communicational (information exchange), interactive (actions exchange) and perceptive ones (during communication partners get acquainted with each other) is proved. The realization prospects of these systems based on the multimodal man-computer interfaces are also shown. The key research and methodology problems of their synthesis are formulated. For their solution based on the suggested approach to effectiveness evaluation, a polymodal infocommunication systems decomposition into multimodal interfaces and data networking has been carried out. A method allowing to synthesize infocommunication systems based on the multimodal man-computer interfaces with required features of sustainability, running speed, efficiency and costs has been suggested and gradually outlined which is proved when bringing into practice new "perceptive" services.

Keywords: Multimodal interface · Polymodal infocommunication system · Data networking · Information-algorithmic structure · System solution · Hardware-software means · Features optimization

1 Introduction

In the modern society the information appears to be an essential component providing full life-sustaining activity of its end users-citizens, as well as of the state on the whole. This fact is reflected in the ongoing tendency of the shift from the industrial society to the informational one. The information provision level of the authorities is regarded as a vital feature of the state governing system in every civilized society. The accumulated experience of the building and improving the state governing system proves that informational provision should be regarded as one of the most strategic directions for increasing the activity efficiency on each level: international, state, regional, field and etc. These regularities demand some active steps towards forming the single information space of the state authorities and securing effective and sustainable state government on this basis.

© Springer International Publishing Switzerland 2015
S. Balandin et al. (Eds.): NEW2AN/ruSMART 2015, LNCS 9247, pp. 128–139, 2015.
DOI: 10.1007/978-3-319-23126-6_12

The single state information space of the public administration system of the Russian Federation should represent a system of organized and technically bound, synchronized in time informational and communicational resources of all kinds of property. The enhancement of the state administration processes on this basis is a complex multifaceted scientific problem, which should be solved gradually by involving experts in system analysis operation study, etc.; field experts in different sectors of science; moreover, scientists engaged in theory and practice of developing new infocommunication technologies. Nowadays, it is a well known fact that without applying the latest achievements in cybernetics, informatics, communication, computing and telecommunication an effective state administration is impossible.

The change in the life-sustaining conditions, expansion of the communicative interaction sphere between subjects of the state administration system and, furthermore, constantly growing loads on their psychological activity make the communication processes between the officials of the administrating authorities more diverse and intense. Moreover, a transformation of the formal-acting communication into businesslike is more easily tractable now, where along with the information exchange the subject's personal qualities, their mood, physiological and psycho emotional states should be taken into account in order to achieve a certain result. For registering the state data, all aspects of communication should be considered: communicational (information exchange), interactive (actions exchange) and perceptive (when communicating partners get to know each other). However, these aspects are revealed only during the direct contact between correspondents. The registering of subjects' psychoemotional state during their interaction with the help of technical communication means appears to be a very difficult task. For example, traditional telecommunication systems, when performing their functions of receiving, processing, transmitting and recovery of the information, fulfill only communicative side of interaction.

In the course of extensive development towards fulfilling multiserviceness (transmitting voice, video and data), telecommunication systems have developed into infocommunication. An infocommunication system will be regarded hereinafter as an intertwined aggregate of processing and storing systems; telecommunication systems which unite them and function under the sole government with the purpose of collecting, processing, storing, protecting, transmitting and reallocating, reflecting and utilizing the information to the benefits of the subscribers (users).

2   Prerequisites for Creating Infocommunication Systems Based on the Multimodal Human-Computer Interfaces

In correspondence to the definition presented above, modern infocommunication systems deliver interactive aspect of communication to some extent. Due to that, in recent years a tendency of separate infocommunication services has been observed, i.e. a subscriber often demands one particular service called "network connection". This service presupposes an opportunity of receiving an accessible or the most suitable interaction form, defined according to personal preferences and physical restriction and also by the environment where the communication is taking place.

During traditional interpersonal communication, people nearly always interact through verbal and nonverbal channels. Accordingly, an objective necessity arises to investigate the ways of subscribes' dialogue polymodality provision in the course of communicative interaction via technical means. Creating polymodal technical communication systems has become possible due to modern achievements of the cognitive science, which studies human perceptive mechanisms and interpersonal dealings. This scientific field has ensured nowadays acquisition of fundamental results in subscribers' behavior modeling and revealed regularities of multimodal man-computer systems constructing and application. Signals of separate modalities (speech, lips movement, eye movement, facial muscles movement, gestures, traditional and pen-based computing) in the following systems appear as the main research objects. They are comprehensively analyzed in terms of compact voice and image presentation and recognition; lip reading; gesture definition; definition of physiological (including tiredness) and psycho-emotional state; authentication (identification, verification) of a subscriber; validity estimation of the transmitted information, voice and image synthesis.

Juxtaposition of the achievements in cognitive science and research results in the field of telecommunications and infocommunications testifies to the necessity and possibility of reallocation and sequence of information from different (traditional and new) sensor systems (analyzers) between subscribers. Thus, we can talk about the viability of the infocommunication systems realization based on the multimodal man-computer interfaces. In this respect, two main tasks appear practically important.

Direct task lies in determining the transport infrastructure resources volume (data network), which is essential for providing the required quality of reception of the transmitted multimodal information.

Dual task - as a result of its solution it is possible at given resources of the data network to transmit the ultimate volume of messages of various modalities with a predetermined quality.

Under such conditions, even a partial refusal from traditional principles of transmitted information division into communication services in favour of multimodal representation demands developing a synthesis (designing) of polymodal infocommunication systems (PICS). While designing infocommunication systems, contradiction inevitably occurs between the desire to introduce the system immediately and achieve necessary results with acceptable costs; the desire to create the system, while following traditional methods, and aspirations to a conceptually new approach resulting from the nature of polymodal systems of infocommunication.

Uniqueness of PICS, defined by state administration specifics, sets restrictions for system technologies application. In other words, there is a finite (and relatively small) aggregate of telecommunication and information technologies; a significantly bigger but finite aggregate of their hardware-software realizations (subscribers and network devices of the access networks; routers and switch-boards of different purposes; means of information cryptographic protection; crypto-routers; multiplexors, etc.); and a nearly infinite aggregate of system solutions.

Given a substantial variety of hardware-software realizations capable of meeting the system requirements with the external efficiency maximum, a problem of their rational choice arises. A number of different markers, which in their turn can affect structures and principles of multimodal information exchange organization, characterize already existing and being developed hardware-software means.

Thus, a problem of choosing rational system solutions arises, while represented considerations require developing a systemic methodological approach to such solutions for multitude forming.

3   System Approach to Polymodal Infocommunication Systems Synthesis

The proposed methodological approach for synthesis of polymodal infocommunication systems includes the following actions:

— reasoning of the requirements list $P = \{p_i\}$, represented in PICS from its users' point;
— defining the total of the premises $Z = \{z_i\}$ for designing PICS including its development prospectives and tendencies;
— a search for optimal (quasioptimal) approaches to forming Δ_α^{rest} of the possible system solutions multitudes;
— choice of the efficiency index E^{PICS} reflecting PICS ability to perform one or several target functions with a stated quality and appearing as a functional $E^{PICS} = F\{P, Z, \Delta_\alpha\}$.

System solutions multitude $\Delta = \{MM, MO, AL, AP\}$ includes multitudes of mathematical models MM, methods MO, algorithms AL and hardware-software implementation AP of the infocommunication services [1-4]. It follows that generally PICS systems synthesis task solution resides in discovering such a variant of $\Delta_\alpha \subset \Delta$ which, when answering the requirements of P considering the premises Z, would possess the highest efficiency:

$$E^{PICS} = \max_{\Delta_\alpha} \left(F\{P, Z, \Delta\} \right) \qquad (1)$$

The formulated task (1) according to its nature is optimization one, so multitude Δ_α from the standpoint of formal operation research theory may be viewed as a system solutions multitude.

It's worth noticing that, as practice shows, the strict solution to the problem stated as optimized on the whole appears to be extremely intricate. The reason for that is different degree of importance of efficiency indices E^{PICS} of the modern infocommunication systems of different purpose along with the complex character of their functional dependence on $\Delta_\alpha \subset \Delta$. Therefore, when choosing basic system integrator solutions for constructing PICS, a well known approach can be used. This approach is as follows: (1) decomposition of PICS into subsystems; (2), optimizing one parameter while restricting the others; (3) a reasonable application of different efficiency indices E^{PICS} with regard to peculiarities of specific tasks under consideration [1].

Requirements documents in the field of infocommunication systems design mostly view *QoS* (Quality of Service) as efficiency criteria indices which reflect the subscriber's degree of satisfaction with the service [5]. In this case, the criteria and parameters QoS should be considered for each specific purpose separately and be defined for an end-to-end communication, the final points of which are the joint points of subscribers' terminals, in terms clear to a subscriber.

In general cases, an infocommunication system provides each pair of subscribers with either a commuting fixed (static multiplexing) or constant virtual (statistical multiplexing) channel. The main aim of the latter is to deliver information as data chunks stream from the subscriber's source-terminal to the subscriber's addressee-terminal. The forming of the channels is carried out by organizing a consistent resource connection of technically associated elements of the guiding systems of the telecommunication and elements of the cross-connect storing devices, both on channel and physical level. So, hereinafter we will use for them a common model concept of "data network" Thus, a data network is a combination of technical means ensuring interaction between remote multimodal terminals of users who are in process of interpersonal communication.

The role and place of the separate data network elements are visually represented in Figure 1, where hierarchical three-tier architecture of the informational infrastructure is depicted. *The service level* unites subscribers' terminals and is responsible for delivering various services to users. *The level of the switched environment* includes means by which information is distributed between the users according to its purpose and type. *The transport network level* is traditionally associated with the function of the frame forwarding (data blocks) between access networks which are distant from each other in space with a stated timing and authenticity [5, 6].

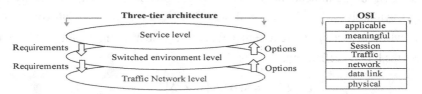

Fig. 1. Three-tire representation of the data network architecture and reference model for Open Systems Interconnection (OSI)

From the standpoint of users, the infocommunication system is required to provide information transmission at the needed time between the stated points in space in the required form, of the definite volume and in a limited time period. Ideally, while exchanging the necessary information, the subscribers do not have to notice the data network existence. If there are no environmental destabilizing factors, the PICS state is characterized by its quality Q_0^{PICS}. As a result of the functioning environment impact, the actual PICS quality will be changed over the time $Q^{PICS}(t)$; at the same time it is important for a subscriber that an absolute deviation of this value from the predicted one does not surpass the given rate

$$\Delta Q^{PICS}(t) = \left| Q_0^{PICS} - Q^{PICS}(t) \right| \le \Delta Q^{EXT}(t). \tag{2}$$

Accordingly, efficiency of PICS should be defined as a maximum value of ΔQ^{PICS} which can be provided under the stated conditions and minding the restrictions placed on the presented expenditures C_{red}^{PICS} on its design and exploitation:

$$\mathfrak{Z}^{PICS} = \Psi(\Delta Q^{PICS},\, C_{st}^{PICS}),\ C_{st}^{PICS} \le C_{st}^{PICS\,ADD}, \tag{3}$$

where $C_{st}^{PICS\,ADD}$ are the allowed stated costs for design and exploitation of PICS.

For assessing PICS efficiency in [7] it is suggested to use a general index which represents its hypothetical volume:

$$V^{PICS} = Pr^{PICS} \cdot RS^{PICS} \cdot Ent^{PICS}, \tag{4}$$

where Pr^{PICS} is data signaling rate (data frames) among all the users (system's productivity); $RS^{PICS} = 1/Tl^{PICS}$ is system's response speed; $Tl^{PICS} = t_{PRO} + (t_D \pm \Delta t_D)$ is PICS timeliness; t_{PRO} is the time of input-output and multimodal information processing; $(t_D \pm \Delta t_D)$ is message delivery delay; $Ent^{PICS} = Ex^{PICS} \cdot Prec^{PICS}$ is PICS entity, a quality which reflects its ability to ensure the required exhaustiveness Ex^{PICS} and precision $Prec^{PICS}$ of the transmitted information (messages) reproduction of the required volume.

On the one hand, the generalized index of the efficiency (4) comprises all the significant indices of the data network functioning qualities; on the other hand, all its components are interconnected with one another but one or two of them are not enough for systematic reflection of the infocommunication system qualities. The level of information infrastructures security and reliability, achieved with the help of the modern technologies, enables one to use the residential efficiency indices as restrictions.

For the purpose of comparing different variants (project alternatives) of the PICS design under restrictions (4), it is reasonable to use the unit cost index [8]:

$$\varsigma^{PICS} = \frac{V^{PICS}}{C_{st}^{PICS}}, \tag{5}$$

the maximum value of which will define the best solution for the conditions under consideration.

Taking into account the accepted hierarchy of PICS functional characteristics, (fig. 1) the index (5) can be decomposed into the unit cost of the information processing means and into data network unit cost:

$$\varsigma^{\mathrm{Inf}} = \frac{V^{\mathrm{Inf}}}{C_{\mathrm{st}}^{\mathrm{Inf}}} = \frac{Ent^{\mathrm{Inf}} \cdot \sum_{j=1}^{J} B_j}{C_{\mathrm{st}}^{\mathrm{Inf}} \cdot t_{\mathrm{PRO}}}; \quad \varsigma^{\mathrm{DN}} = \frac{V^{\mathrm{DN}}}{C_{\mathrm{st}}^{\mathrm{DN}}} = \frac{Ent^{\mathrm{DN}} \cdot \sum_{k=1}^{K} U_k}{C_{\mathrm{st}}^{\mathrm{DN}} \left(t_{\mathrm{D}} \pm \Delta t_{\mathrm{D}} \right)}, \tag{6}$$

where V^{Inf} and V^{DN} is the hypothetical volume of the processed information and the data network accordingly; B_j is the information output speed by the j-source (signals of different nature for traditional infocommunication systems; modalities are for PICS); J is the number of such sources; U_k is the number of the simultaneously transmitted messages (data frame streams).

Bearing in mind that the exhaustiveness of the information transmitted in data frames through data networks $Ex^{\mathrm{DN}} = 1$, entity Ent^{DN} will be fully determined by the precision of the data transmission $Prec^{\mathrm{DN}} = N/t$, where N is the number of the precisely delivered information frames. Then it is reasonable to carry out the task of searching for the best variant for PICS data network construction based on the choice of alternatives, structures and components functional characteristics, which provide unit cost maximization (6). The stated approach is a tested instrument of the estimation of communication networks efficiency [8, 9].

The key index (6) of the information unit cost is its entity $Ent^{\mathrm{Inf}} = Ex^{\mathrm{Inf}} \cdot Prec^{\mathrm{Inf}}$ characterizing the exhaustiveness and precision of the required states reflection of the separately existing accounting items. In traditional telecommunications, it is the information frames, relevant to communication services that appear to be the accounting items on the level of data network. In PICS, however, it is established to consider as accounting items the information frames carrying data about various aspects of communication act (sides of communication). It corresponds with the accepted in this paper refusal from traditional provision of the user with communication services in favour of ensuring communication multimodality via communication technologies.

With regard to what has been already mentioned and based on [8], the multimodal information entity can be determined as

$$Ent^{\mathrm{Inf}} = \frac{1}{M} \sum_{m=1}^{M} x_m R_m, \tag{7}$$

where M is the minimum number of the accounting items required for making a decision, m which is an exhaustiveness index determined as:

$$x_m = \begin{cases} 1, \text{ if it contains accounting item;} \\ 0, \text{ otherwise,} \end{cases} \tag{8}$$

where R_m is the reliability of estimate of the m-accounting item state which should be understood as multimodal information ability to reflect the real or estimated state of the objects and processes (accounting items) of the application domain with a degree of approximation (precision) ensuring an effective use of that information according to the system designated use.

This way, with the given physical structure of the data network and the restrictions upon the presented costs connected with the information processing $C_{st}^{Inf} \le C_{st}^{Inf\,ADD}$, the degree of providing the entity will define PICS efficiency.

4   PICS Synthesis Method

The unit cost index decomposition suggested above (6)-(8) allows one to represent the direct task of PICS synthesis (1) as

$$\varsigma^{Inf} = \max_{\Delta_\alpha}\left(F_1\{R,\Delta\}\right) \tag{9}$$

with $\varsigma^{DN} = \max_{y \in Y}\left(F_2\{Z,Y,H,V,O,X\}\right)$ and dual task as a formula (9) with R=const.

Here F_1 is the information unit cost functional (6); R is the infocommunication resources multitude; F_2 is the data network unit cost functional (6); Y is the controlled variables multitude; H is the noncontrollable variables multitude; $V \supset R$ is the outcome variables multitude; O is the specific tasks multitude (designing operators); X is the graph showing the relations between the elements of the named multitudes.

While realizing the principle of multiserviceness, the PICS synthesis task (fig. 2) resides in [10] forming of the multitude $\Delta_\alpha \subset \Delta$ which, with the unit cost ς^{DN} obtained as a result of data network synthesis and assisting to representation of the given infocommunication resources R and restrictions on the total costs of the information processing, would provide the required information entity (7):

$$Ent^{Inf} = \max_{\Delta_\alpha}\left(F_1\{R,\Delta\}\right). \tag{10}$$

The entity of the multimodal information is influenced by the following restrictions [2,11,12]:

1) From the subscriber (UC), on the ways of the man-computer interaction connected with his subscriber terminal (interface) usage skills, information technologies, personal preferences and physical restrictions;
2) From the subscriber's terminal (DC), on the ways of man-computer interaction connected with its hardware-software abilities;
3) From the man-computer interaction environment (EC), a type of premises and noise level in it; subscribers' number; distance between the subscriber and subscriber's terminal, etc.;
4) The provided services (SC) connected with the subject area, access availability of the infocommunication resources R, their volume and type.

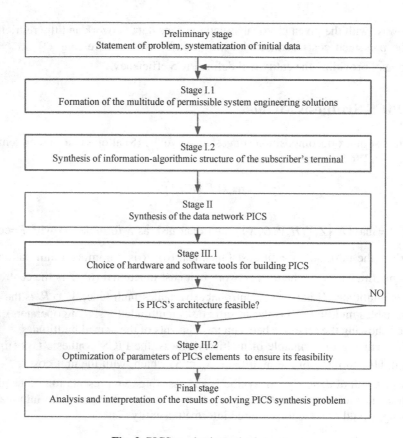

Fig. 2. PICS synthesis method structure

For forming the multitude Δ_α of the acceptable design solutions, a subset of the Cartesian product of the initial design solutions multitude is introduced, presetting synthesis alternatives space:

$$F_{UC}^{(\alpha)} \subseteq MM^{(\alpha)} \times MO^{(\alpha)} \times AL^{(\alpha)} \times AP^{(\alpha)}; \ F_{DC}^{(\alpha)} \subseteq MM^{(\alpha)} \times MO^{(\alpha)} \times AL^{(\alpha)} \times AP^{(\alpha)};$$

$$F_{EC}^{(\alpha)} \subseteq MM^{(\alpha)} \times MO^{(\alpha)} \times AL^{(\alpha)} \times AP^{(\alpha)}; \ F_{SC}^{(\alpha)} \subseteq MM^{(\alpha)} \times MO^{(\alpha)} \times AL^{(\alpha)} \times AP^{(\alpha)};$$

Taking it into consideration, stage I.1 (fig. 2) of the PICS synthesis method is reduced to the search of the optimal (quasioptimal) approaches to forming the multitude Δ_α of the possible design solutions with regard to restrictions *UC, DC, EC, SC* based on the multitude of the input $IM^{(\alpha)} = \left\{ IM_1^{(\alpha)}, IM_2^{(\alpha)}, ..., IM_{N_{IM}}^{(\alpha)} \right\}$ and output $OM^{(\alpha)} = \left\{ OM_1^{(\alpha)}, OM_2^{(\alpha)}, ..., OM_{N_{IM}}^{(\alpha)} \right\}$ modalities:

$$\Delta_\alpha = \begin{cases} \left\langle mm \in MM^{(\alpha)}, mo \in MO^{(\alpha)}, al \in AL^{(\alpha)}, ap \in AP^{(\alpha)} \right\rangle; \\ \Phi : F_{UC}^{(\alpha)} \cap F_{DC}^{(\alpha)} \cap F_{EC}^{(\alpha)} \cap F_{SC}^{(\alpha)} \to B, \end{cases} \tag{11}$$

where Φ is a reflection forming the multitude of the circuit valid values and multitude $B = \{0,1\}$.

Analytical description of the correspondence between elements mm, mo, al, ap of the relevant subsets of the multitude Δ_α and information entity (10) is hindered. Moreover, it is quite often impossible, so the definition (formalization) of Φ reflection appears to be a formidably formalized task, the solution for which lies among methods of fuzzy sets, evolutional calculations, hybrid neural networks and cognitive modeling.

Information algorithmic structure characterizes separate tasks, which are solved by the system, information streams, significant for their solution, tasks solution order and information interaction. Elementary algorithmic items $ne \in AL^{(\alpha)}$ of the information transformation $i \in I$, $I = IM^{(\alpha)} \cup OM^{(\alpha)} \cup AS \cup NS$ ($AS = \{as_q, q \in \mathbb{N}\}$ is the multitude of the artificial signals streams; $NS = \{ns_a, a \in \mathbb{N}\}$ is the multitude of the natural signals streams at the different stages of system functioning, which correspond to a mathematic or a logical operation. Separate groups g (non-empty final subsets $AL^{(\alpha)}$) of the elementary algorithmic items ne and information streams which are incident to them, are described by the corresponding models $MM^{(\alpha)}$. The total of all the links included in the system (subsystem) composes algorithm $al \in AL^{(\alpha)}$ of the system (subsystem) functioning.

Information-algorithmic structure synthesis of the subscriber's PICS terminal (stage I.2 of the synthesis method) presupposes its representation as an oriented acyclic graph at the top of which g elements groups and the optimal hierarchy task solution are placed. In [13] it is shown that the latter can be derived from the search algorithms for optimal trees. What is more, the key element here is differential choice of the structural functional considering the required level of the information-algorithmic structure construction.

Different functioning conditions of the administration system define the space of alternatives synthesis differently. The total of structures, derived as a result of hierarchy synthesis of each of them, determines the complete modalities combination $\bar{\Theta}_\alpha \left(IM^{(\alpha)} \cup OM^{(\alpha)} \right)$ admissible in the synthesized (projected) PICS and design solutions:

$$\bar{\Delta}_\alpha^{lim} = \begin{cases} \left\langle \overline{mm} \in MM^{(\alpha)}, \overline{mo} \in MO^{(\alpha)}, \overline{al} \in AL^{(\alpha)}, \overline{ap} \in AP^{(\alpha)} \right\rangle, \\ \Psi^{(\alpha)} : \bar{\Theta}_\alpha \left(IM^{(\alpha)} \cup OM^{(\alpha)} \right) \times \Delta_\alpha^{lim} \to B. \end{cases} \tag{12}$$

The shift from the information-algorithmic structure to functional one presupposes the selection of corresponding hardware-software realizations $ap \in AP^{(\alpha)}$ (stage III.1 of the synthesis method) based on the developed technique [14-16].

For ensuring subscriber's terminal functional structure realization obtained in the course of applying the technique, the developed synthesis method includes stage III.2 of the PICS elements parameters optimization with the aim of achieving their optimal functional characteristics.

When solving the direct task of the PICS synthesis along with, exceeding of the information source efficiency Pr^{Inf} over the data network efficiency Pr^{DN}, a data network synthesis is carried out (stage II). The corresponding synthesis method (represented in [9]) appears to be the upgrading of the existing apparatus for optimization of data networks functional characteristics at a new level of the development of optic component base above the traditional principles of the capacity resource division into equal parts.

5   Conclusion

On the whole, the suggested methods allow one to synthesize infocommunication systems on the basis of multimodal man-computer interfaces with required qualities of entity, running speed, efficiency and cost. This is verified by the realization of new "perceptive" services [17-19]. The apparatus, developed in terms of each stage, can be used separately for solving the application tasks of optimization of the subscribers' terminals structure, or their functional characteristics. The precision sufficient for practical applications, as well as computational complexity that does not exceed similar characteristics of the existing methods testify in favour of the method's constructability.

Acknowledgments. This work is partially supported by the Russian Foundation for Basic Research (grants № 13-08-0741-a; №15-07-06774-a).

References

1. Ztsarinny, A.A., Ionenkov, Y.S., Kondrashev, V.A.: About the Unified Approach to the Choice of Information-telecommunication Systems Construction System House Solutions. Systems and Means of Informatics **16** (2006). Science
2. Ronzhin, A.L., Karpov, A.A.: Design of Interactive Applications with a Multimodal Interface. TUSUR Reports **1**(21), 124–127 (2010). Part 1
3. Stepanov, S.N.: Teletraffic Basics of Multiservice Networks. Eco-Trends (2010)
4. Saveliev, A.I., Vatamaniuk, I.V., Ronzhin, A.L.: Architecture of data exchange with minimal client-server interaction at multipoint video conferencing. In: Balandin, S., Andreev, S., Koucheryavy, Y. (eds.) NEW2AN/ruSMART 2014. LNCS, vol. 8638, pp. 164–174. Springer, Heidelberg (2014)

5. Gudkova, I.A., Samouylov, K.E.: Modelling a radio admission control scheme for video telephony service in wireless networks. In: Andreev, S., Balandin, S., Koucheryavy, Y. (eds.) NEW2AN/ruSMART 2012. LNCS, vol. 7469, pp. 208–215. Springer, Heidelberg (2012)

6. Popov, S., Kurochkin, M., Kurochkin, L.M., Glazunov, V.: Hardware and software equipment for modeling of telematics components in intelligent transportation systems. In: Balandin, S., Andreev, S., Koucheryavy, Y. (eds.) NEW2AN/ruSMART 2014. LNCS, vol. 8638, pp. 598–608. Springer, Heidelberg (2014)

7. Basov, O.O., Saitov, I.A.: Quality of Functioning and Efficiency of Polymodal Infocommunication Systems. SPIIRAS Proceedings 1(32), 152–170 (2014)

8. Tsybizov, A.A.: Communication Networks Efficiency Evaluation. RSRTU Newsletter 3(29), 19–24 (2009)

9. Saitov, I.A.: Basics of Theory of Constructing Telecommunicational Systems Protected Multiprotocol Optic Transport Networks: monography. Academy of FPS of Russia (2008)

10. Basov, O.O., Karpov, A.A., Saitov, I.A.: Methodological Basics of the State Administration Polymodal Infocommunication Systems Synthesis: Monography. Academy of FPS of Russia (2015)

11. Basov, O.O.: Principles of Construction of Polymodal Info-Communication Systems based on Multimodal Architectures of Subscriber's Terminals. SPIIRAS Proceedings 2(39), 109–122 (2015)

12. Ronzhin, A.L., Saveliev, A.I., Budkov, Victor Yu.: Context-aware mobile applications for communication in intelligent environment. In: Andreev, S., Balandin, S., Koucheryavy, Y. (eds.) NEW2AN/ruSMART 2012. LNCS, vol. 7469, pp. 307–315. Springer, Heidelberg (2012)

13. Voronin, A.A., Mishin, S.P.: Optimal Hierarchic Structures. IPM RAS (2003)

14. Basov, O.O., Bogdanov, S.P., Struev, D.A.: Technique of Choosing Hardware-software Means for Constructing Subscribers' Terminals of Polymodal Infocommunication System. TUSUR Reports 1(35) (2015)

15. Meshcheryakov, R., Bondarenko, V.: Dialogue as a basis for construction of speech systems. Cybernetics and Systems Analysis 44(2), 175–184 (2008)

16. Ronzhin, A., Budkov, V.: Speaker turn detection based on multimodal situation analysis. In: Železný, M., Habernal, I., Ronzhin, A. (eds.) SPECOM 2013. LNCS, vol. 8113, pp. 302–309. Springer, Heidelberg (2013)

17. Ronzhin, A., Vatamaniuk, I., Ronzhin, A., Železný, M.: Algorithms for acceleration of image processing at automatic registration of meeting participants. In: Ronzhin, A., Potapova, R., Delic, V. (eds.) SPECOM 2014. LNCS, vol. 8773, pp. 89–96. Springer, Heidelberg (2014)

18. Ronzhin, A.L., Ronzhin, A.L., Budkov, Victor Yu.: Methodology of facility automation based on audiovisual analysis and space-time structuring of situation in meeting room. In: Stephanidis, C. (ed.) HCII 2013, Part II. CCIS, vol. 374, pp. 524–528. Springer, Heidelberg (2013)

19. Ronzhin, A.L., Karpov, A.A.: A Software System for the Audiovisual Monitoring of an Intelligent Meeting Room in Support of Scientific and Education Activities. Pattern Recognition and Image Analysis 25(2), 237–254 (2015)

Data Mining Algorithms Parallelizing in Functional Programming Language for Execution in Cluster

Ivan Kholod[1(⊠)], Aleksey Malov[2], and Sergey Rodionov[1]

[1] Saint Petersburg Electrotechnical University "LETI", ul. Prof. Popova 5,
Saint Petersburg, Russia
{iiholod,sv-rodion}@mail.ru
[2] Motorola Solutions, Business Centre "T4", Sedova st., 12, 192019 Saint Petersburg, Russia
alexeimal-2@yandex.ru

Abstract. This article describes an approach to parallelizing of data mining algorithms, implemented in functional programming language, for distributed data processing in cluster. Here are provided requirements for the functions which form these algorithms for their conversion into parallel type. As an example we describe Naive Bayes algorithm implementation in Common Lisp language, its conversion into parallel type and execution on cluster with MPI system.

Keywords: Data mining · Distributed data mining · Distributed information processing · Functional language

1 Introduction

Currently, there is a rapid growth of stored information volumes. This brings an urgent problem of effective analysis for extraction of new useful knowledge from accumulated data. The analysis performance improve is possible due to increase of computational clustering resources. This requires adaptation of programs for parallel execution.

Imperative programming languages (Java, C / C ++, Fortran and others.) are not suitable for parallel execution. They are designed for operation in accordance with the model of Turing machine. Its basic idea is to change the state of Turing machine in execution of each program statement. Thus, programs, written in imperative programming languages, presuppose the program state and its change during execution. The main problem of such programs in their parallel execution is requirement for simultaneous access to the program state from parallel branches. This gives rise to such problems as access synchronization, blocking, race and others. Solving these problems is rather laborious intensive process, like for development and for debugging.

Existing expansion imperative languages for parallel execution (Ada, High Performance FORTRAN, High Performance C ++ and others.) allow to parallelize only individual structures such as cycles.

S. Balandin et al. (Eds.): NEW2AN/ruSMART 2015, LNCS 9247, pp. 140–151, 2015.
DOI: 10.1007/978-3-319-23126-6_13

Alternative imperative languages are functional languages (Lisp, Haskell et al.). They are based on theory of λ-calculus proposed by Alonzo Church simultaneously with Turing. However, in contrast with the Turing machine, in the theory λ-calculus program is represented as a function expression where functions are called from each other. All information required for transfer from function to function is transmitted through function arguments and return values. Thus, programs written in functional languages do not use the internal state.

Functions that develop functional program are clean. Hence, these functions can be executed in parallel without requiring additional measures preventing blocking, race and other problems of parallel execution. In theory, functional expression developed from such functions can be parallelized automatically.

The article describes the approach to presentation of Data Mining algorithms as functional programs in Common Lisp language and their conversion from serial to parallel version. As an example, we describe the Naive Bayes algorithm and its conversion into two versions of parallel execution.

The next section is a review of research in the field use of functional programming languages for data mining algorithms and studies in the field of data mining algorithms parallelizing. The third section contains the description of algorithm building of function blocks and the description of blocks common to all data mining algorithms. The fourth chapter describes Apriori algorithms elaboration as a combination of function blocks, including their parallel forms. The fifth chapter describes implementation of function blocks and Apriori family algorithms based on these blocks. The last chapter deals with experiments with the implemented algorithms with citing of the results.

2   Related Work

Research in the field of parallel and distributed data mining have been also carried out for quite a while. As a matter of fact, separate focus areas can be distinguished within the data mining field [1]:

- parallel data mining (PDM): research trend aimed at studying the parallelizing of data mining algorithms for tightly-coupled system applications: systems with common (shared) or distributed memory, clusters (groups) of work stations with shared memory and quick interaction;
- distributed data mining (DDM) research trend aimed at studying the parallelizing of data mining algorithms for applications within loosely-coupled systems consisting of nodes linked into a local network or distributed geographically and united through the global network internet /intranet.

The biggest efforts of the researchers in the field of parallel and distributed data mining are focused towards the elaboration of individual parallel data mining algorithms. However these efforts are aimed at optimization of parallel structure of algorithms for execution under certain conditions. The conditions can be different [2]: shared or distributed memory system, single or multiple of data sources, horizontal or vertical dataset layout (for multilple of data sources) and other.

For this reason the parallel algorithms developed under certain conditions, not will always efficient under other conditions. Examples of data mining algorithms for specific types of computing systems are algorithms:

- decision tree algorithms: Supervised Learning In Quest (SLIQ) [3], Scalable PaRallelizatable INndition of decision Trees (SPRINT) [4], Parallel Decision Tree (PDT) [5];
- association algorithms: Common Candidate Partitioned Database (CCPD) [6], Partitioned Candidate Common Database (PCCD) [6], Asynchronous Parallel Mining (APM) [7] for systems with shared memory and Hash Partitioned Apriori (HPA) [8], Simply Partitioned Apriori (SPA) [8], ParEclat (PE), ParMaxEclat (PME), ParClique (PC), and ParMaxClique (PMC) [9] for systems with distributed memory;
- clustering: Pkmeans [10], MAFIA [11], Distributed Cooperative Clustering in super-peer P2P networks (DCCP2P) [12], P-CLUSTER [13], Collaborative [14], Distributed Information Bottleneck (DIB) [15] and other.

All of these algorithms were implemented by imperative programming languages (Java and C / C ++). With this approach the complexity and effort for developing of parallel algorithms is very high. At that this effort is aimed at adapting the algorithms to execution strictly in the required conditions. The changes to the conditions lead to the necessity of conversion of the algorithm which is in fact a creation of a new algorithm.

In the sphere of implementation of the data mining algorithms in the functional programming languages one should mention, inter alia, the publications devoted to, implementation of algorithms in the Haskell, the creation of the R library.

The tasks of effective analysis for patterns detection in large data sets and production of useful knowledge from accumulated data within machine learning can be solved by means of logic programming. The work [16] proposes using of Prolog and Datalog logic programming languages for relations on large data sets with further classification of these data.

There are a lot of algorithm implementations in functional programming languages. Examples may include algorithms implemented in Haskell for pattern recognition [17-18] and clustering [19]. However, realization data do not suggest parallel execution implementation and specific for certain algorithms.

In contrast, functional language R [20] is specially designed for data analyze. It includes special designs for iterative processing. Furthermore, R library, using the language includes convenient user interface and plurality of package for data analysis. The main disadvantages of the R language include basic single-threaded execution of algorithms and data storage in memory. These disadvantages do not allow effectively process large amounts of data. R language contains tools for parallel execution, but their usage requires significant change in the algorithm code.

Thus, described works for the most part focused on development of individual parallel Data Mining algorithms that requires significant effort. We propose an approach that allows converting sequential Data Mining algorithms, implemented in Common Lisp language, into different parallel versions.

3   Implementation of Data Mining Algorithm in Functional Programming Language

3.1   Data Mining Algorithm Representation as a Function

In order to formally describe a data mining algorithm, first we are going to consider the general concept of their performance (figure 1) [21].

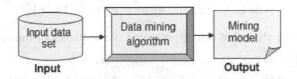

Fig. 1. General concept of data mining algorithm performance

A data mining algorithm takes an input data set and creates an output mining model. Thus, a data mining algorithm can represent as a function with the data set as function argument d and constructed mining model as returned value m:

```
(defun <algorithm_function_name> (d))
```[1]

The simple item of the algorithm is the step (the single operation). The data mining algorithms analyses the data set and builds the mining model on the each step. Builted mining model is passed to the next step. So the each step of the data mining algorithm must take the data set and the mining model as input arguments. The result of step's work is the new mining model [22]. Accordingly the step of the data mining algorithm has the unified interface:

- input: the data set d and the mining model m;
- output: the mining model m.

Functions executed at each step of the algorithm and used for constructing a mining model on the basis of two arguments (analyzed dataset d and mining model m) can be used for introducing a new type of functions:

```
(defun <function_name> (d m))
```

The data set can be represented as a tuple: a list of attributes - *attr_list* and a list of vectors - *vectors_list*:

```
(defstruct d attr_list vectors_list),
```

where

```
(setf attr_list (list attr0, … , attrN, target_attr))
(setf vectors_list (list v0, v1, …, vM))
```

---

[1] Here and below we use Lisp notation.

Each vector vi is list of a value. For example, a vector with 5th attributes:

```
(setf vi (list 0.0 1.0 0.5 0.7 0.8)).
```

A mining model depends on data mining constructed function. In general it can be represented as a tuple: a structure for state parameters - *state* (for example, the number of the current vector, the number of the current attribute, etc.) and a list of rules - *rules_list:*

```
(defstruct m state rules_list)
```

Thus, a data mining algorithm can be presented as a sequence of functions calls:

$$
\begin{array}{ll}
\text{(defun <algorithm\_function\_name> (d)} \\
\quad (fb_n \ d \ (fb_{n-1} \ d \ ... \ (fb_i \ d \ ... \ (fb_1 \ d \ nil)...) \ ...))) & (1)
\end{array}
$$

3.2   Basic Functions for the Implementation of Data Mining Algorithms

Decomposition of any algorithm splits the algorithm into separate logical blocks, cycles, decision, etc. Additionally, data mining algorithms have special blocks: cycle for vectors, cycle for attributes and other. In order to characterize these elements in the form of functional expressions we will add embedded functions and show how they can be used to present the enumerated structural elements in the functional form.

Conditional operator of a data mining algorithm can be expressed as a function:

```
(defun condition_function (d m cf fbₜ fbբ)
              (cond ((cf d m) (fbₜ d m)) (T (fbբ d m)))
```

where
cf – function (3th argument) for calculating a conditional expression;
fb_t – function (4th argument), which is executed if the result of the function cf is true;
fb_f – function (5th argument), which is executed if the result of the function cf is false.

Thus, certain condition function can present as a function of *FB* type. For example, the function of condition which checks weather the current attribute is a target one, and if no will have view:

```
(defun is_curr_attr_target (d m)
   (condition_function
      d
      m
      (eql                                    ;cf function
         (nth m-state-curr_attr d-attr_list)
         m-state-targed_attr)
      nil                                     ;fbₜ function
      for_all_vectors_cycle))                 ;fbբ function
```

The cycle of a data mining algorithm can be presented using a recursive call of a higher-order function:

```
(defun cycle_function (d m cf fb_init fb_pre fb_iter)
               (cycle( d (fb_init d m) cf fb_pre fb_iter))) 
(defun cycle (d m cf fb_pre fb_iter)
(cond ((cf d m)
            (cycle d (fb_iter d (fb_pre d m)) cf fb_pre fb_iter))
  (T (fb_iter d (fb_pre d m))))))))
```
where:

cf is the function (3th argument) determining the condition of a repeated iteration;
fb_{init} is the function (4th argument), which initializes the cycle;
fb_{pre} is the function (5th argument) executed in a cycle prior to execution of the main iteration;
fb_{iter} is the function (6th argument) of the main iteration function.

Thus, certain cycle function can present as a function of *FB* type. For example, data mining algorithms often have the cycle for attributes. We can determine some block of *FB* type as constants for this cycle. Thus, the cycle for attributes can be determined as an embedded higher-order function in the following form:

```
(defun for_all_attributes_cycle (d m fb_iter)
   (cycle_function d m
               (eql nil d-attr_list)   ;cf function
               nil                     ; fb_init function
               (cdr d-attr_list)       ; fb_pre function
               fb_iter))
```

One of the main advantages of developing algorithms from the function blocks is the possibility of their concurrent execution. Data mining algorithms are mostly characterized with data parallelism. In this case explicit conversion of the functional expression is needed with adding of data partitioning function and subsequent aggregation of results.

To make the concurrent execution of a data functional expression, it must be converted into the form in which the function blocks will be invoked as arguments. To this end we should add a higher order function which will allow data-parallelizing in the data mining algorithms for MPI system:

```
(funcname mpi_parallel_function (d m merge split fb)
               (mpi:mpi-init) ;initialize MPI
               (merge d (split d m fb))
               (mpi:mpi-finalize))                           (2)
```

— *split:* function fulfilling the splitting of the data set d (possibly depending on the mining model M) and recovering the list from the n split data sets *[d]* :

— *merge:* function joining the models of knowledge from the list *[m]* and recovering the aggregated model *m*;
— *fb:* function block executed concurrently.

Example of this function will consider below.

4   Implementation of Naive Bayes Algorithm as Functional Expression

As examples, we will consider probability algorithm Naive Bayes [23].

```
for all attributes a
  if a is not target attribute
    for all vectors w
      increment count of vectors for value of attribute a
          equaled value of the attribute a of the vector w
    end for all vectors;
  end if
end for all attributes;
for all classes c
    for all vectors w
      increment count of vectors for class of vector w;
    end for all vectors
end for all classes c
```

The Naive Bayes algorithm can be represented as composition of functions (1):

```
(defun NBAlgorithm (d)
    for_all_classes_cycle (d
        for_all_attributes_cycle (d
            attr_probability_list (d
                attr_val_probability_list (d
                    attr_entr_count (d))))))
```

where

- *for_all_classes_cycle:* is function which adds to the final or partial model row with the count of classes entrances into the data base
- *for_all_attributes_cycle*: is function which executes iteration by attributes
- *attr_probability_list*: is function which executes iteration by possible values of particular attribute
- *attr_val_probability_list*: is function which executes iteration by all classes for the particular value of particular attribute
- *attr_entr_count:* is function which executes iteration by all vectors or by the selected part of the vectors and calculates vectors count having the particular value of particular attribute and particular class

For implementation of Naive Bayes algorithm we have extended as.

```
(defstruct d attr_list vectors_list class_list)
```

where

- *class_list* – list of all possible classes of target attribute.

The mining model for Naive Bayes algorithm can be extended as:

```
(defstruct m status classes_entrances_count
        (attr_1_val_1_entr_count attr_1_val_2_entr_count..)
        (attr_2_val_1_entr_count attr_2_val_2_entr_count..)
        (...))
```

- *classes_entrances_count* – list of each class entrances count in *g_data_base*. Number of the corresponding class determined by position in the list:

```
(setf classes_entrances_count (list 0 0 ... 0))
```

- *attr_N_val_M_entr_count* – list consisting of the count of the vectors in *g_data_base* for each class, where source attribute *N* has value *M*. Number of the corresponding class determined by position in the list. *N* – number of the current attribute, *M* - number of the current value of the current attribute:

```
(setf attr_N_val_M_entr_count (list 0 0 ... 0))
```

Like this algorithm is an expression with nested function calls. In this case nested functions always allow carrying out calculation of one argument - knowledge model m. This means that, in the reduced form Naive Bayes algorithm cannot be executed in parallel (e.g., parallelized on tasks). Parallel execution requires change of the functional expression and add parallelizing function (2). We construct two parallel Naive Bayes algorithms:

- performing parallel processing of vectors

```
(defun NBAlgorithmVectorsParallel (d)
    mpi_vector_parall (d
        for_all_classes_cycle (d
            for_all_attributes_cycle (d
                attr_probability_list (d
                    attr_val_probability_list (d
                        attr_entr_count (d))))))))          (3)
```

- performing parallel processing of attributes

```
(defun NBAlgorithmAttrsParallel (d)
    for_all_classes_cycle ( d
        mpi_attr_parall (d,
            for_all_attributes_cycle (d
```

$$attr\_probability\_list\ (d$$
$$attr\_val\_probability\_list\ (d$$
$$attr\_entr\_count\ (d))))))) \qquad (4)$$

where

- *mpi_vector_parall* : is parallelizing function of vectors for MPI system uses the function (2)

```
(defun mpi_vector_parall (d m)
       mpi_parallel_function (d m
           mpi_merge_vector_parall ;the merge function
           mpi_split_vector_parall ;the split function
           for_all_classes_cycle)) ;the fb function
```

where

— *mpi_merge_vector_parall:* is function performing merge of the list of partial models calculated by each parallel process into final model in the case of parallelization by vectors;
— *mpi_split_vector_parall:* is function performing split of the vectors of data base to the parts intended for each parallel process and performing calculation of the partial model.

- *mpi_attr_parall:* is parallelizing function of attributes for MPI system

```
(defun mpi_attr_parall (d m)
     mpi_parallel_function (d m
         mpi_merge_attr_parall        ;the merge function
         mpi_split_attr_parall        ;the split function
         for_all_attributes_cycle))   ;the fb function
```

where

— *mpi_merge_attr_parall*: is function performing merge of the list of partial models calculated by each parallel process into final model in the case of parallelization by attributes;
— *mpi_split_attr_parall*: is function performing split of the attributes to the parts intended for each parallel process and performing calculation of the partial model.

5   Experiments

We have performed several experiments for the implemented algorithms. The experiments have been performed with various input data sets (Table 1). These data sets contain various numbers of vectors, attributes and the average number of categories for target attribute. Thus, we have used the data sets for which the Naive Bayes algorithm works with various loading.

The experiments have been done on a multicore computer the following configuration: CPU Intel Xeon (4 cores), 2.80 GHz, 4 Gb, Ubuntu 10.04 LTS, Steel Bank Common Lisp (SBCL) v1.0.29.11-1ubuntu1. The parallel algorithms have been executed for the numbers of cores equal to 2 and 4, respectively. The experimental results are provided in Table 2.

The experiments show that parallel execution of the Naive Bayes algorithm for data sets with various parameters is different. The parallel form of algorithm (3) is more efficiently (selected bold font) for data sets with a large number of vectors and a small number of attributes: W1, W5 and W10. It is possible because large count of iteration for all vectors is splitted between a few threads. The parallel form of algorithm (4) is more efficiently (selected bold font) for data sets with a large number of attributes and a small number of vectors (A1, A5 and A10). This is because Naïve Bayes algorithm executes much iteration for all attributes and their values (more than for all vectors) in the case of these data sets. Thus this parallel form of algorithm divides all the iterations between threads. The parallel form of algorithm (3) and (4) have approximately same efficient for data sets with large count of classes: C1, C5 and C10.

Table 1. Experimental results

Input data set	Number of vectors	Number of attributes	Avg. number of classes
W1	10 000	10	5
W5	50 000	10	5
W10	100 000	10	5
A1	100	100	100
A5	100	500	100
A10	100	1 000	100
C1	1 000	100	10
C5	1 000	100	50
C10	1 000	100	100

Table 2. Experimental results (seconds)

Algorithm	Cores	W1	W5	W10	A1	A5	A10	C1	C5	C10
NBAlgorithm	1	0.09	0.45	0.87	0.51	10.44	40.00	0.55	2.61	5.20
NBAlgorithm VectorsParallel	2	**0.05**	**0.24**	**0.45**	0.28	5.33	20.51	0.28	1.33	2.67
	4	**0.03**	**0.13**	**0.28**	0.19	3.39	12.30	0.17	0.74	1.50
NBAlgorithm AttrsParallel	2	0.05	0.25	0.48	**0.28**	**5.27**	**20.34**	0.27	1.32	2.65
	4	0.04	0.19	0.41	**0.16**	**2.96**	**11.47**	0.17	0.74	1.49

6   Conclusion

This article describes an approach to implementation of data mining algorithms in functional programming language. When implementing the algorithm was presented as a functional expression of unified (with the same set of arguments and the output value) functions. For implementation of algorithm structural blocks such as conditional statements, cycles, functions, that have been reduced to a unified type.

For algorithm parallelizing on data we defined the function *mpi_parallel_function*, that can be inserted into call of any unified expression function. This allows you to execute framed function in parallel.

Using the proposed approach we implemented Naive Bayes algorithm as a functional expression. Using *mpi_parallel_function* we got two versions of algorithm parallel execution with simple permutation of function in the expression. An experimental comparison of these versions showed their effectiveness for different data sets.

Thus, using the proposed approach we can easy (by adding a function) obtain various parallel forms of algorithm, that will be effective for different data sets, from sequential data mining algorithm.

Acknowledgments. The work has been performed in Saint Petersburg Electrotechnical University "LETI" within the scope of the contract Board of Education of Russia and science of the Russian Federation under the contract № 02.G25.31.0058 from 12.02.2013. This paper is also supported by the federal project "Organization of scientific research" of the main part of the state plan of the Board of Education of Russia and project part of the state plan of the Board of Education of Russia (task # 2.136.2014/K).

References

1. Paul, S.: Parallel and Distributed Data Mining, New Fundamental Technologies in Data Mining. Funatsu, K. (ed.), pp. 43–54 (2011)
2. Zaki, M.J., Ho, C.-T. (eds.): Large-Scale Parallel Data Mining, pp. 1–23. Springer-Verlag, Heidelberg (2000)
3. Mehta, M., Agrawal, R., Rissanen, J.: SLIQ: a fast scalable classier for data mining. In: Proc. of the Fifth Intl. Conference on Extending Database Technology (EDBT), Avignon, France (1996)
4. Shafer, J., Agrawal, R., Mehta, M.: Sprint: a scalable parallel classier for data mining. In: 22nd VLDB Conference (1996)
5. Kufrin, R.: Decision trees on parallel processors. In: Geller, J., Kitano, H., Suttner, C. (eds.) Parallel Processing for Artiffcial Intelligence 3. Elsevier-Science (1997)
6. Zaki, M.J., Ogihara, M., Parthasarathy, S., Li, W.: Parallel data mining for association rules on shared memory multi-processors. In: Supercomputing 1996 (1996)
7. Cheung, D., Hu, K., Xia, S.: Asynchronous parallel algorithm for mining association rules on shared-memory multi-processors. In: 10th ACM Symp. Parallel Algorithms and Architectures (1998)
8. Shintani, T., Kitsuregawa, M.: Hash based parallel algorithms for mining association rules. In: 4th Intl. Conf. Parallel and Distributed Info. Systems (1996)

9. Zaki, M.J., Parthasarathy, S., Ogihara, M., Li, W.: Parallel algorithms for fast discovery of association rules. Data Mining and Knowledge Discovery: an International Journal 1(4), 343–373 (1997)
10. Johnson, E.L., Kargupta, H.: Collective, hierarchical clustering from distributed, heterogeneous data. In: Zaki, M.J., Ho, C.-T. (eds.) KDD 1999. LNCS (LNAI), vol. 1759, pp. 221–244. Springer, Heidelberg (2000)
11. Goil, S.H.N., Choudhary, A.: MAFIA: Efficient and scalable subspace clustering for very large data sets. Technical Report 9906-010, Center for Parallel and Distributed Computing, Northwestern University (1999)
12. Judd, D., McKinley, P., Jain, A.: Large-scale parallel data clustering. In: Intl Conf. Pattern Recognition (1996)
13. Kashef, R.: Cooperative Clustering Model and Its Applications. PhD thesis, University of Waterloo, Department of Electrical and Computer Enginnering (2008)
14. Hammouda, K.M., Kamel, M.S.: Distributed collaborative web document clustering using cluster keyphrase summaries. Information Fusion 9(4), 465–480 (2008)
15. Deb, D., Angryk, R.A.: Distributed document clustering using word-clusters. In: IEEE Symposium on Computational Intelligenceand Data mining, CIDM 2007, pp. 376–383 (2007)
16. Wrobel, S., Dzeroski, S.: The ILP description learning problem: towards a general model-level definition of data mining in ILP. In: FGML-95 Annual Workshop of the GI Special Interest Group Machine Learning (GI FG 1.1.3) (1995)
17. Kerdprasop, N., Kerdprasop, K.: Mining Frequent Patterns with Functional Programming. International Journal of Computer, Information, Systems and Control Engineering 1(1), 120–125 (2007)
18. Amanda, C., King, R.: Data mining the yeast genome in a lazy functional language. http://users.aber.ac.uk/afc/papers/ClareKingPADL.pdf
19. Aleksovski, D., Erwig, M., Dzeroski, S.: A Functional Programming Approach to Distance-based Machine Learning. http://www.academia.edu/2804496/A_functional_programming_approach_to_distance-based_machine_learning
20. Bloomfield, V.A.: Using R for Numerical Analysis in Science and Engineering. Chapman & Hall/CRC p. 359 (2014)
21. Common Warehouse Metamodel Specification. http://www.omg.org/spec/CWM/1.1/
22. Kholod, I., Karshiyev, Z., Shorov, A.: Formal model of data mining algorithms for algorithm parallelization. The nineteenth international multi-conference on advanced computer systems (ACS 2014). Artificial Intelligence, Software Technologies Biometrics and Information Technology Security (AISBIS 2014), Międzyzdroje, Poland, pp. 385–394, October 22–24, 2014
23. Domingos, P., Pazzani M.: On the optimality of the simple Bayesian classifier under zero-one loss (1997)

Ontology-Based Voice Annotation
of Data Streams in Vehicles

Inna Sosunova[1(✉)], Arkady Zaslavsky[1,2], Theodoros Anagnostopoulos[1,3],
Alexey Medvedev[1], Sergey Khoruzhnikov[1], and Vladimir Grudinin[1]

[1] ITMO University, Kronverkskiy pr., 49, St. Petersburg, Russia
inna_sosunova@corp.ifmo.ru, alexey.medvedev@niuitmo.ru,
xse@vuztc.ru, vlad@digiton.ru
[2] Digital Productivity Flagship, CSIRO, Box 312, Clayton South, VIC 3169, Australia
Arkady.Zaslavsky@csiro.au
[3] Community Imaging Group, University of Oulu, 90570 Oulu, Finland
tanagnos@ee.oulu.fi

Abstract. With proliferation of the Internet of Things, annotation and generation of metadata describing data streams produced by sensors becomes even more urgent and important. This article proposes a method of annotating data streams with voice and extracting semantics from data. The strengths and weaknesses of existing voice recognition systems are discussed and it is argued that ontologies should play important role in making annotations meaningful and useful for various services and applications, including annotating road conditions and traffic situations. The architecture and implementation of the proposed system is discussed and demonstrated.

Keywords: Internet of things · Iot · Ontology · Speech recognition · Speech technology in vehicles · Intelligent transportation systems · ITS

1   Introduction

The main goal of Smart Cities is comfortable human habitation. The integration of information and communication systems into various technical systems and infrastructures of a city are fundamental bases in smart cities. A smart city must combine legacy networks and new communication architectures and configure existing communication networks in order to achieve compatibility and interoperability. Intelligent transportation systems (ITS) are an integral part of smart cities. One of the important tasks of ITS is obtaining data about the situation on the road and road conditions and extracting information from them. In this article we analyze existing approaches and propose a solution based on receiving information directly from the driver.

Our proposed solution is based on speech recognition, which allows getting information without distracting the driver from driving a vehicle. We propose ontology based mobile application – SARRS (Speech-Aware Reporting Road Situation) that recognizes the drivers' speech and annotates traffic situation. For analysis of critical situations on the road various approaches can be used. However information about

© Springer International Publishing Switzerland 2015
S. Balandin et al. (Eds.): NEW2AN/ruSMART 2015, LNCS 9247, pp. 152–162, 2015.
DOI: 10.1007/978-3-319-23126-6_14

many critical situations on the road can be provided mainly by the driver. In this article we suggest an innovative extension of CityWatcher application [2, 3] and propose an approach based on voice recognition and data structuring using ontology. That is, the driver is aware of a problem on the road, will be able to report a problem (annotate a situation) by voice. Then this message and all necessary data will be automatically sent to the associated service.

2   Background and Related Work

We consider some problems such as pits on the road, open road hatches, car passage barriers and existing technologies allowing solve them.

Road Surface Condition Monitoring Systems. Mobile applications, such as the Roads of Russia (Russia) [6], Goodroads (Russia) [7], Street Bump (USA) [8], captures data of the vibration of the vehicle suspension, analyze them and make conclusions about the quality of the road surface. However to trigger the sensor a car have to enter a pit.

Real Time Video Stream Recognition System. It is clear that real time video stream recognition system is a valuable source of information about the situation on the road, although only some of the problems can be detected in this way: recognition of pre-defined objects or object classes and identification - recognition of an individual object instance. Currently transport systems support car number plate recognition and face detection. However, such systems require too much system resources and memory.

Three-Dimensional Sensors can be used to detect holes, bumps, transit impediments, collision avoidance. Main disadvantages of such systems are the high cost and limited coverage.

We offer a solution that allows receiving information directly from the driver and transfer annotations automatically generated using an ontological approach to the control center. To use the proposed solution the driver will need only to download and install application on his/her smartphone. All required data (audio, video) will come from sensors of mobile phone and car DVR.

2.1   Ontologies

Use of the ontological approach allows maximum flexibility in describing the subject area, and ensures logical inferences, i.e. facts not literally present in the ontology, but enabled by the underlying semantics. These entailments may be based on a single document or multiple distributed documents. The ontology consists of a set of subject domain concepts with their attributes and relationships. Concepts are the basic building blocks of ontology. All the rest provides describing of these concepts and relationships between them. SARRS works with RDF/OWL ontologies. Each instance in

OWL [9] is a member of the class owl:Thing and can be perceived as an oriented graph. Figure 1 shows a fragment of our "The Road" ontology.

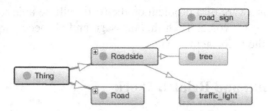

Fig. 1. Fragment of our "The Road" ontology

3   Focus of the Research Approach

Many dangerous situations, such as inoperative traffic light, open hatch, deep hole, remain unsolved. Therefore, the focus of our research is on situations that cannot be detected using existing approaches, and situations, which must be responded to immediately (Fig. 2). In such cases, information can be obtained from the driver. Using the proposed method, information about critical situations that require an immediate

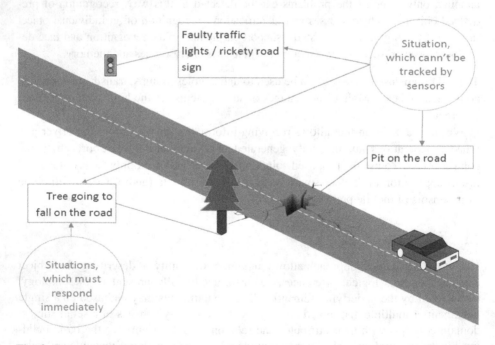

Fig. 2. Cases in which it is planned to use the system: situations that cannot be detected using existing approaches, and situations, which must be responded immediately.

response, will be transmitted to drivers in real time (accident, a tree on the road). An important objective of the system is to obtain all the necessary information without distracting the driver from driving a car and any inconvenience to him/her. One of the promising approaches to address this challenge is speech recognition.

4   The Choice of Speech Recognition System

Speech recognition technology significantly increases the safety of drivers, allowing them to perform operations such as navigation, climate control, picking up the phone without using their hands and at the same time minimizing the need of averting their eyes from the road [16, 17]. In our project, speech recognition is considered as exceedingly effective method of obtaining information from the driver. There are three approaches to the mechanism of speech recognition: speech recognition using cloud services (Google Speech API [18, 19], Yandex SpeechKit [20, 21]); speech recognition using mobile platforms (Android, IOS etc.) built-in solutions; speech recognition based on open source systems and engines which enable custom grammar, custom dictionaries etc. In our project, the basic requirements for a speech recognition system are: flexibility, customizability and speed. Among other things, speech recognition at this stage carried out in accordance with the specified dictionary of key words. Thus rejecting built-in solutions increase the percentage of recognition and reduce the probability of error.

4.1   Comparative Assessment of Speech Recognition Systems

Due to space limitations, the comparative analysis of speech recognition systems cannot be given in its entirety in this article. After a preliminary analysis 4 open source speech recognition systems have been selected for further study, testing and choosing the most suitable for the current task: CMU Sphinx (Pocketsphinx) [22], Julius [23], Kaldi [24], Simon [25].

Following the results of analysis CMU Sphinx was chosen. CMU Sphinx is a speaker-independent continuous speech recognition system developed at Carnegie Mellon University. For our task Pocket Sphinx is most suitable. It is a lightweight speech recognition engine, specifically tuned for handheld and mobile devices and may be built in any other system based on ARM processor.

Table 1. Comparative Assessment of Speech Recognition Systems

	Scope of application	Based on	Advantages	Disadvantages
CMU Sphinx (Pocket-sphinx)	State of art speech recognition algorithms for efficient speech recognition	Hidden markov model and n-gram statistical language model	- Productivity - Speed - Wide range of tools (keyword spotting, alignment, pronunciation evaluation) - Designed specifically for low-resource platform	-
Julius	High-performance, continuous speech recognition	word N-gram and context-dependent HMM	- High productivity - Embeddability	Applies only for the model of Japanese language
Kaldi	Toolkit for speech recognition and research in this area	Subspace Gaussian Mixture Model	- Very customizable - Refactor reliability	-Resource consumption - Speed
Simon	Work on the scenario. Packages are configured for specific tasks	Julius and HTK	No need for training of the system before starting	- Non-configurable - Works only with the ready scenario

5   SARRS Architecture and Implementation

This section describes scenario of system operating, system architecture and options for using the system in the real environment. Currently scenario of system operating is represented as follows: the driver sees an obstacle / accident / some problem, activates the system with voice and describes the situation. The system records and recognizes the driver's voice, searches for keywords in the record and passes information to the control center. Drivers will be able to annotate the traffic situation and to receive alerts about broken traffic lights, traffic jams, accidents, etc.

As a development platform, we chose Java and as database server MySQL. For ontology building, we use OWL (Web Ontology Language). OWL provides the most accurate and full describing of the semantics compared to XML, RDF and RDF Schema.

5.1   Architecture

SARRS operates as follows (Fig. 3): After activation system records a voice message of the driver and extracts keywords in real time using CMU Sphinx. To improve the accuracy of the system, at the current stage recognition is carried out in accordance with specified grammar patterns.

```
grammar road_problems;
public <traffic_light> = (Damaged | Broken | Faulty |
Defective)
(Traffic light;
public <road> = (Pit | Ice | Barrier | Obstacle |
Fallen tree)
(On the road;
```

The speech recognition/ keyword-extracting system now works only with English, but we are planning to implement it in Russian. Thus we need to integrate the Russian-language acoustic and language models. The compilation of Russian language grammar is a separate problem because of the lack of a fixed word order. Potentially the system is multilingual.

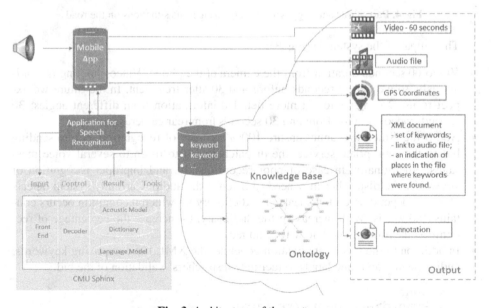

Fig. 3. Architecture of the system.

To implement the understanding and synthesis of information, we use ontological approach. At the moment, the ontology consists of 20 basic terms that describe possible dangerous situations on the road (Fig. 4).

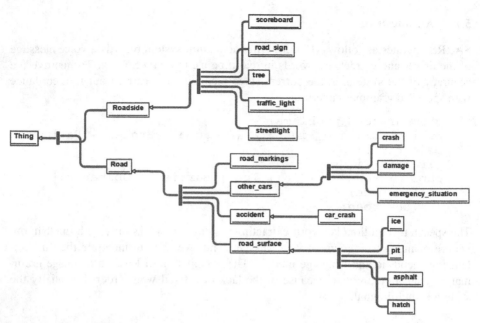

Fig. 4. Part of our ontology with possible dangerous situations on the road.

The output of the system is (Fig. 3):

- Video 60 seconds duration from the camera of the phone. Video shooting is continuously, so we get 30 seconds before and 30 after treatment. In the future we expect to use 2 cameras to get more detailed information from different angles: 30 seconds video from the front and 30 seconds from rear cameras.
- Audio file. Since we cannot ensure 100% accuracy of recognition, before sending bid to the appropriate service, the dispatcher need to listen to several voice messages. Additionally, in case of poor record quality, and improper recognition of keywords, the dispatcher may need to listen to the audio. Currently system detects 60-65% (depending on the speaker), a dictionary by which recognition occurs contains 480 words. With increasing the dictionary, to increase the percentage of recognition the training module will be added.
- In addition to the record application generates the XML file comprising keywords, audio file link and time labels to facilitate dispatchers' validation of the bid.

```
<found>
<keyword>accident</keyword>
<aid>17690458</aid>
<owner_id>664</owner_id>
<audio>k72.waw</audio>
<aduration>195</aduration>
<kwstart>0.097</kwstart>
<kwend>210.000</kwend>
</found>
```

- GPS Coordinates. Location of the smartphone after voice activation. Because drivers can activate the system when on sight a problem, one way or already passed by the problem situation, to determine the position average values will be taken. Currently position error is 12-15 meters. However, based on matching with the map and width the road, it can be improved.

Annotation containing keywords, anticipated incident characteristics, level of danger and level of urgency of response.

```
<Ontology>
    <Declaration>
        <Class IRI="#Crash"/>
    </Declaration>
    <DataPropertyDomain>
        <DataProperty IRI="#degree_of_danger"/>
        <ObjectAllValuesFrom>
            <ObjectProperty abbreviatedI-
RI="owl:topObjectProperty"/>
            <Class IRI="#Crash"/>
        </ObjectAllValuesFrom>
    </DataPropertyDomain>
    <DataPropertyDomain>
        <DataProperty IRI="#urgency_of_response"/>
        <ObjectExactCardinality cardinality="">
            <ObjectProperty abbreviatedI-
RI="owl:topObjectProperty"/>
            <Class IRI="#Crash"/>
        </ObjectExactCardinality>
    </DataPropertyDomain>
</Ontology>
```

At the moment, the use of ontologies allows us to structure annotations by type of incident, there are also two criteria: urgency of response and the degree of danger of the situation. For example, in the situation of car crash it is necessary to respond immediately. On the contrary, the information about slight road pits is not urgent, so it should be processed with the application and then transferred to road services with an indication on which parts of the road the asphalt should be replaced (Fig. 5). In future we are planning extension of ontology and creation of related ontologies. Thus we will be able to add options such as auto-detection of service to which SARRS should pass the information and, perhaps, provide recommendations to the driver in critical situations.

5.2   Implementation

Key indicators of our system are ontology-based annotations (concepts and parameters) and GPS coordinates. According to these parameters, information will be grouped and sorted.

Our proposed application in the long term could be useful for drivers, road services and municipal administration (Fig. 5). In most cases, SARRS sends information about the problem on the road to the dispatch center. Requests are available to dispatchers in a convenient graphical form, grouped by place of handling and type of incident. Depending on the ratio of danger of accidents and the required reaction rate indicator on the map painted in one of three colors: red, yellow, green. For example, a car accident requires ambulance will be displayed on the map as a red indicator. Broken traffic light, which causes of traffic congestion, requiring maximum imminent intervention, but having a low risk rating - the yellow indicator. Non-critical situations that do not require immediate intervention, such as pit on the road are induced by green. It is also assumed to correlate the size and brightness of the indicators with the number of driver appeal. The system is semi-automated. After receiving several requests dispatcher validates information received from the driver, viewing the video, listen to audio (optional) and decides on necessary actions. In the case of hundreds or thousands of requests for the same incident, the proposed system can significantly improve the efficiency and performance of dispatchers by the automated rubrication of drivers' messages. Further information about the problem and prompt for its elimination are transferred to the appropriate service. If necessary, dispatcher also sends a warning to drivers. In the future, system will be able to signal about some the accidents to appropriate services without dispatchers' intervention.

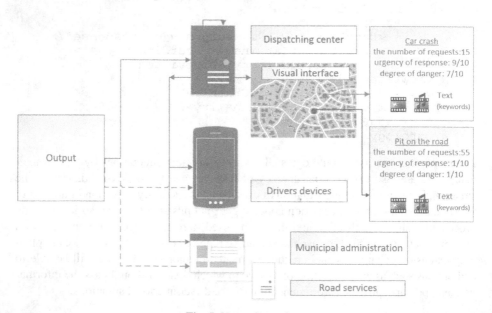

Fig. 5. Usage Scenarios.

6   Conclusions and Future Work

Our approach will allow the following: a). to report problems on the road without being distracted from driving; b). streamline road services; c). facilitate the work of dispatchers by automating the cataloging of information, and providing information in a graphical form, using smart web interface. Also it significantly increases road safety and will allow automating the work of future road and transportation services. In future we are planning to extend the current ontology and to create related ontologies. We also are planning to add attributes and relationships which will lead to safe conclusions about the degree of danger (priority) of the situation, the need for an immediate response, necessity of the immediate warning of other drivers. Next step is to create tools allowing drivers to report the problem to the appropriate service automatically, thus minimizing the participation of dispatchers, giving them ready to use structured information and proposals to resolve the problems. In some situations, drivers will be able to interchange this distributed knowledge between them without the need of a centralized system. In the field of speech recognition, we are planning to add multi-language, configuration options for a particular person (training module) and emotions recognition. The goal is to develop a comprehensive ontology-based voice annotation method instantiated as an app, which will be usable in the real environment and deployed on smartphones.

Acknowledgements. The research has been carried out with the financial support of the Ministry of Education and Science of the Russian Federation under grant agreement #14.575.21.0058.

References

1. Hu, M., Li, C.: Design smart city based on 3s, internet of things, grid computing and cloud computing technology. In: Wang, Y., Zhang, X. (eds.) IOT 2012. CCIS, vol. 312, pp. 466–472. Springer, Heidelberg (2012)
2. Medvedev, A., Zaslavsky, A., Khoruzhnikov, S., Grudinin, V.: Reporting road problems in smart cities using openiot framework. In: Podnar Žarko, I., Pripužić, K., Serrano, M. (eds.) FP7 OpenIoT Project Workshop 2014. LNCS, vol. 9001, pp. 169–182. Springer, Heidelberg (2015)
3. Medvedev, A., Zaslavsky, A., Grudinin, V., Khoruzhnikov, S.: Citywatcher: annotating and searching video data streams for smart cities applications. In: Balandin, S., Andreev, S., Koucheryavy, Y. (eds.) NEW2AN/ruSMART 2014. LNCS, vol. 8638, pp. 144–155. Springer, Heidelberg (2014)
4. Lopez, T., Ranasinghe, D.C., Harrison, M., Mcfarlane, D.: Adding sense to the Internet of Things. Personal and Ubiquitous Computing **16**(3), 291–308 (2012). Springer-Verlag
5. Jara, A.J., Lopez, P., Fernandez, D., Castillo, J.F., Zamora, M.A.: Mobile digcovery: discovering and interacting with the world through the A. F. Internet of Things. Personal and Ubiquitous Computing **18**(2), 323–338 (2014). Springer-Verlag
6. Rusdorogi, April 20, 2015. http://www.rusdorogi.ru/
7. Goodroads, April 20, 2015. http://goodroads.ru/
8. Streetbump, April 20, 2015. http://www.cityofboston.gov/DoIT/apps/streetbump.asp
9. OWL Web Ontology Language Overview, April 20, 2015. http://www.w3.org/TR/owl-features/
10. Yu, L.: Introduction to the Semantic Web and Semantic Web Services. Chapman and Hall/CRC (2007)

11. NigelShadbolt, E.M., Stutt, A., Gibbins, N., Kopsa, J., Mikovec, Z., Slavík, P., Ontology driven voice-based interaction in mobile environment. In: Davies, N., Kirste, T., Schumann, H. (eds.) Mobile Computing and Ambient Intelligence, IBFI, Schloss Dagstuhl, Germany (2005)

12. Hamerich, S.W.: Towards advanced speech driven navigation systems for cars. In: 3rd IET Int. Conf. on Intelligent Environments, IE 2007, Ulm, Germany, pp. 24–25 (2007)

13. Chen, F., Jonsson, I.-M., Villing, J., Larsson, S.: Application of Speech Technology in Vehicles. Speech Technology, pp. 195–219 (2010)

14. Cameron, H.: Speech at the Interface. In: Workshop on Voice Operated Telecom Services, Ghent, Belgium, pp. 1–7 (2000)

15. Barón, A., Green, P.: Safety and Usability of Speech Interfaces for In-Vehicle Tasks while Driving: A Brief Literature Review. Transportation Research Institute (UMTRI) (2006)

16. Chen F., Jonsson, I.-M., Villing, J., Larsson, S., Application of Speech Technology in Vehicles, Speech Technology. Springer US (2010)

17. Huang, X., Acero, A., Hon, H.-W.: Spoken Language Processing. Guide to Algorithms and System Development (2001)

18. Google Speech API Demonstration, April 20, 2015. https://www.google.com/intl/en/chrome/demos/speech.html

19. Google Speech To Text API, April 20, 2015. https://gist.github.com/alotaiba/1730160

20. Yandex Speech Kit for Android, April 20, 2015. https://github.com/yandexmobile/yandex-speechkit-android

21. Yandex Speech Kit, April 20, 2015. http://tech.yandex.ru/speechkit/

22. CMU Sphinx, April 20, 2015. http://cmusphinx.sourceforge.net/wiki/start

23. Julius, April 20, 2015. http://julius.sourceforge.jp/en_index.php

24. Kaldi, April 20, 2015. http://kaldi.sourceforge.net/index.html

25. Simon, April 20, 2015. http://www.simon-listens.org/index.php?id=122&L=1

26. Walker, W., Lamere, P., Kwok, P., Raj, B., Singh, R., Gouvea, E., Wolf, P., Woelfel, J.: Sphinx-4: A flexible open source framework for speech recognition (2004). http://egouvea.users.sourceforge.net/paper/smli_tr-2004-139.pdf

27. Mykowiecka, A., Waszczuk, J.: semantic annotation of City transportation information dialogues using CRF method. In: Matoušek, V., Mautner, P. (eds.) TSD 2009. LNCS, vol. 5729, pp. 411–418. Springer, Heidelberg (2009)

28. Perera, C., Zaslavsky, A., Christen, P., Georgakopoulos, D.: Context Aware Computing for The Internet of Things: A Survey. IEEE Communications Surveys & Tutorials 16(1), 414–454 (2014)

29. Buch, N., Kristl, V.S.A., Orwell, J.: A Review of Computer Vision Techniques for the Analysis of Urban Traffic. IEEE Transactions on Intelligent Transportation Systems 12(3)

Method of Defining Multimodal Information Falsity for Smart Telecommunication Systems

Oleg Basov[1], Andrey Ronzhin[2,3] (✉), Victor Budkov[2], and Igor Saitov[1]

[1] Academy of FAP of Russia, 35, Priborostroitelnaya, Orel 302034, Russia
{oobasov,akramovish}@mail.ru
[2] SPIIRAS, 39, 14th line, St. Petersburg 199178, Russia
{ronzhin,budkov}@iias.spb.su
[3] SUAI, 67, Bolshaya Morskaya, St. Petersburg 190000, Russia

Abstract. In the article, a review of the existing methods of transmitted information falsity diagnostics is presented. A conclusion concerning the purposefulness of this function realization in polymodal infocommunication systems has been drawn. A method of defining the multimodal information falsity transmitted in the course of communication act with the help of these systems has been suggested. Common tendencies concerning the dynamics of subscribers' non-verbal behavior parameters have been formulated. Based on the factor and multiple regressive analysis, the factors depending on such dynamics have been distinguished. Based on the carried out research, a conclusion concerning the possibility of realization of transmitted information falsity in the course of interpersonal communication between subscribers has been drawn and a decisive rule has been formulated.

Keywords: Polymodal infocommunication system · Falsity · Non-verbal behavior · Factor analysis · Factor structure matrix · Multiple regression coefficients

1 Introduction

The constantly growing loads on the psychological activity of subscribers of infocommunication systems make the processes of their communication more and more diverse and emotionally constrained. Despite the constant necessity for solving the problem of the transmitted message falsity (truth) under such conditions, nowadays, the secure methods of the corresponding function realization in the corresponding communication systems do not exist.

During the traditional interpersonal communication, correspondents nearly always interact multimodally, using verbal and non-verbal channels. Application of the existing and expected solutions of the signal processing tasks with different modalities during the polymodal infocommunication systems (PICS) synthesis will provide an opportunity of defining the transmitted information falsity [1]. PICS should be understood as an interconnected aggregate of multimodal interfaces; information processing and storing subsets; telecommunication systems, their unifiers functioning

© Springer International Publishing Switzerland 2015
S. Balandin et al. (Eds.): NEW2AN/ruSMART 2015, LNCS 9247, pp. 163–173, 2015.
DOI: 10.1007/978-3-319-23126-6_15

under the sole management, with the aim of collecting, processing, storing, protecting, transmitting and reallocating, reflecting and using multimodal information to the benefit of the subscribers.

The aim of this work is to develop a method of the transmitted information falsity judging by the dynamics of PICS subscriber's non-verbal behavior parameters.

This paper is organized as follows. Section 2 discusses the methods of transmitted information falsity diagnostics. Section 3 describes the method of defining the transmitted information falsity by the dynamics of subscriber's non-verbal behavior parameters. Section 4 presents the experiments, conditions and results.

2   Review of the Existing Methods of Transmitted Information Falsity Diagnostics

Nowadays for solving the problem of the message falsity determination, the method of instrumental diagnostics with the help of a polygraph detector (lie detector) is mostly used. In terms of this method, the conclusion about the falsity of the information transmitted by a person is drawn upon his psychophysical reaction changes character. Despite its popularity, a number of conditions significantly limit its application in practice. Particularly, the use of polygraph as a contact method is possible only provided meeting the range of requirements concerning the place of study organization (comfortable temperature, optimal humidity, noise insulation etc.) and of personal character (personal consent to studies, absence of somatic diseases, mental disorder, etc.) [2, 3]. All listed make the use of polygraph in the course of PICS subscriber's communicative interaction nearly impossible.

A noncontact method of a person's psychophysiological reaction is also known, and it resides in the person's psychophysical reaction fixation according to his changing electromagnetic field in the process of verbal and nonverbal communication. The disadvantage of this method is in the necessity of instrumental complex implementation and stimulating verbal exposure, which makes its exploitation in infocommunication systems impossible.

The degree of users' replies honesty can be determined with the help of noninvasive video registration of the eye movement parameters during verbal communication [4]. The conclusion concerning the emotional psychophysical excitement, which may appear as a result of information concealing or garbling, is drawn based on the comparison of numbers of blinking acts, figure's area and pupils diameter with the control value. The suggested approach appears to be contact and requires special equipment (video-oculograph) along with abiding to a set of rules, significant for the research procedure performance, which makes this method complicated for subscribers' communication.

The evaluation of the truth (falsity) of the speaker's words can be carried out on the basis of speaker's emotional and psychological features and states by a group of expert observers using one and the same video fragment of the testee with duration not less than 40-60 sec [5]. Experts not less than 10 people should know the basics of the human expressive body movements on the scale of, at least, popular editions [6, 7]. The conclusion concerning the speaker's honesty is drawn with the consideration of the coefficient of visual and audial evaluation correspondence, which is based on the degree of the voice intonation compliance with the whole complex of expressive body movements (facial expressions, posture and gestures). The stated coefficient is determined through the mean estimator for each group of experts as a result of Spearmen's rank-order correlation calculation and is accepted as the psychophysiological measure of the speaker's honesty. The described approach appears to be too subjective, as it requires participation of the group of experts and does not allow identifying individual peculiarities of nonverbal behavior of a particular subscriber, which significantly reduces the reliability of the obtained results.

In addition, there is a great number of medical and special equipment designed for the evaluation of human functional features including his behavior; in this respect, the problem was successfully solved a long time ago. However, the solution to the problem of lie detection in technical, as well as in methodological respect, is far from sample and requires new approaches.

The main way to increase the accuracy of the research during the non-contact information retrieval method resides in creating a method, which helps individualize the approach, perform the adjustment and reveal the most informative indicators of behavior reflecting the transmitted information falsity in each particular case. The prospects of this direction were proved by the results of many scientific studies [2, 8–13]. Realization of this trend presupposes software for determining the most informative parameters of the subscriber's nonverbal behavior during transmitted information polymodal representation.

3   Method of Transmitted Information Falsity Determination

The essence of the developed method for falsity determination of transmitted information based on dynamics of the subscriber's nonverbal behavior parameters consists in the following [14]. After the preliminary study of the subscriber's personality based on different sources of information analysis (personal records data, efficiency report, personnel record list, etc.), the possibility of fulfilling the research based on the structured interview is determined. Its program is developed considering further implementation and assessment of the results directly during the procedure, as well as conducting mathematical processing of the results by an expert with the help of any available and reliable information processing means or without them.

The procedure documenting of the structured interview is carried out on the basis of audio-video recording. The structure of interview consists of fragments (tests), which are standardized according to the categories of various question blocks (neutral and control). At the same time, neutral questions do not relate to the topic under consideration, while the control ones reflect the discussion of socially disapproved behavior. For a nonverbal behavior analysis, 16 indicators are used representing three main types of information (Fig. 1).

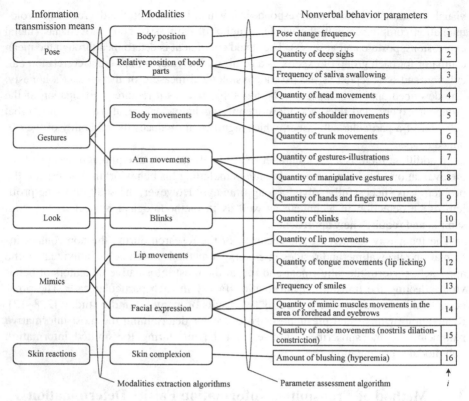

Fig. 1. Subscriber's nonverbal behavior parameters used for transmitted information falsity determination

1. Vegetative reaction, including the following behavior parameters:

— Quantity   of tongue movements (lip licking);
— Quantity of blushing (hyperemia);
— Frequency of saliva   swallowing;
— Quantity of blinks;
— Quantity of deep sighs.

2. Facial reactions, including the following parameters:

— Frequency of smiles;
— Quantity of movements of facial muscles in the area of forehead and eyebrows;
— Quantity of nose movements
— Quantity of lip movements.

3. Pantomimic reactions, including parameters:

— Quantity of head movements;
— Quantity of shoulder movements

— Quantity of body movements;
— Quantity of demonstration gestures;
— Quantity of manipulative gestures;
— Quantity of hand and fingers movements;
— Frequency of body position changes.

For the stated parameters, an average number of reaction for each nonverbal beha-vior parameter per minute is calculated:

$$\overline{r_i} = \frac{r_i}{t_a},$$ (1)

where r_i is the number of reactions in absolute unit for each parameter $(i = 1...16)$; t_a is the analysis time in minutes.

The derived results processing is possible not only based on the number of reactions but on their expressiveness during an interview:

$$\overline{r_i^*} = \omega_i \frac{r_i}{t_a},$$ (2)

where for defining ω_i a system of grading is applied (the maximum grade reflects the highest reaction expressiveness with the account of tendencies).

The structured interview is usually conducted during the preparation and screening tests for employment; promotion to a higher position; professional suitability deter-mination. After such testing, the derived results can be accumulated and processed in a relevant manner to reveal testee's change tendencies for each nonverbal behavior parameter and/or group of parameters.

The tendency in reaction is determined on the ground of juxtaposition of neutral topic indices $\overline{r_i^{(n)}}$ and control blocks questions indices $\overline{r_i^{(c)}}$. If $\overline{r_i^{(c)}} > \overline{r_i^{(n)}}$, then neutral index surpassing the control questions is considered to be a reaction. If $\overline{r_i^{(c)}} < \overline{r_i^{(n)}}$, then it works the other way round.

In the course of experiment a hypothesis has been proved concerning the existing difference between indices of nonverbal behavior of the testees, conveying false in-formation, and those reporting true information. The following common tendencies in the dynamics of the nonverbal behavior of the subscribers reporting false information, compared to those telling the truth, have been outlined.

1. Decrease in head movements (83.3% of the cases, on average by 25.8%);
2. Decrease in leg and feet movements (66.70% of cases, on average by 34.08%);
3. Increase in quantity of smiles (58.3% of cases, on average by 114.47%);
4. Increase in nose movements (nostrils dilatation/striction) (58.3% of cases, on aver-age by 195,89%);
5. Increase in saliva swallowing (58.3% of cases, on average by 90.86%);
6. Increase in a number of blinks (50% of cases, on average by 2.29%).

In other cases, the dynamics direction of the analyzed nonverbal behavior parameters is individualized and can either increase or decrease.

Directly during the communication act this method presupposes registration of parameters r_i, $i = 1...16$ of the non-verbal behavior (fig. 2) of the PICS subscriber (subscribers) under consideration. For such purpose, a video capture and pre-processing of video frames, received from the video camera with image optical resolution 1280x720 pix, progressive scan and image frequency 25 Hz, are carried out. The following modality isolation algorithms are also implemented:

1. Tracking of the body position changes, head, shoulders, trunk and hands movements on the ground of basic Lukas-Kanade algorithm [15];
2. Subscriber's face scan with the help of factorable classifier based on the Viola-Jones algorithm [16], along with targeted areas containing skin complexion, eye, lips, nose areas, mimic muscles in the area of cheekbones, forehead and eyebrows.

The stated algorithms have been realized in OpenCV [17] Library and preliminary trained on pictures of different people.

After assessing r_i parameters of nonverbal behavior, a conclusion is drawn concerning the falsity (truth) of the information transmitted during the stated information act by comparing the resulting reaction $\left(\overline{r_i}\right)$ to the reaction (change tendency for each parameter or/and group of parameters of nonverbal behavior) registered in the course of structured interview.

For the shift from a multitude of initial variables r_i to a substantially smaller number of variables (F_m factors), a factor analysis has been applied [18], which reflects communality assessment for each variable judging by square coefficient of the multiple correlation for the given variable with other variables.

The factor analysis was used for data derived during running the experiment on a group of testees comprising people of both sexes aged from 18 to 32 in total number of 32 participants. The experimental test was carried out on the basis of modeling the situation connected with a breach of work discipline. All the testees were divided into two equal groups. Members of one group possessed the information concerning this fact, while the others were unaware of it. The participants' belonging to a specific group was decided randomly. At the same time, neutral questions dealt with the testee's bibliographic data, whereas control questions were aimed at detecting the person possessing the significant information concerning the concealed fact.

As a result of experiment, a statistically true difference between the indices of the neutral and control questions blocks of the structured interview has been revealed for each of the three groups of parameters (vegetative, mimic and pantomimic reactions) for both testee groups. However, the character of their response was remarkably different. During the examination of the people who possessed the valuable information but denied being informed, the indices $\overline{r_i}^{(n)}$ of a neutral block dominated over the indices $\overline{r_i}^{(c)}$ of the control question block. Regarding the analysis of the experiment results for the second group, the members of which while denying the fact of being aware were completely honest, a distinctive domineering of the nonverbal behavior parameters in the control block of question has been detected.

The derived data showed statistically true difference for all parameters. Based on the hierarchical method of the data grouping, three factors $F_m, m = 1...3$, determining Factor Structure Matrix of the used subscribers' nonverbal behavior parameters were singled out (Table 1). The parameters determining each factor are highlighted in bold in Table 1.

Table 1. Factor Structure Matrix

Nonverbal behavior parameters (*i*)	Factor Load		
	F_1	F_2	F_3
Trunk movement (6)	**0.99**	0.1	0.04
Gestures-manipulations (8)	**0.99**	0.09	0.02
Pose change (1)	**0.98**	0.13	0.08
Deep sigh (2)	**0.97**	0.2	0.05
Head movement (4)	**0.93**	0.2	0.2
Gestures-illustrations (7)	**0.9**	0.3	0.2
Saliva swalllowing (3)	**0.88**	0.24	0.12
Shoulders movements (5)	**0.82**	0.26	0.28
Hyperemia (16)	0.35	**0.79**	0.2
Smile (13)	0.43	**0.75**	0.15
Nose movement (15)	0.4	**0.74**	0.23
Tongue movement (12)	0.5	**0.62**	0.34
Blinking (10)	0.44	**0.58**	0.42
Movements in the area of forehead, eyebrows (14)	0.1	0.38	**0.83**
Hand and finger movement (9)	0.2	0.24	**0.79**
Lip movement (11)	0.1	0.3	**0.63**

The indices division within the designated factors is characterized by stability and is preserved through all stages of a structured interview. The enumerated factors reflect nonverbal behavior exclusively and are viewed as basic, which does not exclude their inclusion into the optional parameters range method with the following determination of additional parameters on their basis.

Values of a_{im} in the cells of Table 1 are factor loads of the variable i for the factor m. Correlation coefficient between any two parameters can be restored using Table 1 as a sum of factor loads products for the following lines:

$$c_{ij} = \sum_{k=1}^{3} a_{ik} a_{jk},$$

where i, j are the parameter numbers in a correlation matrix $C = \{c_{ij}\}$.

Value factors $F_m, m = 1...3$ are used for compact representation of differences between three objects: indices $r_i^{(1)} = \overline{r_i^{(n)}}$ of a neutral block; indices $r_i^{(2)} = \overline{r_i^{(c)}}$ of a control block; and indices $r_i^{(3)} = \overline{r_i^{(a)}}$ registered in a communicative act process. With the known correlations c_{ij} and factor loads a_{im} (Table 1) as an assessment for F_m for the k-th object, linear combinations of the initial variables values are suggested:

$$F_1^{(k)} = \beta_6^{(k)} r_6^{(k)} + \beta_8^{(k)} r_8^{(k)} + \beta_1^{(k)} r_1^{(k)} + \beta_2^{(k)} r_2^{(k)} + \beta_4^{(k)} r_4^{(k)} + \beta_7^{(k)} r_7^{(k)} + \beta_3^{(k)} r_3^{(k)} + \beta_5^{(k)} r_5^{(k)}; \quad (3)$$

$$F_2^{(k)} = \beta_{16}^{(k)} r_{16}^{(k)} + \beta_{13}^{(k)} r_{13}^{(k)} + \beta_{15}^{(k)} r_{15}^{(k)} + \beta_{12}^{(k)} r_{12}^{(k)} + \beta_{10}^{(k)} r_{10}^{(k)}; \quad (4)$$

$$F_3^{(k)} = \beta_{14}^{(k)} r_{14}^{(k)} + \beta_9^{(k)} r_9^{(k)} + \beta_{11}^{(k)} r_{11}^{(k)}, \quad (5)$$

where the multiple regression coefficients [18] act as the factor coefficient $\beta_i^{(k)}$.

When representing (4) – (6), a conclusion, which concerns the falsity (truth) of the information transmitted in a communication act, is drawn on the ground of comparing the reactions $F_m^{(3)} - F_m^{(2)}$ to the reactions $F_m^{(2)} - F_m^{(1)}$ for the detected factors (Table 1).

4   Experimental Results

The results of the research, in the course of which the falsity of information reported by a subscriber was detected, allow speaking about consistency of the suggested approach.

An audio-video recording of a structured interview acquired in the course of subscriber's screening check during his employment was preliminary studied. The research was based on the interview neutral fragment realization, which was followed by control questions. The topics of this conversation were selected with respect to individual peculiarities of the subscriber's personality and were not connected with his further occupation.

As a result of this research, the change tendencies specific to the subscriber under consideration were revealed according to the groups of nonverbal behavior parameters united on the ground of the revealed factors (Table 2).

Table 2. Average indicators of the subscriber's nonverbal behavior assessment

i	$\overline{r}_i^{(n)}$	$\beta_i^{(1)}$	$F_m^{(1)}$	$\overline{r}_i^{(c)}$	$\beta_i^{(2)}$	$F_m^{(2)}$	$\overline{r}_i^{(a)}$	$\beta_i^{(3)}$	$F_m^{(3)}$
6	0	0.125		1.2	0.322		1.6	0.322	
8	1.21	0.524		0.9	0.313		0	0.313	
1	0.1	-0.1		0.3	-0.0001		0.42	-0.0001	
2	0.43	-0.237	0.501	0.12	0.714	0.794	0.3	0.714	0.647
4	10.8	0.009		7.72	0.011		10.4	0.011	
7	7.43	-0.017		10.18	-0.023		4.5	-0.023	
3	2.41	0.011		2.84	0.067		2.1	0.067	
5	6.32	-0.003		6.89	-0.0001		7.2	-0.0001	
16	0	0.307		0.1	0.122		0.4	0.122	
13	0	0.413		0	0.213		0	0.213	
15	1.73	0.094	0.409	0.93	0.134	0.608	1.5	0.134	0.858
12	0	0.098		1.2	0.312		1.7	0.312	
10	5.61	0.044		4.62	0.021		3.7	0.021	
14	4.23	0.022		2.23	0.063		1.8	0.063	
9	11.3	0.001	0.579	8.4	0.001	0.605	5.1	0.001	0.630
11	1.56	0.304		2.14	0.213		2.4	0.213	

After that, there has been carried out a study of the video recording of the conference session in the course of which, during 3 min 25sec, the subscriber communicated some relevant information. The conclusions concerning the transmitted information falsity were formed on the ground of nonverbal behavior parameters $\overline{r}_i,, i = 1...16$ and estimates of factor values $F_m, m = 1...3$.

From the analysis results (Table 2), it is clear that, when stating false information, the subscriber under consideration shows the following parameter dynamics of the nonverbal behavior factor F_1:

1. Increase in quantity of trunk movements;
2. Decrease in quantity of gestures-manipulations;
3. Increase in quantity of the body position changes;
4. Increase in quantity of deep sighs;
5. Increase in quantity of shoulder movements.

As for the second factor F_2, while transmitting false information, a subscriber can be characterized by the following dynamics of nonverbal behavior:

1. Increased amount of blushing;
2. Increase in lip licking quantity;
3. Decrease in blinking;

For the third factor F_3, the following changes in nonverbal behavior parameters are noticed:

1. Reduction in quantity of facial movements in the area of forehead and eyebrows;
2. Reduction in quantity of hand and finger movements;
3. Increase in quantity of lip movements.

The mentioned changes of the parameters resulted in the change of corresponding factors value:

$$F_1^{(2)} - F_1^{(1)} = 0,294 \qquad\qquad F_1^{(3)} - F_1^{(2)} = 0,143 \tag{6}$$

$$F_2^{(2)} - F_2^{(1)} = 0,199 \qquad\qquad F_2^{(3)} - F_2^{(2)} = 0,250 \tag{7}$$

$$F_3^{(2)} - F_3^{(1)} = 0,199 \qquad\qquad F_3^{(3)} - F_3^{(2)} = 0,025 \tag{8}$$

Stable direction of the analyzed parameters (factors) change in the process of communicative interaction compared to the response to neutral and control questions blocks in the course of structured interview indicates the subscriber's insincerity.

Based on reaction (6)(8), a conclusion concerning the transmitted messages falsity can be made if for every $m, (m = 1...3)$:

$$\left(\left(F_m^{(3)} - F_m^{(2)} \right) > 0 \; and \; \left(F_m^{(2)} - F_m^{(1)} \right) > 0 \right) or \left(\left(F_m^{(3)} - F_m^{(2)} \right) < 0 \; and \; \left(F_m^{(2)} - F_m^{(1)} \right) < 0 \right). \tag{9}$$

Under conditions of complete automation of the transmitted multimodal information falsity determination process, convolution of the solutions (9) can be performed with an account of the significance of the corresponding factors $F_m, m = 1...3$.

5   Conclusion

The obtained results testify to the possibility of the realization of the suggested method of transmitted multimodal information falsity determination in the real-time mode and in the process of interpersonal communication between PICS subscribers. Its further development and perfection is due to increasing the accuracy of the algorithms for modality identification and estimation of nonverbal parameters of human behavior.

Acknowledgments. This work is partially supported by the Russian Foundation for Basic Research (grants № 13-08-0741-a; №15-07-06774-a).

References

1. Basov, O.O., Saitov, I.A.: Basic Channels of Interpersonal Communication and their Projection onto Infocommunicational Systems. SPIIRAS Proceedings **30**, 122–140 (2013). (In Russ.)
2. Vrij, A.: Detecting Lies and Deceit: The Psychology of Lying and the Implications for Professional Practice. Wiley (2000). ISBN: 9780471853169
3. Gruzyeva, I.V.: Formal-dynamic and Stylistic Peculiarities of Individuality as Factors of Instrumental Detection of the Concealed Information. Author's Abstract. Cand. Sc. {Psychology} (2006) (in russ.)
4. Usanov, D.A., Romanova, N.M., Skripal, A.V., Rytik, A.P.,Vagarin, A.J., Samokhina, M.A.: Method of estimation of psychophysical condition of person. Patent RU 2337607, Ap-plication 2007111403/14, 28.03.2007, Bull. 31 (2008)
5. Morozov, V.P., Morozov, P.V.: Method for estimating sincerity-insincerity of speaking person. Patent RU 2293518, Application 2005124844/14, 04.08.2005, Bull. 5 (2007)
6. Nierenberg, G.I., Calero, H.H.: How to Read a Person Like a Book. Pocket Books (1990). ISBN: 0671735578
7. Pease, A., Pease, B.: The Definitive Book of Body Language. Bantam (2006). ISBN: 9780553804720
8. Vrij, A.: Detecting lies and deceit: pitfalls and opportunities. Wiley (2008). ISBN: 9780470516249
9. Ekman, P.: Emotions Revealed: Recognizing Faces and Feelings to Improve Communication and Emotional Life. Henry Holt and Co. (2004). ISBN: 080507516X
10. Butovskaya, M.L.: Body Language: Nature and Culture (Evolutionary and Cross-Cultural Bases of the Human Nonverbal Communication). Scientific World (2004) (in russ.)
11. Ronzhin, A., Budkov, V.: Speaker turn detection based on multimodal situation analysis. In: Železný, M., Habernal, I., Ronzhin, A. (eds.) SPECOM 2013. LNCS, vol. 8113, pp. 302–309. Springer, Heidelberg (2013)
12. Ackerl, K., Atzmuller, M., Grammer, K.: The Scent of Fear. Neuroendocrinology Letters **23**, 79–84 (2002)
13. Meshcheryakov, R., Bondarenko, V.: Dialogue as a basis for construction of speech systems. Cybernetics and Systems Analysis **44**(2), 175–184 (2008)
14. Basov, O., Basova, A., Nosov, M.: Human resources management in conditions of operators' psychophysiological state changes. In: Ronzhin, A., Potapova, R., Delic, V. (eds.) SPECOM 2014. LNCS, vol. 8773, pp. 259–267. Springer, Heidelberg (2014)
15. Bouguet, J.-Y.: Pyramidal Implementation of the Lucas-Kanade Feature Tracker Description of the algorithm. Intel Corporation Microprocessor Research Labs (2000)
16. Castrillyn, M., Deniz, O., Hernandez, D., Lorenzo, J.: A Comparison of Face and Facial Feature Detectors based on the Viola-Jones General Object Detection Framework. Machine Vision and Applications **22**(3), 481–494 (2011)
17. Bradsky, G., Kaehler A.: Learning OpenCV. O'Reilly Publisher (2008)
18. Field, A.: Discovering Statistics using IBM SPSS Statistics. SAGE Publications Ltd. (2013)

NEW2AN

Addressing the Influence of Hidden State on Wireless Network Optimizations Using Performance Maps

Kim Højgaard-Hansen[1]([✉]), Tatiana K. Madsen[1], and Hans-Peter Schwefel[1,2]

[1] Network and Security Section, Aalborg University, Aalborg, Denmark
{khh,tatiana}@es.aau.dk, schwefel@ftw.at
[2] The Telecommunications Research Center Vieanna (FTW), Vienna, Austria

Abstract. Performance of wireless connectivity for network client devices is location dependent. It has been shown that it can be beneficial to collect network performance metrics along with location information to generate maps of the location dependent network performance. These performance maps can be used to optimize wireless networks by predicting future network performance and scheduling the network communication for certain applications on mobile devices. However, other important factors influence the performance of the wireless communication such as changes in the propagation environment and resource sharing. In this work we extend the framework of performance maps for wireless networks by introducing network state as an abstraction for all other factors than location that influence the performance. Since network state might not always be directly observable the framework is extended with a network state estimation and prediction function. In the evaluation scenario, resource sharing is used as an example of network state and the framework is applied to the use-case of scheduling TCP-based communication to lower communication overhead. Using extensive simulations for evaluation we show how dynamic network state caused by resource sharing influences the scheduling performance.

1 Introduction

The communication performance experienced by a mobile client in a wireless network is known to be highly correlated to the location of the device. With the widespread adoption of GPS many mobile devices can accurately estimate the current location. If the performance of the wireless network and the device location is sampled a map of the network performance can be generated. A performance map can be used to schedule network communication hereby optimising the use of the network or the application performance as shown in [1][2][3][4].

In [5] a performance map is used to schedule TCP transfers to avoid communication overhead and reduce transfer times. The performance map was generated under ideal conditions where the location of the client device is the only factor influencing the communication performance. However, in a real wireless network many dynamic properties that might not be directly observable will influence

© Springer International Publishing Switzerland 2015
S. Balandin et al. (Eds.): NEW2AN/ruSMART 2015, LNCS 9247, pp. 177–189, 2015.
DOI: 10.1007/978-3-319-23126-6_16

the communication performance. These dynamic properties include changes in the propagation environment, errors/faults in the network and sharing network resources with other users in the wireless network. These dynamic properties cause temporal and spatial variability in the performance measurements.

In an abstract form the influence from all the dynamic properties can be viewed as the state of the network. This paper investigates how information about the network state can be included in the performance map and the performance prediction. In the evaluation scenario, network traffic from other users in the network will act as network state and the term cross traffic will be used to describe it. The application to be optimised is large TCP based transfers with the same optimisation goals as in [5] namely to lower the communication overhead (defined in Section 2.2). We thereby compare different optimization schemes at the same level of 'agressiveness', measured by the total transmission probability(see Section 4), P, averaged over the whole geographic space .

Related Work

Using a map of measured network performance to optimise application performance has been proposed in several previous papers. In [1] and [4] the authors improve the user experience of video streaming by allowing the application to adapt to the predicted available bandwidth. In both cases predictable movement patterns from public transport and vehicles on a freeway make it possible to generate a small scale map of bandwidth and use previous measurements to predict future performance. In [2] the authors demonstrates how energy usage can be lowered significantly by using previous RSSI measurements to schedule background syncing and streaming. The authors of [3] show how just one week of combined sampling of previous mobility and network capabilities can be used to generate connectivity forecasts and lower energy usage by selecting the best network attachment.

In this work we have generalised the concept of recording network performance metrics into a framework where multiple metrics can be collected and used depending on the need of the application like in [3]. The proposed framework also allows for different applications with different optimisation goals to be using the performance map. We do not consider the use of multiple available networks but instead focus on improving the use of one network by mapping the network performance in a much more detailed level than in [3]. We also demonstrate how the state of the network can be included in the performance map and used to improve the scheduling of communication.

Predicting mobility in wireless networks have also been used for other network optimizations than scheduling the communication. In [8] a database of expected throughput is used to optimize handing over from cellular networks to WLAN. In this work we focus solely on optimizing the use of one network instead of making hand-overs to other networks.

Improving the network performance using predictions can also achieved by reserving network resources as demonstrated in [9]. While that works well as an optimization in the infrastructure network our work instead focus on improving the network usage from the client device.

Scheduling TCP communication by predicting future performance can be realised as in our previous work [5], where predictions of TCP throughput are used to plan how the application layer should deliver data to TCP. To achieve this, models for TCP performance were used as defined in [6] and in [7]. In this work, we instead implement a way to start and stop TCP transmission directly in the TCP network layer, and then decide at what times the performance of the network is good enough to use it.

2 Framework for Wireless Network Performance Prediction

The proposed framework for generating and using wireless network performance maps to predict future network performance and enhance application performance is shown in Figure 1. The framework consist of two major parts; the collection part (offline) and the usage/prediction part (online).

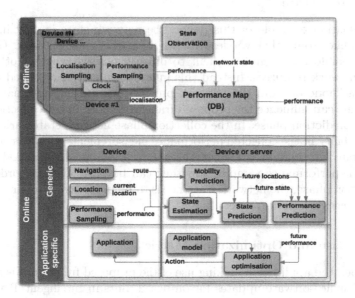

Fig. 1. Overview of the framework

The data collection part consist of a database for storing the collected information, a number of network enabled devices sampling the network performance, and a functionality retrieving information from the network provider about the network state. Each network-enabled device also samples the location of the device and the time, which are added as meta data to the performance samples. The collected data in the database can be viewed as a map of the network performance.

The usage part consist of a network enabled device with a modified network stack making use of the performance map, the performance map (DB) and optionally a server performing the prediction and estimation functions. Since the network performance is correlated to the location a function to predict the mobility of the device is needed. The mobility prediction function receives location samples and optionally route information as input. The network performance is also influenced by the network state, why a function to estimate the network state is needed as well. The state estimation is combined with a model of the network state and the mobility prediction to form a prediction of the network state. Using the mobility prediction, the state prediction and the performance map the network performance can then be predicted. The predicted network performance and a model of the application can then be used to optimise the execution of an application by informing the application to perform different actions.

2.1 Network State

In the evaluation example of this paper, the network state is only influenced by cross traffic from other wireless clients using the same network. Generally it is not possible for a wireless client to directly observe whether other users are using network resources. Instead this information can be obtained from the network backbone or from the network provider monitoring information. The network state can influence the proposed framework both in the collection phase and in the prediction phase. In the collection phase network state variation will create variability in the collected performance measurements. However, when the network state information is known, this information can be included as meta data for the performance samples and they can be partitioned accordingly. In the prediction phase, the network state needs to be predicted in order to predict the correct network performance.

2.2 Use Case and Optimization Metric

In this paper, the application making use of the proposed framework is an application to handle software updates for on-board units in Intelligent Transportation Systems (ITS) scenarios. A software update will often be a transfer of a significant amount of data using a reliable transport protocol like TCP, so the chosen application model is TCP-based file transfer via a WLAN based road-side infrastructure. If the WLAN network is used in areas with poor performance, packets will be dropped and need to be retransmitted adding to the overall cost and causing waste of wireless transmission resources. The optimisation goal here is to lower the cost by scheduling the communication in a way where the communication overhead is decreased. Overhead in the TCP-based use-case is defined by the data volume of retransmissions and duplicate acknowledgements. The TCP analysis filter in Wireshark is used to calculate the precise overhead [10].

3   Scheduling TCP Communication

Our approach to scheduling is to choose a specific performance threshold to decide when the network performance is acceptable for the application performance. Given a static set of performance measurements for a certain area a specific threshold value would mark a part of that area as usable for communication. In this paper the ICMP RTT is used as the performance metric, where a lower RTT means better performance. The performance map is divided into sub areas, which we define as cells and for each cell a number of RTT samples has been collected. The mean value of these RTT samples is defined as the network performance for that cell. Note that these cells refer to geographic areas distinguished by the performance map. Those are typically much smaller than the radio cells of 2G or 3G cellular networks.

If a set of performance samples are collected for an area, it is then possible to decide for each cell, based on the performance threshold, if the network should be used there or not. As an example Figure 2 shows how one setting of the performance threshold divides the map into a set of usable and unusable cells.

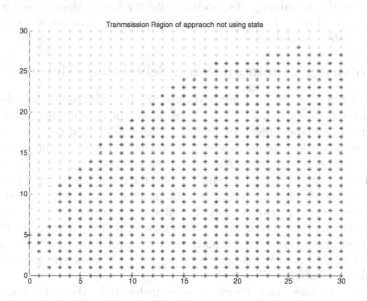

Fig. 2. Transmission region when no network state information is available and performance threshold is selected such that $P = 0.7$.

To limit the evaluation we use a binary network state, indicating the presence or absence of cross-traffic. When collecting the performance metrics, this information will have to be included in the meta-data of the samples, to be able to correlate the performance to the network state. Now assuming a unified collection of performance samples, there will be a number of sample captured when cross traffic was present, and a number of samples captured without cross traffic, for each cell in the map. There are now two mean values for the performance, one with cross traffic present and one with no cross traffic. It is interesting to analyze if a differentiated treatment of network performance based on the hidden state brings some benefit. For that reason, we use two different thresholds here, one that is used during periods that are suspected to be with cross-traffic and one that is used otherwise. The benefits of this differentiation is analyzed later in Section 5.

Prediction of network performance also requires prediction of the device mobility since network performance is correlated to location. This work assumes perfect knowledge about the device mobility to be able to focus on the network state influence. It is clear that a realistic mobility prediction will influence the overall prediction performance, however when working in the ITS context in the presence of navigation systems low mobility prediction errors can be achieved.

An ideal scheduling behaviour has perfect knowledge about the current network state and is able to do perfect prediction of the future conditions. The inverse behaviour is to always use the wrong information, which would serve to show the worst case performance. As a slightly more informative, but still poor, approach, we also include a state estimator which is correct in 60% of the cases and wrong in the remaining 40%, so only slightly better than a random guess. Scheduling can also be done without caring about the network state, we define this as 'no-state' scheduling.

4   Comparing Communication Scheduling Methods

Since the different scheduling methods use different thresholds to select what portion of the network to use for communication, the thresholds cannot be used to compare the different methods directly. This work investigates the influence of including network state awareness in the scheduling, so a comparison across scheduling methods is needed. All scheduling methods are based on selecting how much of the network area should be used for communication. Therefore we introduce a way of selecting a certain amount of the available area to be used for communication that can be used by all scheduling methods. Assuming the probability of being located in each cell is known, and a performance threshold set is selected, it is possible to calculate the total probability for transmission in the whole network area. We define the total probability for transmission as the transmission probability (P).

When no network state information is included in the performance measurement meta-data in the database, all performance measurements for a cell are used to calculate the network performance. Equation (1) shows how P is calculated using the collected performance data with no state information.

$$P^{no-state} = \sum_i^n p_i \cdot 1(\mu_i^{all} \leq T_{all}) \tag{1}$$

The type of P is marked with a superscript. $P^{no-state}$ is the sum of the probabilities of being in the cells that are marked for transmission with a specific threshold T_{all}. The network performance for cell i is the mean value of all recorded performance samples μ_i^{all}. $1()$ is the indicator function that will yield one when the expression in the parentheses is true and zero otherwise. p_i is the probability that the mobile device is in cell i. Such probability is calculated from the mobility model or estimated from empiric mobility data.

Figure 2 illustrates how a certain threshold value marks parts of the area as usable and some as unusable when no network state information is included in the meta data. In this example, a WLAN access point (AP) is placed in bottom right corner of the area and no obstacles are present. Each mark represents one cell and the blue marks represent the cells that are marked as usable. The P for this example is 0.7, while assuming a uniform geographic distribution of the mobile device.

When state information is included in the performance samples meta-data two performance thresholds influence if a cell is selected for transmission or not. With the boolean on/off value of the state information in this work, the performance samples are split into two clusters depending on if cross traffic was present when they were recorded or not. One performance threshold, T_{on}, will decide if the cell should be used when cross traffic is estimated to be on while another threshold, T_{off}, is used otherwise. Equation (2) shows how the P^{state} is calculated when the state information is included.

$$P^{state} = \sum_i^n p_i \cdot (1(\mu_i^{on} \leq T_{on}) \cdot p^{on} + 1(\mu_i^{off} \leq T_{off}) \cdot (1 - p^{on})) \qquad (2)$$

For each state value there is a corresponding performance metric μ^{on} and μ^{off} in the database. The steady-state probability of the network state being on or off is called p^{on} and p^{off} respectively. P^{state} is then the sum off the probabilities of being in the different cells weighted with the probability of using that cell.

To be able to show if including state information into the scheduling influences the application performance, we run simulation experiments for different transmission probabilities: $P \in \{0.3, 0.5, 0.7, 0.9\}$.

Figure 3 shows a plot of the threshold values that can be used to achieve the desired transmission probabilities in the example scenario used later for evaluation. The single stars show the thresholds for the no-state scheduling. The corresponding lines show the tuples (T_{on}, T_{off}) where each point of the curve is a valid choice to achieve the same total transmission probability for $p^{on} = p^{off} = 0.5$ and for uniform p_i in the geometry of Figure 2. Moving to the right along the curves leads to larger T_{off} and correspondingly smaller T_{on} which will give more weight to transmissions during the off periods. The values along the diagonal $(T_{on} = T_{off})$ would correspond to using the same value of the two thresholds. Note that the points on the diagonal are not identical to the star marks, as the star marked scenarios assume that the measurement data is not distinguished according to the network state. Also note that for $P = 0.3$ and

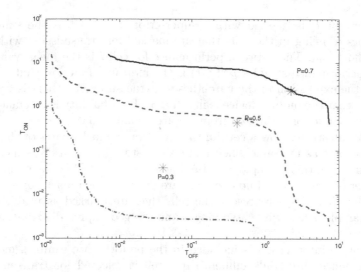

Fig. 3. Threshold choices for different transmission probabilities.

$P = 0.5$ the right and left end of the curves correspond to transmissions only in the estimated off and on states respectively.

5 Scheduling Analysis

The evaluation of the proposed approach is done using an extensive set of simulations in the OMNeT++ network simulator. The simulations are done in two steps; a performance map collection step and a performance map evaluation step. The simulation is configured with a 802.11b network consisting of one mobile client node and one static client node as well as one AP. The AP is connected to a server node using 100Mbit/s Ethernet. The static 802.11b node generates cross traffic using UDP with a packet size of 1000Bytes and a message frequency of 2000Hz resulting in a bandwidth of 16Mbit/s. The cross traffic can be controlled using a time pattern defining on and off periods for one simulation run. The area is defined to be 600x600m and the AP is placed in a corner while the static wireless node is placed 50 meters away from the AP on the area diagonal line. An overview of the simulation configuration is shown in Figure 4.

In the collection step, the mobile node is configured to ping the server node using ICMP, while moving through the area with a constant speed of 10m/s in 60 rows covering the whole area (TractorMobility model). The recorded RTT values collected are added to a DB constituting the performance map. For each sample the location where it was recorded and network state is saved as meta data. One full trip through the area is done without cross traffic and one full trip is done with cross traffic constantly on to have performance samples for both network states for all cells.

In the evaluation step the mobile node is configured with a modified TCP algorithm that can stop/restart an ongoing TCP transfer. This is achieved by

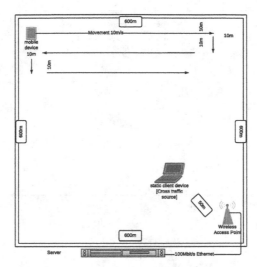

Fig. 4. Overview of simulation configuration. Mobile node is moving using the "Tractor Mobility Pattern".

informing the TCP sender that it should stop transmitting more data by setting the TCP window to 0. To avoid implementing the scheduling algoritms in OMNeT++ a start/stop pattern is pre-calculated and used as simulation configuration. In the evaluation step the cross traffic is on/off according to a exponential time pattern with a mean value of 10s generated once and used for all simulation runs.

For each P, a set of thresholds is chosen covering the full range of possible values from Figure 3. The different scheduling approaches were then used in 100 simulation runs for each threshold-combination, where 10MB application layer data was transmitted using TCP in each run. All results are presented with a 95% confidence interval. For reference, the approach without scheduling is also included, called the 'naive' approach. The naive approach is independent of the transmission probability P and of any threshold choices, while the no-state approach depends only on P.

It is expected that including the network state information in the scheduling will influence the application performance. Cross traffic should affect the network performance, since it will cause packet drops, again causing retransmissions and increased delays. When predicting the network performance without network state information, the performance metric (mean RTT for cell) will be influenced by all recorded samples for both on and off states. For high values of P almost the whole geographic region will be used for communication, and scheduling will have little effect on the application performance. We expect the largest performance gain from using state information for intermediate choices of P.

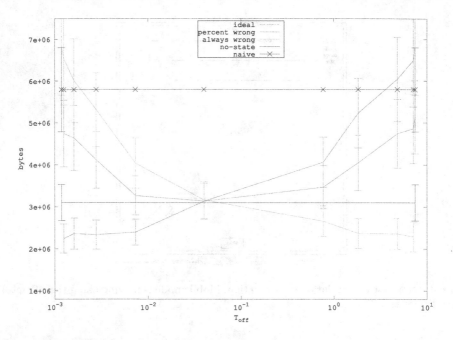

Fig. 5. Comparison of overhead for scheduling approaches for $P = 0.5$

6   Does Network State Influence Scheduling Performance?

The resulting overhead for different schemes and different threshold choices on the x-axis leading to the same transmission probability $P = 0.5$ is shown in Figure 5. When the threshold value for selecting places with no cross traffic (T_{off}) is low, more preference is given to transmissions while cross traffic is assumed to be present. The best performance for the ideal scheduling results, when T_{off} is low, which at first might be surprising. However, the effects of TCP retransmission behavior in a highly congested WLAN are complex and hence lead to this counter intuitive behavior.

The analogue curves for $P = 0.3$ show that here all scheduling implementations perform equally for the best choice of the threshold pair. For $P = 0.7$, the trends looks like Figure 5, but with larger differences in performance (larger gains for state aware scheduling). For $P = 0.9$ the threshold tuple choice no longer makes a difference, since almost all available geographic area is used for communication.

In order to compare the different scheduling methods, we selected for each P the threshold tuple (T_{on}, T_{off}) where the performance of the ideal scheduling algorithm is best in terms of communication overhead. The result for each scheduling algorithm are then obtained based on this threshold choice. Figure 6 shows the difference in communication overhead for the different choices of P. It is clear that the ideal scheduling outperforms the no-state scheduling for $P = 0.5$ and $P = 0.7$. The most interesting area is around $P = 0.7$, where the potential

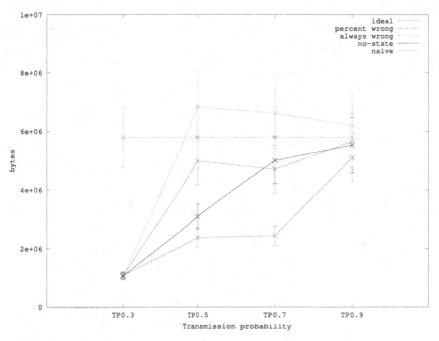

Fig. 6. Comparison overview for all scheduling implementations and all transmission probabilities - overhead.

performance improvement for the expected real-life scheduling reaches 3MB of overhead for a 10MB transfer.

The detailed results show that using network state information when scheduling the communication has a significant influence on the application performance. If the scheduling settings are very constraining ($P = 0.3$), the influence from cross traffic on the communication performance is not very significant and the different scheduling algorithms yield similar results in the best case. When using almost all the geographic space ($P = 0.9$), the scheduling algorithms show no significant difference in performance.

Scheduling the communication to minimize the overhead comes at the cost of increased duration to complete a TCP transfer. To illustrate this trade-off, Figure 7 shows the mean duration of each TCP transfer, measured from when the application request was generated until the transfer is complete. As expected the naive approach outperforms all the scheduling approaches in terms of duration since it never waits to start the transfer. When optimising overhead via scheduling with $P < 1$, the duration will become longer, since less cells are used for communication. The scheduling approaches show an inverse relationship between duration and overhead. No statistical significant difference of duration can be seen between the state and no-state approaches at the threshold value where the minimum value of overhead was achieved. Interestingly the parameter regime around $P = 0.7$, where network state matters, also seems to provide a good trade-off between overhead optimization and duration.

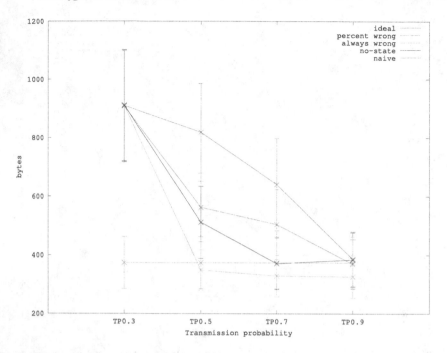

Fig. 7. Comparison overview for all scheduling implementations and all transmission probabilities - duration.

7   Summary and Outlook

We have shown how network state can be used for optimization of wireless communication based on performance maps. The improved optimization comes at a price of increased complexity in the framework. For poor state estimation and prediction functions however, the evaluations show the example of an always-wrong and 40% wrong estimator, the benefit from state-based optimizations however disappears. Although the evaluation results are based on a specific simulation setting, the framework can be applied to any type of wireless network like for example cellular networks.

A number of further improvements to the specific optimization approach are possible: (1) there is not a direct mapping from RTT values to TCP performance; a performanced map based on TCP throuhgput measurements might optimize the specific file-transfer use-case further. (2) The network state is limited to be a boolean value; an interesting extension of this evaluation would be to look a the effect of mapping network state in more detail, with a trade-off in framework complexity.

Acknowledgments. The Telecommunications Research Center Vienna (FTW) is supported by the Austrian Government and by the City of Vienna within the competence center program COMET.

References

1. Yao, J., Kanhere, S., Hassan, M.: Improving QoS in High-speed Mobility Using Bandwidth Maps. IEEE Transactions on Mobile Computing **10**(11) (2011)
2. Schulman, A., Navda, V., Ramjee, R., Spring, N., Deshpande, P., Grunewald, C., Jain, K., Padmanabhan, V.: Bartendr: a practical approach to energy-aware cellular data scheduling. In: Proceedings of MobiCom 2010, pp. 85–96 (2010). doi:10.1145/1859995.1860006
3. Nicholson, A., Noble, B.: BreadCrumbs: forecasting mobile connectivity. In: Proceedings of MobiCom 2008, pp. 46-57. doi:10.1145/1409944.1409952
4. Riiser, H., Endestad, T., Vigmostad, P., Griwodz, C., Halvorson, P.: Video streaming using a location-based bandwidth-lookup service for bitrate planning. ACM TOMCCAP **8**(3) (2012). doi:10.1145/2240136.2240137
5. Højgaard-Hansen, K., Madsen, T., Schwefel, H.P.: Reducing communication overhead by scheduling TCP transfers on mobile devices using wireless network performance maps. In: Proceedings of European Wireless 2012 (2012)
6. Padhye, J., Firoiu, V., Towsley, D., Kurose, J.: Modeling TCP throughput: a simple model and its empirical validation. SIGCOMM Comput. Commun. Rev., **28**(4) (1998). doi:10.1145/285243.285291
7. Cardwell, N., Savage, S., Anderson, T.: Modeling TCP latency. In: Proceedings of IEEE INFOCOM 2000 (2000). doi:10.1109/INFCOM.2000.832574
8. Nielsen, J.J., Madsen, T., Schwefel, H.P.: Location assisted handover optimization for heterogeneous wireless networks. In: Proceedings of European Wireless 2011 (2011)
9. Fei, Y., Leung, V.: Mobility-based predictive call admission control and bandwidth reservation in wireless cellular networks. Elsevier Computer Networks **38**(5), 577–589 (2002). doi:10.1016/S1389-1286(01)00269-9
10. Wireshark Display Filter Reference - TCP. https://www.wireshark.org/docs/dfref/t/tcp.html

Discrete Model of TCP Congestion Control Algorithm with Round Dependent Loss Rate

Olga Bogoiavlenskaia$^{(\boxtimes)}$

Petrozavodsk State University, Lenin St., 33, 185910 Petrozavodsk, Russia
olbgvl@cs.karelia.ru

Abstract. This paper investigates discrete random process describing behavior of the Additive Increase Multiplicative Decrease algorithm of networking Transmission Control Protocol (TCP). We use the sequence of TCP rounds with no data loss events to define the end-to-end path data loss behavior. The Markov chain embedded in the random process is described and the theorem on its ergodic property is proved. Further analysis yields the estimates of stationary first and second moments of the congestion window size which are key performance metrics of TCP protocol.

Keywords: Discrete model · Random process · Markov chain · Networking · AIMD

1 Introduction

The methods of distributed control provide one of the fundamental Internet concepts. Essential part of the methods is responsible for congestion control of the Internet infrastructure (links and routers). A set of highly significant congestion control technologies operate at networking transport layer where data sources follow congestion control algorithms. Thus a source increases the throughput if there is the end-to-end route capacity available and decreases the throughput if it receives a congestion signal. These algorithms are implemented in Transmission Control Protocol (TCP). Since congestion control methods play key role for data communication networks integrity and performance they are the object of an intensive research. Our study is focused on Congestion Avoidance Additive Increase Multiplicative Decease (AIMD) algorithm described by Internet Engineering Task Force (IETF) standard [1], i.e. New Reno TCP version (see also [2]).

Recently AIMD algorithm's disadvantages on the end-to-end paths, that include high-speed, high bandwidth-delay product value or wireless links, are widely discussed as well as its poor ability to meet multi-media flows demands. As a result during last 15 years there were developed more than ten different algorithms and methods of a congestion control targeting particular problems enumerated above. Twelve different versions of TCP congestion control schemes are implemented in Linux operating system (OS) kernel networking component

© Springer International Publishing Switzerland 2015
S. Balandin et al. (Eds.): NEW2AN/ruSMART 2015, LNCS 9247, pp. 190–197, 2015.
DOI: 10.1007/978-3-319-23126-6_17

and they could be used by researchers and system administrators. For example Linux OS since version 2.6.19. sets TCP CUBIC version by default [3], and one of the developments symptomatically is entitled TCP-YeAH, which means Yet Another High Speed TCP protocol [4]. Inspite of significant effort contributed to these developments important performance properties of non standard congestion control schemes are to be investigated and these algorithms has experimental status yet. Henceforth for the most of these schemes their performance, fairness properties and congestion avoidance abilities could not be sufficiently predicted for common networking environments or Internet fragments. Additionally several studies has shown [5] that practical implementations of these schemes vary from the initial algorithms publication. A comparison of standard congestion control with the experimental ones are mainly empirical and scalability of such evaluation varies as well. Therefore further research of the standardized congestion control together with its long deployment history provides foundation for better network administration practices and advanced research and development of congestion control mechanisms.

Currently the area of TCP modeling is rather vast [6]. AIMD algorithm provides congestion control by interpreting data delivery information coming from the receiver as a feedback signal about the state of the end-to-end path. With the aim it uses congestion window size ($cwnd$) which is amount of data that sender may inject in the network without an acknowledgment. If the data are delivered successfully then the sender increases $cwnd$ by adding a constant, if data loss happens then $cwnd$ is multiplied with the value $\alpha < 1$. Generally these approach is supported by the experimental protocols as well.

At the moment there dominate two approaches to the formulation of mathematical and/or performance models of the AIMD algorithm. These are discrete and piecewise linear models. Discrete models are natural, since TCP instance operates with the discrete values, but they are more complicated as well. Piecewise linear models are powerful and fruitful but there are few researches published (e.g. [7], [8]) that study analytically the precision of the piecewise linear models and characterize explicitly the difference between their results and actual behavior of the congestion window stepwise process. Meanwhile most of the publications are concentrated on the evaluation of the stationary expectation of the congestion window size. With the aim those studies apply to Goelder's inequality as the evaluation method. This yields the estimate of the following form $E[W] \leq \sqrt{E[W^2]}$. Here W is $cwnd$ size. Such estimations obviously could not be used for the standard deviation evaluation which is $E[W^2] - E^2[W]$. However $cwnd$ standard deviation is crucial as a performance characteristics for many modern applications.

In this work we consider discrete step-wise stochastic process of AIMD congestion window size. For the process we propose new definition of the segment loss process. Bernoulli loss process used in most of researches of this kind is characterized by the parameter p which represents the segment loss probability. In the case of bulk losses, which is rather typical, New Reno decreases $cwnd$ only once per round inspite of the number of the losses happened during the round [2]. Hence most of the losses do not contribute to the connection's performance and the total number of the lost segment reported by monitors, such as tcpdump, essentially overestimates effective loss rate. Instead of using Bernoulli distribution with independent losses, we introduce distribution of the number of the rounds without losses observed by TCP sender at the end-to-end path. This approach simplifies analytical transformations and provides reliable results of the loss process identification for practical purposes by the real data. Although the last one demands more monitoring efforts. For this model we derive lower bound of the *cwnd* stationary expectation and lower and upper bounds of its second moment. Since the expectation estimate is obtained without Goelder's inequality then further analysis lets ones characterize *cwnd* stationary standard deviation. Explicit terms of the estimates applicability as well as their precision characteristics are presented as well.

The rest of the paper is organized as follows. Section 2 provides detailed problem statement, the stochastic model formulation and the conditions of the ergodic properties of the Markov chain embedded. Section 3 presents estimate of the stationary *cwnd* expectation, section 4 is devoted to *cwnd* second moment and section 5 contains conclusion.

2   Discrete Stochastic Process of the Congestion Window Size

Let us define the stochastic process which describes a behavior of AIMD congestion window size. Let's assume that multiplicative decrease ratio α is rational number and hence $\alpha = \dfrac{n}{m}$, where n, m are natural numbers and $n < m$.

Let t_n are the moments of AIMD-rounds end-points. Hence $[t_{n-1}, t_n]$ are round trip intervals, and RTT length is $\xi_n = t_n - t_{n-1}$. Let us denote $w(t)$ a congestion windows size under AIMD control at the time moment t. Then $\{w(t)\}_{t>0}$ is a stepwise process such that

$$
w(t_n + 0) = \begin{cases} \left\lfloor \dfrac{w(t_n)n}{m} \right\rfloor, & \text{if during the interval (TCP round) } [t_{n-1}, t_n] \\ & \text{one or more TCP segment losses has happened,} \\ w(t_n) + 1, & \text{if all data delivery in the round was acknowledged.} \end{cases}
$$

Between the moments t_n the process $\{w(t)\}_{t>0}$ stays constant.

We define that data *loss event* has happened if at the time interval $[t_{n-1}, t_n]$ more than one TCP segments were lost. Let τ_k be equal to the first moment t_n, arrived after kth data loss event, i.e.

$$
\tau_k = t_n : \; w(t_n + 0) = \left\lfloor \frac{w(t_n)n}{m} \right\rfloor, \; k = 1, 2 \ldots.
$$

and $\tau_{k_1} < \tau_{k_2}$, if $k_1 < k_2$. Now we assume that segment loss pattern is defined by discrete distribution $\{f_j\}_{j \geq 0}$, i. e. the distribution of the number of the consequential rounds at the end of which $w(t)$ size increased. In other words f_j is

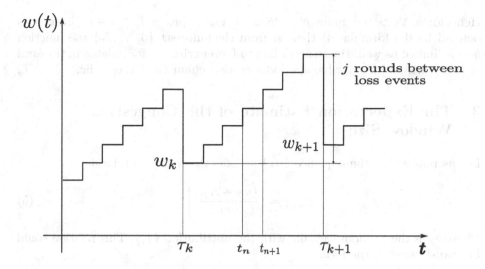

Fig. 1. Evolution of TCP congestion window size

the probability that a sequence of j rounds without losses took place after a segment loss event occurred. Then the sequence $\{w_k = w(\tau_k + 0)\}_{k>0}$ is the Markov chain. The example of the random process $\{w(t)\}_{t>0}$ trajectory is presented at Fig. 1.

Let's define the expectation determined by the distribution $\{f_j\}$ as $R = \sum_{j=1}^{\infty} j f_j$. Following theorem defines ergodic properties of the Markov chain $\{w_k\}$.

Theorem 1. *If R is finite then the Markov chain $\{w_k\}$ has steady state distribution.*

Proof. Let us compute $g(i) = \mathsf{E}[w_{k+1}|w_k = i]$ as follows

$$g(i) = \sum_{k=0}^{\infty} \left(\left\lfloor \frac{in}{m} \right\rfloor + k \right) f_k = \left\lfloor \frac{in}{m} \right\rfloor + R. \tag{1}$$

Henceforth if the condition of the theorem holds then

$$\mathsf{E}\{w_{k+1}|w_k = i\} = \left\lfloor \frac{in}{m} \right\rfloor + R, \tag{2}$$

where R is finite. Now we consider following condition

$$\left\lfloor \frac{in}{m} \right\rfloor + R \le i - \epsilon, \tag{3}$$

or

$$i > \frac{m}{m-n}(R + \epsilon - 1). \tag{4}$$

Henceforth, $\forall \epsilon > 0 \; \exists$ finite N : $\forall i > N \; \mathsf{E}[w_{k+1}|w_k = i] \leq i - \epsilon$. Since N is defined by the formula (4) then $\forall i$ from the finite set $\{0, \ldots, N\}$ the function $g(i)$ is limited as well. Hence according to Foster criterion formulated in the form of [10, § 3.1, p.2] irreducible aperiodic Markov chain $\{w_k\}$ is ergodic.   □

3   The Expectation Estimate of the Congestion Window Size

Let us notice that the sequence $\{w_k\}$ satisfies the following relation

$$w_{k+1} = \left\lfloor \frac{(w_k + r_k)n}{m} \right\rfloor, \tag{5}$$

where r_k is the random variable with the distribution $\{f_j\}$. This relation could be transformed to the form

$$w_{k+1} = \frac{(w_k + r_k)n - i_k}{m}, \tag{6}$$

where i_k is the integer random variable and $0 \leq i_k < m$. Using the results of the work [9] one concludes that difference equation (6) has a stationary solution

$$w_k^* = \sum_{j=0}^{\infty} \left(\frac{n}{m}\right)^{j+1} r_{k-j} - \frac{1}{m} \sum_{j=0}^{\infty} \left(\frac{n}{m}\right)^j i_{k-j}. \tag{7}$$

Theorem 2. *If the inequality $R > \dfrac{n}{m}$ holds then the stationary expectation of the congestion window size can be estimated as follows*

$$\mathsf{E}[w_k^*] \geq R\frac{n}{m-n} - \frac{m-1}{m-n}. \tag{8}$$

Proof. Computing the expectation according to the expression (7) yields the following

$$\mathsf{E}[w_k^*] = \sum_{j=0}^{\infty} \left(\frac{n}{m}\right)^{j+1} \mathsf{E}[r_{k-j}] - \frac{1}{m} \sum_{j=0}^{\infty} \left(\frac{n}{m}\right)^j \mathsf{E}[i_{k-j}]. \tag{9}$$

By our definitions $\forall k, \; j \; \mathsf{E}[r_{k-j}] = R$ and $\mathsf{E}[i_{k-j}] \leq m - 1$. Hence

$$\mathsf{E}[w_k^*] \geq R \sum_{j=0}^{\infty} \left(\frac{n}{m}\right)^{j+1} - \frac{m-1}{m} \sum_{j=0}^{\infty} \left(\frac{n}{m}\right)^j = \tag{10}$$

$$= R\frac{n}{m-n} - \frac{m-1}{m-n}.$$

The last expression is positive if $Rn > m$. This proves condition of the theorem.   □

According to the definition above R is expectation of the number of rounds between two consecutive loss events. In practice R typically takes values of several hundreds for wired networks and at least several tens for wireless ones. Meanwhile too little n/m ratio means long recovery period for TCP connection and hence such values typically are not used by AIMD implementations. IETF standard [1] defines the ratio equal to 0.5. In the case negative term of the estimate is equal to 1 segment which corresponds to the result presented in [8].

4   The Interval Estimation for the Second Moment

Let us calculate second moment of w_k^*.

$$E\left[w_k^{*2}\right] = E\left[\left(\sum_{j=0}^{\infty}\left(\frac{n}{m}\right)^{j+1} r_{k-j}\right)^2\right] - \tag{11}$$

$$\frac{2}{m}E\left[\sum_{j=0}^{\infty}\left(\frac{n}{m}\right)^{j+1} r_{k-j} \sum_{s=0}^{\infty}\left(\frac{n}{m}\right)^{s} i_{k-s}\right] + \frac{1}{m^2}E\left[\left(\sum_{j=0}^{\infty}\left(\frac{n}{m}\right)^{j} i_{k-j}\right)^2\right]$$

Further analysis of each term of the expression (11) yields

$$W_1 = E\left[\left(\sum_{j=0}^{\infty}\left(\frac{n}{m}\right)^{j+1} r_{k-j}\right)^2\right] =$$

$$E\left[\sum_{j=0}^{\infty}\left[\left(\frac{n}{m}\right)^{j+1} r_{k-j}\right]^2\right] + E\left[\sum_{j=0}^{\infty}\sum_{\substack{s=0\\s\neq j}}^{\infty}\left(\frac{n}{m}\right)^{j+s+2} r_{k-j}r_{k-s}\right] =$$

$$\sum_{j=1}^{\infty}\left(\frac{n}{m}\right)^{2j} E\left[r_{k-j-1}^2\right] + 2\sum_{j=0}^{\infty}\sum_{s=j+1}^{\infty}\left(\frac{n}{m}\right)^{j+s+2} E\left[r_{k-j}r_{k-s}\right]. \tag{12}$$

Let's denote $\forall k, j$ $R^{(2)} = E[r_{k-j}^2]$ and according to our assumptions the elements of the random sequence $\{r_k\}$ are independent. Hence $\forall k, j \neq s$ $R^2 = E\left[r_{k-j}r_{k-s}\right]$. Then

$$W_1 = R^{(2)}\sum_{j=1}^{\infty}\left(\frac{n}{m}\right)^{2j} + 2R^2\sum_{j=0}^{\infty}\sum_{s=j+1}^{\infty}\left(\frac{n}{m}\right)^{j+s+2} =$$

$$R^{(2)}\frac{n^2}{m^2 - n^2} + 2R^2\frac{n^3}{(m-n)(m^2 - n^2)} \tag{13}$$

Now for the second term one obtains

$$W_2 = \frac{2}{m} \sum_{j=0}^{\infty} \sum_{s=0}^{\infty} \left(\frac{n}{m}\right)^{j+s+1} \mathsf{E}[r_{k-j}i_{k-s}] <$$

$$2R \sum_{j=0}^{\infty} \left(\frac{n}{m}\right)^{j+1} \left(1 - \frac{n}{m}\right)^{-1} = \frac{2Rmn}{(m-n)^2} = E_l. \tag{14}$$

Finally for the third term one obtains

$$W_3 = \frac{1}{m^2} \mathsf{E}\left[\sum_{j=0}^{\infty} \left(\frac{n}{m}\right)^{2j} i_{k-j}^2 + 2 \sum_{j=0}^{\infty} \sum_{s=j+1}^{\infty} \left(\frac{n}{m}\right)^{j+s} i_{k-j}i_{k-s}\right] =$$

$$\frac{1}{m^2} \sum_{j=0}^{\infty} \left(\frac{n}{m}\right)^{2j} \mathsf{E}\left[i_{k-j}^2\right] + \frac{2}{m^2} \sum_{j=0}^{\infty} \sum_{s=j+1}^{\infty} \left(\frac{n}{m}\right)^{j+s} \mathsf{E}\left[i_{k-j}i_{k-s}\right] \tag{15}$$

Furthermore one notices that $\forall k, \ j \ \mathsf{E}[i_{k-j}^2] < m^2$. Also $\forall k, \ j \neq s \ \mathsf{E}[i_{k-j}i_{k-s}] < m^2$ as well. Therefore the following estimate holds

$$W_3 < \sum_{j=0}^{\infty} \left(\frac{n}{m}\right)^{2j} + 2 \sum_{j=0}^{\infty} \sum_{s=j+1}^{\infty} \left(\frac{n}{m}\right)^{j+s} =$$

$$\frac{m^2}{m^2 - n^2} + \frac{2nm^2}{(m-n)(m^2 - n^2)} = E_g. \tag{16}$$

Now using terms W_1, E_l, and E_g the stationary second moment for the congestion window size could be evaluated as

$$W_1 - E_l < \mathsf{E}\left[w_k^{*2}\right] < W_1 + E_g \tag{17}$$

The following analysis derives the applicability terms for the expression (17). The upper bound is the unconditioned one. For the lower bound the sufficient condition is

$$\frac{2n^3 R^2}{(m-n)^2(n+n)} > \frac{2Rmn}{(m-n)^2} \tag{18}$$

or

$$\frac{n^2 R}{m+n} > m. \tag{19}$$

The parameters defined by IETF standard [1] are $n = 1$, $m = 2$. These values transform the condition (19) into $R > 6$. The last one holds for overwhelming majority of the end-to-end connections under AIMD control. The considerations presented above prove the following theorem

Theorem 3. *If*

$$\frac{n^2 R}{m+n} > m. \tag{20}$$

then the following interval estimation holds

$$W_1 - E_l < \mathsf{E}\left[w_k^{*2}\right] < W_1 + E_g. \tag{21}$$

A comparison of the estimation (17) with the results of the piecewise linear approximation (see e.g. [11]) shows that the term W_1 coincides up to notation and the terms E_l and E_g appear due to the discrete nature of the random process considered and due to the using of the *floor* operation in the AIMD control. Further analysis yields bounds of the standard deviation which could not be obtained through expectation estimates constructed using Goelder's inequality.

5 Conclusion

Actual TCP implementations support discrete arithmetics for the congestion window size control. Its most important performance metrics are *cwnd* expectation and standard deviation. In this work we formulate stochastic process which describes *cwnd* evaluation and derive for the embedded Markov chain the estimates of its first and second moments. Since expectation bounds are derived without addressing to Goelder's inequality the estimates obtained could be used for evaluation of the standard deviation. Applicability terms and precision characteristics for all estimates obtained are formulated as well.

References

1. Allman, M., Paxon, V.: TCP Congestion Control. RFC 5681 (2009)
2. Floyd, S., Henderson, T.: The NewReno Modification to TCP's Fast Recovery Algorithm. RFC 2582 (1999)
3. Ha, S., Rhee, I., Xu, L.: CUBIC: A New TCP-Friendly High-Speed TCP Variant. ACM SIGOPS Operating System Review (2008)
4. Baiocchi, A., Castellani, A.P., Vacirca, F.: Yeah-tcp: yet another highspeed tcp. In: 5th International Workshop on Protocols for Fast Long-Distance Networks (PFLDnet), March 2007
5. Leith, D.J., Shorten, R.N., McCullagh, G.: Experimental evaluation of cubic TCP. In: Proc. Protocols for Fast Long Distance Networks, Los Angeles (2007)
6. Afanasyev, A., Tilley, N., Reiher, P., Kleinrock, L.: Host-to-host congestion control for TCP. IEEE Communications Surveys & Tutorials, **12**(3) (2010)
7. Mathis, M., Semke, J., Mahdavi, J., Ott, T.: The macroscopic behavior of the TCP congestion avoidance algorithm. ACM SIGCOMM Computer Communication Review **27**(3), 67–82 (1997)
8. Bogoiavlenskaia, O.: Comparison of stepwise and piecewise linear models of congestion avoidance algorithm. In: Balandin, S., Andreev, S., Koucheryavy, Y. (eds.) NEW2AN/ruSMART 2014. LNCS, vol. 8638, pp. 609–618. Springer, Heidelberg (2014)
9. Brandt, A.: The Stochastic Equation $Y_{n+1} = A_n Y_n + B_n$ with stationary coefficients. Adv. Appl. Prob. **18**, 211–220 (1986)
10. Gnedenko, B.V., Kovalenko, I.N.: Introduction to Queuing Theory. Radio i Sviaz, Moscow (1987)
11. Altman, E., Avrachenkov, K., Barakat, C.: A stochastic model of TCP/IP with stationary random losses. In: Proceedings of ACM SIGCOMM 2000, Stockholm, pp. 231-242 (2000)

Chunked-Swarm: Divide and Conquer for Real-Time Bounds in Video Streaming

Christopher Probst, Andreas Disterhöft, and Kalman Graffi(✉)

Technology of Social Networks, University of Düsseldorf, Düsseldorf, Germany
{christopher.probst,disterhoeft,graffi}@hhu.de
http://www.tsn.hhu.de/en.html

Abstract. Live user-generated video streaming platforms like Twitch.tv generate a large portion of the Internet traffic. Millions of viewers daily watch user channels, although roughly 85 % of all channels have less than 200 views during one session. Due to latency, Twitch.tv provides one or more servers for each of Twitch.tv's supported countries. An alternative approach could enable peer-to-peer communication in order to utilize the capacities of the user devices. Solutions up to now, mainly offer only best effort delay guarantees on the distribution speed from initial seeders to all peers. In this paper, we present Chunked-Swarm, a swarm-based approach, which aims to offer predictable streaming delays, independently of the number of peers. Evaluation shows the various impact of the number of peers, number of video parts and chunks on the streaming delay. Being able to hold specific deadlines for up to 200 peers, predestines our solution to be suitable for the majority of Twitch.tv's channels.

Keywords: Real-time streaming · Twitch.tv · P2p swarm · Qos

1   Introduction

Traditional video streaming applications are mostly based on the client / server model, where the server only responds to client requests and the clients do not know each other. In case of real-time video streaming, content delivery deadlines are crucial to meet in order to provide a satisfying quality of experience. Naturally, single servers or server clusters are limited in their bandwidth and thus application providers have to upgrade their servers in order to cope with increasing bandwidth requirements leading to high operational costs. Instead of continuously upgrading servers, unutilized resources of their clients could be used to improve this imbalance.

Twitch.tv is a good example for a growing streaming application for user-generated content with hundreds of thousands of streamers and millions of spectators requiring high operational investments. Other examples are the streaming

This work was partially supported by the German Research Foundation (DFG) Grant "OverlayMeter" (GR 4498/1-1)

S. Balandin et al. (Eds.): NEW2AN/ruSMART 2015, LNCS 9247, pp. 198–210, 2015.
DOI: 10.1007/978-3-319-23126-6_18

of live-lectures in universities, which are becoming increasingly important as students are used to online resources and e-learning [2]. In case of Twitch.tv, Zhang and Liu present in their measurement study [20] a model for the distribution from viewer to streamer using a Weibull and / or Gamma function. The measurements show that around 85 % of all streamers have less than 200 viewers during their live streaming session. Although viewers are free to start and to switch to other channels, Nascimento et al. [12] modeled the behavior of viewers in Twitch.tv and concluded that 'the content is mainly consumed by a small fraction of very assiduous streamers'.

Based on these observations of these stream characteristics, namely being user-generated, having mainly less than 200 viewers and very low churn, we propose our fully decentralized streaming solution in which the viewers use their own upload capacity to help the streamer to distribute streaming content within a given transmission deadline. In specific, we present a peer-to-peer (p2p) system which provides guarantees on the delivery time and performs best on networks with up to 200 peers. All peers (viewers) predictably receive continuous video parts, while the seeder (streamer) uploads parts only once (in case there is no churn). Our solution, in contrast to solutions from literature, emphasizes on the real-time requirements of the content distribution and meets specific distribution deadlines to achieve a certain level of quality of service (QoS).

In Section 2 we discuss related work and point out that our approach in aiming at clear content delivery timings is in contrast to the majority of solutions in literature. In Section 3, we elaborate the theory and concept of our approach. In this section, we also model the transmission time using our approach. We conduct an extensive simulation study of the proposed system. Section 4 introduces our evaluation setup and the obtained results. In specific, we highlight four scenarios that we used to analyze the impact of the number of peers, the number of chunks and number of parts on the performance of our content delivery strategy and its overhead. Evaluation shows that the desired real-time characteristic is reached. In Section 5, we conclude our work.

2   Related Work

In this paper, we propose a novel real-time streaming protocol for p2p networks, which provides delivery delay guarantees in relation to the slowest node's bandwidth. Therefore, we review related work in the field of p2p (live) video streaming and highlight first, that this desired guarantee on the delivery delay is novel. Three main solution categories have emerged for live video streaming: content delivery networks, p2p networks and IP multicast. Content delivery networks assist in the delivery of media content through a worldwide network of data centers. IP multicast is limited to a few Internet Service Providers (ISP) only and typically is not reaching beyond their networks. P2P networks offer the option to stream video to a large number of users with small operational costs, thus taking a serious impact on the media industry [9]. The content delivery is mainly operated by the end devices of the users, sharing storage, computing time and bandwidth.

For p2p live video streaming three main solution classes have been identified: centralized schemes, clustering recursive schemes and full-distributed schemes, which are further sub-subcategorized in swarm-based schemes and tree-based schemes. Tree-based schemes can be further sub-categorized in structured, network-driven and data-driven. To ease the p2p exchange of parts, each part is split in chunks or pieces. Typical chunk sizes range from 8 KB, over 256 Kb, to 1 Mb.

In centralized approaches, a central element is coordinating the transmissions between the peers. In ALMI [15] a central element defines an explicit bi-directional fat-tree topology, which is optimized for low delays and high bandwidth. The information towards the root lists the position of the nodes in the tree and their free capacities. The information flow towards the leafs contains the video content. CoopNet [13] is a more flexible tree and uses a delay-based metric space, in which nodes are placed according to their delay distances to each other. In Graffi et al. [6], the DHT supports to match ideal transmissions between the peers based on a variety of node characteristics. Centralized approaches can reach a low delay at the costs of high load at the essential central element. Due to the tree structure of height $O(\log(N))$ with N being the number of peers, the leafs have to wait $O(\log(N))$ submission rounds in order to receive the full video part. A submission round is hereby the time needed to download the video part once. Here, we aim at a maximum of two submission rounds.

Clustered approaches organize nodes in multi-layered topologies. On each layer, the nodes are organized in clusters or mini-swarms in which they exchange data and elect a cluster-head which is representing them at the higher layer. While this layering scheme is similar to a tree structure, the topology is more complex as nodes on the same 'tree-level' also share data among each other. Examples for this are NICE [1] and THAG [17]. Due to the hierarchical layering, a tree of height $O(\log(N))$ emerges which involves each chunk of the video to be transmitted at least $O(\log(N))$ times. So, $O(\log(N))$ submission rounds are needed to transfer the video part, we aim at two rounds.

The two main categories for fully distributed schemes are tree-based solutions and swarm-based solutions. Tree-based solutions aim at creating a tree structure which allows to share the video content from the seeding root to all nodes in a hierarchical manner. One option for trees is to rely on a distributed hash table, reusing the routing table or the capability to have responsible nodes for given IDs. Examples for this are SplitStream [3] and Bayeux [23]. Using a DHT involves besides the advantages also a considerable amount of maintenance overhead. Trees can be network-driven or data-driven. Network-driven trees are optimizing for having peers with high bandwidth capacities in high layers of the tree, and thus avoiding bottlenecks in the early steps of the video dissemination. Examples are mTreebone [18] and TreeClimber [21]. They support the dynamic rearrangement of nodes in the tree in order to support changes in the nodes' network conditions. Finally, data-driven tree-based solutions organize the nodes according to their current position in the video. Parent nodes are chosen in a way that their are ahead of the playback position of the considered node and

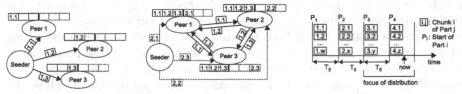

(a) Initial distribution of chunks

(b) Parallel distribution of chunks in full mesh

(c) Live video streaming: Distribution of max. 2 parts

Fig. 1. Chunked-Swarm Distribution

thus are guaranteed to have the desired video chunks. As a results each layer in the tree is a time step behind in the video position than its parent peer. Examples are CoolStreaming [19], Substream Trading [10] and SPANC [4]. Also here, $\mathcal{O}(\log(N))$ submission rounds are needed to transfer the video part, while we aim at a maximum of two submission rounds.

Swarm-based schemes are the second large group of solution for p2p-based live video streaming. In these, peers exchange information on the chunks they share with their neighbors and either push or pull then further chunks to complete their video part. Swarms use the network capacities of all nodes in contrast to several tree based solutions which omit the upload potential of the leaf nodes. Examples are BTLive [16], Chainsaw [14] and LayerP2P [11]. Nodes exchange small bitmaps on their chunk offerings and use a small share set with typically four nodes to exchange individual chunks. While most large-scale p2p-based live video streaming solutions use a swarm-based scheme, they are reported [22] to have a playback delay between 5s and 20s, which allows a lost packet to be re-transmitted once or twice in case of errors. The playback delays are by this significantly longer than in tree-based schemes.

In this paper, we present a p2p-based live video streaming scheme which is fully-distributed, swarm- and pull-based. Our approach aims to distribute a video stream with a fixed and predictable start-up and streaming delay for a set of roughly 200 peers in a low churn environment like 85 % of Twitch.tv's streams. For the delay we aim at a maximum of twice the transmission time of the seeder to one peer, which is not supported by previous work due to their tree-based core with a minimum delay of $\mathcal{O}(\log(N))$ or swarm-based core with arbitrary delays between 5s and 20s. In contrast to other (swarm-based) p2p solutions, our approach manages that the seeder uploads the video part only once, thus even weak peers like smartphones are perfectly suitable to be a seeder. Next, we present our approach.

3   Our Approach: Chunked-Swarm

In Chunked-Swarm we use a full mesh topology, in which all peers (viewers) in the swarm are connected to each other, thus the suggested limit of 200 peers. The seeder (streamer) splits the video stream into parts, which are then split into chunks and announced to all connected peers, see Fig. 1(a). Therefore, the chunk count should always be equal to or greater than the number of peers. Peers request a random chunk from the seeder, but are rejected when requesting a chunk, which has already been requested by another peer. Those peers remove this chunk from their interest set and start a new request. This procedure is repeated until all peers have found an unique chunk to download. Thus, the seeder delegates the responsibility for distributing chunks to the peers. If a peer goes offline, the initial chunk received from the seeder might also leave the swarm. The seeder detects those leaving peers and announces the lost chunk again. After downloading a chunk from the seeder, peers announce their own chunk to all other peers in form of a space-efficient bitmap. After this step, each peer owns a chunk, which is of interest for the other peers. This strategy gives every peer the ability to participate equally in the distribution of each part. Therefore, the seeder does not have to participate anymore in sharing, unless chunks are lost, e.g. due to churn in the rather short period of a part. In the following, all peers download in parallel from all other peers in a pull-based manner, see Fig. 1(b).

To inform peers about available chunks, an accumulative and space-efficient bitmap-state is announced regularly to all peers. While this involves $O(\log N^2)$ messages in the network, for small and medium sized swarms the overhead is manageable and the start-up delay is below $2 * T_0$, where T_0 is the time needed to transfer an entire video part from the seeder to the slowest peer. This limit is fixed and independent of the number of peers. Therefore, our solution can theoretically distribute a video part between an arbitrary number of peers in the same time as it would take a server, to serve only two clients. If a new peer connects to the seeder, it gets an address list of all other peers in the swarm to which it connects then. The joining peer only participates for the exchange of the next part, to not induce churn in the current part exchanging clique.

3.1   Analysis of the Transmission Time of One Part

Fig. 1(a) and 1(b) show an example of the Chunked-Swarm model with one seeder and three peers distributing two video parts. Each part is of size (s) and consists of three chunks, but for now, we only concentrate on the first part, whose chunks are visualized with solid arrows. For reasoning, we assume the seeder and the peers have the same upload and download bandwidth (b). Each of the remaining peers request one unique chunk of the first part, which is a third of the whole video part ($\frac{1}{3} * s$), from the seeder in parallel. Therefore, each peer also gets $\frac{1}{3}$ of the seeder's upload bandwidth, so it takes $3 * (\frac{1}{3} * s) * \frac{1}{b} = \frac{s}{b} = T_0$ seconds to upload all chunks once to the swarm, in specific one distinct chunk to each peer in parallel.

This is the same time span it would take a server to upload the video part to one client. At this point, the entire first video part is present in the swarm, where each peer has exactly one unique chunk ($\frac{1}{3}$ of the part). Therefore, the chunks can be distributed by the peers among themselves in parallel. Since each peer has to upload one third of the first video part ($\frac{1}{3} * s$) to 2 peers, it takes each peer $2 * (\frac{1}{3} * s) * \frac{1}{b} = \frac{2}{3} * T_0$ seconds, to upload its own chunk to the two other peers. Now, every peer contains the first video part after $T = T_0 + \frac{2}{3} * T_0$ seconds. Analogously, a variable number of peers (n) needs $T(n) = T_0 + \frac{n-1}{n} * T_0$ seconds to finish a single video part. Since $0 \leq \frac{n-1}{n} \leq 1$ is always true, T never exceeds $2 * T_0$. When the chunk count (c) is doubled, the model behaves better, as the peers can start to upload their own chunks while they are downloading the next unique chunk from the seeder. In theory, following formula applies:

$$T(n, c) = T_0 + \frac{n}{c} * \frac{n-1}{n} * T_0 = (1 + \frac{n-1}{c}) * T_0 < 2 * T_0 \qquad (1)$$

with $c = a * n, a = 2^i \in \mathbb{N}_0$ as chunk count and n the peer count.

3.2   Live Video Streaming: Distribution of Several Parts

So far, we have only discussed the distribution of a single video part, which can only be watched by viewers after completion. For live video streaming, we use multiple video parts, each containing a consecutive time interval of the live stream. Naturally, the peers have to download and distribute the first video part completely, in order to start watching the live video stream continuously. The time needed to distribute the first video part across all peers, termed start-up streaming delay, is predictable and can be calculated using Equation 1. To provide a continuous flow of video parts, the seeder publishes the next video, while the peers distribute the last part. Therefore, the seeder and peers are never halted. Fig. 1(b) shows the collection and distribution phases, which run in parallel, after the first video part has been published.

To guarantee that multiple video parts do not interfere with each other, we evaluate the usage of 20 video parts in Scenario 4, which is presented in Section 4.4. Please note, that this scenario implements a form of video on-demand streaming, since the video parts remain in the network; late peers are able to watch the video stream from the beginning. However, as presented in Fig. 1(c), live video streaming only requires two video parts to be available at any given time. While the seeder is distributing the video part P_i, peers are busy distributing part P_{i-1} and finish their job before the seeder announces part P_{i+1}, since the $2 * T_0$ limit is respected. Older parts are not meant to be distributed, so they can be dropped by both, the seeder and peers. Therefore, the overhead caused by live video streaming based on Chunked-Swarm is very manageable. A realistic live video streaming use case, which also elaborates the relation between payload and announcements overhead, is presented in Section 4.6.

3.3   Churn

The time model in Section 3.1 disregards the influence of churn. A streaming start-up delay of $2 * T_0$, with T_0 being the time to transfer one video part, is only guaranteed if all peers participate in the chunk distribution. When peers leave during the transmission of a part, the time limit might be violated. In which extent depends on the number of concurrent leaving peers during the short time T_0. As T_0 is considered to be short, only very few peers are expected to leave within this time. Joining peers do not participate in the distribution of the current, but only of the next part. The seeder adjusts the number of chunks to the current number of peers, when a new video part is published.

4   Evaluation

An analysis of the efficiency of the Chunked-Swarm model with its live streaming option is given in Section 3.1. Here, we aim to stepwise evaluate the performance of live video streaming while looking at whether the $2 * T_0$ start-up delay limit is fulfilled in practice. The first step applies the distribution performance with only one video part and a variable number of viewers, while the second step evaluates the live video streaming with more than two parts. The second step is mandatory as two adjacent video parts may interfere with each other as mentioned in Section 3.2. We implemented and evaluated our approach in a standalone application, which is more accurate than the evaluation in a p2p simulator [5,8].

4.1   Evaluation Setup

We performed four scenarios in total using our Java implementation and compared their results and overhead caused by announcement messages. Each scenario is run ten times. The results are merged by calculating the mean and the 95 % confidence interval. The first scenario is considered the default, in the remaining scenarios we vary one parameter at a time to measure the influence of this parameter. All scenarios ran in real-time on a server with 14 *Intel Xeon* 2.1 GHz cores and 38 GB main memory. For benchmarking, we reduced the upload bandwidth of each peer to $16\,384\,\frac{byte}{sec}$, which might seem unrealistic. However, the upload bandwidth does not matter in our case, since we calculate the video part size according to the upload bandwidth and the desired start-up delay. This gives us the freedom to pick an arbitrary upload bandwidth for the evaluation, without being unrealistic.

4.2   Scenario 1: Default

Scenario 1, also called default scenario, simulates 63 peers and 1 seeder owning only one video part, performing our first step towards live video stream. This scenario is used to evaluate the distribution of only one video part during a stream of user-generated content. The size (s) of the video part is calculated from the simulated upload bandwidth (b) of $16\,384\,\frac{byte}{sec}$, such that a single transfer from the seeder to one peer takes exactly 10 minutes ($T_0 = 600\,sec$):

$s = 600\,sec * 16\,384\frac{byte}{sec} = 9\,830\,400\,byte$. This implicates a target start-up delay of 20 minutes ($2 * T_0 = 1200\,sec$). While video streams should usually start within seconds, such a long start-up delay can help to evaluate the behavior of each peer during the start-up phase in more detail. Also, a higher upload bandwidth or smaller part would certainly decrease the start-up delay, but it

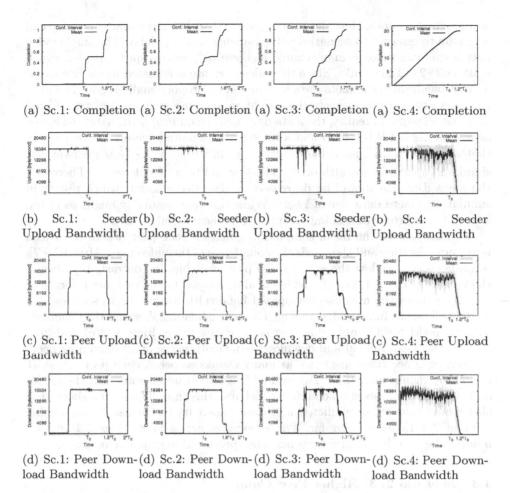

(a) Sc.1: Completion (a) Sc.2: Completion (a) Sc.3: Completion (a) Sc.4: Completion

(b) Sc.1: Seeder (b) Sc.2: Seeder (b) Sc.3: Seeder (b) Sc.4: Seeder
Upload Bandwidth Upload Bandwidth Upload Bandwidth Upload Bandwidth

(c) Sc.1: Peer Upload (c) Sc.2: Peer Upload (c) Sc.3: Peer Upload (c) Sc.4: Peer Upload
Bandwidth   Bandwidth   Bandwidth   Bandwidth

(d) Sc.1: Peer Down- (d) Sc.2: Peer Down- (d) Sc.3: Peer Down- (d) Sc.4: Peer Download Bandwidth   load Bandwidth   load Bandwidth   load Bandwidth

Fig. 2. Scenario 1 - **Fig. 3.** Scenario 2 - **Fig. 4.** Scenario 3 - **Fig. 5.** Scenario 4 -
63 (+1) Peers   127 (+1) Peers   191 (+1) Peers   20 Parts

Table 1. Overhead tradeoff (download)

Scenario	Payload [KiB]	Chunks in total	Size per chunk [KiB]	Overhead [KiB]	Overhead share [%]
1 (63 peers)	9600	126	76.2	36.4	0.38
2 (127 peers)	9600	254	37.8	168.8	1.73
3 (191 peers)	9600	382	25.1	451.1	4.49
4 (63 peers)	9600	2520	3.8	1172.8	10.89
5 (31 peers)	9600	62	154.8	11.9	0.12
6 (255 peers)	9600	510	18.8	1212.4	11.2
7 (63 peers - chunk x4)	9600	252	38.1	79.1	0.8
8 (63 peers - chunk x8)	9600	504	19.0	153.6	1.6
9 (63 peers - chunk x16)	9600	1008	9.5	320.5	3.2

is important to note, that the results gathered with a large start-up delay can easily be transferred to scenarios with a smaller start-up delay. The single video part is split into twice as many chunks as there are peers except the seeder, which makes $63 * 2 = 126$ chunks. Since the video part size is adjusted to the decreased upload bandwidth, the meta data size should be proportionally reduced as well. The meta data size represents the size of an announcement packet, transferred between all peers. In reality, the meta data would require approximately 64 *bytes*, as it contains a *SHA-1* hash and a bit set or differential vectors. If we assume, that a peer with an upload bandwidth of $1\,048\,576\frac{byte}{sec} = 1\frac{MiB}{sec}$ is realistic, our simulated upload bandwidth of $16\,384\frac{byte}{sec}$ would be a 64× decrease. Therefore, the meta data size should be decreased by the same amount, which results in simulating a meta data size of 1 *byte*. While this value arguably might seem too low, only the proportion is important to gain meaningful results.

Fig. 2(a) shows the mean completion graph for each peer, where the x-axis represents the time and the y-axis the completion of the video part. After $1.5 * T_0$ seconds every peer has the entire video part available. In contrast, Equation 1 shows: $T(63, 126) = 1.49 * T_0$, which is almost equal to the measured duration. Fig. 2(b) shows the mean seeder upload bandwidth usage. For T_0 seconds, the seeder uploads at full speed after which it stops uploading, because the Chunked-Swarm model forbids uploading the same chunk twice. Figures 2(c) and 2(d) present the mean upload and download bandwidth usage of all remaining peers in average. Since there are twice as many chunks as peers, each peer can start uploading chunks after $0.5 * T_0$ seconds. It is important to note, that the seeder and the remaining peers upload in parallel after this time. The results show, that this model behaves as predicted and even undercuts the initial start-up delay limit of $2 * T_0$ seconds by far. The ensuing scenarios in Sections 4.3 and 4.4 modify the default scenario by using more peers and multiple video parts.

4.3   Scenario 2, 3: Higher Peer Count

These scenarios are used to observe the performance impact of using 127 and 191 peers, instead of 63 peers in a live video stream scenario with one part. There is still only one seeder. The results are shown in the vertically aligned Figures 3 and 4, which show a minor decrease in performance. While the default scenario takes about $1.5 * T_0$ seconds, these scenarios take $1.6 * T_0$ (with 127 peers) and $1.7 * T_0$ seconds (with 191 peers) respectively. Though both scenarios do not exceed $2 * T_0$ seconds, the performance drops when using more peers.

If a peer downloads its chunk too fast, it might download a second chunk, which should have been distributed by another peer. Thus it distributes two chunks instead of one, as shown in Fig. 3(c), 3(d) and 4(c), 4(d). Some peers start distributing their own chunks before $0.5 * T_0$ seconds, while others seem to upload chunks even after $1.5 * T_0$ seconds. While these small effects lead to an increased download time, the aim of $2 * T_0$ is met.

If n is the number of peers, the whole p2p network consists of n^2 connections. Simulating in real-time on one machine might induce this effect. If the peer count is doubled, the chunk count is doubled as well, so the number of announcements actually quadruples, because every peer notifies other peers about available chunks while downloading them. Theoretically, the Chunked-Swarm model works for any number of clients, in practice the overhead introduced by the growing number of chunks, and thus announcements, is just too high at some point. In reality, every announcement would also increase the latency caused by the *RTT (Round Trip Time)*. In the considered user-generated live streaming scenario, such as Twitch.tv, 85% of all streams have less than 200 users. We simulated up to 191 peers. The benchmark server actually runs in that case $191 * 192 = 36.672$ connections at once, which has a great impact on CPU and main memory usage. However, in reality the overhead is distributed evenly among all peers, so a single peer should be able to connect to more peers. We aim to prove the last statement with a setup of real peers in future work.

4.4   Scenario 4: Multiple Video Parts

In our second step towards live video streaming, which involves the distribution of two parts in parallel as shown in Fig. 1(c), we evaluate whether two adjacent video parts interfere with each other. For this reason we performed Scenario 4 which uses 20 video parts to multiply the effects of parallel part distributions. Please note, that this scenario represents more than a live video stream as the whole video, from the begin to the live time, is distributed by the peers. To maintain consistent notation, the meaning of $T_0 = 600s$ remains, now representing the time needed to distribute all video parts across all peers. Therefore, we introduce a new variable $T_{part} = 30s$, which is the time needed to transfer a single video part from the seeder to a peer and the rate at which the seeder publishes the video parts, so the streaming delay limit is now $2 * T_{part} = 60s$.

Fig. 5(a) shows the average video part completion graph of all peers. Interestingly, this graph almost represents a bisector, which indicates, that all peers finish each video part roughly at the same time. The reason is, that the download and the distribution phases of each video part ran in parallel. So the seeder starts uploading the next part while peers are still distributing the last part but they do not interfere with each other, because every peer is uploading one chunk to $n - 1$ peers and downloading one chunk from $n - 1$ peers plus one new chunk from the seeder. After an improved $1.2 * T_0$ seconds, each peer has all 20 parts available, although each video part still only has a chunk count factor of two as in the previous scenarios. This means that our delay limit of $2 * T_{part} = 60s$ has been undercut, thus the average delay for each video part is $\frac{1}{20} * 1.2 * T_0 = 36s$. The higher confidence intervals in Figures 5(b), 5(c) and 5(d) are caused by the large amount of chunks, which arise from using multiple video parts. When simulating ten runs, it is very unlikely, that all results are equal. As the confidence intervals show, it is still very likely that a further run will be similar. Therefore, we can conclude that a live video stream can be achieved by distributing and announcing only the latest two parts of the video stream. Next, we discuss the overhead in Section 4.5 and in Section 4.6 our results with a real live video streaming application in mind: Twitch.tv.

4.5 Overhead

Compared to other approaches, where the overhead partially depends on the actual payload size, the overhead of our Chunked-Swarm approach only depends on the number of chunks and peers. To fulfill the $2 * T_0$ start-up delay limit reliably, we recommend to use twice as many chunks as peers, so an increase of peers always causes an increase of chunks. The overhead of using more peers grows quadratic, while more chunks cause only a linear growth of overhead.

To reduce protocol overhead we use accumulated announcements, but we still encounter an asymptotically quadratic relationship. To determine the influence of using more peers and chunks more reliably, further scenarios were taken into account. The performance in these scenarios was not considered during the evaluation, as their results did not deliver more insight into the performance characteristics than the other four scenarios did.

Table 1 shows the overhead for these scenarios. When comparing Scenario 1 to 7, 8 and 9, where the chunk count was modified, the overhead seems to grow linearly, as shown in the *Overhead* column, which is what the model predicted. Scenarios 2, 3, 5 and 6 increase the peer count comparing to Scenario 1, thus the overhead should grow quadratically. Since the chunk count is increased as well due to the coupling with the peer count, additionally the overhead should be further increased. Interestingly, the results certainly show an exponentially growth, but it is not quite quadratic. This is due to the internal optimizations, so the quadratic growth can be seen as an upper limit, it is far less in practice.

4.6 Realistic Live Video Streaming Use Case: Twitch.tv

To conclude, we transfer our work's results to the well-known live video streaming platform Twitch.tv and calculate the needed bandwidth of a streamer for a certain start-up delay. To provide similar delays, as occurring in Twitch.tv [20], we choose a delay limit of 10s, so $2 * T_{part} = 10s$ and $T_{part} = 5s$. In a swarm of 191 nodes with one video part, 382 announcements (see Table 1) has been sent, resulting in 764 packets for two video parts distributed at the same time. The size of the 764 announcements (each 64 byte) with network headers (20+20+18 bytes for TCP+IP+Ethernet header) is $764 * (64 + 58)$ byte $= 93.2$ kByte.

The maximum supported bitrate streamers are allowed to stream via Twitch.tv is $3500\frac{kBit}{s} = 438\frac{kByte}{s}$ currently – as of 26th February 2015. Thus, viewers send each 5s traffic with $438 * 5 = 2190$ kByte video payload and 93.2 kByte announcement overhead, which results to 2283.2 kByte, thus our announcements introduce a 4.3 % overhead. Finally, all participants need an upload bandwidth of $\frac{2283.2*8}{5} \frac{kByte}{s} = 3.65\frac{MBit}{s}$, which is below the global broadband upload bandwidth with $10\frac{MBit}{s}$ (see Ookla – http://www.netindex.com/upload/). Thus the Twitch.tv scenario is well supported and our solution would allow private users to stream their content to up to 200 viewers with a small local bandwidth consumption and with reliable delay guarantees.

5 Conclusion

In this paper, we presented our approach for live video streaming, termed Chunked-Swarm. It uses a full mesh topology and a p2p chunk distribution algorithm that guarantees the distribution of the video to all connected peers under $2 * T_0$, where T_0 is the time needed to transfer the desired data from the seeder to the slowest peer. If the number of chunks is chosen wisely, which is coupled with the number of peers, we are able to undercut the $2 * T_0$ mark, which is novel as related work introduces a delay of at least $\mathcal{O}(\log(N) * T_0)$. With a small modification, this model is also capable for streaming applications with a predictable start-up delay. In the evaluation we analyzed the implementation of our model, confirmed our expectations on the analysis of the distribution time mentioned in Equation 1. We evaluated the overhead and calculated a realistic use case. The injected overhead, caused by announcement messages, depends on the chunk count and also peer count, which are fixed for some distribution scenarios but may vary during a live video streaming. If the number of peers is doubled, the chunk count is also doubled and leads to an quadratic upper bound for the overhead. Nevertheless, the overhead share is still reasonable for file sharing and live video streaming applications for small and medium sized scenarios up to hundreds of nodes. As long as the transmission time aim is reached and the seeder has to upload the video only once, the overhead for the peers is considered bearable. Thus, we summarize that our solution is suitable for 85 % of Twitch.tv's streamers and would decrease the operational costs of Twitch.tv by utilizing unused resources of the spectators.

References

1. Banerjee, S., Bhattacharjee, B., Kommareddy, C.: Scalable application layer multicast. In: ACM SIGCOMM 2002 (2002)
2. Cardall, S., Krupat, E., Ulrich, M.: Live Lecture versus Video-recorded Lecture: Are Students Voting with Their Feet? Academic Medicine 83(12) (2008)
3. Castro, M., Druschel, P., Kermarrec, A., Nandi, A., Rowstron, A., Singh, A.: SplitStream: high-bandwidth multicast in cooperative environments. In: ACM SOSP 2003
4. Chan, T.K., Chan, S.G., Begen, A.C.: SPANC: Optimizing Scheduling Delay for Peer-to-Peer Live Streaming. IEEE Transactions on Multimedia 12(7) (2010)
5. Graffi, K.: PeerfactSim.KOM: A P2P system simulator - experiences and lessons learned. In: IEEE P2P 2011 (2011)
6. Graffi, K., Kaune, S., Pussep, K., Kovacevic, A., Steinmetz, R.: Load balancing for multimedia streaming in heterogeneous peer-to-peer systems. In: ACM NOSSDAV 2008 (2008)
7. Kovacevic, A., Graffi, K., Kaune, S., Leng, C., Steinmetz, R.: Towards benchmarking of structured peer-to-peer overlays for network virtual environments. In: IEEE ICPADS 2008 (2008)
8. Kovacevic, A., Kaune, S., Heckel, H., Mink, A., Graffi, K., Heckmann, O., Steinmetz, R.: PeerfactSim.KOM - A Simulator for Large-Scale Peer-to-Peer Networks. Technical Report Tr-2006-06, TU Darmstadt (2006)

9. Liebau, N., Pussep, K., Graffi, K., Kaune, S., Jahn, E., Beyer, A., Steinmetz, R.: The impact of the P2P paradigm on the new media industries. In: AMCIS (2007)

10. Liu, Z., Shen, Y., Ross, K., Panwar, S., Wang, Y.: Substream trading: towards an open P2P live streaming system. In: IEEE ICNP 2008 (2008)

11. Liu, Z., Shen, Y., Ross, K., Panwar, S., Wang, Y.: LayerP2P: using layered video chunks in P2P live streaming. IEEE Trans. on Multimedia 11(7) (2009)

12. Nascimento, G., Ribeiro, M., Cerf, L., Cesario, N., Kaytoue, M., Raissi, C., Vasconcelos, T., Meira, W.: Modeling and analyzing the video game live-streaming community. In: LA-WEB 2014 (2014)

13. Padmanabhan, V., Wang, H., Chou, P., Sripanidkulchai, K.: Distributing streaming media content using cooperative networking. In: ACM NOSSDAV 2012 (2002)

14. Pai, V., Kumar, K., Tamilmani, K., Sambamurthy, V., Mohr, A.E.: Chainsaw: eliminating trees from overlay multicast. In: van Renesse, R. (ed.) IPTPS 2005. LNCS, vol. 3640, pp. 127–140. Springer, Heidelberg (2005)

15. Pendarakis, D., Shi, S., Verma, D., Waldvogel, M.: ALMI: an application level multicast infrastructure. In: USENIX USITS 2001 (2001)

16. Rückert, J., Knierim, T., Hausheer, D.: Clubbing with the peers: a measurement study of BitTorrent live. In: IEEE P2P 2014 (2014)

17. Tian, R., Xiong, Y., Zhang, Q., Li, B., Zhao, B.Y., Li, X.: Hybrid overlay structure based on random walks. In: van Renesse, R. (ed.) IPTPS 2005. LNCS, vol. 3640, pp. 152–162. Springer, Heidelberg (2005)

18. Wang, F., Xiong, Y., Liu, J.: mTreebone: A Collaborative Tree-Mesh Overlay Network for Multicast Video Streaming. IEEE Transactions on Parallel and Distributed Systems 21(3) (2010)

19. Xie, S., Li, B., Keung, G.Y., Zhang, X.: Coolstreaming: Design, Theory, and Practice. IEEE Trans. on Multimedia 9(8) (2007)

20. Zhang, C., Liu, J.: On Crowdsourced Interactive Live Streaming: A Twitch. TV-Based Measurement Study (2015). arXiv preprint arXiv:1502.04666

21. Zhang, X., Hassanein, H.S.: TreeClimber: a network-driven push-pull hybrid scheme for peer-to-peer video live streaming. In: IEEE LCN 2010 (2010)

22. Zhang, X., Hassanein, H.S.: A Survey of Peer-to-Peer Live Video Streaming Schemes - An Algorithmic Perspective. Comp. Networks 56(15) (2012)

23. Zhuang, S., Zhao, B.Y., Joseph, A.D., Katz, R.H., Kubiatowicz, J.: Bayeux: an architecture for scalable and fault-tolerant wide-area data dissemination. In: ACM NOSSDAV 2001 (2001)

Adaptive Mobile P2P Streaming System for Wireless LAN

Geun-Hyung Kim(✉)

Department of Game and Visual Image Engineering, Dong-Eui University,
176 Eomgwang-no, BusanJin-Gu, Busan 614-714, Korea
geunkim@deu.ac.kr

Abstract. Recent advance in digital technologies, such as broadband wireless networking, media compression technologies, and embedded system technologies, has enabled it possible to provide a real-time video streaming service over the Internet. Specially, IEEE 802.11 standards have been developing to handle new application demands like a video streaming. The mobile video traffic has been already dominating network usage, as the number of mobile multimedia devices is increasing. Therefore, mobile peer-to-peer(P2P) streaming system can be considered as a promising way to distribute media streams over the wireless LAN because of its scalability and efficiency. However, the bandwidth of wireless links changes dynamically depending on the mobility of peers and symmetric allocation. The link bandwidth fluctuation can affect negatively the mobile P2P streaming performance. In order to guarantee overall service quality of P2P streaming system in wireless environment, an adaptive mechanism, that considers the timing-varying wireless link conditions, is required. In this paper, we investigate the effect of wireless LAN on a P2P streaming and propose the adaptive P2P streaming system considering the wireless network conditions.

Keywords: Mobile P2P streaming · Adaptive P2P system · Wireless LANs

1 Introduction

Recent advances in wireless Internet technologies and proliferation of wireless devices have changed both the way users enjoy contents and the devices they utilize. According to the latest Cisco Visual Networking Index [1], global mobile traffic growth will outpace global fixed traffic growth by a factor of three and mobile video will represent 72% of global data traffic by 2019 (up from 55% in 2014).

The IEEE 802.11 standards based wireless LANs(WLANs) are used almost everywhere, from home to public places like hotels, airports, and cafes, etc. The WiFi offload traffic will surpass cellular traffic and 54% of total mobile data traffic will be offloaded by 2019, even though broadband cellular technologies have been steadily deployed. Specifically, streaming among smartphones,

© Springer International Publishing Switzerland 2015
S. Balandin et al. (Eds.): NEW2AN/ruSMART 2015, LNCS 9247, pp. 211–219, 2015.
DOI: 10.1007/978-3-319-23126-6_19

tablets, smart TVs, and other smart devices over the WLAN has become an important enabling technology for home entertainment and multi-screen applications. Therefore, an efficient dissemination of video streams over the WLAN is an important issue.

Several video streaming architectures, such as client/server architecture, content delivery network(CDN), and P2P architecture, have been discussed over few years to disseminate video contents efficiently. The traditional client/server architecture provides good performance and high availability when the number of clients is limited. The number of clients is proportional to the required bandwidth on the server. In short, the bandwidth required for a stream and the bandwidth of network interface in the server limit the number of concurrent clients served by the server. Therefore, the client/server architecture is insufficient to support large scale video streaming applications.

In CDN architecture, various dedicated servers located geographically different locations are used and they are parts of an overlay content delivery network. Whenever the CDN encounters a request, it forwards the request to the closest server to reduce latency issues and scalability issues of client/server architecture. However, the biggest drawback of CDN is the cost issue, such as expensive deployment and maintenance.

Among those architectures, P2P architecture has been considered as a very promising way to disseminate contents efficiently and extensively, because it has the features of off-server, scalability, high performance-cost ratio, robustness and high load balancing. P2P architecture has several advantages, such as higher utilization of network resources and the elimination of a single-point failure. Recently, various P2P streaming systems have been deployed for live and video on demand (VoD) streaming services over the Internet. Compared to the conventional approaches, each peer contributes its own resources to the participating streaming session. Hence, administration, maintenance, and responsibility for operations are dispersed among several clients(peers) instead of focusing on few servers [2].

However, because of random peer selection and ignorance of the different abilities among peers, the P2P streaming systems have the problems of low resource utilization and unreasonable resource distribution. Long start-up latency, interpeer playback lag, playback discontinuity, and total dissemination rate in P2P systems are critical issues for high quality of Experience(QoE) requirements [3]. Even though previous researches have been performed to solve these problems for the P2P streaming, they select the highest service peers based on peer's inherent abilities without considering the dynamic changes of the network. Eventually, they can not adapt the change of service ability caused by network state changes, such as, sudden network congestion and wireless link rate change by peer's movement. Since P2P video streaming has become more popular in fixed network, extension of P2P video streaming to a mobile environment is necessary.

Mobile nodes in WLANs share radio resources and connect with each other through access point(AP). Therefore, it is hard to migrate fixed P2P video streaming services into mobile environment, because of the features of WLANs, such as mobility, symmetric resource allocation, link rate adaptation, congestion, and dynamics, and sharing. For the mobile P2P streaming in WLANs, we should consider the following problems, such as an efficient dissemination of video stream to all peers with good experience quality, adaptive mechanism to mitigate the impact of peer's mobility and dynamics, and the mechanism for congestion mitigation in WLAN. The effective bandwidth on mobile peers in WLAN is non-deterministic and time-varying because of the peers' mobility, symmetric wireless resource allocation, and wireless resource sharing. Therefore, it is hard to guarantee the bandwidth consistently compared with fixed P2P video streaming systems. In order to ensure overall service quality in mobile P2P streaming system, an adaptive algorithm that reflects the time-varying wireless channel status, is indispensable to ensure a consistent service quality. In this paper, we investigate the effect of the peer's mobility and rate adaptation on overall service quality and propose an adaptive algorithm. This proposed algorithm determines serving bandwidth and the number of serving peers depending on the link status and selects the peers with that each peer exchanges chunks.

The rest of this paper is organized as follows. We discuss the related works in Section 2. In Section 3, we discuss the problems in mobile P2P streaming architecture. We describe the proposed algorithm and performance analysis in Section 4. In Section 5, we summarize this paper and show the future work.

2 Related Works

In terms of the overlay network topology, the P2P streaming architectures are classified in three categories: *tree-based*, *mesh-based*, and *hybrid* topology. BitTorrent(BT) [4] is one of the most successful mesh-based P2P architecture to disseminate contents over the Internet. It has been considered as a scalable file sharing protocol, incorporating swarming data transfer mechanism. However, it is inappropriate for a time sensitive content distribution, since the chunk selection algorithm focuses only on minimizing the probability for rare pieces rather than guaranting the playback deadline.

BitToS system [5] extends the current BitTorrent system, dividing the missing chunks into *high priority set* and *remaining piece set* and requesting with higher probability chunks from the *high priority set*. In the BitToS system, the requested chunks are not always available for playback on time, due to asymmetric nature of the Internet connections and heterogeneity of the peers [6].

BitMax [7] is an uplink allocation mechanism, that determines the number of unchoked peers during each unchoking interval and the percentage on the available uplink bandwidth for an unchoked peers. In the BitMax, a peer allocates maximum rate for each unchoked peer. It specifically focuses on the uplink utilization and download period reducing by uplink scheduling, but it does not consider mobile P2P streaming and link bandwidth dynamics. The utility-based rate allocation framework [8] optimizes the data flow rates in a P2P streaming architecture for wireless networks. It focuses only on maximizing the aggregate utility throughout the wireless P2P overlay. The optimistic unchoking algorithm for BitTorrent [9] prevents free-riding and improves the efficiency of optimistic unchoking and activeness-based seed choking algorithm [10] defines activeness values of requesting peers as the ratio of the average download bandwidth to the available upload bandwidth. These two works [9] [10] focus on restraining the free-riding of selfish peers using adaptive unchoking algorithms. Smart Caching in Access Points(SCAP) [11] is an efficient caching mechanism, reducing the amount of traffic between AP and peers by exploiting temporal locality. Once a peer receives the *interested message* from other peers and the interested chunks was in its buffer, the peer sends an information about the destination of this chunk to the AP. The AP simply assembles packets with the chunks stored in its local buffer and sends them out. We consider the more realistic adaptive algorithm to get overall high performance of mobile P2P streaming in WLANs.

3 Problems in Mobile P2P Streaming

Nowaday, IEEE 802.11 based WLANs have widely deployed for ubiquitous environments and they will be prominent technologies to be used for streaming service. Next generation WLAN technologies also endeavor to provide broadband download capacity.

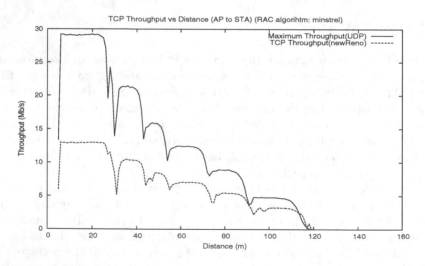

Fig. 1. TCP throughput vs. Distance between an AP and a mobile node.

The wireless network performance depends on signal-to-noise ratio(SNR). The higher the SNR, the better the performance. As a mobile node moves away from an AP while background noise remains constant, the SNR decreases progressively and a mobile node operates at successively lower link rates. Eventually, wireless communication is going to be no longer supported at a certain distance where the signal is lost in the background noise [12].

In mobile P2P streaming systems, exchanging chunks among peers utilizes a transmission control protocol (TCP) as a transport protocol. The mobile P2P streaming service performance in WLAN depends on the TCP throughput on a mobile node. We investigate the maximum throughput and the TCP newReno throughput in IEEE 802.11a WLAN through the ns-3 [13] simulation, as a mobile node moves away from an AP. We show the TCP throughput on a mobile node at a given distance from an AP in Fig. 1. The TCP throughput on a mobile node is influenced by where other mobile node is, even though a mobile node does not move.

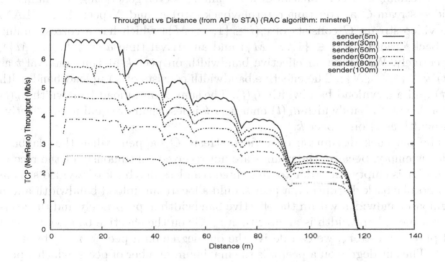

Fig. 2. TCP throughput vs. Distance between two nodes.

In addition, we investigate the effect of a distance between a sender and a receiver on a TCP throughput between two nodes. In this investigation, we measure the TCP throughput by a ns-3 simulation, as a receiver node moves away from an AP while a sender is at 5m, 30m, 50m, 60m, 80m, and 100m from an AP. Fig. 2 shows the simulation results, that the TCP throughputs on a receiver node are affected by not only its position but also the sender's position. From this investigation, we can find the link capacity of a mobile node changes dynamically according to the position of the node and the corresponding node. Therefore, we should consider the node's mobility to determine mobile node's effective bandwidth in a P2P streaming system.

4   Proposed Mechanism

In a BitTorrent-based P2P streaming system, the peers, having more advanced chunks, may receive more chunk request messages than others and they have a dominant role in a chunk dissemination. As we discussed previously, the link bandwidth of a mobile node changes dynamically according to mobile node's movement. To guarantee a seamless playback with no playback disruption, a peer adjusts its operations adaptively, reflecting the peer's status. In this paper, we propose an adaptive upload bandwidth allocation and adaptive peer selection algorithm based on the streaming service capability of peers. Therefore, proposed algorithm can adapt to the change of peer's service capability.

In a P2P streaming system, video stream data are divided into small pieces with the same size for the dissemination of a video stream. Each piece is called as a chunk and is assigned by an unique identifier. We assume that a mobile P2P streaming system consists of a streaming server generating M chunks of a video stream C at constant rate(playback rate) r_s and N peers in a WLAN. The video stream C consists of $c_i(i \in \{1, \ldots, M\})$ which has a corresponding playback deadline $p_i(i \in \{1, \ldots, M\})$ and an arrival time $a_i(i \in \{1, \ldots, M\})$, respectively. We define an effective bandwidth on peer k at time interval t as $b_k(t)(k \in \{1, \ldots, N\})$. The effective bandwidth consists of an upload bandwidth $u_k(t)$ and a download bandwidth $d_k(t)$. The sum of an upload bandwidth $u_k(t)$ and a download bandwidth $d_k(t)$ on a peer k is less than or equal to an effective bandwidth $b_k(t)$ on a peer k.

The playback disruption of a video stream C happens when the playback buffer is empty, because the chunks are not received continuously or the receiving buffer is empty. In order to guarantee seamless playback of a video stream with no playback disruption, a peer should allocate an upload bandwidth and a download bandwidth within the effective bandwidth appropriately and the average download bandwidth is greater than r_s. Given the effective bandwidth $b_k(t)$ and playback rate r_s, we can derive the out-degree on a peer k, $n_k(t)$ by using Eq. 1. The out-degree on a peer k is the maximum number of peers, which a peer k supports. We define the set of peers to which peer k uploads its chunks as S_k. The members in a set S_k are dynamically selected according to the changes of network condition and the number of members in a set S_k ($|S_k|$) also changes dynamically.

$$n_k(t) = \lfloor (b_k(t) - r_s)/r_s \rfloor = \lfloor b_k(t)/r_s - 1 \rfloor. \quad k \in \{1, .., N\}. \tag{1}$$

As discussed in previous section, the maximum effective bandwidth is show in Fig. 1. When the mobile peers communicate each other via an AP, the throughput is less than the maximum effective bandwidth and depends on the sender's position. We define the throughput between a peer i and a peer j as $b_{i,j}$. The $b_{i,j}$ is the upload bandwidth of a peer i for a peer j. The total upload bandwidth of peer i is $\sum_{k \in S_i} b_{i,k}$.

For seamless playback of a video stream based on the network conditions, following constrains are defined in our algorithm. Eq. 2 defines that the sum of throughputs from a peer i to the members of S_i should be less than or equal to an upload bandwidth of a peer i which considers the playback rate as a download bandwidth. Eq. 3 defines the constraint for the number of supporting peers. The MAX_OUT_DEGREE defines the maximum number of supporting peers and is usually six as the system parameter. Eq. 4 defines the constraint which a chunk m should arrive before its playback deadline (p_m).

$$\sum_{k \in S_i} b_{i,k} \leq u_i(t) = b_i(t) - r_s \tag{2}$$

$$|S_i| \leq min(n_i(t), MAX\_OUT\_DEGREE) \tag{3}$$

$$a_m < p_m \ \ for \ \ m \in \{1, \ldots, M\} \tag{4}$$

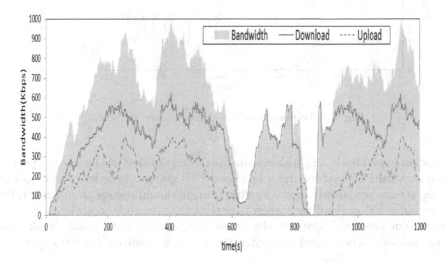

Fig. 3. The bandwidth usage in a conventional P2P streaming algorithm.

Fig. 3 shows the upload bandwidth, the download bandwidth, and the sum of the upload bandwidth and download bandwidth in mobile peer that participates in mobile P2P streaming service over an WLAN. In this measurement the peer moves away from an AP and comes back to the AP. Between 800 sec to 900 sec, the peer uses the effective bandwidth to upload chunk instead of downloading chunks. So, the playback disruption occurs because playback buffer is empty.

When a peer is close to an AP, its effective bandwidth is high and so it may have an advanced chunks. If a peer has lots of advanced chunks, it receives many request messages from other peers and selects the peers that it will support. In the conventional algorithm, the peer has a constant out-degree, so it will send chunks to peers as many as its fixed out-degree. When it moves away from an AP its effective bandwidth becomes low, it is sending chunks to peers despite it has no enough effective bandwidth and so its buffer becomes empty. Consequently, the playback disruption happens.

The proposed adaptive algorithm constraints an upload bandwidth in order to guarantee a seamless playback of a video stream. The buffer in the proposed adaptive algorithm does not become empty. it gives the insight that there is no

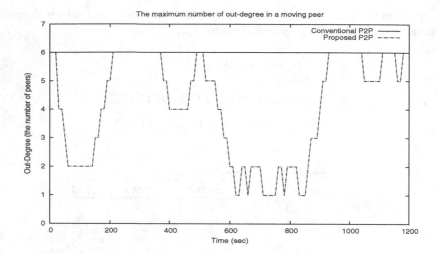

Fig. 4. The maximum out-degree in moving peer.

disruption in playback. Fig. 4 shows the maximum out-degree in a moving peer of a conventional P2P system and a proposed P2P system. In order to playback of a video stream seamlessly in the moving peer, it should reduce an upload traffic to guarantee a download rate when it notices that network condition is getting poor. In our proposed adaptive algorithm, the peers do not upload chunks and try to maintain a download rate when the network condition is getting poor.

5   Conclusion

Mobile P2P streaming architecture will be considered as a promising way to disseminate a video stream over the WLAN, due to its scalability and efficiency. In the WLAN, the effective bandwidth of a mobile peer varies in time, because of its flexible resource sharing and its mobility. Hence, mobile P2P streaming service in the WLANs is a challenge. In this paper, we propose an adaptive algorithm that the peer manages upload bandwidth and utilizes its effective bandwidth in order to guarantee a seamless playback. The major objective of the proposed algorithm is to reflect the variation of effective bandwidth of maintaining overall system performance. We show the proposed adaptive algorithm outperforms the conventional algorithm in terms of seamless playback. As the future work, we will extend the adaptive algorithm into 802.11n WLAN environment.

Acknowledgments. This research was supported by Basic Science Research Program through the National Research Found (NRF) funded by Ministry of Education, Science and Technology (NRF-2010-0025069) and the Dong-Eui University Research Grant of 2015(2015AA185).

References

1. Cisco White Paper: Cisco VNI: Global Mobile Data Traffic Forecast Update 2014–2019 White Paper (2015)
2. Thampi, S.M. : A Review on P2P Video Streaming (2013). arXiv preprint arXiv:1304.1235
3. Wu, X., Chan, X., Wang, H.: An Effective Scheme for Performance Improvement of P2P Live Streaming Systems. Journal of Networks, 9(4) (2014)
4. BitTorrent Protocol Specification v1.0 (2005). http://wiki.theory.org/BitTorrentSpecification
5. Vlavianos, A., Iliofotou, M., Faloutsos, M.: BiToS: enhancing BitTorrent for supporting streaming applications. In: INFOCOM 2006, vol. 1, no. 6, pp. 23-29, April 2006
6. Pandey, R.R., Patil, K.K.: Study of BitTorrent based Video on Demand Systems. International Journal of Computer Applications 1(11), 29–33 (2010)
7. Laoutaris, N., Carra, D., Michiardi, P.: Uplink Allocation Beyond Choke/Unchoke, ACM CoNEXT 2008, Madrid, Spain (2008)
8. Qiu, F., Bai, J., Cui, Y., Xue, Y.: Optimal rate allocation in peer-to-peer streaming over wireless networks. In: International Conference on Collaboration Technologies and System (CTS) 2011, pp. 23–27. IEEE Press, NJ (2011)
9. Ma, Z., Qiu, D.: A novel optimistic unchoking algorithm for BitTorrent. In: Proc. of the 6th IEEE Conference on Consumer Communications and Networking Conference, CCNC 2009, pp. 327–330. IEEE Press, NJ (2009)
10. Huang, K., Zhang, D., Wang, L.: An Activeness-based seed choking algorithm for enhancing BitTorrent's robustness. In: Abdennadher, N., Petcu, D. (eds.) GPC 2009. LNCS, vol. 5529, pp. 376–387. Springer, Heidelberg (2009)
11. Tan, E., Guo, L., Chen, S., Zhang, X.: SCAP: smart caching inwireless access points to improve P2P streaming. In: International Conference on Distributed Computing Systems, ICDCS 2007, pp. 61–69. IEEE Press, NJ (2007)
12. Jupiter Networks White Paper: Coverage or capacity making the best use of 802.11n (2011)
13. NS-3 Network Simulator. http://nsnam.org

Application for Selective Streaming of Video Components

Humera Siddiqua[(⊠)] and Hamid Shahnasser[(⊠)]

Electrical Engineering Deparment, San Francisco State University,
San Francisco, USA
humera@mail.sfsu.edu, hamid@sfsu.edu

Abstract. Streaming of video components has become a popular form of media service on the Internet. With the advancement in recent computer and networking technologies audio-video streaming has given rise to interesting applications. But, the variable nature of the wireless medium and limited bandwidth will not lend itself easily to support audio-video streaming. In this paper, we propose an application to provide a standard interface for selective streaming of different components of the video from the server within the desired bandwidth allocation. Specifically, the application is designed for capabilities of searching and selecting the best combination of audio, and/or video files with varying resolutions to deliver a scalable, high-performance audio and/or video seamless streaming experience over the internet. This avoids undesirable throttling of data, meets desired needs, saves user bandwidth and power consumption.

Keywords: Adaptive bitrate streaming · Bandwidth allocation · Power consumption · Selective streaming

1 Introduction

The expeditious advancement in computing and networking technologies, especially with the emergence of Internet and mobile devices, the audio-video streaming applications are becoming very popular. The propagation of mobile devices is generating a new wave of applications that enable users to lean on smart phones in their daily activities [1]. People prefer mobile devices to search and browse video and or audio content on the move and stream it over the network. When a streaming is done across the devices within a local network the download allowances of Internet service providers (ISPs) is not used. However the mobile devices often go beyond the local network when they are on the go and end up using ISPs allocated data allowances. An hour's video play could well use up over 1GB of data. Some ISPs also have online broadband monitors and after certain amount of data usage they impose throttling of the download speeds to the devices. Video streaming on mobile devices over the ISP network consumes more bandwidth and run in to buffering issues due to slower speeds

© Springer International Publishing Switzerland 2015
S. Balandin et al. (Eds.): NEW2AN/ruSMART 2015, LNCS 9247, pp. 220–228, 2015.
DOI: 10.1007/978-3-319-23126-6_20

(compared to local Wi-Fi networks) and ends up in the throttling of the speed by the ISP. This leads to streaming video getting chocked in playing as it keeps retrying to get the content. And this also leads to another problem of draining the battery of mobile device in the process of retrying and buffering. There is a difference between progressive download and streaming. Progressive download is getting a file into memory and playing it back. For progressive download, we set the source of the media element and its just downloads and plays. In streaming, we dont have to deal with the complexities of streaming inside of our application. Our proposed approach, presented in this paper can also be useful with smart phones with a smart stay sensor such that, if the user is not watching the video, the application would switch off or pause the video streaming and just load the audio streaming based on the user configuration. Streaming only the audio component, which is lower in size compared to the complete video content size, helps in reduced power consumption. Also, if the streaming device is operated with smartphone screens getting turned off, for just listening the stream, it further reduces the power consumption for the display.

The rest of the paper is organized as follows: Section 2 discusses the motivating use cases for better utilization of bandwidth allocation. Section 3 discusses the related work. Section 4 describes the proposed application. Section 5 shows the client-server interface. Section 6 shows the sequence of steps that are involved between the client interface and streaming server. Section 7 evaluates the amount of bandwidth saved with this application. Finally, section 8 concludes the paper.

2   Background and Motivating Use Cases

With the limitation of bandwidth, a mobile device user with limited data plan from the carrier may be interested When a user is interested only in listening to the streaming content, then downloading the video stream wastes the bandwidth of the network and also makes the streaming slower. In such a case, the application that can selectively retrieve and stream just the audio component is desirable. Some of the uses cases are given below:

A mobile device user with limited data plan from the carrier may be interested in just listening to the content of a video either while driving or in public commute and not interested in watching the video. For example, the videos involving convocation speech, motivation speech, interviews, success stories where the audio content is of more importance than its associated video, which may need to be just listened to without wasting the bandwidth in streaming its video when we are not looking at it. Also Wi-Fi is not available during driving and on most of public transports; hence saving mobile data usage becomes more significant [3]. In places where mobile carriers do not have enough infrastructures to provide high bandwidths, the videos would not load and / or get stuck in buffering frequently even though user may be willing to pay for the data usage. In that case, users could just get an option to at least listen to the video on slower network with an un-interrupted audio streaming. Even for the users who have the devices in well-developed telecom infrastructure, the careers would still do throttling of the data if the users tend to download beyond certain data limit in a billing cycle.

In Such cases, even an unlimited data plan would not be able to load and run the videos without getting stuck. Having an audio streaming application would help in listening to the desired videos even at throttled network speed [12]. Even if the network is too slow to stream the audio content, the application would provide the options to selectively pick the available audio formats with different quality. The lower quality would stream better at lower speeds. The best-fit approach thus helps user to get the content seamlessly. The application in itself is a platform independent such that it should be portable to run on any mobile device platform.

3   Related Work

Earlier work has been done in scalable video coding (SVC) such as H.26/MPEG-4 SVC. HaechulChoi et al. [16] has evaluated SVC performance and concludes that compared to H.264/MPEG-4AVC, SVC had negligible overhead bit- rate for temporal scalability and around 10% overhead bit rate for special/quality scalability. Garrido-Cantos et al. [15] shows a technique of scalable video transcoding for mobile communications such that, it converts a H.264/AVC bit stream without scalability to a scalable bit stream with temporal scalability. Xinying Liu [17] has studied the decoding efficiency and packet transmission of audio and video. Luis Herranz [18] proposes the integration of content analysis with scalable video adaptation model. Christopher Muller et al. [7] have integrated SVC to MPEG DASH (Motion Picture Expert Groups-Dynamic Adaptive Streaming over HTTP) for a higher average bit-rate. Our proposed application is to make a standard interface for selectively streaming the video components by incorporating user preferences and requirements for a better utilization of the available bandwidth.

4   System Design and Overview

4.1   System Overview

The proposed model is to make an application platform, where the users will be able to select and play files as they desire based on their bandwidth allocation, data and battery usage. The application is responsible for retrieving all of the video and/or audio segments and put them into the media pipeline through the media source we create. The application features are to stream data as per user needs. Stream the audio component of the video when the user just wants to listen, dynamically adapting video bit rate [4]. It provides different modes of video qualities for users of different bandwidth. Stream videos when users want to watch videos and provide the platform, where the user can select an option of audio or video or both. As in most of the smart phones Smart stay can also be a feature, finally we opt to make a universal app that works on PCs, mobile phones and tablets. In this app project, one of the design features will be to figure out the current bandwidth speed and adapt to the audio and or video file format based on the available bandwidth allocation. For example, when a user selects a video for streaming, the applications extracts the available formats from the server and allows the users with an option to select for best quality or best utilization of bandwidth [2]. For each such option there is an associated format code to request the required stream from the server. If a user selects best quality (that is mp4 with 1280x720) resolution, the application would use the corresponding format code to request that stream, from the server. However since that could take more bandwidth, the user may choose a different lower quality option or chose to just request one of the available audio only options and save the bandwidth. Also on top of user selection, the application can dynamically select and adapt to seamless streaming by selecting next best of selected option in order to avoid buffering and loading issues with the stream [8]. Thus, it dynamically adapts to video bit rate and all of the adaptive streaming starts after the selection of file format.

4.2   System Architecture

In order for an application to selectively retrieve the audio component, the corresponding content provider interface at the server should allow the required interfacing APIs. Since there are no standard interface APIs for negotiating the components of the content across different service provides, the application would need to run through an abstraction layer, which does different interfacing requests to different servers while providing the single interface to the user [5, 6]. Having such abstraction layer software in itself is a different scope in parallel to the objective of this project; the other option is to have different one to one application to different content provider interfaces. In such case one application will operate with only one provider. In the scope of fulfilling the objective of this project, we would select the YouTube.com as content provider and interface our application with it. Since YouTube is the single widely used and it holds the highest content database, the usability of the application in conjunction with You Tube is more desirable than other services. However the concept as such would work with interfacing any content provider and not just YouTube.

There are five major components involved in serving the objective of the application. They are the application user interface, portable concept on any mobile platform, the interface layer that would communicate with the content provider for given user request and get the results. There is an optional abstraction layer to communicate across different content providers. However if the application is targeted to single content provider, this is not required. We then have an interface layer at the content provider, which would respond to the service requests from user application and a video content library database of the content provider. While scoping the objective of the project to YouTube, we would use the YouTube interface APIs to request and retrieve the contents from YouTube database directly over the network [11]. In that case the first two components (platform independent Application and its YouTube interface layer) need to be implemented. With such application if a user is able to selectively retrieve and stream the audio of a video present in YouTube library, then the objective is achieved and would serve the above mentioned use cases.

5   Client-Server Interface

With the purpose of searching and selecting the available format as per user needs and bandwidth allocation, we are analyzing the clientserver communication. We have carried out an experiment using YouTube APIs, writing a source code in C language and by executing the code in Ubuntu Server [14]. The client is connected to the Internet through the wireless connection and receives the video components of the media file from the YouTube server [13].

Table 1. Set of Available Formats

File Format	Bit rate (kb/s)	Resolution	Note
webm	112	audio only	DASH audio
m4a	132	audio only	DASH audio
m4a	260	audio only	DASH audio
m4a	134	256x144	DASH video, video only
webm	168	426x240	DASH video, video only
m4a	276	426x240	DASH video, video only
webm	319	640x360	DASH video, video only
mp4	603	640x360	DASH video, video only
webm	641	854x480	DASH video, video only
mp4	1105	854x480	DASH video, video only
webm	1318	1280x720	DASH video, video only
mp4	2202	1280x720	DASH video, video only

From the client interface, the interface API is used to retrieve the available formats of the video file from the YouTube server. Table 1 shows the list of available formats, received from the server. Based on the user-input, preferences and available bandwidth, one of the available formats is selected and requested to stream. In case if a user is interested in just listening to the audio, the client interface can just select one of the available audio formats. Apart from audio-only selection it also helps in selecting the lower quality videos to avoid buffering and help seamless adaptive streaming of the video when the network is slow. For Mobile devices, this not only helps in optimal usage of the allocated bandwidth but also reduce the battery usage of device with selective format of videos [10].

6   Client-Server Time Sequence Graph

Below is the timing sequence of a Client-Server graph, showing the communication that takes place between client and server.

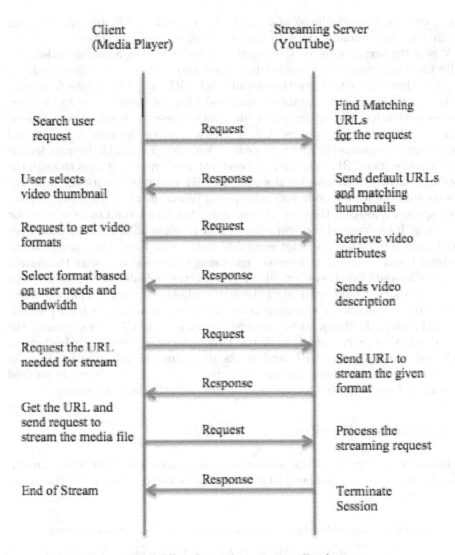

Fig 1. Client-Server time sequence Graph

Fig. 1. Client Server time sequence Graph

As shown in Fig. 1, the following sequence of steps are involved between the client interface and streaming server in order to fetch and stream different formats or components of the video. A user interface running on any platform (mobile application or a Linux or other OS terminal) would interface with the client interface library and trigger a search on specific keywords to look for the desired video for its streaming options. User can also provide the search filters like maximum views, upload dates [14]. Client interface would take the given search key words (and filters) and invoke corresponding API function through http request to communicate with the streaming server (YouTube).

When the server receives the request, it looks for the available videos (or audio files) for given search keywords / filters and replies to the client with the list of options along with its thumbnails and URLs [7]. Client interface picks the response and present the thumbnails and URLs of these videos to the user. The user can select the desired video file from the results shown, based on users need and the available options. Client interface picks the user selection and further again, requests the server to provide with list of available formats for the selected video files URL. When server receives these requests, it goes through the attributes of the video associated with that URL and replies with the available options with the format code for each option. When the client interface receives this response, it looks at the user criteria to decide which format to be selected for streaming. If the desired configuration of user is audio only, the client interface would pick the format code for available audio only files. Also based on the bandwidth and speed, client interface can prompt the user to go with the desired selection for optimal streaming [9]. In general, the default option, to save the users bandwidth, would be to select the audio only best quality streaming, unless user overrides this option. Client interface selects the format code from previous step and makes a final request to the server to retrieve the URL for streaming the desired selection. Server sends back the URL associated with desired selection. Client interface gets the URL and starts streaming the selection on a media player [13]. When streaming is done or interrupted by the user, it would proceed to continue from step one, for the next search, selection and streaming

7   Evaluation Results

In this section, we evaluate the measurement of bandwidth saved, when streaming only the audio component of the video file as listed in Table 2. The above

Table 2. Bandwidth Saved in Streaming only the Audio Components

Video Index	Mp4 size (MiB)	Mp4 Resolution	Webm size (MiB)	Webm Resolution	BandWidth Saved	% BandWidth Saved
V1	727.9	1280x720	38.53	Audio only	689.37	94.707
V2	798.01	1280x720	56.28	Audio only	741.73	92.947
V3	426.1	1280x720	20.26	Audio only	405.84	95.245
V4	534.82	1280x720	34.07	Audio only	500.75	93.630
V5	281.66	1280x720	17.31	Audio only	264.35	93.854
V6	370.73	1280x720	17.96	Audio only	352.77	95.156
V7	253.83	1280x720	11.53	Audio only	242.3	95.458
V8	257.47	1280x720	12.66	Audio only	244.81	95.083
V9	215.39	1280x720	12.51	Audio only	202.88	94.192
V10	461.55	1280x720	30.04	Audio only	431.51	93.491
V11	233.27	1280x720	10.11	Audio only	223.16	95.666

evaluation shows the result that, an average of 95% of the bandwidth can be saved by streaming the audio component of the video file rather than streaming both the audio and video components. The files V1, V2, V3 and V4 till V11 are the set of YouTube video files. The mp4 file is the audio plus video file while the webm extension file is the audio only file. For each file, we retrieved the attributes of both these formats and checked their individual sizes and resolutions. The overall bandwidth saved is obtained by calculating the difference between the file sizes of the mp4 and webm file formats. For example, for the video file V1, compared to the default high quality mp4 selection of size 727.9 MiB, the desired audio format selection of size 38.53 MiB helps save the bandwidth of about 689.37 MiB. From the difference, the percentage of bandwidth saved while using the audio file is computed. We observed that a significant amount of bandwidth could be saved by streaming the audio component of the video file for the users who are interested in audio only streaming.

8   Conclusion

In this paper, we presented an application for selective streaming of different components of the video from the server within the desired bandwidth allocation, incorporating user preferences and needs. Our evaluation shows that on an average, more than 95% of the bandwidth can be saved by selectively streaming only the audio component of the video, when user wants to just listen to the video. This meets the user requirements, avoids the unnecessary wastage of bandwidth and provides fast seamless streaming even at very low bandwidths. Streaming only the audio component, which is over 90% lower in size, compared to the size of the complete video content, helps in reduced power consumption. The evaluation is performed through streaming different video files using the YouTube server as the source for the video files. Test results showed that streaming just the audio component of a video file, compared to streaming the whole video file, have the best usage as per user requirements and available network bandwidth for different media files evaluated. This application is platform independent and apart from YouTube, it can also be used to stream the media files along with other servers and selectively stream components of a video

References

1. Talaat, M.A., Koutb, M.A., Kelash, H.M., Aboelez, R.H.: Content-aware adaptive video streaming system, information and communications technology. In: 3rd ITI International Conference on Enabling Technologies for the New Knowledge Society, pp. 265–276 (2005)
2. Omerasevic, D., Behlilovic, N., Mrdovic S., Sarajlic, A.: Comparing randomness on various video and audio media file types. In: 21st IEEE Telecommunications Forum (TELFOR), pp. 381–384 (2013)
3. Protzmann, R., Massow, K., Radusch, I.: On performance estimation of prefetching algorithms for streaming content in automotive environments. In: 11th Wireless On-Demand Network Systems and Services (WONS) Annual Conference, p. 147 (2014)

4. Zhao, X., Tian, D.: The architecture design of streaming media applications for android OS. In: 3rd IEEE International Conference Software Engineering and Service Science (ICSESS), pp. 280–283 (2012)

5. De Cicco, L., Mascolo, S.: An adaptive video streaming control system: modeling, validation, and performance evaluation. IEEE/ACM Transactions on Networking **22**(2), 526–539 (2014)

6. Zambelli, A.: IIS smooth streaming. Technical overview, Microsoft Corporation (2009)

7. Muüller, C., Renzi, D., Lederer, S., Battista, S., Timmerer, C.: Using scalable video coding for dynamic adaptive streaming over HTTP in mobile environments. In: 20th IEEE European Signal Processing Conference (EUSIPCO), pp. 2208–2212 (2012)

8. Cisco, San Jose, CA, USA.: Cisco Visual Networking Index: Forecast and methodology 2013–2018. White Paper (2014)

9. YouTube, San Bruno, CA, USA.: Getting Started with YouTube APIs and Tools

10. Akhshabi, S., Begen, A., Dovrolis, C.: An experimental evaluation of rate-adaptation algorithms in adaptive streaming over HTTP. In: Proc. ACM MMSys., pp. 157–168 (2011)

11. Sodagar, I.: The MPEG-DASH standard for multimedia streaming over the internet. IEEE Multimedia **18**(4), 62–67 (2011)

12. Vanam, R., Kerofsky, L.J., Reznik, Y.A.: Perceptual pre-processing filter for adaptive video on demand content delivery. In: IEEE International Conference on Image Processing (ICIP), pp. 2537–2541 (2014)

13. Youtube Search Engine. https://www.youtube.com

14. Ubuntu Sever. http://www.ubuntu.com

15. Garrido-Cantos, R., De Cock, J., Martinez, J., Van Leuven, S., Cuenca, P., Garrido, A.: Scalable video transcoding for mobile communications. Web of Science on Telecommunication Systems **55**(2), 173–184 (2014)

16. Choi, H., Lee, II. K., Bae, S.-J., Kang, J.W., Yoo, J.-J.: Performance evaluation of the emerging scalable video coding. In: International Conference Digest on Technical Papers on Consumer Electronics, pp. 1–2 (2008)

17. Liu, X.: Study on the packet transmission method of audio and video. In: International Conference on Electronic and Mechanical Engineering and Information Technology, pp. 189–192 (2011)

18. Herranz, L.: Integrating semantic analysis and scalable video coding for efficient content-based adaptation. Multimedia Systems **13**(2), 103–118 (2007)

Modeling and Monitoring of RTP Link on the Receiver Side

Andrey Borisov$^{(\boxtimes)}$, Alexey Bosov, and Gregory Miller

Department of Information Technologies in Control, Federal Research Center
"Computer Science and Control" of Russian Academy of Sciences,
44/2, Vavilova Str., 119333 Moscow, Russia
{ABorisov,ABosov,GMiller}@ipiran.ru
http://www.ipiran.ru

Abstract. The paper presents a new mathematical model of a link carrying by the Real Time Transport Protocol. The model attempts to meet the key features of the real link functioning like the frame delays, losses, bursting reception etc. The proposed approach is based on the Hidden Markov concept. The unobservable state is assumed to be a finite-dimensional Markov process. The observation is a non-Markovian multivariate point process that indicates heterogenous frames reception. The paper also contains the formulation and solution to the filtering problem of the hidden link state given the observable multivariate point process. Proposed link model validity and filtering algorithm performance are illustrated by processing of captured real video streams delivered via 3G mobile network.

Keywords: RTP link · Packet delay variation · Hidden markov model · Multivariate point process · Optimal state filtering

1 Introduction

The mathematical models of communication channels functioning are the subjects of extensive research during the last half of the century. On the one hand, a consistent model of a real phenomenon serves a powerful tool for its detailed analysis and further hardware and software development/optimization. On the other hand, new communication devices and transport protocols appearance leads to mathematical models design and improvement.

In general, any mathematical model developed should fit the nature of the link functionality, including the non-stationary packet sending/receiving processes and the random sequences of the packet delays, permutations and losses.

The classical models [1] and [2] had been initially developed to describe bit losses in the bursting phone network channels, but later the model applicability was extended rather successfully to the lossy computer networks governed by the TCP/IP protocols. The models were ideally simple, nevertheless they reflected few essential features of the TCP/IP networks.

© Springer International Publishing Switzerland 2015
S. Balandin et al. (Eds.): NEW2AN/ruSMART 2015, LNCS 9247, pp. 229–241, 2015.
DOI: 10.1007/978-3-319-23126-6_21

To take into account more subtle network phenomena the authors attempted to involve the structure and functioning discipline of the real physical queueing systems/networks, processing the packet flows. These investigations led to utilization of the finite-state Markov chain framework [3], [4], [6] and [7]. Enlargement of the dimensionality in these models and investigation of the heavy traffic conditions caused appearance of the fluid and diffusion approximations [8], [9], [10] and [11]. The attempts of partial or complete rejection of the Markov property led to the usage of selfsimilar processes or hidden Markov models (HMM) [12], [13], [14], [15] and [16].

In the paper we present a new mathematical model of the link functioning under the Real-Time Transport Protocol (RTP), which is a basebone protocol of VoIP and IPTV services. In our opinion, it has some specificities which should be taken into account.

First, the RTP is an unreliable protocol, i.e. the receiver gets the packet stream with possible waste, bursting reception and disordering/permutation.

Second, logged packet reception instants represent important statistical information concerning the packet delay characteristics corresponding to the different channel states.

Third, the real data indicate that the inter-arrival times of the packet receptions are non-exponentially distributed in any link state, i.e. the packet flow is not Markovian one [21] and [20]. It is also obvious, the current link state can not be observed exactly, because it is a function of the current status of hard- and software forming and serving the channel, corresponding queues lengths and so on. At the same time non-anticipation of the link functioning looks rather natural. So, in the paper we offer to use the HMM ideology. Namely, the current link state is supposed to be unobservable (hidden) finite-state Markov process. The available observation is a sequence of pairs "the reception instant — the packet header". However, the "raw" observations are too redundant and low informative for the estimation and/or identification purposes. Nevertheless, they can be preprocessed in the on-line manner to extract the information concerning the link state. The unit of a video stream is a single frame, so in our opinion the "refining" procedure should aggregate "raw" logging data into the frame logging. Further, we suggest to consider the resulted observable sequence of pairs "frame reception instant — frame status" as a multivariate point process (MPP) [17]. Thus, the proposed link model belongs to the class of partially observable stochastic dynamic systems with the obvious mathematical framework of the stochastic analysis to solve the observation/identification and control problems.

The paper is organized as follows. Section 2 contains both the verbal description of the RTP stream receiving process and its mathematical model in terms of the partially observable stochastic differential systems. The preprocessing procedure, transforming the packet reception stream into the frame reception sequence, is also described. Section 3 presents an on-line monitoring problem of the link state as a filtering problem. The latter could be interesting as an auxiliary routine in a parameter identification procedure or for the purposes of the control with incomplete information. The general case of the filtering algorithm including the detailed proof is given in [18], and the discussion of the model and filtering algorithm usage in the queueing networks is proposed in [19]. Section 4 illustrates the presented results. It demonstrates good fitness of suggested model to the real statistical data obtained from the Linphone VoIP service communicated via the 3G mobile network. The concluding remarks are given in Section 5.

2   Verbal Description and Mathematical Model

We suppose the RTP link functioning can be described by a stochastic dynamic observation system having two components. The first one is the link state itself. It is a function of multitude of random parameters: physical topology of the channel, statuses and characteristics of the communication hardware supporting the channel, firmware states and embodied algorithms of the data transmission. This is a base of network queueing management and congestion control. The most important factor is the link resource sharing: the communication hardware is utilized simultaneously by great number of users. Hence the current "external" channel workload significantly affects the link state and serves as an additional source of uncertainty. All the mentioned parameters and factors, which determine the link state, are unobservable, hence the link can be treated as an unobservable (hidden) random process. At the same time, a great number of independent homogenous users, simultaneously accessing the link resources, provide aftereffect absence for the link state process. So, the hypothese of Markovianity for the state process is rather likely.

In this particular case we suppose that the link state is a homogenous Markov process with the initial distribution p_0, intensity matrix Λ and three possible states $\{e_1, e_2, e_3\} \triangleq \mathbb{S}^3$. Here the notion e_j stands for the j-th unit vector of the Euclidean space $\mathbb{R}^3$. The value e_1 is assigned to the link state *"free"*, e_2 — for the state *"moderate load"* and e_3 — for the state *"heavy traffic and/or link failure"*. The choice of unit vectors for the formal state designation gives a possibility to represent X_t as a solution to the stochastic equation

$$X_t = \xi + \int_0^t \Lambda^\top X_s ds + M_t^X, \tag{1}$$

where M_t^X is a martingale. The objects (p_0, Λ) are supposed to be known or could be identified given the statistical data as described in the next section.

The second component of the system represents the observation scheme. The initial form of statistical information is the sequence of pairs "instant of packet reception — received packet header". The statistics has some distinctive features. First, by contrast with the TCP the RTP packet loss can not be registered directly. Second, the RTP packet flow is nonordinary: most of the VoIP and IPTV server software assigns the same timestamp to all of packets forming a frame, and send them simultaneously. This property is inherited somehow by the reception flow. Additionally, the sending device uses some low-level protocols to get additional information concerning the link quality and/or failure for subsequent adaptation of the sending procedure. During the disconnect or low-quality connection periods the RTP packets are accumulated in the buffer to be sent just after the link improvement. In this situation packets can be registered on the receiver simultaneously, and even in wrong order.

Finally, we can conclude that the link behaviour towards the playback performance is determined by the uniformity of the frames arrival and their quality. So, the statistical information in the initial form is something redundant and could be preprocessed to extract the frame events. The result of the preprocessing is an MPP, i.e. a sequence of pairs $\{(\tau(n), Y(n))\}_{n \in \mathbb{N}}$, where $\tau(n)$ is a random instant of the n-th observable event, and $Y(n)$ is the type of the event. The "alphabet" of feasible events is $\{f_1, f_2, f_3, f_4\} \triangleq \mathbb{S}^4$. Here the notion f_k stands for the k-th unit vector of the Euclidean space $\mathbb{R}^4$. The value f_1 is assigned to the event *"large frame received"*, f_2 — for the event *"small frame received"*, f_3 — for the event *"two or more received frames registered simultaneously"* and f_4 — for the event *"lossy frame received"*. The events of the first three types are registered in the obvious on-line manner at the last moment of the corresponding frame packet reception. The loss in the specific frame is detected, if at least one packet of the frame has not been received until the next frame marked packet reception. Nevertheless, the moment of the lossy frame registration is determined by the instant of the last frame packet received.

The observation in the form of the MPP $\{(\tau(n), Y(n))\}_{n \in \mathbb{N}}$ is equivalent to the following continuous-time process with the counting components

$$Y_t \triangleq \sum_{n \in \mathbb{N}} Y(n)\mathbf{I}(t - \tau(n)). \tag{2}$$

Here $\mathbf{I}(\cdot)$ is the Heaviside step function continuous from the right. The components of the 4-dimensional process Y_t are counting ones, and each component indicates the total number of specific events, occurred during the interval $[0, t]$.

The following conditional probability density functions $\pi_{jk\ell}(\cdot)$ are supposed to be known or could be identified preliminarily:

$$\mathbf{P}\{\tau(n) \leqslant t, Y(n) = f_j | \tau(n-1) = s, Y(n-1) = f_k, X_u \equiv e_\ell, u \in [s, t]\} =$$

$$= \mathbf{I}(t-s) \int_s^t \pi_{jk\ell}(u-s)du. \tag{3}$$

It turns out that the process Y_t admits the following representation

$$Y_t = \int_0^t \phi(\omega, u)X_{u-}du + M_t^Y, \tag{4}$$

where M_t^Y is a martingale and $\phi(\omega, u) : \Omega \times \mathbb{R}_+ \to \mathbb{R}^{4 \times 3}$ is a predictable matrix-valued function of the instantaneous intensity. The components of the matrix $\phi(\cdot) = \|\phi_{j\ell}(\cdot)\|$ have nice probabilistic interpretation: for any $n \in \mathbb{N}$

$$\phi_{j\ell}(\omega, t)dt =$$

$$= \mathbf{P}\left\{\tau(n) \in [t, t+dt), Y(n) = f_j \big| \{(\tau(i), Y(i))\}_{i=1}^{n-1} \vee \{X_{t-} = e_\ell\} \vee \{\tau(n) \geqslant s\}\right\}.$$

The intensity matrix can be calculated via the pdfs $\pi_{jk\ell}(\cdot)$ (here and further we set $\tau_0 \triangleq 0$):

$$\phi(\omega, t) = \sum_{n \in \mathbb{N}} \mathbf{I}_{(\tau(n-1), \tau(n)]}(t) \varphi(\tau(n-1), Y(n-1), t), \tag{5}$$

where $\varphi(\cdot) = \|\varphi_{j\ell}(\cdot)\|$ is the matrix-valued function of appropriate dimensionality with the components

$$\varphi_{j\ell}(\tau(n-1), Y(n-1), t) \triangleq$$

$$\triangleq \sum_{k=1}^{4} f_k^{\mathsf{T}} Y(n-1) \pi_{jk\ell}(t - \tau(n-1)) \left(\sum_{i=1}^{4} \int_t^{+\infty} \pi_{ik\ell}(u - \tau(n-1)) du \right)^+ .$$

The adequacy of the presented model to the real RTP link functioning will be illustrated in Section 4. The next section contains the statement and solution to the on-line state estimation problem.

3   State Filtering Problem

The mathematical model of the link state in the form of stochastic observation system (1), (4) allows to solve a variety of the system analysis problems. Moreover, the effective and powerful martingale framework gives a possibility to approach the optimal or robust control problem given the incomplete noisy observations. The first step to do this is to solve the state filtering problem.

Let us remind that the state filtering problem is to find the conditional mathematical expectation (CME) of the state given all the observations available till the moment t: $\widehat{X}_t \triangleq \mathbf{E}\{X_t | Y_{[0,t]}\}$.

In fact the CME is a key to the solution of the whole class of different state filtering problems, since it is optimal with respect to various criteria. First of all,

$$\widehat{X}_t = \mathbf{E}\{X_t | Y_{[0,t]}\} = \sum_{j=1}^{3} e_j \mathbf{P}\{X_t = e_j | Y_{[0,t]}\},$$

i.e. $\widehat{X}_t$ defines the conditional distribution of X_t explicitly: j-th component of $\widehat{X}_t$ equals to the conditional probability that $X_t = e_j$ given the observations $Y_{[0,t]}$.

Furthermore, let $\overline{X}_t = \overline{X}_t(Y_{[0,t]})$ be an arbitrary estimate of the state X_t, such that $\overline{X}_t(Y_{[0,t]})$ is a measurable function of the available observations, and the conditions of non-negativity $e_j^{\mathsf{T}} \overline{X}_t \geqslant 0$ ($j = 1, 2, 3$) and normalisation $1_3 \overline{X}_t = 1$ hold $\mathbf{P}$-a.s. Here 1_3 is a three-dimensional row-vector formed by units. We denote the class of similar estimates by $\mathfrak{X}_t$ and consider the nonnegative lower-semicontinuous risk function $\rho(\cdot) : \mathbb{R}^3 \to \mathbb{R}$, such that $\rho(0) = 0$.

The estimate $\widetilde{X}_t \in \mathfrak{X}_t$ is optimal with respect to the risk function ρ, if $\widetilde{X}_t \in \underset{\overline{X}_t \in \mathfrak{X}_t}{\mathrm{Argmin}} J(\overline{X}_t)$, where $J(\overline{X}_t) \triangleq \mathbf{E}\{\rho(\overline{X}_t - X_t)\}$ is an optimality criterion.

Using the properties of the CME we can obtain that

$$J(\overline{X}_t) = \mathbf{E}\left\{\mathbf{E}\left\{\rho(\overline{X}_t - X_t)|Y_{[0,t]}\right\}\right\} = \mathbf{E}\left\{\sum_{k=1}^{3}\rho(\overline{X}_t - e_k)\mathbf{P}\{X_t = e_k|Y_{[0,t]}\}\right\}$$

$$= \mathbf{E}\left\{\sum_{k=1}^{3}\rho(\overline{X}_t - e_k)e_k^\top \widehat{X}_t\right\} = \mathbf{E}\left\{r(\overline{X}_t)\widehat{X}_t\right\},$$

where $r(\overline{X}_t) \triangleq \mathrm{row}(\rho(\overline{X}_t - e_1), \ldots, \rho(\overline{X}_t - e_3))$. Hence,

$$\min_{\overline{X}_t \in \mathfrak{X}_t} J(\overline{X}_t) = \mathbf{E}\left\{\min_{\overline{X}_t \in \mathfrak{X}_t} r(\overline{X}_t)\widehat{X}_t\right\} = \mathbf{E}\left\{\min_{\gamma \in \mathfrak{G}_t}\gamma\widehat{X}_t\right\},$$

where $\gamma \in \mathbb{R}^3$ is a row-vector belonging to the set $\mathfrak{G}_t \triangleq \{\gamma \in \mathbb{R}^3 : \exists\ x \in \mathfrak{X}_t, r(x) = \gamma\}$.

The optimal estimate $\widetilde{X}_t$ is, in fact, a solution to the generalized linear programming problem, such that the coefficients of the objective function are determined via $\widehat{X}_t$. Hence we can conclude that $\widetilde{X}_t$ should be a measurable function of the CME $\widehat{X}_t$. For example, if the risk function is the squared Euclidean norm of the estimate error i.e. $\rho(\Delta) = \|\Delta\|^2_{\mathcal{L}_2} = \Delta^\top\Delta$, or it is the $\mathcal{L}_1$-norm of Δ i.e.

$$\rho(\Delta) = \|\Delta\|_{\mathcal{L}_1} = \sum_{k=1}^{3}|e_k^\top\Delta|,$$ then one can verify easily, that $\widetilde{X}_t = \widehat{X}_t$ $\mathbf{P}$-a.s. If

the risk function is simple, i.e. $\rho(\Delta) = \mathbf{I}_{\{0\}}(\Delta)$, then $\widetilde{X}_t \in \mathrm{Argmax}_{e_1,\ldots,e_3} e_k^\top \widehat{X}_t$.

Theorem 1. *The CME $\widehat{X}_t$ is determined as the unique solution to the following stochastic system*

$$\widehat{X}_t = p_0 + \int_0^t \Lambda^\top \widehat{X}_{s-}ds + \int_0^t \left[\mathrm{diag}\,\widehat{X}_{s-} - \widehat{X}_{s-}\widehat{X}_{s-}^\top\right]\phi^\top(\omega, s-)\times$$

$$\times \left[\mathrm{diag}\left(\phi(\omega, s-)\widehat{X}_{s-}\right)\right]^{-1}\left(dY_s - \phi(\omega, s-)\widehat{X}_{s-}ds\right). \quad (6)$$

4   Numerical Experiments

The proposed numerical experiment consists of several stages described below.

1. We organize a video-conference using a VoIP service and register the corresponding video-stream receiving on some participating node. This statistical data is then used to identify the parameters of the observation system (1) and (4). It should be noted, the identification is implemented only after the preprocessing of the initial statistical data as it is mentioned in Section 2. We show good fitting of the real statistical data by the suggested model. Further, the identification procedure itself presumes the link state estimation in the off-line smoothing manner. This estimates are considered as the "ideal pattern" for the subsequent comparison.

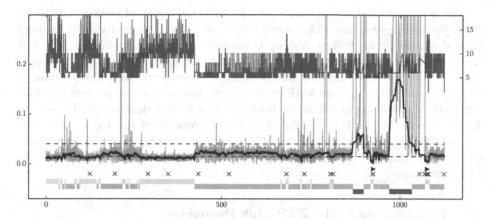

Fig. 1. The RTP session data used for the link parameters identification. Legend (from top to bottom):

- frame size F_t (thin dark gray line);
- reception speed S_t and its median $\mathrm{med}(S_t)$ (thin gray and thick black line);
- event "two or more frames received simultaneously" (triangle marks);
- event "lossy frame received" (cross marks);
- link states e_1, e_2, e_3 (solid light gray, gray and dark gray lines);
- thresholds 0.015 and 0.04 (dashed black lines).

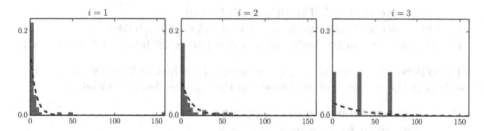

Fig. 2. Histograms for times spent in states e_i and their fitted exponential pdfs.

2. We apply the optimal filtering algorithm (6) to the data previously used for the model identification. The obtained estimates can be treated as the result of the RTP link monitoring. The quality of this on-line estimation procedure is compared with the "ideal pattern" calculated via the off-line smoothing.
3. To verify the statistical stability of the model (1) and (4), and preservation of the identified parameter values, we repeat registration of another video-stream under the similar experiment conditions. Using the same pre-processing procedure, we find the smoothed "ideal pattern" estimates.
4. Finally, we solve the state filtering problem for the new statistical data using the parameter values calculated by the previous statistics recorded on Stage 1. Good performance of the filtering estimates approves both the model (1), (4) quality and high accuracy of the filtering estimate (6).

For our numerical experiments we had collected series of packet data from video streaming sessions. Video sequences were being streamed via 3G channel from a mobile device to the desktop computer, where the incoming traffic was captured, parsed in order to extract the particular video session and then processed with respect to the mathematical model developed. As a video streamer we had chosen the Linphone VoIP phone, since this open source project provides a variety of client applications for both the mobile and desktop platforms and does not encrypt the RTP traffic. The latter property is the most important for our studies, since it makes the RTP traffic processing clear: we can easily separate the particular RTP sessions from each other and distinguish the audio and video sequences.

4.1   Identification of the RTP Link Parameters

In order to identify the RTP link parameters we take the data of approximately 20 minutes long RTP session (Fig. 1). As the indicator of the link state we choose the reception speed S_t, i.e. the mean time between packets reception within a frame. This process still has rapid oscillations caused by the high-frequency nature of the digital communication channels. To extract the long-term component of the state responding to the model (1) we have to smooth the speed process, calculating its median $\mathrm{med}(S_t)$, and then use the following state selection procedure:

- e_1 – link state is *"free"*, if $\mathrm{med}(S_t) \in [0, 0.015)$;
- e_2 – link state is *"moderate load"*, if $\mathrm{med}(S_t) \in [0.015, 0.04)$;
- e_3 – link state is *"heavy traffic and/or link failure"*, if $\mathrm{med}(S_t) \in [0.04, +\infty)$.

The observations depend on the frame size F_t, which is literally the number of packets in the frame, and are chosen in the manner described earlier:

- f_1 – event *"large frame received"*: $F_t \geq 8$;
- f_2 – event *"small frame received"*: $F_t < 8$;
- f_3 – event *"two or more frames received simultaneously"*,
- f_4 – event *"lossy frame received"*.

The observation f_1 refers more likely to the state e_1, f_2 does to e_2, meanwhile the observations f_3 and f_4 appear mostly when the link is in the state e_3. The point is, the version of the sender software for the mobile devices uses a low-level hardware protocol to identify temporary mobile link failures. The software acts to prevent the packet losses if possible, and accumulates the frame packets in its buffer to send them simultaneously as soon as the link restores.

With respect to these data the intensity matrix can be identified as follows:

$$\Lambda = \begin{pmatrix} -2.971 \cdot 10^{-3} & 2.971 \cdot 10^{-3} & 0 & 0 \\ 1.364 \cdot 10^{-3} & -1.442 \cdot 10^{-3} & 7.867 \cdot 10^{-5} & 0 \\ 0 & 1.11 \cdot 10^{-3} & -1.11 \cdot 10^{-3} & 0 \\ 0 & 9.11 \cdot 10^{-3} & 0 & -9.11 \cdot 10^{-3} \end{pmatrix}$$

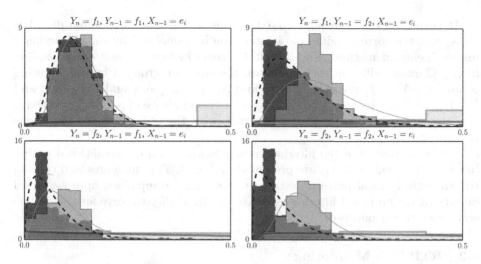

Fig. 3. Estimated conditional pdfs $\pi_{jk\ell}(\cdot)$ (3). Legend:

- $X_{n-1} = e_1$: histogram inner color dark gray, estimated pdf dashed line;
- $X_{n-1} = e_2$: histogram inner color gray, estimated pdf thin line;
- $X_{n-1} = e_3$: histogram inner color light gray, estimated pdf thick line.

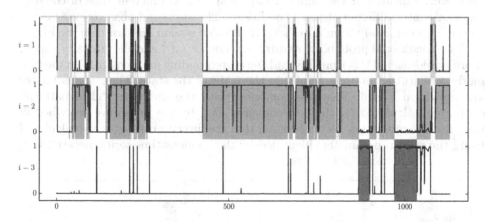

Fig. 4. The filtering estimates $\mathbf{P}\{X_t = e_j | Y_{[0,t]}\}$ for $j = 1, 2, 3$ (thin black line) in comparison with the smoothed "ideal pattern" estimates (different gray shades).

The histograms of the duration for each state are shown on Fig. 2 in comparison with the corresponding exponential approximations. Visually, the exponential pdfs explain the real data rather successfully on the 1st and 2nd state. The failure of the 3rd state duration time identification can be explained by too short sample available. Small discrepancy of the 1st histogram and its exponential approximation is not surprising: similar "tail lacks" often take place in various applied pdf fitting problems.

It turns out that the pdfs $\pi_{jk\ell}(\cdot)$ are visually similar to Gamma distributions, and the corresponding parameters can be calculated by the classic maximum likelihood method. Fig. 3 contains some histograms and corresponding fitting Gamma pdfs. On each subfigure the plots are grouped by the following conditions: $Y_n = f_j$ given $Y_{n-1} = f_k$ and $X_{n-1} = e_\ell$ (i.e. k and j are fixed and $\ell = 1, 2, 3$). The plots allow to check good correspondence of pdfs estimates with the histograms, and compare various pdfs with the different state conditions $X_{n-1} = e_\ell$.

The components of the filtering estimates which are the conditional probability of the event $X_t = e_\ell$, are presented on Fig. 4. The subfigures also contain the smoothed "ideal pattern" estimate. The visual comparison approves good quality of the proposed filtering estimate and its ability to serve for the on-line state monitoring purposes.

4.2   RTP Link Monitoring

The identified model can further be used for the online RTP link monitoring. In this section we provide the results of another RTP session state estimation. In order to eliminate the device and software peculiarities the session data was obtained in the similar setting as the previous one. Nevertheless, since the sessions were captured at the different days and even at different time of the day, one can expect different channel conditions. The state and observations for the second approximately 30 minutes long RTP video session are shown on the Fig. 5.

The conditional probability density functions $\pi_{jk\ell}(\cdot)$ and the intensity matrix Λ are not identified this time. Instead the corresponding parameters from the first model are utilized in order to obtain the CME of the state. The state estimates are shown on Fig. 6. They are compared with the smoothed "ideal pattern" estimates, calculated for the new captured data by the procedure described in the previous subsection. We note again that the filtering estimates are calculated using the system parameters identified by the previous time series, nevertheless,

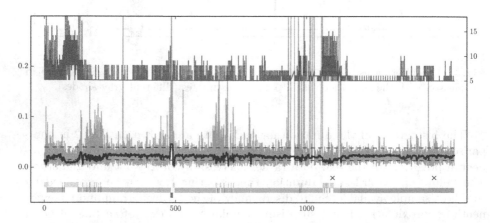

Fig. 5. The RTP session data used for channel state estimation.

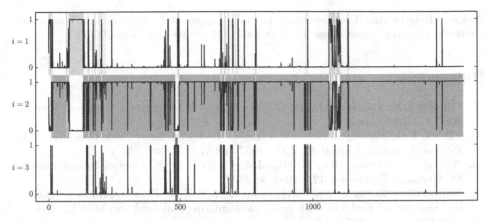

Fig. 6. The filtering estimates $\mathbf{P}\{X_t = e_j | Y_{[0,t]}\}$ for $j = 1, 2, 3$ (thin black line) in comparison with the smoothed "ideal pattern" estimates (different gray shades).

the comparison with the "ideal pattern" demonstrates rather high performance of the filtering estimate.

5 Conclusion

We consider the paper contribution as following. First, we propose a new phenomenological description of the RTP link functioning. The model is based on the partially observable dynamic system concept. The unobservable (hidden) link state is treated as a finite-dimensional Markov jump process, meanwhile the observable frame reception flow is supposed to be some MPP. The suggested model has obvious advantages and is supported by the powerful mathematical framework of the stochastic analysis. Second, the proposed preprocessing procedure, transforming the "raw" packet flow into the "refined" frame one, turns out to be successful both for the subsequent off-line identification of the link parameters and for the on-line state filtering. Third, the suggested choice of both the state and observation "alphabets" demonstrates the good quality of real data fitting: the corresponding histograms are unimodal and can be approximated by the well-known Gamma distribution.

Nevertheless, in our opinion, the investigation is not completed. The presented model needs to be verified comprehensively. Further, the model should be developed to take into account additional available information concerning the network topology, hard- and software algorithms functioning in the link, available RTCP packet information etc. On the other hand, the stable robustified versions of the model are encouraged. It is also necessary to design the unified procedure of simultaneous model parameters identification/tuning and the state filtering. Finally, the statement and solution to the optimal and/or robust control problems in these models with incomplete information are appraised traditionally as the "crown" of each investigation.

Acknowledgments. The research is partially supported by the Russian Foundation for Basic Research (grants Nos. 13-01-00406, 13-07-00408 and 15-37-20611).

References

1. Gilbert, E.N.: Capacity of a Burst-Noise Channel. Bell System Technical Journal **39**, 1253–1265 (1960)
2. Elliott, E.O.: Estimates of Error Rates for Codes on Burst-Noise Channels. Bell System Technical Journal **42**, 1977–1997 (1963)
3. Altman, E., Avrachenkov, K., Barakat, C.: TCP in Presence of Bursty Losses. Performance Evaluation **42**, 129–147 (2000)
4. Miller, B., Avrachenkov, K., Stepanyan, K., Miller, G.: Flow control as stochastic optimal control problem with incomplete information. In: Proc. of IEEE INFO-COM, pp. 1328–1337 (2005)
5. Bruno, R., Conti, M., Gregori, E.: Throughput Analysis and Measurements in IEEE 802.11 WLANs with TCP and UDP Traffic Flows. IEEE Trans. on Mob. Comput. **7**(2), 171–186 (2008)
6. Haßlinger, G., Hohlfeld, O.: The Gilbert-Elliott model for packet loss in real time services on the Internet. In: Proc. of the 14th GI/ITG Conference on Measurement, Modelling and Evaluation of Computer and Communication Systems (MMB), pp. 269–283 (2008)
7. Malik, M., Aydin, M., Shah, Z., Hussain, S.: Stochastic Model of TCP and UDP traffic in IEEE 802.11b/g. In: Proc. of the IEEE 9th Conference on Industrial Electronics and Applications (ICIEA), pp. 2170–2175 (2014)
8. Whitt, W.: Stochastic-Process Limits: An Introduction to Stochastic-Process Limits and their Application to Queues. Springer, New York (2002)
9. Bohacek, S.: A stochastic model of TCP and fair video transmission. In: Proc. of IEEE INFOCOM, vol. 2, 1134–1144 (2003)
10. Domańska, J., Domański, A., Czachórski, T., Klamka, J.: Fluid Flow Approximation of Time-Limited TCP/UDP/XCP Streams. Bulletin of the Polish Academy of Sciences: Technical Sciences **62**(2), 217–225 (2014)
11. Liu, Y., Gong, W.: On Fluid Queueing Systems with Strict Priority. IEEE Trans. Autom. Contr. **48**(12), 2079–2088 (2003)
12. Yariv, E., Merhav, N.: Hidden Markov Processes. IEEE Trans. Inform. Th. **48**(6), 1518–1569 (2002)
13. Anisimov, V.: Switching Processes in Queueing Models. Wiley, New York (2008)
14. Ellis, M., Pezaros, D.P., Kypraios, T., Perkins, C.: Modelling packet loss in RTP-based streaming video for residential users. In: Proc. of the 37th Conference on Local Computer Networks, pp. 220–223 (2012)
15. Borisov, A., Miller, G.: Hidden markov model approach to TCP link state tracking. In: Proc. of the 43th CDC, vol. 4, pp. 3726–3731 (2004)
16. Tsybakov, B., Georganas, N.: Overflow and Losses in a Network Queue with a Self-similar Input. Queueing Systems **35**(1–4), 201–235 (2000)
17. Liptser, R., Shiryayev, A.: Theory of Martingales. Kluwer, London (1989)
18. Borisov, A.: Monitoring Remote Server Accesability: the Optimal Filtering Approach. Informatics and Applications **8**(3), 53–69 (2014)
19. Borisov, A.: Partially observable multivariate point processes with linear random compensators: analysis and filtering with applications to queueing networks. In Proc. of the 1st IFAC Conference on Modelling, Identification and Control of Nonlinear Systems (MICNON), pp. 1119–1124 (2015)

20. Paxson, V., Floyd, S.: Wide-Area Traffic: The Failure of Poisson Modelling. IEEE/ACM Transactions on Networking **3**(3), 226–244 (1995)
21. Kaj, I., Marsh, I.: Modelling the arrival process for packet audio. In: Proc. of Quality of Service in Multiservice IP Networks, pp. 35–49 (2003)

Suitability of MANET Routing Protocols for the Next-Generation National Security and Public Safety Systems

Pavel Masek[1](✉), Ammar Muthanna[2], and Jiri Hosek[1]

[1] Brno University of Technology, Technicka 3082/12, 61600 Brno, Czech Republic
xmasek12@phd.feec.vutbr.cz, hosek@feec.vutbr.cz
[2] State University of Telecommunication, Pr. Bolshevikov, 22, St. Petersburg, Russia
ammarexpress@gmail.com

Abstract. One of new domains where the D2D communication can be applied is the Public Protection and Disaster Relief (PPDR) and National Security and Public Safety (NSPS) services. The key requirement for employing the PPDR and NSPS services is to provide the access to the communication services even if cellular network is highly overloaded or become fully dysfunctional due to some public disaster or emergency situations. In such case, the most important task is to share emergency information between citizens even when most of them do not have infrastructural connectivity. Since D2D is by default a single-hop communication, it is crucial to find the mechanism how to transmit any data via randomly assembled network even for longer distances. Therefore, this paper focuses on the utilization of MANET routing protocols for D2D communication in case that the connection to the cellular network is down. The selected routing protocols used in MANET are described and their suitability is verified in the simulation environment Network Simulator 3 (NS-3).

Keywords: AdHoc · Cellular network · D2D · LTE · MANET · WiFi-Direct

1 Introduction

The original idea of Device-to-device (D2D) communication was to fulfill the commercial use cases, in which real-time information needs to be exchanged between two (or more) mobile devices in proximity to each other [1], [2]. D2D communication can be implemented in two modes. First, the so-called *overlaid* mode describes the case where D2D is used in licensed spectrum and is transparent for the end users [1]. On the other side, when D2D operates in *underlay* mode, the eNodeB (evolved NodeB) controls the operations between end users using D2D communication by the maintaining and controlling plane association [2]. The advantages of D2D communication in comparison with the traditional cellular method (direct connection / communication between the eNodeB and end

© Springer International Publishing Switzerland 2015
S. Balandin et al. (Eds.): NEW2AN/ruSMART 2015, LNCS 9247, pp. 242–253, 2015.
DOI: 10.1007/978-3-319-23126-6_22

device) are better proximity usage, and improved spectral and energy efficiency of the whole communication system [3].

In recent years, several research works have proposed the integration of the IEEE 802.11 standard (short-range communication) and AdHoc / Mobile AdHoc NETwork (MANET) networking in cellular network [4], [5], [6]. From these papers could be seen that the short-range communication can take advantage of the cellular control layer in spreading of multimedia content between end users relying on unlicensed spectrum [4]. As a result, AdHoc mobile nodes can increase the system capacity, reduce the transmission power (for mobile nodes) and improve the system coverage at the same time [5]. Also the spectrum sharing schemes can allow D2D users to opportunistically transmit data via the alternative radio interface; keep the interference level within the specified boundaries can result in significant power saving [6].

Based on above mentioned facts, direct D2D communication is expected to be main driver for the future communication between devices. Recognizing the importance of public safety and the need for the NSPS and PPDR types of services, the 3rd Generation Partnership Project (3GPP) started to study the scenarios, the requirements and the technology enablers for given emergency applications [7]. Communication for NSPS and PPDR has a number of specific requirements; one of the most important of them is an ability to set up the communication between devices regardless of the presence / absence of a fixed cellular infrastructure [8], see Fig. 1.

In this paper we review the MANET routing protocols and their usability as drivers for direct communication between end devices to extend the communication range beyond the standard one-hop D2D. As the first, we implemented two scenarios in simulation environment Network Simulator 3 (NS-3) where the attention is paid to the situation with and without the infrastructural connection to the cellular network. Then we simulate the data transmission between end devices using the MANET routing protocols. Namely, we used the AdHoc On-Demand Distance Vector (AODV) and Destination-Sequenced Distance-Vector Routing (DSDV) routing protocols as frequently used mechanisms.

The rest of this paper is organized as follows. Section 2 describes the common routing protocols in AdHoc networks as a way how to communicate directly between end devices. In the Section 3 we introduce the created simulation scenarios (with and without the connection to the eNodeB). Section 4 presents the obtained results and finally, in Section 5, we draw conclusion together with our future plans in this research area.

2   D2D Communication

Standardization of 5th generation (5G) mobile networks will be completed around 2020. Compared to other networks such as 4G, 5G network will provide larger bandwidth which is approximately 1000 times larger and its spectral efficiency will be 10 times higher than that of 4G (LTE-A) networks. The purpose of 5G systems is to bring a ubiquitous connectivity for people

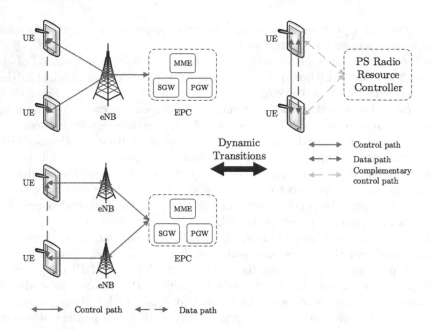

Fig. 1. The main goal for NSPS and PPDR is to provide dynamic transitions between situations where the cellular network connection is (left) / is not available (right). If the cellular connection is not available, the UEs should be able to perform Radio Resource Management (RRM) and ensure the lower service degradation if the cellular network becomes nonfunctional. In case, when the public safety radio resource controller (RRC) is implemented (right) an complementary control path is established to manage radio resources for the D2D communication [9].

and machines: e.g Human-to-Human (H2H), Human-to-Machine (H2M) and Machine-to-Machine (M2M). There is a number of advanced technologies which can provide new possibilities for 5G networks such as D2D communication, M2M communication in millimeter (MMW) frequency range, Multiple Input Multiple Output (MIMO) signal transmission, program-controlled Software Defined Networks (SDN), Heterogeneous Networks (HetNets), etc. A lot of these technologies are based on the ideas of self-organizing networks. D2D communication, fundamentally changed the composition architecture of the base station zone from the existing radial to mesh.

LTE technology and D2D together enable an improved state of affairs with the acute problems facing information security. In the light of the aforementioned, in LTE-A Rel. 12 introduces the concept of proximity based services (ProSe) to users located close to each other [10] using the D2D standard. The recently introduced Rel. 13 considers two ProSe scenarios: the direct transfer of data when two devices communicate without any network element and local communication when D2D user equipment communicates with local networks using relay through the control unit without the basic elements of the network [10]. It is obvious that in the second case the information security problems can be solved at the level of local network. The first scenario, of course, requires further research in the subject area.

Based on the considered business models, operator can have different levels of control of communication link or prefers not to have any. When having control, operator can either exercise full/partial control over the resource allocation among source, destination, and relaying devices. In total, there are four main types of D2D architectures, presented in Fig. 2 [3]:

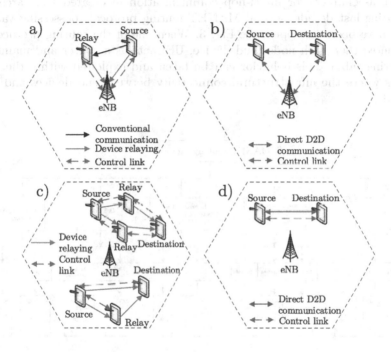

Fig. 2. D2D communication types [3].

- Fig. 2 a) - A device at the edge of a cell or in a poor coverage area can communicate with the BS by relaying its information via other devices. This allows the device to achieve a higher QoS or more battery life. The operator communicates with the relaying devices for partial or full control link establishment.
- Fig. 2 b) - The source and destination devices talk and exchange data with each other without the need for a BS, though they are assisted by the operator for link establishment.
- Fig. 2 c) - The operator is not involved in the process of link establishment. Therefore, source and destination devices are responsible for coordinating communication, using relays between each other.

– Fig. 2 d) - The source and destination devices have direct communication with each other without any operator control. As such, source and destination devices use the resource in such a way as to ensure limited interference with other devices in the same tier and the macro-cell tier.

2.1   MANET Routing Protocols

Following the information given in the Section 2, the MANET routing protocols can serve as enablers for multi-hop communication in created D2D architecture. During last decade, several MANET routing protocols possessing various features have been developed, see Fig. 3. When one of the routing protocols is implemented then each node (end device; UE) acts as a router and maintains own routing table. This behavior can be taken and exploited within the D2D scenarios where the infrastructural connectivity between end devices and eNB is limited.

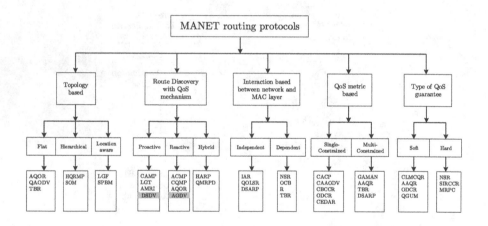

Fig. 3. Overview of available routing protocols introduced during last decade.

From the available routing protocols, the AODV and DSDV were chosen as well-recognized reliable solutions. The AODV represents the *reactive / on-demand* routing protocol. In this case, the information about the possible paths between end devices is not available until there is any transmission request in the network. The second routing protocol (DSDV) maintains routing table on each node for all the time and the updates of routing tables are done regularly. Therefore, this proactive approach performs better in *high-mobility* networks where the links are changing dynamically. In this paper, we implemented and evaluated representatives of both groups of MANET routing protocols (reactive and proactive) in the scenario where the connectivity to the LTE network is limited and propagation of important message is done in local network using the selected routing protocol.

3   Created Model

As it was described in previous Sections 1 and 2, we focus in this paper on
the D2D communication using the routing protocols AODV and DSDV (originally introduced for MANET networks); these protocols were chosen as popular
representatives of routing protocols in MANET networks. DSDV represents the
protocol suitable for the continuous data transfers (higher throughput, higher
delay in comparison with the AODV). AODV shows the less packet drops (in
comparison with the DSDV) [11]. As a simulation environment, the NS-3 in version 3.22 [13] was used together with the LENA framework [14] [1]. We initially
focused on the scenario with fully functional cellular network. This scenario is
depicted in Fig. 4a) where the end devices (UEs) can reach the cellular (LTE)
network and communicate with the remote hosts located in the Internet.

The second implemented scenario, see Fig. 4b), depicts the situation of public
disaster or emergency situations. In this case, the infrastructural network (concretely the Random Access Network (RAN) as part of LTE network) is highly
overloaded or completely out of order and so not able to manage connection
requests from the UEs [2]. The UEs in the second scenario were implemented
with the following parameters:

- UE maximum transmit power = 10 dBm,
- UE antenna model = Isotropic Antenna Model,
- Number of UEs = 10, 20, 30,
- UE mobility model = RandomWalk2dMobilityModel,
- Boundaries of space = 1000 x 1000 meters,
- Velocity = 1.5 m/s,
- Radio technology = IEEE 802.11 b (WiFi Direct),
- WiFi MAC mode = AdhocWifiMac (Algorithm AARF-CD).

Key parameters of the LTE network can be summarized in these points:

- Cell layout = 1 eNodeB, 1 sector,
- Duplex format = LTE-FDD,
- Maximum transmit power = 30 dBm,
- System bandwidth = 3 MHz (15 PRBs),
- Scheduler = Pf Df Mac Scheduler,
- Path loss model = Friis Spectrum Propagation Loss Model,
- eNB antenna model = Isotropic antenna model,
- Frequency reuse factor = 1.

---

[1] The simplification of the data plane protocol stack which is implemented in LENA
framework is the merge of the SGW and PGW functionality within a single
SGW/PGW node. This removes the need for the S5 a S8 interfaces specified by
3GPP. The S1-U protocol stack and the LTE radio protocol stack, all specified by
3GPP, are described in [15].

[2] We performed the simulation focused on the overloading of the RAN part in our
previous paper [16].

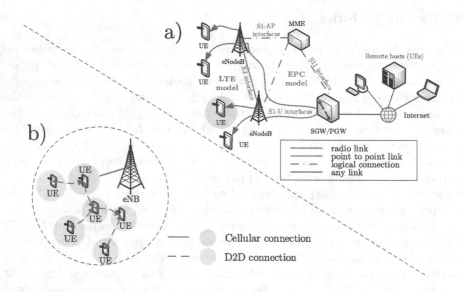

Fig. 4. a) First scenario - default (fully functional) topology of created LTE network. b) Second scenario - D2D communication in case of emergency situations.

4   Simulation Results

This paper covers two simulation scenarios. The first scenario 4.1, so called *Default LTE*, imitates the situation when the UE is connected directly to the LTE network and transmit data to the remote host (located in the Internet) via standard infra link. In comparison, the utilization of MANET protocols for D2D is evaluated in the second scenario. The results from the second scenario are described in 4.2; the latency for two routing protocols (AODV and DSDV) is discussed in relation to varying number of end devices.

4.1   Communication in LTE network

In this scenario, see Fig. 4a), the UE is connected directly to the LTE network. Data is generated using the On/Off application in NS-3 (transport protocol UDP; File size 500 B; Data rate 100 Kb/s). In this case, the static routing is used between the UE and eNB [17]. The latency measured during the data transmission is shown in Fig. 5 with the average value 22.8 ms. The obtained latency results are in line with the theoretical assumptions for LTE-TDD [18].

To evaluate the correct behavior of created traffic from UE, the trace files were created. In LTE network, the data is encapsulated by the GPRS Tunneling Protocol (GTP) which is used over the S1-U, X2, S4, S5 and S8 interfaces of the Evolved Packet System (EPS); note that the S5 and S8 interfaces are not implemented in LENA framework yet. The correct handling of the data traffic within the LTE network was verified by an analysis performed in Wireshark.

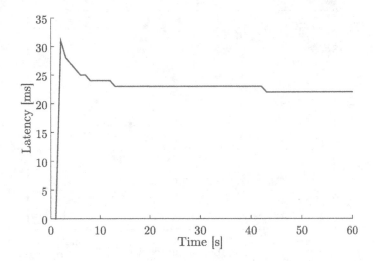

Fig. 5. Latency of transmitting data from end device to remote host

4.2   Emergency D2D Scenario

In this scenario, see Fig. 4b), the usability of MANET routing protocols for D2D communication is modeled and assessed. We performed the simulation where only one of the end devices is able to connect to the cellular network (to eNB); the rest of end devices fail to connect during the initial procedure. In this case, data is transmitted between users without assistance from cellular network. As mentioned above, AODV and DSDV routing protocols were implemented [3].

For each of both selected routing protocols we performed three different simulation runs for 10, 20 and 30 UEs. The data transmission was always set up between two the outermost UEs; following that fact, the simulation results give us the overview of the time which is needed to deliver the information to the farthest end device. The remaining nodes act as relays.

From Fig. 6 it can be seen that the average value of latency for three different scenarios with AODV protocol (containing 10, 20 and 30 end devices) is in the range from 9.54 ms to 16.20 ms. In comparison with the results for the DSDV routing protocol, see Fig. 7, the average latency for 10 end devices is 21.20 ms. In case of 20 end devices, the average latency is 40.48 ms and the average latency for last scenario (30 end devices) is 77.45 ms [4].

The obtained results show that the transmission delay of AODV protocol is lower and more stable for all scenarios which supports theoretical hypothesis

---

[3] In case of both implemented routing protocols, the default configuration in NS-3 was used.

[4] In this scenario the outlier represented by the latency 845 ms (in simulation time 0.5 second) was not take into account due to the fact, that this is caused by the behavior of simulation environment NS-3.

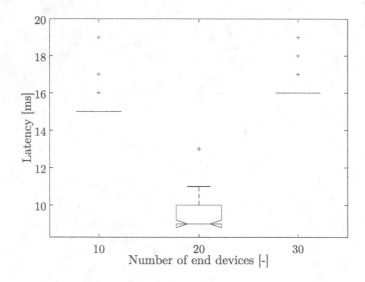

Fig. 6. AODV - Latency of D2D communication between end devices

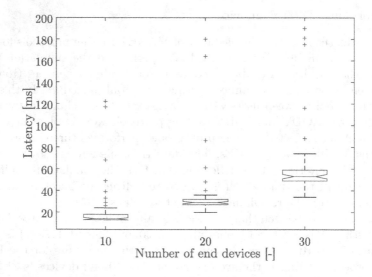

Fig. 7. DSDV - Latency of D2D communication between end devices

that AODV is more suitable for the networks with higher mobility (especially in case of large number of mobile nodes) comparing to DSDV. The main reason is a special technique to maintain routing information implemented by AODV. Despite the fact that AODV belongs to the reactive protocols, it utilizes both an on-demand and a table-driven routing approaches. It adopts flat routing tables containing one entry per destination [19]. Therefore the AODV is more efficient during the path discovery procedure and so more suitable for the applications where the attributes as end-to-end delay and packet loss are critical.

Besides the latency, we evaluated also the hop count as key metric influencing the overall efficiency of D2D transmission, especially in terms of energy consumption. Therefore we implemented into the NS-3 function for counting hops between end devices during the communication. The hop counts are shown in Fig. 8. Results show that in case of AODV, from 3 to 4 hops are required to deliver the particular message to the destination. In case of DSDV, the number of hops increases from 3 (for 10 end devices) through 4 (for 20 end devices) to 6 (for 30 end devices). The observed difference is caused mostly by different routing algorithms utilized by both protocols. The general trend of growing number of hops (evident especially in DSDV scenario) is due to an increasing number of UEs which extends the distance between source and destination node. However, the greater efficiency of AODV protocol is noticeable.

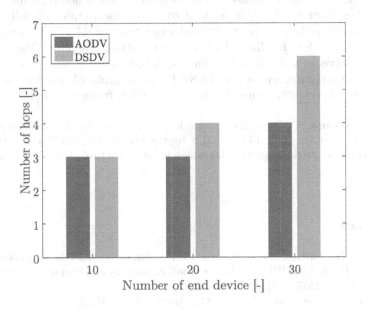

Fig. 8. Hop count for used routing protocols

5 Conclusion

Today, there is a growing interest in applying the commercial cellular networks to the field of public safety (i.e. in USA, the National Public Safety Telecommunications Council is interested in defining interoperable national standard for the NSPS; in EU, the European agencies work with the Electronic Communications Committee of the European Conference of Postal and Telecommunications Administrations to establish frequency band for PPDR).

In this paper we address the question of direct communication between end devices in situation when the connectivity to the cellular network (RAN part of LTE network) is restricted due to the public disaster or emergency situations. We performed simulations where two routing protocols originally coming from MANETs were used to route data traffic between end devices. Namely the AODV and DSDV protocols were implemented in simulation environment NS-3.

The obtained results, see Section 4, confirm that the propagation latency in case of both routing protocols is at maximum several tens of milliseconds. The overall latency in combined scenario (D2D + LTE connection to a remote host located in the Internet) is up to 450 ms for DSDV and even better results were obtained while using AODV protocol, where overall delay is up to 83 ms (both measured in 20-node scenarios). Along with the lower number of required hops, AODV protocol was recognized as suitable candidate for D2D emergency systems.

These initial results show that using the MANET routing protocols in case when the cellular network / infrastructure becomes malfunction is promising approach. On the other hand it is important to notice that obtained results give us only very rough picture of possible implementations of MANET routing protocols as enablers for the direct D2D communication. Therefore, the obtained results will serve as a background for the future more extensive simulations where we want to implement other MANET routing protocols like DSR or OLSR together with closer interconnection with the LENA framework in NS-3.

Acknowledgments. The described research was supported by the National Sustainability Program under grant LO1401 and by the project CZ.1.07/2.3.00/30.0005 of Brno University of Technology. For the research, infrastructure of the SIX Center was used.

References

1. Kaufman, B., Aazhang, B.: Cellular networks with an overlaid device to device network. In: Proc. 42nd IEEE Asilomar Conf. Signals, Syst. Comput., Pacific Grove, CA, USA, pp. 1537–1541, October 2008
2. Doppler, K., Rinne, M., Wijting, C., Ribeiro, C.B., Hugl, K.: Device-to-device communication as an underlay to LTE-advanced networks. IEEE Commun. Mag. **47**(12), 42–49 (2009)
3. Fodor, G., et al.: Design aspects of network assisted device-to-device communications. IEEE Commun. Mag. **50**(3), 170–177 (2012)
4. Fitzek, F.H.P.: Cellular controlled short-range communication for cooperative p2p networking. In: Proc. Wireless World Res. Forum (WWRF) Contrib., pp. 141–155, November 2006
5. Luo, H., Ramjee, R., Ramjee, R., Sinha, P., Li, L.E., Lu, S.: UCAN: A unified cellular and ad-hoc network architecture. In: Proc. MobiCom, San Diego, CA, USA, pp. 353–367, September 2003
6. Vinel, A., Vishnevsky, V., Koucheryavy, Y.: A simple analytical model for the periodic broadcasting in vehicular ad-hoc networks. In: 2008 IEEE Globecom Workshops, GLOBECOM 2008 (2008)

7. Brahmi, N., Venkatasubramanian, V.: Summary on preliminary trade-off investigations and first set of potential network-level solutions. In: Proc. Eur. 7th Framework Res. Project METIS, October 2013. http://bit.ly/1LpmVhr
8. Delivering Public Safety Communications with LTE, 3GPP White Paper. http://3gpp.org/Public-Safety
9. Hovey, R. (ed.): Scenarios and requirements for general use cases and national security and public safety, 3rd Generation Partnership Project (3GPP), 650 Route des Lucioles-Sophia Antipolis, France, Tech. Rep. 22.803, May 2013
10. Gerasimenko, M., Petrov, V., Galinina, O., Andreev, S., Koucheryavy, Y.: Impact of machine-type communications on energy and delay performance of random access channel in LTE-advanced. European Transactions on Telecommunications **24**(4), 366–377 (2013)
11. Chauhan, P., Vijay, S., Arya, P.: Comparative Analysis of Routing Protocols in AD-HOC Network: AODV, DSDV, DSR. International Journal of Current Engineering and Technology. ISSN 2277–4106
12. Muthanna, A., Prokopiev, A., Paramonov, A., Koucheryavy, A.: Comparison of protocols for Ubiquitous wireless sensor network. In: 6th International Congress on Ultra Modern In: Telecommunications and Control Systems and Workshops (ICUMT 2014), pp. 334–337. October 2014
13. Network Simulator 3: Discrete-event network simulator. NSNAM. http://www.nsnam.org
14. LENA: LTE-EPC Network simulAtor. CTTC. http://networks.cttc.es/mobile-networks/software-tools/lena/
15. Masek, P., Zeman, K., Hosek, J., Tinka, Z., Makhlouf, N., Muthanna, A., Novotny, V.: User performance gains by data offloading of LTE mobile traffic onto unlicensed IEEE 802. 11 links. In: Proceedings of the 38th International Conference on Telecommunication and Signal Processing, TSP 2015, pp. 1–5. Asszisztencia Szervezo Kft., Prague (2015). ISBN: 978-1-4799-8497-8
16. Masek, P., Hosek, J., Dubrava, M.: Influence of M2M communication on LTE networks. In: Sbornik Prispevku Studentske Konference Zvule 2014, pp. 53–56, January 2014. ISBN: 978-80-214-5005-9
17. Masek, P., Uhlir, D., Zeman, K., Masek, J., Bougiouklis, C., Hosek, J.: Multi-Radio Mobile Device in Role of Hybrid Node Between WiFi and LTE networks. International Journal of Advances in Telecommunications, Electrotechnics, Signals and Systems **4**(2), 1–6 (2015). ISSN: 1805–5443
18. Yang, L., Liu, L., Mingju, L., Lan, C.: Uplink control for low latency HARQ in TDD carrier aggregation. In: 2012 IEEE 75th Vehicular Technology Conference (VTC Spring), pp. 1–5, May 6–9, 2012
19. Shivahare, B., Wahi, C., Shivhare, S.: Comparison Of Proactive And Reactive Routing Protocols In Mobile Adhoc Network Using Routing Protocol Property. International Journal of Emerging Technology and Advanced Engineering **2**(3), 356–359 (2012). ISSN: 2250–2459

Analysis of Approaches to Internet Traffic Generation for Cyber Security Research and Exercise

Tero Kokkonen[1,2(✉)], Timo Hämäläinen[2], Marko Silokunnas[1], Jarmo Siltanen[1],
Mikhail Zolotukhin[2], and Mikko Neijonen[1]

[1] Institute of Information Technology,
JAMK University of Applied Sciences, Jyväskylä, Finland
{tero.kokkonen,marko.silokunnas,jarmo.siltanen,
mikko.neijonen}@jamk.fi
[2] Department of Mathematical Information Technology,
University of Jyväskylä, Jyväskylä, Finland
{timo.t.hamalainen,mikhail.m.zolotukhin}@jyu.fi
tero.t.kokkonen@student.jyu.fi

Abstract. Because of the severe global security threat of malwares, vulnerabilities and attacks against networked systems cyber-security research, training and exercises are required for achieving cyber resilience of organizations. Especially requirement for organizing cyber security exercises has become more and more relevant for companies or government agencies. Cyber security research, training and exercise require closed Internet like environment and generated Internet traffic. JAMK University of Applied Sciences has built a closed Internet-like network called Realistic Global Cyber Environment (RGCE). The traffic generation software for the RGCE is introduced in this paper. This paper describes different approaches and use cases to Internet traffic generation. Specific software for traffic generation is created, to which no existing traffic generation solutions were suitable.

Keywords: Internet traffic generation · Cyber security research and exercise · Cyber security · Network security

1 Introduction

The JAMK University of Applied Sciences has built a closed Internet-like network called Realistic Global Cyber Environment (RGCE). RGCE mimics the real Internet as closely as possible and contains most services found within the real Internet, from tier 1 Internet Service Providers (ISP) to small local ISPs and even individual home and corporate ISP clients. The fact that RGCE is completely isolated from the Internet allows RGCE to use accurate GeoIP information for all IP addresses within RGCE. This allows the creation of exercises or research cases where the attackers and the defenders are seemingly in different parts of the world and any device (real or virtual) will assume that it is actually operating within the real Internet. RGCE also contains various web services found in the real Internet [1, 2].

© Springer International Publishing Switzerland 2015
S. Balandin et al. (Eds.): NEW2AN/ruSMART 2015, LNCS 9247, pp. 254–267, 2015.
DOI: 10.1007/978-3-319-23126-6_23

Due to the fact that RGCE is isolated from the real Internet, RGCE does not contain background user traffic of its own. This poses a problem: how can you realistically train for a scenario where your public services are being attacked by an unknown party, and the attack traffic is concealed within normal user traffic if there is no normal user traffic? This is the basic problem to be solved in order to efficiently use RGCE for cyber security exercises or research.

Traffic generation has an important role when characterizing behaviour of the Internet. Behaviour of the real Internet consists of the rapid changes of the network, network traffic and user behaviour as well as the variables of characterization vary from the traffic links and protocols to different users or applications [3]. In addition changing nature of connections in Internet is influenced by the behaviour of the users, which determines the page level, and the connection level correlation that should be included to the traffic generation models [4]. According to the study [4] this is neglected by the scientific literature.

There are two fundamental approaches to Internet-like traffic generation, trace-based generation and analytical model-based generation. In trace-based generation the content and the timings of the captured real traffic are retransmitted and in analytical model-based generation the traffic is generated based on the statistical models [5, 6].

Due to increasing amount of traffic, applications and users deep analysis of real Internet traffic is essential for planning and managing networks [7]. Deep analysis of real Internet traffic also gives an efficient viewpoint for realizing the extensive processes of the Internet [8]. Thus the deep analysis of the real network traffic can be used for developing Internet traffic generation software using realistic traffic patterns from both humans and machines.

In this paper the Internet traffic generation software is introduced. First, the requirements, existing solutions and different approaches for traffic generation are presented. Then the developed solution is introduced and evaluated.

2   Found Requirements

The main purpose of developed Internet traffic generation software is to generate user traffic for the cyber security exercises conducted within RGCE. To meet the requirements for cyber security exercises the Internet traffic generation software was implemented according to the following self-generated requirements:

- Centralized control; the system shall have a single point of control and the control mechanism shall enable the generation of a large volume of traffic with minimal user interaction.
- Ability to generate legitimate traffic; the generated traffic shall adhere to the generated protocol.
- Ability to generate meaningful traffic on several layers of the OSI model; the system shall be able to generate meaningful traffic on OSI layers 3-6. This shall include IP, TCP, HTTP and other application layer protocols.

- Ability to generate attack traffic; the system shall be able to generate traffic for various attacks commonly encountered on the Internet. Examples of such attacks include SYN flood, NTP and DNS amplification DDoS attacks.
- Generated traffic shall look like real Internet traffic; the traffic shall be as indistinguishable as possible from real traffic for both humans and machines.
- Ability to make the traffic look like it is coming from anywhere within RGCE; it shall be possible to deploy parts of the system to various parts of RGCE to make the geolocation information look realistic.
- Generated traffic shall not be a replay; replaying previously recorded traffic would make it easy to distinguish generated traffic from normal user traffic, unless the recorded captures are of significant length.
- Generated traffic shall work with existing servers; the system shall be able to use normal, non-modified servers as targets for traffic generation. A simple example would be HTTP: the system shall be able to generate legitimate non-identical requests to a given HTTP server, with varying HTTP headers and make those requests at human-like intervals.
- The system shall be highly autonomous; the system shall be able to recover from errors without human intervention as much as possible. The system shall be able to generate traffic without human intervention for extended periods of time.

3   Existing Solutions for Traffic Generation

There are a number of proprietary and open source tools available for Internet-like traffic generation, such as TG Traffic Generator [9], NetSpec [10], Netperf [11], Packet Shell [12] and D-ITG [13, 14]. A detailed listing and analysis of available tools can be found from the study [5]. Those mentioned tools approach the problem from the viewpoint of workload generation through statistical models. Their goal is to generate repeatable workloads for networks and monitoring tools.

Such tools suffer from the fact that they are often implemented on top of non-real-time operating systems (OS). This causes their behaviour to be un-deterministic due to various scheduling decisions made by the OS as introduced in study [15]. Performance of D-ITG is also analysed in [14]. Netbed has a different viewpoint, it is a tool for integrating three experimental environments: network emulator, network simulator and real networks [16].

Developed Internet traffic generation software avoids many of above-mentioned problems, mainly because the goal is not in the generation of realistic workloads, but rather in meaningful payloads and good integration with existing off-the-shelf products with minimal customization.

4   Approaches

There are different approaches to Internet traffic generation with their pros and cons. These described approaches were analysed for the development of Internet traffic generation software.

4.1   Network Layer Traffic Generation

Generating traffic on the network layer is a simple approach to traffic generation. It is trivial to implement using, for example, Linux raw sockets [17], and can be implemented for both IPv4 and IPv6.

This approach works by generating a large number of IP packets with randomized payloads. The use of Linux raw sockets also allows the source IP address of the packet to be spoofed, which allows a single machine to simulate a huge number of individual hosts. The machine sending the IP packets could be considered to be the default gateway for a large organization, such as university or a company.

An example system could work by requiring a definition of a range of source IP addresses to use (e.g. 10.0.0.1-10.0.0.255 for IPv4) and then generating a large number of IP packets with the source field set to one of the IP addresses within the source range.

The generated packets are only meaningful when analysed on the IP layer. If the requirements for the generated traffic are such that the traffic has to be meaningful on higher layers (e.g. TCP), this approach is not suitable without a considerable amount of effort. This means that implementing a custom TCP stack on top of Linux raw sockets is required. The only benefit over regular Linux TCP sockets [18] is the ability to spoof source IP addresses for individual TCP segments [19].

Various analytical model-based network traffic generation tools utilize this approach. Such tools put emphasis on IP traffic characteristics (e.g. packet size and timing), rather than the transmitted data itself [13, 9, 10].

This is not feasible for the purposes of Internet traffic generation within RGCE (see Section 2). But it is relevant for testing various other aspects of network performance.

4.2   Transport Layer Traffic Generation

Using existing TCP stacks found in operating systems to handle the TCP connections significantly reduces the complexity of the implementation but makes IP spoofing [20] difficult. It is still possible to use a single machine to simulate a larger amount of hosts by using IP aliasing.

An elementary approach to traffic generation on the transport layer would be to utilize TCP stack provided by the operating system. This greatly simplifies the implementation of the traffic generator, as the OS TCP stack will take care of retransmission and other TCP details. As a downside, this approach does not allow for much control over the generated traffic characteristics.

As with the network layer approach (Section 4.1) this method works as long as meaningful exchanges on higher layers of the OSI model are not required (e.g. HTTP). It is possible to overcome this problem for the simplest of cases, such as creating multiple identical HTTP requests and always expecting an identical reply. More complex transmissions are also possible to implement but in most cases it would be more straightforward to just implement the approach described in Section 4.6

4.3   Replaying Traffic

When considering approaches to Internet traffic generation, replaying PCAP files [21] is a rather natural option. Typically, traffic replay aims to generate repeatable workloads for systems under test [6]. This is achieved by replaying recorded data [22] or synthesizing [23] traffic traces and then replaying them through the network. Tcpreplay [22] is existing software solution that is able to replay captured TCP traffic from files.

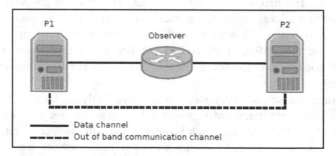

Fig. 1. Simple traffic replaying environment

It is necessary to use Out of band communication channel (Fig. 1) if the orchestration should not interfere with the system under test. Orchestration could include communicating the roles, timing, and bandwidth quotas associated with the replay [5].

In order to make this approach work, some processing is required for the PCAP file:

- Filtering out all unnecessary data streams. Unless the PCAP file is captured with the intention of replaying it, it is likely that the file contains a lot of unnecessary packets.
- Compiling the payload bytes from the TCP segments. Sending individual TCP segments from the PCAP file is not a feasible approach because network conditions are very likely to differ between the recording environment and the replaying environment. When the bytes are properly extracted and sent over the operating system's TCP stack the implementation does not have to concern itself with TCP details. It also makes it easier to detect and handle networking problems in the replaying environment.
- Constructing an intermediate presentation of the PCAP file that describes what to send and what to receive for each participant of the conversation.

It is worth noting that replaying PCAP files is only feasible for reliable transport layer protocols (e.g. TCP and SCTP). While it is possible to just extract the sent UDP (or other unreliable transport level protocols) packets and resend them, it will require extra steps to ensure that the packets get to their destination due to the nature of UDP [24]. There are two approaches to overcome this problem:

- Protocol awareness. The system needs to be aware of the protocol it is replaying and in case of lost packets mimic the simulated protocol's behaviour in such situations (if any). This requires considerable effort to duplicate the protocol's functionality and the solution starts to resemble the approach detailed in section 4.4.
- Out of band communication channel. An out of band communication could be utilized to transfer information about sent and received packets between participants. While this approach makes sure that all packets get delivered, it does not reliably reproduce the simulated protocol, because it is acceptable to lose packets in some UDP based protocols.

The following subsections will detail the out of band communication channel approach and its limitations. It is worth noting that the out of band communication channel must use a reliable transport layer protocol, such as TCP. The out of band communication channel also introduces additional latency to the replaying caused by TCP.

4.4   Replaying in a Reliable Network

It is assumed in Fig. 2 that the replaying environment does not suffer from packet loss, thus introducing no unexpected side effects.

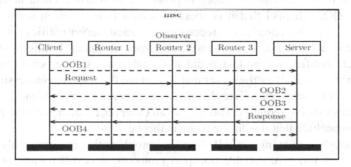

Fig. 2. No network problems

Replaying in a reliable network is processed as follows: client notifies Server through the OOB channel that it is about to send a request, client sends the actual request, server notifies client that it received the request, server notifies client that it is about to send a response, server sends the response and finally client notifies server that it received the response successfully.

This scenario works as expected and does not introduce any additional side effects; the observer sees a single request and a single response and thus cannot tell the traffic apart from the real traffic.

4.5   Replaying in an Unreliable Network

The fact that the out of band communication channel introduces some reliability features to the system can cause the observer to see responses without requests, this can be seen from Fig. 3.

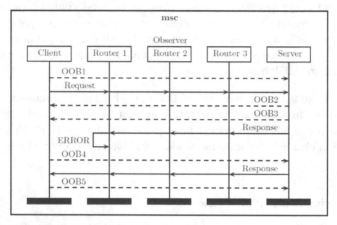

Fig. 3. Network problems

Replaying in an unreliable network is processed as follows: client notifies server through the OOB channel that it is about to send a request, client sends the actual request, server notifies client that it received the request, server notifies client that it is about to send a response, server sends the response, router 1 fails to deliver the packet, client notifies Server that it did not receive a response, server resends the response and Client notifies the server that it received the response successfully.

In the case of a network failure the observer observes multiple identical responses without corresponding requests. This allows an observer familiar with the protocol in question, to conclude that this traffic is not authentic.

Even though replaying PCAP files is problematic for protocols that are implemented on top of unreliable transport protocols, it is still robust for protocols utilizing reliable transport protocols. But still this approach cannot be used with existing servers and will end up repeating the same conversation over and over again, thus not fulfilling the requirements listed in section 2.

4.6   Simulating Clients

Simulating full clients for Internet traffic generation offers a flexible solution to traffic generation, as it allows fine-grained control over the generated traffic and the depth of simulation. This approach does not allow precise traffic generation on packet level or control of the various packet characteristics that are available in other traffic generation solutions, such as Inter Departure Time (IDT) and Packet Size (PS).

This approach can fulfil the requirements listed in section 2; it can be used with any server and the generated traffic is sufficiently diverse. It is also possible to implement very specific types of traffic (e.g. deliberately broken TCP traffic). If the client simulation is sufficiently sophisticated, it is very difficult for the server and observers to distinguish it from the realistic clients.

It is worth noting, that this approach requires a significant amount of effort, as it is difficult to create a single solution that could simulate multiple clients and protocols in a convincing manner. This means that the system will require multiple protocol specific modules.

5   Implemented Solution

Implemented solution aims to generate traffic that looks meaningful to a human observer. It was decided to implement Internet traffic generation software using the full client simulation approach. Solution consists of a hierarchical network of nodes. The network forms a tree like structure. The network forms an opt-in botnet, where each individual node is a host.

5.1   Terminology

The network consists of three different node types: King, Slavemaster and Botmaster. Bots are not nodes (hosts), but are run on the same host as the Botmaster.

King is the root node of the tree. King acts as a bridge between the UI and the rest of the network. The UI is running on a webserver on this node. Both Slavemasters and Botmasters can connect to King. Every message sent into the network by the user passes through King.

Slavemaster connects to the King or another Slavemaster and acts as a router between nodes. Slavemasters can connect to other Slavemasters and thus the depth of the tree representing the network can be arbitrary. Slavemasters have full knowledge of the tree underneath themselves. When a Slavemaster receives a message it checks the message recipient. If the recipient is the Slavemaster it broadcasts that message to all of its children, who then broadcast it to their children and so on. If the recipient is one of the descendants of the Slavemaster message is forwarded towards it.

Botmaster is a leaf node of the tree. Botmasters are charged with performing the actual traffic generation. Botmasters run one or more Bots. Multiple Bots can be running simultaneously. Botmaster receives messages and status updates from its Bots and forwards them to King, which will then update the UI accordingly. In the current implementation Bots are ran in the same process as the Botmaster.

Bot handles the actual traffic generation. Each Bot is tasked with generating traffic of a certain type (e.g. HTTPBot generates HTTP traffic). If a Bot encounters an error it sends a notification to the UI about it.

5.2   Implementation

Current implementation of the system contains traffic generation profiles for various protocols and services, such as HTTP, SMTP, DNS, FTP, NTP, IRC, Telnet, SSH, CHARGEN and ICMP. Each protocol or service is capable of containing different profiles. For example there are five different bots for HTTP protocol: HTTPBot mimics an user that is browsing the internet, SlowlorisBot performs the slowloris HTTP DoS attack, SlowPOSTBot performs the slow POST HTTP DoS attack, HTTPAuthBot repeatedly attempts to authenticate using HTTP Basic Auth and HTTPDDoSBot repeats the same HTTP request continuously.

Implementation is done for GNU/Linux using the Go programming language [25]. Each node of the system (King, Slavemaster and Botmaster) has its own binaries. In the current implementation Bots are ran in the same process as the Botmaster, but this is required to change in the future development. Go was chosen as the implementation language due to the fact that it has native support for coroutines (called goroutines in Go) and easy interfacing with C programming language.

Go also provides a way to facilitate communication between goroutines using channels, which are derived from Hoare's CSP [26]. The first few versions of our traffic generation software utilized the C interface significantly, but the current version contains no C code. The need for interfacing with C reduced as the Go ecosystem grew and more libraries became available. Most of the C code in the early versions was related to utilizing raw sockets to conduct IP spoofing, which can now be achieved using Go.

Bots are run inside the Botmaster as goroutines. A Bot can contain multiple goroutines. Naturally, the Botmaster contains goroutines that are not Bots as well, such as goroutines related to communication with the rest of the network.

Bots communicate with the Botmaster using Go's channels. Botmasters, Slavemasters and King communicate between each other using a custom text based protocol implemented on top of TCP.

The UI is implemented as a single-page web application served by a web server running on the King. The UI communicates with the King using a custom protocol on top of WebSockets [27]. The King acts as a translator between the WebSocket protocol and the protocol used by the rest of the network. User interface (UI) of developed Internet traffic generation software can be seen in Fig. 4.

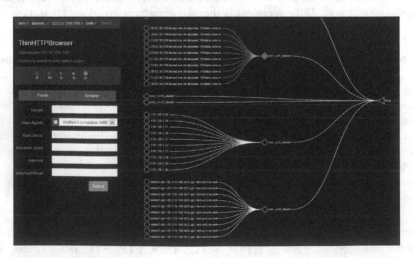

Fig. 4. UI of developed Internet traffic generation software

5.3   Evaluation

The evaluation for the developed solution is studied as follows. There were three different networks and three different amount of data generation Botmasters chosen. The different networks chosen were Localhost, Local Area Network (LAN) and Internet. The webserver with webpage including text pictures and links to 30 sub-pages was installed and HTTPBots browsed the contents of that webserver on each network cases. The LAN was in the JAMK University of Applied Sciences and the Internet scenario was between Netherlands and Finland (the webserver located in the Netherlands). The amount of Botmasters was 5, 25 and 125. The network data was captured on both sides, server side and client side. Time period of every single capture is 30 minutes long.

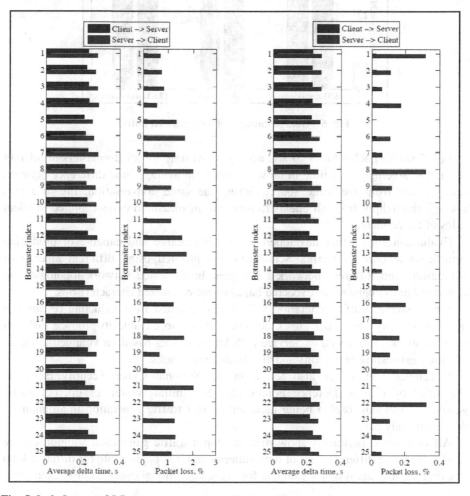

Fig. 5. Left: Internet 25 Botmasters, average delta time (s) and packet loss (%) Right: LAN 25 Botmasters, average delta time (s) and packet loss (%)

Evaluation data was generated using the HTTPBot, which mimics a browser by first downloading an HTML page from the targeted HTTP server. HTTPBot downloads all images, JavaScripts and CSS files referenced in the HTML document. Once all of the files are downloaded, the HTTPBot searches for a link to another page within the same domain or another domain. A link is chosen at random and the same process is repeated again.

The network traffic (PCAP data) was captured from the client and server side of the connection and also log data from the server was collected.

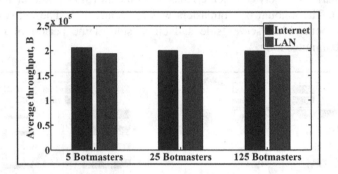

Fig. 6. Average throughput for generated traffic

Fig. 5 shows packet loss (%) and average delta time for 25 Botmasters in Internet and LAN. Average delta time in those figures is an average time difference between sent packets from the same source during the same conversation. Fig. 6 shows average throughput from all clients to server in all measured cases captured in client sides of the connections.

Evaluation shows that developed system is scalable and capable of producing significant amount of traffic. Scalability was proofed using different amount of Botmasters and different network topologies. In all tested network topologies the generated traffic behaves as expected based on the calculated characteristics.

Since Internet traffic generation software is designed for conducting research in network security and cyber attack detection, it is also capable to produce different sorts of attacks e.g. Denial of Service DoS/DDoS attacks based on volumetric traffic, resource exhausting or exploits and also bruteforce attacks.

Developed solution is also tested in the National Cyber Security Exercises organized by Finnish Defence Forces [28, 29]. Initial version of Internet traffic generation software is also being used for Internet traffic generation in an anomaly detection study [2].

All of those experiments show that developed traffic generation solution can be used to generate different kind of data patterns, and it is appropriate for different kind of cyber security analysis for example for big scale National cyber security exercises.

5.4   Lessons Learned

The generation of the Internet traffic is extremely important for research and development of cyber security. For example research and development of Anomaly Detection algorithms or Intrusion Detection Systems requires an environment with realistic legitimate background traffic and design made attacks [30, 2]. Generation of Internet traffic has an important role in cyber security exercises and training.

There were some lessons learned from the use cases that caused extra development for the data generation software. OOB communication for control traffic is very important when generating lot of traffic (e.g. HTTPDDoSBot). Generated data might block the outgoing data and if the command communication data uses the same interface it is also blocked. That causes situation where one Bot blocks the Botmaster out from the network. Another lessons learned that required changes for development was CPU bound meaning that if there is Bot doing resource intensive processing it might harm the whole process and block the Botmaster out of communication. Bots that are CPU resource intensive (e.g. SYN flood Bot) cannot have permission to generate traffic as fast as they are capable of processing, thus there must be a limit e.g. 5000 packets/second/Bot.

6   Conclusion

In this study, approaches to realistic Internet traffic generation for cyber security research and exercise were considered. First requirements for traffic generation were analysed. After that different solutions and approaches were described. Finally suitable approach was chosen and developed Internet traffic generation software was introduced. As a conclusion it can be said that the developed Internet traffic generation software met the requirements and it is suitable for modelling the Internet traffic as a part of the cyber security research and exercise. Requirements for next phase were found and in future the development of those requirements are planned to execute. The deployment of the several Botmasters shall be automated and replaying of PCAP data (limited only for TCP) between Botmasters shall also be developed.

References

1. JAMK University of Applied Sciences, Jyväskylä Security Technology (JYVSECTEC), Realistic Global Cyber Environment (RGCE). http://www.jyvsectec.fi/en/rgce/
2. Zolotukhin, M., Hämäläinen, T., Kokkonen, T., Siltanen, J.: Analysis of HTTP requests for anomaly detection of web attacks. In: 2014 IEEE 12th International Conference on Dependable, Autonomic and Secure Computing, pp. 406–411, August 2014
3. Floyd, S., Paxson, V.: Difficulties in Simulating the Internet. IEEE/ACM Trans. Netw. 9(4), 392–403 (2001)
4. Casilari, E., Gonzblez, F.J., Sandoval, F.: Modeling of HTTP traffic. Communications Letters, IEEE 5(6), 272–274 (2001)

5. Botta, A., Dainotti, A., Pescapè, A.: A tool for the generation of realistic network work-load for emerging networking scenarios. Computer Networks (Elsevier) **14**(15), 3531–3547 (2012)
6. Hong, S.-S., Wu, S.: On interactive internet traffic replay. In: Valdes, A., Zamboni, D. (eds.) RAID 2005. LNCS, vol. 3858, pp. 247–264. Springer, Heidelberg (2006)
7. Pries, R., Wamser, F., Staehle, D., Heck, K., Tran-Gia, P.: On traffic characteristics of a broadband wireless internet access. In: Next Generation Internet Networks, NGI 2009, pp. 1–7, July 2009
8. Li, T., Liu, J., Lei, Z., Xie, Y.: Characterizing service providers traffic of mobile internet services in cellular data network. In: 2013 5th International Conference on Intelligent Human-Machine Systems and Cybernetics (IHMSC), vol. 1, pp. 134–139, August 2013
9. The University of Southern California (USC-ISI), The Information Sciences Institute, TG Traffic Generation Tool. http://www.postel.org/tg/
10. The University of Kansas, The Information and Telecommunication Technology Center (ITTC), NetSpec Tool. http://www.ittc.ku.edu/netspec/
11. Netperf. http://www.netperf.org/netperf/
12. pksh -the Packet Shell. http://tecsiel.it/pksh/index.html
13. Universita' degli Studi di Napoli "Federico II", D-ITG, Distributed Internet Traffic Generator. http://traffic.comics.unina.it/software/ITG/
14. Angrisani, L., Botta, A., Miele, G., Vadursi, M.: An experimental characterization of the internal generation cycle of an open-source software traffic generator. In: 2013 IEEE International Workshop on Measurements and Networking Proceedings (M N), pp. 74–78, October 2013
15. Botta, A., Dainotti, A., Pescapè, A.: Do You Trust Your Software-Based Traffic Generator. IEEE Communications Magazine, 158–165 (2010)
16. White, B., et al.: An integrated experimental environment for distributed systems and networks. In: Proceedings of the 5th Symposium on Operating Systems Design and Implementation, 2002. USENIX Association, December 2002
17. Kleen, A.: Linux Programmer's Manual RAW(7). http://www.manpages.info/linux/raw.7.html
18. Kleen, A., Singhvi, N., Kuznetsov's, A.: Linux Programmer's Manual TCP(7). http://www.manpages.info/linux/tcp.7.html
19. Postel, J.: Transmission Control Protocol. RFC 793 (INTERNET STANDARD). Updated by RFCs 1122, 3168, 6093, 6528. Internet Engineering Task Force, September 1981. http://www.ietf.org/rfc/rfc793.txt
20. Tanase, M.: IP Spoofing: An Introduction. The Security Blog, March 2003. http://www.symantec.com/connect/articles/ip-spoofing-introduction
21. WireShark Wiki, Libpcap File Format. http://wiki.wireshark.org/Development/LibpcapFileFormat/
22. Tcpreplay. http://tcpreplay.synfin.net/
23. Khayari, R.E.A., Rücker, M., Lehman, A., Musovic, A.: ParaSynTG: a parameterized synthetic trace generator for representation of WWW traffic. In: SPECTS (2008), pp. 317–323, June 16-18, 2008
24. Postel, J.: User Datagram Protocol. RFC 768 (INTERNET STANDARD). Internet Engineering Task Force, August 1980. http://www.ietf.org/rfc/rfc768.txt
25. The GO Programming Language. https://golang.org/
26. Hoare, C.A.R.: Communicating Sequential Processess. Communications of the ACM **21**(8), 666–677 (1978)

27. Fette, I., Melnikov, A.: WebSocket Protocol. RFC 6455 (INTERNET STANDARD). Internet Engineering Task Force, December 2011. http://www.ietf.org/rfc/rfc6455.txt
28. Ministry of defense press release May 8, 2013. Cyber Security Exercise in Jyväskylä, May 13–17, 2013. Kyberturvallisuusharjoitus Jyväskylässä, May 13–17, 2013. http://www.defmin.fi/ajankohtaista/tiedotteet/2013/kyberturvallisuusharjoitus_jyvaskylassa_13.-17.5.2013.5502.news
29. Finnish Defence Forces Press Release, June 3, 2014, Performance of Cyber Security is developed by co-operation between Government Authorities and University. Kyber-suorituskykyä hiotaan viranomaisten ja korkeakoulujen yhteistyöllä. http://www.fdf.fi/wcm/su+puolustusvoimat.fi/pv.fi+staattinen+sivusto+su/puolustusvoimat/tiedotteet/kybers uorituskykya+hiotaan+viranomaisten+ja+korkeakoulujen+yhteistyolla
30. Luo, S., Marin, G.A.: Realistic Internet traffic simulation through mixture modeling and a case study. In: Proceedings of the 2005 Winter Simulation Conference, pp. 2408–2416, December 4–7, 2005

CloudWave Smart Middleware for DevOps in Clouds

Boris Moltchanov[1] and Oscar Rodríguez Rocha[2(✉)]

[1] Research and Prototyping/Strategy and Innovation, Telecom Italia, Turin, Italy
`boris.moltchanov@telecomitalia.it`
[2] Politecnico Di Torino, Corso Duca Degli Abruzzi, 24, 10129 Turin, Italy
`oscar.rodriguezrocha@polito.it`

Abstract. Nowadays the IT Clouds are used to provide services and run applications (in a flexible and scalable way) where the entire lifecycle of the application or service is under the control of the *Cloud Management Subsystem* that is able to automatically and autonomously start, stop, restart a server or an application. Additionally, it is able to determine when a certain application or service is running under overload conditions and reduce the load by adding additional virtual machines hosting the affected software or releasing the resources not used anymore. These are classic operations performed by nowadays Clouds. However little effort has been dedicated to support a developer during the application development process or during the service execution by changing the application behavior dependently on the service's components response and availability. The CloudWave project founded by the European Commission within the FP7 Research Program aims to contribute on these underdeveloped concepts. It contributes to the development process of a service or application in a feedback-driven way and also to the service execution by adapting the service accordingly to the execution environment and condition changes.

1 Introduction

This paper describes the main goals of the CloudWave project [1] and emphasizes the key concepts that it addresses, such as feedback-driven development (FDD) and the coordinated adaptation (CA) during the development or execution of a service or application.

The research leading to these results has received funding from the European Community's Seventh Framework Programme (FP7/2007-2013) under grant agreement n. 610802.

The feedback used for the FDD is collected from all the levels of the application runtime environment such as: physical, virtualization and application execution tiers. Then specific execution analytics, decide when and which tools and better patterns to suggest to the developer.

On the other hand, the CA intervenes during the service execution when the load of the application changes or when something abnormal occurs in the execution environment, thus the application should change either its behavior and possibly its user interface, or detach certain components and their respective functionalities if unavailable. This implies that the runtime environment provides feedback and

S. Balandin et al. (Eds.): NEW2AN/ruSMART 2015, LNCS 9247, pp. 268–273, 2015.
DOI: 10.1007/978-3-319-23126-6_24

analytics able to make decisions and perform changes in the environment in such a way that the application is able to undertake the changes required by the changing environment and taken decisions.

This work aims to be used in an Open Source environment, to accomplish that, Open Stack [2] has been chosen as cloud virtualization ecosystem.

2   The Innovative Concept

CloudWave is going to refine the modern cloud infrastructures and tools by enabling agile development and delivery of adaptive cloud services, which dynamically adjust to changes in their environment in order to optimize service quality and resource utilization. To do so, CloudWave provides a powerful foundation (based on open standards) for creating innovative cloud services, and for realizing the full promise of the cloud computing.

Moreover, CloudWave aims to tangibly deliver an open architecture and standards-based reference implementation of an advanced cloud software stack, with novel capabilities for adaptation across all cloud layers, and tools and methods for agile development of reliable and adaptable cloud services, facilitated by the new stack.

Technologically, CloudWave surpasses the state-of-art along three areas:

- **Execution Analytics**: A new framework where specialized algorithms dynamically analyze the cloud infrastructure and the application behaviors, seamlessly integrate data pertaining to physical and virtual resources and IoT elements, and provide consolidated feedback to drive service evolution and adaptation;
- **Coordinated Adaptation**: A new software technology where cloud services, cloud infrastructure and end-user devices exploit Execution Analytics to collaboratively and automatically undertake complex adaptation actions across the cloud stack, ensuring quality of service and effective utilization of ICT re sources;
- **Feedback-Driven Development**: A new agile approach for developing cloud applications, where developers exploit Execution Analytics to incrementally determine and evolve application features, extensions and optimizations, based on observed user needs.

These three innovations pillars are show in the Fig. 1.

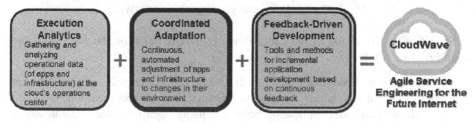

Fig. 1. CloudWave innovation pillars

CloudWave empowers developers of Cloud applications by exploiting the fact that these applications, throughout their lifecycle, are hosted and deployed by Cloud prviders from Cloud operation centers, and therefore can be also controlled at that point of delivery. In contrast to on-premises software, Cloud service delivery models allow for an iterative cycle of application evolution and improvement through continuous analysis of application-centric operational data gathered at the Cloud operations center; and applying adjustments during operations, if applications, the infrastructure or platforms deviate from their expected service quality and behavior. This opportunity can be exploited in order to deliver continuously improving applications, which rapidly evolve and dynamically adapt to their execution environment, connected devices, and user requirements. Enabling this revolutionary development and management approach across all Cloud delivery models and management mechanisms is the challenge addressed by CloudWave.

In fact, acknowledging the increasing importance of better aligning software development and IT operations, CloudWave takes inspiration from the emerging DevOps paradigm [3], that promotes communication, collaboration and integration between these two disciplines and their stakeholders such as to contribute to software quality. CloudWave is the first technology-based, holistic solution leveraging DevOps' principles as shown in Fig. 2.

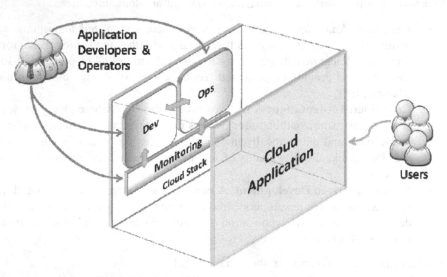

Fig. 2. The DevOps paradigm adopted in CloudWave

Specifically, the vision of the CloudWave project is based on the following three main innovation pillars:

- **Execution Analytics** (in Fig. 2), where programmable mechanisms and specialized algorithms are used to dynamically introspect and analyze Cloud infrastructure and application behavior, using open interfaces, seamlessly integrating data pertaining to physical resources, virtualized resources and IoT elements, thereby supporting *application adaptation* and *agile incremental development*;

- **Coordinated Adaptation** ("Ops" in Fig. 1), where new distributed algorithms and data models enable cloud infrastructures and applications to take coordinated adaptation actions across the system stack in response to dynamic changes in their execution environment, aiming to optimize the quality of service and resource utilization, while providing a feedback on runtime adaptations to developers;
- **Feedback-Driven Development** ("Dev" in Fig. 1), representing a novel engineering paradigm and set of modular platform services for agile development of Cloud applications, where developers are provided with easy access to runtime and contextual data; this allows them to enhance business applications for the sake of superior customer experience, based on observed user needs and context as well as application/infrastructure behavior.

Based on those novel ideas, CloudWave goes significantly beyond the state-of-art in cloud computing, and investigates into emerging capabilities and mechanisms such as cloud-aware applications; dynamic offloading of application code between cloud, on-premise servers and mobile devices; dynamic binary optimization of application code to support migration across the cloud, and more. CloudWave thereby goes over and beyond the scalability aspect of QoS, and complementarily addresses aspects such as service stability, availability, reliability, fail-safe provisioning, security and privacy.

3   CloudWave Architecture

In order to technically realize the vision of CloudWave, the project defines and implements a novel architecture, which incorporates the three main innovation pillars shown in Fig. 1. As described above, these pillars are: Execution Analytics, Coordinated Adaptation, and Feedback-Driven Development. This architecture is the foundation upon which the CloudWave software stack is built.

Fig. 3 presents a conceptual view of the CloudWave architecture. At this highest level, the architecture consists of three major building blocks, each providing the capabilities of one of the three key pillars. These building blocks are interconnected through feedback and refinement interfaces (arrows in Fig. 3). The CloudWave architecture overall includes components, which in part complement the application runtime environment (upper part of the Fig. 3), and in part extend the cloud infrastructure (bottom part), which also has an interface to the Internet of Things (IoT).

Fundamentally, the CloudWave architecture introduces two related feedback cycles: the first is automated and deals with self-adaptation of applications and infrastructure, and the second involves "humans in the loop" who, based on feedback from cloud operations, manually evolve the application. The realization of these two feedback cycles and the interaction between them is at the heart of the CloudWave innovative architecture – which exploits the interdependence between software development and IT Operations.

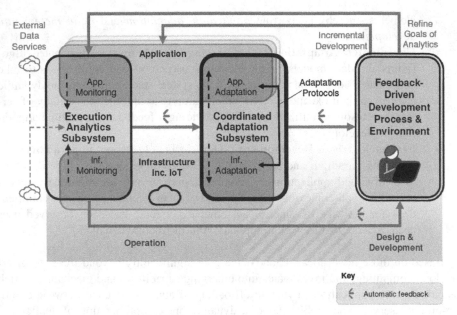

Fig. 3. The CloudWave Architecture

Architecturally, a cloud application executing in the CloudWave environment includes an application-specific monitoring component as well as accompanying software modules that contain filtering and analytics instructions pertaining to the application's operational data. When the application is deployed over the cloud infrastructure, application operational data is monitored and fed to the Execution Analytics subsystem through the application-monitoring component. In addition, application-related infrastructure operational data is gathered and filtered with respect to the monitoring goals (by the data filtering module provided with the application). The combined application-centric operational data, from the application and the infrastructure, is then processed and analyzed by a Feedback Generator, based on the analytics instruction provided with the application. The result of this data analysis, referred to as feedback, can be exploited to optimize application performance via two following two feedback cycles:

- **Automated Adaptation Feedback Cycle**. Feedback compiled from application-centric operational data is continuously fed to the Coordinated Adaptation subsystem. This subsystem is responsible for triggering and regulating adaptation actions, based on an *Adaptation Model* that specifies runtime conditions which the developer identifies as triggers for possible adaptation of the application or the infrastructure, alongside the actions to be taken when conditions are triggered. Adaptation decisions are taken through negotiation among the relevant layers (applications, cloud infrastructure, end-user devices), using the Adaptation Protocol. Based on the agreed adaptation decisions, actions are taken to best adapt the application and, possibly, the ICT infrastructure. This, in turn, affects the application (and infrastructure) behavior, and is reflected and followed-up through the monitoring components.

- **Manual Adaptation Feedback Cycle.** When requirements are continuously changing or are unknown, the feedback from cloud operations – stemming from the actual use of the applications by end-users – can be exploited as a basis for another adaptation scheme, where application developers are explicitly engaged. In this scheme, the feedback from the Analytics subsystem, along with insights about adaptation feasibility from the Coordinated Adaptation subsystem, are fed to the developer and used to create the next version of the application.

Figuratively, one can think of the automated adaptation feedback cycle as an online control mechanism that tries to position the application at the center of its operational envelope. This is done either by adapting the application itself, or by adapting the infrastructure (i.e., changing the envelope). However, a key innovation of CloudWave is that this control mechanism not only tries to perform best, but it also knows its own limits, and it can provide feedback to the manual adaptation cycle so the developer can create a new and improved version of the application.

4   Conclusions and Future Work

The CloudWave project has just started a few months ago and its overall execution duration is 36 month. The very high-level architecture fitting to the Fig. 3 is already defined, but its detailed elaboration is still missing. The project is going to start with implementation of three service scenarios based on the first draft of the architecture, which are following:

- Telco application development, when many software components are running in the CloudWave platform and the developers combine many features and run them within different environments such as functional testing and prototyping, preproduction then production monitoring vital application parameters and moving the application very fast across different aforementioned environments, improving and tuning the application. This drastically reduces the development cycle and time-to-market when run in production;

- VoIP application where the relay nodes and voice codecs are located in the cloud and dynamically switching depending on required Quality of Service, Network availability or device capability;

- Enterprise application development based on the enterprise grade environment where professional developers needs quickly create business supporting and operation-supporting applications in customized way for their business, enterprise data and structure.

References

1. CloudWave FFP7 Project: Agile Service Engineering for the Future Iternet. http://cloudwave-fp7.eu
2. OpenStack Open Source Cloud Computing Software. http://www.openstack.org
3. Debois, P.: Devops: A software revolution in the making? Journal of Information Technology Management (2011)

Data Mining Approach for Detection of DDoS Attacks Utilizing SSL/TLS Protocol

Mikhail Zolotukhin[1]([✉]), Timo Hämäläinen[1], Tero Kokkonen[2],
Antti Niemelä[2], and Jarmo Siltanen[2]

[1] Department of Mathematical Information Technology, University of Jyväskylä,
Jyväskylä, Finland
{mikhail.m.zolotukhin,timo.t.hamalainen}@jyu.fi
[2] JAMK University of Applied Sciences, Jyväskylä, Finland
{tero.kokkonen,antti.niemela,jarmo.siltanen}@jamk.fi,
tero.t.kokkonen@student.jyu.fi

Abstract. Denial of Service attacks remain one of the most serious
threats to the Internet nowadays. In this study, we propose an algorithm
for detection of Denial of Service attacks that utilize SSL/TLS protocol.
These protocols encrypt the data of network connections on the application layer which makes it impossible to detect attackers activity based on
the analysis of packet payload. For this reason, we concentrate on statistics that can be extracted from packet headers. Based on these statistics,
we build a model of normal user behavior by using several data mining algorithms. Once the model has been built, it is used to detect DoS
attacks. The proposed framework is tested on the data obtained with the
help of a realistic cyber environment that enables one to construct real
attack vectors. The simulations show that the proposed method results
in a higher accuracy rate when compared to other intrusion detection
techniques.

Keywords: Network security · Intrusion detection · DoS attack · Data
mining · Anomaly detection

1 Introduction

Due to the fact that Internet has become the major universal communication
infrastructure, it is also subject to attacks in growing numbers and varieties.
One of the most serious threats to the Internet nowadays is Denial of Service
(DoS) attacks [2]. This kind of attack disables the network servers using lots of
messages which need response and consumes the bandwidth of the network or
the resource of the system. Because it is difficult for an attacker to overload the
targets resources from a single computer, modern DoS attacks are launched via
a large number of distributed attacking hosts in the Internet. Such distributed
DoS (DDoS) attacks can force the victim to significantly downgrade its service

© Springer International Publishing Switzerland 2015
S. Balandin et al. (Eds.): NEW2AN/ruSMART 2015, LNCS 9247, pp. 274–285, 2015.
DOI: 10.1007/978-3-319-23126-6_25

performance or even stop delivering any service [1]. Moreover, DDoS attacks are more complex and harder to prevent compared to conventional DoS attacks.

Traditional DDoS attacks are carried out at the network layer, e.g. ICMP flooding, SYN flooding, and UDP flooding. The purpose of these attacks is to consume the network bandwidth and deny service to legitimate users of the victim systems. This type of attack has been well studied recently and different schemes have been proposed to protect the network and equipment from such bandwidth attacks [3–7]. For this reason, attackers shift their offensive strategies to application-layer attacks. Application-layer DoS attacks may focus on exhausting the server resources such as Sockets, CPU, memory, disk bandwidth, and I/O bandwidth. Unlike network-layer DoS attacks, application-layer attacks do not necessarily rely on inadequacies in the underlying protocols or operating systems. They can be performed by using legitimate requests from legitimately connected network machines. The most popular application-layer DoS attacks are HTTP page flooding and low-rate DoS attacks [1].

The problem of application-layer DoS and DDoS attacks detection is of great interest nowadays. For example, in [8], an anomaly detector based on hidden semi-Markov model is proposed to describe the spatial-temporal patterns of normal users and to detect application-layer DDoS attacks for popular websites. Study [9] proposes an advanced entropy-based scheme, which divides variable rate DDOS attacks into different fields and treats each field with different methods. Paper [10] considers detection of slow DoS attacks by analyzing specific spectral features of network traffic over small time horizons. In [11], authors detect application-layer DDoS attacks by constructing a random walk graph based on sequences of web pages requested by each user. Paper [12] shows a novel detection technique against HTTP-GET attacks, based on Bayes factor analysis and using entropy-minimization for clustering.

Despite the rising interest to the detection of application DDoS attacks, most of the current researches concentrate on various HTTP DDoS attacks. In this study, we propose an algorithm for the detection of DDoS attacks utilizing SSL/TLS protocol [13]. These protocols encrypt the data of network connections in the application layer which makes it impossible to detect attacker's activity based on the analysis of packet's payload. For this reason, we concentrate on statistics that can be extracted from packet headers. Based on these statistics, we build a model of normal user behavior by using several data mining algorithms. Once the model has been built, it is used to detect DoS attacks.

The rest of the paper is organized as follows. Extraction of feature vectors from network packets is considered in Section 2. Section 3 introduces a scheme that uses data mining techniques to build a model of normal user behavior and subsequently detects DoS attacks. In Section 4, we present some numerical results to evaluate the algorithm proposed and compare it with certain analogues. Finally, Section 5 draws the conclusions and outlines future work.

2   Problem Formulation

We consider a web server that provides several services working in two application layer protocols: HTTP and HTTPS. Outgoing and incoming traffic of this server is captured during some time period $[T_s, T_e]$. There is no guarantee that the traffic captured is free of attacks. For this reason, we aim to investigate captured traffic and discover behavior patterns of normal users. In this study, it is assumed that the most part of the traffic captured is normal. In real world, this can be achieved by filtering the traffic with the help of a signature-based intrusion detection system [14]. Once normal behavior patterns have been discovered, these patterns can be used to analyze network traffic and detect DoS and DDoS attacks against the web server in online mode.

We concentrate on attack detection based on the analysis of statistics that can be extracted from packet headers. For this purpose, for each packet, we extract the following information: time stamp, IP address and port of the source, IP address and port of the destination, protocol, packet size, window size, sequence number, acknowledgment number, time to live and TCP flags.

Packets with some common properties passing a monitoring point in a specified time interval can be combined into flows. As a rule, these common properties include IP address and port of the source and IP address and port of the destination. Thus, for each packet we also extract the index of the network traffic flow this packet belongs to.

3   Algorithm

In order to detect network flows related to DoS and DDoS attacks we apply an algorithm that can be divided into two main steps. First, we apply an anomaly detection algorithm in order to find time intervals when an attack takes place. After that, the traffic sent during these time intervals is analyzed in more details to detect flows related to the attack.

3.1   Detection of Network Anomalies

In order to find time intervals which contain traffic anomalies we consider the network traffic as time series. For this reason, the analyzed time period $[T_s, T_e]$ is divided into equal overlapping time bins of length ΔT by points $T_s + \frac{t}{w}\Delta T$, where $t = \{w, w+1, \dots, w\frac{T_e - T_s}{\Delta T} - 1\}$. The length of each time bin ΔT should be picked in such a way that it contains enough information to detect anomalies. Moreover, the value of w should be big enough for the earlier detection of attacks.

For each resulting time interval the following features are calculated:

1. Source IP address sample entropy,
2. Source port sample entropy,
3. Destination IP address sample entropy,
4. Destination port sample entropy,

5. Total number of flows,
6. Average flow duration,
7. Average number of packets in one flow,
8. Average size of packets,
9. Average size of TCP window for packets.

Features $7 - 9$ are extracted separately for packets sent from source to the destination and packets sent from the destination to the source.

Sample entropy allows one to capture the degree of dispersal or concentration of the parameter's distribution. Let us assume that in the t-th time interval the i-th parameter has n_i^t unique values which appear with frequencies $p_{i1}^t, \ldots, p_{in_i^t}^t$. In this case, sample entropy E_i^t for the i-th parameter in the t-th time interval is defined as follows:

$$E_i^t = -\sum_{k=1}^{n_i^t} p_{ik}^t \log_2 p_{ik}^t. \tag{1}$$

The model of normal user behavior is built by calculating chi-square values. Chi-square values are used to find out how much the observed values of a particular given sample are different from the expected values of the distribution [15,16]. Chi-square value for the t-th time interval is calculated as follows:

$$\chi_t^2 = \sum_{i=1}^{n_x} \frac{(x_i^t - \mu_i)^2}{\mu_i}, \tag{2}$$

where n^x is the number of features (in our case it is equal to 12), x_i^t is the value of the i-th feature for the t-th time interval and μ_i is the mean value of the i-th feature during the analyzed time period $[T_s, T_e]$. Chi-square value in (2) is similar to the traditional chi-square statistic for independence tests.

Once chi-square values have been calculated, some filtering can be applied to remove outliers and noise to build the model of normal user behavior. As proposed in study [17], we define the following distance function:

$$d(\chi_{t_1}^2, \chi_{t_2}^2) = p(\chi_{t_1}^2 \text{ is normal}) - p(\chi_{t_2}^2 \text{ is normal}), \tag{3}$$

where $p(x \text{ is normal})$ is the probability that value x is normal and can be found as follows:

$$p(x \text{ is normal}) = \begin{cases} \frac{\sigma_{\chi^2}^2}{(x-\mu_{\chi^2})^2}, & \text{if } x \geq \mu_{\chi^2} + \sigma_{\chi^2}, \\ 1, & \text{if } x < \mu_{\chi^2} + \sigma_{\chi^2}, \end{cases} \tag{4}$$

where μ_{χ^2} and $\sigma_{\chi^2}^2$ are the mean and the variance of chi-square values, respectively. In this case, the distance d between a normal pattern and an outlier pattern is expected to be higher than the distance d between two normal patterns or two outlier patterns.

It is easy to divide all the chi-square values into two clusters, i.e. normal values and outliers, by using single-linkage clustering algorithm. This algorithm belongs to a class of agglomerative hierarchical clustering methods. In the beginning of the algorithm, each chi-square value forms a cluster, i.e. the number of clusters is equal to the number of chi-square values, and every cluster consists of only one element. At each iteration, the algorithm combines those two clusters which are the least distant from each other. The distance $d(C_i, C_j)$ between two clusters C_i and C_j is defined as the minimal distance between two chi-square values of these clusters, such that one value is taken from each cluster:

$$d(C_i, C_j) = \min_{\chi_t^2 \in C_i, \chi_\tau^2 \in C_j} (d(\chi_t^2, \chi_\tau^2)). \tag{5}$$

The algorithm stops when the required number of clusters is formed. In our case, this number is equal to two: one cluster for normal values C_n and another one for outliers C_o. All outliers are removed from the model and all normal values are used for detecting anomalies.

The χ^2 statistic approximately follows a normal distribution according to the central limit theorem [18], regardless of the distribution that each of the extracted features follows. If we make the assumption that values of cluster C_n resemble a normal distribution, approximately 99.7% of all chi-square values should fall within three standard deviations of the mean value of cluster C_n.

In order to classify network traffic during the recent time interval, we discover necessary features for this time interval and calculate chi-square value χ^2. Network traffic at this time interval is classified as anomalous if

$$\chi^2 > \bar{\mu}_{\chi^2} + \alpha \bar{\sigma}_{\chi^2}, \tag{6}$$

where $\bar{\mu}_{\chi^2}$ and $\bar{\sigma}_{\chi^2}$ are the mean and the standard deviation of chi-square values from cluster C_n, and parameter $\alpha \geq 3$. Thus, we can find the time interval when an attack takes place.

3.2   Detection of Intrusive Flows

In order to build a model for detection of network flows related to a DDoS attack we consider time intervals during which traffic is classified as normal. For each flow in these time intervals, the following information is extracted:

1. Average, minimal and maximal size of packets,
2. Average, minimal and maximal size of TCP window,
3. Average, minimal and maximal time since the previous packet,
4. Average, minimal and maximal time to live,
5. Percentage of packets that have TCP flag SYN,
6. Percentage of packets that have TCP flag ACK,
7. Percentage of packets that have TCP flag PSH,
8. Percentage of packets that have TCP flag RST,
9. Percentage of packets that have TCP flag FIN.

We extract these features separately for both directions: from the source to the destination and from the destination to the source. Thus, the i-th flow is presented as as a feature vector y_i of length $n = 34$.

Values of vectors y_i can have different scales. In order to standardize the feature vectors of the training set max-min normalization is used. Max-min normalization performs a linear alteration on the original data so that the values are normalized within the given range [19]. In this paper, we map vectors y_i to range $[0, 1]$. To map a value y_{ij} of an attribute $(y_{1j}, y_{2j}, \ldots, y_{nj})$ from range $[\min_{1 \leq i \leq n'} y_{ij}, \max_{1 \leq i \leq n} y_{ij}]$ to range $[0, 1]$, the computation is carried out as follows

$$z_{ij} = \frac{y_{ij} - \min_{1 \leq i \leq n} y_{ij}}{\max_{1 \leq i \leq n} y_{ij} - \min_{1 \leq i \leq n} y_{ij}}, \tag{7}$$

where z_{ij} is the new value of y_{ij} in the required range.

The model of normal behavior can be found with the help of density-based spatial clustering of applications with noise (DBSCAN). DBSCAN is a powerful density-based clustering algorithm, which is often used for detecting outliers. It discovers clusters in the training dataset starting from the estimated density distribution of feature vectors [20].

DBSCAN requires two parameters: the size of neighborhood ε and the minimum number of points required to form a cluster N_{min}. The algorithm starts with an arbitrary feature vector z that has not been checked. The number of feature vectors $N_\varepsilon(z)$ contained in the ε-neighborhood of z is found and compared to N_{min}:

$$\begin{cases} \text{If } N_\varepsilon(x) < N_{min}, \text{ then } z \text{ is labeled as noise,} \\ \text{If } N_\varepsilon(x) \geq N_{min}, \text{ then } z \text{ is a part of a cluster.} \end{cases} \tag{8}$$

Vectors marked as noise might later be discovered as a part of another vector ε-environment and hence be made a part of a cluster. If a vector is found to be a part of a cluster, its ε-neighborhood is also part of that cluster. After that, each point $\bar{z}$ contained in the ε-neighborhood is checked. If $\bar{z}$ is density-reached from z with respect to ε and N_{min}, it is added to the cluster. Vector $\bar{z}$ is density-reachable from z with respect to ε and N_{min}, if there is a chain of points $z_1, z_2, \ldots, z_m$, where $z_1 = z$ and $z_m = \bar{z}$, such that $\forall i \in \{1, 2, \ldots, m - 1\}$ the two following conditions are satisfied:

$$\begin{cases} d(z_i, z_{i+1}) \leq \varepsilon, \\ N_\varepsilon(z_i) \geq N_{min}, \end{cases} \tag{9}$$

where $d(z_i, z_{i+1})$ is the Euclidean distance between z_i and z_{i+1}. The cluster is built when all vectors density-reachable from z have been found. Then, a new unvisited vector is processed, leading to a discovery of a further cluster or noise. As a rule, all points which remain cluster-less after the algorithm is finished are classified as anomalies. Since, we assume that all flows in time intervals considered are normal we consider all cluster-less points as clusters which contain only one point. Figure 1 shows an example of the application of DBSCAN, with $N_{min} = 3$ and $\varepsilon = 0.25$ (radius of each circle).

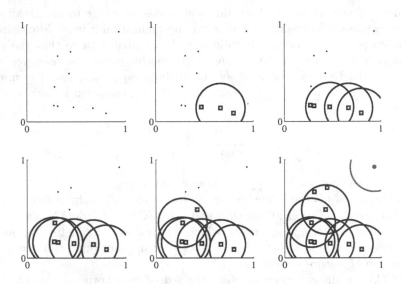

Fig. 1. An example of the application of DBSCAN

Thus, DBSCAN can find arbitrarily-shaped clusters and does not require to know the number of clusters in the dataset a priori. DBSCAN requires just two parameters that should be optimally chosen: the size of neighborhood ε and the minimum number of points required to form a cluster N_{min}. There are several studies devoted to this problem [21,22]. In this study, N_{min} is selected to be equal to 10% of the total number of flows in time intervals that have been classified as normal, and ε is equal to the average Euclidean distance between feature vectors corresponding to these flows.

Once the clustering has been completed, the maximal pairwise Euclidean distance is calculated for each cluster. Let us denote the maximal pairwise distance of elements of the i-th cluster C_i as m_i:

$$m_i = \max_{z_j, z_k \in C_i} d(z_j, z_k). \tag{10}$$

For each cluster-less point, m_i is selected to be equal to the minimal value of maximal pairwise distances of elements of all clusters. In order to classify a new flow in the recent time interval, all necessary features are extracted from this flow into vector z. After this, the cluster (or cluster-less point) which contains the vector which is the least distant from z is found:

$$i^* = \arg \min_{z_j \in C_i} d(z, z_j). \tag{11}$$

This flow is classified as an attack if the distance between z and the least distant vector is greater than the corresponding maximal pairwise distance:

$$\min_{z_j \in C_{i^*}} (d(z, z_j)) > m_{i^*}. \tag{12}$$

After this, the traffic that corresponds to intrusive flows can be blocked to prevent the development of the attack.

4   Numerical Simulations

Realistic Global Cyber Environment (RGCE) was used for the DoS/DDoS data generation. RGCE is closed Internet-like environment developed and hosted by JAMK University of Applied Sciences. As one of the main features, RGCE executes real IP-addresses and geolocations [23,24]. A web server serving a static main page with some text and a picture through HTTPS (SSL/TLS) was installed as a part of the RGCE infrastructure. After that, RGCE was used to generate both legitimate user traffic and DoS/DDoS traffic to this web server. RGCE data generation software uses botnet architecture where the bots are controlled by botmasters [23]. The bots were distributed in RGCE with different global IP-addresses that also simulates global geological distribution inside the RGCE. Two different types of bots were used for the data generation: bots that generated legitimate HTTPS traffic and bots that generated non-legitimate HTTPS traffic (DOS/DDoS) to the test web server. The bot that generated legitimate traffic crawled through the targeted web-site and generated requests for found content and followed found links. The bot that generated the non-legitimate traffic created multiple requests to the targeted single web-site page. The resulting data set contains mainly HTTPS traffic as the test web server communicated with the clients only through encrypted protocol. The non-encrypted traffic (HTTP) in the dataset are the initial handshakes between the clients (bots) and the test web server before the encrypted channel was created. All the traffic was captured as PCAP-files [25] for the numerical analysis.

To test attack detection algorithms we consider one of such PCAP-files. This file contains 80 minutes of traffic or 429202 traffic flows. To build a model of normal user behavior we use the training set which contains 8 minutes or 10% of traffic. This traffic is free of attacks but it contains few outliers. The testing set contains remaining 72 minutes or 90% of the traffic. This set contains DDoS attack from two different subnets. The attack starts at 8:53.19 and ends at 68:37.2. The length of time bin ΔT was selected to be equal to five seconds, whereas parameter w is equal to 5. In this case, each time interval contained enough information to detect anomalies every second.

Figure 4 shows features extracted for each second of the time period considered. On this figure, blue line represents normal traffic whereas red line corresponds to the DDoS attack. Chi-square values calculated for each second for the features extracted are presented on Figure 4. As one can see, chi-square values corresponding to time intervals when the attack takes place are higher. In order to find time intervals that contain anomalous traffic we apply the method described in the previous section. Parameter α is selected to be equal to 5. As a result, time of the start and time of the end of the attack can be defined up to the size of time window which is equal to 1 second in our simulation.

We compare the performance of DBSCAN with other clustering and outlier detection techniques: K-means, K-Nearest Neighbors (KNN), Support Vector

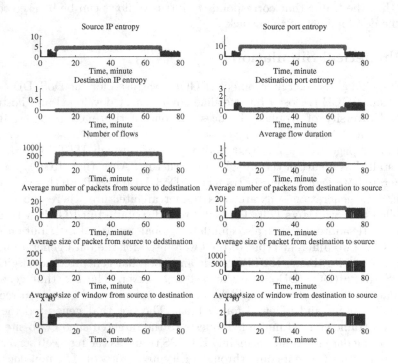

Fig. 2. Features extracted for each time interval to find anomalous network traffic (blue line - normal traffic, red line - traffic during the attack).

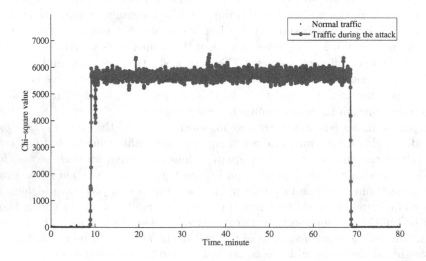

Fig. 3. Chi-square values of features extracted.

Data Description (SVDD) and Self-organizing Map (SOM). To evaluate the performance of each algorithm, the following characteristics are calculated in our tests:

- True positive rate – the ratio of the number of correctly detected anomalous samples to the total number of anomalous samples in the testing set
- False positive rate – the ratio of the number of normal samples classified as anomalous to the total number of normal samples in the testing set
- Detection accuracy – the ratio of the total number of normal samples detected as normal and anomalies detected as anomalies to the total number of samples in the testing set.

Figure 4 shows the dependence between false positive and true positive rates for different detection methods and different parameters. To compare the accuracy of the methods, we select optimal parameters of the methods based on the training set. The comparison results are listed in Table 1 As one can notice, all methods are able to detect all intrusive conversations (TPR = 100 %). However, SVDD gives the worst results with the highest number of false alarms (FPR = 6.0627 %). DBSCAN outperforms other methods in terms of accuracy (Accuracy = 99.9993 %) and number of false alarms (FPR = 0.0697 %). DBSCAN's FPR equal to 0.0697 % corresponds to only one flow in the testing set which is normal but is classified as an attack. This false alarm can be explained by the fact that

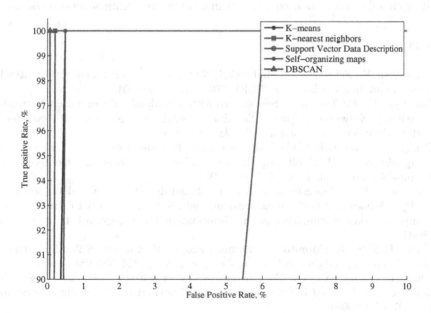

Fig. 4. Dependence between false positive and true positive rates for different detection methods and different parameters.

Table 1. Intrusion detection accuracy of different detection methods.

Algorithm	TPR	FPR	Accuracy
K-means	100 %	0.4878 %	99.9951 %
KNN	100 %	0.2091 %	99.9979 %
SVDD	100 %	6.0627 %	99.9390 %
SOM	100 %	0.4878 %	99.9951 %
DBSCAN	100 %	**0.0697 %**	**99.9993 %**

the number of conversations in the training set is very low and probably not enough for building the accurate model of normal user behavior.

5   Conclusion

In this paper, the scheme for detection of Denial of Service attacks that utilize SSL/TLS protocol is proposed. The scheme is based on the analysis of statistics extracted from packet headers. Based on these statistics, we build a model of normal user behavior by calculating Chi-square values and clustering flows with DBSCAN. Once the model has been built, it is used to detect DDoS attacks. The method is tested on the data obtained with the help of a realistic cyber environment. The simulation results show that the scheme proposed allows to detect all intrusive flows with very low number of false alarms. In the future, we are planning to improve the algorithm in terms of the detection accuracy, and test it with a dataset which contains traffic captured during several days.

References

1. Durcekova, V., Schwartz, L., Shahmehri, N.: Sophisticated denial of service attacks aimed at application layer. In: ELEKTRO, pp. 55–60 (2012)
2. Gu, Q., Liu, P.: Denial of Service Attacks. Handbook of Computer Networks: Distributed Networks, Network Planning, Control, Management, and New Trends and Applications, vol. 3. John Wiley & Sons (2008)
3. Peng, T., Leckie, K.R.M.C.: Protection from distributed denial of service attacks using history-based IP filtering. In: Proc. of IEEE International Conference on Communications, vol. 1, pp. 482–486 (2003)
4. Limwiwatkul, L., Rungsawangr, A.: Distributed denial of service detection using TCP/IP header and traffic measurement analysis. In: Proc. of IEEE International Symposium on Communications and Information Technology, vol. 1, pp. 605–610 (2004)
5. Yuan, J., Mills, K.: Monitoring the macroscopic effect of DDoS flooding attacks. IEEE Tran. Dependable and Secure Computing 2(4), 324–335 (2005)
6. Chen, R., Wei, J.-Y., Yu, H.: An improved grey self-organizing map based dos detection. In: Proc. of IEEE Conference on Cybernetics and Intelligent Systems, pp. 497–502 (2008)
7. Ke-Xin, Y., Jian-Qi, Z.: A novel DoS detection mechanism. In: Proc. of International Conference on Mechatronic Science, Electric Engineering and Computer (MEC), pp. 296–298 (2011)

8. Xie, Y., Yu, S.-Z.: Monitoring the Application-Layer DDoS Attacks for Popular Websites. IEEE/ACM Transactions on Networking **17**(1), 15–25 (2008)
9. Zhang, J., Qin, Z., Ou, L., Jiang, P., Liu, J., Liu, A.: An advanced entropy-based DDOS detection scheme. In: Proc. of International Conference on Information Networking and Automation (ICINA), vol. 2, pp. 67–71 (2010)
10. Aiello, M., Cambiaso, E., Mongelli, M., Papaleo, G.: An on-line intrusion detection approach to identify low-rate DoS attacks. In: Proc. of International Carnahan Conference on Security Technology (ICCST), pp. 1–6 (2014)
11. Xu, C., Zhao, G., Xie, G., Yu, S.: Detection on application layer DDoS using random walk model. In: Proc. of IEEE International Conference on Communications (ICC), pp. 707–712 (2014)
12. Chwalinski, P., Belavkin, R., Cheng, X.: Detection of application layer DDoS Attacks with clustering and bayes factors. In: Proc. of IEEE International Conference on Systems, Man, and Cybernetics (SMC), pp. 156–161 (2013)
13. Dierks, T., Rescorla, E.: The transport layer security (TLS) protocol. IETF RFC 4346 (2006)
14. Gollmann, D.: Computer Security, 2nd edn. Wiley (2006)
15. Ye, N., Borror, C.M., Parmar, D.: Scalable Chi-Squae Distance versus Conventional Statistical Distance for Process Monotoring with Uncorrelated Data Variables. Quality and Reliability Engineering International **19**(6), 505–515 (2003)
16. Muraleedharan, N., Parmar, A., Kumar, M.: A flow based anomaly detection system using chi-square technique. In: Proc. of the 2nd IEEE International Advance Computing Conference (IACC), pp. 285–289 (2010)
17. Corona, I., Giacinto, G.: Detection of server-side web attacks. In: Proc of JMLR: Workshop on Applications of Pattern Analysis, pp. 160–166 (2010)
18. Johnson, R., Wichern, D.: Applied Multivariate Statistical Analysis. Prentice-Hall, Upper Saddle River (1998)
19. Saranya, C., Manikandan, G.: A Study on Normalization Techniques for Privacy Preserving Data Mining. International Journal of Engineering and Technology (IJET) **5**(3), 2701–2704 (2013)
20. Ester, M., Kriegel, H., Jörg, S., Xu, X.: A density-based algorithm for discovering clusters in large spatial databases with noise, pp. 226–231. AAAI Press (1996)
21. Kim, J.: The anomaly detection by using DBSCAN clustering with multiple parameters. In: Proc. of the ICISA, pp. 1–5 (2011)
22. Smiti, A.: DBSCAN-GM: an improved clustering method based on gaussian means and DBSCAN techniques. In: Proc. of the IEEE 16th International Conference on Intelligent Engineering Systems (INES), pp. 573–578 (2012)
23. Jyvsectec-rgce - homepage. http://www.jyvsectec.fi/en/rgce/
24. Zolotukhin, M., Hämäläinen, T., Kokkonen, T., Siltanen, J.: Analysis, of http requests for anomaly detection of web attacks. In: Proc. of the 12th IEEE International Conference on Dependable, Autonomic and Secure Computing, pp. 406–411 (2014)
25. WireShark Wiki, Libpcap File Format. http://wiki.wireshark.org/Development/LibpcapFileFormat/

An Applicability of AODV and OLSR Protocols on IEEE 802.11p for City Road in VANET

Raj K. Jaiswal$^{(\boxtimes)}$ and C.D. Jaidhar

Department of Information Technology,
National Institute of Technology Karnataka, Surathkal, India
jaiswal.raaj@gmail.com, jaidharcd@nitk.edu.in

Abstract. Vehicular Ad-hoc Network (VANET) improves, makes more safe and comfortable road transportation by using vehicular communication and the Internet. VANET is the subset of Mobile Ad-hoc Network (MANET). Thus, due to their similar characteristics, MANET routing protocols may also be applicable into VANET. Hence, the performance of MANET routing protocols should be evaluated only on IEEE 802.11p communication standard, which is specifically designed for VANET communication, with urban and non-urban vehicular traffic. This work compares the performance of Ad-hoc On-Demand Distance Vector (AODV) routing protocol with Optimized Link State Routing protocol (OLSR) on two different road network scenarios, particularly a complex road network, which represents the city road network, having multiple crossroad and an intersection of two roads. We used two distinct simulators such as Vehicular Ad-hoc Networks Mobility Simulator (VANETMOBISIM), to simulate the city road network and vehicular traffic in an area of 700mx700m and NS-2.35 network simulator to simulate the communication network. AODV and OLSR performances are assessed on different transmission range, i.e. 250m and 500m with four different data generation rate of 512, 1024, 1536 and 2048 Kbps.

The primary goal of this work is to do an assessment to scrutinize the applicability of AODV and OLSR protocols in VANET with different traffic scenario and transmission ranges of IEEE 802.11p standard.

Keywords: AODV · IEEE 802.11p · OLSR · VANETMOBISIM

1 Introduction

Goal of Intelligent Transportation System (ITS) is to provide services to users while driving the vehicle. ITS integrates sensors, computer, communication, location finding device and control management approach to reduce the traffic and accidents. Inter Vehicular Communication (IVC) is an important component of ITS, it initiates the formation of Vehicle to Vehicle (V2V) and Vehicle to Roadside Unit (V2R) communication network [1]. In V2V communication, vehicle communicates with other vehicle either directly or using intermediate vehicles (hop). In the latter one, vehicle communicates directly to fixed Roadside Unit (RSU) to exchange the messages.

© Springer International Publishing Switzerland 2015
S. Balandin et al. (Eds.): NEW2AN/ruSMART 2015, LNCS 9247, pp. 286–298, 2015.
DOI: 10.1007/978-3-319-23126-6_26

Vehicular Ad-hoc Network renders its services to ITS to achieve the desired goals by exploiting communication between vehicle to vehicle and vehicles to RSU.

VANET employs wireless communication between vehicle to vehicle and vehicle to RSU through Dedicated Short Range Communication (DSRC) Standard especially IEEE 802.11p [7]. The IEEE 802.11a standard is altered to meet IEEE 802.11p standard, specifically to reduce the overhead operations and frequent signal drop [2].

Robust routing and forwarding mechanism are the important key issues in VANET communication, to disseminate safety, infotainment services and traffic information quickly in time. Due to the specific characteristics of VANET such as speed, mobility, intermittent link and delay in packet delivery, makes conventional routing protocol not to be used in VANET communication. Nevertheless, due to similar characteristics of MANET and VANET such as self configuring node and multi-hop routing, routing protocols applicable to MANET such as reactive, proactive and position based may also be applied in VANET. Thus, the performance of MANET routing protocols should be evaluated in the VANET environment on IEEE 802.11p standard for the typical and non-typical city road network, which consist of multiple crossroad and single crossroad networks respectively. This work is restricted to test the applicability of MANET topology based routing protocols such as AODV, OLSR in VANET. In this work, we have used vehicle and node interchangeably.

AODV [3] and OLSR [5] are the efficient and most used protocols in MANET. Hence, it is essential to evaluate the AODV and OLSR protocols behaviour in VANET environment on IEEE 802.11p standard. To simulate the city road network for vehicular traffic, VANETMOBISIM [13] and for communication network, ns-2.35 [14] are used.

The remaining section of this paper is organized as follows: section 2 briefly discusses the related work. Section 3 explains the working of AODV and OLSR protocols, simulation parameters used in the simulation are described in section 4. Section 5 discusses the simulation results followed by conclusion and future work in section 6.

2   Related Work

In the literature, AODV and OLSR performance are compared either on IEEE 802.11a or IEEE 802.11DCF standards, which are not congruent with VANET characteristics. Thus, it is a prime requirement to evaluate these protocols on IEEE 802.11p standard, which is specifically designed to meet the VANET requirement, for instance signal drop rate and quality is improved compared to IEEE 802.11a. Research conducted so far to test the applicability of MANET protocols in VANET on IEEE 802.11p standard is limited and discussed as follow:

Clausen et al. [9], compared the performance of AODV and OLSR protocols specifically for MANET. In their work, IEEE 802.11 standard and *Random*

Waypoint mobility are used. In *Random Waypoint* mobility node decides the destination and speed randomly, hence it does not provide vehicular characteristics. In addition, node moves into a region where obstacles are not considered on their path. Hence, the parameters chosen for the performance evaluation are not sufficient to test the applicability in VANET. For instance, in VANET environment, routing performance depends on several factors such as mobility, speed and consistency in network connectivity. Hence, applicability of MANET routing protocols for VANET should be evaluated only with mobility generated with VANET characteristics and communication standards.

Jerome Haerri et al. [8], compared the performance of AODV & OLSR protocols for VANET on IEEE 802.11DCF standard. However, they did not mention any signal propagation loss model in their work, which impacts the routing performance due to signal fading in the city. Hence, parameters are inappropriate to conclude the applicability of AODV and OLSR protocols to be used in VANET based on their measured performance.

Imran Khan et al. [10], assessed the performance of AODV and OLSR protocols using IEEE 802.11p with Nakagami fading model. In their work, they considered uniform speed of 40Kmph, which may be insufficient to represent realistic vehicular traffic for a city. Further, they did not discuss about the parameters used for micro mobility in simulation, which also affect the routing performance.

Evjola Spaho et al. [11], compared the AODV and OLSR performance on IEEE 802.11p standard for VANET. In their work, CAVENET simulator is used to generate vehicular mobility, which does not support many essential features of VANET such as accelaration and decelaration, politeness factor and intersection management [12]. It is found that vehicles are distributed around crossroad, which is insufficient to assess the applicability of AODV and OLSR protocols in VANET. In addition, Vehicles cover each other in their transmission range as the simulation area is restricted to 200m X 200m and the vehicle have a transmission range of 250 meters. Consequently, the route discovery procedure is required only when the vehicle needs to communicate with the diagonal vehicle. Accordingly, communication links do not get disconnected due to their transmission range which envelopes the entire simulation area.

In general, simulation parameters and scenario so far used in the experiments are inadequate to test the applicability of AODV and OLSR protocols in VANET.

Hence, it is crucial to re-evaluate the performance of AODV and OLSR on IEEE 802.11p standard with the mobility generated specifically considering the VANET characteristics. In addition, vehicles must be deployed unevenly and randomly around the crossroad and on the road segment wherein vehicle can communicate with the other vehicle moving in different location of the city as depicted in fig. 1 in which vehicles are represented by their node Id's.

3   Routing Protocols

Generally, routing protocols proposed in MANET for topology based routing are classified as reactive and proactive protocols. In reactive routing, routes are created on demand, when a node need to communicate with the other node. Whereas in proactive routing, routes are precomputed and stored in the route table. The following subsection explains briefly about AODV & OLSR protocols.

3.1 Ad-hoc On-Demand Distance Vector Routing

In AODV, routes are created on demand. Each node periodically broadcasts *HELLO* packets to maintain the topology. It does not store precomputed route in the route table until it has to communicate with the other node. The source node discovers the route for the destination through a route discovery mechanism which follows *Route_Request*, *Route_Reply* and *Route_Maintenance* procedures. During route search procedure, source node broadcasts *Route_Request* packet to its neighbor nodes, which consist *source_addr*, *source_sequence#*, *broadcast_id*, *dest_addr*, *dest_sequence#* & *hop_cnt* information in which source sequence number represents the freshness of the route. Upon receiving of *Route_Request* packet, the node sends *Route_Reply* packet in response to the sender node, if it finds a route to the queried destination, else rebroadcast the *Route_Reply* packet to neighbor nodes by increasing *hop_cnt*.

Route_Reply packet contains *source_addr*, *dest_addr*, *dest_sequence#*, *hop_cnt* & *lifetime* information. Node discards the *Route_Reply* packet, if it has received previously from other node.

Frequent movement and speed of the node, leads to link failure. To address these issues, *Route_Maintenance* procedure is opted by the node in which it sends *Route_Error* packet back to the node from where it has received [3] [4].

3.2 Optimized Link State Routing

OLSR is a proactive routing protocol, which is designed on the basis of link state routing. Basically, It optimizes the routing control overhead by appointing a subset of *MultiPoint Relay* (MPR) nodes among the neighbor nodes. The MPR nodes are only authorized to forward the packets to the queried destination, if it is in reach, else forward to the next MPR. Thus, a node multicast the packets only to the MPR node to reduce the control overhead. OLSR follows two basic functions, to complete the route procedure as follows:

- *Neighbor Sensing*: Each node periodically broadcasts *HELLO* packet to the neighbor node to maintain the link status. *HELLO* packet contains the information of neighbor nodes and their link status.
- *Multipoint Relay Selection*: Each node appoints a set of MPR nodes among the neighbor nodes. The MPR set may contain a subset of up to one hop neighbor node, which covers up to two hop neighbors. The node broadcasts the *Topology control* (TC) packets periodically to declare the set of the MPR selection list. A TC packet contains the list of neighbors, who is selected by the sender node as an MPR node [5] [6].

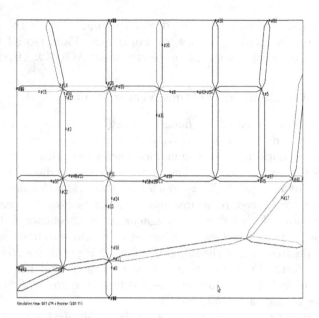

Fig. 1. Entire city Road Network

4 Simulation Parameters

Primarily, simulation is conducted for two different city road network, multiple crossroad and single crossroad as depicted in fig. 1 and 2 respectively. Multiple crossroad scenario depicts the typical city road topology, on which vehicle can move on any road segment and each road segment has speed limit. Whereas single crossroad has chosen to observe the AODV and OLSR performance in non-urban area over the entire city, in which nodes are located nearby. Further, each scenario is sub-divided into different transmission range, 250m and 500m. Vehicles are assigned different speed between 0 to 35 m/s randomly. In addition, routing protocols are also evaluated considering uniform and constant speed of 20 m/s of the vehicles.

To fix the transmission range of the vehicle, required transmission power is computed using *Two-Ray Ground Propagation* path loss model for the range 250m and 500m as depicted in the equation 1.

$$P_r = \left(\frac{P_t * G_t * G_r (h_t^2 * h_r^2)}{d^4 * L} \right) \tag{1}$$

$P_r = Receiving\ Power,\ P_t = Transmitting\ Power,\ G_t = Transmitting$
$Antenna\ Gain,\ G_r = Receiving\ Antenna\ Gain,\ d = distance,$
$L = Loss\ Factor,$
$h_t = Transmitting\ Antenna\ Height,\ h_r = Receiving\ Antenna\ Height.$

Height of antenna is set for 1.5m for both transmitter and receiver and loss factor is assumed to be 1 in the equation 1.

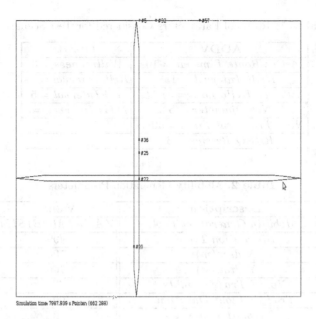

Simulation time: 7997.939 s Pointer (662 288)

Fig. 2. Single Crossroad

Mobilities are generated for 15, 30, 45 and 60 nodes using VANETMOBISIM. Nodes are deployed randomly in 700mx700m simulation area for each scenario. In first scenario, some crossroads are equipped with traffic lights with the duration of 60 seconds as depicted in fig. 1, whereas in second scenario only single crossroad is used which is also equipped with traffic light as shown in 2. Each road segment is divided into two lanes, which are categorized into the fast and slow moving lane.

IDM-IM model controls the movement of the vehicle such as acceleration and deceleration according to the front moving vehicle or traffic lights, including driver behaviour during overtaking using politeness factor. It also includes a safety gap between two vehicles to keep the safe distance and safe headway time in mobility. Table 2 represents the mobility parameters considered in the simulation.

We used AODV protocol available in NS-2.35, whereas OLSR protocol is implemented using UM-OLSR patches in NS-2.35. Parameters for both the protocols are kept as per standards defined in the RFC as depicted in the Table 1. We conducted the simulation on different data generation rate of 512, 1024, 1536 and 2048 Kbps for each category to observe the protocol behaviour w.r.t. data rate. Table 3 contains the list of parameters used in NS-2.35.

5   Simulation Results

AODV and OLSR performances are compared on performance matrix of *Packet Delivery Ratio* (PDR), *Routing Overhead, Throughput* and *Average Delay.*

Table 1. Protocol Parameters Considered for Test Scenario

AODV	OLSR
Active Route Timeout = 10*s*	*Willingness* = 3
Hello Interval = 1*s*	*HelloInterval* = 2*s*
Allowed Hello Loss = 3*Pkts*	*TC Interval* = 5
Net Diameter = 5	*MID Interval* = 5
Node Traversal Time = 30*ms*	
RREQ Retries = 3	

Table 2. Mobility Generation Parameters

Description	Values
Mobility Generation Tool	*VANETMOBISIM*
Simulation Time	499
X dim.(m)	700
Y dim.(m)	700
No. of Traffic lights	05
Traffic Light Duration	60*s*
No. of Lanes	2
Min. Speed (m/s)	0.5
Max. Speed (m/s)	35
Constant Speed	20(*m/s*)
Politeness Factor for Road Segment	0.2, 0.5, 0.8
Maximum Acceleration	0.9(m/s^2)
Maximum Deceleration	0.6(m/s^2)
Minimum Congestion Distance	2*m*
Safe Headway Time	2*s*
Length of Vehicle	5*m*

The results are obtained for two different scenarios as aforementioned for 250m and 500m transmission range with variable speed and constant speed. We have considered only the best results and abbreviation A and O are used for AODV and OLSR respectively in the figures. Which are discussed as follows:

5.1   Packet Delivery Ratio

PDR is a ratio of total number of packets received and send by destination and source, respectively. Fig. 3 shows the PDR performance w.r.t.[1] vehicle density. From results, it is clear that AODV outperform compared to OLSR in 250m transmission range for the entire city road network. It achieves 78% PDR for 250m transmission range at 512 Kbps data generation rate. PDR decreases as the vehicle density and data transmission rate increases. Whereas OLSR does better compared to AODV with 500m transmission range and it achieves 82% PDR

---

[1] w.r.t. is an acronym for "with respect to"

Table 3. Network Simulator Parameters

Parameters	Values
Network Simulator	$NS - 2.35$
Simulation Time(S)	499
Routing Protocols	$AODV$ & $UM - OLSR$
Antenna Model	$Omni - Directional\ Antenna$
Modulation Technique	$BPSK$
Radio Propagation Model	$Two - Ray\ Ground$
Transmission Range(m)	$250, 500$
MAC Type	$IEEE\ 802.11p$
MAC Rate	$2\ Mbps$
Interface Queue Type	50
Transport Protocol	UDP
Data Type	CBR
CBR Generation Rate (Kbps)	$512, 1024, 1536, 2048$
Packet Size	$512\ Bytes$
No. of Connections	$50\%\ of\ Number\ of\ Vehicles$
No. of Vehicles	$15, 30, 45, 60$
Node Density	$\#nodes.(\pi.range^2/x_{dim}.y_{dim})$

w.r.t. 512 Kbps data transmission rate. However, PDR of both the protocols decreases as the vehicle density and data generation rate increases.

PDR marginally increases in single crossroad scenario to 84% w.r.t. OLSR on 512Kbps whereas AODV reaches 81% for 250m transmission range. However, their performances are much closer to each other w.r.t 500m transmission range as shown in fig. 4. Nevertheless, OLSR has slight stable performance compared to AODV.

Though, AODV & OLSR achieves acceptable PDR with 512Kbps data rate. However, PDR of both the protocols decreases as the vehicle density and data generation rate increases. Hence, OLSR and AODV could not achieve acceptable PDR for high vehicle density and data transmission rate, which is a prime requirement to disseminate safety messages in a critical situation in VANET.

5.2 Routing Overhead

Routing overhead represents the number of bytes required to construct and maintain routing table. In fig. 5 and 6 routing overhead increases exponentially w.r.t vehicle density for both the protocols. However, it decreases as data generation rate increases in all scenarios. OLSR has more routing overhead unlike AODV, due to route table maintenance.

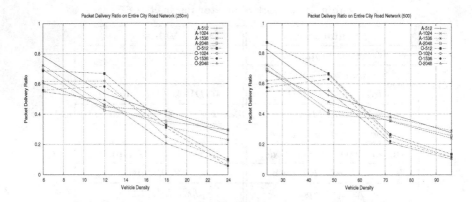

Fig. 3. Packet Delivery Ratio for City Road Network

Hence, intuitively, transmission range and road topology do not affect routing overhead. From results, it can be concluded that data transmission rate, vehicle density and routing mechanism affect the routing overhead. Hence, protocols which maintain the route table has more routing overhead in VANET and makes it less applicable.

5.3   Throughput

Throughput of the network is the average number of bytes transmitted per second. Both the protocols AODV & OLSR attain 525 Kbps & 580 Kbps highest throughput on 250m and 500m transmission range, respectively w.r.t. vehicle density as shown in fig. 7 for the entire city road network. However, in both the case throughput decreases as data transmission rate and vehicle density increases.

In case of single crossroad scenario throughput increases as the vehicle density increases due to movement of nodes near to each other as the nodes are deployed only on road segments and its intersection point rather than a city road network. It decreases as data transmission rate increases as shown fig. 8.

Hence, throughput of the protocols decreases w.r.t. node density and data transmission rate for the City road network, whereas it increases with single crossroad. Which makes it less applicable in VANET w.r.t throughput.

5.4   Average Delay

Average Delay is the measurement of end to end transmission delay between sender and receiver only for the correctly received packets. Delay increases as the vehicle density increases and decreases as the data generation rate increases in all scenarios as depicted in fig. 9 and 10. However, in all the scenarios AODV has a more average delay compared to OLSR due to on demand route discovery

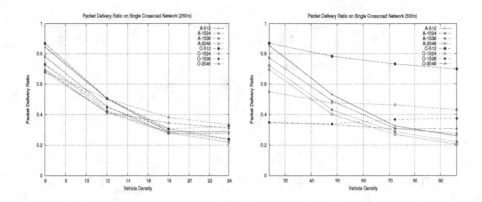

Fig. 4. Packet Delivery Ratio for Single Crossroad Network

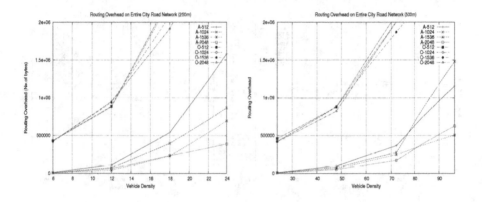

Fig. 5. Routing Overhead for City Road Network

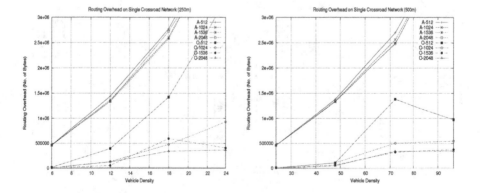

Fig. 6. Routing Overhead for Single Crossroad Network

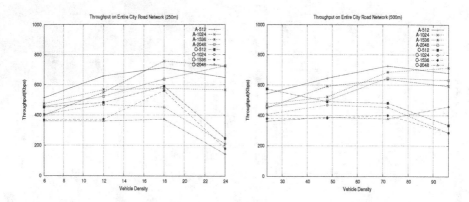

Fig. 7. Throughput for City Road Network

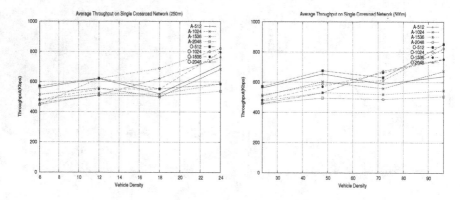

Fig. 8. Throughput for Single Crossroad Network

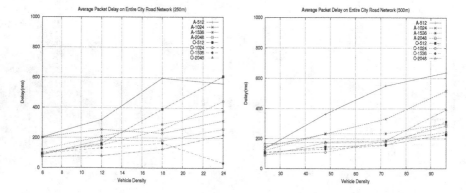

Fig. 9. Average Delay for City Road Network

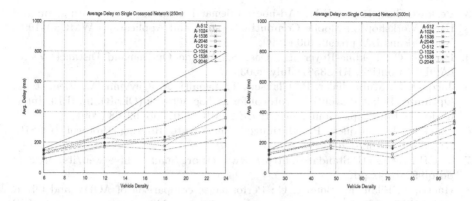

Fig. 10. Average Delay for Single Crossroad Network

procedure. We also observe that transmission range and road topology have less effect on packet delay.

Hence, based on results both the AODV and OLSR protocols are less feasible to be used in VANET due to more average delay w.r.t. vehicle density.

6 Conclusion and Future Work

AODV and OLSR protocols do not have stable PDR and throughput w.r.t to vehicle density and data generation rate. However, Routing overhead and Average delay increase with vehicle density and decreases when data generation rate increases. We also observed that performance is better on single crossroad compared to multiple crossroad. Hence, performance computed considering single crossroad is not appropriate to decide the applicability of MANET routing protocols in VANET. In addition, performance with constant speed is better than variable speed as it does not affect the topology compared to real scenarios. Hence, based on results obtained for the entire city road network scenario, AODV and OLSR protocols are not feasible for VANET as endurance with vehicle density and data generation rate are not satisfactory with VANET characteristics.

As a future work, topology and position based routing protocols of MANET will be re-evaluated on Nakagami channel fading model using single crossroad and multiple crossroad scenarios.

References

1. Boukerche, A.: Algorithm and Protocols for Wireless and Mobile Ad-hoc Network. Wiley Series on Parallel and Distributed Computing
2. Al-Sultan, S., Al-Doori, M.M.: A Comprehensive Survey on Vehicular Ad-hoc Network. Journal of Network and Computer Applications **37**, 380–392 (2014)

3. Perkins, C.E., Royer, E.M.: Ad-hoc on-demand distance vector routing. In: 2nd IEEE Workshop on Mobile Computing System and Applications WMCSA-1999, New Orleans LA, pp. 90–100 (1999)
4. Perkins, C., Belding-Royer, E., Das, S.: Ad-hoc On-Demand Distance Vector (AODV) Routing, RFC3561, July 2003
5. Jacquet, P., Muhlethaler, P., Clausen, T., Laouiti, A., Qayyum, A., Viennot, L.: Optimized link state routing protocol for ad-hoc network (OLSR). In: IEEE International Multi Topic Conference, IEEE INMIC-2001, pp. 62–68 (2001)
6. Clausen, T., Jacquet, P.: Optimized Link State Routing Protocol (OLSR), RFC3626, October 2003
7. The IEEE 802.11P Standards. http://www.ietf.org/mail-archive/web/its/current/pdfqf992dHy9x.pdf
8. Haerri, J., Filali, F., Bonnet, C.: Performance comparison of AODV and OLSR in VANETs urban environments under realistic mobility patterns. In: 5th IFIP Networking Workshop, Italy, June 14–17 (2006)
9. Clausen, T.H., Jacquet, P., Viennot, L.: Comparative Study of Routing Protocols for Mobile Ad-hoc NETworks (2002)
10. Khan, I., Qayyum, A.: Performance evaluation of AODV and OLSR in highly fading vehicular ad-hoc network environments. In: IEEE 13th International Conference on Multitopic, INMIC 2009, pp. 1–5, December 2009
11. Spaho, E., Ikeda, M., Barolli, L., Xhafa, F., Younas, M., Takizawa, M.: Performance evaluation of OLSR and AODV protocols in a VANET crossroad scenario. In: IEEE 27th International Conference on Advanced Information Networking and Applications (AINA), pp. 577–582, March 25–28, 2013
12. De Marco, G., Tadauchi, M., Barolli, L.: CAVENET: description and analysis of a toolbox for vehicular networks simulation. In: 2007 International Conference on Parallel and Distributed Systems, vol. 2, pp. 1–6, December 2007
13. Härri, J., Filali, F., Bonnet, C., Fiore, M.: VanetMobiSim: generating realistic mobility patterns for VANETs. In: Proceedings of the 3rd International Workshop on Vehicular Ad hoc Networks (VANET 2006), pp. 96–97. ACM, New York (2006)
14. The ns-2 Network Simulator. http://www.isi.edu/nsnam/ns/

State of the Art and Research Challenges for Public Flying Ubiquitous Sensor Networks

Andrey Koucheryavy, Andrey Vladyko, and Ruslan Kirichek[✉]

The Bonch-Bruevich Saint-Petersburg State University of Telecommunications,
Saint-Petersburg, Russia
akouch@mail.ru, vladyko@bk.ru, kirichek@sut.ru

Abstract. The article deals with theoretical and practical directions of Public Flying Ubiquitous Sensor Networks (FUSN-P) research. Considered the distinctive features of this type of networks from the existing ones. A wide range of issues is covered: from the methods of calculation FUSN to the new types of testing and model network structure for such networks. Presented a model network for full-scale experiment and solutions for the Internet of Things.

Keywords: Public flying ubiquitous sensor networks · Public unmanned aerial vehicles · Delay-Tolerant networks · New type of testing

1 Introduction

The invention of unmanned aerial vehicles (UAV) has led to the new challenges in networking and communications systems, and to possibility of providing new services supported by Flying Ad Hoc Networks FANET (Flying Ad Hoc Networks) [1,2,3]. FANET networks have already been used for military purposes [6], for surveillance [7], for locating purposes [8] etc. The research tasks for FANET are to study the questions of protocols' choice for rational functioning of FANET [1,9], dynamic routing [4], the usage of Non Beacon-enabled protocols to improve the quality of transmission video [7], data synchronization on Delay-Tolerant Networks [10] and so on.

It should be noted that almost all of these works are developing the idea of MANET networks in the flying network. At the same time, Ubiquitous Sensor Networks (USN) has taken the first place among the existing networks not only in terms of implementation, but in the provided services as well. It brings up the question about possibility and feasibility of flying ubiquitous sensor networks, the main purpose of which will be to collect data from terrestrial sensor fields and their maintenance. Such networks were called flying ubiquitous sensor networks (FUSN) [24]. The mobile sinks for USN can be UAV which increase connectivity to the ground sensor nodes [5]. Furthermore, the UAV can be temporary cluster head for ground USN clusters [21] and so on. A non-hierarchical structure of flying ubiquitous sensor networks with one or a plurality of public unmanned aerial vehicles (UAV-P) e.g. drons, quadracopters flying approximately in one plane will be considered in this paper. We have named this network as public flying ubiquitous networks (FUSN-P).

© Springer International Publishing Switzerland 2015
S. Balandin et al. (Eds.): NEW2AN/ruSMART 2015, LNCS 9247, pp. 299–308, 2015.
DOI: 10.1007/978-3-319-23126-6_27

2   FUSN-P Features

The FUSN is the network which consists of two segments (flying and terrestrial), and the communication between these segments is supported by ubiquitous sensor network protocols ZigBee, 6LoWPAN, RPL, Bluetooth Low Energy (BLE) and so on. The public unmanned aerial vehicles (UAV-P) can be used for FUSN creation and FUSN-P is the network abbreviation in this case. The FUSN-P architecture example with some UAV-P in the flying segment is shown on the Figure 1.

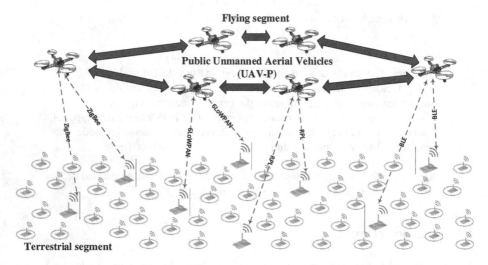

Fig. 1. The FUSN-P architecture example

The UAV-P's are controlled by people who usually have no professional education in this area. They can be farmers, hunters, fishermen, sportsmen, students and so on. Hereby the UAV-P should fly according to the fixed route. Thus the information from the UAV-P can be transmitted either by existing mobile communications networks, or be delivered by UAV-P to the base (Fig. 2). When sending information to the network, database can be regarded as Delay-Tolerant Network. It requires some research in the field of permissible delay information that is directly related to the problem of choosing the optimal route to fly around the sensor fields.

Example of incremental algorithm of download UAV flight path to collect data is presented on the Figure 3. Open source software, such as Mission Planner, can be used to plan a route. The area in which one needs to collect data from the sensor nodes is given on Google Maps (step 1). In the output of the program the flight mission which is loaded into the UAV-P is formed (step 2). UAV-P is equipped with sensors and obstacle detection algorithm that enables one to fly around obstacles encountered on the way, for example quadrocopter DJI Matrix 100 with Guidance (step 3). It is worth noting that the UAV is actually a self-contained during the flight, but the control of its movement is realized through the telemetry channel. The infrared sensors enable to measure the distance to bend around obstacles in line with its target algorithms. Also the battery charge and discharge are monitored in the case

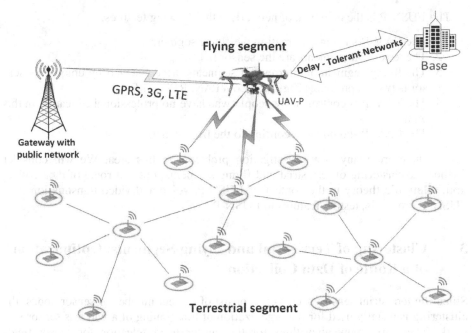

Fig. 2. Transmission of information from the UAV-P

of the UAV-P returns to base on its own. Furthermore UAV P can be equipped with cameras and other tools if needed. In step 4, the data collected is realized from the sensor nodes using the optimal algorithms, which are calculated on the location of the sensors. After collecting data UAV-P shall return to the base, and deliver data for later analysis or sending to the Internet through a gateway (step 5).

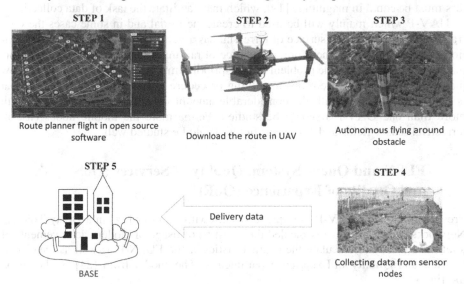

Fig. 3. Collecting data from sensor nodes using UAV-P

The FUSN-P is the new type of network with following features:

1. The UAV-P's using for creation the flying segments.
2. The terrestrial segments are the sensor fields.
3. The flying segment and terrestrial segments are connected by ubiquitous sensor networks protocols ZigBee, 6LoWPAN, RPL, BLE.
4. The UAV-P is controlled by people who have no professional education in this area.
5. The UAV-P should fly according to the fixed route.

So, there are many new investigation problems in this area. We will consider further the clustering of terrestrial and flying segments, optimal route of data collection, teletraffic theory in the context of FUSN-P, voice and video transmission over FUSN-P protocols, testing features to FUSN-P.

3   Clustering of Terrestrial and Flying Segments. Optimization of a Route of Data Collection

Since the terrestrial sensory fields can consist of a great number of sensor nodes, the clustering is usually used for the organization of functioning of a wireless sensor network. There are many algorithms for the cluster head selection for sensor fields located in the plane for the stationary [12, 13] and mobile sensor nodes [14,15]. However, the works in this direction are still in progress, and the appearance of new algorithms for terrestrial sensor fields [16] can improve performance FUSN-P. There are not so many algorithms for the cluster head selection for the ubiquitous sensor networks in three-dimensional space [17, 18]. Considering the complexity of collecting information using PUAV in three-dimensional space, it seems efficient that clusters must be equal in magnitude [19], which may facilitate the task of data collection.

UAV-P which mainly will be used to create the initial and in some cases the only fragment that flies to the surface of terresrial has a very limited capacity for the duration of being in flight. Therefore, the choice of rational route of collecting the information is considered as the problem of first-priority importance. The first results show that it is reasonable to use problem-solving procedure of travelling salesman[20] that is naturally connected with the considerable amount of calculations. The cases with more than one UAV-P's should be studied. Furthermore the mobile nodes case in terrestrial sensor fields and 3D sensor fields should be studied well.

4   FUSN and Queue System. Quality of Services (QoS) and Quality of Experience (QoE)

Presentation of the UAV-P as a queue system with a buffer [24] in a Delay-Tolerant Networks [11], has basically tackled the question of usage of modern achievements of teletraffic theory to calculate the characteristics of the FUSN-P as well as Quality of Service and Quality of Experience parameters. The model from [24] is shown on the fig. 4.

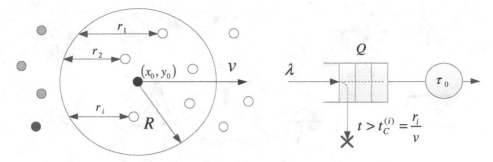

Fig. 4. UAV-P as a queuing system

R - radius of the circle in which the UAV-P can collect information from the sensor nodes;

r_i-a specific distance from the sensor assembly to the border of the circle with a radius R;

V-a velocity of the UAV-P;

λ – a parameter of the message flow of sensor nodes on the UAV-P;

τ_0 - a service entity;

Q – a queue.

However, the presence of multiple UAV-P for solving FUSN-P and a plurality of terrestrial sensory fields in conditions a Delay-Tolerant Networks require an extensive theoretical research in this field. Furthermore the presentation FUSN-P as a queue network can be very interesting.

The voice and video transmission from terrestrial segment to UAV-P is the next investigation task because often it may be the only chance to pass the necessary information to the area of terresrial sensor fields. There is a positive experience of transfer of voice data over the protocol ZigBee [25]. In connection with this problem the comparison of voice quality and video quality of experience during their transmission from the terrestrial segment for years using different protocols (ZigBee, 6LoWPAN, RPL), and different languages is of great interest for FUSN-P.

5   Model Network and Testing of the FUSN-P

In practical terms, it is appropriate to establish a model network for conformance and interoperability testing to ensure the implementation of FUSN-P, as well as other networks [22]. Since FUSN-P is one of the components of the Internet of Things, such a model network is created in the SUT in the Internet of Things laboratory. The structure of model network is given in fig. 5.

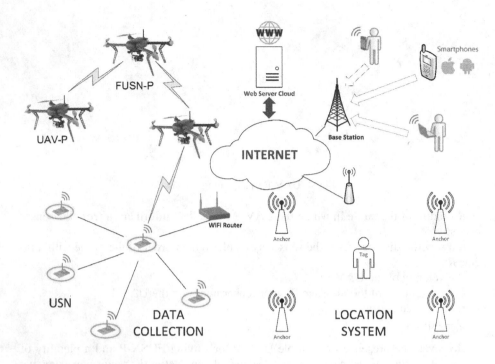

Fig. 5. Model network for full-scale experiment and solutions for the Internet of Things

The model network consists of 4 segments:

- The first segment: Data Collection is used to test the full life cycle of a wireless sensor network, the study of the principles of self-organizing sensor nodes and the study of protocols ZigBee, 6LoWPAN, RPL. SBCs Intel Edison, Intel Galileo # 2, Raspberry Pi, Arduino Yun, VoCore and others are used as the sensor nodes. In addition to the single-board computers there are radios with microcontrollers: ESP8266, EM357, nRF51822. Interaction with the Internet is realized through a gateway.

- The second segment is used for testing problems related to positioning indoors. At the moment, the Internet of Things is using two positioning systems in the lab: one which is based on IEEE 802.15.4 and another one which is based on IEEE 802.11. The accuracy of determining the position of the label is +/- 1.5 m.

- The third segment presents the flying ubiquitous sensor network. On the basis of public quadrocopters there is working off the tasks connected with data collection from the sensor fields and its transfer to the Internet. In addition quadrocopters are used to build the swarm structures that considerably accelerate the process of data collection.

- The fourth segment is implemented as a virtual entity of real Internet of things, such as: applications for smart phones, web applications and other means of control devices.

The laboratory has a server with special software that enables to track the delay, loss, jitter delivery of data from the endpoints.

The existing model enables the network to test any script of the Internet of Things interaction in the shortest time and to obtain quantitative and qualitative assessments.

One type of test is to determine the quality of the communication channel functioning between the terresrial and the flying segment. Modern UAV-P's aboard have several radio modules operating in different frequency bands (Figure 1). To pass parameters to telemetry (sending up to 500 parameters at once) there is a radio channel 433 MHz for Europe and 915 MHz for the US, to broadcast streaming video from cameras UAV, for example, for the First Person View-flight, 5.8 GHz radio is used. There is the likelihood of errors due to the interference signals. For this reason, data corruption can occur in the physical layer, resulting in delays and losses during data transmission [26,27,28].

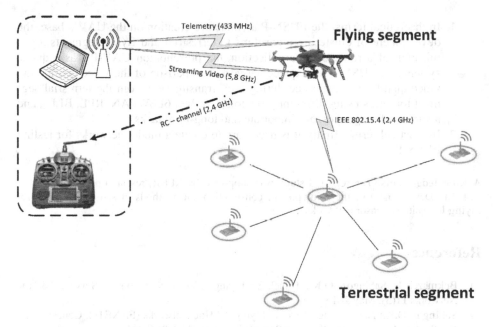

Fig. 6. Testing the quality of the communication channel between the terrestrial and the flying segment FUSN

The series of tests have shown that when testing the impact of radio drones on IEEE 802.15.4 channel by increasing the distance between the flying and terrestrial segment FUSN-P, there is a significant increase in the percentage of packets received with errors. Thus, at a distance of 20 meters, this value reaches 63% of the total number of the sent packets. Thus, collecting data from the terrestrial component is not possible with such an amount of errors. Flight altitude UAV-P's above 10 meters is critical for FUSN-P, in all tests, as losses in excess of 5% to 10% lead to significant delays in the delivery of packages.

In addition to traditional forms of testing, such as compliance, compatibility, benchmarking testing FUSN-P has added two new species. This is Delay-tolerant testing and test of legality.

In addition to collecting information in the moment of UAV-P above the touch field, it can test the terrestrial network segment or part of the residual energy parameters. This type of testing is named Delay-Tolerant Testing by analogy with Delay-Tolerant.

The second new type of testing is related to the fact that in ubiquitous sensor networks one of the common types of intrusion is the cloning of sensor nodes [23]. Therefore, the function of testing the legality of sensor nodes on a flying segment of FUSN-P could be entrusted, and it makes up a current terrestrial network segment of Public Flying Ubiquitous Networks.

6   Conclusions

1. In theoretical terms, the FUSN-P architecture creation on the UAV's base, the development of clustering algorithms to terrestrial and flying segments, optimization of a route of data collection, the presentation UAV-P's as a queue system and FUSN-P as a queue network, comparison of the voice quality and video quality of experience during their transmission from the terresrial segment for years using different protocols (ZigBee, 6LoWPAN, RPL, BLE), and a variety of languages, are important and long-term tasks.
2. In practical terms, firstly it is necessary to create a model networks for testing FUSN-P.

Acknowledgements. The reported study was supported by RFBR, research project No15 07-09431a "Development of the principles of construction and methods of self-organization for Flying Ubiquitous Sensor Networks".

References

1. Bekmezci, I., Sahingoz, O.K., Temel, S.: Flying Ad-Hoc Networks: A Survey. Ad Hoc Networks **11**(3), May 2013
2. Sahingoz, O.K.: Networking Model in Flying Ad Hoc Networks (FANETs): Concepts and Challenges. Journal of Intelligent &Robotics Systems **74**(1-2) (2014)
3. Singh, S.K.: A Comprehensive Survey on Fanet: Challenges and Advancements. International Journal of Computer Science and Information Technologies **6**(3) (2015)
4. Rosati, S., Kruzelecki, K.; Heitz, G.; Floreano, D: Dynamic Routing for Flying Ad Hoc Networks. http://ieeexplore.ieee.org/xpl/RecentIssue.jsp?punumber=25 arXiv preprint arXiv:1406.4399
5. de Freitas, E.P., Heimfarth, T., Netto, I.F., Lino, C.E., Pereira, C.E., Ferreira, A.M., Wagner, F.R., Larsson, T.: UAV relay network to support WSN connectivity. Proceedings of the ICUMT 2010
6. Orfanus, D., Eliassen, F., de Freitas, E.P.: Self-organizing relay network supporting remotely deployed sensor nodes in military operations. In: Proceedings of the 6 th ICUMT, St. Petersburg, Russia, 6-8 October

7. Rosário, D., Zhao, Z., Braun, T., Cerqueira, E., Santos, A.: A comparative analysis of beaconless opportunistic routing protocols for video dissemination over flying ad-hoc networks. In: Balandin, S., Andreev, S., Koucheryavy, Y. (eds.) NEW2AN/ruSMART 2014. LNCS, vol. 8638, pp. 253–265. Springer, Heidelberg (2014)
8. DeLima, P., York, G., Pack, D.: Localization of ground targets using a flying sensor network. In: Proceedings of the IEEE International Conference on Sensor Networks, Ubiquitous, and Trustworthy Computing, 2006, Taichung, Taiwan, vol. 1, pp. 5–7, June 2006
9. Vasiliev, D.S., Meitis, D.S., Abilov, A.: Simulation-based comparison of AODV, OLSR and HWMP protocols for flying Ad Hoc networks. In: Balandin, S., Andreev, S., Koucheryavy, Y. (eds.) NEW2AN/ruSMART 2014. LNCS, vol. 8638, pp. 245–252. Springer, Heidelberg (2014)
10. Phuong, H., Yamamoto, H., Yamazaki, K.: Data synchronization method in DTN sensor network using autonomous air vehicle. In: Proceedings, International Conference on Advanced Communication Technology, ICACT 2014, Phoenix Park, Korea (2014)
11. Akyildiz, I.F., Akan, O.B., Chen, C., Fang, J., Su, W.: InterPlaNetary Internet: state-of-the-art and research challenges. Computer Networks **43** (2003)
12. Heinzelman, W., Chandrakasan, A., Balakrishnan, H.: Energy-efficient communication protocol for wireless microsensor networks. In: Proceedings 33rd Hawaii International Conference on System Sciences (HICSS), Wailea Maui, Hawaii, USA, January 2000
13. Koucheryavy, A., Salim, A.: Cluster head selection for homogeneous wireless sensor networks. In: Proceedings, International Conference on Advanced Communication Technology, ICACT 2009, Phoenix Park, Korea (2009)
14. Kim, D., Chung, Y.: Self-organization routing protocol supporting mobile nodes for wireless sensor network. In: Proceedings of the First International Multi-Symposiums on Computer and Computational Sciences, vol. 2 (2006)
15. Koucheryavy, A., Salim, A.: Prediction-based clustering algorithm for mobile wireless sensor networks. In: Proceedings, International Conference on Advanced Communication Technology, ICACT 2010, Phoenix Park, Korea (2010)
16. Al-Qadami, N., Laila, I., Koucheryavy, A., Saker Ahmad, A.: Mobility adaptive clustering algorithm for wireless sensor networks with mobile nodes. In: Proceedings, International Conference on Advanced Communication Technology, ICACT 2015, Phoenix Park, Korea (2015)
17. Attarzadeh, N., Mehrani, M.: A New Three Dimensional Clustering Method for Wireless Sensor Networks. Global Journal of Computer Science and Technology **11**(6), version 1.0, April 2011
18. Abakumov, P., Koucheryavy, A.: The cluster head selection algorithm in the 3D USN. In: Proceedings, International Conference on Advanced Communication Technology, ICACT 2014, Phoenix Park, Korea (2014)
19. Abakumov, P., Koucheryavy, A.: Clustering algorithm for 3D wireless mobile sensor network. In: The 15th International Conference on Internet of Things, Smart Spaces, and Next Generation Networks and Systems, NEW2AN 2015. LNCS. Springer, Heidelberg (2015 accepted)
20. Kirichek, R., Paramonov, A., Vareldzhyan, K.: Optimization of the UAV's motion trajectory in flying ubiquitous sensor networks (FUSN). In: The 15th International Conference on Internet of Things, Smart Spaces, and Next Generation Networks and Systems, NEW2AN 2015. LNCS. Springer, Heidelberg (2015 accepted)
21. Futahi, A., Koucheryavy, A., Paramonov, A., Prokopiev, A.: Ubiquitous sensor networks in the heterogeneous LTE network. In: Proceedings, International Conference on Advanced Communication Technology, ICACT 2015, Phoenix Park, Korea (2015)

22. Recommendation Q.3900. Methods of testing and model network architecture for NGN technical means testing as applied to public telecommunication networks. ITU-T, Geneva, July (2006)
23. Koucheryavy, A., Bogdanov, I., Paramonov, A.: The mobile sensor network life-time under different spurious flows intrusion. In: Balandin, S., Andreev, S., Koucheryavy, Y. (eds.) NEW2AN 2013 and ruSMART 2013. LNCS, vol. 8121, pp. 312–317. Springer, Heidelberg (2013)
24. Kirichek, R., Paramonov, A., Koucheryavy, A.: Flying ubiquitous sensor networks as a quening system. In: Proceedings, International Conference on Advanced Communication Technology, ICACT 2015, Phoenix Park, Korea (2015)
25. Touloupis, E., Meliones, A., Apostolacos, S.: Speech codes for high-quality voice over ZigBee applications: Evaluation and implementation challenges. IEEE Communications Magazine 50(4), April 2012
26. Moltchanov, D., Koucheryavy, Y., Harju, J.: Cross-layer modeling of wireless channels for data-link and IP layer performance evaluation. Computer Communications 29(7), April 24, 2006
27. Dunaytsev, R., Moltchanov, D., Koucheryavy, Y., Harju, J.: Modeling TCP SACK performance over wireless channels with completely reliable ARQ/FEC. International Journal of Communication Systems 24(12), December 2011
28. Dunaytsev, R., Koucheryavy, Y., Harju, J.: TCP NewReno throughput in the presence of correlated losses: the slow-but-steady variant. In: Proceedings of the 25th IEEE International Conference on Computer Communications, INFOCOM 2006

A Practical Implementation of a Cooperative Antenna Array for Wireless Sensor Networks

Edison Pignaton de Freitas[1,2](✉), Ricardo Kehrle Miranda[3], Marco Marinho[3],
João Paulo Carvalho Lustosa da Costa[3], and Carlos Eduardo Pereira[1,2]

[1] Institute of Informatics, Federal University of Rio Grande Do Sul, Porto Alegre, Brazil
epfreitas@inf.ufrgs.br, cpereira@ece.ufrgs.br
[2] School of Engineering, Federal University of Rio Grande Do Sul, Porto Alegre, Brazil
[3] Department of Electrical Engineering, University of Brasília, Brasília, Brazil
{rickehrle,jpdacosta}@unb.br, marco.marinho@ieee.org

Abstract. Energy consumption is a key issue to be handled in Wireless Sensor Networks, especially considering low-end sensor nodes, i.e. sensor with severe energy resources limitations. When sensor nodes have their energy resources depleted, they stop working which can compromise the whole network functioning, thus its lifetime. As communication is the most energy-consumption task, enhancements in communication that diminish the amount of messages lost and the need for retransmissions are very important to preserve energy resources and extend the network lifetime. Considering the impact of the energy preservation and the opportunity to exploit it in terms of communication, this paper discusses the practical implementation of a cooperative MIMO scheme based on virtual antenna array using sensor nodes in order to enhance data communication in wireless sensor networks. The conducted experiments present evidence of the feasibility of the proposed approach highlighting performance aspects.

Keywords: Wireless sensor networks · Cooperative antenna array · Cooperative multiple input multiple output · Energy efficiency

1 Introduction

The use of wireless sensor networks (WSN) is considered a key enabling feature to a number of emerging applications in many areas, from precision agriculture to military and defense systems [1]. Despite the number of applications that can benefit of the usage of WSN, their energy resource limitation is a practical issue in their employment, which still hinders their massive usage. Wireless sensor nodes are usually tiny resource constrained platforms, driven by batteries with limited energy budget. In most of cases, they are deployed in places difficult to be accessed, such as large areas mixed with all kind of hazards or even inside building structures. This particularity in the WSN deployment makes impracticable the replacement of their energy resources. To overcome this problem, a smart energy resource management must take place.

Studies demonstrate that the most energy consuming task in wireless sensor nodes is communication [2]. Thus, efficient communication mechanisms are highly

© Springer International Publishing Switzerland 2015
S. Balandin et al. (Eds.): NEW2AN/ruSMART 2015, LNCS 9247, pp. 309–318, 2015.
DOI: 10.1007/978-3-319-23126-6_28

desirable to reduce the sensor nodes energy depletion and consequently, enlarge the whole network lifetime. Observing this aspect, a number of proposals try to address the problem with alternative routing protocols [3], energy aware broadcast [4], energy aware MAC protocols [5] and Multiple Input Multiple Output (MIMO) based systems [6]. Among these approaches, the three first are extensively explored in several different ways to optimize respectively the network and data link layers. The last one is a promising technology that tries to complement the energy saving effort in the upper layer addressing the problem in the physical layer.

Conventional MIMO techniques explore antenna arrays installed in a single node, as how it is used in WIFI access-points and routers. Due to the resource scarcity of the wireless sensor nodes, the same strategy used in WIFI equipment is not feasible. However, WSN are inherently cooperative distributed systems. Observing this aspect, an opportunity can be explored, which is the composition of a virtual MIMO system, or a cooperative MIMO system. In this alternative the antenna array is formed by the antennas of different nodes, which cooperate to compose this virtual antenna array [6]. Despite the promising advantage of this approach, there are technical issues that might hinder its feasibility, which is mainly related to the synchronization of the sender nodes, as discussed in [7]. Observing these practical issues, the goal of this project is present the results of a feasibility test of a cooperative MIMO approach application to WSN by the evaluation of a practical implementation.

The remaining text is structured as follows: Section 2 presents a brief discussion about cooperative MIMO and its usage in WSN. Section 3 provides an overview about the synchronization problem. Section 4 describes the framework for the feasibility tests, while Section 5 reports and discusses the acquired results. Section 6 concludes the paper providing directions for future work.

2   Cooperative MIMO and Its Usage in WSN

Wireless sensor networks are cooperative by nature. Observing this aspect, a cooperative MIMO approach can be implemented in order to minimize the energy consumption due to communication. As opposed to traditional MIMO systems, in which a set of antenna is present at the transmitter and at the receiver nodes, the cooperative MIMO utilizes a virtual MIMO scheme. In this virtual MIMO scheme, the multiple antennas involved are present at different nodes. This avoids the increased hardware complexity, which is especially important due to the limitations of the hardware platform of the sensor nodes. The additional complexity is transferred to the communication protocol.

Figure 1 presents an example in which two clusters of sensors establish a communication as a MIMO system. Notice that in this text the term "cluster" refers to groups of cooperating nodes in a cooperative MIMO arrangement, it is not the same meaning as used in hierarchical WSN. The figure also presents another possible situation in which a cluster of sensors establish communication with a single sensor similar to a single input multiple output/ multiple input single output (SIMO/MISO) system. Using this approach, if two sensors near to each other cooperate to transmit information, and two sensors on a far cluster cooperate to receive data, the efficiency is effectively doubled, as two symbols can be transmitted over the same time slot [8].

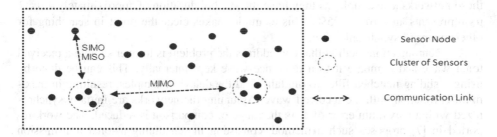

Fig. 1. Cooperative MIMO Applied to WSN: communication between clusters of sensors nodes and SIMO/MISO communication between a cluster an individual sensor.

Figure 2 presents the steps involved in a cooperative MIMO communication.

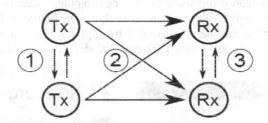

Fig. 2. Cooperative MIMO Transmission Steps.

In the 2 by 2 MIMO system example presented in Figure 1 the numbers represent the steps that are described as follows: (1) The transmit sensors exchange the information that needs to be transmitted; (2) Both sensors transmit different symbols at the same time slot; (3) The receive sensors exchange the received information so that the original symbol sequence can be obtained. If the data is destined to only one sensor of the receiving cluster this exchange becomes uni directional. Another option is to exchange only a portion of the received information so that every sensor is responsible for part of the decoding, alleviating the computational burden of a single node.

3 The Synchronization Problem in Cooperative MIMO Communications

The synchronization problem is widely studied in the WSN research area, with a number of different solutions proposed in the literature [9]. Some of these solutions rely on GPS synchronization, which despite the great accuracy (with variations being kept as small as 200 ns), has the severe drawback of the energy consumption overhead and the problem of WSN dependence to the GPS system. Other approaches proposed broadcast synchronization schemes, which are capable to achieve 1 μs accuracy, which is a good result, but not sufficient considering, for instance, conventional sensor networks operating at a 256 kbps rate and using BPSK modulation. Taking these networks as example, as they have the symbol duration of approximately 4 μs, 1 μs represents an error of 25%. This example makes clear the point in searching for other means of synchronization.

A synchronization method that can address the problem is to over sample a received tonal wave and compare it to a reference wave kept internally. This can be done by using a sliding matched filter to digitally find the delay, in samples, resulting in maximum correlation with the received wave. Assuming the networks are initially synchronized with a maximum error of 1 μs the range of comparison is reduced. The work reported in [7] proposes such a method, consisting in scheduling a tonal transmission between a pair or tonal broadcast to a group of nodes, the sampling on the receiving nodes will start at the scheduled time, and the clock error can be compensated.

In order to avoid problems created by sampling with a difference of more than a period of the tonal wave, the time length of the tonal transmission needs to be known to the receiving nodes. With this information, if a signal with less than the expected length is received, the receiver can compensate by starting sampling earlier or later, adjusting its internal clock accordingly. This mechanism is shown in Figure 3, in which sampling synchronization error d can be compensated by applying the sliding correlator. For a complete discussion about this method, interested readers are referred to [7].

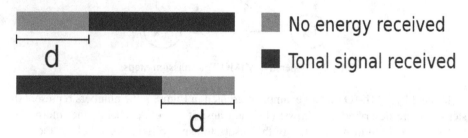

Fig. 3. Cooperative MIMO Transmission Steps.

4   Feasibility Test Framework

As introduced in Sections 2 and 3, despite the usefulness of the cooperative MIMO approach in WSN, there are practical implementation issues that have to be addressed in order to effectively apply cooperative MIMO. Motivated by these issues, the following feasibility test framework was designed.

The framework is composed of two sensor nodes, whose description is kit freescale MC1322x Sensor Node, instrumented with a developed firmware to control the communications tests and a software-programmable radio transceiver equipment from Texas Instruments (USRP 2932).

The developed firmware was generated by the *Beekit* based on the firmware *Connectivity Test* provided by Freescale. It basically controls the transmission and reception of synchronization messages. Its block diagram is depicted in Figure 4.

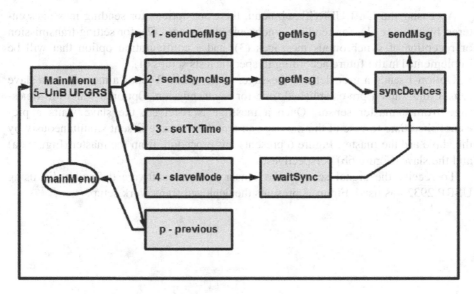

Fig. 4. Block Diagram of the Developed Firmware.

The UART serial interface of the sensor nodes was used for data input and output. To handle this interface, it is necessary to use a serial communication software, as the HyperTerminal, which was the one used in this work. An option labeled *UnB UFRGS* was introduced in the start menu of the *Connectivity Test* firmware (Figure 5a) to access the sub-menu with functionalities developed to perform the planned feasibility tests (Figure 5b).

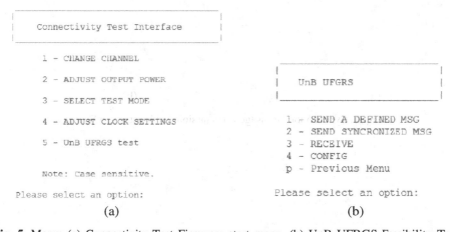

Fig. 5. Menu: (a) Connectivity Test Firmware start menu. (b) UnB UFRGS Feasibility Test Submenu.

Accessing the UnB UFRGS submenu, there are options for sending messages entered by the user (1), for sending synchronized messages (2), for setting transmission or reception of synchronous messages (3), and a configuration option that will be implemented in the future according to specific tests setups (4).

Option 1 sends a user defined message, while option 2 sends a message to a slave sensor and waits a pre-established time for retransmission. Option 3 waits for a message from a master sensor. Once a message is received, the slave waits a pre-established time to resend the message so that the message is sent simultaneously by the slave and the master. Figure 6 presents this procedure from the master (Figure 6a) and the slave (Figure 6b) perspectives.

To receive the signal sent by the sensor nodes, a software defined radio using USRP 2932 was used. Figure 7 presents the deployed framework setup.

(a)   (b)

Fig. 6. Synchronous message transmission: (a) Master. (b) Slave

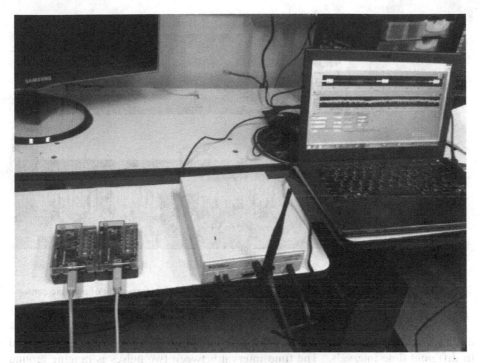

Fig. 7. Deployed Test Framework Setup

5 Experimental Results

Using the USRP 2932, a constructive interference between signals from the two sensor nodes was generated to increase the communication range using a cooperative MIMO virtual array.

The performed tests used the channel 21 of the IEEE 802.15.4 standard with central frequency of 2455 MHz, which is located between the channels 9 and 10 of the IEEE 802.11 standard (WIFI) due to the fact that this was the choice that receives less interference in the place where the tests were performed. Figures 8 to 10 show the received signals.

In Figure 8, observe the signal transmitted by one single radio in time and frequency without perceptible interferences. The signal amplitude is approximately 0.0005 V according to the display Amplitude [V] vs Time [s]. The time interval between two pulses is approximately 0.01 s, while the pulse width is approximately 0.0033 s.

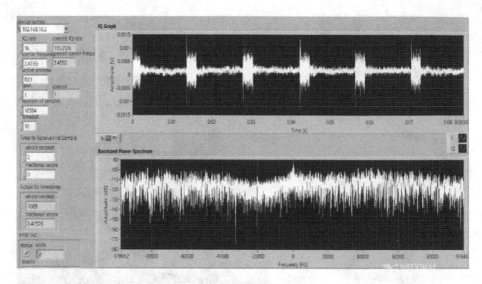

Fig. 8. Results acquired by the Spectrum Analyzer during the transmission of a single sensor

In Figure 9 the results of the radios of the two sensor nodes performing an unsynchronized transmission are shown. Note that the pulses from each radio are received in different time intervals. The time interval between two pulses is ranging around 0.005 s.

Figure 10 show the constructive interference of the signals from the two sensor nodes, which are almost completely overlapped. Note the increased signal power due to this overlap.

Performing these feasibility tests, we show the gain obtained due to the synchronization by means of a software implementation. Note that when the transmissions of the two sensors were synchronized, the amplitude of the signal approximately doubles. Moreover, during the experiments, it was possible to notice that there is sufficient channel diversity to assure the MIMO transmission, at least for the tested case of two sensor transmitting at the same time.

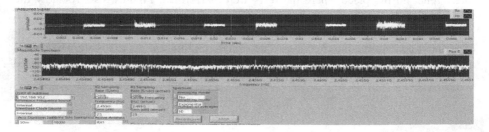

Fig. 9. Results acquired by the Spectrum Analyzer for non-synchronized sensors

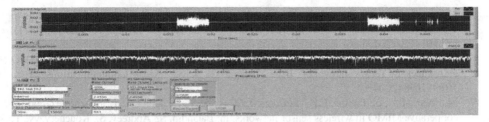

Fig. 10. Results acquired by the Spectrum Analyzer when the sensors are synchronized.

6 Conclusions and Future Work

Radio transmissions are responsible for the highest amount of consumed power in wireless sensor networks. This work provided a practical experiment for testing a promising approach for saving power on data transmissions, the cooperative MIMO. To take full advantage of cooperative MIMO in WSNs, synchronization is the key aspect to provide increased data rates or transmission range. In this work a firmware was created to check the feasibility of cooperative MIMO by only adapting the higher layers of communication. The tests showed that it is possible to merge packets and increase transmission power. However, achieving complete and precise synchronization is still challenging.

The next step in this type of experiment is to access the waveform directly from the physical layer. Once the physical layer is accessed, the first task it to synchronize carriers. Once that is done, WSNs can take full advantage of the increased performance and efficiency provided by cooperative MIMO. These techniques can then be explored in order to save power in wireless sensor networks.

Acknowledgements. The authors thank the Research Financing Agency of Federal District in Brazil (FAPDF), the Research Financing Agency of Rio Grande do Sul in Brazil (FAPERGS) and the Brazilian National Counsel of Technological and Scientific Development (CNPq) for the provided support to develop this research.

References

1. Business Week: 21 ideas for the 21st century, pp. 78–167, August 30, 1999
2. Mini, R.A.F., Loureiro, A.A.F.: Energy in wireless sensor networks. In: Garbinato, B., Miranda, H., Rodrigues, L. (eds.) Middleware for Network Eccentric and Mobile Applications. Springer, pp. 3–24 (2009)
3. Goyal, D., Tripathy, M.R.: Routing Protocols in Wireless Sensor Networks: A Survey. Second International Conference on Advanced Computing & Communication Technologies, pp. 474–480 (2012)
4. Durresi, A., Paruchuri, V., Barolli, L., Raj, J.: QoS-energy aware broadcast for sensor networks. In: Proceedings of 8th ISPAN, pp. 6 (2005)

5. Huang, P., Xiao, L., Soltani, S., Mutka, M.W., Xi, N.: The Evolution of MAC Protocols in Wireless Sensor Networks: A Survey. IEEE Communications Surveys & Tutorials **15**(1), pp. 101–120 (2013)
6. Cui, S., Goldsmith, A.J., Bahai, A.: Energy-efficiency of MIMO and cooperative MIMO techniques in sensor networks. IEEE Journal on Select. Areas Communications **22**(6), 1089–1098 (2004)
7. Marinho, M.A., de Freitas, E.P., da Costa, J.P.C.L., de Sousa Júnior, R.T.: Synchronization for cooperative MIMO in wireless sensor networks. In: Balandin, S., Andreev, S., Koucheryavy, Y. (eds.) NEW2AN 2013 and ruSMART 2013. LNCS, vol. 8121, pp. 298–311. Springer, Heidelberg (2013)
8. Wolniansky, P.W., Foschini, G.J., Golden, G.D., Valenzuela, R.A.: V-BLAST: an architecture for realizing very high data rates over the rich-scattering wireless channel. In: Signals, Systems and Electronics (1998)
9. Lasassmeh, S.M., Conrad, J.M.: Time synchronization in wireless sensor networks: a survey. In: Proceedings of the IEEE SoutheastCon 2010 (SoutheastCon), pp. 242–245 (2010)

Coverage and Connectivity and Density Criteria in 2D and 3D Wireless Sensor Networks

Nasser Al-Qadami and Andrey Koucheryavy[✉]

Saint-Petersburg State University of Telecommunications,
Pr. Bolshevikov, 22, Saint Petersburg, Russia
alkadami@spbgut.ru, akouch@mail.ru

Abstract. Sensor networks have implemented in many resent interest, such as remote tracking, atmospheric, flying sensor networks, space communications and environmental monitoring in real-time, etc. Although most wireless sensor networks are designed to work on two-dimensional (2D) plane, in reality, sensor networks usually deployed in three dimensions (3D) space. Expanding wireless sensor networks from 2D to appropriate 3D spaces facing many difficulties and the network topology is getting more complex in 3D than the topology of 2D. This paper focuses on major issues in designing WSNs: coverage, connectivity, density criteria and the ratio of transmission range to sensing range in 2D and 3D WSNs, where the goal is to improve the accuracy of wireless sensor networks and the results of this paper could be used for extending the design of WSNs from 2D to 3D sensor networks.

Keywords: Network topology · Deployment · Coverage · Connectivity · Network density

1 Introduction

Wireless sensor networks are made up of ultra-low-power consuming, low cost, distributed devices called sensor nodes that combine sensing, computation and communication [1]. Features such as low cost, low power, compact size and robustness of these sensor nodes aid them in being used for various applications [2].

Sensors are deployed in a region, they wake up, organize themselves as a network, and start sensing the environment and communicating the information to a data collection center [3]. Although most wireless sensor networks are based on two dimensional (2D) design, in reality, such networks operate in three dimensions (3D) areas in real applications, for example on multiple floors of a building, in a forest monitoring (on trees of different heights), underwater monitoring, atmospheric and flying sensor network in the air for tracking chemical plumes etc. Transition from 2D to 3D is not always easy, since many problems in 3D are significantly harder than their 2D counterparts. Many problems that are trivial to solve in 2D turn out to be major research challenges in 3D that remain unsolved even for centuries [4].

© Springer International Publishing Switzerland 2015
S. Balandin et al. (Eds.): NEW2AN/ruSMART 2015, LNCS 9247, pp. 319–328, 2015.
DOI: 10.1007/978-3-319-23126-6_29

The deployment strategy (random, regular, planned, etc.), sensing range, communication rang of sensor nodes and density of the WSN are important issues in designing WSNs [5]. Coverage and connectivity are two of the most fundamental issues in WSNs, which have a great impact on the performance of WSNs. Optimized deployment strategy, sleep scheduling mechanism, and coverage radius cannot only reduce cost, but also extend the network lifetime [6]. Maximizing the sensing coverage is a fundamental requirement for many critical applications of sensor networks. For example, the military applications and the medical applications require a high degree of coverage, as the failure of even a single node can lead to disasters. On the other hand, applications targeted for commercial use do not require much fault tolerance and can have a low coverage degree which reduces the cost factor also [7].

The network in which every point within sensor field (or every target in the target set) achieves the required coverage degree called as a 100% coverage area (complete coverage).

As a matter of fact, there are several problems in the geometry structures that are solved in 2D, but are very complex in 3D. Examples include the art gallery and sphere-packing problems that are closely tied to sensing coverage optimization [8].

Figure. 1 illustrates the geometric virtual structure for 2D and 3D sensor networks.

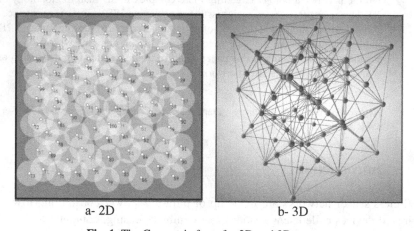

a- 2D   b- 3D

Fig. 1. The Geometric form for 2D and 3D

In spite of the difference between the normal 2D and the more realistic 3D scenario is only one extra dimension, extending the design of WSNs from 2D to 3D sensor networks, the sensing area of sensor node and the network topology are getting more complex, therefore when talking about the relative position of sensor nodes, we need 3D sense of space to construct the model.

2   Related Works

Several works has been done that address one way or other main issues of designing WSNs such as, deployment density, coverage and connectivity. The problem of providing sensing coverage has received significant attention in the context of two-dimensional networks [4]. How to achieve better deployment of sensors network is addressed by [9]. In [10] the authors studied the complex surface coverage problem in sensor networks. The critical radius for connectivity Ad Hoc network was analyzed in [11]. In [12] the authors discussed the expected coverage ratio which belongs to the surface coverage, by considering different 3D terrain models. In [13] the authors studied the surface coverage problem which is related to the optimal sensor deployment on the three-dimensional terrains.

In this paper, we focus on major issues in WSNs designing where the goal is to guarantee at least 90% area coverage of 2D and 3D WSNs, where all the node have the same sensing range and the same transmission range. In particular, we want to answer the following questions:

- What should be the minimum sensing range and sensor density to get between 90...99% of the field coverage for 2D and3D WSNs?
- What is the number of nodes required for surveillance of a 2D area and 3D space is minimized, while guaranteeing 90...99% coverage?
- What is the suitable SR/CR ratio that could lead to improve the performance of the network, such as network life time, stability period and throughput of the network by using the ratio of R_c/R_s.

The remainder of this paper is organized as follows: in section 3, we introduce a short description about deployment challenges in 2D and 3D WSNs .In section 5, the relationship between sensing range (R_s), coverage fraction (c) and sensor density (ρ) in 2D WSNs and 3D WSNs. In section 6, the critical transmission radius and the ratio of the communication radius to the sensing radius in section 7. Finally, in section 8, we draw the main conclusions.

3   Deployment Challenges in 2D and 3D WSNs

The area coverage and network connectivity are very important issues in designing wireless sensor networks and depend on node deployment strategy. Deployment method concerns with how a sensor network is constructed. In general sensor nodes can be deployed in primarily two ways: (a) deterministic deployment and (b) random deployment [7]. In deterministic deployment methods, sensor nodes are deployed only in specific locations determined by the user and this method of deployment guarantee that the target area is entirely covered and this method can be applied to a small or medium sensor network in a friend environment. However, sometimes the deterministic placement of sensor nodes is not possible because of high densities of nodes and deployment in unfriendly or hazardous environments, then the random sensor deployment might be the only choice.

In random deployment methods, the nodes can be scattered over a vast region for the purpose of detecting or monitoring some special events. In this case the first challenge encountered in WSNs is how to guarantee that the monitoring region is covered perfectly. To make sure of that, we should densely deploy the sensor node.

However, in this case, there may be redundancy in the network in the sense that two or more sensors may have an overlapping sensing area. Accordingly, the question is how to increase the coverage area at a time in which we use the least possible number of active nodes? One may want to know how many sensor nodes are needed before the nodes deployment; this is known as sensor density(ρ) and defined as the number of nodes per unit area. The minimal number of nodes required for 100% area coverage defined as the critical sensor density (CSD) of the network.

4   The Relationship Between Sensing Range (R_s)Coverage Fraction (c) and Sensor Density (ρ) in 2D WSNs and 3D WSNs

In this section, we study the relationship between sensing range, coverage fraction and sensor density in 2D and 3D WSNs. Where sensors are randomly distributed in a vast geographical area. R_s- represents the common sensing range of the sensors, c-represents the coverage fraction and ρ- represents the density of the underlying Poisson point process.

4.1   Two-Dimensional Case

Suppose n total sensor nodes are randomly distributed in 2D plane, a sensor's sensing ability is Omni-directional; its coverage range is a disk whose radius is R_s and whose area is:

$$S = \pi R_s^2 \tag{1}$$

With network density

$$\rho = \frac{n}{S} \tag{2}$$

The coverage fraction (c) defined as the ratio of covered area (A) or volume (V) to the total area or volume of interest at time t > 0 is shown in [15] and is given by:

$$c = 1 - e^{\rho S} \tag{3}$$

In order to achieve desired target area coverage($0.9 \le c \le 1$) approximately, the node density should be

When $c \ge 0.99$ ... .$0.90 \Leftrightarrow \rho S \ge 4.6$........2.3 respectively as follow:

$c = 0.98 \Leftrightarrow \rho S \ge 3.912, \quad c = 0.97 \Leftrightarrow \rho S \ge 3.5, \quad c = 0.96 \Leftrightarrow \rho S \ge 3.21$
$c = 0.95 \Leftrightarrow \rho S \ge 2.995, \quad c = 0.94 \Leftrightarrow \rho S \ge 2.813, \quad c = 0.93 \Leftrightarrow \rho S \ge 2.66$
$c = 0.92 \Leftrightarrow \rho S \ge 2.526, \quad c = 0.91 \Leftrightarrow \rho S \ge 2.408, \quad c = 0.90 \Leftrightarrow \rho S \ge 2.30$

Effectively detect any crossing object; sensors should be deployed at a density higher than the critical value.

$$\rho = \frac{-\log(1-c)}{\pi r^2} \Leftrightarrow R_s = \sqrt{\frac{-\log(1-c)}{\pi \rho}} = \sqrt{\frac{-\log(1-c)S}{\pi n}} \tag{4}$$

Figure. 2 shows the relationship between sensing range (R_s), the percentage of area coverage and how many sensors should be deployed in every square meter in 2D plane WSNs to get a certain degree of coverage.

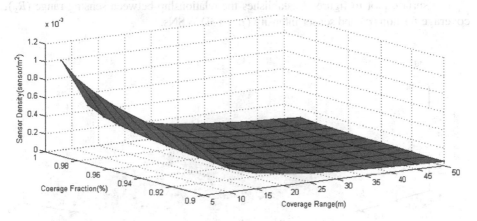

Fig. 2. The relationship between sensing range (R_s), coverage fraction (c) and sensor density (ρ) in 2D WSNs.

4.2   Three-Dimensional Case

We are given a set of sensors, $S = \{S_1, S_2, \ldots, S_n\}$, in a three-dimensional cuboid sensing field A. Each sensor S_i, $i=1\ldots n$, is located at coordinate (x_i, y_i, z_i) inside V and has a sensing range of R_s. So the sensing area for each sensor S_i can be represented by a sphere centered at (x_i, y_i, z_i) with radius R_s.

$$V_s = \tfrac{4}{3}\pi R_s^3 \tag{5}$$

The number of sensors located in V are given by:
$$n = V * \rho \tag{6}$$
The fraction of the volume being covered is hence given by:
$$c = 1 - e^{\rho V_s} \tag{7}$$

This formula can be used in the network planning to determine the required sensor density.

Similarly, as in the previous case when $c \geq 0.99 \ldots 0.90 \Leftrightarrow \rho V_s \geq 4.6 \ldots 2.3$ respectively.

In sensor network planning or deployment, the designer may want to determine the minimal number of nodes required to ensure complete area coverage. Using the above area coverage result, we can derive the required sensor density (ρ) in order to achieve a desired area coverage (c).

In general: the sensor density can be calculated by the following form:

$$\rho = \frac{-\log(1-c)}{\tfrac{4}{3}\pi R_s^3} \Leftrightarrow R_s = \left[\frac{-\log(1-c)}{\tfrac{4}{3}\pi \rho}\right]^{1/3} = \left[\frac{-\log(1-c)V_s}{\tfrac{4}{3}\pi n}\right]^{1/3} \tag{8}$$

Where:

c- The coverage fraction, ρ - Represents the expected number of nodes per unit volume, R_s - The sensing range.

The surface plot in figure. 3 establishes the relationship between sensing range (R_s), coverage fraction (c) and sensor intensity (ρ) in 3D WSNs.

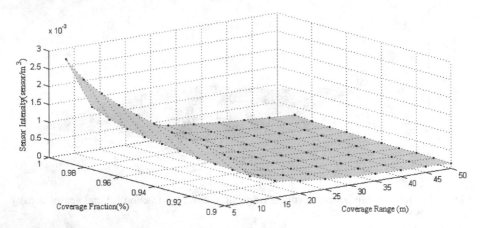

Fig. 3. The relationship between sensing range (R_s), coverage fraction (c) and sensor density (ρ) in 3D WSNs.

The figure shows the relationship between sensing range (R_s), the rea coverage and how many sensors should be deployed per cubic meter in three-dimensional space to get a certain degree of coverage.

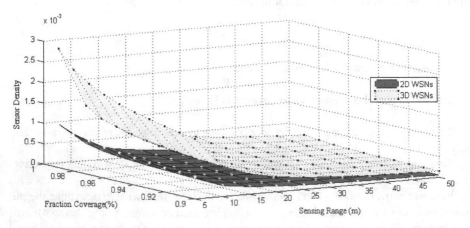

Fig. 4. The relationship between sensing range (R_s), coverage fraction (c) and sensor density (ρ) in 2D and 3D WSNs.

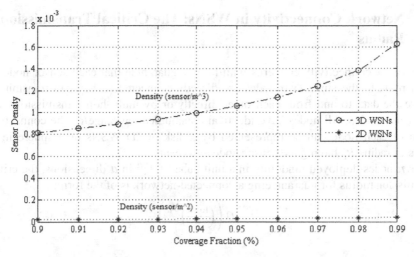

Fig. 5. Coverage fraction (c)as a function of sensor density (ρ) in 2D WSNs and 3D WSNs when $R_s = 30m$.

For the domain to be covered with high probability, say $C \geq 0.90$ we must have $\rho \geq \frac{2.30}{V_s}$ Vs. The sensing region of a node will intersect with the sensing regions of all nodes that are less than $2R_s$ away. The number of such nodes, on an average, willbe $\lambda(\frac{4}{3}\pi.(2R_s)^3) = 2^3 * \rho V_s \geq 2.30 * 2^3 \approx 19$ in 3D. The corresponding number in 2D is ≈ 9 and so on; figure 6.Shows the relationship between coverage fraction and the number of neighbors that are less than $2R_s$ in a random 2D and 3D WSNs.

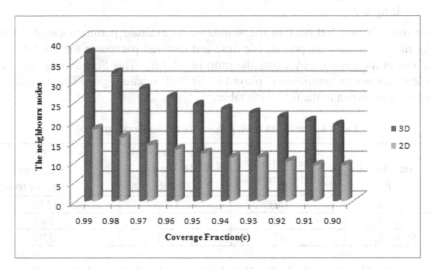

Fig. 6. The relationship between coverage fraction (c)and the number of neighbors that are less than $2R_s$ in a random 2D and 3D WSNs

5   Network Connectivity in WSNs: The Critical Transmission Radius

The network connectivity concerns with how to guarantee that each sensor node can find a route to the sink. Two nodes are directly connected if they can transmit and receive the data to and from each other directly or by multi-hop transmissions with some other nodes. All nodes have identical communication range R_c; the connectivity region of each node can be represented by a disk in 2D WSNs or a sphere in 3D WSNs of radius R_c having the sensor node at its center.

For n nodes deployed randomly in a unit cube, $[0, 1]^d$ in d dimensions, the critical transmission radius for guaranteeing a connected-network is of the form:

$$O\left(\left(\frac{\log n}{n}\right)^{1/d}\right) \tag{9}$$

6   The Ratio of the Communication Radius to the Sensing Radius

Using a suitable R_c / R_s ratio is a very important factor to guarantee the continuous working for the network until the death of the last node. In our research we have simulated the following cases for 2D WSNs in order to study the effect of R_c / R_s on the performance of the network.

- If the communication radius is at least twice of the sensing radius $R_c \geq 2R_s$
- If $R_c = R_s = 30$ m
- If $R_c < R_s$

We used C# and VB.NET in the simulation to evaluate performance of the networks in term of stability period, life time and received packets to the BS which located out of the network by using the ratio of R_c / R_s. This Scenario assumes 200 mobile sensor nodes, uniformly deployed in 400*400 square region. Three cases have been tested as shown in the following table.

Table 1. The effect of R_c / R_s on the network performance

The ratio of R_c / R_s	The first node died at round	The 40% of the nodes died at round	The received packet to BS
$R_c \geq 2R_s$	140	1200	1525
$R_c = R_s$	70	460	790
$R_c < R_s$	67	370	770

From the table above, it is clear that the period of stability and the number of received packets to the base station for the first case about doubled compared to the second and third cases. Furthermore, in the second and third cases we observed during the simulation process that although the network continued to operate, the work was unproductive compared to the first situation because most of the nodes had lost connectivity with each other, due to the small communication radius. Accordingly, in order to avoid network fault-tolerance and enhance the stability period of WSNs, we suggested using a radio station (radio translation) to guarantee the connectivity between nodes in the network after a period of time or using a suitable SR/CR ratio.

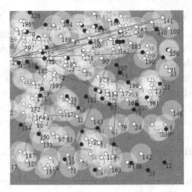

Fig. 7. Losing connectivity between nodes after dying 40% of all nodes

As a result we can make conclusion that to guarantee network-connected and long life working, the communication radius should be at least twice the sensing radius in 2D WSNs. In 3D WSNs there are many strategies for nodes placement for example, Hexagonal Prism, Dodecahedron or Octahedron. For a Cube where the distance between any two neighboring nodes is $2R_s/\sqrt{3}$, the transmission radius must be at least $1.1547R_s$.

7   Conclusion

The paper addressed the problem of coverage, connectivity, deployment and the ratio of transmission range to sensing range in 2D and 3D WSNs; we studied the geometric structure for the wireless sensor networks on 2D plane and in 3D space. We focused on the relationship between the sensing range and sensor density where the goal is to find the number of nodes required for surveillance of 2D area and 3D space is minimized, while guaranteeing 90.99% sensing coverage area for 2D and 3D WSNs, We evaluated the performance of the networks using metrics, such as network life time, stability period and throughput of the network in order to find the appropriate ratio of R_c/R_s which have a great impact on the performance of WSNs. It is clear from the simulation results that the transmission radius that is required for 3d WSNs is much more compared with 2D WSNs. The results of this paper could be used for the transformation strategy from 2D to 3D space and the necessity to analyze the real-world applications of WSNs before deployment of sensor nodes.

Acknowledgement. This work was supported by RFBR under the research project №15-07-09431a "Development of principles of construction and methods of self-organizing for flying sensor network".

References

1. May, M.: Design of a Wireless Sensor Node Platform. University of Waikato (2012)
2. Misread, S., Kumara, M.P., Obaidatb, M.S.: Connectivity preserving localized coverage algorithm for area monitoring using wireless sensor networks. Kharagpur, India, Monmouth University, New Jersey, USA (2011)
3. Bowing. Coverage Control in Sensor Networks, Computer Communications and Networks. doi:10.1007/978-1-84996-059-5_2, ©. Springer-Verlag London Limited (2010)
4. Alam, S.M., Zygmunt, J.H.: Coverage and Connectivity in Three-Dimensional Networks. In: MobiCom 2006. Cornell University, Los Angeles, California, USA, September 23–29, 2006
5. Onur, C., Ersoy, H.: Delic, and L. Akarun: Surveillance wireless sensor networks: Deployment quality analysis. IEEE Network 21(6), 48–53 (2007)
6. Zhu, C., Zheng, C., Shu, L., Han, G.: A survey on coverage and connectivity issues in wireless sensor networks. Hohai University, Changzhou, China. Osaka University, Japan (2011)
7. See, C.H., Abd-Alhameed, R.A., Hu, Y.F., Horoshenkov, K.V.: Wireless sensor transmission range measurement within the ground level, in: Antennas and Propagation Conference 2008, Loughborough, UK, pp. 225–228, March 2008
8. Poduri, S., Pattem, S., Krishnamachari, B., Sukhatme, G.S.: Sensor Network Configuration and the Curse of Dimensionality. University of Southern California, Los Angeles, CA 90089-0781, USA (2006)
9. Unaldi, N., Temel, S., Vijayan, K.A.: Method for Optimal Sensor Deployment on 3D Terrains Utilizing a Steady State Genetic Algorithm with a Guided Walk Mutation Operator Based on the Wavelet Transform. Sensors 12, 5116–5133 (2012). doi:10.3390/s120405116-19
10. Kong, L., Zhao, M., Liu, X., Liu, Y., Wu, M., Shu, W.: Surface coverage in sensor networks. IEEE Transaction on Parallel and Distributed Systems 25(1), 234–243 (2014)
11. Radwan, A.A.A., Mohamed, M.H., Mofaddel, M.A., El-Sayed, H.: A Study of Critical Transmission Range for Connectivity in Ad Hoc Network. Minia University, Assiut Univesity, Sohag University, Egypt, May 1, 2013
12. Liu, L., Ma, H.: On coverage of wireless sensor networks for rolling terrains. IEEE Transactions on Parallel and Distributed Systems 23(1), 118–125 (2012)
13. Jin, M., Rong, G., Wu, H., Guo, X. and Shuai, L.: Optimal surface deployment problem in wireless sensor networks. In: Proceedings IEEE INFOCOM, USA, pp. 2345–2353 (2012)

Internet Connection Sharing Through NFC for Connection Loss Problem in Internet-of-Things Devices

Ismail Turk[1,2]($\boxtimes$) and Ahmet Cosar[1]

[1] Computer Engineering Department, Middle East Technical University, Ankara, Turkey
[2] Field Application Engineer, NXP Semiconductors, Istanbul, Turkey
ismturk@gmail.com

Abstract. Contactless devices and smart cards have been widely in use in daily life transactions for a long time. At first, those systems were designed to work fully offline for both the reader and the card side. With technological improvements, Internet connection can be available even in very small embedded devices (IoT devices). As a result, current systems have connected devices as a part of the transaction design and so keeping the system operational all the time relies on the availability of continuous Internet connectivity of such devices. After the invention of NFC technology, contactless cards in our wallets are being replaced with virtual cards in mobile handsets. Regular contactless cards never had the capability to connect the Internet, whereas almost all of the modern mobile handsets routinely access the Internet. This has resulted in a trend shift in contactless transaction designs. Offline verification operations are being replaced with online operation where the system uses the connection capabilities of the mobile handset. However, the connection capability either in the mobile handset or in the transaction acceptance reader is not completely stable yet to rely on. Therefore, in this paper, we examine such connection loss problems in IoT devices and we offer a connection sharing mechanism through NFC so that the transaction can be completed if either one of these two entities has the Internet connection. Our proposed mechanism covers both reader-card communication and also peer-to-peer communication of NFC.

Keywords: NFC · Internet of Things · Connection sharing · Online transaction

1 Introduction

Smart Cards have been in use for a long time and offer easy and secure identification with a compact form which is comfortable to carry. After a while, smart cards had been improved and equipped with contactless communication capability. This improvement also let the technology to be used in mobile handsets. This is called, NFC.

The first mobile handsets were dedicated to make phone calls only however the convenience of carrying an embedded device let the technology fly and now billions of people are carrying a smart phone which is more functional than a PC for many aspects, and most of them are having Internet Connection. Internet connectivity let people to connect to online resources anywhere they are. According to [1] about half billion of mobile phones with full IP connectivity will be sold every year by 2015.

© Springer International Publishing Switzerland 2015
S. Balandin et al. (Eds.): NEW2AN/ruSMART 2015, LNCS 9247, pp. 329–342, 2015.
DOI: 10.1007/978-3-319-23126-6_30

For the point where this paper is focusing on, mobile handsets are now used for many online transactions such as payment, identification, social networking and more. And in parallel, ease of enabling Internet connectivity in portable devices reflected to transaction acceptance devices and so the regular transaction devices are equipped with connectivity modules and now they are also capable to make online clearance for their transactions.

A regular NFC payment between a transaction acceptance device (i.e. POS terminal) and a transaction media (i.e. contactless card, NFC phone) today is enriched with Internet connectivity and therefore, there is the trend shift in transaction designs to rely on online accounts. This is the convenient way of having the transaction but the main problem is ensuring the Internet connectivity on the entities used for the transaction.

This paper is focusing on this problem and so offering a solution that can keep the system operational either one entity of the transaction has the Internet connection by sharing this connectivity with the other entity which needs to have the connection in order to complete the transaction.

2   Technology Background

2.1   NFC Domain Information

Since 2004 we have NFC in our Mobile devices. Nowadays, more than 250 phone models are equipped with this technology. NFC is the technology which lets a contactless device to communicate with other contactless device within a range of a few centimeters. Using this technology a mobile device can have following NFC Modes; Reader Mode, Card Emulation Mode and Peer-to-Peer Mode [3].

Reader Mode enables an NFC enabled mobile handset to read all other contactless smart cards and also other NFC phones which has Card Emulation Mode.

Card Emulation Mode of NFC was of interest for payment industry for a long time because using this mode an NFC phone can be used as a mobile wallet.

Peer-to-Peer mode is the mode that was not in contactless cards before because it requires embedded software that is driving the execution. In this mode, two entities first make the handshake to start communication, and each entity switches between roles of sending the content and listening to the content. By this way, both entities can exchange any content in between. Since this mode is not available in contactless smart cards it is introduced after having NFC technology in mobile handsets with a different ISO standard [3].

2.2   Transaction Components

Transaction Acceptance Devices (TAD) are everywhere in our lives and they perform the transaction verification related tasks and so decides either to accept it or not. For examples all the POS terminals that are used for credit card payment are transaction acceptance devices. Another example could be the reader that is amount to the wall of an office entrance that people tap their badges to let the gate open.

Transaction media (TM) can either be a contact only smart card or a contactless card depending on the system design. If contactless cards are used in the system, it is also possible to use NFC phones as the Transaction Media.

2.3   Offline Transaction versus Online Transaction

TAD devices can perform either offline or online transaction; each has its own advantages and disadvantages. In an offline transaction all the transaction flow is executed within the TAD without any need to connect to any online resource. After the execution of the transaction flow, TAD decides either to allow the action or not. In an online transaction, at least a part of the transaction flow needs to be executed online therefore the TAD needs to be connected all the time to make a transaction clearance.

Considering today's Internet connection capabilities and the connection bandwidths, an offline transaction system is faster than an online transaction system. Therefore, systems requiring fast transactions like public transport payments choose offline transaction designs. An offline transaction system simply has the TAD and a transaction media without any connection to online systems.

For some systems it is not always possible to put all the transaction flow execution in an offline device because of security concerns. Therefore, online transaction design is chosen, although it is slower than an offline transaction. An online transaction design can have two possibilities for the connectivity, which is either requiring to have the TAD online as shown in Fig. 1, or to have the transaction media online as shown in Fig. 2.

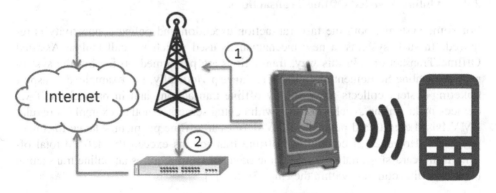

Fig. 1. TAD Online design of Online Transaction.

In TAD Online design of the Online Transaction, the TAD can have the connectivity either through Ethernet by being connected to a network switch (marked as "2" in Fig. 1) or through GSM network by having a GSM SIM inserted into the TAD (marked as "1" in Fig. 1). During this transaction, TAD receives required information from the transaction media and completes the transaction flow in conjunction of inside and offside execution.

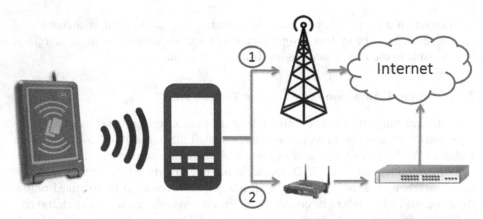

Fig. 2. Transaction media online design of Online Transaction.

In Transaction Media online design of Online Transaction, the TAD performs the transaction flow execution offline but within the transaction flow it retrieves some information from transaction media which is gathered from Online Systems. TAD performs the validity check of the information provided by transaction media which is in fact retrieved from online systems. In this design, if the transaction media is a mobile handset, it can be connected to the Internet either using the GSM network (marked as "1" in Fig. 2) or using the WiFi (marked as "2" in Fig. 2).

2.4   Online Assisted Offline Transactions

For some systems, both the fast transaction execution and online connectivity is required. In such systems, a new mechanism is used which we call Online Assisted Offline Transactions. In this way, transactions are performed offline but the system requires Online Settlement or Online Clearance periodically. For example, a transport ticketing system collects payments by offline transactions and in parallel the TAD devices make online synchronization with central servers. Another example is regular EMV based credit card payment. EMV defines an offline payment scheme for micro payments. However, if consecutive offline transactions exceed the defined total offline limit, card stops authorizing offline payments and requires an online transaction to reset offline counters within the chip [5], [6], [7].

3   Problem Definition

In today's system designs online transaction methods are commonly chosen because of its advantages like ensuring higher security, ease of online account management, etc. "TAD Online", "Transaction Media Online" or "Online Assisted Offline Transaction" methods are chosen according to the system requirements. However, for all of them the system relies on the continuous Internet connectivity.

3.1   Problems in TAD Online Method

In TAD Online design, all infrastructure problems, network issues, overloaded networks, limited bandwidth issues can cause connection loss problem. Whenever it happens, it automatically means for a TAD Online system to stop accepting transactions. In such case, the system operator needs to make an on-site visit and investigate the issue. The connectionless period can cause several problems and financial loss if the system is a payment system.

3.2   Problems in Transaction Media Online Method

If the Transaction Media is a mobile handset, connectivity is completely depending on the location of the customer during the transaction. For example, a coffee vending machine is accepting mobile wallet payments which are using online accounts of the mobile wallet provider and so requires mobile handset to be online for the payment. In this case, users will not be able to use coffee vending machines placed in the metro stations because of having no GSM network available under the ground.

Another connection loss reason for Transaction Media is the usage of limited data package in user's GSM account. Data packages (which let mobile handset to reach Internet using GSM network) are generally offered with an upper limit of the usage within the month, like 500MB, 1GB, etc. If the limit is reached, the Internet connection is not available any more until the month ends. Considering such a customer is using his/her phone for online transaction, they will not be able to make transactions whenever their data package limit is reached.

4   Related Work

Besides the connection loss problem that we mention in this paper, our research relates to the resource sharing between entities and also relates to the context awareness to detect the connection loss problem and to automatically setup the required connection. In this approach, there are several academic studies that are targeting to solve problems in this domain.

Considering a user having several IoT devices that are connected to Internet using an IP infrastructure and as well containing NFC feature available in those devices, then the same user will have several Secure Elements that reside in different devices. Available services within those secure elements in fact targeting the same user but there is no inter-connectivity between them. Therefore, it turns into a resource sharing issue between the devices. In [4] a cooperative architecture is introduced to solve this problem. Paper focuses on using the Android devices to share these resources in between within the same intranet.

Another resource sharing scheme introduced in [11]. The paper is introducing IoT based mHealth applications that shares sensor resources and uses sensor information in a real-time distributed way.

Context awareness in NFC to improve the advantages gained from NFC technology is crucial also. NFC technology is a well-defined technology containing very generic standards of communication and so lets a lot of different applications to work on the same infrastructure. However, to make some operations detected automatically and so taking the related action quickly the technology needs context awareness improvements. Our work is one of them as it is to automatically detect the online connection need and quickly performing the setup to provide this connection. Another context aware approach for NFC technology is introduced in [8]. It is a learning system to determine the transactions made by the NFC phone and so offers faster transactions after detecting the context.

Besides having a TAD and a transaction media in a transaction, sometimes it is also possible to have a mobile handset to execute a transaction flow by just reading a tag to retrieve transaction information. Reference [12] offers a generic solution for transaction flow detection from an NFC tag by combining Semantic Web Architectures interrelated with the NFC tag content.

5   Solution

As explained in Section 2, having an online transaction system has several advantages but ensuring the connectivity all the time is a challenge because of the reasons explained in Section 3. A regular transaction system has two components and having connection loss problems in one of them is likely to happen but having connection loss problem at both entities at the same time is more unlikely to happen. That's why our main focus in this paper is to share this connectivity in between the entities during the transaction.

This paper is focusing on the NFC transactions in our daily lives, thus, we developed a solution which is used to share the Internet connection using the regular NFC protocol. An NFC transaction can work in two possible ways;

- Reader-Card Mode: In this transaction mode, TAD is in the Reader Mode and the Transaction Media is either a contactless card or an NFC device having Card Emulation mode.
- Peer-to-Peer Mode: In this transaction mode, both entities performing NFC Peer-to-Peer mode to communicate and execute the transaction.

Reader-Card mode is the most common way of doing NFC transactions today because of the interoperability between regular contactless cards and NFC enabled card emulations. Peer-to-Peer mode is generally used for media sharing between two NFC devices, but it is the convenient way of doing any data exchange. Therefore, we expect to see more transaction designs depending on the Peer-to-Peer mode in future. Only pre-requisite for this is to have more NFC enabled phones in the field. Our solution is covering both modes of the transaction to offer a solution for today's use cases and as well to be ready to the future improvements.

In our proposed solution we tried to keep the command structure common for both Reader-Card and Peer-to-Peer modes. The main aim is to share the Internet connectivity in between; therefore, having the structure common for both types makes it easier to develop an inter-operable system. As a result, proposed solution only differs at underlying NFC protocol level, uses the protocol communication channel in the way it is defined but the actual commands exchanged between entities to perform connection sharing is common.

5.1   Reader-Card Mode

In this mode, reader side (TAD device) is the active side that is performing the transaction flow execution. Reader sends commands to the transaction media and gets response back from it. At any time, transaction media cannot send a message to reader side unless reader sends a command to it. Thus, transaction flow execution has several command and reply pairs in between the TAD and the transaction media.

As a result, any connection sharing protocol requires an initial handshake between two entities before starting the transaction execution. In this initial messaging both entities tells to each other if they need an online connection to perform the transaction or not.

After this messaging, if connection sharing is necessary and if resources are able to share this connectivity, the reader creates a new transaction flow setup automatically to have connectivity sharing within the transaction. At this stage we can have two possibilities of connection sharing; either TAD needs to share TM's connectivity or TM needs to share TAD's connectivity. Pseudo code of this execution is shown below.

1. TAD connects to TM
2. Connection Sharing Handshake

        -> TAD CSSB
        <- TM CSSB

3. Perform Status Analysis

        TAD = OK
        TM = OK
        if (TAD needs online connection) {
                if (TAD is Offline) {
                        TAD = NOK
                }
        }
        if (TM needs online connection) {
                if (TM is Offline) {
                        TM = NOK
                }
        }
        if (TAD = OK and TM = OK) {
                Continue Regular Transaction.
        }
        else if (TAD = NOK and TM = OK) {

```
                if (TM connection sharable)
                        Ask TM to share con.
            else
                        Reject Transaction
        }
        else if (TAD = OK and TM = NOK) {
                if (TAD connection sharable)
                        Share connection
            else
                        Reject Transaction
        }
        else if (TAD = NOK and TM = NOK) {
                Reject Transaction
        }
    4. Setup Connection Sharing
        if (TAD will share TM connection)
                -> Sharing request
                <- Sharing approval
        if (TM will share TAD connection)
                -> Sharing notification
                <- ACK
    5. Execute Transaction
```

The first step is the regular connection handling of TAD to the TM. The second step is to exchange Connection Sharing Status Bytes (CSSB) between entities. First, TAD sends the command to TM including its CSSB byte and TM replies back with its CSSB byte. Coding of CSSB values are shown in Fig. 3.

Connection Sharing State Coding										
b7	b6	b5	b4	b3	b2	b1	b0	Meaning for TAD	Meaning for TM	Meaning for Peer in P2P Mode
0								TAD is offline	TM is offline	Peer is offline
1								TAD is online	TM is online	Peer is online
	0			if b7=1				TAD connection is not sharable	TM connection is not sharable	Peer connection is not sharable
	1							TAD connection is sharable	TM connection is sharable	Peer connection is sharable
		0	0					TAD can work offline	TM can work offline	Peer can work offline
		0	1	if b7=0				Online Connection may be used	Online Connection may be used	Online Connection may be used
		1	0					Online Connection is a Must!	Online Connection is a Must!	Online Connection is a Must!
		1	1					RFU	RFU	RFU
				x	x	x	x	RFU	RFU	RFU

Fig. 3. CSSB Coding for Transaction Acceptance Device, Transaction Media and Peer Device

The third step is to perform status analysis based on the shared CSSB values. TAD checks if entities need to share other entities connectivity. TAD rejects the transaction if one entity needs the connectivity but cannot share it from the other.

The fourth step is to setup the connection sharing if it was decided in third step. If TAD will share the connection of TM, then TAD requests it from TM. TM updates its commands dispatcher to expect connection request commands and replies back to TAD with the approval of connection sharing. If TM will share the connection of TAD, the TAD first enables connection sharing and notifies TM about the availability of the shared connection. TM replies back with an ACK.

In the fourth step, besides the need of connection sharing, it can also be a good-to-have to share the connectivity in between entities. Therefore, CSSB contains flag for "Online Connection may be used" state. This is to indicate that entity does not need to have the connectivity for completing the transaction but having the connectivity is useful for its current state. To clarify the situation with an example; in online assisted offline transaction systems, the TAD device keeps an offline cache for the transactions and periodically synchronizes them with the central system and so clears the offline cache. In such systems, TAD device has a limit for the maximum offline transaction to store inside. When this limit is reached, TAD stops transaction acceptance until it can reach to the Internet. However, when TAD's offline cache is very close to be full, TAD can indicate "Online connection may be used" flag to request connection sharing and therefore blocking a possible cache-full problem before it occurs.

The last step is to execute the regular or auto-updated transaction flow to complete the transaction. This step contains several commands that are related to transaction execution and also connection sharing.

Detailed descriptions of these commands are provided in Section 6.

5.2   Peer-to-Peer Mode

Peer-to-Peer mode is very similar to Reader-Card Mode execution but in this case both parties are using the same scheme and one of them can be the active side depending on the "Active Initiator" of the transaction [3].

If connection sharing is necessary and if resources are able to share it, the Active Initiator creates a new transaction flow to have the connection sharing. In P2P mode, the active initiator is playing the role of Reader in Reader-Card mode. Execution flow is the same but only the protocol details are different because P2P mode devices are directly driven by the host system of the TM.   Connection sharing status byte coding for P2P mode is shown in Fig. 3.

6   Protocol

In this section we provide the detailed command specification for connection sharing implementation that is explained in this paper. Commands explained in Reader-Card mode are in ISO 7816 APDU frame format. It contains CLA, INS, P1, P2 bytes in the header and contains LC, CDATA, Le optional bytes afterwards. CDATA is the array of data that reader wants to send to the card, LC is the length of the sent CDATA [2]. The card responds with DATA and SW1, SW2 bytes. SW1 and SW2 bytes are the status words which indicate execution result. 0x90 0x00 is the success status word [2].

Commands used in P2P mode are using SNEP protocol to exchange data in between and NDEF messages are used with NDEF Record that carry the byte data inside [3]. In this paper, our focus is to provide the protocol details for connection sharing, therefore, we do not mention about the NDEF details but only the protocol values exchanged between the Peers.

6.1   Reader-Card Mode Connection Sharing Handshake

In order to exchange connection sharing needs between the entities, TAD sends the command shown command (1) and gets back 1 byte from TM which is indicating TM's CSSB byte.

Command (1): TAD-TM Connection Sharing Handshake;

CLA=0x1A, INS=0x00, P1=P2=0x00, Lc=0x01, CDATA=<1 byte TAD CSSB>   (1)

6.2   Reader-Card Mode Connection Sharing

If TAD to share TM's connectivity, TAD first issues Connection Request command as shown in Command (2). If TAD receives a success code in reply, it prepares the package to be sent to the online system and sends all the parts of the package to the TM. First, it issues a package start command as shown in Command (3). Connection sharing method only supports TCP/IP socket connection to the online system for the sake of simplicity. IP address of the online system is coded in 4 bytes (0xFF means 255), and the port is encoded in two bytes as a short value representation. Last byte in CDATA is indicating the number of data blocks that will be sent for the package. Each data block contains 250 bytes and therefore before issuing online connection command TAD prepares the data blocks first. NDB value is the ceiling value of [total # of bytes]/250.   After receiving a success reply to this command, TAD starts a loop to deliver all the data blocks of the package using the commands explained in Command (4). Data block number is indicated as P2 value of the command.

Command (2): Connection Sharing Request;

$$CLA=0x1A, INS=0x01, P1=P2=0x00 \qquad (2)$$

Command (3): TAD Package Start Command;

CLA=0x1A, INS=0x02, P1=P2=0x00, Lc=0x07, CDATA= IP, Port and NDB   (3)

Command (4): Package Data Blocks;

CLA=0x1A, INS=0x02, P1=0x01, P2=DB#, Lc=DBL, CDATA=Data Block   (4)

After completing these commands, TM terminates the NFC connection from TAD. Host application goes online to deliver the message from TAD using the IP and Port values retrieved from TAD. After this execution, online system can have a reply to the TAD. TM's host application writes back the reply into its local module and activates the NFC again. Whenever TAD determines the NFC availability of TM, it starts execution again to retrieve the reply from its online system. TAD first issues the command in Command (5) to understand the total number of blocks available in TM. At this step, TAD starts a loop to retrieve data blocks from TM that will loop NDB times. TAD issues the command shown in Command (6) to retrieve each block.

Command (5): GetData Block Size From TM

$$CLA=0x1A, INS=0x03, P1=P2=0x00 \tag{5}$$

Command (6): Get Reply from TM

$$CLA=0x1A, INS=0x04, P1=0x00, P2=DB\# \tag{6}$$

Connection sharing commands can be repeated as many as TAD needs. And then, TAD continues with the regular transaction execution flow.

If TM to share TAD's connectivity, TAD first issues connection sharing notification command as CLA=0x1B, INS=0x01, P1 and P2 = 0x00. TM replies back to this command with the connection details. 7 bytes of data is returned coded as "ip1 ip2 ip3 ip4 port1 port2 NDB". First 6 bytes are coding the TCP/IP socket information and the last byte indicates the total number of data blocks of the package. Each data block can contain 256 byte of data. Thus the NDB value is the ceiling value of [Total # of bytes]/256. After receiving this command reply, TAD starts data block retrieval loop using the command structure in Command (7).

Command (7): Get Data Block Size from TM

$$CLA=0x1B, INS=0x02, P1=0x00, P2=DB\# \tag{7}$$

After completing these commands, TAD goes online. Online system can have a reply to TM; TAD retrieves this information and prepares commands to send it to the TM using the command structure shown in Command (8). P2 value in the command indicates if it is the last block or not (LBI = Last Block Indication). 0x01 indicates the last block and all other values indicate that there will be following commands of data blocks. When TM retrieves the last block it unpacks the information and executes necessary updates internally. For each command the TM replies back with a Loop Indication Byte. For continues data block retrieval commands, this value is sent as 0x00, only for the last block retrieval it may have different value. When TM executes the reply from online system, it decides if it needs to send another package to the online system or not. If yes, then tells this using the Loop Indication Byte by setting it to 0x01. When TAD retrieves 0x01 value it again starts connection sharing sequence to let TM to make another connection through it.

Command (8): Write Reply from TM

$$CLA=0x1B, INS=0x03, P1=0x00, P2=LBI, Lc=Length, CDATA=Data \tag{8}$$

After completing all the connection sharing commands TAD goes into the regular transaction execution flow.

6.3   Peer-to-Peer Mode Connection Sharing Handshake

In order to exchange the connection needs between the Peers, Active Initiator of the transaction sends the CSSB value to other Peer and gets back other Peer's CSSB value. Both Peers use 0x2a tag header to indicate this value is a CSSB value and as data CSSB value is sent.

6.4   Peer-to-Peer Mode Connection Sharing

If Active Initiator side needs to share other Peer's connectivity it sends a command indicating this request. This request is indicated with 0x2b tag header and followed by 6 data bytes coded as ip1 ip2 ip3 ip4 port1 port2. If Active Initiator will share its connectivity with other Peer then it sends the sharing notification command with 0x2c tag header without data bytes. In reply to this command, other Peer replies back with 0x2c tag header and 6 data bytes coded as ip1 ip2 ip3 ip4 port1 port2.

After this sharing setup, at any step of transaction execution the Peer can request other Peer which is sharing its connectivity to go online. For this purpose the command with 0x2d tag header is used together with the data bytes. The Peer connecting to the online system retrieves back the information and sends back to other Peer using the command with 0x2e tag header and the data bytes retrieved from the online system.

7   Experimental Work

In order to show our work we developed applications that can fall into the connection loss situations that we explained in this paper. In our experimental work, we intentionally let our devices to lose connectivity and we examined how they share the connectivity in between to continue transaction clearance. During this work we developed following applications;

Chip Application: A basic payment application that is running on JCOP platform. Application allows 5 consequent offline payments and requires an online counter reset operation when consequent payments reach to 5. With this application we tried to simulate EMV offline payment.

Counter Reset PC Application: We developed a PC application which is used to reset offline counter of the chip application. With this application we tried to simulate an online transaction of EMV which eventually reset offline counters.

Counter Reset Online Application: We developed another PC application which listens to a TCP port and retrieves messages from the chip and replies back with counter reset approval. We used this application to check if our chip application can use connection sharing to reset its counters.

Transaction PC Application: We developed another PC application which performs offline transaction with our chip application and also it makes connection sharing handshake at the beginning of each transaction by indicating that it can be used for connection sharing.

Test Scenario;

Step 1: We deployed our chip application to a JCOP card.

Step 2: We made 5 offline payments using our PC Application.

Step 3: We used Counter Reset Application to reset counters.

Step 4: We tried offline payment and it is approved. And we again made transactions until reaching offline transaction limit.

Step 5: We turned off connection sharing flag of our Transaction Application.

Step 6: We tried another transaction using our chip. Chip requested connection from TAD but Transaction Application rejected the sharing, and so transaction failed.

Step 7: We turned on connection sharing flag of our Transaction Application.

Step 8: We tried another transaction. Chip requested connection from TAD, and it is shared. TAD collected information from chip and forwarded it to our Counter Reset Online Application. Execution Information is retrieved from our application and forwarded to chip application. TAD automatically continued to regular transaction execution and completed the transaction successfully.

8   Conclusions

Considering the current implementations, we have Internet of Things devices everywhere and they are being used for many purposes including financial transactions. Up to now, limited functionality devices were in use for such operations and so there were no resource sharing functionalities in between the entities. As we started to use more complicated and more functional devices for our daily life operations and transactions, we think it is worth to consider resource sharing to leverage better from this trend shift. As a result of our work, we have shown that Internet connection sharing between transaction entities is possible. Also, we think our work will be the very first example of many other resource sharing schemes that will be introduced in the near future.

9   Future Work

As a clear outcome of our work we see several different academic and commercial researches that can be undertaken to continue and advance this work.

Our work focus on the NFC transactions and according to the current technology that is in use the connection baud rate between two entities is limited to a maximum of 848 Kbps. This communication speed maybe sufficient for regular transaction related requirements, however for connection sharing we may have bigger packages that needs to be delivered to the online systems. This NFC communication in between entities can become a bottleneck for the whole operation execution time. ISO/IEC 14443 standard defined higher bit rates called very high bit rates (VHBR) for the contactless communication between the NFC entities and after this implementation we will have bit rates between 10.17 and 27.12 Mbps [9]. Therefore, Integrating VHBR with connection sharing schemes can be an interesting future work.

We introduced our work based on NFC Transactions, but there is a new trend in IoT device communication which is a new Bluetooth technology that was very recently introduced. It is called Bluetooth Low Energy (BLE) and it offers a convenient Bluetooth connection and data sharing when it is compared to regular Bluetooth technology. Because of this fact, there have been some applications of using this technology for communication between transaction entities and creating transaction schemes using this technology. As a future work, our Internet connection sharing scheme can be extended for BLE based transaction systems also.

In our work, during the Internet connection sharing between the entities we assumed that one entity can create a secured package while transferring it through the other entity. However, security concerns can still be valid. In [10] a secure communication channel creation in between NFC entities is introduced. As a future work, Internet connection sharing schemes can be enhanced with security related improvements that can make package communication more secure.

References

1. http://www.gartner.com/it/page.jsp?id=1622614 (accessed February 5, 2015)
2. ISO 7816. Cards Identification - Integrated Circuit Cards with Contacts. The International Organization for Standardization (ISO)
3. NFC Forum Protocol Technical Specification. http://www.nfc-forum.org (accessed February 5, 2015)
4. Pascal, U., Christophe, K.: A new cooperative architecture for sharing services managed by secure elements controlled by android phones with IP objects. In: The International Conference on Collaboration Technologies and Systems (CTS 2012), pp. 404–409
5. EMV Integrated Circuit Card Specifications for Payment Systems. Book 1 – Application Independent ICC to Terminal Interface Requirements, November 2011
6. EMV Integrated Circuit Card Specifications for Payment Systems. Book 2 – Security and Key Management, November 2011
7. EMV Integrated Circuit Card Specifications for Payment Systems. Book 3 – Application Specification, November 2011
8. Karan, K., Harsh, G., Abhishek, G.: Integrating contactless near field communication and context-aware systems: improved internet-of-things and cyberphysical systems. In: 5th International Conference- Confluence The Next Generation Information Technology Summit, pp. 365–372 (2014)
9. Saminger, C., Stark, M., Gebhart, M., Grünberger, S., Langer, J.: Introduction of very high bit rates for NFC and RFID. e&i elektrotechnik und informationstechnik, 218–223 (2013). Springer Verlag Wien
10. Ceipidor, U.B., Medaglia, C.M., Marino, A., Sposato, S., Moroni, A.: KerNeeS A protocol for mutual authentication between NFC phones and POS terminals for secure payment transactions. In: 2012 9th International ISC Conference on Information Security and Cryptology, pp. 115–120 (2012)
11. Stefan, F., Theo, K., Olle, J.: Real-time distributed sensor-assisted mHealth. Applications on the internet-of-things. In: 2012 IEEE 11th International Conference on Trust, Security and Privacy in Computing and Communications, pp. 1844–1849 (2012)
12. Broll, G., Siorpaes, S., Rukzio, E., Paolucci, M., Hamard, J., Wagner, M., Schmidt, A.: Supporting mobile service usage through physical mobile interaction. In: 5th Annual IEEE International Conference on Pervasive Computing and Communications, White Plains, NY, USA (2007)

Clustering Algorithm
for 3D Wireless Mobile Sensor Network

Pavel Abakumov[(✉)] and Andrey Koucheryavy

Saint-Petersburg State University of Telecommunications, Saint Petersburg, Russia
pvl.abakumov@gmail.com, akouch@mail.ru

Abstract. Scope of sensor networks is increasing every year, finding practical application in many areas of our lives. There are a lot of algorithms for organizing stationary sensor networks where nodes location does not change. Nowadays network nodes mobility is one of the basic properties for wireless sensor network. In this regard, there are some new problems, such as connecting nodes and optimizing their energy consumption, network coverage and others, which require a new approach to organizing sensor network algorithms. The appearance of flying sensor networks actualizes this task even more. This paper discusses problems associated with mobile wireless sensor networks implementation, proposes a new clustering algorithm for three-dimensional space, simulation and comparison with the LEACH-M algorithm.

Keywords: Mobile WSN · LEACH-M · Flying sensor networks · Energy efficiency

1 Introduction

Over the past 15 years, a whole class of new types of networks has appeared, consisting of tiny wireless devices. These devices are called sensors and generally consist of a battery, the processor and the wireless communication module with low power consumption and a unit which allows to implement functions such as measurement of necessary parameters, or control than either [1,2]. The main features of such networks are self-organizing [3,4], energy efficiency and maximum coverage of sensor field. Initially, the organization sensor network algorithms assumed stationary location sensor nodes, node position did not change throughout the life-time [5,6]. As a result of the development of sensor networks their scope has expanded considerably. There were even flying sensor networks consisting of ground and flying segments [7].

New problems appear as a result application of wireless sensor networks under the impact of external factors where the sensor could change their position, moving relative to the previous location. Such applications include mobile sensor networks, for example, the measurement of air pollution, where the nodes can change their location by the wind or the measurement of parameters in the water under the influence of the watercourse.

S. Balandin et al. (Eds.): NEW2AN/ruSMART 2015, LNCS 9247, pp. 343–351, 2015.
DOI: 10.1007/978-3-319-23126-6_31

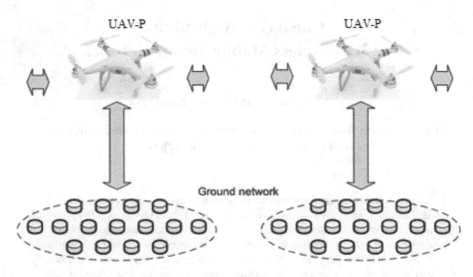

Fig. 1. One PUAV flying sensor network

At present time appears a new class of Ubiquitous Sensor Networks (USN) - Flying Ubiquitous Sensor Networks (FUSN) (Fig. 1). In our case FUSN is a data collection network that consists of several sensors fields and Public Unmanned Air Vehicle (UAV-P) which collects data from sensor fields. UAV-P flight path and data collection methods are the subject of current research. UAV-P needs to change flight path every time if the head nodes are not located near the center of the cluster. The proposed algorithm makes this task easier, because it uses a static clusters and cluster head node selection consider its location within the cluster. So we know the head nodes cluster locations and this locations do not change significantly during life cycle, in this situation UAV-P can fly the same path all time.

The algorithm organizing mobile sensor network implementation imposes additional problems compared with the networks with fixed position of nodes that can not be ignored [8,9]. Among them we should highlight the following issues:

Initial location of sensors and their life cycle movement can affect the appearance of empty regions without nodes or only nodes with sufficient energy to send data to the base station. But these regions may be important for us. In the most cases it is impossible or impractical to change the unit or the battery. In this case, appears "the dead clusters" and holes in the network coverage appear.

As a rule, these algorithms applied cluster organization, allowing reducing energy consumption. Head cluster nodes aggregate the data and transmit them to the operator directly, or using head nodes of neighboring clusters in case of failing to direct communication. In some situations data cannot be transferred because neighboring clusters are "dead".

Sensor node can move from cluster to cluster because of their mobility (Fig. 2). Also the energy consumption depends on the data transfer distance. Transmission data to the closest head node is a rational idea.

There are a lot of algorithms for organizing static sensor networks where nodes cannot move. Some of these algorithms cannot be applied for mobile sensor nodes, because they do not supply changing of cluster by nodes per round. Some algorithms can be used but without significant success.

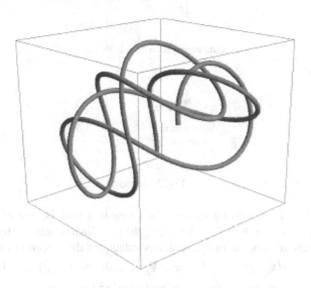

Fig. 2. Sensor node movement

This paper proposes a new algorithm for clustering wireless sensor network SCA (Static Cluster Algorithm), which providing efficient coverage, taking into account the terms of the mobility of sensor nodes. The algorithm involves working with homogeneous mobile wireless sensor network with a random arrangement of nodes.

2   Network Model

Cluster networking involves two roles that can play sensor nodes – the head node and the cluster member. Homogeneous sensor network consists of nodes with the same characteristics, each node that can take on any role. The head node of the cluster has been receiving data from the cluster members, data aggregation and transmission to the base station, creating communication schedules for cluster members. The proposed algorithm is focused on sensor networks, operating in three-dimensional space [10,11,12], but can be successfully used for planar sensor fields.

The life time of a sensor network is divided into rounds, in each round the head node receives information from members of the cluster. After that gateway transmits the aggregate information and, if necessary, passes the head node to another cluster member with the appropriate parameters (Fig. 3). Two nodes can communicate, if the distance between them is less than the maximum radius of connection, in other case the node is not involved in the current round, but will be able to participate in the next round, if the distance is reduced. The sensor is "dead" if its energy supply has expired. The cluster is "dead" if it does not contain nodes with sufficient energy.

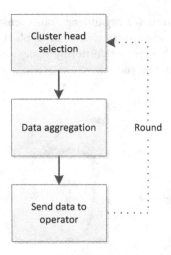

Fig. 3. Round

Each node has a positioning system, for example it may be one of a global positioning systems, or a system of calculating the coordinates relative to multiple base stations. The law of motion in space nodes calculates the direction of movement in three dimensions (dx_n, dy_n, dz_n), speed V_n, coordinates (x_n, y_n, z_n), where n the number of rounds and can be calculated by following formulas:

Direction:

$$(dx_n, dy_n, dz_n) = (dx_{n-1}k_1 + rand_x k_2, dy_{n-1}k_1 + rand_y k_2, dz_{n-1}k_1 + rand_z k_2) \tag{1}$$

Speed:

$$V_n = V_{n-1}k_1 + rand_V k_2 \tag{2}$$

Coordinates:

$$(x_n, y_n, z_n) = (x_{n-1} + V_n dx_n, y_{n-1} + V_n dy_n, z_{n-1} + V_n dz_n,) \tag{3}$$

Parameters k_1 and k_2 used for changing the motion smoothness, which the into account movement style in the previous round, *rand* - a random number.

3 Clustering Algorithm

One of the main features of the algorithm is a fixed location and sizes of clusters. Determination of the coordinates and size of the clusters network performed during the network initialization, which occurs once in the beginning of its operation. Each node knows the coordinates of the centers of clusters and their serial number from the base station, which transmits them to broadcast packets. Thereafter, each sensor knows own location, determines its rating Rtg, which represents the distance to the nearest cluster center and determines own cluster. Sensor that has minimum distance to cluster center and sufficient energy gets "the cluster head" role. Other nodes located in this cluster are members and transmit data to its head node once or several times per round. The cluster head node aggregates data and transmits data through a gateway to operator (Fig. 4).

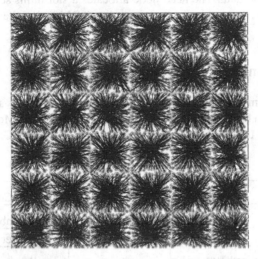

Fig. 4. Clustering projection of sensor field on the XY plane

Communication between nodes of mobile wireless sensor network bases on the TDMA (Time Division Multiple Access) approach. Each head node makes the schedule for its cluster, reserved slots in the schedule for the connection of new nodes that have moved from one cluster to another. The algorithm can provide for the transfer of data from the head node to the base station through head nodes of neighboring clusters, knowing the location of all clusters in the network, if the distance exceeds the range of the base station radio unit. TDMA schedule has additional slots for this communication.

A member of the cluster sends a message to the head node, which consists of a cluster address, the sender's address and data of the current rating Rtg. The head node sends a reply confirming receipt, in which the address of the cluster, the recipient's address and own current rating Rtg_h, when $Rtg_h <= Rtg$. Otherwise, the role of the head node transmitted to a member of the cluster. If the energy unit E is less than the threshold E_{min}, the rating is maximum. In this case sensor cannot be cluster head node.

$$\begin{cases} Rtg = d, E > E_{min} \\ Rtg = 0xFFFF; E <= E_{min} \end{cases} \qquad (4)$$

The head node is assigned to the sensor with the required energy reserve, located as close as possible to the center of the cluster. The frequent reassignment of head node role involves additional energy consumption for transmitting overhead information, including TDMA schedules.

Sensor nodes can move within a cluster and between them. The transition from one cluster to another sensor node determines on its location. When node moves between clusters, sensor waits a slot in the schedule which reserved for new members of the cluster and sends the data. The head node allocates a slot in his schedule for the new node. If there is no message in the corresponding TDMA slot, the head node releases the slot.

4   Simulation

LEACH-M algorithm was chosen for results compeering, which has all the advantages of the LEACH, but can be used for mobile wireless sensor networks. Modeling and displays the results produced using the C++ programming language and MFC.

Table 1. Initial parameters

Name	Value	Dimension	Comment
N	4000	K	Number of nodes
P	5	%	Parameter for LEACH-M
x	100	m	Sensor field length
y	100	m	Sensor field width
z	100	m	Sensor field height
E_s	1	J	Initial energy
E_{elec}	50	nJ	Energy admission
E_{amp}	10	pJ/bit/m^2	Amplification
m	4000/80	bit	Message length
E_{ag}	5	nJ	Aggregation energy
R_{max}	30	m	Maximum range

Energy consumption for transmission and reception of messages:

$$E_{tx} = E_{elec}m + E_{amp}md^2 \qquad (5)$$

$$E_{rx} = (E_{elec} + E_{ag})m \text{ - For head node} \qquad (6)$$

$$E_{rx} = E_{elec}m \text{ - For cluster member} \qquad (7)$$

Message length may be 4000 bits in the case of the transition of head node role and 80 bits (cluster number, nodes address and data) in the case of confirmation messages. Also take into account the energy consumption on the formation of TDMA schedules and routing table for the data transmission over the head nodes of neighboring clusters.

The sensor field is a cube and is divided into clusters with the distance between the centers is 17 meters, the number of clusters for the proposed algorithm is 216. The optimum value of the parameter P for the algorithm LEACH-M is 5%, in case of 4000 nodes the average number of head nodes for the algorithm LEACH-M is 200.

Parameters k_1 and k_2 are equal to 0.9 and 0.1 respectively, which smooth the sensors movement for simulation. In general, it can be random from 0 to 1.

5   Simulation Results

The size of clusters using the algorithm LEACH-M may be different due to the random selection of head nodes. The proposed algorithm forms clusters of comparable size range, selection of head nodes depends on the distance to the cluster center, which reduces the distance between the members of the cluster head node, and reduces power consumption for data transfer. Another advantage of this algorithm is simple routing of messages between adjacent head nodes towards the parent base station.

The main performance parameters of the algorithms are the total lifetime of the sensor network, the number of live nodes in the network in each round, the coverage and the residual energy of nodes. Comparison of the parameters of the algorithm is shown in figure 5.

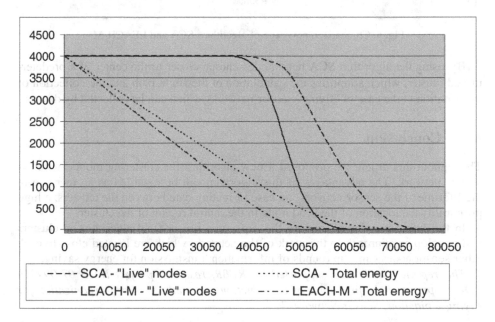

Fig. 5. A chart of "live" nodes and the total energy of nodes for SCA algorithm and LEACH-M

The abscissa denotes the current round, and the ordinate is the amount of nodes in the network. The simulation results showed that using the algorithm SCA increases the lifetime of the sensor network and increases the energy flow efficiency of nodes that can be seen from a comparison of the parameters "total energy supply." Due to the fact that the communication range is limited in this simulation, not all the nodes can be connected in each round, so when using the algorithm LEACH-M coverage reaches 100% only in rare cases. To estimate that, we assume that the coverage in the current round cannot be more than the minimum for all previous rounds.

The results of the comparison coverage for algorithms are shown in figure 6.

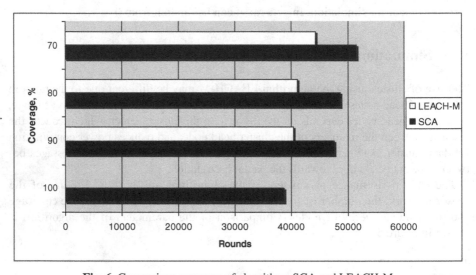

Fig. 6. Comparison coverage of algorithms SCA and LEACH-M

By using the algorithm SCA head nodes chosen closer to the center of a predetermined cluster, which simplifies the calculation of the flight path and the collection of information in the case of "flying" sensor networks which can be collected by drones.

6   Conclusion

Throughout this article we proposed a new algorithm for clustering mobile wireless sensor network. The advantages of this algorithm can be regarded as an increase in the lifetime of the sensor network, an ease of routing data between the clusters, a high probability the location of the head node in the central region of the cluster.

In the future we plan to explore the separation sensor field algorithm on clusters and develop an approach to the work of the network, which are located close to each other sensor nodes can skip rounds of information transmission for energy saving.

The reported study was supported by RFBR, research project No15 07-09431a "Development of the principles of construction and methods of self-organization for Flying Ubiquitous Sensor Networks".

References

1. Akyildiz, I.F., Vuran, M.C., Akan, O.B., Su, W.: Wireless Sensor Networks: A Survey revisited. Computer Networks Journal (2005)
2. Gerasimenko, M., Petrov, V., Galinina, O., Andreev, S., Koucheryavy, Y.: Impact of machine-type communications on energy and delay performance of random access channel in LTE-advanced. European Transactions on Telecommunications 24(4), June 2013
3. Khan, S., Pathan, A.-S.K., Alrajech, N.A.: Wireless Sensor Networks: Current Status and Future Trends. CRC Press (2012)
4. Vinel, A., Vishnevsky, V., Koucheryavy, Y.: A simple analytical model for the periodic broadcasting in vehicular ad-hoc networks. In: 2008 IEEE Globecom Workshops, GLOBECOM (2008)
5. Heinzelman, W., Chandrakasan, A., Balakrishnan, H.: Energy-efficient communication protocol for wireless microsensor networks. In: Proceedings 33rd Hawaii International Conference on System Sciences (HICSS), Wailea Maui, Hawaii, USA, January 2000
6. Koucheryavy, A., Salim, A.: Cluster head selection for homogeneous wireless sensor networks. In: Proceedings, International Conference on Advanced Communication Technology, ICACT 2009, Phoenix Park, Korea
7. Kirichek, R., Paramonov, A., Koucheryavy, A.: Flying ubiquitous sensor networks as a quening system. In: Proceedings, International Conference on Advanced Communication Technology, ICACT 2015, Phoenix Park, Korea, July 01–03, 2015
8. Kim, D.S., Chung, Y.J.: Self-organization routing protocol supporting mobile nodes for wireless sensor network. In: Proceedings First International Multi Symposium on Computer and Computational Sciences, Hangzhou, China, June 2006
9. Koucheryavy, A., Salim, A.: Prediction-based clustering algorithm for mobile wireless sensor networks. In: Proceedings, International Conference on Advanced Communication Technology, ICACT 2010, Phoenix Park, Korea
10. Attarzadeh, N., Mehrani, M.: A New Thre Dimensinal Clustering Method for Wireless Sensor Networks. Global Journal of Computer Science and Technology 11(6), April 2011. version 1.0
11. Hooggar, M., Mehrani, M., Attarzadeh, N., Azimifar, M.: An Energy Efficient Three Dimensional Coverage Method for Wireless Sensor Networks. Journal of Academic and Applied Studies 3(3), March 2013
12. Abakumov, P., Koucheryavy, A.: The cluster head selection algorithm in the 3D USN. In: Proceedings, International Conference on Advanced Communication Technology, ICACT 2014, Phoenix Park, Korea

Optimization of the UAV-P's Motion Trajectory in Public Flying Ubiquitous Sensor Networks (FUSN-P)

Ruslan Kirichek[✉], Alexander Paramonov, and Karine Vareldzhyan

The Bonch-Bruevich Saint-Petersburg State University of Telecommunications,
Saint Petersburg, Russia
kirichek@sut.ru, alex-in-spb@yandex.ru, karisha.var@gmail.com

Abstract. In this article we propose a method of interaction with terrestrial USN using public unmanned aerial vehicles (UAV-P). This method optimizes the UAV-P's motion trajectory in order to minimize the information delivery time to the user. In this article we also present the comparative evaluation of different selection algorithms of UAV-P's motion trajectory. The purpose of the UAV-P's motion is to collect information from terrestrial ubiquitous sensor network.

Keywords: Ubiquitous Sensor Networks (USN) · The Public Unmanned Aerial Vehicles (UAV-P) · Delay-Tolerant Networks (DTN) · Clustering · Traveling salesman problem · The penalty algorithm

1 Introduction

Ubiquitous sensor Networks (USN) [1,2] are the technological basis for the concept of the Internet of Things [3]. Intensive development of such networks, allows them to be integrated in numerous areas of everyday life. In many cases, the task of such network is to monitor some of the environmental parameters or certain objects. Self-organization [4,5], simplicity and low cost of construction of the USN can significantly improve the efficiency of such tasks. A wide variety of sensor networks implementations [6] is also expressed in a wide range of technical solutions for their implementation [7]. In some cases, a sensor network can be located over a large landscape, situated in remote areas, and can be set up just for a short period of time. In some of such cases, building infrastructure, which ensures the permanent communication channels with USN using wired and wireless technologies, is economically unreasonable or impossible. A compromise is necessary where the required parameters for the network are provided with the minimal use of technical and other means. One option is to use the public unmanned aerial vehicles (UAV-Ps) as communication and maintenance elements. This method provides rather effective solution due to affordability and diversity of UAV-Ps. Such networks have been proposed [8] and called the Public Flying Ubiquitous sensor Networks (FUSN-P). FUSN-P include terrestrial and flying segments. In this paper, we discuss one of the possible ways of the interactions between terrestrial USN and UAV-P, which would solve the data transmission

© Springer International Publishing Switzerland 2015
S. Balandin et al. (Eds.): NEW2AN/ruSMART 2015, LNCS 9247, pp. 352–366, 2015.
DOI: 10.1007/978-3-319-23126-6_32

problem between network nodes and their users. We propose the method for data collection from the terrestrial USN using FUSN-P. The goal of the proposed method is to optimize the FUSN-P resources spending for this purpose. Proposed method is based on the optimal UAV-P's route selection.

2   Problem Statement

We assume that the USN is placed on a flat surface. The network topology (nodes coordinates) can be obtained, for example, from constructing a network (if the fixed coordinates of each node are at its location) or by exploring the network using a UAV-P.

Once the network topology is known, the interaction between UAV-P and USN nodes collects the acquired information. The role of UAV-P in USN-UAV-P-Gateway system may be different depending on the method of organizing the data delivering process, Figure 1.

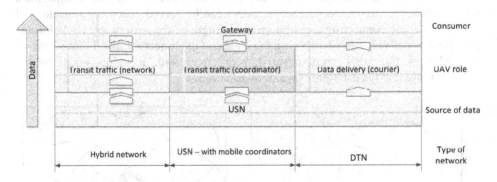

Fig. 1. The role of UAV-Ps in the USN – UAV-P – Gateway system

For example, the following options are possible:

-Transit Traffic through a network of UAV-Ps. In this case, the UAV-P is equipped with USN unit, which acts as a transit node (possibly the coordinator) [9,10]. There can also be UAV-P-UAV-P type of communications which forms a network between the mobile units providing route traffic to the gateway.

-Transit Traffic through the UAV-P node. In this case, the UAV-P is equipped with USN unit which acts as a transit and / or the coordinator. UAV-P also communicates with the gateway and is located in the zone within the gateway;

-Data delivering. In this case, the UAV-P is equipped with USN unit which acts as the coordinator or the gateway. The UAV-P receives data and stores it in its memory for the time required to move the UAV-P to a base station where the data is read. In this case there is a data delivery delay equal to the travel time of the UAV-P.

Let's consider the last option of using UAV-Ps. In this case the system can be regarded as a Delay-Tolerant Network (DTN) [11]. There are two main ways of the data transmission (transport) in this network: using wireless communication technologies (USN-UAV-Ps) and moving the UAV-P in space (mechanical). It is likely that in this case the second method requires significantly more time. One of the main indicators of quality of DTN functioning is data delivery time. Further we propose a method to minimize this data delivery time.

3   The Solution

Data delivery (transportation) involves data collection and delivery to the point in space where it can be transferred to a gateway. Data is collected using the technology of the USN communication nodes. USN nodes have a limited communication range. Therefore, depending on the size of the network and the requirements of the data delivery time, data collection is a movement of the UAV-P between the nodes or groups of nodes and actual data collection. The trajectory of the UAV-P in this case depends on the placement of nodes in the service area. If the communication range of the UAV-P can provide interaction with a number of USN nodes, it is advisable to find a point, where the maximum number of nodes can be served. Then the trajectory of UAV-P is a line connecting these points. The length of the path of the UAV-P route and the data delivery time depends on the choice of the trajectory line. Figure 2 shows examples of the USN units covered with circle with a radius equal to the UAV-P range and the different types of nodes location.

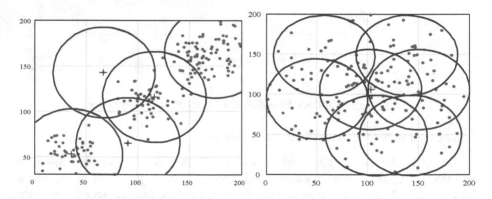

Fig. 2. Various types of the units placement in service area

In order to find the path which provides UAV-P the ability to receive data from all of the network nodes we should find the coverage of all USN nodes. The time used should be minimal. Assuming that the speed of the UAV-P is constant, it is required to find the path of a minimal length.

We assume that UAV-P's communication range is a circle with a radius R and UAV-P can read information from USN nodes at arbitrary points. We divide this problem into two components:

1. Search for the minimum number of points at which UAV-P need to read the information (minimizing);
2. Determination of the optimal UAV-Ps route between these points.

The key of finding the minimum number of points for reading information is to cover initial set of points (nodes) of the network by minimum number of circles with radius R. In order to solve this problem the method of cluster analysis, in particular the method FOREL, can be applied [12, 13-14].

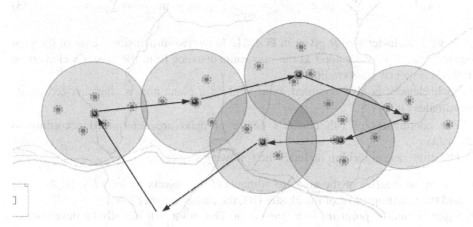

Fig. 3. An example of the UAV-P path at a known network topology

FOREL algorithm solves the clustering problem. It minimizes the total squared distance between the elements of the clusters (USN nodes) and the centers of mass of these clusters. Minimized function can be written as:

$$M = \sum_{i=1}^{k} \sum_{x_j \in S_i} (x_j - \mu_i)^2 \qquad (1)$$

where k - the number of clusters;

S_i - A number of elements of the i-th cluster;
μ_i - Coordinates of the center of mass of the i-th cluster;
x_j - Coordinates of the center of mass of the i-th element of the cluster.

The value $(x_j - \mu_i)$ is the Euclidean distance between the element of the cluster and the center of mass of the cluster.

Consider the algorithm for the two-dimensional space (the plane). In this case, each element is regarded as a point on a plane and is characterized by its coordinates (x_j, y_j).

The coordinates of the center of mass of i-th cluster are defined as

$$x_i^{(\mu)} = \frac{1}{n_i} \sum_{j=1}^{n_i} x_j \, , y_i^{(\mu)} = \frac{1}{n_i} \sum_{j=1}^{n_i} y_j \tag{2}$$

In addition, each object (point) other than the coordinates may be characterized by some parameter ("weight") m_j. Then the center of mass is defined as (the center of mass of a plane figure):

$$x_i^{(\mu)} = \frac{1}{m_i^{(\Sigma)}} \sum_{j=1}^{n_i} m_j x_j \, , y_i^{(\mu)} = \frac{1}{m_i^{(\Sigma)}} \sum_{j=1}^{n_i} m_j y_j \, , m_i^{(\Sigma)} = \sum_{j \in S_i} m_j \tag{3}$$

There is a cluster size R given in FOREL. In the two-dimensional case of the geometric plane, R is understood as the maximum distance from the cluster's element to cluster's center of mass (radius).

Each element is also regarded as a point on a plane and is characterized by its coordinates (x_j, y_j).

The coordinates of i-th cluster's center of mass are determined according to (2) or (3).

The clustering algorithm includes the following steps:

1. Set the boundaries of the area, coordinates of the objects (points) $I = \{I_1, I_2, \dots I_n\}$ and the maximum size of the cluster (R), the cluster number $i = 1$.
2. Select a random point m_i in a given area. This point will initially be the center of mass of the cluster.
3. Add all objects located at a distance equal to or less than R to this cluster.
4. Calculate the resulting cluster's center of mass according to formulas (2) or (3) $\widehat{m_i}$. If the calculated coordinates of the center of mass coincide with the point m_i, it is believed that the cluster i is defined. Thus, all assigned to this cluster points are marked with cluster number and are excluded from further consideration. Go to step 5 (search for the next cluster).
   If the calculated coordinates of the center of mass do not coincide with the point, the search process continues, i.e. we accept $m_i = \widehat{m_i}$, go to step 3.
5. Check if there are any objects left that are not assigned to any cluster. If not, then all the clusters are defined, the search is complete (go to step 6.).
   If there are such objects, then go to step 2 (search next cluster).
6. Obtain the results of the formed clusters with a) objects within them and b) centers of their mass. The end of the search.

Note. During the search of the next cluster there can be no object at a distance of less than R from the selected center of mass. In that case, make a selection of a new center of mass (random selection). The general algorithm is shown in Figure 4 FOREL.

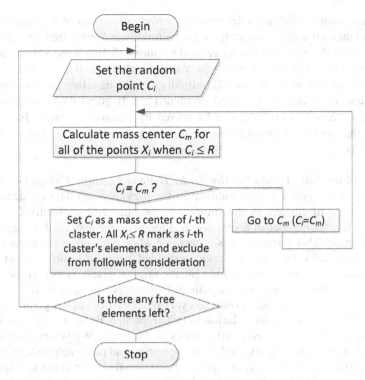

Fig. 4. FOREL Clustering Algorithm

It should be noted that similar to clustering in two dimensions, this problem can be solved for the three-dimensional space when the network elements are not located on a flat surface.

When the problem of choosing the data collection points is solved, it is necessary to find the trajectory of UAV-P. There can be different requirements imposed for the properties of the desired path such as the length, the number of data collecting points, angle of direction change, flight altitude, etc. In this paper we take the length of the path as an only criterion. We present the found data collection points as a complete graph's vertices. We believe that the UAV-P's path can start and end at any data collection point within the service provided territory. These points may be located both in the vertices of the obtained graph and outside the graph. In the second case the graph should be complemented by two vertices corresponding to those points. We can show that if the UAV-P's flight should start and finish in the same point, this point can be represented by two points with zero distance between them. In general, the desired trajectory is a route connecting all the vertices of the graph with the minimal traveled path length.

If the length of the route is greater than the maximum permissible (possible UAV-P flight range or data delivery time) you should find several routes which provide a parallel solution of the data collection problem. In general, the task of finding the solution that would "pull" a predetermined number of vertices of the graph by edges with minimum total weight associates with the minimum Steiner tree problem (SMT). It is known that this problem has no general solution [15,16]. If we add a requirement for the route such as passing through each vertex only once, this problem is reduced to a problem of finding a minimal Hamiltonian cycle (traveling salesman problem). This problem is defined as NP-hard problem [15]. If graph has a relatively small number of vertices this problem can be solved by exhaustive search. However, the growth of the number of nodes (network size) increases the complexity of this issue exponentially.

Definition of the Initial Data for the Task. There is a complete graph G. We define the vertices of the graph as the obtained in the previous step data collection points (centers of mass of clusters in the solution of the clustering problem) $X = \{x_i\}, i = 1..k$, where k – the number of found points. Moreover, we assume that graph G is an undirected graph. This assumption is acceptable when there are no restrictions on the direction of the UAV-Ps over the service provided territory. In order to set the start and end points of the route we supplement the original graph by two vertices x_S and x_t respectively. We assume that coordinates of the points x_S and x_t are known. Therefore, distances from these points to every other vertices of the graph can also be calculated. We assume that the solution to this problem is the route between the given vertices x_S and x_t which passes all vertices of the graph. We believe that the route should start at the vertex x_S and end at the vertex x_t and pass each node of the graph only once. The length of the route should be minimal. Under these restrictions the solution of the problem is the shortest Hamiltonian cycle (traveling salesman route) [15]. A variety of methods can be used in order to solve this problem such as precise methods (exhaustive search) or approximate methods (for example, a decision tree algorithm or penalty algorithm) [15]. Precise methods require too much computing. Therefore, we give preference to the penalty algorithm (Christofides algorithm) for the ease of its implementation and meeting the existing requirements. [15]. This algorithm is relatively simple in implementation and in most cases guarantees convergence to the optimal solution. As shown in [15], in case of nonconvergence of the algorithm to the optimal solution the near-optimal solution can be obtained. Penalty method is based on sequential search for the shortest spanning trees (SST) and "penalizing" its vertices the degree of which exceeds 2. The result of the algorithm is the shortest spanning tree, all vertices of which except vertices x_S and x_t have degree 2, i .e. the desired route with length close to the length of a Hamiltonian cycle.

If the task is to build several routes, the original complete graph can be divided into the required number of complete subgraphs. In that case the problem of finding optimal routes for UAV-P has to be solved for each of them.

The Route Search Using Penalty Algorithm. Let the graph be a matrix of lengths of its edges

$$C_0 = \|c_{ij}\|; \quad c_{ij} < \infty; \quad i, j = 1..k \tag{4}$$

where k - number of vertices.

Let's substitute values in rows or columns i = s and i = t or j = s and j = t with values $c_{sj} = c_{sj} + N; j = 1..k$ or $c_{is} = c_{is} + N; i = 1..k$, where N - sufficiently large number, i.e. $N \gg c_{ij}; i, j = 1..k$. The resulting matrix:

$$C_0 = \|c_{ij}\|; \ i, j = 1..k \tag{5}$$

Next we find the shortest spanning tree using the matrix C. If SST is a Hamiltonian cycle, i.e. each vertex of the graph except s and t has a degree equal to 2, then the problem is solved. If not, we "penalize" vertices with a degree greater than 2. If the vertex r has a degree greater than 2, then

$$c_{ir} = c_{ir} + \eta \cdot (c_{ir} - 2); \ i, r = 1..k \tag{6}$$

where η- the size of the penalty.

Next we find the shortest spanning tree for the updated matrix C and check for Hamiltonian path. We repeat these iterations until the next shortest spanning tree is not a Hamiltonian path, which is a solution. In general, if the number of iterations is too large and the shortest spanning tree is still not found, the process can be stopped. In that case a different search method should be used. Note, the nonzero probability of the search not converging to an optimal solution is typical for many known methods. This problem requires further investigation, however, as a practical solution we can suggest to solve it by modifying the original graph. The original graph can be modified, for example, by adding a new vertex with random coordinates or by changing coordinates of the vertices of degree greater than 2 to a random value lying within their evaluation precision. The second case is in fact equivalent to "penalizing" vertices on a random amount. In this case and in case of nonconvergence of the algorithm, the heuristic method of nearest neighbor, for example, will give a less accurate but guaranteed solution.

Finding the solution of the problem depends on whether there is a required Hamiltonian cycle or not and on the convergence of the algorithm. Definition of a Hamiltonian cycle is the fact that all vertices of the original graph should have a degree of no less than 2. This is also a required condition for the existence of the Hamiltonian cycle. Pausch condition[17] is yet another sufficient condition. With such condition, the graph contains a Hamiltonian cycle if the number of vertices k, $0 < k < \frac{n-1}{2}$ and if for any k, $0 < k < \frac{n-1}{2}$ the number of vertices of degree non exceeding k does not exceed k. The peculiarity of this problem is that the original graph is complete, i.e. all vertices are associated to one another, so that each vertex has a degree $n - 1$. In this case, the Pausch condition is fulfilled, therefore any possible graph is a Hamiltonian graph. In this case, the possibility of finding the solution depends only on convergence of the route searching algorithm (a Hamiltonian path).

The convergence and runtime of the penalty algorithm strongly depend on the penalty η. As it is shown in [15], if the values of penalty are too low there is a need of a greater number of iterations. On the other side, if the values of penalties are high the algorithm may not converge. The same paper suggested several approaches for choosing the penalty η.

In order to analyze the convergence and a runtime of the algorithm we run tests on a sufficient number of random complete graphs of various dimensions. In those tests we used fixed positive penalties. The amount η was chosen based on the distribution of lengths of the original graph edges. The results of testing penalty algorithm, its random penalty modifications and the nearest neighbor method, are shown below. For the testing we used C# program which implements described in this article algorithms. Testing was performed on graphs of different dimensions from 10 to 60, for each of the dimensions 20 tests on random graphs were performed. A Poisson field of points in a plane of a square bounded area was used as vertices of a random graph. The length of its edges was the distance between the points. Figure 5 shows an example of the route found using the fixed penalty method in the graph of 60 vertices.

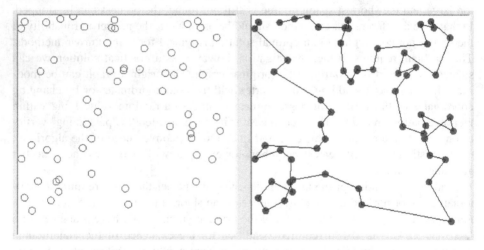

Fig. 5. Example of found Hamiltonian path in the graph of 60 vertices

Examples of the results obtained by different search methods are shown in Figure 6 which shows: initial field of points, the route found using the fixed penalty method. There are two options shown for the random penalty method. There are two options shown for the random penalty method and one option for the nearest neighbor method. The length of the path ("L") is also shown in the Figure 6.

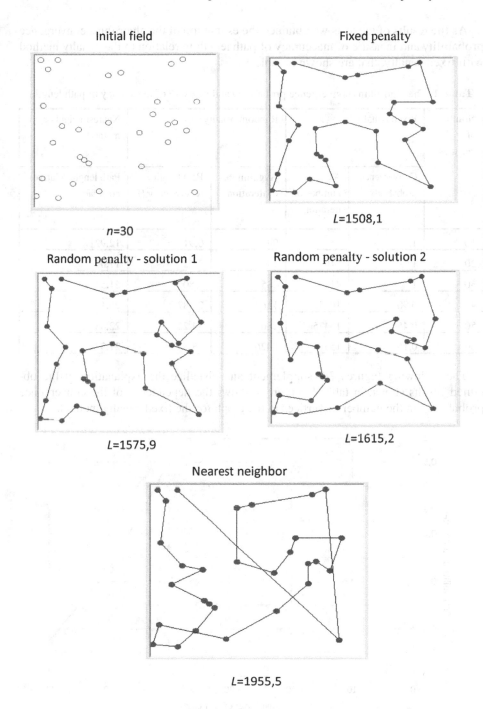

Fig. 6. Examples of the route search results obtained by different methods

As the results of the tests we obtained the estimation of the algorithms convergence probability and measure of inaccuracy of path length in relation to the penalty method with fixed penalty, that are shown in table 1.

Table 1. The algorithms convergence probability and measure of inaccuracy of path length.

Number of vertices	Fixed penalty		Random penalty		Nearest neighbor method
	Convergence probability	Avg. number of iteration	Avg. number of iteration	Path length relative error %	Path length relative error %
10	1	95	94	0,91	12,92
20	1	610	147	3,72	17,05
30	1	4106	235	7,80	21,52
40	0,90	10426	1168	20,87	26,14
50	0,45	13475	1861	34,22	22,68
60	0,15	15573	2426	38,33	21,57

The following Figures 7-9 supplement and visualize the explanation of the obtained results shown in table 1. Figure 7 shows the dependence of the convergence probability on the number of vertices in the graph for the fixed penalty method.

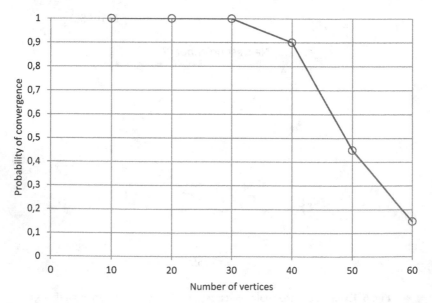

Fig. 7. Dependence of the convergence probability on the number of vertices in the graph for the fixed penalty method

According to the results the convergence probability begins to decline significantly when the graph has more than 30 vertices, with more than 40 vertices we can see a sharp decrease of the algorithm convergence probability. It should be noted that the random penalty and nearest neighbor algorithms provide 100% convergence throughout all range of the graph sizes.

Figure 8 shows the relative error for the length of the found route of the random penalty and nearest neighbor methods in relation to the fixed penalty method.

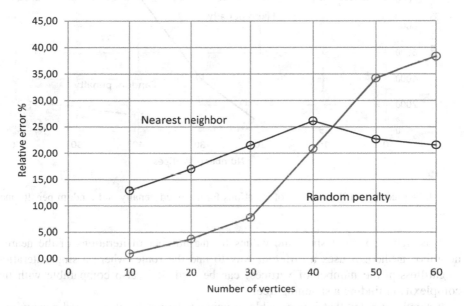

Fig. 8. Relative error for the length of the route found using the random penalty and nearest neighbor methods

As can be seen from the graphs, when the number of vertices is small, the value of the relative error for random penalty method is lower than for nearest neighbor method. However, increasing the number of vertices to 45..50 we can see an increase of errors for random penalty method compared with the nearest neighbor method. It should be noted that the computational complexity of these methods differs significantly in favor of the nearest neighbor method.

Fig. 9 shows the number of iterations for the fixed penalty and random penalty methods. In this case we define iteration as a single iteration of the minimum spanning tree search algorithm. As can be seen from the figure, the number of iterations needed to obtain a solution using the random penalty method is considerably less, although it is commensurate with the number of iterations for the fixed penalty method.

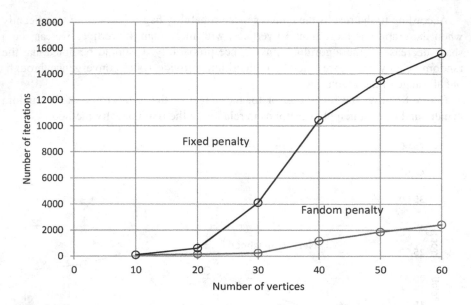

Fig. 9. Dependence of the number of iterations for the fixed penalty and random penalty methods

This figure does not show the values for the number of iterations of the nearest neighbor method. It uses a different way to find the route, where a single iteration (regardless of the number of vertices) can be used. It is also comparable with the complexity of finding a spanning tree.

It should be noted that it is possible to improve the results of the fixed penalty method using a more effective strategy of "penalizing" vertices, as suggested in [15]. In this paper such modifications have not been considered. These modifications would probably improve the convergence of the method, but would not significantly reduce its computational complexity compared to heuristically nearest neighbor method.

4   Conclusions

Comparing the accuracy and computing complexity of the methods listed below, we can conclude the following:

1. Search for a UAV-P route should be divided into two tasks: finding points of interaction with the network nodes and laying a route with the shortest length through these points. In order to solve the first problem the clustering method FOREL can be used, while for the second task we can use the methods for solving the Traveling Salesman Problem.
2. The Traveling Salesman Problem is an NP-hard problem. Therefore, to solve this problem, depending on the conditions, both methods of search precision (optimal) solution and heuristics methods of search close to the optimal solution may be used.

3. The accuracy of the results obtained by the methods discussed above depends on the dimension of the graph (number of readout data points). Size of the relative error of the proposed heuristic fixed penalty method and heuristic nearest neighbor method decreases with the number of vertices.
4. The computational complexity of the proposed random penalty method is significantly lower than complexity of the penalty method, but it has a similar growth trend with an increase of the number of vertices in the graph.
5. The analysis of the penalty methods and of the proposed random penalty modifications has shown that the first of them can be used (in terms of current research) in case of a relatively small number of data readout points (up to 30 points). The proposed random penalty method can be applied if number of points is from 30 to 50. For a larger number of points it is appropriate to use heuristic nearest neighbor method.

Acknowledgements. The reported study was supported by RFBR, research project No15 07-09431a "Development of the principles of construction and methods of self-organization for Flying Ubiquitous Sensor Networks".

References

1. Akyildiz, I.F., Vuran, M.C., Akan, O.B., Su, W.: Wireless Sensor Networks: A Survey revisited. Computer Networks Journal (2005)
2. Dargie, W.W., Poellabauer, C.: Fundamentals of Wireless Sensor Networks: Theory and Practice. John Wiley & Sons Ltd. (2010)
3. Recommendation Y.2060: Overview of Internet of Things. ITU-T, Geneva, June 2012
4. Stojmenovic, I.: Handbook of sensor networks. Algorithms and architectures. John Wiley & Sons (2005)
5. Kim, D.S., Chung, Y.J.: Self-organization routing protocol supporting mobile nodes for wireless sensor network. In: Proceedings First International MultiSymposium on Computer and Computational Sciences, Hangzhou, China, June 2006
6. Li, Y., Thai, M.T., Wu, W.: Wireless Sensor Networks and Applications. Springer (2008)
7. El Emery, I.M.M., Ramakrishnan, S.: Wireless Sensor Networks: From Theory to Applications. CRC Press, Taylor & Francis Group (2013)
8. Koucheryavy, A., Vladyko, A., Kirichek, R.: Flying Sensor Networks. Electrosvyaz, no. 9 (2014) (in Russian)
9. Koucheryavy, A., Salim, A.: Cluster head selection for homogeneous wireless sensor networks. In: Proceedings, International Conference on Advanced Communication Technology, ICACT 2009, Phoenix Park, Korea
10. Koucheryavy, A., Salim, A.: Prediction-based clustering algorithm for mobile wireless sensor networks. In: Proceedings, International Conference on Advanced Communication Technology, ICACT 2010, Phoenix Park, Korea
11. Akyildiz, I.F., Akan, O.B., Chen, C., Fang, J., Su, W.: InterPlaNetary Internet: state-of-the-art and research challenges. Computer Networks **43** (2003)
12. Paramonov, A.: Investigation of the Traffic Models for Public Networks. D.Sc. Thesis. SPbSUT, St.Petersburg (2014) (in Russian)

13. Mandel, I.D.: Cluster analysis. Finance and Statistics, Moscow (1988) (in Russian)
14. Enyukov, I.S. (ed.) Factor, discriminant and cluster analysis, Moscow (1989). Translation of: Kim, J.-O., Mueller, C.U., Klecka C. (in Russian)
15. Christofides, N.: Graph Theory. An Algorithmic Approach. Hardcover (1975)
16. Berge, C.: The Theory of Graphs and its Applications. Wiley (1962)
17. Ore, O.: Theory of Graphs. American Mathematical Society, Colloquium Publications, vol. 38 (1967)

Reliable and Scalable Architecture for Internet of Things for Sensors Using Soft-Core Processor

U.V. Rane, V.R. Gad, R.S. Gad$^{(\boxtimes)}$, and G.M. Naik

ALTERA SoC laboratory, Department of Electronics, Goa University, Goa 403206, India
{udaysingvrane,gadvinaya}@gmail.com, {rsgad,gmnaik}@unigoa.ac.in

Abstract. With significant technological developments as well as advances in sensors, wireless communications and Internet – a lot of research areas have emerged, such as wearable computing, context-aware homes, mobile phone sensing and smart vehicle systems. From those emerging areas, there is a clear trend to augment the physical devices/objects with sensing, computing & communication capabilities, connect them together to form a network and make use of the collective effect of networked smart things – the Internet of Things(IoT). This paper proposes the IoT architecture "Kaivalyam"[1] for sensors/actuator. The configurable interface generates the identity of the 'Thing' and stores in 32-bit datagram. The various configuration of the 32-bit datagram from 4,8,12 and 16-bit device data and respective 14, 12, 10 and 8 bits duplicate identifier can scale the number of devices connected to design platform. These datagram's read into FIFO of the Triple Speed Ethernet (TSE) for transmission using Low Density Parity Check (LDPC) encoder. Simulation studies for the system were performed for block length 512 bits, which is the minimum Ethernet frame length of 64 bytes. AWGN(Additive White Gaussian Noise) is introduced in the channel and BER(Bit Error Rate) is computed for different (1dB to 6dB) SNR(Signal to Noise Ratio) showing BER of 10-4 -4 to 10-5 can be achieved for SNR of 2.5 dB indicating the secured and reliable data transmission.

Keywords: Internet of Things · Architecture · Scheduling · Ethernet and LDPC

1 Introduction

"All things appear and disappear because of the concurrence of causes and conditions. Nothing ever exists entirely alone; everything is in relation to everything else" is what said by Prince Gautama Siddharta[1], which has direct relevance with the idea of

---

[1] Kaivalyam, is the ultimate goal of yoga; and means "Mukti" or "detachment" for absolute freedom. It is said in Vedas / Upanishads that there re 4 types of Mukthi/Kaivalyam. Namely: Saalokya, Saaroopya, Saameepya, & Saayujya. Which can be explain as when person worships a Deity he attains to the world of that Deity called 'Saalokya'; further he attains 'Saaroopya' i.e. form of that Deity; then he attains 'Saameepya' i.e. proximity to that Deity and finally he becomes one with that Deity i.e. 'Saayujya' (http://www.namadwaar.org/nibbles/?p=150). This way the person attains the absolute freedom which is similar to 'Thing' in IoT getting freedom in space.

© Springer International Publishing Switzerland 2015
S. Balandin et al. (Eds.): NEW2AN/ruSMART 2015, LNCS 9247, pp. 367–382, 2015.
DOI: 10.1007/978-3-319-23126-6_33

Internet of Things (IoT) introduced in the PCANS Model[2]. The PCANS model suggests that "all systems are structured along these three domains, Individuals, Tasks, and Resources" and also introduces the concept that networks occur across multiple domains and that they are interrelated. Network science studies complex networks such as engineered, information, biological, cognitive, semantic and social networks. The field draws on theories and methods including graph theory from mathematics, statistical mechanics from physics, data mining and information visualization from computer science, inferential modeling from statistics and social structure from sociology. A telecommunications network is a collection of terminals, links and nodes which connect to enable telecommunication between users of the terminals. Each terminal in the network has a unique address so messages or connections can be routed to the correct recipients. The links connect the nodes together and are themselves built upon an underlying transmission network which physically pushes the message across the link; using circuit switched, message switched or packet switched routing. Examples of telecommunications networks are: computer networks, the Internet, the telephone network, the global Telex network, the aeronautical 'ACARS' network.

IoT works has been proposed in several application scenarios, such as environmental monitoring, e-health, intelligent transportation systems, military, and industrial plant monitoring. Technically this requires embedding sensing, actuation, processing, securing, and reliable networking into common objects. Sensors device availability and decreased cost of hardware has triggered the era of new computing systems to integrate the vehicles, devices, goods and everyday's object to be a part of IoT[3].

This paper proposes IoT architecture based on the extended Ethernet MAC approach providing various communication front-ends interfaces to reconfigure them for the propose common 'Kaivalyam1' interface which explore novel way for person to object and object to object communication like the sensors or actuator network interface on IoT platform sensors/actuator. Further it elaborates the capabilities of architectures for scalability for nodes and reliability of data using Error Correction Coding with LDPC coding schemes.

2   Architecture of IoT and Related Enabling Technologies and Standardizations

There are three IoT components which enables seamless ubicomp: (a) Hardware—made up of sensors, actuators and embedded communication hardware (b) Middleware—on demand storage and computing tools for data analytics and (c)Presentation—novel easy to understand visualization and interpretation tools which can be widely accessed on different platforms and which can be designed for different applications. Gartner 2012 Hype Cycle of expected emerging technologies indicates the IoT will take 10-12 years to reach to plateau of productivity[4] . The claims of the Gartner are justified by the ongoing status of standardization activities of EPCglobal, GRIFS, M2M, 6LoWPAN, ROLL and other stakeholders of IoT. Most of them are supporting the data rate of 10^2 kbps over 1- 100 meters range of communication.

2.1   Enabling Technologies IoT

Development of certain enabling technologies such as nano-electronics, communications, sensors, smart phones, embedded systems, cloud computing and software technologies will be essential to support important future IoT product innovations affecting the different industrial sectors. In addition, systems and network infrastructure (Future Internet) are becoming critical due to the fast growth and advanced nature of communication services as well as the integration with the healthcare systems, transport, energy efficient buildings, smart grid, smart cities, and electric vehicles initiatives.

2.1.1 More Than Moore (MtM)

Since 2007, the ITRS (The International Technology Roadmap For Semiconductors: 2012 Update) has addressed the concept of functional diversification under the title "More than Moore" (MtM). The MtM approach typically allows for the non-digital functionalities which don't scale as per 'Moor's Law' (e.g., RF communication, power control, passive components, sensors, actuators) to migrate from the system board-level into a particular package-level (SiP) or chip-level (SoC) system solution. The basic idea of MtM is the pervasive presence around us of a variety of things or objects which, through unique addressing schemes, are able to interact with each other and cooperate with their neighbors to reach common goals [5]. With regards to the IoT paradigm at large, a very interesting standardization effort is now starting in ETSI [6] (the European Telecommunications Standards Institute), that has purpose of conducting standardization activities relevant to M2M systems and sensor networks (in the view of the IoT). The goals of the ETSI M2M committee include: the development and the maintenance of an end-to-end architecture for M2M (with end-to-end IP philosophy behind it), strengthening the standardization efforts on M2M, including sensor network integration, naming, addressing, location, QoS, security, charging, management, application, and hardware interfaces [7]. Presently, Micro-electro-mechanical systems (MEMS) technologies can fabricate micrometer-sized mechanical structures (suspended bridges, cantilevers, membranes, fluid channels, etc.) that are often integrated with analog and digital circuitry. MEMS can act as sensors, receiving information from their environment, or as actuators, responding to a decision from a control system to change the environment. It also reviews emerging MEMS applications, including optical filters, picoprojectors, the electronic nose, microspeakers, and ultrasound devices.

2.1.2 Communication Technologies

Internet Protocol is used in network technology for connecting smart objects around the world. According to the Internet Protocol for Smart Objects (IPSO) vision, the IP stack is a light protocol that already connects a huge amount of communicating devices and runs on tiny and battery operated embedded devices. This guarantees that IP has all the qualities to make IoT a reality. By reading IPSO whitepapers, it seems that through a wise IP adaptation and by incorporating IEEE 802.15.4 into the IP architecture, in the view of 6LoWPAN [8], the full deployment of the IoT paradigm will be automatically enabled. We have proved the same for smart system control platform for Ethernet enabled devices [9]. As this represents only a partial functional requirement in the IoT, similar to the role of communication technology in the Internet and equaling communication technologies such as WiFi, Bluetooth, ZigBee, I2C, CAN, 6LoWPAN, ISA 100, WirelessHart /802.15.4, 18000-7, LTE to the Internet of Things is too simplistic. The 6LoWPAN concept originated from the idea that low-power devices with limited processing capabilities should be able to participate in the Internet of Things. Wireless sensor network and MANET is capable of performing various mechanisms [10] such as self-configuration, multi hop communication, energy efficient operations, in network processing, data centric and content-based networking, exploiting location and activity pattern, positioning, scheduling, time synchronization topology control and routing. However, we can say that these technologies certainly might be part of Internet of Things.

2.1.3 Addressing and Networking Issues

The IoT will include an incredibly high number of Nodes. Currently, the IPv4 protocol identifies each node through a 4-byte address and these addresses are depleting rapidly and will soon reach zero. IPv6 addressing has been proposed for low-power wireless communication nodes within the 6LoWPAN context. IPv6 addresses are expressed by means of 16 bytes to define 1038 addresses, which should be enough to identify any object which is worth to be addressed. IPv6 addresses are assigned to organizations in much larger blocks as compared to IPv4 address assignments—the recommended allocation is a /48 block which contains 280 addresses, being 248 or about 2.8×1014 times larger than the entire IPv4 address space of 232 addresses and about 7.2×1016 times larger than the /8 blocks of IPv4 addresses, which are the largest allocations of IPv4 addresses. The total pool, however, is sufficient for the foreseeable future, because there are 2128 or about 3.4×1038 (340 trillion trillion trillion) unique IPv6 addresses. This address space is many times that of the world population of 7 billion which will accommodate the generation of huge quantities of data over IoT, between 1.000 and 10.000 per person per day [11].

2.1.4 Embedded Devices

RFID or wireless sensor networks (WSN), may be part of the Internet of Things, but as standalone applications (intranets) they miss the back-end information infrastructures necessary to create new services. The IoT has come to mean much more than just networked RFID systems. While RFID systems have at least certain standardized information architectures to which all the Internet community could refer, global WSN infrastructures have not yet been standardized. An IoT vision statement, which goes well beyond a mere "RFID centric" approach, is also proposed by the consortium CASAGRAS [12]. With regards to the RFID technology, it is currently slowed down by fragmented efforts towards standardization, which is focusing on a couple of principal areas: RFID frequency and readers-tags (tags-reader) communication protocols, data format placed on tags and labels. The major standardization bodies dealing with RFID systems are EPCglobal, ETSI, and ISO.

2.2   Internet of Things Architecture Technology

RFID-installations in production and logistics today can be considered as an Intranet of Things or Extranet of Things. Traditional communication means, such as EDIFACT (Electronic Data Interchange For Administration, Commerce and Transport), are used to communicate with a limited number of preferred partners. These early approaches need to be extended to support open Internet architectures. Most of the RFID installations introduce a novel read-out method for a hierarchical wireless master slave RFID reader architecture of multi standard Near Field Communication (NFC) and Ultra High Frequency (UHF) technologies. This can be used to build a smart home service system those benefits in terms of cost, energy consumption and complexity [13].

There are several projects and standardization initiatives on sensor networks, which may eventually converge with the IoT. The core objective of the COBIS project was to provide the technical foundation for embedded and wireless sensor network technology in industrial environments. SENSEI creates an open, business-driven architecture that fundamentally addresses the scalability problems for a large number of globally distributed wireless sensors and actuator devices. It provides network and information management services to enable reliable and accurate contextual information retrieval and interaction with the physical environment.

Likewise, other smaller research projects exist, such as GSN[14], SARIF[15], and MoCoSo, that combine concepts of object identification, sensor data and the Internet. Sensor networks can be integrated in the IoT for example, by integration with the EPCglobal Architecture Framework. Although the EPCglobal Network does not yet provide adequate support for the inclusion of sensor values in the streams of data, the Action Groups inside the GS1/EPCglobal community are actively researching issues such as 'Active Tagging' and 'Sensor and Battery Assisted Passive Tags'. The EPC Sensor Network[16] is an effort of the Auto-ID Lab in Korea to incorporate Wireless Sensor Networks (WSN) and sensor data into the EPCglobal Network architecture and standards. While identification, sensing and actuator integration are core functionalities in an IoT, there are further requirements such as scalability and robustness that need to be addressed.

3   Architecture 'Kaivalyam' for IoT Using Ethernet MAC Backbone

Open standards are required to use and extend its functionality. It will be a huge network, considering that every object has its virtual representation. Therefore, scalability is required. The Internet of Things will need to be flexible enough to adapt to changing requirements and technological developments. Proposed architecture (Fig. 1.) support the flexibility as one can add as many communication interfaces developing in near future. Also the protocols can be coded decoded at the Kaivalyam platform. The architecture for the sensor integration should address issue like Unique Identity, Integration of dynamic data, Support for non-IP devices, Integration of an actuator interface, Data synchronization for offline support, Optional Interface for software agents etc. Our architecture supports almost all these features except the last. The Optional interface (Fig. 2.) can be also provided using the soft-core NIOS processor System on Chip (SoC) solution by adding the respective lightweight IP protocols in the system software of the SoC[9]. We have proposed here the 4-port switch having soft-core processor for monitoring and routing the packets.

The data is encoded using LDPC (Low Density Parity Check) codes which are transmitted over usually the Gaussian channel after modulation. The data transmitted is demodulated and then decoded for errors corrections, if any.

3.1   Extended MAC Network Interface Card Architecture 'Kaivalyam' for IoT

Wireless sensor network solutions are based on the IEEE 802.15.4 standard, which defines the physical and MAC layers for low-power, low bit rate communications in wireless personal area networks(WPAN) [17].WSN is capable of performing various mechanisms such as self-configuration, multi-hop communication, energy efficient operations, in-network processing, data centric and content based networking, scheduling, time synchronization topology control and routing. IEEE 802.15.4 does not include specifications on the higher layers of the protocol stack, which is necessary for the seamless integration of sensor nodes into the Internet. This is a difficult task for several reasons, the most important are given below:

Sensor networks may consist of a very large number of nodes. This would result in obvious problems as today there is a scarce availability of IP addresses. The largest physical layer packet in IEEE 802.15.4 has 127 bytes; the resulting maximum frame size at the media access control layer is 102 octets, which may further decrease based on the link layer security algorithm utilized. Such sizes are too small when compared to typical IP packet sizes.In many scenarios sensor nodes spend a large part of their time in a sleep mode to save energy and cannot communicate during these periods. This is absolutely anomalous for IP networks.

In other words, between two different objects communicating, the communication path may be broken into different sections [18-19]. And how will all these products manage to talk to each other? The 'language' will be based on a type of protocol, similar to the built-in formula that enables our mobile phones to talk using WiFi, ISA100, ZigBee or BlueTooth[20]. Hence such an intelligent, configurable network interface is an effective solution. A reconfigurable NIC (Network Interface Card) allows rapid prototyping of new system architectures for network interfaces [21].

The architectures can be verified in real environment, and potential implementation bottlenecks can be identified. Thus, what is needed is a platform, which combines the performance and efficiency of special-purpose hardware with the versatility of a programmable device. Hence we have proposed the platform which is processor-based implemented using a configurable hardware [22]. An FPGA (Field Programmable gate array) with an embedded processor is a natural fit with this requirement. Also, the reconfigurable NIC must have different memory interfaces including high capacity memory and high speed memory for adding new networking services. This is SGMII interface (Fig. 2.) to support the various speed over Ethernet and Physical media.

3.2   Scalability for Nodes with 'Kaivalyam' Packet

The 'Kaivalyam' packet is optimized for the smaller size of 256- bits due to limited resources on node. The design is such that the 128-bits of IPv6 address are allocated for Things unique universal address. It is desirable to program the node for smart features which could be exercise through proper control register. The status of the Thing could be read through the status register for sleeping, hibernation etc. for energy minimization of node. Here small overhead of 16-bit is kept for the parity information with proper hamming distance windows for data error checking. The remaining 96-bits corresponding to 12-bits of data payloads are kept for the node. Such small size packets are very much possible on MtM sensors nodes.

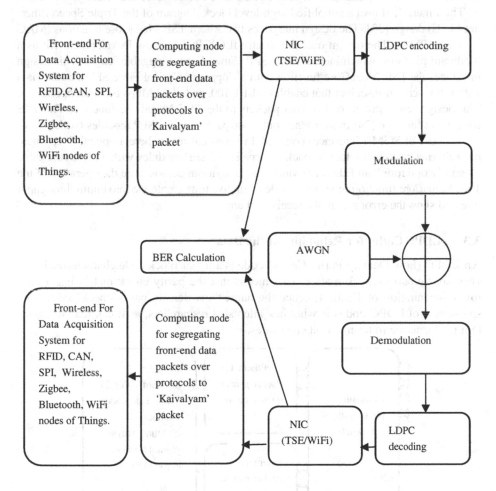

Fig. 1. The system details of the 'Kaivalyam' Architecture of IoT.

The unique universal ID's, data, status and parity information of the node transmitted over like RFID, CAN, SPI, Zigbee, Bluetooth, WiFi, Wireless etc.; could be extracted over front-end interfaces during the segregation with help of configurable computing node having integrated N x 4 switch. Thus one can generate the 256-bits of 'Kaivalyam' packets as shown in Figure 3. Data-synchronization for offline support feature can be program in the soft-core processor (NIOS) built on SoC using FPGAs to store the data of the devices over banks in the limited Flash memory (or may be over SAN's) in the offline mode and the same data can be transmitted during active mode. Also the optional integration of software agents feature can be programmed in the scenarios like complex global supply networks requiring more decentralized and automated decision making. Software-agents have been researched broadly.

The Figure 4, shows a simplified high-level block diagram of the Triple Speed Ethernet (TSE) design [23]. The design integrates two Altera TSE MegaCore functions (MAC + PCS + PMA). The design uses the Stratix II GX PCI Express Development Kit as a hardware platform, which includes two SFP (Small Form Pluggable) cages. This design interfaces the TSE MegaCore function with a Copper or Optical Fibre SFP module via a 1.25 Gbps serial transceiver that enables all 10, 100, and 1000 Mbps Ethernet operations. The design sends stream of Ethernet packets to the TSE MegaCore function. The TSE MegaCore function [24] in turn sends out those packets to the SFP modules (which serve the purpose of SGMII interface) connected to physical media here a optical fiber where the Ethernet packets are looped back externally via SFP modules with an Ethernet cable assembly or through an Ethernet switch. The design can demonstrate the operation of the TSE MegaCore function in various modes with live traffic upto the maximum throughput rate and show the error rate in the receiver, if any.

3.3   LDPC Codes for Reliability of the Data

An LDPC (Low Density Parity Check) code is a linear block code characterized by a very sparse parity-check matrix. This means that the parity check matrix has a very low concentration of 1's in it, hence the name "low-density parity-check" code. The sparseness of LDPC codes is what has interested researchers, as it can lead to excellent performance in terms of bit error rates.

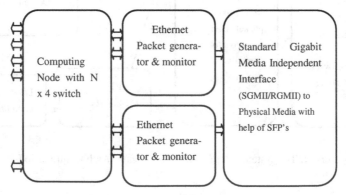

Fig. 2. Computing node for segregating data packets over protocols

Table 1. Protocols of popular physical communication interfaces exploited by communication-enabled objects.

Physical Communication interface type	Communication type	Protocols	OSI Layers
Zigbee,Bluetooth, RFID,etc.	Wireless	NWK/APS/API defined by each standardization body	Network/Transport/ Upper
WiFi	Wireless	IP/TCP-UDP	Network /Transport/Upper
UWB	Wireless		Network /Transport /Upper
Sensor network busses (CAN, Profibus)	Fixed	Upto Data link	Data link
Serial	Fixed	Upto Data link	Data link
USB	Fixed,Wireless	Upto Data link	Data link
DeviceNet	Fixed	DeviceNet network and transport	Network /Transport Upper
ControlNet	Fixed	ControlNet network and transport	Network /Transport /Upper
Ethernet/IP	Fixed	IP/TCP-UDP	Network /Transport /Upper
Power line (KNX, LonWorks)	Fixed	Network /transport layers according Network layer/Transport to KNX and Lonworks specifications	Network /Transport/Upper

128-bit address parity of things(IPv6)	96-bits data	16-bit status	16-bit Control Register	16-bit Parity

Fig. 3. 256-bits of 'Kaivalyam' packet datagram details for 'Thing' on IoT.

LDPC Codes are characterized by the sparseness of ones in the parity-check matrix. This low number of one's allows for a large minimum distance of the code, resulting in improved performance. Although proposed in the early 1960's by Gallager, it has not been since recently that codes have emerged as a promising area of research in achieving channel capacity. This is part due to the large amount of processing power required to simulate the code. In the case of any coding scheme larger block length

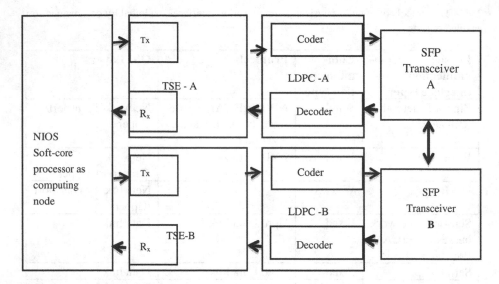

Fig. 4. Simplified block diagram of Triple Speed Ethernet (TSE) reference design.

codes provide better performance, but require more computing power. Performance of a code is measured through its bit error rate (BER) vs. signal to noise ratio Eb/No *in* dB. The curve of a good code will show a dramatic drop in BER as SNR improves. The best codes have a cliff drop at an SNR slightly higher than the Shannon's limit (0.18dB).In addition to presenting his seminal work in 1960, Gallager also provided a decoding algorithm that is effectively optimal. The algorithm iteratively computes the distributions of variables in graph-based models and comes under different names, such as 'Message Passing Algorithm", "Sum-Product Algorithm (SPA)" or "belief propagation Algorithm". SPA-Logdomain and SPA-Min Sum Algorithm are the simplified algorithms and easy to implement in an embedded environment[25] .

4   Results and Discussion

We have conceptualized the computing node using the NIOS-II soft-core processor as a processing element having 8-bit interface with 4x4 I/O switch (Figure 2). The design is implemented with ALTERA Inc. Quartus-II software (Ver. 8.0). The computing node is a switch embedded with soft-core IP NIOS-II processor having processing capabilities. Presently the features of generating and receiving ATM packets programs are tested on the NIOS IDE of Quartus-II environment.

Table 2. Details of Input packets

Input data byte stream					
1st byte	2nd byte	3rd byte	4th byte	5th byte	6-53 bytes
72	75	6C	95	76	11
A9	A5	BE	DC	AD	22
DF	EB	9E	02	E3	33
15	10	20	48	1A	44

Table 3. Details of Output packets

Output data byte stream					
1st byte	2nd byte	3rd byte	4th byte	5th byte	6-53 bytes
72	71	23	45	76	11
A9	AA	BC	DC	AD	22
DF	E8	19	32	E3	33
15	1D	EB	A8	1A	44

Functional simulations were performed using QUARTUS II software. Here the computing node use the router static look-up table Virtual Channel Identifier (VCI) information obtained for the shortest path optimization computation. This VCI information can be altered for the   Nx4 switch in the proposed 'Kaivalyam' Architecture to route the Kaivalyam packets on to duplex mode Ethernet Packets Generator and Monitor for transmission over Ethernet backbone. Hence , depending on the look-up table information the packets VCI information is altered for the specific route. The functional simulation results are demonstrated in the input 'Table 2' and output 'Table 3' for routing of the packets over 4x4 I/O port switch. The scheduler circuit in the switch schedules the packets available in the 'VOQ_input_x'to the appropriate output through crossbar fabric(Voq_fabric). The scheduling of the packets is implemented using combination of priority based round robin diagonal propagation over the five I/P ports and four O/P ports of the computing node design. Diagonal propagation has advantage of dependencies over crossover fabric.

Simulation studies were performed (using MATLAB Simulink) for block length 512 bits, which is the minimum Ethernet frame length of 64 bytes. AWGN(Additive White Gaussian Noise) is introduced in the channel and BER(Bit Error Rate) is computed for different SNR(Signal to Noise Ratio). The SNR is increased from 1dB to 6dB. Figures 6(a) to 6(f) indicate that a BER of 10^{-4} to 10^{-5} can be achieved for SNR of 2.5 dB. Also increasing the number of iterations decreases the BER. However for considerable improvement in BER, you need to increase the block length. Also SPA-Logdomain Algorithm is showing better performance than SPA-Min Sum Algorithm. LDPC can be used for LANs since the normal noise level involved in optic fiber cables would range from 3 to 4 dB.

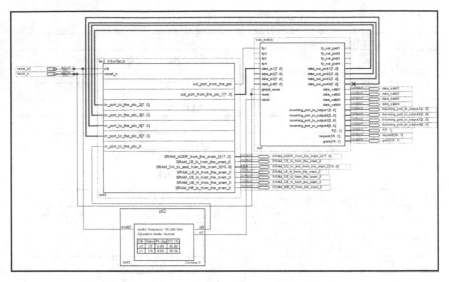

Fig. 5. Computing Node embedded with 4x4 I/O switch.

Concept that IoT has primarily to be focused on the "Things" and that the road to its full deployment has to start from the augmentation in the Things' intelligence. This is possible through proposed 'Kaivalyam' packets which is in line of 'spime'. The spime are defined as object that can be tracked through space and time throughout its lifetime and that will be sustainable, enhanceable, and uniquely identifiable [26]. Although quite theoretical, the spime definition finds some real-world implementations in so called Smart Items. These are a sort of sensors not only equipped with usual wireless communication, memory, and elaboration capabilities, but also with new potentials.

Inputs are particularly expected from the Machine-to-Machine Workgroup of the European Telecommunications Standards Institute (ETSI) and from some Internet Engineering Task Force (IETF) Working Groups. 6LoWPAN [27], aiming at making the IPv6 protocol compatible with low capacity devices, and ROLL [28], more interested in the routing issue for Internet of the Future scenarios, are the best candidates. The multi sensor fusion algorithms like estimation (Non recursive, Recursive) classification (parametric, cluster, K-means), inference (Baysian, Dempster-shafer) and ANN (Expert, Adaptive, Fuzzy) methods [29] integrated in the MtM could give diver scope and potential application in the areas of co-operative, community sensing applications for IoT in areas of self-assembly and self-organization system. IEEE 802.11 is a set of media access control (MAC) and physical layer (PHY) specifications for implementing wireless local area network (WLAN) computer communication in the 2.4, 3.6, 5 and 60 GHz frequency bands. The IEEE 802.11 can be extended for higher data rates with the multiple-antenna also known as spatial multiplexing with multiple input multiple-output (MIMO) system design, wherein data for transmission is divided into independent data streams to be transmitted through multiple antennas. In a multi-antenna system the adjacent antennas must be separated by a minimum distance, around half a wavelength (27 mm for 802.11ac), to reduce the coupling between antennas as well as correlation between streams. For applications where size matters, this requirement limits the number of antennas and consequently the number of streams and maximum bit rate. At 60 GHz the carrier wavelength is only 5 mm, so relatively high gain antennas can be implemented in a small package with MtM technology in place. For example, a 13 dB patch array antenna printed on Duroid substrate ($r = 2.2$) occupies an area of 5 mm × 6 mm [30].

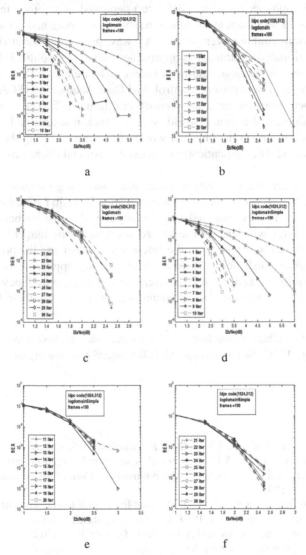

Fig. 6. BER vs SNR for 512 bytes block size for 100 frame over 1 to 30 iteration with logdoman(a-c) and logdomainSimple(d-f) algorithms.

On web front, we moved from www (static pages web) to web2 (social networking web) to web3 (ubiquitous computing web), the need for data-on-demand using sophisticated intuitive queries increases significantly. Another interesting paradigm which is emerging in the Internet of the Future context is the so called Web Squared, which is an evolution of the Web 2.0. It is aimed at integrating web and sensing technologies [31] together so as to enrich the content provided to users. Presently, this is obtained by taking into account the information about the user context collected by the sensors (microphone, cameras, GPS, etc.) deployed in the user terminals. Such Web Squared could be enhances for functionalities using more nodes for better virtualization applications running over the IoT. Anyway, proprietary industrial approaches ignoring international standardization approaches as well as political discussion will try to set their own de-facto-standards. A recent malware attack (Stuxnet), aiming to spy on and reprogram Supervisory Control And Data Acquisition (SCADA) systems, has revealed once more the need for security in a future IoT. The Internet has been misused to manipulate the virtual world, such as stock markets; and hence IoT will have direct implications on the physical world. Measures ensuring the architecture's resilience to attacks, data authentication, access control and client privacy need to be established[32].

Nevertheless, there are also certain threats and issues of governance, security, and privacy that need to be considered. Open governance in an IoT remains an important issue. However, it may be assumed that the ongoing discussions between different regions and countries will lead to a federated structure in the longer term, similar to the domain structures we know from the Internet today. Among the possible applications, we may distinguish between those either directly applicable or closer to our current living habitudes and those futuristic, which we can only fancy of at the moment, since the technologies and/or our societies are not ready for their deployment.

Acknowledgment. Authors would like to acknowledge financial assistance from University Grant Commission (UGC, New Delhi) and ALTERA Inc.USA for the support under University Program.

References

1. Carrithers, M.: The Buddha: A Very Short Introduction. Oxford University Press (2001)
2. Krackhardt, D., Carley, K: A PCANS model of structure in organizations. In: Proceedings of the 1998 International Symposium on Command and Control Research and Technology, California, June 1998
3. Kranz, M., Holleis, P., Schmidt, A.: Embedded Interaction Interacting with the Internet of Things, Internet of Things Track. IEEE Internet Computing (2010)
4. Gartner's hype cycle special report for 2011, Gartner Inc. (2012). http://www.gartner.com/technology/research/hype-cycles/
5. Giusto, D., Iera, A., Morabito, G., Atzori, L. (eds.) The Internet of Things. Springer (2010). ISBN: 978-1-4419-1673-0
6. http://www.etsi.org
7. Shelby, Z.: ETSI M2M Standardization, March 16, 2009. http://zachshelby.org

8. Hui, J., Culler, D., Chakrabarti, S.: 6LoWPAN: Incorporating IEEE 802.15.4 Into the IP Architecture – Internet Protocol for Smart Objects (IPSO) Alliance, White Paper #3, January 2009. http://www.ipso-alliance.org
9. Fernandes, F.F., Gad, R.S., Gad, V.R., Lobo, S., Naik, G.M., Parab, J.S.: Embedded System Communication: A Control Platform for Ethernet –Enabled Systems. Circuit Cellar (255), October 2011
10. Hill, Jason, Szewczyk, Robert, Woo, Alec, Hollar, Seth, Culler, David, Pister, Kristofer: System architecture directions for networked sensors. ACM SIGPLAN Notices **35**(11), 93–104 (2000). doi:10.1145/356989.356998
11. Internet 3.0: The Internet of Things. © Analysys Mason Limited (2010)
12. Dunkels, A., Vasseur, J.P.: IP for Smart Objects, Internet Protocol for Smart Objects (IPSO) Alliance, White Paper #1, September 2008. http://www.ipso-alliance.org
13. Darianian, M., Michael, M.P.: Smart home mobile RFID-based internet-of-things systems and services. In: 2008 International Conference on Advanced Computer Theory and Engineering
14. Aberer, K., Hauswirth, M., Salehi, A.: Global Sensor Networks, Technical report LSIR-REPORT- 2006-001 (2006)
15. Shim, Y., Kwon, T., Choi, Y.: SARIF: A novel framework for integrating wireless sensors and RFID networks. IEEE Wirel. Commun. **14**, 50–56 (2007). doi:10.1109/MWC.2007.4407227
16. Sung, J., Sánchez López, T., Kim, D.: The EPC sensor network for RFID and WSN integration infrastructure. In: Pervasive Computing and Communications Workshops, Fifth IEEE International Conference on Pervasive Computing and Communications Workshops (PerComW 2007) (2007)
17. http://ieee802.org/15
18. van Kranenburg, R., et al.: The Internet of Things. In: 1st Berlin Symposium on Internet and Society (2011)
19. Internet of Things in 2020 ROADMAP FOR THE FUTURE, European Commission Information Society and media & European Technology Platform on Small Systems Integration, May 2008
20. Sundmaeker, H., et al.: Vision and Challenges for Realising the Internet of Things European Commission Information Society and media, March 2010
21. Tinoosh Mohsenin, Rice University: Design and Evaluation of FPGA-Based Gigabit-Ethernet/PCI Network Interface Card Thesis (2004)
22. Gad, V.R., Gad, R.S., Naik, G.M.: Performance analysis of gigabit ethernet standard for various physical media using triple speed ethernet IP core on FPGA. In: International Conference on Network Communications held in Delhi, India on May 25–27, 2012, AIRCC, ICCSEA & Springer and published in International Journal of Advances in Intelligent Computing
23. Triple Speed Ethernet Data Path Reference Design AN-483-June 2009 ver. 1.1 Altera Corporation (2009)
24. Triple-Speed Ethernet MegaCore Function User Guide © December 2010 Altera Corporation Application Note 483 (2010)
25. Ryan, W.E.: An introduction to LDPC codes, August 2003
26. Sterling, B.: Shaping Things – Mediawork Pamphlets. The MIT Press (2005)
27. Kushalnagar, N., Montenegro, G., Schumacher, C.: IPv6 Over Low- Power Wireless Personal Area Networks (6LoWPANs): Overview, Assumptions, Problem Statement, and Goals, IETF RFC 4919, August 2007

28. Weiser, M.: The computer for the 21st century. ACM Mobile Computing and Communications Review **3**(3), 3–11 (1999)
29. IEEE Sensors Journal **2**(2), April 2002
30. Biglarbegian, B., et al.: Optimized Microstrip AntennaArrays for Emerging Millimeter-Wave Wireless Applications. IEEE Trans. Antennas Propagation **59**(5), 1742–1747 (2011)
31. O'Reilly, T., Pahlka, J.: The 'Web Squared' Era, Forbes, September 2009
32. Computer Law & Security Review **24**(26), 23–30 (2010)

Modelling and Performance Analysis of Multicast File Repair in 3GPP LTE Networks

Konstantin E. Samouylov, Irina A. Gudkova, and Darya Y. Ostrikova[✉]

Department of Applied Probability and Informatics, Peoples' Friendship University of Russia,
Ordzhonikidze str. 3, 115419 Moscow, Russia
{ksam,igudkova,dyostrikova}@sci.pfu.edu.ru
http://api.sci.pfu.edu.ru

Abstract. As the demand for higher bit rates in LTE networks increases, new technologies to enhance the effectiveness of radio resource utilization are introduced. One of them is Multimedia Broadcast Multicast Service (MBMS) that enables the usage of multicast technology for multiple resource demanding services, e.g. file repair procedure, intended for resending lost or damaged file segments to users. An important question is to find the periodicity of resending file segment, i.e. forthcoming interval for collecting user requests for repair the same file segment. To address this problem, we propose a Markov model of file repair procedure and formulas for calculating the performance measures – blocking probability, mean delay of file repair to start, and radio channel downtime probability. We also formulate and numerically solve a non-linear optimization problem for mean value of the mentioned polling interval.

Keywords: LTE · MBMS · File repair · Polling interval · Non-linear optimization

1 Introduction

Currently, fourth-generation LTE wireless networks supporting high bit rates continue to expand throughout the world, allowing operators to provide a wide range of multimedia services. Increasing demand for high-speed services caused a need for effective use of network resources. To resolve this problem, a multicast mode of data transmission or multicast technology is used. In modern cellular networks starting with the third-generation (UMTS, 3GPP Release 6) the MBMS subsystem has been developed for this purpose. In fourth-generation networks (LTE, 3GPP Release 8), this subsystem is called enhanced MBMS (E-MBMS) [11].

There are four types of services provided with the use of MBMS subsystem: streaming services (audio, video), file download services, and the so-called "carousel" services, which are the combination of streaming services (text and still images) and file download services as well as TV services [13]. Often file downloading is accompanied by losses or damages of the file segments, that is unacceptable taking into

The reported study was partially supported by RFBR, research project No. 15-07-03608.

S. Balandin et al. (Eds.): NEW2AN/ruSMART 2015, LNCS 9247, pp. 383–392, 2015.
DOI: 10.1007/978-3-319-23126-6_34

account high requirements to the quality of services in LTE network. To solve this problem, file repair procedure [12] is used in unicast, broadcast and, multicast modes. In unicast mode, the same lost or damaged file segments are transmitted to different users at different frequencies. Broadcast mode involves transmission of the same file segments at the same frequency to all network users. Finally, the multicast mode allows the file segments to be transmitted at same frequency only to those network users who sent a request for repair these segments. Evidently, the multicast file repair mode is the most effective one because it allows to save network resources and to avoid transmission of file segments to those users who do not require them.

Currently, extensive research on file repair procedure is being carried out [7, 13]. 3GPP specifications for MBMS are mainly focused on unicast file repair mode. However, for multicast mode the step by step algorithms of multicast forming are not given. So research groups offer a variety of such algorithms. Research is mainly based on simulation modelling [1, 2, 5]. For example, in paper [5] the start of multicast forming occurs at the moment when the transmission of damaged file is complete. In this situation, the group may be empty, but the server will be occupied. This problem is solved in paper [6], where the start of multicast forming occurs at the moment when the first request for file repair arrives. Authors of [6], in addition to simulation modelling use methods of queuing theory and describe the multicast forming as a queuing system $M / G / \infty$.

The algorithm proposed in this paper is a combination of two approaches described above. As in the first approach, the start of multicast forming is independent on the arrival of request. As in the second approach, there is at least one request in multicast group. The idea of the algorithm is as follows: after the timer Δ expires, if there are no requests for file repair, the timer value is increased and becomes equal to 2Δ. This procedure continues until the first request is received. Thereby, this algorithm reduces the mean time of multicast forming.

Thus, in this paper, we used the apparatus of queuing theory to construct a model of multicast file repair with random batch forming time. The model proposed in [10] describes the case when batch forming time is determined by number of users. However, we consider the case when the batch forming time is made up of fixed polling intervals. Note that the multicast technology was analysed in [4, 9], models with elastic traffic were introduced in [3, 8].

The remainder of this paper is organized as follows. In Sect. 2, we propose the Markov model for analysing multicast file repair in LTE network and formulas for calculating main performance measures. In Sect. 3, we illustrate a numerical example as well as formulate and numerically solve a non-linear optimization problem for mean value of polling intervals. Finally, we conclude the paper in Sect. 4.

2   Single-Server Bulk Service Queuing System with Random Batch Forming Time

2.1   Problem Formulation and Batch Forming Procedure

In this paper, we study the multicast file repair procedure. Here we consider the case when the time of multicast forming is random and made up of polling intervals.

Such polling within a few cells is provided by a functional element of MBMS architecture – the controller that monitors users who have received damaged file at small intervals (polling intervals) [11]. Multicast group is considered to be formed if during the polling interval at least one user sent a request for file repair and there are available network resources. Then file segments will be transmitted at the same frequency to all users of a given group. If there are no available resources at the end of the polling interval then the time of multicast forming is increased by one interval more.

As a mathematical model of multicast file repair procedure in LTE networks, a single-server bulk service queue with random batch forming time can be considered. In this paper, we assume that the requests for transmission of different files segments are the requests of the same type. The main performance measures of the proposed model are the blocking probability B, mean delay W of file repair to start, and also radio channel downtime probability P_0.

2.2 Markov Process and State Transition Diagram

We consider a queuing system with one server and a buffer of volume r, $r < \infty$. Requests arrive according to Poisson process with rate λ. Let t_i, $i > 0$ be the moment of arrival of i-th request. We assume that in the moment $t_0 = 0$ the system is empty. $\xi_i = t_i - t_{i-1}$, $i > 0 -$ is the random variable of the time between the arrivals of $(i$-1)-th and i-th requests. Requests are served in groups. Service of the group starts only at the moments τ_j, $j > 0$ under two conditions. The first condition is that there is at least one request in a queue. The second condition is that the server is not occupied by service of previous group. The group forming time is composed of random variables $\Delta_j = \tau_j - \tau_{j-1}$, $j > 0$ distributed exponentially with mean γ^{-1}. Service of the group begins at moment τ_j, $j > 0$ of the end of group formation, and completes at moment $\tilde{\tau}_k$, $k > 0$. $\eta_k = \tilde{\tau}_k - \tau_j$, $k > 0 -$ is random variable of service duration of k-th group of requests. This variable is exponentially distributed with mean μ^{-1}.

Let $N(t)$ be the number of requests in the buffer at the moment t, $M(t)$ be the state of server at the moment t. Wherein:

$$M(t) = \begin{cases} 0, & \text{if the server is free at moment } t, \\ 1, & \text{if the server is busy at moment } t. \end{cases} \tag{1}$$

The Markov process $(N(t), M(t))$ represents the system states. The system state space $\mathbf{X}$ satisfies the relation (2):

$$\mathbf{X} = \{(n, m) : n = 0, 1, \ldots, r; \; m = 0, 1\}. \tag{2}$$

The various scenarios of group formation depending on the state of server and received requests are shown in Fig. 1. In the first scenario the group is formed by one interval (Δ_1, Δ_2). In the second scenario the group is formed by two intervals: because of busy server $(\Delta_3 + \Delta_4)$ and because of no requests in the buffer $(\Delta_5 + \Delta_6)$.

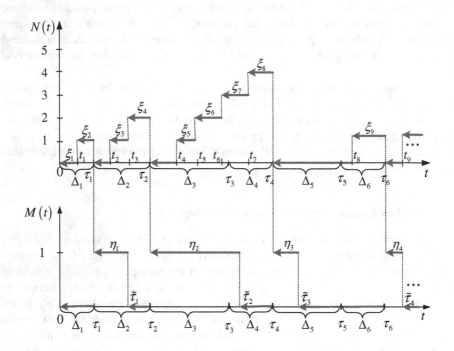

Fig. 1. Number of requests in the buffer and the state of the server

2.3   System of Equilibrium Equations

Fig. 2 shows a *state transition diagram* for the Markov process $(N(t), M(t))$.

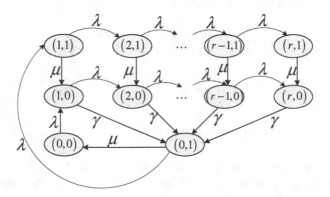

Fig. 2. State transition diagram

System of equilibrium equations can be represented as follows:

$$\lambda p_{00} = \mu p_{01},$$
$$(\lambda + \mu)\, p_{01} = \gamma \sum_{n=1}^{r} p_{n0},$$
$$(\lambda + \mu)\, p_{n1} = \lambda p_{n-1,1},\ n = 1,\ldots,r-1,$$
$$(\lambda + \gamma)\, p_{n0} = \lambda p_{n-1,0} + \mu p_{n1},\ n = 1,\ldots,r-1, \qquad (3)$$
$$\mu p_{r1} = \lambda p_{r-1,1},$$
$$\gamma p_{r0} = \lambda p_{r-1,0} + \mu p_{r1},$$
$$\sum_{n=0}^{r} p_{n0} + \sum_{n=0}^{r} p_{n1} = 1.$$

We can find stationary probability distribution from the system of equilibrium equations by linear transformations.

2.4 Stationary Probability Distribution

Stationary probabilities $p(n,m)$ of system states can be calculated by the formula (4):

$$p_{n1} = \left(\frac{\rho}{\rho+1}\right)^{n} \rho\, p_{00},\ n = 0,\ldots,r-1,$$

$$p_{r1} = \left(\frac{\rho}{\rho+1}\right)^{r-1} \rho^{2} p_{00},$$

$$p_{n0} = \left[\left(\frac{a}{a+1}\right)^{n} + \sum_{k=1}^{n}\left(\frac{\rho}{\rho+1}\right)^{k}\left(\frac{a}{a+1}\right)^{n-k+1}\right] p_{00},\ n = 1,\ldots,r-1,$$

$$p_{r0} = \left[a\left(\frac{a}{a+1}\right)^{r-1} + a\sum_{k=1}^{r-1}\left(\frac{\rho}{\rho+1}\right)^{k}\left(\frac{a}{a+1}\right)^{r-k} + \rho a\left(\frac{\rho}{\rho+1}\right)^{r-1}\right] p_{00}, \qquad (4)$$

$$p_{00} = \left(1 + \sum_{n=1}^{r-1}\left[\left(\frac{a}{a+1}\right)^{n} + \sum_{k=1}^{n}\left(\frac{\rho}{\rho+1}\right)^{k}\left(\frac{a}{a+1}\right)^{n-k+1}\right] + a\left(\frac{a}{a+1}\right)^{r-1} + \right.$$
$$\left. a\sum_{k=1}^{r-1}\left(\frac{\rho}{\rho+1}\right)^{k}\left(\frac{a}{a+1}\right)^{r-k} + \rho a\left(\frac{\rho}{\rho+1}\right)^{r-1} + \right.$$
$$\left. + \sum_{n=0}^{r-1}\left(\frac{\rho}{\rho+1}\right)^{n}\rho + \left(\frac{\rho}{\rho+1}\right)^{r-1}\rho^{2}\right)^{-1},$$

where $\rho = \dfrac{\lambda}{\mu}$ and $a = \dfrac{\lambda}{\gamma}$.

It should be noted that ρ is the intensity of the offered load. This value can be interpreted as a mean number of requests received during the server operation. By analogy, the value a can be interpreted as a mean number of requests received within one polling interval.

2.5   Performance Measures

Having found the probability distribution $p(n,m)$, $(n,m) \in \mathbf{X}$ of the mathematical model of multicast file repair procedure in LTE networks, one may compute its performance measures, notable blocking probability B of user requests for file repair, mean W delay of file repair to start, and also the radio channel downtime probability P_0 :

$$B = p_{r1} + p_{r0}, \tag{5}$$

$$W = \frac{\sum_{n=1}^{r} n(p_{n0} + p_{n1})}{\lambda(1-B)}, \tag{6}$$

$$P_0 = \sum_{i=0}^{r} p_{i0}. \tag{7}$$

3   Performance Analysis of Multicast File Repair Procedure

3.1   Numerical Example: Blocking Probability and Mean Delay

We present an example of a single cell supporting multicast file repair procedure to illustrate the performance measures defined above. Let us consider a cell with peak throughput rate of 50 Mbps. 38% of total network resources are allocated for file repair procedure, thus file segments are transmitted to users at a rate of $C = 19$ Mbps,

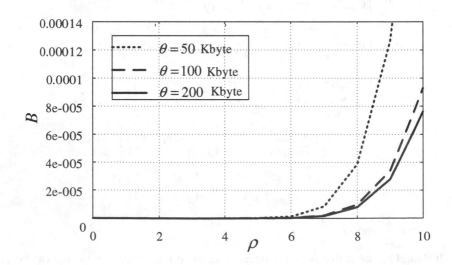

Fig. 3. Blocking probability

buffer size $r = 100$, and mean polling interval γ^{-1} equals to 20 ms. By changing the offered load ρ from 0 to 10, we compute the performance measures (5), (6) for different values of mean time of file repair $\mu^{-1} = \theta / C$, where file sizes are $\theta = 50, 100, 200$ Kbyte [7]. The results are plotted in Fig. 3 and Fig. 4.

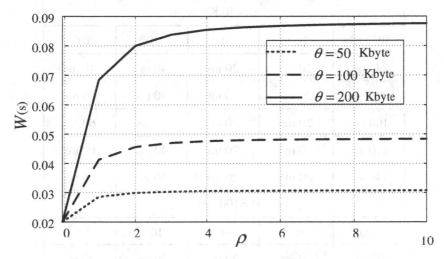

Fig. 4. Mean delay of the file repair to start

3.2   Non-linear Optimization of Polling Interval

It is important for mobile operator to determine the optimal length of polling interval. For the numerical example described above, we have chosen the minimum acceptable length of polling interval equal to 20 ms. It means that the signal information is transmitted by the controller every 20 ms. Such periodicity may cause excessive load on the network. However, the controller can monitor users who have received damaged file at longer intervals. These intervals must be multiples of 20 ms, i.e., 40 ms, 60 ms, 80 ms, etc. Solving a non-linear optimization problem numerically, one can obtain recommended length of polling interval.

It is necessary to maximize the radio channel downtime probability P_0. In this case, according to Service Level Agreement, blocking probability B of user requests for file repair and mean W delay of file repair to start should not exceed the values B^* and W^* respectively. Then a non-linear optimization problem can be formulated as follows:

$$
\begin{cases}
P_0\left(\gamma^{-1}\right) \to \max, \\
W\left(\gamma^{-1}\right) \le W^*, \\
B\left(\gamma^{-1}\right) \le B^*.
\end{cases}
\tag{8}
$$

Table 1 shows the examples of recommended mean length of polling interval at different values of B^* and W^* and initial data of $C = 19$ Mbps, $r = 250$ and $\theta = 50,\ 100,\ 150$ Kbyte.

Table 1. Recommended mean length of polling interval (ms)

$\theta = 50$ Kbyte				
W^* / B^*	10^{-5}	10^{-4}	10^{-3}	10^{-2}
60 ms	20 ms	20 ms	40 ms	40 ms
80 ms	20 ms	20 ms	40 ms	60 ms
100 ms	20 ms	20 ms	40 ms	60 ms
120 ms	20 ms	20 ms	40 ms	60 ms
140 ms	20 ms	20 ms	40 ms	60 ms
$\theta = 100$ Kbyte				
W^* / B^*	10^{-5}	10^{-4}	10^{-3}	10^{-2}
60 ms	20 ms	20 ms	20 ms	20 ms
80 ms	40 ms	60 ms	60 ms	60 ms
100 ms	40 ms	60 ms	80 ms	80 ms
120 ms	40 ms	60 ms	80 ms	100 ms
140 ms	40 ms	60 ms	80 ms	120 ms
$\theta = 150$ Kbyte				
W^* / B^*	10^{-5}	10^{-4}	10^{-3}	10^{-2}
60 ms	–	–	–	–
80 ms	40 ms	40 ms	40 ms	40 ms
100 ms	60 ms	60 ms	60 ms	60 ms
120 ms	80 ms	80 ms	80 ms	80 ms
140 ms	80 ms	100 ms	100 ms	100 ms

Fig. 5 illustrates the solution of non-linear optimization problem (8) under constraints $B^* = 10^{-4}$ and $W^* = 100$ ms. Analysing this figure, we can conclude that the optimum length of polling interval equals to 60 ms. On reaching this value the radio channel downtime probability is maximum and constraints for B^* and W^* are valid in the entire offered load range.

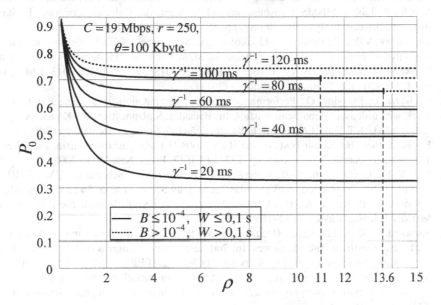

Fig. 5. Radio channel downtime probability

4   Conclusion

In this paper, we addressed a multicast forming problem for file repair procedure in LTE wireless networks, and we presented a mathematical model of multicast file repair procedure with random batch forming time. We propose a solution of a non-linear optimization problem for mean length of polling interval. An interesting task for future studies is to develop a simulation model for verification of our obtained results as well as to construct a mathematical model of multicast file repair procedure in LTE network as a single-channel bulk service queuing system with an arbitrary distribution of the length of polling interval.

Acknowledgments. The authors are grateful to B.Sc. graduate Peter Kharin for performing the numerical experiment.

References

1. Alexiou, A., Asimakis, K., Bouras, C., Kokkinos, V., Papazois, A., Tseliou, G.: Reliable multicasting over LTE: a performance study. In: IEEE Symposium on Computers and Communications, pp. 603–608. IEEE Press, New York (2011)
2. Alexiou, A., Bouras, C., Kokkinos, V., Papazois, A., Tseliou, G.: Forward Error Correction for Reliable e-MBMS Transmissions in LTE Networks. Cellular Networks - Positioning, Performance Analysis, Reliability, pp. 353–374. InTech (2011)
3. Borodakiy, V.Y., Buturlin, I.A., Gudkova, I.A., Samouylov, K.E.: Modelling and analysing a dynamic resource allocation scheme for M2M traffic in LTE networks. In: Balandin, S., Andreev, S., Koucheryavy, Y. (eds.) NEW2AN 2013 and ruSMART 2013. LNCS, vol. 8121, pp. 420–426. Springer, Heidelberg (2013)
4. Gudkova, I., Plaksina, O.: Performance measures computation for a single link loss network with unicast and multicast traffics. In: Balandin, S., Dunaytsev, R., Koucheryavy, Y. (eds.) ruSMART 2010. LNCS, vol. 6294, pp. 256–265. Springer, Heidelberg (2010)
5. Hechenleitner, B.: Repair costs of the IPDC/DVB-H file repair mechanism. In: Wireless Telecommunications Symposium, pp. 137–144. IEEE Press, New York (2008)
6. Lai, Y.-C., Lin, P., Lin, Y.-B., Chang, L.-T.: A File Repair Scheme for UMTS MBMS Service. IEEE Transactions on Vehicular Technology 57(6), 3746–3756 (2008). Florida
7. Lohmar, T., Ibanez, J.-A., Blockstrand, M., Zanin, A.: Scalable push file delivery with MBMS. Ericsson Review 1, 12–16 (2009)
8. Samouylov, K., Gudkova, I.: Recursive computation for a multi-rate model with elastic traffic and minimum rate guarantees. In: 2nd International Congress on Ultra Modern Telecommunications and Control Systems, pp. 1065–1072. IEEE Press, New York (2010)
9. Samouylov, K., Gudkova, I.: Analysis of an admission model in a fourth generation mobile network with triple play traffic. Automatic Control and Computer Sciences 47(4), 202–210 (2010). Latvia
10. Chukarin, A.V., Pershakov, N.V., Samouylov, K.E.: Performance of Sigtran-based signaling links deployed in mobile network. In: 9th International Conference on Telecommunications, pp. 163–166. IEEE Press, New York (2007)
11. GPP TS 22.246: Multimedia Broadcast/Multicast Service (MBMS) user services; Stage 1 (Release 12). 3GPP, Valbonne, France (2014)
12. GPP TS 23.246: Multimedia Broadcast/Multicast Service (MBMS); Architecture and functional description (Release 12). 3GPP, Valbonne, France (2014)
13. GPP TS 26.346: Multimedia Broadcast/Multicast Service (MBMS); Protocols and codecs (Release 12). 3GPP, Valbonne, France (2014)

LTE Positioning Accuracy Performance Evaluation

Mstislav Sivers[1] and Grigoriy Fokin[2(✉)]

[1] St. Petersburg State Polytechnic University, St. Petersburg, Russia
m.sivers@mail.ru
[2] The Bonch-Bruevich St. Petersburg State University of Telecommunications,
St. Petersburg, Russia
grihafokin@gmail.com

Abstract. In this paper we investigate the positioning accuracy of user equipment (UE) with observed time difference of arrival (OTDoA) technique in Long
Term Evolution (LTE) networks using dedicated positioning reference signal
(PRS) by means of comprehensive simulation model. System-level model includes typical cellular network layout with spatially distributed eNodeB (eNB)
and link-level model simulates PRS transmission and reception with LTE System Toolbox. Matlab "TDOA Positioning Using PRS example" was developed
to include known linear least squares (LLS), nonlinear Gauss-Newton (GN) and
Levenberg-Marquardt (LM) positioning algorithms and compare it with Cramer-Rao lower bound (CRLB). Resulting estimates clarify known results and
reveal that simple LLS achieves close to LM positioning accuracy when the
number of eNB is 6, while further increasing the number of eNB degrades
positioning algorithms accuracy.

Keywords: LTE · UE · eNB · OTDoA · PRS · Cramer-Rao lower bound ·
Levenberg-Marquardt · Gauss-Newton · Linear least squares · Positioning
accuracy

1 Introduction

Positioning of user equipment (UE) in LTE wireless networks is an important trend in
next-generation mobile communications for delivering location based services (LBS),
such as emergency services and location-based social network. Traditional Global
Navigation Satellite Systems (GNSS) such as Global Positioning System (GPS) and
GLObal NAvigation Satellite System (GLONASS) provide rather accurate position
estimates outdoor, however this is not always the case and indoor these systems have
limitations because of non-line of sight (NLOS) conditions as a result of reduced satellite visibility. To enhance GNSS positioning capabilities in wireless networks 3rd
Generation Partnership Project introduced LTE Positioning Protocol (LPP) [1] which
provides three positioning schemes: Assisted Global Navigation Satellite System (A-
GNSS) positioning, Observed Time Difference of Arrival (OTDoA) positioning and
Enhanced Cell ID (E-Cell ID) positioning.

A-GNSS positioning is based on cellular network assistance of GNSS receiver by
providing assistance data for the visible satellites [2].

© Springer International Publishing Switzerland 2015
S. Balandin et al. (Eds.): NEW2AN/ruSMART 2015, LNCS 9247, pp. 393–406, 2015.
DOI: 10.1007/978-3-319-23126-6_35

E-CID positioning is based on the knowledge of the geographical coordinates of UE serving eNB, which can be obtained executing tracking area update or by paging. The position accuracy is in that case is rather coarse and depends on the cell size, which of course would not fulfill the accuracy requirements [2].

OTDoA is a downlink positioning method in LTE which is based on hyperbolic trilateration by measuring the difference in arrival times of dedicated downlink signals from multiple eNB at the UE [3]. These special dedicated Positioning Reference Signals [4] were introduced with 3GPP Release 9 to provide Reference Signal Time Difference (RSTD) measurement, specified in [5]. Thus, OTDoA positioning scheme based on RSTD measurements performed on PRS can be regard as not only GNSS enhancing, but as self-sufficing LTE wireless network technique for providing UE positioning by itself and is an object of research in presented paper.

Analysis of publications about the area of time difference of arrival positioning accuracy performance evaluation has shown us that this problem had already been solved before [6, 7]. In work [6] the achievable localization accuracy of the LTE positioning reference signal (PRS) has been analyzed only for maximum likelihood (ML) estimation algorithm, and the RMSE performance of the ML method was evaluated based on the Cramer-Rao lower bound (CRLB) for time delay estimation. In work [7] localization accuracy has been analyzed for iterative Gauss-Newton and proposed iterative Levenberg-Marquardt (LM) algorithms [8, 9], but without taking into consideration special features of the LTE positioning reference signal (PRS).

The aim of this paper is to evaluate the positioning accuracy of UE with OTDoA technique in LTE networks using dedicated PRS by solving hyperbolic trilateration problem with linear least squares (LLS) and nonlinear GN and LM algorithms and compare it with CRLB computed for RSTD measurements. Following an approach, proposed in [10], we develop Matlab example "Time Difference Of Arrival Positioning Using PRS" [11] to include linear least squares (LLS) [12] and GN and LM positioning algorithms into resulting simulation model. Thus system-level simulation model includes typical cellular network layout with spatially distributed eNodeB which enables us to compare RSTD measurements and compute CRLB, while link-level model carefully simulates PRS transmission and reception with LTE System Toolbox.

The material in the paper organized in the following order. LTE positioning support, including architecture, protocols and PRSs, essential for performance evaluation, presented in the second part. System-level positioning model considering OTDoA method formalized in the third part. LLS, GN, LM positioning algorithms presented in the fourth part. Simulation model is described in the fifth part. Finally, we draw the conclusions in the sixth part.

2   LTE Positioning Support

LTE Positioning Protocol (LPP) [1] supports several TDOA positioning methods, depending on the measurement and position estimation entities: depending on the measurement entity, UE or eNB, LPP supports downlink or uplink TDOA positioning respectively. Depending on the position estimation entity, UE or eNB, LPP supports UE-assisted or eNB-assisted TDOA positioning respectively, where term "assisted" is used to classify the measurements [2]. OTDoA is a downlink TDOA position estimation method where UE serves as a measurement entity and Location Server (LS) is an estimation entity. LS is represented in the fig. 1 by Evolved Serving Mobile Location Center (E-SMLC) or SUPL (Secure User Plane Location) Location Platform (SLP) that manages positioning for UE by obtaining measurements and providing assistance data to the UE.

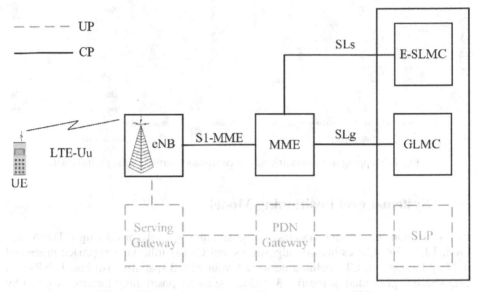

Fig. 1. LTE location services architecture overview ([3])

OTDoA positioning procedure include following signaling in control plane (CP). First, Mobility Management Entity (MME) receives a request for location service from UE; then MME sends a location service request to an E-SMLC. E-SMLC after processing location service request transfers OTDoA assistance data to the target UE. UE performs RSTD measurements and sends them back to E-SMLC. E-SMLC performs hyperbolic trilateration computing and sends positioning result back to the UE [3].

For UE to perform proper Reference Signal Time Difference timing measurements on downlink signals eNBs should be synchronized when transmitting positioning reference signals. In [5] RSTD is defined as the relative timing difference between two cells – the reference cell and a measured cell – calculated as the smallest time difference between two subframe boundaries receiving from the two different cells.

To provide reliable RSTD measurements 3GPP Release 9 introduced dedicated PRS for eNBs to be transmitted on antenna port 6 [4]. The mapping of positioning reference signals to resource elements is shown in fig. 2 for normal cyclic prefix and one-or-two transmit antenna ports (left), and four transmit antenna ports (right). Each square in fig. 2 indicates a resource element with frequency-domain index k and time-domain index l.

The squares labelled R_6 indicate PRS resource elements within a block of 12 subcarriers over 14 or 12 OFDM symbols, respectively. Comparing with Cell-specific Reference Signal (CRS) PRS signals provides less inter-cell interference.

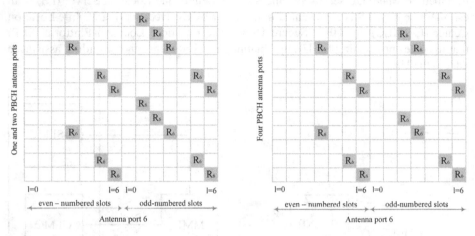

Fig. 2. Mapping of positioning reference signals (normal cyclic prefix) ([4])

3 System-Level Positioning Model

In this section we present system-level positioning model considering OTDoA method, LLS, GN, LM estimation algorithms and CRLB following approach presented in [10]. Example of LTE cellular network layout with 3 spatially distributed eNBs in a 2-D space is presented in the fig. 3. UE as the target positioning location is given by intersection of at least two hyperbolas. eNBs located at the distances d_i from UE are time synchronized and transmit PRSs. UE receives positioning reference signals and measures the time of arrival (TOA) of PRSs $\tau_1=d_1/c$,$\tau_2= d_2/c$ and $\tau_3= d_3/c$ received from eNB_1, eNB_2 and eNB_3 respectively, where $c=3\cdot10^8$ m/s is the speed of light. For L eNBs there are (L-1) non-redundant TDOAs from all eNB pairs denoted by $\tau_{ij}= \tau_i-\tau_j$, i, j=1,2...L with i>j [12]. Location server then computes TDOA by subtracting TOA of a reference eNB from several TOAs of neighbor eNBs. For example presented in the fig. 3 with eNB_1 as a reference we get two distinct TDOAs: $\tau_{21}= \tau_2-\tau_1$ and $\tau_{31}= \tau_3-\tau_1$. These are the so called real time differences $RTD_i= \tau_i-\tau_j$, i=2,3,...L. Because of the different errors and synchronization inaccuracy represented by the TDOA measurement error term $\Delta_i = \delta| \tau_i-\tau_j |$ there is measurement uncertainty, which is illustrated in fig. 3 with hyperbolas having a certain width [3]. Reference signal time difference measurement $RSTD_i$ takes RTD_i and TDOA measurement error Δ_i into account and gives OTDoA model expressed as $RSTD_i = RTD_i + \Delta_i$.

Following notations of [12] lets denote known coordinates of eNBs as $\mathbf{x_i}=[x_i,y_i]^T$, i=1,2...L, where L≥3 is the number of eNBs, and unknown coordinates of UE as $\mathbf{x}=[x,y]^T$. The distance between the UE and the ith eNB denoted by d_i and is given by

$$d_i = \left\| \mathbf{x} - \mathbf{x_i} \right\|_2 = \sqrt{\left(x - x_i\right)^2 + \left(y - y_i\right)^2}, \; i = 1,2,...L . \qquad (1)$$

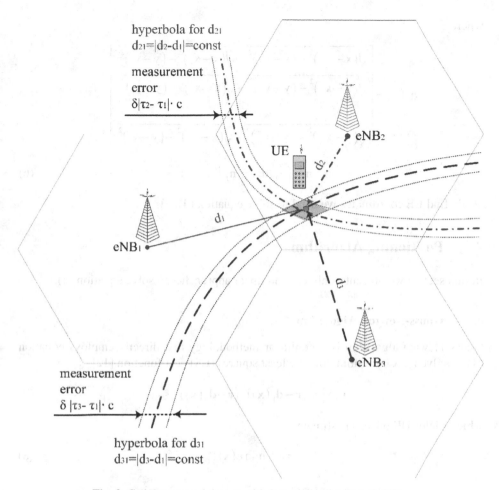

Fig. 3. Cellular network layout with 3 spatially distributed eNBs

Non-redundant real time differences RTD_i with eNB_1 as a reference point define following range differences $d_{i,1}$

$$d_{i,1} = d_i - d_1. \tag{2}$$

Taking TDOA measurement error Δ_i into account defines the following received signal range difference measurement model r_i

$$r_i = d_{i,1} + n_i, \tag{3}$$

where range difference measurement is given by $n_i = \Delta_i \cdot c$.

In matrix form observed range difference measurement model is

$$\mathbf{r} = \mathbf{d}_1(\mathbf{x}) + \mathbf{n}, \tag{4}$$

where

$$\mathbf{d}_1(\mathbf{x}) = \begin{bmatrix} \sqrt{(x-x_2)^2-(y-y_2)^2} - \sqrt{(x-x_1)^2-(y-y_1)^2} \\ \sqrt{(x-x_3)^2-(y-y_3)^2} - \sqrt{(x-x_1)^2-(y-y_1)^2} \\ \vdots \\ \sqrt{(x-x_L)^2-(y-y_L)^2} - \sqrt{(x-x_1)^2-(y-y_1)^2} \end{bmatrix}, \tag{5}$$

$$\mathbf{n} = [n_2, n_3, ..., n_L]^T. \tag{6}$$

To find UE coordinates we need to solve equation (4).

4   Positioning Algorithms

In this section we present nonlinear and linear approaches to solve equation (4).

4.1   Gauss-Newton Algorithm

Gauss-Newton algorithm is a nonlinear methodology that directly employs equation (4) to solve for $\mathbf{x}$ by minimizing the least squares (LS) cost function [12]

$$\varepsilon(\mathbf{x}) = (\mathbf{r} - \mathbf{d}_1(\mathbf{x}))^T (\mathbf{r} - \mathbf{d}_1(\mathbf{x})), \tag{7}$$

which yields UE position estimate

$$\hat{\mathbf{x}} = \arg\min_{\mathbf{x}} \varepsilon(\mathbf{x}). \tag{8}$$

The iterative Gauss-Newton algorithm procedure for $\hat{\mathbf{x}}$ is [12]

$$\hat{\mathbf{x}}^{k+1} = \hat{\mathbf{x}}^k + \left(\mathbf{G}^T(\hat{\mathbf{x}}^k)\mathbf{G}(\hat{\mathbf{x}}^k) \right)^{-1} \mathbf{G}^T(\hat{\mathbf{x}}^k)\left(\mathbf{r} - \mathbf{d}_1(\hat{\mathbf{x}}^k) \right), \tag{9}$$

where $\mathbf{G}(\hat{\mathbf{x}}^k)$ is the Jacobian matrix of $\mathbf{d}_1(\hat{\mathbf{x}}^k)$ computed at $\hat{\mathbf{x}}^k$

$$\mathbf{G}(\hat{\mathbf{x}}^k) = \begin{bmatrix} \dfrac{x-x_2}{d_2} - \dfrac{x-x_1}{d_1} & \dfrac{y-y_2}{d_2} - \dfrac{y-y_1}{d_1} \\ \dfrac{x-x_3}{d_3} - \dfrac{x-x_1}{d_1} & \dfrac{y-y_3}{d_3} - \dfrac{y-y_1}{d_1} \\ \vdots & \vdots \\ \dfrac{x-x_L}{d_L} - \dfrac{x-x_1}{d_1} & \dfrac{y-y_L}{d_L} - \dfrac{y-y_1}{d_1} \end{bmatrix}. \tag{10}$$

Gauss-Newton algorithm provides fast convergence, but coarse initial values lead to noninvertible matrix.

4.2   Levenberg-Marquardt Algorithm

Levenberg-Marquardt algorithm for TDOA positioning was introduced in [7] and is based on damped GN procedure given by

$$\hat{x}^{k+1} = \hat{x}^{k} + \left(G^{T}\left(\hat{x}^{k}\right)G\left(\hat{x}^{k}\right) + \lambda^{k}I_{2} \right)^{-1} G^{T}\left(\hat{x}^{k}\right)\left(r - d_{1}\left(\hat{x}^{k}\right)\right), \qquad (11)$$

where I_{n} is the $n \times n$ identity matrix. The dumping parameter leads to invertible matrix implementation in comparison to GN and considered to be more robust for inaccurate initial values.

4.3   Linear Least Squares Algorithm

Linear least squares algorithm is presented in [12] and converts equation (4) into a set of linear equations in x. In matrix form solution is given as

$$AQ + q = b, \qquad (12)$$

where

$$A = 2\begin{bmatrix} x_{1} - x_{2} & y_{1} - y_{2} & -r_{2} \\ x_{1} - x_{2} & y_{1} - y_{3} & -r_{3} \\ \vdots & \vdots & \vdots \\ x_{1} - x_{L} & y_{1} - y_{L} & -r_{L} \end{bmatrix}, \qquad (13)$$

$$Q = \begin{bmatrix} x - x_{1} & y - y_{1} & R_{1} \end{bmatrix}^{T}, \qquad (14)$$

$$q = \begin{bmatrix} m_{2} & m_{3} \dots m_{L} \end{bmatrix}^{T}, \qquad (15)$$

$$b = \begin{bmatrix} r_{2}^{2} - \left(x_{1} - x_{2}\right)^{2} - \left(y_{1} - y_{2}\right)^{2} \\ r_{3}^{2} - \left(x_{1} - x_{3}\right)^{2} - \left(y_{1} - y_{3}\right)^{2} \\ \vdots \\ r_{L}^{2} - \left(x_{1} - x_{L}\right)^{2} - \left(y_{1} - y_{L}\right)^{2} \end{bmatrix}, \qquad (16)$$

$$m_{i} = n_{i}^{2} + 2n_{i}\sqrt{\left(x - x_{i}\right)^{2} + \left(y - y_{i}\right)^{2}}, \qquad (17)$$

$$R_{1} = \sqrt{\left(x - x_{i}\right)^{2} + \left(y - y_{i}\right)^{2}}. \qquad (18)$$

LLS position estimate is obtained from the first and second entries of $\mathbf{Q}$ by

$$\hat{\mathbf{x}} = \left[\left[\mathbf{Q} \right]_1 + x_1 \ \left[\mathbf{Q} \right]_2 + y_1 \right]^{\mathrm{T}}. \tag{19}$$

5   Simulation Model

In this section we describe link and system level simulation model developed for LTE positioning accuracy performance evaluation. We developed Matlab example "Time Difference Of Arrival Positioning Using PRS" [11] to include LLS, GN [12] and LM [7] positioning algorithms into resulting simulation model. Thus system-level simulation model includes typical cellular network layout with spatially distributed eNodeB which enables us to compare RSTD measurements and compute CRLB, while link-level model carefully simulates PRS transmission and reception with LTE System Toolbox.

Link-level model includes eNBs transmitter configuration which is created using Reference Measurement Channel (RMC) function on the basis of unique eNB cell identity. Positioning reference signals are generated and mapped onto the resource grid which is then OFDM modulated according to [4] for 3 MHz bandwidth downlink Shared Channel (PDSCH) transmission.

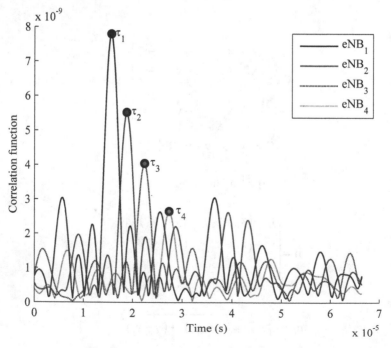

Fig. 4. Delay estimation by correlation peaks

System-level model includes network layout where several eNBs are randomly distributed around UE which is placed in the origin of coordinates. Distances between eNBs and UE are calculated using (1). To model the reception of eNBs waveforms by UE, appropriate delays and attenuation are applied to eNBs PRS transmissions according to calculated distances d_i. Attenuations are calculated using Urban Macro Line Of Sight (LOS) path loss model [13]. Received waveform delays are computed with LTE basic time unit which $T_s=1/(15000 \cdot 2048)=32$ ns.

Absolute TOAs cannot be estimated because UE and eNBs are not synchronized, however relative arrival times estimation is based on UE correlation of received positioning reference signal with a local PRS generated with the cell identity of each eNodeB; relative delay estimates are identified by the peak correlations and are presented in fig. 4. Then delay differences are calculated assuming than eNB_1 as a reference which produces hyperbolas of constant delay difference. Intersection of hyperbolas defines the position of the UE and is presented in fig. 5.

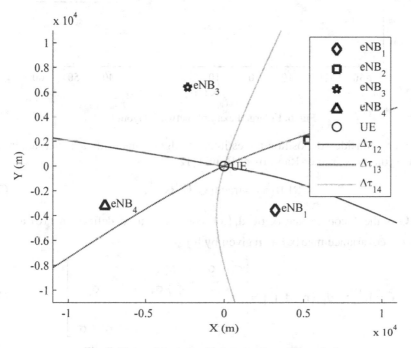

Fig. 5. Network layout with intersection of hyperbolas

Enlarged example network layout with intersection of hyperbolas and position estimations computed according to considered algorithms is presented in fig. 6.

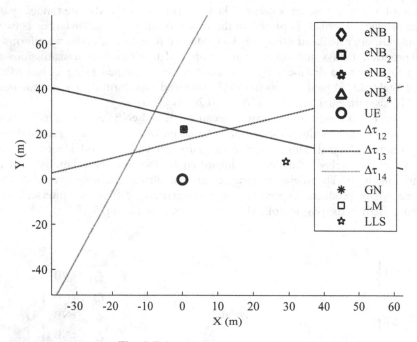

Fig. 6. Enlarged example network layout

To compare considered positioning estimation algorithms with a lower attainable bound we compute Cramer-Rao Lower Bound [12]

$$CRLB(\mathbf{x}) = \text{trace}\left(\mathbf{G}^{\text{T}}\mathbf{C}^{-1}\mathbf{G}\right)^{-1},\tag{20}$$

where $\mathbf{G}$ is the Jacobian matrix of $\mathbf{d}_1(\mathbf{x})$ computed at $\mathbf{x}$ defined in (10) and $\mathbf{C}$ denotes the covariance matrix for $\mathbf{n}$ given by [6]

$$\mathbf{C} = E\left\{(\mathbf{r}-\mathbf{d}_1)(\mathbf{r}-\mathbf{d}_1)^{\text{T}}\right\} = \begin{bmatrix} \sigma_1^2 + \sigma_2^2 & \sigma_1^2 & \cdots & \sigma_1^2 \\ \sigma_1^2 & \sigma_1^2 + \sigma_3^2 & \cdots & \sigma_1^2 \\ \vdots & \vdots & \ddots & \vdots \\ \sigma_1^2 & \sigma_1^2 & \cdots & \sigma_1^2 + \sigma_L^2 \end{bmatrix},\tag{21}$$

where σ_i is the standard deviation, which is defined by measurement error. Since all RSTDs are determined with respect to the reference eNB$_1$, n_i, i=2,3,...,L in (3) are correlated and hence $\mathbf{C}$ is not a diagonal matrix and computed during simulation as a multiple of basic time unit $T_s=32$ ns [10].

Root Mean Square Error (RMSE) position in meters with respect to true position $\mathbf{x}$ for comparing considered positioning algorithms estimates $\hat{\mathbf{x}}$ is computed as [7]

$$RMSE = \sqrt{E\left\{\left\|\mathbf{x} - \hat{\mathbf{x}}\right\|_2\right\}},\tag{22}$$

and averaged over a thousand   simulation trials.

Initial value for the iterative GN and LM positioning algorithms (9) and (11) is calculated as the mean value of the positions of all eNBs as

$$\hat{\mathbf{x}}^0 = \frac{1}{L}\sum_{i=1}^{L}\mathbf{x}_i .$$
(23)

Fig. 7 shows a performance of positioning algorithms in terms of RMSE in dB versus the measurement error σ_i computed during simulation as a multiple of basic time unit T_s for the case of L=3 eNB.

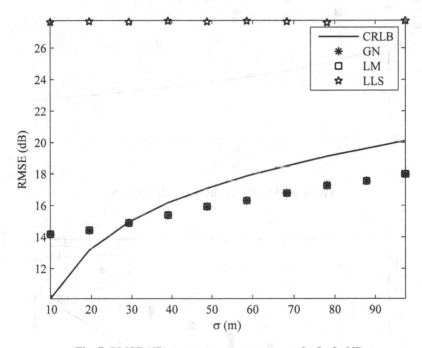

Fig. 7. RMSE (dB) versus measurement error for L=3 eNB

It can be seen from the fig. 7, LLS algorithm positioning performance is poor, compared to nonlinear algorithms. One more interesting fact is, that RMSE for GN and LM algorithms outperform CRLB, however, as stated in [12], it is possible in the case of biased estimator.

Fig. 8 shows a performance of positioning algorithms in terms of RMSE in dB versus the number L of eNB for measurement error σ_i computed for basic time unit T_s.

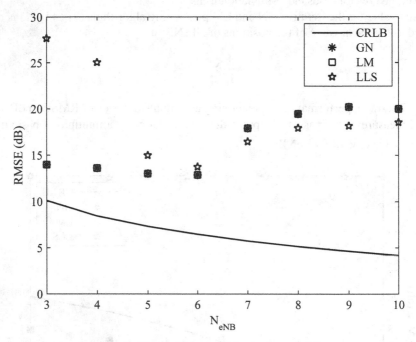

Fig. 8. RMSE (dB) versus the number of eNBs

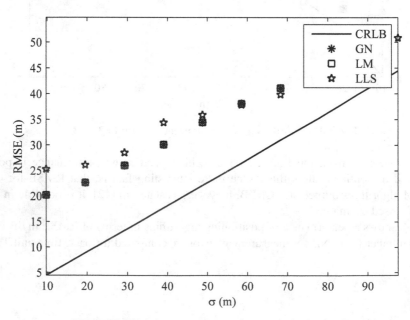

Fig. 9. RMSE (m) versus measurement error for L=6 eNBs

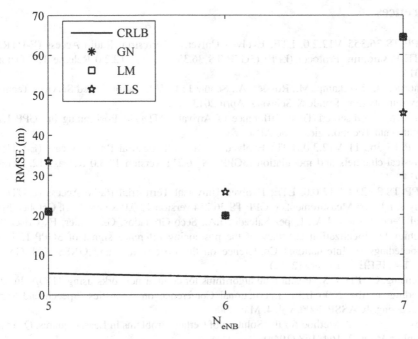

Fig. 10. RMSE (m) versus the number of eNBs

RMSE in fig. 7 and fig. 9 is computed in dB because of substantial difference in error for LLS and nonlinear algorithms. However, for L=6 fig. 8 reveal that simple LLS achieves close to LM   positioning accuracy, while further increasing the number of eNB degrades positioning algorithms accuracy.

To validate result of fig. 8 we plot RMSE in meters for the case of L=[5,6,7] eNBs. Fig. 9 demonstrates, that RMSE in meters for the case of L=6 eNBs has comparable performance in the case of simple LLS and iterative nonlinear GN and LM algorithms.

6   Conclusion

In this paper we evaluated the positioning accuracy of UE with OTDoA technique in LTE networks using dedicated PRS by solving hyperbolic trilateration problem with linear least squares (LLS) and nonlinear GN and LM algorithms and compared it with CRLB computed for RSTD measurements. Performed system-level simulation clarify known results and reveal that simple LLS achieves close to LM   positioning accuracy when the number of eNB is 6, while further increasing the number of eNB degrades positioning algorithms accuracy.

References

1. GPP TS 36.355 V12.2.0. LTE; Evolved Universal Terrestrial Radio Access (E-UTRA); LTE Positioning Protocol (LPP) (3GPP TS 36.355 version 12.2.0 Release 12), October 2014
2. Thorpe, M., Kottkamp, M., Rössler, A., Schütz J.: LTE Location Based Services Technology Introduction. Rohde & Schwarz, April 2013
3. Fischer, S.: Observed Time Difference of Arrival (OTDOA) Positioning in 3GPP LTE. Qualcomm Technologies, June 2014
4. GPP TS 36.211 V12.3.0. LTE; Evolved Universal Terrestrial Radio Access (E-UTRA); Physical channels and modulation (3GPP TS 36.211 version 12.3.0 Release 12, October 2014
5. GPP TS 36.214 V12.0.0. LTE; Evolved Universal Terrestrial Radio Access (E-UTRA); Physical layer; Measurements (3GPP TS 36.214 version 12.0.0 Release 12), October 2014
6. Del Peral-Rosado, J.A., Lopez-Salcedo, J.A., Seco-Granados, G., Zanier, F., Crisci, M.: Achievable localization accuracy of the positioning reference signal of 3GPP LTE. In: Proceedings of International Conference on the Localization and GNSS (ICL-GNSS), pp. 1–6. IEEE, Starnberg (2012)
7. Mensing, C., Plass, S.: Positioning algorithms for cellular networks using TDOA. In: Proceedings of the 2006 IEEE International Conference on Acoustics, Speech and Signal Processing, ICASSP 2006, vol. 4, May 2006
8. Levenberg, K.: A Method for the Solution of Certain Problems in Least-Squares. Quarterly Applied Math. **2**, 164–168 (1944)
9. Marquardt, D.: An Algorithm for Least-Squares Estimation of Nonlinear Parameters. SIAM Journal Applied Math. **11**, 431–441 (1963)
10. Sivers, M., Fokin, G., Duxovnickiy, O.: LTE Mobile Station Positioning Using Time Difference of Arrival. Control systems and Information Technologies **1**(59), 55–61 (2015). Voronezh State Technical University,V.A. Trapeznikov Institute of Control Sciences, RAS, ISSN: 1729-5068
11. LTE System Toolbox™. The MathWorks, Inc. http://www.mathworks.com/products/lte-system/
12. Zekavat, R., Buehrer, M.R.: Handbook of Position Location: Theory, Practice and Advances. Wiley-IEEE Press (2011)
13. GPP TR 36.814 V9.0.0. 3rd Generation Partnership Project; Technical Specification Group Radio Access Network; Evolved Universal Terrestrial Radio Access (E-UTRA); Further advancements for E-UTRA physical layer aspects (GPP TR 36.814 Release 9 2 V9.0.0), March 2010

On Capturing Spatial Diversity of Joint M2M/H2H Dynamic Uplink Transmissions in 3GPP LTE Cellular System

Amir Ahmadian[1], Olga Galinina[1], Irina A. Gudkova[2], Sergey Andreev[1（✉）],
Sergey Shorgin[3], and Konstantin Samouylov[2]

[1] Tampere University of Technology (TUT), Tampere, Finland
{amir.ahmadian,olga.galinina,sergey.andreev}@tut.fi
http://www.tut.fi/en
[2] Peoples' Friendship University of Russia (PFUR), Moskva, Russia
{igudkova,ksam}@sci.pfu.edu.ru
http://www.rudn.ru/en
[3] Federal Research Center, Computer Science and Control
of the Russian Academy of Sciences, Moscow, Russia
sshorgin@ipiran.ru
http://www.ipiran.ru

Abstract. While queuing theory has indeed been instrumental to various communication problems for over half a century, the unprecedented proliferation of wireless technology in the last decades brought along novel research challenges, where user location has become a crucial factor in determining the respective system performance. This recent shift turned important to characterize large cellular macrocells, as well as the emerging effects of network densification. However, the latter trend also called for increased attention to the actual user loading and uplink (UL) traffic dynamics, accentuating again the necessity of queuing analysis. Hence, by combining queuing theory and stochastic geometry in a feasible manner, we may quantify the dependence of system-level performance on the traffic loading.

As performance of both session- and file-based UL transmissions has already been investigated recently in the context of heterogeneous networks, this paper explores a possibility of combining these two applications to provide a first-order evaluation of joint machine-to-machine (M2M) and human-to-human (H2H) transmissions in 3GPP LTE cellular systems. Employing a two-dimensional Markov chain for the aggregated process, we provide an approximation for the state transitions and, finally, arrive at a system-level approximation for the steady-state mode, which allows estimating a variety of system parameters averaged across space and time.

The reported study was partially supported by GETA, Finnish Academy of Sciences, and RFBR, research project No. 15-07-03051.

S. Balandin et al. (Eds.): NEW2AN/ruSMART 2015, LNCS 9247, pp. 407–421, 2015.
DOI: 10.1007/978-3-319-23126-6_36

1   Introduction

The emergence of novel technological paradigms supported by ubiquitous connectivity will likely bring us to the vision of networked society in foreseeable future. Targeting massive connectivity of identifiable objects [1], the world of heterogeneous unattended devices referred to as the *Internet of Things (IoT)* has been attractive for both industry and academia, which are investing large amounts of resources into developing it. In turn, *machine-to-machine (M2M)* communication is predicted to account for a considerable portion of the IoT market by 2020 [2], [3], [4].

Along with the legacy cellular technologies, the recent 3GPP Long Term Evolution (LTE) system has indeed been helpful to enable M2M services and applications [5]. However, the available capacity of the corresponding LTE signaling channels may deteriorate quickly in case of massive M2M connectivity on the one hand and negatively affect the expected QoS requirements for conventional *human-to-human (H2H)* devices on the other hand [6]. While developing the M2M-aware LTE scheduling mechanisms that handle the overloaded traffic efficiently, there have been many partial solutions proposed in [7], [8], and [9], focused mostly on simulation studies.

At the same time, from the analytical point of view, recent theoretical advances in wireless networking have been widely exploiting the powerful tools coming from stochastic geometry [10]. However, network densification, as one of the major trends in the emerging 5G networks, encourages the research community to look back at the user loading and the corresponding UL traffic dynamics. This trend calls for re-employing the queuing theoretical methods, which have been extensively investigated over the last century [11], [12], [13]. Combining these two important methodologies – queuing theory and stochastic geometry – in a simple fashion, we may build understanding of the *dynamic* system-level performance, which helps us characterize the practical system-level behavior.

While the respective performance of spatially-distributed session- and file-based UL transmissions has indeed been captured in [14] and [15], this paper explores a possibility of combining these two distinct QoS classes and suggests a *first-order evaluation* of the joint M2M/H2H operation under a particular LTE scheduler, which in turn has been analyzed for the fixed-location scenario in [16] and [17]. Using the two-dimensional Markov chain for the aggregated process, we provide an approximation for the state transitions and, finally, arrive at the system-level approximation for the steady-state mode, which allows estimating a variety of operational parameters averaged in both space and time, while accounting for M2M/H2H QoS requirements with a joint scheduling scheme.

We emphasize that the intended *main contribution* of this paper is not in investigating a particular scheduling discipline as such, but rather in offering a new approach to combining the characteristic features of M2M and H2H transmissions, where all of the devices share the available 3GPP LTE resource. Similarly, the proposed approach could be applied to any specific scheduler, which remains tractable from the queuing theoretical perspective. The remainder of this text is organized as follows. In Section 2, we describe the system model, emphasizing both geometric and dynamic components. Section 3 summarizes our analytical method of building a combined M2M/H2H system-level abstraction. Finally, we present selected numerical results and conclusions in Section 4.

2 System Model

This section describes the important system entities under consideration and introduces our main assumptions associated with them. Importantly, combining space and time components implies (i) a geometrical model for the M2M/H2H device locations on the plane and, independently, (ii) a queuing model, which would reflect the system evolution in time due to the scheduler and admission control features. Hence, below we begin with introducing the geometrical component and then, after explaining the channel model, arrive at the description of our joint scheduler, respective QoS requirements, and admission control settings for both classes of traffic individually.

2.1 General Description

Let us consider a centralized wireless architecture (illustrated in Fig. 1) consisting of (i) a single base station (BS) located at the center of a circular cell and (ii) a number of identical M2M and H2H devices spatially distributed within the cell and associated with the BS.

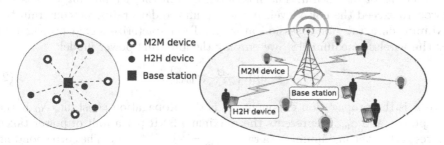

Fig. 1. Topology of the envisioned system: connected M2M and H2H devices.

Together with the spatial component, we introduce traffic dynamics, when the connection between the M2M/H2H device and the BS (termed a *session*) exists for a certain random time interval and also depends on the connection type. In particular, M2M devices associated with the BS transmit data "files", whereas H2H devices require voice or web streaming sessions with predefined rate requirements. The data transmissions of both classes share the same LTE uplink channel, and scheduling requests are managed centrally at the BS. The scheduling procedure as part of the radio resource management mechanism is performed based on the connection type and the available channel resources. It results in either fulfilling the QoS requirements or declining the session. In what follows, we detail our key assumptions prior to explaining the mathematical derivations behind the proposed model.

2.2 Topology Distribution

Both M2M and H2H devices are assumed to be spatially distributed within the cell of radius R, so that the probability density function (PDF) for the distance to the BS is given by $f_d(d)$. In particular, we consider a special case when the devices are distributed according to a Poisson Point Process (PPP) on the plane and thus are uniformly deployed within the area of interest. In this case, the distribution of distances within the circle of radius R is defined by the expression $f_d(d) = \frac{2d}{R^2}$. However, we emphasize here that the geometric component irrespectively of its complexity delivers as an ultimate result the distribution of distances $f_d(d)$ between all the transmitters and the receiving BS, which is further used as one of the input parameters for the queueing model below and defines the transitions between the corresponding states of the Markov process.

2.3 Power-Rate Mapping and Signal Propagation

We assume that the instantaneous data rate for the M2M/H2H device obeys the Shannon's formula:

$$r = w \log (1 + \gamma p), \tag{1}$$

where w denotes the spectral bandwidth, p is the transmit power, and γ is the signal-to-noise ratio (SNR) per a unit of power. Note that γ and p are the functions of distance d. The interference from the neighboring cells is not allowed to exceed the noise level. Avoiding infinite data rates, we constrain the maximum data rate for a session at $r_{\max}$. To characterize signal propagation over the wireless medium [18], we employ the following power model:

$$\gamma = \min \left[\frac{G}{d^\kappa N_0}, \gamma_{\max} \right], \tag{2}$$

where κ is the propagation exponent, G is the propagation constant, N_0 is the noise power, and $\gamma_{\max}$ represents the maximum SNR per a unit of power taken with respect to the maximum data rate $\gamma_{\max} = \frac{1}{p} \left(e^{\frac{r_{\max}}{w}} - 1 \right)$. The corresponding equation for γ implies that if the distance to the BS becomes less than a certain threshold $d_{\min} = \left(\frac{G}{\gamma_{\max} N_0} \right)^{\frac{1}{\kappa}}$, the instantaneous data rate as well as the SNR remain constant. Thereby, knowing the maximum data rate $r_{\max}$, the maximum value $\gamma_{\max}$, and the distance threshold $d_{\min}$ by the above expressions, we may define γ as well as the instantaneous data rate r as follows:

$$
\begin{aligned}
&\text{(i)}\ r = r_{\max}, \quad \gamma = \gamma_{\max}, \quad 0 \le d \le d_{\min}, \\
&\text{(ii)}\ r = w \log \left(1 + \frac{pG}{d^\kappa N_0} \right), \quad \gamma = \frac{G}{d^\kappa N_0}, \quad d_{\min} < d \le R.
\end{aligned}
\tag{3}
$$

2.4 Arrivals and QoS Requirements

Both M2M and H2H arrivals into the system follow a Poisson process of intensity λ_m and λ_h, respectively. A new M2M device targets to transmit a single data file of size s, which is distributed exponentially with the average θ. Importantly, any M2M data transmission requires the *minimal bit-rate* b_m. However, the actual achieved bit-rate may exceed this threshold. If admitted, the BS serves the M2M device until the entire file is transmitted. In contrast, each H2H device requires a *fixed bit-rate* b_h during the time interval of exponential duration (with the average μ^{-1}). Once the H2H device arrives into the system, the BS allocates the needed resource, and the device is served continuously until its session ends.

2.5 Scheduler Properties

Naturally, the BS is assumed to schedule the requests generated by both M2M and H2H devices within a specified resource pool, which we assume being of volume 1 without loss of generality. As mentioned above, the individual shares of the total resource have to be assigned such that they satisfy both M2M and H2H QoS requirements. The *actual data rate* may be calculated as a product of the random *instantaneous rate* r^m (M2M) or r^h (H2H), and a dedicated share of the total resource.

Both instantaneous rates r^m and r^h are obtained based on the respective power policy, which in our study (i) for M2M translates into the fixed power p_m for simplicity, and (ii) for H2H leads to *SNR-threshold* power control, when the power is either set to achieve the SNR η (and the respective rate r_η) within a circle of radius $d_\eta = \left(\frac{G}{p_{\max} N_0} \right)^{\frac{1}{\kappa}}$ or to the maximum level $p_{\max}$. Therefore, the system settings may be summarized as follows:

$$
\begin{aligned}
&\text{(i) } p_m = const, \quad r^m = \min[w \log\left(1 + \gamma p_m\right), r_{\max}], \\
&\text{(ii) } p_h = \min\left[\frac{\eta}{\gamma}, p_{\max}\right], \quad r^h = \min\left[w \log\left(1 + p_{\max}\gamma\right), r_\eta\right].
\end{aligned}
\tag{4}
$$

M2M Request Scheduling. For all the data sessions generated by M2M devices, the actual time spent in the system until service completion depends on the variable *actual data rate*. Naturally, this actual data rate for one device is a function of the allocated resource, which is assigned to a *set of devices* as an *integer number* of identical chunks of size $c < 1$. All j devices in service *share* the allocated resource so that the individual share for one device equals $\frac{1}{j} z\left(j\right) c$, where $z\left(j\right)$ is the minimum possible number of chunks to guarantee b_m for everyone. Clearly, if $B\left(i\right)$ is the share of the total resource occupied by all ongoing H2H sessions, then:

$$
z\left(j\right) c + B\left(i\right) \leq 1.
\tag{5}
$$

Moreover, each new M2M device requires the rate of at least b_m, or, equivalently, the share $\frac{b_m}{r^m}$ of the total resource. Therefore, the condition $z\left(j\right) \leq \frac{b_m j}{r_{\max} c}$ should be satisfied for any device out of $j \leq N_m^{\max}$ (the maximum number of M2M devices). Otherwise, this new device cannot be served and is considered *blocked* by the system.

H2H Request Scheduling. The H2H devices are guaranteed to obtain the exact rate b_h, so as to maintain their target QoS. Additionally, all H2H sessions in service together with the M2M traffic utilize not more than the total available resource 1, i.e., (5) is satisfied. Correspondingly, the BS is able to accept a new H2H request if it does not violate the minimum data rate requirement on the one hand, and if there are still resources available to the new requests as well as their number does not exceed the maximum number of H2H devices in service $N_h^{\max}$ on the other hand. The new request is *blocked* and considered lost permanently if any of the above conditions is not met.

3   Performance Analysis

In this section, we provide the description of our analytical approach to evaluate the system-level performance metrics. As it has been mentioned above, the analysis comprises two core components: network geometry and queuing model. While the geometrical component results in distribution $f_d(d)$ of distances between the transmitters and the BS, the aforementioned scheduling and admission control mechanisms comprise the queuing model, which is detailed below. The geometrical component is incorporated into the queuing model via the transition rates between the states and impacts directly on the stationary distribution together with the arrival rates.

3.1   General Remarks

We remind that according to our assumptions the inter-arrival times for both M2M and H2H traffic patterns, as well as the M2M file size and the duration of the H2H session, are distributed exponentially. Exploiting this fact, we may describe the system behavior by means of a Markov process $X(t)$, the state of which is defined by all currently running M2M and H2H sessions. Specifically, as the key parameter we select the individual distance between the transmitter (either M2M or H2H device) and the BS. We note that a set of such distances for all ongoing sessions determines the behavior of the entire system. Consequently, we describe the state of the process $X(t)$ as $x = \left(\left(d_1^h, \ldots, d_i^h \right), \left(d_1^m, \ldots, d_j^m \right) \right)$, where j and i are the current numbers of M2M and H2H devices, correspondingly, while d_i^h (d_i^m) are the distances between the M2M (H2H) device i and the BS.

3.2   Aggregated Markov Process

We note that the process $X(t)$ features the uncountable number of states, which makes it rather complex to analyze straightforwardly. In order to tackle the problem at hand and decrease the number of states, we employ the *state aggregation* technique. Aggregating the states of the initial process $X(t)$ by the number of M2M and H2H, we eventually arrive at the aggregated Markov process $Y(t)$, where the state is defined as $y = (i, j)$. The state space of this process is:

$$\mathcal{Y} \subset \{(i, j) \geq (0, 0) : i = 0, \ldots, N_h^{\max}; j = 0, \ldots, N_m^{\max}\}, \tag{6}$$

where $N_h^{\max}$ and $N_m^{\max}$ are the maximum possible numbers of devices in service.

To this end, Fig. 2 illustrates the general structure of the discussed process and, in particular, the selected state (i, j) together with the corresponding transitions $Q_{(i,j) \to (\bullet, \bullet)}$. We note that not all the states in (6) are possible due to restrictions on the transmission rate. Otherwise, we would have had an opportunity to allocate infinitely-small shares of resource and observe all possible combinations of (i, j).

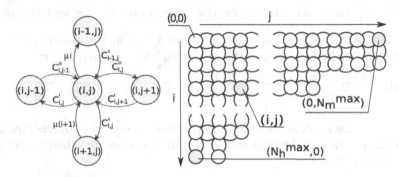

Fig. 2. State transition diagram for the aggregated Markov process $Y(t)$: a particular state (i, j) and the structure of the state space $\mathcal{Y}$. Transitions $C_{i,j}^I, C_{i,j}^{II}, and C_{i,j}^{III}$ correspond to Propositions 1, 2, and 3, respectively.

Hence, we construct the process such that the state transitions would reflect the spatial nature of the system under consideration and provide a spatially-averaged estimate of the system performance. In order to obtain the steady-state distribution of the process $Y(t)$, we derive the infinitesimal generator matrix Q. While the backward transition from the state (i, j) to the state $(i - 1, j)$ is near-trivial and equals $Q_{(i,j) \to (i-1,j)} = i\mu$, the rest of the transitions are more involved and deserve separate attention (given in Propositions 1-3). The following Lemma provides subsidiary expressions:

Lemma 1. *The first two moments of the random variables $\frac{1}{r^m}$ and $\frac{1}{r^h}$ are given by:*

$$E\left[\tfrac{1}{r}\right] = \tfrac{1}{r_{\max}} F_d(d) + \int\limits_d^R \frac{1}{w \log\left(1 + \frac{Gp}{x^\kappa N_0}\right)} f_d(x) dx, \ and \qquad (7)$$

$$E\left[\tfrac{1}{r^2}\right] = \tfrac{1}{r_{\max}^2} F_d(d) + \int\limits_d^R \frac{1}{w^2 \log^2\left(1 + \frac{Gp}{x^\kappa N_0}\right)} f_d(x) dx, \qquad (8)$$

where $d = d_{\min}$, $p = p_m$ for M2M devices and $d = d_\eta$, $r_{\max} = r_\eta$, $p = p_{\max}$ for H2H devices, and $F_d(d)$ is the cumulative distribution function (CDF) of distances d.

Proof. The proof is straightforward and is omitted here.

We note that distributions of the instantaneous rates have the same shape for both H2H and M2M transmissions, but H2H power allocation leads to a certain distribution of powers, while M2M power is kept on the fixed level.

Therefore, any urther calculation of energy consumption and its derivatives has to take into account the stochastic nature of transmit power values.

Further, the transition to the state $(i, j-1)$ may be obtained by calculating the average service time spent at the state. Let $Z(j) = z(j)c$ stand for the proportion of resource occupied and equally shared by j M2M devices in service, where $z(j)$ is a discrete random variable representing the number of allocated chunks. Given the current state (i, j), the average service time may be expressed as $E\left[\frac{s}{\frac{Z(j)}{j}r^m}\right]$, where s is the exponentially-distributed file size with the average θ, and $\frac{Z(j)}{j}$ is the resource allocated per one M2M device. Hence, due to the independence of the file size and the rest of variables, we may write:

$$Q_{(i,j)\to(i,j-1)} = \frac{j}{E\left[\frac{s}{\frac{Z(j)}{j}r^m}\right]} = \frac{1}{E\left[\frac{s}{Z(j)r^m}\right]} = \frac{1}{\theta E\left[\frac{1}{Z(j)r^m}\right]} = \frac{1}{\theta E\left[\frac{1}{Z(j)}\right]E\left[\frac{1}{r^m}\right]}, \quad (9)$$

where $Z(j)$ is the random variable reflecting the amount of resources consumed by all j M2M devices and represented as a number of allocated chunks multiplied by the size of one chunk.

Proposition 1. *The transitions $Q_{(i,j)\to(i,j-1)}$ may be approximated by the following expression:*

$$Q_{(i,j)\to(i,j-1)} = \frac{1}{\theta E\left[\frac{1}{Z(j)r^m}\right]} \approx \frac{\left\lceil\frac{b_m j}{cr^m_{\min}}\right\rceil c}{\theta E\left[\frac{1}{r^m}\right]}, \quad (10)$$

where $E\left[\frac{1}{r^m}\right]$ is given by (7) for $d = d_{\min}$, $p = p_m$.

Proof. We note that the variable $Z(j)$ denotes the amount of resources occupied by j M2M devices and consists of the variable number of chunks $z(j)$ of size c. As it has been mentioned above, to serve a set of M2M devices, for every device k the value of $z(j)$ has to exceed $\frac{b_m j}{r^m c}$, which translates into the following expression:

$$z(j) = \left\lceil\frac{b_m j}{c\min_{k=\overline{1,j}}\{r^m_k\}}\right\rceil. \quad (11)$$

For the fixed j, the distribution of $z(j)$ may be obtained through the functional transform discretization of the random variable $r^m_{\min}(j) = \min_{k=\overline{1,j}}\{r^m_k\}$. In turn, the distribution of the random variable $r^m_{\min}(j)$ may be calculated via a product of j equal components, i.e., $F_{r^m_{\min}(j)}(r) = 1 - (1 - F_r(r))^j$, where $F_r(r)$ is the CDF of the random variable r^m. However, this would lead to cumbersome expressions, and to simplify further calculations we replace the random variable $r^m_{\min}(j)$ with the minimum value of the rate $r^m_{\min}$, thus arriving at the approximation:

$$Z(j) \leq \tilde{Z}(j) = \left\lceil\frac{b_m j}{cr^m_{\min}}\right\rceil c, \quad \forall j : \tilde{Z}(j) < 1, \quad \text{i.e. } Z(j) \approx \min\left(1, \left\lceil\frac{b_m j}{cr^m_{\min}}\right\rceil c\right), \quad (12)$$

where $r^m_{\min} = \max\{r_{\max}, w\log\left(1 + p_m\frac{G}{R^\kappa N_0}\right)\}$ is the minimum achievable instantaneous rate at the maximum distance, and b_m is the minimum requirement for the M2M device. Therefore, due to independence of $\tilde{Z}(j)$ and the current

r^m, we may write the following:

$$Q_{(i,j)\to(i,j-1)} = \frac{1}{\theta E\left[\frac{1}{Z(j)r^m}\right]} \approx \frac{\left[\frac{b_m j}{cr^m_{\min}}\right]c}{\theta E\left[\frac{1}{r^m}\right]}, \tag{13}$$

where $E\left[\frac{1}{r^m}\right]$ is given by (7). At this step, we assume the worst-case resource allocation, i.e., when all the M2M devices are located at the edge of the cell.

The transitions from the state (i, j) to the state $(i+1, j)$ are obtained by taking into account the share of the actually accepted H2H devices as the following proposition shows. We remind here that the geometrical component is included into the transitions owing to the aggregation of the initial process. Focusing on the forward transitions, we explicitly consider not only the probability to arrive at the next state, but also the condition that the previous state has been achieved. Keeping this in mind, we formulate the following two propositions and the corresponding auxiliary corollaries.

Proposition 2. *The transitions $Q_{(i,j)\to(i+1,j)}$ may be approximated as follows:*

$$Q_{(i,j)\to(i+1,j)} \approx \lambda_h \frac{\Phi\left(\frac{1-\tilde{Z}(j)-m_{i+1}}{\sigma_{i+1}}\right)}{\Phi\left(\frac{1-\tilde{Z}(j)-m_i}{\sigma_i}\right)}, \tag{14}$$

where $\tilde{Z}(j)$ is given by equation (12), $\Phi(x)$ is the standard function $\frac{1}{\sqrt{2\pi}}\int_{-\infty}^{x} e^{-\frac{t^2}{2}}dt$, whereas the parameters m_i and σ_i are:

$$m_i = i b_h E\left[\frac{1}{r^h}\right], \quad \sigma_i^2 = i b_h^2 \left(E\left[\left(\frac{1}{r^h}\right)^2\right] - \left(E\left[\frac{1}{r^h}\right]\right)^2\right). \tag{15}$$

The transition from the zero state is given by:

$$Q_{(0,j)\to(1,j)} \approx \lambda_h F_d\left(\min\left\{R, \left(\frac{Gp}{N_0}\right)^{\frac{1}{\kappa}}\left(e^{\frac{b_h}{(1-\tilde{Z}(j))w}}-1\right)^{-\frac{1}{\kappa}}\right\}\right), \tag{16}$$

where $F_d(d)$ is the CDF of the random distance $d \geq d_{\min}$.

Proof. Calculation of the transition rates automatically translates into the estimation of the conditional probability that the H2H device would be accepted by the system:

$$\Pr\{i+1, j \text{ can be served}|i, j \text{ are served}\} = \frac{\Pr\{i+1, j \text{ can be served}; i,j \text{ are served}\}}{\Pr\{i,j \text{ are served}\}} =$$

$$\frac{\Pr\{i+1, j \text{ are served}\}}{\Pr\{i, j \text{ are served}\}} = \frac{\Pr\left\{\sum\limits_{k=0}^{i+1}\frac{b_h}{r^h}+Z(j)<1\right\}}{\Pr\left\{\sum\limits_{k=0}^{i}\frac{b_h}{r^h}+Z(j)<1\right\}}, \tag{17}$$

where $Z(j) = z(j)c$ is the amount of resources already allocated to j M2M devices and is represented by a discrete number $z(j)$ of resource chunks.

In order to obtain the above probability, one needs to calculate the distribution of the sum of independent random variables $\frac{b_h}{r^h}$ as well as the distribution of the discrete random variable $z(j)$, and finally find the convolution of both distributions. Even though it is doable, we instead focus on a computationally simpler expression, which results in being an approximation. Therefore, at this step we suggest replacing the random variable $Z(j)$ by its estimate $\tilde{Z}(j)$ (12). At the same time, the sum of two and more random variables we may substitute with the random variable following the normal distribution $\mathcal{N}(m_i, \sigma_i^2)$ with the parameters $m_i = i b_h E\left[\frac{1}{r^h}\right]$, $\sigma_i^2 = b_h^2 i \left(E\left[\frac{1}{(r^h)^2}\right] - \left(E\left[\frac{1}{r^h}\right]\right)^2\right)$, where the corresponding moments are given by (7) and (8). Hence, the final transition rate may be rewritten as:

$$Q_{(i,j)\to(i+1,j)} \approx \lambda_h \frac{\Phi\left(\frac{1-\tilde{Z}(j)-m_{i+1}}{\sigma_{i+1}}\right)}{\Phi\left(\frac{1-\tilde{Z}(j)-m_i}{\sigma_i}\right)}, \tag{18}$$

where $\tilde{Z}(j)$ is given by (12), $\tilde{Z}(j) < 1$, and $\Phi(x)$ is an integral corresponding to the Gaussian distribution.

For the transition from the state 0 to the state 1, due to the absence of any conditionality we have:

$$Q_{(0,j)\to(1,j)} \approx \lambda_h \Pr\left\{\frac{b_h}{r^h} \leq 1 - \tilde{Z}(j)\right\} = \lambda_h \Pr\left\{\frac{b_h}{w \log\left(1+\frac{Gp}{d^\kappa N_0}\right)} \leq 1 - \tilde{Z}(j)\right\} =$$

$$\lambda_h \Pr\left\{d \leq \left(\frac{Gp}{N_0}\right)^{\frac{1}{\kappa}} \left(e^{\frac{b_h}{(1-\tilde{Z}(j))w}} - 1\right)^{-\frac{1}{\kappa}}\right\} =$$

$$\lambda_h F_d\left(\min\left\{R, \left(\frac{Gp}{N_0}\right)^{\frac{1}{\kappa}} \left(e^{\frac{b_h}{(1-\tilde{Z}(j))w}} - 1\right)^{-\frac{1}{\kappa}}\right\}\right), \tag{19}$$

where $F_d(d)$ is the CDF of the random distance d.

We note that this approximation for the fraction in (17) may be exploited only for those values i, j, for which the following inequality holds:

$$i > \left\lfloor \frac{1-\tilde{Z}(j)}{b_h/r_{\min}^h} \right\rfloor, \tag{20}$$

where $r_{\min}^h = \min\left[r_\eta, w \log(1 + p_{\max}\frac{G}{R^\kappa N_0})\right]$ is the minimum rate for the H2H device.

Corollary 1. *The probability for the H2H device to be blocked at the state (i, j) may be obtained by means of the same approximation as:*

$$P_b^h(i, j) \approx 1 - \frac{\Phi\left(\frac{1-\tilde{Z}(j)-m_{i+1}}{\sigma_{i+1}}\right)}{\Phi\left(\frac{1-\tilde{Z}(j)-m_i}{\sigma_i}\right)}, \tag{21}$$

and by analogy for the state $(0, j)$.

Finally, we formulate our third proposition explaining the calculation of the forward transition from the state (i, j) to the state $(i, j + 1)$ and corresponding to the M2M traffic arrival.

Proposition 3. *The transitions* $Q_{(i,j)\to(i,j+1)}$ *may be approximated by the following expression:*

$$Q_{(i,j)\to(i,j+1)} \approx \lambda_m \frac{[F_d(d_{j+1})]^{j+1}}{[F_d(d_j)]^j}, \tag{22}$$

where $d_j = \left(\frac{Gp_m}{N_0}\right)^{\frac{1}{\kappa}} \left(exp\left(\frac{b_m j}{\left\lceil\frac{1-B(i)}{c}\right\rceil cw}\right) - 1\right)^{-\frac{1}{\kappa}}$ *and* $B(i) = ib_h E\left[\frac{1}{r^h}\right].$

Proof. Similarly, the probability to accept a new M2M device may be calculated as:

$$\Pr\{i,j+1 \text{ can be served}|i,j \text{ served}\} = \frac{\Pr\{i,j+1 \text{ can be served}\}}{\Pr\{i,j \text{ can be served}\}}.$$

The probability in the latter expression translates into the probability that for all j devices the rate requirements are satisfied given the maximum possible room $1 - B(i)$ shared between them, where $B(i)$ is the random variable reflecting the amount of resources consumed by all i H2H devices. In other words, the inequality $\frac{b_m}{r^m} \leq \frac{1}{j}\left\lfloor\frac{1-B(i)}{c}\right\rfloor c$ has to hold for every $k = \overline{1,j}$. We note that the straightforward consideration of the distribution of the random variable $B(i)$ would numerically complicate the solution. Hence, aiming at a simpler approximation, we replace $B(i)$ with the average value of the resource share occupied by the H2H devices $ib_h E\left[\frac{1}{r^h}\right]$, which brings us to the final expression for the transition rate:

$$Q_{(i,j)\to(i,j+1)} \approx \lambda_m \frac{\left(\Pr\left\{r_i \geq \frac{\frac{b_m(j+1)}{1-B(i)}}{c}\right\}\right)^{j+1}}{\left(\Pr\left\{r_i \geq \frac{\frac{b_m j}{1-B(i)}}{c}\right\}\right)^j} = \lambda_m \frac{\left(\Pr\left\{w \log(1+p_m \frac{G}{d_i^\kappa N_0}) \geq \frac{\frac{b_m(j+1)}{1-B(i)}}{c}\right\}\right)^{j+1}}{\left(\Pr\left\{w \log(1+p_m \frac{G}{d_i^\kappa N_0}) \geq \frac{\frac{b_m j}{1-B(i)}}{c}\right\}\right)^j} =$$

$$\lambda_m \frac{\left(\Pr\left\{d \leq \left(\frac{Gp_m}{N_0}\right)^{\frac{1}{\kappa}}\left(exp\left(\frac{\frac{b_m(j+1)}{1-B(i)}}{cw}\right)-1\right)^{-\frac{1}{\kappa}}\right\}\right)^{j+1}}{\left(\Pr\left\{d \leq \left(\frac{Gp_m}{N_0}\right)^{\frac{1}{\kappa}}\left(exp\left(\frac{\frac{b_m j}{1-B(i)}}{cw}\right)-1\right)^{-\frac{1}{\kappa}}\right\}\right)^j} = \lambda_m \frac{[F_d(d_{j+1})]^{j+1}}{[F_d(d_j)]^j}, \tag{23}$$

where $d_j = \left(\frac{Gp_m}{N_0}\right)^{\frac{1}{\kappa}}\left(exp\left(\frac{b_m j}{\left\lceil\frac{1-B(i)}{c}\right\rceil cw}\right)-1\right)^{-\frac{1}{\kappa}}.$

Corollary 2. *The probability for the M2M device to be blocked at the state* (i,j) *may be obtained similarly as:*

$$P_b^m(i,j) \approx 1 - \frac{[F_d(d_{j+1})]^{j+1}}{[F_d(d_j)]^j}, \tag{24}$$

where the parameters are given in the statement of Proposition 3.

3.3   Performance Measures of Interest

Having obtained the transition intensities $Q_{(\bullet,\bullet)\to(\bullet,\bullet)}$, we compose the infinitesimal matrix $\mathbf{Q}$. Solving the corresponding system of equilibrium equations, we may finally establish the steady-state probability distribution $\{P(i,j),(i,j) \in \mathcal{Y}\}$. Based on it, we can derive any stationary metrics of interest, such as, e.g., the average number of M2M and H2H devices:

$$E[N_m] = \sum_{(i,j)\in\mathcal{Y}} jP(i,j), \quad E[N_h] = \sum_{(i,j)\in\mathcal{Y}} iP(i,j).$$

Delay and, hence, the average energy consumption per M2M device may be easily estimated using the above result together with the Little's law, while the energy consumption of the H2H device requires additional calculations of the transmission power distribution (due to the corresponding power control mechanism). However, in this paper we primarily focus on another important system parameter, which is the system blocking probability for either M2M or H2H type of devices:

$$P_b^m = \sum_{(i,j)\in\mathcal{Y}} P_b^m(i,j)P(i,j), \quad P_b^h = \sum_{(i,j)\in\mathcal{Y}} P_b^h(i,j)P(i,j).$$

4   Numerical Results and Conclusions

In order to illustrate our analysis with some numerical results, we consider the practical scenario of a small cell, where the human users (H2H devices) are sharing the available system resource jointly with, e.g., smart home gateways [19] (M2M devices), which collect data from all home sensors and send the aggregated files to the network whenever ready.

Importantly, the proposed first-order estimation provides useful insights into the scheduler operation, allowing, for example, the optimization of its parameters as shown in Fig. 3. By varying the input parameter c for both M2M and H2H devices, we may select sub-optimal settings (e.g., $c = 0.3$) that minimize the resulting blocking probability. However, while the said sub-optimal parameter balances the service rate for the files, keeping at the same time sufficient free room for the constant-bit-rate traffic, the transmission delay is expectedly rising with the reduction in the allocated resource c. Such trade-offs naturally open the possibility to optimizing this or other scheduler settings mindful of practical delay constraints.

Additionally, Fig. 4 illustrates the blocking probability surface against the changes in the system loading (the M2M and H2H arrival rates, λ_m and λ_h, respectively). These numerical results confirm our practical intuition on that the overall system performance degrades when the offered M2M/H2H loading grows, which results in having more devices at service.

Summarizing, this paper provides a novel framework for the exploration of joint operation between two distinct device classes with different QoS requirements (in particular, session- and file-based UL transmissions). Importantly, such joint operation captures not only the spatial diversity, which is crucial due to the wireless nature of the channel [20], but also the dynamic traffic loading, which is expected to become a key issue in emerging wireless systems. Going further, any

Table 1. Primary system parameters

Notation	Parameter description	Value
R	Cell radius	100 m
w	Total UL bandwidth	10 MHz
N_0	Noise level	-102 dBm
η	Target SNR	20 dB
$\text{SNR}_{\max}$	Maximum SNR, changing rate	25 dBm
p_m	Power level for M2M	6 dBm
$p_{\max}$	Maximum power level for H2H	23 dBm
b_m	Minimal bit-rate for H2H	50 Kbps
b_h	Target bit-rate for H2H	100 Kbps
θ	Average M2M file size	1 Mbit
$N_h^{\max}$	Maximum number of H2H users	30
$N_m^{\max}$	Maximum number of M2M users	30

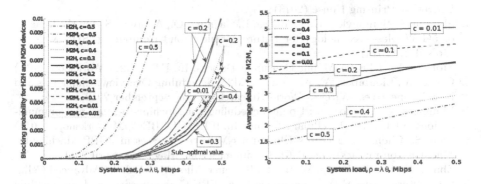

Fig. 3. Dependence of the system blocking probability (a) and the M2M transmission delay (b) on the M2M loading for different values of the chunk size c.

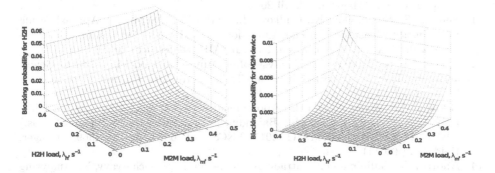

Fig. 4. Illustration of the impact of the system loading (both M2M and H2H arrival rates) on the system blocking probability for H2H (a) and M2M (b) devices.

more detailed information on the joint scheduler for the M2M/H2H users would help establish a higher degree of modeling accuracy thus becoming a fruitful area for our future work.

References

1. Gerasimenko, M., Petrov, V., Galinina, O., Andreev, S., Koucheryavy, Y.: Impact of Machine-Type Communications on Energy and Delay Performance of Random Access Channel in LTE-Advanced. European Transactions on Telecommunications, June 2013
2. Ahmadian Tehrani, A.M.: Modeling Contention Behavior of Machine-Type Devices over Multiple Wireless Channels, Master of Science Thesis, Tampere University of Technology, January 2015
3. Mazhelis, O., Warma, H., Leminen, S., Ahokangeas, P., Pusinen, P., Rajahonka, M., Siuruainen, R., Okonen, H., Shveykovskiy, A., Mylykoski, J.: Internet-of-Things Market, Value Networks, and Business Models: State of the art report, Jyväskylä University Printing House (2013)
4. Shorgin, S., Samouylov, K., Gudkova, I., Galinina, O., Andreev, S.: On the Benefits of 5G Wireless Technology for Future Mobile Cloud Computing. SDN & NFV, MoNeTec. (2014)
5. Morioka, Y.: LTE for Mobile Consumer Devices. In: ETSI M2M Workshop (2011)
6. Gotsis, A., Lioumpas, A., Alexiou, A.: M2M Scheduling Overview: Challenges and New Perspectives, Wireless World Research Forum, September 2012
7. Lioumpas, A., Alexiou, A.: Uplink Scheduling For Machine-to-Machine Communications in LTE-based Cellular Systems. In: IEEE GLOBECOM Workshops (2011)
8. Lien, S., Chen, K.: Towards Ubiquitous Massive Accesses in 3GPP Machine-to-Machine Communications. IEEE Communications Magazine, April 2011
9. Zhang, K., Hu, F., Wang, W.: Radio Resource Allocation in LTE-Advanced Cellular Networks with M2M Communications. IEEE Communications Magazine, July 2012
10. Andreev, S., Pyattaev, A., Johnsson, K., Galinina, O., Koucheryavy, Y.: Cellular Traffic Offloading onto Network-Assisted Device-to-Device Connections. IEEE Communications Magazine, April 2014
11. Shorgin, S., Samouylov, K., Gudkova, I., Markova, E., Sopin, E.: Approximating Performance Measures of Radio Admission Control Model for Non real-time Services with Maximum Bit Rates in LTE. In: AIP Conference (2015)
12. Samouylov, K., Gudkova, I.: Analysis of an Admission Model in a Fourth Generation Mobile Network with Triple Play Traffic, Automatic Control and Computer Sciences (2013)
13. Gudkova, I.A., Samouylov, K.E.: Approximating Performance Measures of a Triple Play Loss Network Model. In: Balandin, S., Koucheryavy, Y., Hu, H. (eds.) NEW2AN 2011 and ruSMART 2011. LNCS, vol. 6869, pp. 360–369. Springer, Heidelberg (2011)
14. Andreev, S., Galinina, O., Pyattaev, A., Johnsson, K., Koucheryavy, Y.: Analyzing Assisted Offloading of Cellular User Sessions onto D2D Links in Unlicensed Bands. IEEE Journal on Selected Areas in Communications, January 2015
15. Galinina, O., Pyattaev, A., Andreev, S., Dohler, M., Koucheryavy, Y.: 5G Multi-RAT LTE-WiFi Ultra-Dense Small Cells: Performance Dynamics, Architecture, and Trends. IEEE Journal on Selected Areas in Communications, June 2015

16. Gudkova, I., Samouylov, K., Buturlin, I., Borodakiy, V., Gerasimenko, M., Galinina, O., Andreev, S.: Analyzing of Coexistence Between M2M and H2H Communication in 3GPP LTE System. In: IEEE WWIC (2014)
17. Borodakiy, V.Y., Buturlin, I.A., Gudkova, I.A., Samouylov, K.E.: Modelling and Analysing a Dynamic Resource Allocation Scheme for M2M Traffic in LTE Networks. In: Balandin, S., Andreev, S., Koucheryavy, Y. (eds.) NEW2AN 2013 and ruSMART 2013. LNCS, vol. 8121, pp. 420–426. Springer, Heidelberg (2013)
18. Andreev, S., Koucheryavy, Y., Himayat, N., Gonchukov, P., Turlikov, A.: Active-mode Power Optimization in OFDMA-based Wireless Networks. In: IEEE Globecom Workshops (2010)
19. Galinina, O., Mikhaylov, K., Andreev, S., Turlikov, A., Koucheryavy, Y.: Smart Home Gateway System over Bluetooth Low Energy with Wireless Energy Transfer Capability. EURASIP Journal on Wireless Communications and Networking (2015)
20. Moltchanov, D., Koucheryavy, Y., Harju, J.: Simple, Accurate and Computationally Efficient Wireless Channel Modeling Algorithm. In: Braun, T., Carle, G., Koucheryavy, Y., Tsaoussidis, V. (eds.) WWIC 2005. LNCS, vol. 3510, pp. 234–245. Springer, Heidelberg (2005)

Performance Evaluation of Methods for Estimating Achievable Throughput on Cellular Connections

Lars M. Mikkelsen$^{(\boxtimes)}$, Nikolaj B. Højholt, and Tatiana K. Madsen

University of Aalborg, Wireless Communication Networks, Fredrik Bajers Vej 7,
9220 Aalborg East, Denmark
{lmm,tatiana}@es.aau.dk, nhajho11@student.aau.dk

Abstract. The continuous increase in always connected devices and the advance in capabilities of networks and services offered via the network is evident. The target group of most of these devices and services is users, so how users perceive the network performance is of great importance. Estimating achievable throughput (AT) is the main focus of this paper, which can be expressed as the data rate that users experience. We establish the Bulk Transfer Capacity (BTC) method as the ground truth of the AT. We choose to evaluate the Trains of Packet-Pair (TOPP) method as an alternative to BTC in estimating AT, due to its much reduced resource consumption. Based on real-life measurements of the two methods we conclude that TOPP is a good candidate to estimate AT, based on similarity in results with BTC.

Keywords: BTC · TOPP · Achievable throughput · Cellular networks

1 Introduction

The usage of mobile data has increased dramatically in recent years and it will continue to increase in the coming years[1]. This is due to the introduction of always connected smart devices, due to the capabilities of the networks, and due to service types being offered to users. A result is users getting used to having services available everywhere and all the time via networks. For this reason it is critical that the networks deliver continuously high quality service and only rarely drop in performance. Generally, network behavior can be captured via network monitoring, and the information about the connection quality that the network provides can be obtained through network performance measurements, performed actively from devices connected to the network.

When discussing network performance one of the important things is estimating the data rate that can be achieved for connected devices. In measuring data rates on networks it is important to be clear on what the goal is, i.e. what you want to measure, and how this is measured. Overall there are 3 different types of results related to data rates: capacity, available bandwidth (AB), and achievable throughput (AT).

© Springer International Publishing Switzerland 2015
S. Balandin et al. (Eds.): NEW2AN/ruSMART 2015, LNCS 9247, pp. 422–435, 2015.
DOI: 10.1007/978-3-319-23126-6_37

- Capacity, often called minimum capacity, is a measure of the overall capacity of the link.
- AB is a measure of the unused capacity of a link. As there typically is traffic from many users on a link, AB is the upper bound for data rate of a transmission on the link for a user.
- AT is a measure of what portion of the link that the user actually is assigned when performing a transmission. So this can be seen as a measure of what is experienced by a user.

Our goal is to find a good method to deploy in measurement systems aimed at providing end users with information about the connection quality they get. From the perspective of a user it is not that relevant to know what the capacity of the connection is, because he will most likely never experience this. The same argument could be applied to AB as it is an upper bound of what can be experienced. AB could potentially be much higher than what the user will ever experience. For this reason the focus of this paper will be on estimating AT as experienced from the end user.

We consider a case of mobile cellular networks. It is known that bandwidth estimation is challenging for this type of networks [2], and only very limited attempts exist to adapt measurement techniques for this use case [4].

Estimating data rates is a field where a great amount of research has been done, and where many different approaches exists [9]. Traditionally most approaches aim to either estimate capacity or AB for wired networks with first-come first-served store-and-forward links. Recently estimations of AT started to get some attention, as this metric is conceptually different from AB for cellular links. The majority of AB estimation methods are developed for traditional LAN and WLAN networks and not specifically for mobile networks. But as user devices typically are connected to the Internet via mobile networks, this is a big concern. Mobile networks are managed much more dynamically than LAN and WLAN networks due to high mobility of the connected devices. This should be taken into account when selecting and measuring the network performance. So in selecting the method to use, one has to consider the network type, how connected devices act, and how the network is managed.

Another issue in estimating AT on mobile networks is that the bandwidth assigned to a user varies much more rapidly than for other network types. Consequently the challenge is to estimate the throughput instantaneously, to counter the rapid changes in connection performance. Additionally when doing active measurements the goal is to reduce overhead and put as little traffic on the network as possible. But when doing estimations fast, the amount of data on which to base the estimate is reduced, making the estimate less confident. So the trade-off is the time spent and the amount of data collected during a measurement versus the correctness of the result.

In practice there is a great need for AT estimation methods aimed at mobile networks from tools such as Open Signal, NETRADAR, and NetMap.

Open Signal[8] uses crowd sourcing to perform signal strength measurements used to generate a network coverage map, and to localize base stations.

Furthermore, Open Signal also allows the user to perform individual measurements such as round trip times (RTT) and throughput estimation.

NETRADAR[7] also applies crowd sourcing to gather signal strength measurements and perform network performance measurements such as RTT and AT. The main focus of this system is to analyze the correlation between signal strength measurements and network performance measurements.

In NetMap[6] crowd sourcing is used to performance measurements such as RTT and throughput estimation. The goal of this system is to make a network performance map (NPM) that should replace the network performance maps provided by ISPs. The NPM provided from ISPs are based on signal strength measurements and theoretical models of these, and from this the network performance metrics such as RTT and throughput are estimated.

In the present work it is chosen to look at active measurement methods that attempt to estimate the AT. Within this area two methods have been chosen that are very different in resource consumption; Bulk Transfer Capacity (BTC) which is a traditional method that applies a very simplistic brute force approach in estimating the AT; and Trains Of Packet Pairs (TOPP) that applies a more delicate approach utilizing information of packet interarrival times in estimating the AT. These two methods will be compared based on the results from a number of measurements. The measurements will be done on a commercial mobile network in Denmark with devices connected via 4G (LTE).

The rest of the paper is organized as follows: Section 2 will describe what we will measure and the concept of ground truth used in this paper. Section 3, 4 and 5 will describe measurement setup and the measurement methods used. In Section 6 results from the method will be compared and evaluated. Lastly in Section 7 the results will be discussed and conclusions will be drawn.

2   What Will Be Measured

When estimating anything it is important to know if your estimation is close to the actual value of what is being estimated. The actual value can be denoted as the ground truth - the true value of what is being estimated. In our case we are estimating the AT, i.e. how much data the user is allowed to put through his connection. The challenge occurs when we are performing estimations based on real-life measurements. The question is then which of the measured values is the ground truth. That is if any of the obtained values is the ground truth. This problem could be overcome by using simulations instead. In simulations it is possible to extract all information about users and network states. This information is often not available when making real-life measurements, and in many cases they are not observable.

There will be fluctuations of the AT for the user due to different factors. These factors include cross-traffic (load sharing among multiple users), channel conditions and interference, and mobility and geographical location of users. Only the first aspect related to cross-traffic is present in wired networks, while all the other aspects are specific to wireless access networks. Additionally, cellular networks employ a proportional fair scheduler that allocates resources to users, taking into account change in channel conditions and interference, e.g. providing some compensation for users near the cell edge.

The fluctuations of AT can be roughly divided into fast and slow. The slow fluctuations will be caused by the general user load of the network changing according to location and over time. The fast fluctuations are related to when other users connect to or leave the network, and when other users start or stop transmitting on the network. Additionally, under loaded network conditions a user with bad signal conditions will not be scheduled until the scheduler predicts that the user will experience good signal to noise ratios. On cellular networks resource allocation to the users are done every transmission time interval, which is as low as 2 ms for HSPA and 1 ms for LTE. It is not feasible to measure with such high resolution, and the measurements would be required to be done on the physical layer. Furthermore, as we are interested in the user experience, which also relates to higher layers, we will measure no lower than the transport layer. On the transport layer the measurement frequency is however limited by the scheduler of the device, and by the chosen size of the samples. Despite these limitations we will still be able to observe fluctuations in the throughput, only less rapid than what potentially could be observed on the physical layer.

As AT is a measure of how much data the user actually is allowed to put through a connection, we will approximate the ground truth using the method Bulk Transfer Capacity (BTC). BTC is measured by putting as much data as possible through a connection over a duration of time, using TCP. This represents a typical user file transfer, as TCP is the most widely used transport protocol on the Internet. Furthermore, by using TCP, which employ flow control and congestion avoidance, the data rate is controlled and adjusted according to the capabilities of the user connection, just as any other file transfer would be.

3   Measurement Setup and Methodology

Implementation: BTC (see Section 4) is implemented using TCP and TOPP (see Section 5) implemented using UDP. The client implementation is done as an Android application in Java, and the server implementation is also done in Java. For this reason the samples are done on the application layer, why packet size is containing payload + IP header + UDP/TCP header. The measurements are performed between a mobile device and a server acting as a measurement point for all the measurements.

Client Device: All the measurements are performed with the same Android device (Samsung Galaxy S5 (SM-G900F)) running Android 5.0 (Samsung stock Lollipop). The device is connected to the Internet using a LTE (4G) connection to a large danish Internet Service Provider (ISP). All measurements are done while being stationary at the same location near (<500m) the cell tower. The location of the client device is indoor at the Aalborg University campus.

Measurement Server: The measurement server is placed at Aalborg University and is connected to the Internet using a Gbit connection to the danish research network, making the shortest path for data from the device to the server through the ISP network, via the Danish Internet Exchange point onto the danish research network, to the server. The server is a stationary PC running Ubuntu 14.04, where no other software than the measurement software is running during the test.

4   Bulk Transfer Capacity (BTC)

The BTC method, as defined by the Internet Engineering Task Force (IETF)[10], is the average amount of data that can be transmitted through a link per second, $BTC = data\_sent/elapsed\_time$. Here $data\_sent$ is the amount of unique bits transmitted excluding headers. $elapsed\_time$ is the time between receiving the first bit and the last. A way to implement BTC is to produce samples and to stop the measurements when a predefined number of samples is received, or when a certain time to do measurements has elapsed. The size of a sample is fixed pr measurement and the value for the sample size should be selected in a clever way, e.g. adjusting it according to the currently used connection technology. This will help to achieve more comparable results and additionally to some extend limit the amount of data used per test. Figure 1 shows a sequence diagram of BTC for the downlink case.

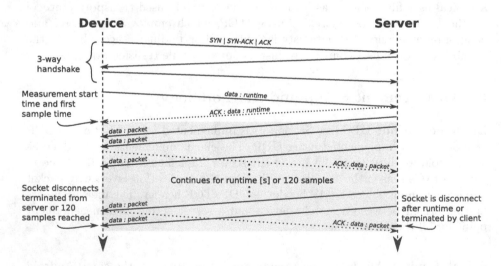

Fig. 1. Sequence diagram of a BTC download measurement.

BTC is implemented using TCP which applies flow control mechanisms in terms of slow start and congestion avoidance that throttles the data flow according to the capabilities of the connection. This causes the samples of the BTC

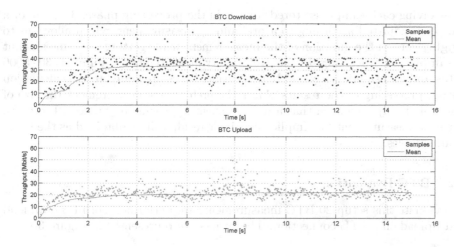

Fig. 2. Example of samples obtained via a BTC measurement on a LTE connection.

measurement to fluctuate a lot, as TCP attempts to adjust the rate according to the capabilities of the connection. This can be seen from Figure 2.

From Figure 2 it can be seen that there is a big spread on the individual samples, why some processing must be performed to get an idea of the tendency of the samples. This is done by calculating a mean of the samples, which is the red line on the plots. The mean is calculated for sample 1 to sample n, where n runs from 1 to the total number of samples. This reveals the impact that the flow control of TCP has on the data transfer, in that the mean rises in the beginning and settles after approximately 3 seconds.

The slow start should be taken into account when performing the measurements, or rather when processing the results. One approach could be to only calculate the mean of samples after the initial 3 seconds, and discard the initial samples. However, as TCP is chosen as the transport protocol, the rise time of the throughput is an integrate part of the result, as it says something about the stability of the connection.

Another thing to consider is for how long to measure, or alternatively how much data to transmit. With a long measurement the result will be more stable, but at the same time more expensive in terms of data usage. And conversely, with a short measurement the result will be more affected by fluctuations, but it will be cheaper in data usage.

4.1   BTC Setup

In our implementation when performing a BTC measurement the bytes are counted at the receiver side and sampled. We have selected the following parameters: Each measurement consists at most of 120 samples, or a duration of 15 sec. Thus, a single measurement round might include less than 120 samples if the connection is slow at the time of measurement (e.g. below 1 Mbit/s). The time of receiving each sample is stored for use in the processing phase. The size of a sample is adjusted according to a preliminary measurement, which allows us to roughly control the duration of the measurement while getting the same amount of samples for processing. In the preliminary measurement a sample is $30 \cdot 5000$ bytes, only considering payload bytes. Header bytes are added to the calculation during processing of the results. After the preliminary measurement the size of a sample is adjusted such that a measurement approximately takes 10 seconds. For each measurement for simplicity a single result will be obtained as the mean of all samples.

4.2   BTC Evaluation

To evaluate this setup 20 BTC measurements are performed both for download and upload on a LTE connection. The results are represented in Figure 3.

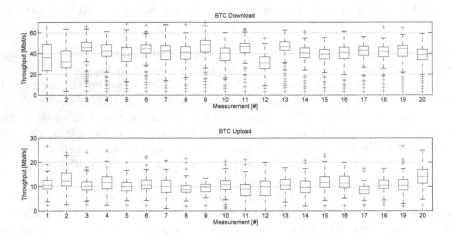

Fig. 3. Boxplot of 20 BTC download and upload measurements. The box corresponds to the center samples between the 25th and 75th percentile, the red line in the box is the mean, the whiskers extend to the most extreme points that are not considered outliers, and the red pluses are outliers. Note: download plot is limited to 70 Mbit/s and upload to 30 Mbit/s.

From Figure 3 it can be seen that the means of the measurements, both for upload and download, are rather steady around the same value.

Furthermore, from Table 1 it can be seen that the data usage is quite extensive. This is due to the size of sample being fixed such that each measurement approximately lasts for 10 seconds, to ensure that the rate will have time to settle.

Table 1. Statistics based on 20 BTC download and upload measurements.

	Mean Throughput [Mbit/s]	Variance of Means [Mbit/s]	Mean Data Usage [MB]	Mean Time [s]
Download	39.9049	13.5181	55.1067	12.91
Upload	10.8834	2.3749	8.3098	7.07

5   Trains Of Packet-Pair (TOPP)

5.1   BTC Alternatives

Based on the results in Section 4.2, it is apparent that the resource usage of BTC is very high. So another method is needed that preferably achieves similar results, but significantly reduces the resource consumption both regarding data and time. There exists a wide range of methods developed for estimating AB. These methods generally fall into one of two categories: packet rate methods and packet gap methods[4]. Packet gap methods pose high requirements to measurement equipment and for that reason is a bit more tricky to implement in our setup. Our setup consist of standard mobile phones, why the necessary time resolution of measurements is not supported. Therefore we choose a method from the packet rate method class, namely Trains of Packet-Pairs (TOPP). TOPP is a well known technique for estimating AB, and some studies have indicated that using AB techniques for measuring in cellular environments would result in AT estimations[4]. This motivates our choice and calls for more extensive studies of its applicability for AT estimation.

5.2   TOPP Description

In order to describe TOPP[5] the principle of packet-pair will be described first. The idea behind packet-pair is sending two packets back-to-back with the same payload L. When the packets arrive at the receiver the packets will arrive with a dispersion δ as illustrated in Figure 4.

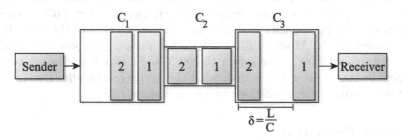

Fig. 4. The height of C_1, C_2 and C_3 simply illustrates the capacity of the link between sender and receiver. The packets leave the sender back to back, and they arrive at the receiver with a dispersion that is determined by the narrow link.

The UDP protocol is used as it does not implement flow control. The dispersion δ is given by the payload L over the capacity of the link C. From this it can easily be deduced that $C = L/\delta$. As proposed by [3] the larger the payload L the greater the impact the cross traffic will have on the measurements. Therefore, for big L the measurements are trending towards the tight link instead of the narrow link. Whereas if L is small the result trends towards the narrow link capacity or post-narrow link capacity.

TOPP is based on the packet-pair principle with an increase in the number of packets transmitted ($N > 2$). The packets are sent to a receiver with a fixed packet size. When the number of packets N is sufficiently large the dispersion δ between all packets becomes an average packet delay. The total dispersion $\Delta(N)$ can be found by:

$$\Delta(N) = \sum_{n=1}^{N-1} \delta_n \tag{1}$$

The measured AT can thereby be calculated by:

$$A = \frac{(N-1) \cdot L}{\Delta(N)} \tag{2}$$

5.3   TOPP Setup

When performing a TOPP measurement the transmitted packets are counted at the receiver side, and the time of arrival of each packet is noted and used during the processing phase. For an initial test of the method we have choosen the following parameters: Each measurement consists of a maximum of 800 packets. The amount of packets counted at the receiving end can potentially be lower due to packet loss during the measurement. The packet size is set to 1500 bytes, based on the standard MTU size. However, we are aware that due to fragmentation and packetization done by link layer protocols in cellular networks, chunks of data of 1500 bytes will end up being transmitted in different physical layer packets, and thus, other packet sizes might be preferable.

Furthermore, it was discovered that for the download case there was a high amount of packet loss, while none for the upload case. We believe this is caused by the server having a network connection with high capacity (1Gbit/s), while the client connection capacity is significantly lower (LTE Cat4 theoretically 150 Mbit/s downlink, 50 Mbit/s uplink). For this reason we choose to limit the download transmit rate at the server to 200 Mbit/s, which still leaves room for packets to be transmitted faster than what the device can handle. This approach significantly reduces the packetloss for the download case.

5.4   TOPP Evaluation

In Figure 5 the results of 20 TOPP measurements performed on a LTE connection can be seen.

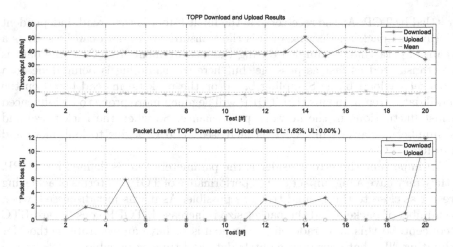

Fig. 5. Top plot: The results of 20 TOPP download and upload measurements performed on a LTE connection, and the mean of the download and upload results respectively. Bottom plot: The packet loss registered in the TOPP measurements.

From Figure 5 it can be seen that both for download and upload the results are mostly very consistent, only fluctuating very little. Furthermore it can be seen that there still is some packet loss on the download measurements, however much less than what previously was the case, and the packet loss that is present does not seem to have any impact on the results.

To be able to compare TOPP with BTC the statistics of these TOPP measurements are presented in Table 2. From this table it can be seen that the results in terms of mean values and variance of means are similar. The variance for the TOPP means is slightly lower, than for BTC. TOPP only uses approximately 2% and 14% of the data that BTC uses for download and upload respectively. Another thing to notice is that TOPP uses significantly less time pr measurement, why the 20 measurements can be performed much quicker than BTC, leaving less time for the network to change.

Table 2. Statistics based on 20 TOPP download and upload measurements.

	Mean Throughput [Mbit/s]	Variance of Means [Mbit/s]	Mean Data Usage [MB]	Mean Time [s]
Download	39.0834	12.2029	1.1444	0.23
Upload	8.4744	1.2309	1.1444	1.09

5.5 Considerations

When comparing TOPP to BTC, they apparently apply two significantly different methods. In BTC data is transmitted as fast as possible, where the rate is throttled by TCP. At the receiver side the input buffer is then read, independent of individual packets. Because of this the transmission of data can be seen as a stream of data that is sampled during the measurement. In TOPP the data is also transmitted as fast as possible, but here the sampling is being done on a pr packet payload basis. Sampling based on the packets, instead based on the overall data stream, means that TOPP will be much more prone to be influenced by fast fluctuations in the network performance. So when the packet size and the number of packets in TOPP is increased, TOPP begins to look more and more like BTC.

An important thing to consider is the parameters that is defined for TOPP, as they may have a big impact on the performance of TOPP in terms of achieving a result as close to the ground truth as possible. As it was just discussed, when the number of packets and the packet size is increased, TOPP goes toward BTC in concept. So this is a trade off as the goal is to have an accurate method for estimating AT, that consumes as little data and time as possible.

6   Comparison of BTC and TOPP

To evaluate whether TOPP is sufficiently accurate in estimating the AT, a number of tests will be done where the results will be compared to the ground truth that will be obtained using BTC. In the comparison the parameters of TOPP will be changed to find the best trade off between data usage and accuracy of the results. The different combinations of TOPP settings will furthermore be evaluated in different scenarios. The network load will be different in the different scenarios, so it can be examined what impact this has on the results of TOPP.

6.1   Measurement Setup

In Table 3 the different combinations of TOPP settings can be seen. 3 different number of packets and 3 different packet sizes have been chosen. For each set of 3 TOPP measurements with the same number of packets, a BTC measurement will be performed, which will serve as the ground truth for that set of TOPP measurements. The BTC measurements will be performed as described in Section 4. For each unique setup of TOPP, and for each BTC, the measurement will be repeated 20 times. Based on this, a full schedule of measurements is as follows:

Table 3. TOPP setttings matrix.

	300 bytes	1500 bytes	3000 bytes
200 packets	a1	a2	a3
800 packets	b1	b2	b3
1500 packets	c1	c2	c3

BTC; TOPPa1; TOPPa2; TOPPa3; BTC; TOPPb1; TOPPb2; TOPPb3; BTC; TOPPc1; TOPPc2; TOPPc3.

A full schedule of the measurements covering all settings are done both at noon and at night, but at the same location as described in Section 3. This will allow for evaluation of how the setups are influenced by cross traffic on the network, as during night there are only very few users in the area.

6.2   Results Evaluation

In Figures 6 and 7 the results are presented for download and upload measurements respectively.

From Figure 6 it can be seen that the download rates measured at noon are slightly higher than during night. Furthermore, for most of the cases the night results seem to be less fluctuating than the day rates, indicated by slightly smaller confidence intervals. For TOPP the best setup, i.e. the results closest to the BTC results, seem to be with the bigger packet sizes (1500 and 3000 bytes). The number of packets pr measurement does not seem to have any impact on the mean, but there is a slight tendency that the confidence intervals are lower for bigger number of packets.

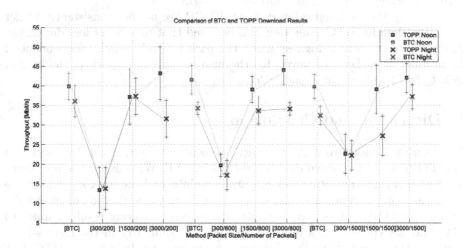

Fig. 6. Comparison of BTC and TOPP download measurements performed at noon and at night. Each result is based on 20 measurements where the square or x marks the mean. The whiskers indicate the 95% confidence interval of the means. Whether the result is for BTC or TOPP with a certain setup is indicated in the x-axis label. From left to right each BTC result is linked with the following 3 TOPP results, according to time of day.

From Figure 7 it can be seen that the upload rates almost consistently are higher at night than at noon. Also here the bigger packet sizes for TOPP seem

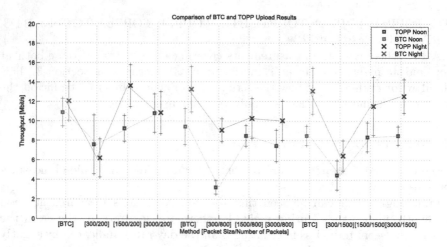

Fig. 7. Comparison of BTC and TOPP upload measurements performed at noon and at night. See Figure 6 for description.

to achieve the results closest to BTC. But the confidence intervals for upload seem to be slightly smaller at noon than at night.

Generally the confidence intervals for TOPP are neither consistently bigger or smaller than for BTC. Finally it can be noted that the packet loss during the TOPP measurements is highest when the packet size is small, independent of the number of packets. However, for the highest number of packets the biggest packet size also shows high packet loss.

7   Discussion and Conclusion

Based on the results from the two measurement methods, it can be concluded that TOPP is a good candidate for estimating AT. What is interesting about this result is that the measurements are done in a real life network, where the cross-traffic and varying user load is very real. From the evaluation of the results it seems as TOPP offers a good trade-off between delivering results similar to BTC and not consuming a lot of resources. The similarity in the results both relates to the means of the measurements being close to the means of BTC. But also the fluctuations in the means are similar to those of BTC. Furthermore, TOPP seems to deliver good results both in the heavily and the lightly loaded network scenario.

From the presented results the recommendation for the settings of TOPP would be a packet size 1500 bytes, and 800 packets pr measurement. However, further measurements and evaluations based on more packet sizes and different number of packets could reveal setups that consume even less resources while still being stable in results. Also, to better be able to compare the methods it should be attempted to perform the measurements simultaneously on several devices. Other future work should include a deeper analysis of the impact of the cellular network scheduler on the measurements should be performed. Furthermore, there should be performed further evaluation of the validity of BTC as provider of the ground truth. And lastly, other alternative measurement methods should be tested and evaluated along with TOPP.

The results presented in this work are obtained in a very limited scenario. The scenario could be expanded both measuring at more times and at more locations relative to the base station. This would yield stronger results and a better general image of the performance of the methods. Furthermore, to get a complete understanding of the dynamics in cellular networks, the measurements should also include cellular specific parameters such as signal strength and parameters related to mobility.

References

1. Cisco: Cisco visual networking index: Forecast and methodology, 2012–2017, May 2014. http://www.cisco.com/c/en/us/solutions/collateral/service-provider/ip-ngn-ip-next-generation-network/white_paper_c11-481360.html
2. Devi, U., Viswanathan, H., Kokku, R., Pichapati, V., Kalyanaraman, S.: On the estimation of available bandwidth in broadband cellular networks. In: 2014 Eleventh Annual IEEE International Conference on Sensing, Communication, and Networking (SECON), pp. 19–27, June 2014
3. Dovrolis, C., Ramanathan, P., Moore, D.: Packet-dispersion techniques and a capacity-estimation methodology. IEEE/ACM Transactions on Networking 12(6), 963–977 (2004)
4. Koutsonikolas, D., Hu, Y.: On the feasibility of bandwidth estimation in wireless access networks. Wireless Networks 17(6), 1561–1580 (2011). doi:10.1007/s11276-011-0364-5
5. Melander, B., Bjorkman, M., Gunningberg, P.: A new end-to-end probing and analysis method for estimating bandwidth bottlenecks. In: Global Telecommunications Conference, GLOBECOM 2000, vol. 1, pp. 415–420. IEEE (2000)
6. Mikkelsen, L.M., Thomsen, S.R., Pedersen, M.S., Madsen, T.K.: NetMap - creating a map of application layer qos metrics of mobile networks using crowd sourcing. In: Balandin, S., Andreev, S., Koucheryavy, Y. (eds.) NEW2AN/ruSMART 2014. LNCS, vol. 8638, pp. 544–555. Springer, Heidelberg (2014)
7. NETRADAR: Netradar, September 2014. https://www.netradar.org/en
8. Open Signal: Open signal, May 2014. http://opensignal.com/
9. Prasad, R., Dovrolis, C., Murray, M., Claffy, K.: Bandwidth estimation: metrics, measurement techniques, and tools. IEEE Network 17(6), 27–35 (2003)
10. The Internet Engineering Task Force: Ip performance metrics, May 2014. http://datatracker.ietf.org/wg/ippm/charter/

Alternating Priorities Queueing System with Randomized Push-Out Mechanism

Alexander Ilyashenko, Oleg Zayats$^{(\boxtimes)}$, Vladimir Muliukha, and Alexey Lukashin

Peter the Great Saint-Petersburg Polytechnic University, Saint Petersburg, Russia
{ilyashenko.alex,zay.oleg}@gmail.com,
vladimir@mail.neva.ru, lukash@neva.ru

Abstract. This paper is written about a priority queueing models in combination with randomized push-out mechanism. Considered model has two incoming flows having different intensities, limited buffer size (k) and two types of priorities for each flow. Priority in this model is realized as alternating priority and it can be switched from one flow to another in moment when system will complete serving all packets from high priority flow and will have in buffer only low-priority packets. In this paper presented algorithm for computing characteristics of the model like loss probability for both types of packets. For getting solution is used generating functions method. This method allowed to reduce size of Kolmogorov's linear equations system from $k(k+1)$ to $(4k-2)$, where k is a model buffer size. Also authors obtained (using this method) areas of "closing" borders for non-priority packets and areas of "linear behavior" where can be used linear law depending on push-out probability α for approximating loss probabilities.

Keywords: Priority queueing · Alternating priority · Randomized push-out mechanism

1 Introduction

Queueing systems can be used for modelling different kinds of real processes, like data transfer, production lines, electronic queues in various institutions and others. Using such models allows to improve performance and to reduce waiting time for being serviced.

In queueing theory are used two main ways to change the behavior of queueing models: adding a priority and including push-out mechanism. Prioritizing as a method of bandwidth control is effective only in two cases: when system has infinite buffer size or it's weakly loaded. If the buffer is limited, and intensity of low priority flow is high, system can get filled by low priority packets and the effect of prioritization will get nullified. To avoid this problem usually introduce a push-out mechanism, which allows high-priority packets to push out of the system buffer low-priority packets. In literature analyzed in detail deterministic push-out mechanism, which is only effective when high-priority packets flow has low intensity. When it reaches certain critical level the model have buffer filled only by high priority packets.

© Springer International Publishing Switzerland 2015
S. Balandin et al. (Eds.): NEW2AN/ruSMART 2015, LNCS 9247, pp. 436–445, 2015.
DOI: 10.1007/978-3-319-23126-6_38

For avoiding such situations can be used randomized push-out mechanism, introduced in [2, 3] and actively studied in [1, 5-7]. For such type of mechanism pushing out of the buffer does not occur every time when priority packets coming to the system, but only with a provided probability α. It can be considered as a parameter of this mechanism. Also this mechanism generalizes two special cases: deterministic push-out ($\alpha = 1$) and its absence ($\alpha = 0$).

In queueing theory considered three main types of priorities [8]: preemptive, relative and alternating. Preemptive and relative priorities were studies in [1,2,3]. In this article will be studied model with alternating priority. Using this type of priority allows to switch priority between incoming flows. Assume that at first flow has high priority and second flow has low. When system will complete servicing all packets from first flow and will have only low priority packets in the buffer it'll set high priority to second flow. Also from this moment first flow packets will have low priority. Later it can be switched back, when system will have no packets from second flow in buffer.

This type of priority guarantees minimal amount of switches between priorities [8]. This feature can be useful in systems where changing of priority can take a lot of time, like production lines where operator has to reconfigure service device or reconfigurable network devices.

In [2, 3] provided classification for queueing systems with push out mechanisms where has been used a symbol $f_i^{\,j}$. Originally, i takes the values: $i = 0$ (no priority), $i = 1$ (relative priority), $i = 2$ (absolute priority). The superscript j chosen out of the two values: $j = 0$ (no push-out), $j = 2$ (deterministic push-out). Following recommendations in [2], let's use $j = 1$ for randomized push-out. Furthermore, in this paper we propose to reserve $i = 3$ for alternating priority. Model which will be studied in this article can be classified using described notation as $\overrightarrow{M_2}/M/1/k/f_i^1$ where $i=3$.

2   Building Kolmogorov's System of Linear Equations

Consider a model with two incoming flows, alternating priority and randomized push-out mechanism. At first we have to build phase space of this model and notation to all states of the system. States of this model are characterized by three-dimensional phase vector

$$\{N_0(t), N_1(t), N_2(t)\}, \tag{1}$$

where N_1 and N_2 meaning amount of packets from flow with appropriate number which are in buffer of the system, and N_0 specifies number of flow which packet is on service at the moment t. The phase space is given in the form

$$\Omega = \{\varnothing\} \cup \left\{(n, i, j) : n = \overline{1,2}; i = \overline{0, k-1}; j = \overline{0, k-1-i}\right\}, \tag{2}$$

and state graph consists of two subgraphs shown in Fig. 1.

Consider the study only the steady state, denoting the final probabilities through $P_{i,j}^{(n)}$, where n is the type of packet which is on service (like N_0 in phase vector (1)), and i and j - number of packets in buffer of each type. Kolmogorov's system of linear equations for this state graph provided in Theorem 1. For building of this system were used standard approach [4]. On the left side of all equations in Theorem 1 probability of being system in specified state (n, i, j) multiplied by total intensity of all outgoing transitions from it. On the right side the sum of products of the transition intensities and probabilities of states from which you can get in (n, i, j) state. By equating left and right part for all states from (2) getting balance equations (3).

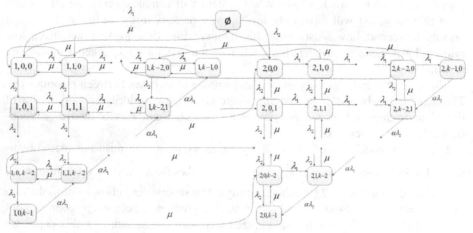

Fig. 1. The state graph of system $\overrightarrow{M}_2 / M / 1 / k / f_3^1$

Theorem 1. *Kolmogorov's system of linear equations for model* $\overrightarrow{M}_2 / M / 1 / k / f_3^1$

$$
\begin{cases}
(\mu + (\lambda_1 + \lambda_2)(1 - \delta_{i+j,k-1}) + \alpha\lambda_1\delta_{i+j,k-1}(1 - \delta_{j,0}))P_{i,j}^{(1)} = \\
= \lambda_1(1 - \delta_{i,0})P_{i-1,j}^{(1)} + \lambda_2(1 - \delta_{j,0})P_{i,j-1}^{(1)} + \\
+ \mu\delta_{j,0}(1 - \delta_{i,k-1})P_{i+1,0}^{(2)} + \mu(1 - \delta_{i+j,k-1})P_{i+1,j}^{(1)} + \\
\alpha\lambda_1\delta_{i+j,k-1}(1 - \delta_{i,0})P_{i-1,j+1}^{(1)} + \lambda_1\delta_{i,0}\delta_{j,o}P_{\varnothing}, \\
(\mu + (\lambda_1 + \lambda_2)(1 - \delta_{i+j,k}) + \alpha\lambda_2\delta_{i+j,k-1}(1 - \delta_{i,0}))P_{i,j}^{(2)} = \\
= \lambda_2\delta_{i,0}\delta_{j,0}P_{\varnothing} + \lambda_1(1 - \delta_{i,0})P_{i-1,j}^{(2)} + \\
+ \lambda_2(1 - \delta_{j,0})P_{i,j-1}^{(2)} + \mu(1 - \delta_{i+j,k-1})P_{i,j+1}^{(2)} + \\
+ \mu\delta_{i,0}(1 - \delta_{j,k-1})P_{0,j+1}^{(1)} + \alpha\lambda_2\delta_{i+j,k-1}(1 - \delta_{j,0})P_{i+1,j-1}^{(2)},
\end{cases}
\tag{3}
$$

where $\delta_{i,j}$ *is a Kronecker delta and indices i and j have following limitations:*
$i = \overline{0, k-1}; \, j = \overline{0, k-1-i}$.

To make matrix of this system of linear equations not singular replace any equation by following normalization condition:

$$\sum_{w=1}^{2}\sum_{i=0}^{k}\sum_{j=0}^{k-i}P_{i,j}^{(w)}=1.$$

This system has size equal to $k(k+1)$. In the next paragraph method from [1] will be applied to reduce size of matrix from $k(k+1)$ to $4k-2$.

3   Generating Function Method Application

Let's introduce generating function for probabilities $P_{i,j}^{(n)}$ as:

$$G(u,v,w)=\sum_{n=1}^{2}\sum_{i=0}^{k-1}\sum_{j=0}^{k-1-i}P_{i,j}^{(n)}u^iv^jw^{n-1}=G^{(1)}(u,v)+wG^{(2)}(u,v),\qquad(4)$$

where $G^{(1)}(u,v)$ and $G^{(2)}(u,v)$ are generating functions for left and right subgraph of state graph from Fig. 1.

Using approach from [1] obtain equations for generation functions $G^{(1)}(u,v)$ and $G^{(2)}(u,v)$ given in Theorem 2. These equations include only four sets of unknown probabilities: $\{P_{k-1-i,i}^{(1)}\}_{i=0}^{k-1},\{P_{i,k-1-i}^{(2)}\}_{i=0}^{k-1},\{P_{0,i}^{(2)}\}_{i=0}^{k-1},\{P_{i,0}^{(1)}\}_{i=0}^{k-1}$.

Theorem 2. *For model with alternating priorities* $\overrightarrow{M_2}/M/1/k/f_3^1$ *generating functions* $G^{(1)}(u,v)$ *and* $G^{(2)}(u,v)$ *satisfy following equations:*

$$[(1+\rho)-(\rho_1u+\rho_2v)-1/u]G^{(1)}(u,v)=\rho_1P_\varnothing+((\rho-\alpha\rho_1)-$$

$$-(\rho_1u+\rho_2v)+\alpha\rho_1u/v)\sum_{i=0}^{k-1}P_{i,k-1-i}^{(1)}u^iv^{k-1-i}+\alpha\rho_1P_{k-1,0}^{(1)}u^{k-1}(1-u/v)+$$

$$+[G^{(2)}(u,0)-P_{0,0}^{(2)}-G^{(1)}(0,v)]/u,$$

$$[(1+\rho)-(\rho_1u+\rho_2v)-1/v]G^{(2)}(u,v)=\rho_2P_\varnothing+((\rho-\alpha\rho_2)-\qquad(5)$$

$$-(\rho_1u+\rho_2v)+\alpha\rho_2v/u)\sum_{i=0}^{k-1}P_{i,k-1-i}^{(2)}u^iv^{k-1-i}+$$

$$+[-G^{(2)}(u,0)+G^{(1)}(0,v)-P_{0,0}^{(1)}]/v+\alpha\rho_2P_{0,k-1}^{(2)}v^{k-1}(1-v/u).$$

Using equation (5) and technique from [1] can express probabilities of all system states (6) through probabilities $p_i^{(1)}=P_{k-1-i,i}^{(1)}$, $p_i^{(2)}=P_{i,k-1-i}^{(2)}$ and $s_i^{(1)}=P_{0,i}^{(1)}$, $s_i^{(2)}=P_{i,0}^{(2)}$.

Theorem 3. *Probabilities of all states for model* $\overrightarrow{M_2} / M / 1 / k / f_3^1$ *are expressed through "diagonal" probabilities* $p_i^{(1)} = P_{k-1-i,i}^{(1)}, p_i^{(2)} = P_{i,k-1-i}^{(2)}$ *and "border" probabilities* $s_i^{(1)} = P_{0,i}^{(1)}, s_i^{(2)} = P_{i,0}^{(2)}$ *as*

$$P_{k-1-j,i}^{(1)} = \sum_{S=1}^{i} p_S^{(1)}((\rho_1^{-1}+\alpha)\rho_1^{(1-j+S)/2}c_{j-i-1}^{i-S+1}(t_0^{(1)}) -$$

$$-\rho_1^{(-j+S)/2}c_{j-i-2}^{i-S+1}(t_0^{(1)}))\beta_1^{i-S} - \alpha\sum_{S=0}^{i}\rho_1^{(1-j+S)/2}p_{S+1}^{(1)}c_{j-i-1}^{i-S+1}(t_0^{(1)})\beta_1^{i-S} -$$

$$-\beta_1^i\sum_{S=0}^{j-2-i}s_{k-1-S}^{(2)}\rho_1^{(S-j)/2}c_{j-S-i-2}^{i+1}(t_0^{(1)}) +$$

$$+p_0^{(1)}\rho_1^{(-1-j)/2}[c_{j-i-1}^{i+1}(t_0^{(1)}) - \rho_1^{1/2}c_{j-i-2}^{i+1}(t_0^{(1)})]\beta_1^i,$$

$$\tag{6}$$

$$P_{i,k-1-j}^{(2)} = \sum_{S=1}^{i} p_S^{(2)}((\rho_2^{-1}+\alpha)\rho_2^{(1-j+S)/2}c_{j-i-1}^{i-S+1}(t_0^{(2)}) -$$

$$-\rho_2^{(-j+S)/2}c_{j-i-2}^{i-S+1}(t_0^{(2)}))\beta_2^{i-S} - \alpha\sum_{S=0}^{i}\rho_2^{(1-j+S)/2}p_{S+1}^{(2)}c_{j-i-1}^{i-S+1}(t_0^{(2)})\beta_2^{i-S} -$$

$$-\beta_2^i\sum_{S=0}^{j-2-i}s_{k-1-S}^{(1)}\rho_2^{(S-j)/2}c_{j-S-i-2}^{i+1}(t_0^{(2)}) +$$

$$+p_0^{(2)}\rho_2^{(-1-j)/2}[c_{j-i-1}^{i+1}(t_0^{(2)}) - \rho_2^{1/2}c_{j-i-2}^{i+1}(t_0^{(2)})]\beta_2^i,$$

where

$$t_0^{(1)} = \rho_1^{-1/2}(1+\rho_1+\rho_2);$$
$$t_0^{(2)} = \rho_2^{-1/2}(1+\rho_1+\rho_2);$$
$$\beta_1 = -\rho_2\rho_1^{-1/2}; \beta_2 = -\rho_1\rho_2^{-1/2}$$

The last step to reduce size of linear equations system (3) is to build system of linear equations for four sets of unknown probabilities $p_i^{(1)} = P_{k-1-i,i}^{(1)}, p_i^{(2)} = P_{i,k-1-i}^{(2)}$ and $s_i^{(1)} = P_{0,i}^{(1)}, s_i^{(2)} = P_{i,0}^{(2)}$, which are used in (6). For doing this we have to get $4k-2$ equations. First $2k$ equations can be built by substitution of $j=k-1$ into (5). To get $2k-2$ more equations find $\lim_{u\to 0} G^{(1)}(u,v)$ and $\lim_{v\to 0} G^{(2)}(u,v)$ and equate to zero coefficients of all powers of argument v for limit of $G^{(1)}$ and u for limit of $G^{(2)}$ in (5).

4   Computational Results

Most interesting characteristic of considered queueing model is probability of losing packets of both types. Losing probability for each type can be obtained as

$$P_{loss}^{(1)} = p_0^{(1)} + (1-\alpha)\sum_{i=1}^{k-1} p_i^{(1)} + \alpha\frac{\rho_2}{\rho_1}\sum_{i=1}^{k-1} p_i^{(2)} + \sum_{i=0}^{k-1} p_i^{(2)},$$

$$P_{loss}^{(2)} = p_0^{(2)} + (1-\alpha)\sum_{i=1}^{k-1} p_i^{(2)} + \alpha\frac{\rho_1}{\rho_2}\sum_{i=1}^{k-1} p_i^{(1)} + \sum_{i=0}^{k-1} p_i^{(1)}.$$

$$(7)$$

On Fig. 2 shown loss probability for model with flows intensities $\rho_1 = 0.2, \rho_2 = 0.9$. From this figure can be seen that loss probability is getting close to 1 just by changing parameter of randomized push-out mechanism α from 0 to 1. Loss probability of second type packets is changing from by almost 70%. This effect can be considered as "closing" of a model for low priority packets, because with so high loss probability these packets have really small chance to get serviced.

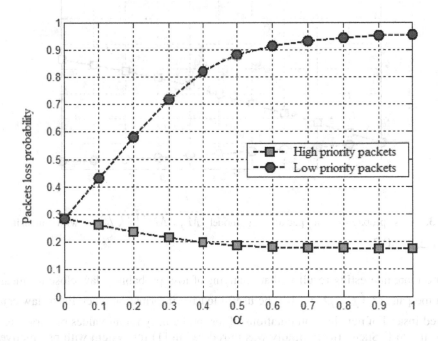

Fig. 2. Loss probability of packets in model $\overrightarrow{M_2}/M/1/k/f_3^1$ with intensities $\rho_1 = 0.2, \rho_2 = 0.9$.

On Figure 3 we see that changing of α parameter has a significant effect on the loss probability in case when first flow has higher intensity. It's not getting close to 1, but changes from 10% to almost 45%. As we can see that on Fig. 2 loss probability is getting almost close to 1, but on Fig. 3 it's getting only up to value close to 0,45. Such effect was introduced in [1] and it was called "closing" effect. For building such areas we have to find pairs of (ρ_1, ρ_2) for which loss probability will be close to 1 when parameter α is equal to 1.

On Fig.4 presented such areas for model with alternating priorities. As in [1] these borders were computed for intensities from 0 to 4.0 with a 0.01 step and computed losing probability of non-priority packets for push-out probability α equal to 1. On Figure 4 presented a plot of areas built using levels for value $(1 - P_{loss}\big|_{\alpha=1})$ [0, 0.5, 0.9, 0.95, 0.99, 0.999, 0.9999].

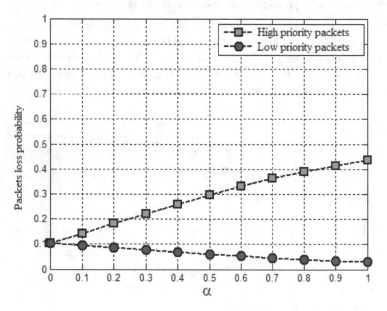

Fig. 3. Loss probability of packets in model $\overrightarrow{M_2}/M/1/k/f_3^1$ with intensities $\rho_1 = 1.2, \rho_2 = 0.2$.

One more interesting result is that changing of loss probability law close to linear for some pairs of (ρ_1, ρ_2). It'll be useful to have information when linear law can be used instead of detailed computations of loss probability for all values of parameter α from 0 to 1. Such effect initially was introduced in [1] for system with preemptive priority and for this model were obtained areas of "linear behavior" too.

On Fig. 5-6 presented borders of areas where "linear behavior" effect can be seen. For checking linear behavior was used following condition: calculated relative deviation of losing probability curve and its linear approximation (8) has to be less that specified level.

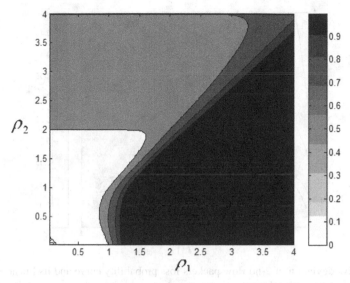

Fig. 4. Areas of system "closing" for levels [0,0.5,0.9,0.95,0.99,0.999,0.9999].

$$\delta P_{loss}^{(i)} = \max_{0 \leq \alpha \leq 1} \left\{ \left| P_{loss}^{(i)}(\alpha) - (P_{loss}^{(i)}(0)(1-\alpha) + \alpha P_{loss}^{(i)}(1)) \right| / \left| P_{loss}^{(i)}(\alpha) \right| \right\} \qquad (8)$$

As a one more expected result, areas of linear behavior have to be symmetrical for pairs of (ρ_1, ρ_2) and (ρ_2, ρ_1), because in considered model with alternating priorities changing flows numbers is not significant for such characteristics. Fig. 5-6 confirm this feature of model.

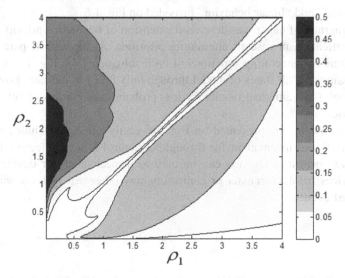

Fig. 5. Relative deviation of 1-st flow packets loss probability curve and its linear approximation from 0% to 50% with 1% step.

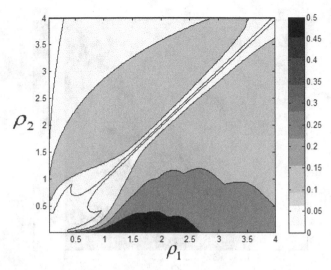

Fig. 6. Relative deviation of 2-nd flow packets loss probability curve and its linear approximation from 0% to 50% with 1% step

5   Conclusion

In this article was studied model with two incoming flows, alternating priorities and randomized push-out mechanism. For this model was applied generating function method and it allowed to reduce size of matrix in Kolmogorov's system of linear equations from $k(k+1)$ to $4k-2$. Using this reduced system were calculated borders for areas of "closing" and "linear behavior" provided on Fig. 4-6.

In first paragraph of paper was described extension of Basharin's priority queueing systems classification in case of alternating priorities. At the second paragraph obtained equations for generating functions of each subgraph from Fig. 1 and were obtained expressions for all states of model through only four sets of states probabilities. Third paragraph has numerical results for loss probabilities, borders of "closing" and "linear behavior" areas.

From numerical results presented on Fig. 2-6 easily can be concluded that using randomized push-out mechanism for throughput control is very effective for model with alternating priorities too and can be used as a mechanism for developing new technical devices for data transfer or controlling any other real systems with limited buffer size and queues.

References

1. Ilyashenko, A., Zayats, O., Muliukha, V., Laboshin, L.: Further investigations of the priority queuing system with preemptive priority and randomized push-out mechanism. In: Balandin, S., Andreev, S., Koucheryavy, Y. (eds.) NEW2AN/ruSMART 2014. LNCS, vol. 8638, pp. 433–443. Springer, Heidelberg (2014)
2. Avrachenkov, K.E., Shevlyakov, G.L., Vilchevsky, N.O.: Randomized push-out disciplines in priority queueing. Journal of Mathematical Sciences **122**(4), 3336–3342 (2004)
3. Avrachenkov, K.E., Vilchevsky, N.O., Shevlyakov, G.L.: Priority queueing with finite buffer size randomized push-out mechanism. Performance Evaluation **61**(1), 1–16 (2005)
4. Kleinrock, L.: Queueing systems. John Wiley and sons, NY (1975)
5. Zaborovsky, V., Mulukha, V., Ilyashenko, A., Zayats, O.: Access control in a form of queueing management in multipurpose operation networks. International Journal on Advances in Networks and Services **4**(3/4), 363–374 (2011)
6. Zayats, O., Zaborovsky, V., Muliukha, V., Verbenko, A.S.: Control packet flows in telematics devices with limited buffer, absolute priority and probabilistic push-out mechanism. Part 1. Programmnaya Ingeneriya (2), 22–28 (2012). (in Russian)
7. Zayats, O., Zaborovsky, V., Muliukha, V., Verbenko, A.S.: Control packet flows in telematics devices with limited buffer, absolute priority and probabilistic push-out mechanism. Part 2. Programmnaya Ingeneriya (3), 21–29 (2012). (in Russian)
8. Gnedenko, B.V., Danielyan, E.A., Dimitrov, B.N., Klimov, G.P., Matveev, V.F.: Priority queueing systems. Moscow State University, Moscow (1973). (in Russian)

An Analytical Approach to SINR Estimation in Adjacent Rectangular Cells

Vyacheslav Begishev[1], Roman Kovalchukov[1], Andrey Samuylov[1],
Aleksandr Ometov[2]($\boxtimes$), Dmitri Moltchanov[2], Yuliya Gaidamaka[1],
and Sergey Andreev[2]

[1] Peoples' Friendship University of Russia, Ordzhonikidze Str. 3,
115419 Moscow, Russia
{vobegishev,rnkovalchukov,ygaidamaka}@sci.pfu.edu.ru,
aksamuylov@gmail.com
[2] Tampere University of Technology, Korkeakoulunkatu 10, 33720 Tampere, Finland
{aleksandr.ometov,dmitri.moltchanov,sergey.andreev}@tut.fi

Abstract. Signal-to-interference-plus-noise ratio (SINR) experienced by a mobile user is the key metric determining the throughput it receives over a certain wireless technology of interest. In this paper, we propose an analytical model for SINR estimation in rectangular-shaped cells for cases when noise can be considered negligible. This scenario is common for offices, shopping malls, dormitories, etc. We address uplink and downlink communications showing that in both cases the integral expressions can be derived. Various extensions to the proposed model are discussed. We also elaborate on the numerical results highlighting the important dependencies for the chosen scenario.

1 Introduction

The signal-to-interference-plus-noise ratio (SINR) is one of the most important metrics characterizing the quality of communication and describing the performance of a wireless system [1]. Effectively, via the Shannon's law, it determines the maximum throughput that a user receives over the wireless technology of interest. Introducing additional technology-specific factors pertaining to the type of modulation and coding used at the air interface, one could also derive the instantaneous throughput delivered to the user.

SINR characterizes the wireless channel between a transmitting and a receiving device. Particularly, SINR can be considered as a ratio between the useful energy and the interference-and-noise. Due to the users' mobility in various cases like direct [2], vehicular [3], and machine-to-machine [4] communications, SINR is often a function of the current location of the user and thus can be considered as a random variable [5]. To this end, SINR depends on the following factors [6]: the distance between the transmitter and the receiver, the set of active transmitters operating over a channel of interest [7], and the noise power. In other words,

---

The reported study was partially supported by RFBR, research projects No. 14-07-00090, 15-07-03051, 15-07-03608.

S. Balandin et al. (Eds.): NEW2AN/ruSMART 2015, LNCS 9247, pp. 446–458, 2015.
DOI: 10.1007/978-3-319-23126-6_39

SINR at the receiving device indicates the extent of how much the effective signal is superior over various detrimental effects.

In this work, we analyze the SINR in a characteristic cellular scenario under assumption of zero noise power, that is the signal-to-interference (SIR) ratio. The environment of interest consists of adjacent rectangular cells of certain dimensions, with wireless base stations or access points (APs) deployed in their geometrical centers (one AP per cell). The positions of active mobile users are uniformly distributed over the area of interest. Concentrating on two adjacent cells, and assuming that at most one entity could be transmitting at any time inside a single cell, we apply the methods from stochastic geometry to develop a mathematical framework for SIR estimation in uplink and downlink channels. Our framework allows to obtain integral expressions for SIR values and could be further extended to characterize the throughput received by mobile users or APs in the considered scenario.

The rest of the paper is organized as follows. In Section 2, we describe the system model and introduce our main assumptions. In Section 3, we develop the analytical model and derive expressions for SIR in the uplink channel. The case of the downlink channel is briefly addressed as well. Section 4 illustrates some numerical results and offers a discussion on the key properties of SIR ratio in the considered environment. Conclusions are drawn in the last section.

2   System Model

2.1   Environment of Interest

We assume that the mobile users are deployed across the area composed of adjacent rectangular cells (e.g., rooms) with the sides of certain length, as shown in Fig. 1. The APs (access points) are located in the geometrical centers of these cells. In each cell, mobile users are assumed to be uniformly distributed over its area.

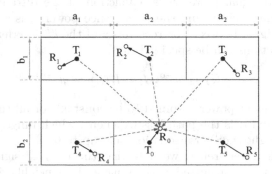

Fig. 1. System operation in scenario with rectangular cells

Even though we do not focus on any particular wireless technology, we adopt a number of assumptions on its characteristic details. First, mobile users operating in adjacent cells are assumed to utilize the common set of frequencies to communicate. This means that all transmitting entities interfere with each other. This is the case, for example, in Wi-Fi systems operating on the same channel or in micro-LTE systems that schedule the same set of resource blocks for users in adjacent cells. In the latter case, the model proposed in this paper corresponds to the worst-case interference scenario. Further, at each time instant t, we assume that at most one entity per cell is allowed to use the channel.

For the described scenario, we focus on both uplink and downlink channels. Whereas there are some differences between these options, the framework developed in this paper may generally address both of them. Due to the space limitations, below we provide full analysis for the uplink operation. In turn, the downlink channel is briefly addressed as an extension to the uplink case.

2.2 SINR Details

Formally, SINR can be expressed as

$$SINR = \frac{S}{\sum_{i=1}^{N} I_i + \sigma^2},$$ (1)

where S is the received signal power, N is the number of interfering sources, I_i is the interference power from the ith source, and σ^2 is the noise power.

Generally, we can classify the detrimental noises onto (i) the natural noises, including the Johnson–Nyquist noise, the atmospheric noise, the solar noise, etc. and (ii) the artificial noises, e.g. the industrial noises [8]. Occasionally, however, the noise may be assumed to be near-zero and we thus need to investigate SIR

$$SIR = \frac{S}{\sum_{i=1}^{N} I_i},$$ (2)

where the received signal power S is a function of the distance between the transmitter and the receiver, and the interference power I_i is a function of the distance and the signal between the receivers and the ith interfering user.

The above functions can be specified as

$$S = S(l) = gl^{-\alpha}, I_i = I_i(l_i) = gl_i^{-\alpha_i},$$ (3)

where g is the transmit power assumed to be constant for all the transmitters, l is the distance, and α is the path loss exponent, which ranges from 2, in the case of a Line-of-Sight (LoS) transmission, and up to 6 [9].

Focusing on the interference, we note that the received signal is primarily affected by the signals transmitted at the same and the neighboring frequencies. In today's cellular systems, the interfering users are, on the one hand, mobile devices being in proximity and, on the other hand, the cellular base stations that utilize the common frequency spectrum. The latter is typical for high-density urban deployments.

Here, we analyze the formulation conceptually similar to that in [10], by focusing on a pair of users in adjacent rectangular cells with the side lengths of (a_1, b_1) and (a_2, b_2), respectively (see Fig. 1). The receiver of interest, denoted as Rx_0, is located at the geometrical center of one tagged cell and the coordinates of the corresponding receiver are uniformly distributed over the area of this cell. In Fig. 1, black solid arrows represent logical downlink channels from the transmitter to its respective receiver. The adjacent cells are associated with the transmitters $T_i, i = 1, 2, \ldots, 5$ and the receivers $Rx_i, i = 1, 2, \ldots, 5$ that all use the same wireless technology and the common set of frequencies. Then, red dash lines indicate the interfering signals from the neighboring transmitters.

Further, we *tag* a pair of communicating users. For this pair, we denote $TR_0 = < Tx_0, Rx_0 >$. Similarly, for all the interfering pairs we impose $TR_i = < Tx_i, Rx_i >, i = 1, 2, \ldots, 5$. Let us then denote the distance between the transmitter Tx_0 and the receiver Rx_0 as R_0. Note that the received power is a function of distance between the interfering users, i.e. between Tx_i and Rx_0. We denote this distance as $D_i, i = 1, 2, \ldots, 5$.

Under the above assumptions, (2) can be simplified as

$$SIR = \frac{S(R_0)}{\sum_{i=1}^{5} I_i(D_i)}, \tag{4}$$

where

$$S(R_0) = gR_0^{-\alpha_0}, \qquad \sum_{i=1}^{5} I_i(D_i) = g\sum_{i=1}^{5} D_i^{-\alpha_i}. \tag{5}$$

Assuming constant transmit power, (4) reads as

$$SIR = \frac{gR_0^{-\alpha_0}}{g\sum_{i=1}^{5} D_i^{-\alpha_i}} = \frac{R_0^{-\alpha_0}}{\sum_{i=1}^{5} D_i^{-\alpha_i}}. \tag{6}$$

3   Proposed Analytical Model

In this section, due to the space limitations, we concentrate primarily on the uplink communication. The downlink channel can be analyzed similarly. In addition, to simplify the resulting expressions, we consider the square cells of the same size.

3.1   Signal-to-Interference Ratio for Uplink

In the uplink case, the interference is effective at the AP side and Fig. 2 reflects the considered scenario, where the tagged cell incorporates the receiver of interest, whereas the interfering user is located in the adjacent cell named the *interfering cell*. Let both cells be of square shape with the side length of c. The receivers denoted by Rx_0 and Rx_1 are positioned in the geometrical centers of their respective cells, while the transmitters (e.g., mobile users) are uniformly distributed in the corresponding cells. In this configuration, Rx_0 is the tagged receiver and Tx_1 is the interfering transmitter. The distance between Tx_1 and Rx_0 is denoted by D_1 and the distance between Tx_0 and Rx_0 is denoted by R_0. A Cartesian coordinate system is chosen, such that the coordinates of the tagged receiver are $(c/2, c/2)$.

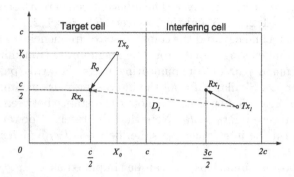

Fig. 2. Uplink SIR estimation details.

For the considered case, SIR takes the following form

$$SIR = \frac{gR_0^{-\alpha_1}}{gD_1^{-\alpha_2}} = \frac{R_0^{-\alpha_1}}{D_1^{-\alpha_2}} = \frac{D_1^{\alpha_2}}{R_0^{\alpha_1}}. \tag{7}$$

According to (7), SIR is proportional to the distance D_1 and inversely proportional to R_0. Note that these distances are random variables and they are independent from each other, as the coordinates of the transmitters Tx_0 and Tx_1 are chosen independently. There are two ways of how to obtain the SIR distribution. First, we may determine the distributions of D_1 and R_0, then obtain the distributions of their powers, and finally calculate their ratio. This is completely feasible as the distances D_1 and R_0 are independent from each other and the joint distribution of their powers is simply a product of the individual distributions of their powers. However, this technique is not applicable to the downlink case, as D_1 and R_0 are not independent any more. An alternative approach is to directly employ the coordinates of the transmitter and the interferer. We follow this path for our proposed analysis below.

Let us first introduce the following coordinates

$$\chi_1 = X_0, \; \chi_2 := Y_0, \; \chi_3 := X_1, \; \chi_4 := Y_1, \tag{8}$$

where (X_0, Y_0) and (X_1, Y_1) are the coordinates of the transmitters Tx_0 and Tx_1, respectively, whereas (χ_1, χ_2) and (χ_3, χ_4) are the particular values that they take. Since the coordinates of each transmitter are uniformly distributed within the square of side length c, the joint probability density function (pdf) of their coordinates is

$$w_{\chi_1,\chi_2}(x_1, x_2) = f_{X_0,Y_0}(x_1, x_2) = 1/c^2,$$
$$w_{\chi_3,\chi_4}(x_3, x_4) = f_{X_1,Y_1}(x_3, x_4) = 1/c^2. \tag{9}$$

Thereby, our goal is to determine the pdf of the random variables ξ, $W_\xi(y_1)$, denoting the SINR based on the joint densities $w_{\chi_1,\chi_2}(x_1, x_2)$ and $w_{\chi_3,\chi_4}(x_3, x_4)$ of random variables $\chi_1, \chi_2, \chi_3, \chi_4$. Using the Cartesian coordinates, we can write (7) as

$$\xi = \frac{D_1^{\alpha_2}}{R_0^{\alpha_1}} = \frac{\left(\sqrt{(\chi_3 - c/2)^2 + (\chi_4 - c/2)^2}\right)^{\alpha_2}}{\left(\sqrt{(\chi_1 - c/2)^2 + (\chi_2 - c/2)^2}\right)^{\alpha_1}}, \tag{10}$$

where c is the square side length.

Following [11, 12], in order to establish the pdf $W_\xi(y_1)$ we need to evaluate the densities of the numerator and the denominator of (10) and then integrate out over auxiliary variables y_2, y_3, y_4. To this end, let us determine the pdf $W_{\eta_1}(y_3)$. Recalling the form of the numerator in (10), we establish

$$y_3 := f(x_3, x_4) = \sqrt{(x_3 - c/2)^2 + (x_4 - c/2)^2}, \tag{11}$$
$$y_4 = x_4,$$

where y_4 is an auxiliary variable.

The inverse of (11) takes the following form

$$x_3 = \varphi(y_3, y_4) = \frac{c \pm \sqrt{-c^2 + 4y_3^2 + 4c\, y_4 - 4y_4^2}}{2}, \tag{12}$$

which is two-valued with the branches $x_{1,i} = \varphi_i(y_1, y_2, y_3, y_4), i = 1, 2$.

The pdf we are looking for can be determined as in [11]

$$W_{\eta_1,\eta_2}(y_3, y_4) = \sum_{i=1}^{2} w_{\chi_3,\chi_4}(\varphi_i(y_3, y_4), y_4) \left| \frac{\partial \varphi_i(y_3, y_4)}{\partial y_3} \right|. \tag{13}$$

To deliver $W_\xi(y_1) = W_{\eta_1}(y_1)$, we need to integrate out y_4 in (13) obtaining

$$W_{\eta_1}(y_3) = \sum_{i=1}^{2} \int_{Y_{2,i}} w_{\chi_3,\chi_4}(\varphi_i(y_3, y_4), y_4) \left| \frac{\partial \varphi_i(y_3, y_4)}{\partial y_3} \right| dy_4, \tag{14}$$

where $Y_{2,i}$ is the region of y_4 for the ith branch of φ_i in (12).

For the branches defined in (12), we have

$$\frac{\partial \varphi_i(y_3, y_4)}{\partial y_3} = (-1)^i \frac{2y_3}{\sqrt{-c^2 + 4y_3^2 + 4cy_4 - 4y_4^2}}, \quad i = 1, 2. \tag{15}$$

Recalling (9), the integrands in (14) take the following form

$$I_1(y_3, y_4) = w_{\chi_3, \chi_4}(\varphi_1(y_3, y_4), y_4) \left| \frac{\partial \varphi_1(y_3, y_4)}{\partial y_3} \right| = \frac{2y_3}{c^2 \sqrt{-c^2 + 4y_3^2 + 4cy_4 - 4y_4^2}},$$
$$I_2(y_3, y_4) = w_{\chi_3, \chi_4}(\varphi_1(y_3, y_4), y_4) \left| \frac{\partial \varphi_1(y_3, y_4)}{\partial y_3} \right| = \frac{2y_3}{c^2 \sqrt{-c^2 + 4y_3^2 + 4cy_4 - 4y_4^2}}. \tag{16}$$

The ranges $Y_{2,i}$, $i = 1, 2$, can be obtained by observing $x_3 \in [c, 2c]$ and $x_4 \in [0, c]$. For example, to calculate $Y_{2,1}$ we need to solve

$$\begin{cases} c \leq \frac{c + \sqrt{-c^2 + 4y_3^2 + 4c\,y_4 - 4y_4^2}}{2} \leq 2c, \\ 0 \leq y_4 \leq c, \\ y_3 \geq 0. \end{cases} \tag{17}$$

The solution of (17) is in the form $Y_{2,1} = Y_{2,1}^1 \cup Y_{2,1}^2 \cup Y_{2,1}^3$, where

$$Y_{2,1}^1 = \begin{cases} \frac{c}{2} < y_3 \leq \frac{c}{\sqrt{2}}, \\ \frac{c}{2} - \frac{1}{2}\sqrt{-c^2 + 4y_3^2} \leq y_4 \leq \frac{c}{2} + \frac{1}{2}\sqrt{-c^2 + 4y_3^2}. \end{cases}$$
$$Y_{2,1}^2 = \begin{cases} \frac{c}{\sqrt{2}} < y_3 \leq \frac{3c}{2}, \\ 0 < y_4 \leq c. \end{cases} \tag{18}$$
$$Y_{2,1}^3 = \begin{cases} \frac{3c}{2} < y_3 \leq c\sqrt{\frac{5}{2}}, \\ \left[\begin{matrix} 0 < y_4 \leq \frac{c}{2} - \frac{1}{2}\sqrt{-9c^2 + 4y_3^2} \\ \frac{c}{2} + \frac{1}{2}\sqrt{-9c^2 + 4y_3^2} < y_4 \leq c. \end{matrix} \right. \end{cases}$$

Evaluating the integral (14), we establish for $W_{\eta_1}(y_3)$

$$\begin{cases} 0, y_3 \leq \frac{c}{2} \\ -\frac{y_3}{c^2}\left(\arcsin\left[\frac{-\sqrt{-c^2 + 4y_3^2}}{2y_3} \right] - \arcsin\left[\frac{\sqrt{-c^2 + 4y_3^2}}{2y_3} \right] \right), \frac{c}{2} < y_3 \leq \frac{c}{\sqrt{2}} \\ -\frac{y_3}{c^2}\left(\arcsin\left[\frac{-c}{2y_3} \right] - \arcsin\left[\frac{c}{2y_3} \right] \right), \frac{c}{\sqrt{2}} < y_3 \leq \frac{3c}{2} \\ -\frac{y_3}{c^2}\left(\arcsin\left[\frac{\sqrt{-9c^2 + 4y_3^2}}{2y_3} \right] - \arcsin\left[\frac{c}{2y_3} \right] \right) - \\ \quad -\frac{y_3}{c^2}\left(\arcsin\left[\frac{-c}{2y_3} \right] - \arcsin\left[\frac{-\sqrt{-9c^2 + 4y_3^2}}{2y_3} \right] \right), \frac{3c}{2} < y_3 \leq c\sqrt{\frac{5}{2}} \end{cases} \tag{19}$$

Further, we estimate the pdf $W_{\eta_3}(y_5)$ of D^{α_2} arriving at

$$\begin{cases} 0, y_5 \leq \left(\frac{c}{2} \right)^{\alpha_2} \\ \frac{2}{\alpha_2 c^2} y_5^{\frac{2}{\alpha_2} - 1}\left(\arcsin\left[\frac{c}{2y_5^{1/\alpha_2}} \right] - \arcsin\left[\frac{\sqrt{-9c^2 + 4y_5^{2/\alpha_2}}}{2y_5^{1/\alpha_2}} \right] \right), \left(\frac{3c}{2} \right)^{\alpha_2} < y_5 \leq \left(c\sqrt{\frac{5}{2}} \right)^{\alpha_2} \\ \frac{2}{\alpha_2 c^2} y_5^{\frac{2}{\alpha_2} - 1} \arcsin\left[\frac{\sqrt{-c^2 + 4y_5^{2/\alpha_2}}}{2y_5^{1/\alpha_2}} \right], \left(\frac{c}{2} \right)^{\alpha_2} < y_5 \leq \left(\frac{c}{\sqrt{2}} \right)^{\alpha_2} \\ \frac{2}{\alpha_2 c^2} y_5^{\frac{2}{\alpha_2} - 1} \arcsin\left[\frac{c}{2y_5^{1/\alpha_2}} \right], \left(\frac{c}{\sqrt{2}} \right)^{\alpha_2} < y_5 \leq \left(\frac{3c}{2} \right)^{\alpha_2} \end{cases} \tag{20}$$

The density of R^{α_1}, $W_{\eta_4}(y_6)$ is obtained by following the same procedure

$$\left(\frac{2}{\alpha_1}y_6^{\frac{1}{\alpha_1}-1}\right)\begin{cases} \pi\frac{y_6^{1/\alpha_1}}{c^2}, 0<y_6 \le \left(\frac{c}{2}\right)^{\alpha_1} \\ 2\frac{y_6^{1/\alpha_1}}{c^2}\left(\arcsin\left[\frac{c}{2y_6^{1/\alpha_1}}\right] - \arcsin\left[\frac{\sqrt{-c^2+4y_6^{2/\alpha_1}}}{2y_6^{1/\alpha_1}}\right]\right), \left(\frac{c}{2}\right)^{\alpha_1} < y_6 \le \left(\frac{c}{\sqrt{2}}\right)^{\alpha_1} \end{cases}$$

$$\left(\frac{2}{\alpha_1 c^2}y_6^{\frac{2}{\alpha_1}-1}\right)\begin{cases} \pi, 0<y_6 \le \left(\frac{c}{2}\right)^{\alpha_1} \\ 2\left(\arcsin\left[\frac{c}{2y_6^{1/\alpha_1}}\right] - \arcsin\left[\frac{\sqrt{-c^2+4y_6^{2/\alpha_1}}}{2y_6^{1/\alpha_1}}\right]\right), \left(\frac{c}{2}\right)^{\alpha_1} < y_6 \le \left(\frac{c}{\sqrt{2}}\right)^{\alpha_1} \end{cases}$$

Evaluating the ratio at hand similarly, we produce

$$W(y_1) := \int_0^\infty \sum_{i=1}^3 I_i(y_1,y_2)dy_2, \tag{21}$$

where the components I_1, I_2, I_3, respectively, are given by

$$I_1(y_1,y_2) := \begin{cases} y_2\frac{4}{\alpha_2 c^2}(y_1 y_2)^{\frac{2}{\alpha_2}-1}\arcsin\left[\frac{\sqrt{-c^2+4(y_1 y_2)^{\frac{2}{\alpha_2}}}}{2(y_1 y_2)^{\frac{1}{\alpha_2}}}\right]\left(\frac{2}{\alpha_1 c^2}y_2^{\frac{2}{\alpha_1}-1}\right) \times \\ \quad \times \left(\arcsin\left[\frac{c}{2y_2^{1/\alpha_1}}\right] - \arcsin\left[\frac{\sqrt{-c^2+4y_2^{2/\alpha_1}}}{2y_2^{1/\alpha_1}}\right]\right), \\ if\left[\frac{c^{\alpha_2-\alpha_1}}{(\sqrt{2})^{2\alpha_2-\alpha_1}} \le y_1 \le \left(\frac{c}{2}\right)^{\alpha_2-\alpha_1} \cap \left(\frac{c}{2}\right)^{\alpha_2}\frac{1}{y_1} \le y_2 \le \left(\frac{c}{\sqrt{2}}\right)^{\alpha_1}\right] \cup \\ \cup\left[\left(\frac{c}{2}\right)^{\alpha_2-\alpha_1} < y_1 \le \frac{c^{\alpha_2-\alpha_1}}{(\sqrt{2})^{\alpha_2-\alpha_1}} \cap \left(\frac{c}{2}\right)^{\alpha_1} \le y_2 \le \left(\frac{c}{\sqrt{2}}\right)^{\alpha_1}\right] \cup \\ \cup\left[\left(\frac{c}{\sqrt{2}}\right)^{\alpha_2-\alpha_1} < y_1 \le \left(\frac{c}{\sqrt{2}}\right)^{\alpha_2}\left(\frac{c}{2}\right)^{-\alpha_1} \cap \left(\frac{c}{2}\right)^{\alpha_1} \le y_2 \le \left(\frac{c}{\sqrt{2}}\right)^{\alpha_2}\frac{1}{y_1}\right]; \\ \\ y_2\frac{2\pi}{\alpha_2 c^2}(y_1 y_2)^{\frac{2}{\alpha_2}-1}\arcsin\left[\frac{\sqrt{-c^2+4(y_7 y_6)^{\frac{2}{\alpha_2}}}}{2(y_1 y_2)^{\frac{1}{\alpha_2}}}\right]\left(\frac{2}{\alpha_1 c^2}y_2^{\frac{2}{\alpha_1}-1}\right), \\ if\left[\left(\frac{c}{2}\right)^{\alpha_2-\alpha_1} \le y_1 \le \left(\frac{c}{\sqrt{2}}\right)^{\alpha_2}\left(\frac{c}{2}\right)^{-\alpha_1} \cap \left(\frac{c}{2}\right)^{\alpha_2}\frac{1}{y_1} \le y_8 \le \left(\frac{c}{2}\right)^{\alpha_1}\right] \cup \\ \cup\left[y_1 > \left(\frac{c}{\sqrt{2}}\right)^{\alpha_2}\left(\frac{c}{2}\right)^{-\alpha_1} \cap \left(\frac{c}{2}\right)^{\alpha_2}\frac{1}{y_1} \le y_2 \le \left(\frac{c}{\sqrt{2}}\right)^{\alpha_2}\frac{1}{y_1}\right]; \\ 0, otherwise \end{cases}$$

$$
I_2(y_1,y_2) := \begin{cases}
y_2 \frac{4}{\alpha_2 c^2}(y_1 y_2)^{\frac{2}{\alpha_2}-1} \arcsin\left[\frac{c}{2(y_1 y_2)^{1/\alpha_2}}\right] \left(\frac{2}{\alpha_1 c^2} y_2^{\frac{2}{\alpha_1}-1}\right) \times \\
\quad \times \left(\arcsin\left[\frac{c}{2y_2^{1/\alpha_1}}\right] - \arcsin\left[\frac{\sqrt{-c^2+4y_2^{2/\alpha_1}}}{2y_2^{1/\alpha_1}}\right]\right), \\
if \left[\left(\frac{c}{\sqrt{2}}\right)^{\alpha_2-\alpha_1} \le y_1 \le \left(\frac{c}{\sqrt{2}}\right)^{\alpha_2}\left(\frac{c}{2}\right)^{-\alpha_1} \cap \left(\frac{c}{\sqrt{2}}\right)^{\alpha_2}\frac{1}{y_1} \le y_2 \le \left(\frac{c}{\sqrt{2}}\right)^{\alpha_1}\right] \cup \\
\quad \cup \left[\left(\frac{c}{\sqrt{2}}\right)^{\alpha_2}\left(\frac{c}{2}\right)^{-\alpha_1} < y_1 \le \frac{3^{\alpha_2}c^{\alpha_2-\alpha_1}}{(\sqrt{2})^{2\alpha_2-\alpha_1}} \cap \left(\frac{c}{2}\right)^{\alpha_1} \le y_2 \le \left(\frac{c}{\sqrt{2}}\right)^{\alpha_1}\right] \cup \\
\quad \cup \left[\frac{3^{\alpha_2}c^{\alpha_2-\alpha_1}}{(\sqrt{2})^{2\alpha_2-\alpha_1}} < y_1 \le \left(\frac{3c}{2}\right)^{\alpha_2}\left(\frac{c}{2}\right)^{-\alpha_1} \cap \left(\frac{c}{2}\right)^{\alpha_1} \le y_2 \le \left(\frac{3c}{2}\right)^{\alpha_2}\frac{1}{y_1}\right]; \\
y_2 \frac{2\pi}{\alpha_2 c^2}(y_1 y_2)^{\frac{2}{\alpha_2}-1} \arcsin\left[\frac{c}{2(y_1 y_2)^{1/\alpha_2}}\right]\left(\frac{2}{\alpha_1 c^2}y_2^{\frac{2}{\alpha_1}-1}\right), \\
if \left[\left(\frac{c}{\sqrt{2}}\right)^{\alpha_2}\left(\frac{c}{2}\right)^{-\alpha_1} \le y_1 \le \left(\frac{3c}{2}\right)^{\alpha_2}\left(\frac{c}{2}\right)^{-\alpha_1} \cap \left(\frac{c}{\sqrt{2}}\right)^{\alpha_2}\frac{1}{y_1} \le y_2 \le \left(\frac{c}{2}\right)^{\alpha_1}\right] \cup \\
\quad \cup \left[\left(\frac{3c}{2}\right)^{\alpha_2}\left(\frac{c}{2}\right)^{-\alpha_1} < y_1 \cap \left(\frac{c}{\sqrt{2}}\right)^{\alpha_2}\frac{1}{y_1} \le y_2 \le \left(\frac{3c}{2}\right)^{\alpha_2}\frac{1}{y_1}\right]; \\
0, otherwise
\end{cases}
$$

$$
I_3(y_1,y_2) := \begin{cases}
y_2 \frac{4}{\alpha_2 c^2}(y_7 y_8)^{\frac{2}{\alpha_2}-1}\left(\arcsin\left[\frac{c}{2(y_1 y_2)^{1/\alpha_2}}\right] - \arcsin\left[\frac{\sqrt{-9c^2+4(y_1 y_2)^{2/\alpha_2}}}{2(y_1 y_2)^{1/\alpha_2}}\right]\right) \times \\
\quad \times \left(\frac{2}{\alpha_1 c^2}y_2^{\frac{2}{\alpha_1}-1}\right)\left(\arcsin\left[\frac{c}{2y_2^{1/\alpha_1}}\right] - \arcsin\left[\frac{\sqrt{-c^2+4y_2^{2/\alpha_1}}}{2y_2^{1/\alpha_1}}\right]\right), \\
if \left[\left(\frac{3c}{2}\right)^{\alpha_2}\left(\frac{c}{\sqrt{2}}\right)^{-\alpha_1} \le y_1 \le c^{\alpha_2-\alpha_1}\sqrt{\frac{5^{\alpha_2}}{2^{\alpha_2}-\alpha_1}} \cap \left(\frac{3c}{2}\right)^{\alpha_2}\frac{1}{y_1} \le y_2 \le \left(\frac{c}{\sqrt{2}}\right)^{\alpha_1}\right] \cup \\
\quad \cup \left[c^{\alpha_2-\alpha_1}\sqrt{\frac{5^{\alpha_2}}{2^{\alpha_2}-\alpha_1}} < y_1 \le \left(\frac{3c}{2}\right)^{\alpha_2}\left(\frac{c}{2}\right)^{-\alpha_1} \cap \left(\frac{3c}{2}\right)^{\alpha_2}\frac{1}{y_1} \le y_2 \le c^{\alpha_2}\left(\sqrt{\frac{5}{2}}\right)^{\alpha_2}\frac{1}{y_1}\right] \cup \\
\quad \cup \left[\left(\frac{3c}{2}\right)^{\alpha_2}\left(\frac{c}{2}\right)^{-\alpha_1} < y_1 \le c^{\alpha_2}\left(\sqrt{\frac{5}{2}}\right)^{\alpha_2}\left(\frac{c}{2}\right)^{-\alpha_1} \cap \left(\frac{c}{2}\right)^{\alpha_1} \le y_2 \le c^{\alpha_2}\left(\sqrt{\frac{5}{2}}\right)^{\alpha_2}\frac{1}{y_1}\right]; \\
y_2 \frac{2\pi}{\alpha_2 c^2}(y_1 y_2)^{\frac{2}{\alpha_2}-1}\left(\arcsin\left[\frac{c}{2(y_1 y_2)^{1/\alpha_2}}\right] - \arcsin\left[\frac{\sqrt{-9c^2+4(y_1 y_2)^{2/\alpha_2}}}{2(y_1 y_2)^{1/\alpha_2}}\right]\right) \times \\
\quad \times \left(\frac{2}{\alpha_1 c^2}y_2^{\frac{2}{\alpha_1}-1}\right), \\
if \left[\left(\frac{3c}{2}\right)^{\alpha_2}\left(\frac{c}{2}\right)^{-\alpha_1} \le y_1 \le c^{\alpha_2}\left(\sqrt{\frac{5}{2}}\right)^{\alpha_2}\left(\frac{c}{2}\right)^{-\alpha_1} \cap \left(\frac{3c}{2}\right)^{\alpha_2}\frac{1}{y_1} \le y_2 \le \left(\frac{c}{2}\right)^{\alpha_1}\right] \cup \\
\quad \cup \left[c^{\alpha_2}\left(\sqrt{\frac{5}{2}}\right)^{\alpha_2}\left(\frac{c}{2}\right)^{-\alpha_1} < y_1 \cap \left(\frac{3c}{2}\right)^{\alpha_2}\frac{1}{y_1} \le y_2 \le c^{\alpha_2}\left(\sqrt{\frac{5}{2}}\right)^{\alpha_2}\frac{1}{y_1}\right]; \\
0, otherwise.
\end{cases}
$$

3.2   Discussion and Extensions

While performing our above analysis, we adopted a number of assumptions, including the one on the square cells. Extending these results for the case of rectangular cells is rather straightforward. The most restrictive component is the use of the same propagation exponent α on both paths, D_1 and R_0. For the uplink scenario, the extension is simple as we can obtain the distributions of the numerator and the denominators individually. Instead of working with the coordinates, we may as well obtain the distributions of the distances D_1 and R_0, and then proceed with deriving the distribution of the ratio of their powers.

The downlink case is illustrated in Fig. 3 and is more involved compared to the uplink. The reason is that the distances R_0 and D_1 are no longer independent, as the coordinates of the receiver Rx_0 determine both of them. Hence,

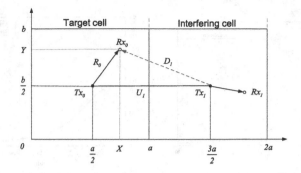

Fig. 3. An illustration of the model for downlink SINR analysis.

the approach based on the distributions of the distances will not lead to the solution, as we will not be able to determine the joint density of the powers of these distances. However, working with the coordinates similarly to what we did in the case of the uplink scenario is still feasible.

4   Numerical Results

We first verify the proposed analytical framework with our own simulator implemented in R [13]. Then, we demonstrate the impact of different input parameters by highlighting various dependencies. We provide the results for both uplink and downlink cases.

4.1   Verification of the Model

The developed simulator allows to vary all the parameters of interest used in the system model (see Fig. 2 and Fig. 3), including the dimensions of the cells a and b, the path loss exponent α between the transmitter and the receiver, the number of experiments, and the sampling frequency. At the output, our tool provides the empirical density functions, the sample means, the sample standard deviations, and the estimates for the quantiles of random variables R_0, D_1, SIR, and $10 \lg SIR$.

Along these lines, Fig. 4 compares the analytical results against those obtained with simulations for the uplink and downlink scenarios. For both cases, the dimensions of the cells have been set as $a = b = c$, where $c = 1$, while the same path loss exponent $\alpha = 2$ is utilized on both paths. The number of samples used to construct the empirical density has been $10e6$. The simulation data shows a perfect match with the analytical results. In both cases, the Kolmogorov-Smirnov goodness-of-fit statistical test [14] has been performed with the level of significance set to 0.05. It shows that the sample data belongs to the analytical distribution, which allows us to rely on the analytical model in the rest of this section.

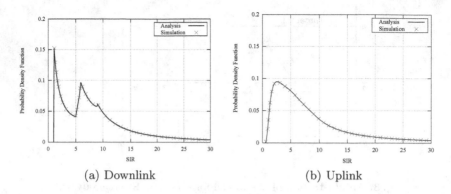

(a) Downlink   (b) Uplink

Fig. 4. Validation of the proposed model, $\alpha = 2$, $a = b = 1$.

4.2 Performance Evaluation

The considered scenario has a number of interesting properties that could be demonstrated by employing the proposed model. In particular, the results are not very sensitive to the choice of some input parameters. Therefore, Fig. 5(a) illustrates the effect of the different values of α for the downlink scenario with the cell dimensions set as $a = b = 1$. Note that the same α is again used for both directions: the interferer to the receiver and the transmitter to the receiver. As one may observe, the value of α provides the scaling effect on the resulting SIR density. Further, Fig. 5(b) highlights the insensitivity of the model to the dimensions of interest. The results are demonstrated for two dimensions, $a = b = c$, where $c \in [1, 2, 3]$, while the path loss propagation exponent α is set to 2. It is important to note that these properties hold for the downlink scenario as well, across different a and b, as well as various path loss exponents (describing the path from the transmitter to the receiver and from the interferer to the receiver). However, we do not consider these effects in detail here due to the space limitations.

So far, we provided our results for the same path loss exponent at the propagation paths between the transmitter and the receiver, as well as the interferer and the receiver. These assumptions are only likely to hold in the open-space environments. Whenever there are walls separating the cells, the propagation path between the interferer and the received is characterized by a more severe attenuation. Accordingly, Fig. 6 highlights the effect of different values of α in the uplink scenario with α_1 corresponding to the propagation path between the transmitter and the receiver, and α_2 characterizing the propagation path between the interferer and the receiver. As we observe, this effect is fairly straightforward and self-explanatory: the larger the α_2 is, the better the interference picture at the receiver becomes. Recall that the value of α_2 is primarily determined by the material of walls. Our proposed model thus allows to compute the SIR for different wall materials.

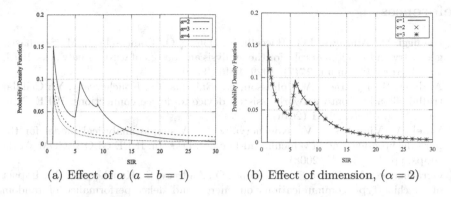

(a) Effect of α $(a = b = 1)$ (b) Effect of dimension, $(\alpha = 2)$

Fig. 5. Insensitivity of the model to the input parameters for downlink scenario

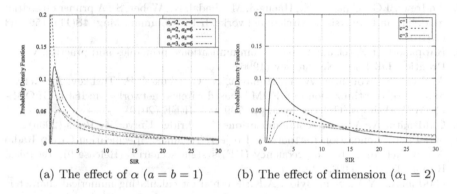

(a) The effect of α $(a = b = 1)$ (b) The effect of dimension $(\alpha_1 = 2)$

Fig. 6. Insensitivity of the model to the input parameters for uplink scenario

5 Conclusions and Future Work

In this paper, we analyzed the SINR behavior in adjacent rectangular-shaped cells. We demonstrated that for both uplink and downlink channels we can esti-mate the SINR distribution analytically without the need for time-consuming simulations. Our model allows for a number of useful extensions, including the one taking into account more than one adjacent room. Our numerical inves-tigation reveals that for the square configurations of cells the SINR value is insensitive to the side length of the square for both uplink and downlink scenar-ios. Further, the impact of the wall material affecting the propagation exponent could be significant. Finally, in our setup the SINR is positive at all times as the transmitting entity is always closer to the receiver compared to the interfering entity. However, this may not be the case if more than one adjacent cell (room) is considered. Developing an analytical model for this case is the subject of our ongoing work.

References

1. Haenggi, M., Andrews, J.G., Baccelli, F., Dousse, O., Franceschetti, M.: Stochastic geometry and random graphs for the analysis and design of wireless networks. IEEE J. Sel. Areas Commun. **27**(7), 1029–1046 (2009)
2. Andreev, S., Pyattaev, A., Johnsson, K., Galinina, O., Koucheryavy, Y.: Cellular traffic offloading onto network-assisted device-to-device connections. IEEE Commun. Mag. **52**(4), 20–31 (2014)
3. Vinel, A., Vishnevsky, V., Koucheryavy, Y.: A simple analytical model for the periodic broadcasting in vehicular ad-hoc networks. In: IEEE GLOBECOM Workshops, pp. 1–5. IEEE (2008)
4. Gerasimenko, M., Petrov, V., Galinina, O., Andreev, S., Koucheryavy, Y.: Impact of machine-type communications on energy and delay performance of random access channel in LTE-advanced. Transactions on Emerging Telecommunications Technologies **24**(4), 366–377 (2013)
5. Andrews, J.G., Ganti, R.K., Haenggi, M., Jindal, N., Weber, S.: A primer on spatial modeling and analysis in wireless networks. IEEE Commun. Mag. **48**(11), 156–163 (2010)
6. Rappaport, T.S., et al.: Wireless communications: principles and practice, vol. 2. Prentice Hall PTR, New Jersey (1996)
7. Andreev, S., Koucheryavy, Y., Himayat, N., Gonchukov, P., Turlikov, A.: Active-mode power optimization in OFDMA-based wireless networks. In: IEEE GLOBECOM Workshops (GC Wkshps), pp. 799–803. IEEE (2010)
8. Goldsmith, A.: Wireless communications. Cambridge University Press (2005)
9. 3GPP, 3rd Generation Partnership Project; Technical Specification Group Radio Access Networks; Radio Frequency (RF) system scenarios (Release 9). Technical Report 25.942 v9.0.0, 12–2009
10. Gaidamaka, Y.V., Samouylov, A.K.: Method for calculating numerical characteristics of two devices interference for device-to-device communications in a wireless heterogeneous network. Computer Science and its Applications **1**(9), 10–15 (2015)
11. Ross, S.M.: Introduction to probability models. Academic Press (2014)
12. Levin, B.: Theoretical principles of statistical radiophysics. Sovetskoe Radio, Moscow (1969)
13. Team, R.C.: R Language Definition (2000)
14. Massey Jr, F.J.: The Kolmogorov-Smirnov test for goodness of fit. Journal of the American statistical Association **46**(253), 68–78 (1951)

Improving BER Performance of Uplink LTE by Using Turbo Equalizer

Aleksandr Gelgor[1(✉)], Anton Gorlov[1], Pavel Ivanov[2], Evgenii Popov[1], Andrey Arkhipkin[2], and Tatiana Gelgor[1]

[1] Peter the Great St. Petersburg Polytechnic University, St. Petersburg, Russia
a_gelgor@mail.ru, {anton.gorlov,tanya.gelgor}@yandex.ru,
eugapop@gmail.com
[2] ZAO SBT, Moscow, Russia
{ivanov,ava}@sbtcom.ru

Abstract. The potential of turbo-equalization technique applied to uplink (UL) LTE signals detection is analyzed in this paper. The turbo equalizer, which is also called iterative receiver, represents a popular approach for detection of signals passed through a fading channel. The receiver performs equalization and decoding of error-correcting code in a loop. For implementation of the iterative receiver we performed two frequency-domain equalizers: the approximate MMSE SISO-equalizer and the soft interference canceller (SIC) SISO-equalizer. During the simulation, we analyzed several configurations of UL LTE with QPSK, 16-QAM, 64-QAM signal constellations and allocation of 25 and 100 resource blocks. All considered modes used rate 2/3 parallel concatenated convolutional code and single input single output antennas pattern. Bit error rate (BER) performance was estimated during the simulation with the extended vehicular A (EVA) model of multipath fading channel.

Keywords: Uplink LTE · Turbo equalizer · SISO-equalizer · Approximate MMSE · Soft interference canceller

1 Introduction

European specifications of 4th generation systems for mobile data transmission LTE define the physical layer which uses OFDM [1] in downlink channel and SC-FDMA [2] in uplink channel; a cyclic prefix is used in both of these cases. The cyclic-prefixed OFDM is already known as effective technique to be used in multipath fading channels. At first, it performs a channel frequency response estimation and equalization with lower complexity with respect to conventional time-domain methods; at second, resource allocation between multiple users becomes easy and users individual channel responses can be taken into account. The only significant disadvantage of OFDM is a high value of peak to average power ratio (PAPR) of transmitted signal, and this can be crucial for user equipment powered by a battery. Due to this disadvantage LTE physical layer defines the use of SC-FDMA for the data

© Springer International Publishing Switzerland 2015
S. Balandin et al. (Eds.): NEW2AN/ruSMART 2015, LNCS 9247, pp. 459–472, 2015.
DOI: 10.1007/978-3-319-23126-6_40

transmission in the uplink channel; the SC-FDMA scheme actually generates single-carrier signals with lower value of PAPR. Research and development of new detection algorithms including equalizers, which can deal with SC-FDMA signals, formulate a relevant problem.

A conventional receiver for digital data transmission system consist of separate blocks: equalizer, demodulator, deinterleaver, channel decoder (of error-correcting code). The demodulator block inside pioneer digital receivers provided hard decisions to be processed by the decoder, but increase of computational capabilities of mobile devices led to the soft decisions processing, and the decoding became more accurate. In particular, SISO algorithms were used for the iterative decoding of turbo codes. The next stage of receiver evolution is extension of the iterative decoder concept to the entire receiver chain – implementation of so-called turbo equalizer [3]. The turbo equalizer offers the iterative execution of all receiver blocks and each next iteration uses the result of the previous one. We should note that some researchers use the term "iterative receiver" instead of the "turbo equalizer".

The objective of this paper is research and comparison of various turbo equalizer algorithms applied to detection of SC-FDMA signals generated in accordance with LTE specifications.

2   From Turbo Code to Turbo Equalizer

Shannon's works defined the channel capacity – the upper bound for data transmission rate [4]. It was shown that the data transmission rate could reach the channel capacity by using of error correcting codes, but the codes were not specified. For a long time all decoders used the maximum likelihood sequence estimation (MLSE) criterion. For instance, the Viterbi algorithm [5] was widely used for decoding of convolutional codes.

In 1974 Bahl et al. firstly proposed the algorithm based on the maximum a-posteriori symbol probability criterion for the decoding of convolutional code [6]. The algorithm is usually called BCJR in view of list of its originators, or just MAP (maximum a-posteriori probability) symbol estimation. Studies of the BCJR algorithm showed that in terms of BER performance it is almost equal to the Viterbi algorithm, but the former is more computationally complex and sensitive to the noise variance estimation. Due to these features the BCJR algorithm has not been used for a long time.

A new rise of interest to the MAP symbol decoding occurred in the beginning of 90th years of XX century, when Berrou et al. proposed turbo codes and the iterative decoding algorithm [7]. First turbo codes in [7] were constructed by parallel concatenation of two systematic convolutional codes, and data bits were randomly interleaved before the one of two component encoding procedures. These turbo codes are also called parallel concatenated convolutional codes (PCCCs) in accordance with their construction principles. The BCJR algorithm was used for decoding of each component code providing a-posteriori probabilities or log-likelihood ratios (LLRs) for transmitted data bits. The key principle of the iterative decoder is alternate decoding of two component codes, when each consequent decoding procedure uses the result of previous one as a-priori information. After some number of iterations, the iterative scheme is stopped and hard decisions about information bits are obtained.

A significant contribution to development of turbo codes was made by the research group headed by Benedetto. In particular, they proposed serial concatenation of convolutional codes with transitional bit interleaver [8, 9]. These serial concatenated convolutional codes (SCCCs) exceeded PCCCs but only in the range of very low (less than 10^{-5}) bit error probabilities.

In 1995, the group of researchers headed by Douillard proposed the iterative receiver performing equalization and decoding procedures closed in a loop [3]. If the chain including a convolutional encoder, interleaver and modulator is used for the signal generation, then the output signal passed through a fading channel can be interpreted as signal with SCCC where inner code is formed by the fading channel. However, it should be noticed that the fading channel cannot provide energy gain since it does not introduce any redundancy. Hence, an iterative receiver performing cyclic equalization-decoding procedures at best can compensate the ISI and provide BER performance corresponding to the convolutional code gain in AWGN channel. The iterative receiver with both blocks of equalizer and decoder based on the soft output Viterbi algorithm (SOVA, [10]) was considered in [3]. For the convolutional code with the constraint length 5 and for the bit error probability 10^{-5} this SOVA turbo equalizer provided a 3–5 dB energy gain over the conventional receiver with single-executed blocks.

Further development of iterative receivers is closely associated with the adoption of conventional equalization algorithms in order to be used in turbo equalizer. The adopted block must be able to perform the equalization taking bits a-priori information into account. Due to relative simplicity of linear equalizers, they have become more relevant for this application. The linear equalizers can be constructed with the minimum maximal error criterion (zero-forcing, ZF-equalizers) or in view of the minimum mean squared error criterion (MMSE-equalizers). Decision feedback equalizers (DFE) are also considered for application in iterative receivers. The adopted linear equalizers are usually applied for implementation of the simplified turbo equalizer. Tuchler's works [11, 12] and also [13–16] investigate these approaches in detail. It is shown that simplified iterative receivers still exceed conventional schemes in energy efficiency, although, the simplification leads to a reduction of its energy gain.

The linear equalization algorithms including DFE-equalizers were examined in [17–19] for application to SC-FDMA signals. The key conclusion of these works claims that the DFE equalizer with frequency-domain feedforward and time-domain feedback filters provides better BER performance than frequency-domain linear equalizers. However, a complexity of DFE equalizers remains higher than for linear algorithms. We should note that considered papers analyze receiver schemes with independent single-executed blocks of equalizer and decoder. Despite the simplicity of conventional receivers, improving of LTE uplink energy efficiency by using more complex turbo equalizer formulates a relevant subject for research.

A SC-FDMA signal generated by user equipment for high-rate data transmission is affected by deep ISI during the propagation through urban wireless channels. Because of the deep ISI, implementation of the optimal iterative receiver employing a time-domain equalizer becomes practically impossible. Hence, for application of the simplified iterative receiver to UL LTE signals we chose two adopted frequency-domain equalizers. The first algorithm is described in part IV.D of [20], it represents the approximate frequency-domain MMSE equalizer. The same MMSE SISO-equalizer for BPSK signals was considered in [21]. The second chosen algorithm implements the soft interference cancellation (SIC) approach, which is described by Jar in [22].

3   Simulation Model, SISO-Equalizers

The simulation model described below was made in accordance with technical specifications [23, 24] and was used for efficiency estimation and comparison of receiver algorithms in several UL LTE modes.

In LTE system, a signal transmission from a user equipment to a base station is performed in subframes carrying some integer number of encoded blocks. Our simulation model also performs signal generation, channel simulation and detection in subframe basis. Each subframe carries two slots, each slot consists of seven SC-FDMA symbols in accordance with normal cyclic prefix mode. In addition, the model is developed under number of assumptions: the number of subbands $N_{sb} = 1$; frequency hopping is disabled; the system allocates all UL resource blocks for the sole user; physical uplink control channel (PUCCH) and control information on physical uplink shared channel (PUSCH) are not simulated and are assumed non-existent during the mapping user data.

Basic procedures of encoding, interleaving and bits allocation are performed in accordance with LTE specifications. The rate of the code was chosen equal to 2/3, also we chose the coded block length equal to the maximal permitted coded block length which fits to one subframe (table 1).

Table 1. Simulation model parameters

Signal constellation	Resource blocks allocated	Information block length	Coded block length	Number of blocks in subframe	Number and percentage of dummy bits in subframe
QPSK	25	4736	7110	1	90 / 1,25 %
16-QAM	25	6144	9222	1	5178 / 35,96 %
64-QAM	25	6144	9222	2	3156 / 14,61 %
QPSK	100	6144	9222	3	1134 / 3,94 %
16-QAM	100	6144	9222	6	2268 / 3,94 %
64-QAM	100	6144	9222	9	3402 / 3,94 %

The signal generation and iterative receiver schemes are shown in fig. 1. In the detection procedure log-likelihood ratios (LLRs) L_{ext} for coded bits given by the SISO-equalizer go to the inverse rate matching and deinterleaving block and after to the iterative decoder of the turbo code. Updated coded bits LLRs L' from the decoder output go to the interleaving and rate matching block, and after are used by the SISO-equalizer as a-priori information. After some number of iterations the receiver is stopped, and hard decisions about information bits are given by the decoder.

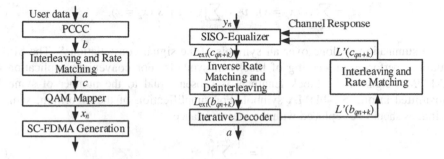

Fig. 1. Signal generation and iterative receiver for UL LTE

We should note that turbo codes provide good energy characteristics leaving 1–3 dB up to the Shannon limit [1]. However, some of corrupted blocks may require high number of decoder iterations (more than 10) to correct all bit errors. For decoding of these blocks, the turbo equalizer can demonstrate a high efficiency. For instance, two (outer) iterations of the receiver each comprising three (inner) iterations of the decoder can be sufficient for correction of all bit errors.

The SISO-equalizer block built in the iterative receiver represents a subject of interest, it primarily performs the equalization and the calculation of coded bits LLRs; a-priori information about coded bits is involved in each stage of the equalization. In this paper, two algorithms for the SISO-equalizer are analyzed. The first algorithm uses the approximate MMSE criterion, and the second implements the soft interference cancellation approach. Both of the algorithms perform the signal processing in frequency domain.

3.1 Approximate MMSE SISO-Equalizer

The first stage of SISO-equalization is the processing of a-priori information which is represented in form of coded bits LLRs L'. Based on these values, a-priori symbol probabilities s' are computed as follows:

$$s'(x_n = x) = \prod_{k=0}^{q-1} \frac{\exp\left(-m_k L'(c_{qn+k})\right)}{1+\exp\left(-L'(c_{qn+k})\right)}, \quad x = \text{map}(m_0,...,m_{q-1}), \quad (1)$$

where x is an ideal symbol from a signal constellation of size 2^q and q bits are mapped to one symbol. We should note that both considered algorithms operate with normalized signal constellations, such that the mean symbol energy is equal to one. Symbol expectation values μ_n and variances υ_n are computed from symbol probabilities as follows:

$$\mu_n = \sum_{x \in S} xs'(x_n = x), \quad \upsilon_n = \sum_{x \in S} |x - \mu_n|^2 s'(x_n = x), \tag{2}$$

where summation is done over all symbols of the signal constellation S. The SISO-equalizer performs processing of blocks of symbols. For convenient application to SC-FDMA signals, the block length N is chosen equal to the number of symbols transmitted in one SC-FDMA symbol. For simplification of the algorithm, symbol variances should be replaced with the mean variance:

$$\bar{\upsilon} = \frac{1}{N} \sum_{n=0}^{N-1} \upsilon_n. \tag{3}$$

At the end of the a-priori information processing stage, symbol expectations values are transformed to frequency domain:

$$\mu_{F,n} = \frac{1}{\sqrt{N}} \sum_{k=0}^{N-1} \mu_k \exp\left(-j\frac{2\pi kn}{N}\right). \tag{4}$$

For implementation of the approximate MMSE criterion, inner parameters of the SISO-equalizer are computed as

$$s = \frac{1}{N} \sum_{n=0}^{N-1} \frac{|h_{F,n}|^2}{\sigma^2 + \bar{\upsilon}|h_{F,n}|^2}, \quad \kappa = \frac{1}{1 + (1 - \bar{\upsilon})s}, \tag{5}$$

where σ^2 is the variance of complex-valued additive white Gaussian noise (AWGN), $h_{F,n}$ are samples of the channel frequency response. The received block of complex symbols y_n, which are SC-FDMA symbol samples, also must be transformed to frequency domain:

$$y_{F,n} = \frac{1}{\sqrt{N}} \sum_{k=0}^{N-1} y_k \exp\left(-j\frac{2\pi kn}{N}\right), \tag{6}$$

where the $1/\sqrt{N}$ factor retains an average signal power after the transformation. It is required in order to keep the noise variance σ^2 during the transformation.

Estimated symbol values in frequency domain are computed with the use of channel frequency response samples $h_{F,n}$ as follows:

$$\hat{x}_{F,n} = \kappa\left(\frac{\left(h_{F,n}^* y_{F,n} - |h_{F,n}|^2 \mu_{F,n}\right)}{\sigma^2 + \bar{\upsilon}|h_{F,n}|^2} + s\mu_{F,n}\right), \quad n = 0, 1, \ldots N{-}1, \tag{7}$$

and after this step time-domain symbol estimations are computed as:

$$\hat{x}_n = \frac{1}{\sqrt{N}} \sum_{k=0}^{N-1} \hat{x}_{F,k} \exp\left(j\frac{2\pi kn}{N} \right). \tag{8}$$

The final step of the SISO-equalizer is the calculation of coded bits log-likelihood ratios:

$$L_{ext}(c_{qn+k}) = \ln \frac{\displaystyle\sum_{\mathbf{m}:m_k=0} \exp\left(-|\hat{x}_n - \mathrm{map}(\mathbf{m})|^2 / (1-\kappa s)\right) \prod_{j\neq k} \exp\left(-m_j L'(qn+j)\right)}{\displaystyle\sum_{\mathbf{m}:m_k=1} \exp\left(-|\hat{x}_n - \mathrm{map}(\mathbf{m})|^2 / (1-\kappa s)\right) \prod_{j\neq k} \exp\left(-m_j L'(qn+j)\right)}, \tag{9}$$

where $n = 0, \ldots, N-1$, $k = 0, \ldots, q-1$, the numerator summation is done over all symbols corresponding to "0" bit transmission on the kth position, and the denominator summation covers all other signal constellation symbols corresponding to transmission of "1" bit. All terms in sums are multiplied by exponents taking a-priori LLRs as arguments except the one LLR corresponding to the kth position, for which a new LLR L_{ext} is being calculated. Thus the a-priori L' value corresponding to the kth position is eliminated from the calculation, this is done in order to avoid the "errors accumulation" effect and to compute so called extrinsic LLR value [20].

On the first iteration of the SISO-equalizer there is no any a-priori information about coded bits, all symbol expectations μ_n are equal to zero, and variances v_n are equal to one. The SISO-equalizer performs the computation of frequency-domain symbol estimations like a conventional MMSE equalizer for OFDM. From the second iteration a-priori LLRs become available, and symbol expectations μ_n are no longer equal to zero. In case of convergence of the equalizer, these symbol expectations are concentrated near complex values corresponding to signal constellation points, and every subsequent iteration makes the variety of μ_n visually similar to the signal constellation (fig. 2).

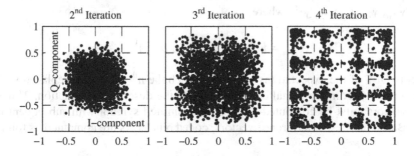

Fig. 2. Symbol expectation values μ_n at some iterations of SISO-equalizer

3.2 Soft Interference Canceller SISO-Equalizer

The first stage of the SIC SISO-equalization algorithm is also the a-priori information processing, which calculates the symbol expectations vector $\overline{X} = (\mu_0, \mu_1, ..., \mu_{N-1})^{\mathrm{T}}$ and the mean symbol variance $\overline{v}$.

Before the equalization of block of N symbols, the diagonal matrix must be computed:

$$\tilde{\Sigma} = \overline{v}\tilde{\mathbf{H}}\tilde{\mathbf{H}}^H + \sigma^2\mathbf{I}, \tag{10}$$

where $\tilde{\mathbf{H}}$ contains frequency channel response samples on the main diagonal:

$$\tilde{\mathbf{H}} = \begin{pmatrix} \tilde{h}_{f,0} & 0 & 0 \\ 0 & \ddots & 0 \\ 0 & 0 & \tilde{h}_{f,N-1} \end{pmatrix}, \tag{11}$$

$\mathbf{I}$ is identity matrix, σ^2 is also the variance of AWGN, and samples of the channel frequency response are indicated as $\tilde{h}_{f,n}$ to retain author's notation used in [22]. The inverse covariance matrix is computed from the inverse matrix $\tilde{\Sigma}^{-1}$ in accordance with the Sherman-Morrison formula [25]:

$$\tilde{\Sigma}[n]^{-1} = \tilde{\Sigma}^{-1} + \overline{v}\frac{\tilde{\Sigma}^{-1}\tilde{\mathbf{h}}_{F_1}[n]\left(\tilde{\mathbf{h}}_{F_1}[n]\right)^H \tilde{\Sigma}^{-1}}{1 - \overline{v}\left(\tilde{\mathbf{h}}_{F_1}[n]\right)^H \tilde{\Sigma}^{-1}\tilde{\mathbf{h}}_{F_1}[n]}, \quad n = 0, 1, ..., N-1, \tag{12}$$

where $\tilde{\mathbf{h}}_{F_1}[n]$ is the nth column of the matrix product $\tilde{\mathbf{H}}\mathbf{F}_1$, and $\mathbf{F}_1$ is discrete Fourier transform matrix of size N. The interference component is computed by subtraction of the product $\tilde{\mathbf{H}}\overline{X}'_{x,n}$ from the frequency-domain vector of received symbols $Y' = (y_{f,0}, y_{f,1}, ..., y_{f,N-1})^{\mathrm{T}}$, where

$$\overline{X}'_{x,n} = \mathbf{F}_1\left(\mu_0, \mu_1, ..., x, ..., \mu_{N-1}\right)^{\mathrm{T}}, \tag{13}$$

which is DFT of the expectations vector, in which the nth position is replaced with the ideal symbol x from the signal constellation. By the described subtraction, 2^q interference vectors must be computed for every transmitted symbol with subsequent number n. Based on these interference vectors, coded bits LLRs are computed as follows:

$$L_{\text{ext}}(qn + k) =$$

$$= \ln \left(\frac{\displaystyle\sum_{\mathbf{m}:\, m_k=0} \exp\left(-\frac{1}{2}\left(Y' - \tilde{\mathbf{H}}\bar{X}'_{\text{map}(\mathbf{m}),\, n}\right)^H \tilde{\Sigma}[n]^{-1}\left(Y' - \tilde{\mathbf{H}}\bar{X}'_{\text{map}(\mathbf{m}),\, n}\right) \right) L_0(\mathbf{m}, qn)}{\displaystyle\sum_{\mathbf{m}:\, m_k=1} \exp\left(-\frac{1}{2}\left(Y' - \tilde{\mathbf{H}}\bar{X}'_{\text{map}(\mathbf{m}),\, n}\right)^H \tilde{\Sigma}[n]^{-1}\left(Y' - \tilde{\mathbf{H}}\bar{X}'_{\text{map}(\mathbf{m}),\, n}\right) \right) L_0(\mathbf{m}, qn)} \right), \quad (14)$$

$$L_0(\mathbf{m}, qn) = \prod_{j \neq k} \exp\left(-m_j L'(qn + j)\right), \quad n = 0, 1, \ldots, N-1, k = 0, \ldots, q-1.$$

In this expression the numerator summation is done over all signal constellation symbols map($\mathbf{m}$) corresponding to the transmission of "0" bit on the kth position, and the denominator summation contains terms corresponding to "1" bit on the kth position. As we can see the idea of SIC algorithm is based on the assumption that interference vector components have N-dimensional Gaussian distribution with the covariance matrix equal to $\tilde{\Sigma}[n]$.

4   Numerical Results

This section describes numerical results obtained by the simulation. Also, the comparison of conventional receiver and turbo equalizer is shown below. The turbo equalizer was built on the MMSE or SIC SISO-equalization algorithm and the iterative decoder of turbo code, which were closed in a loop. The conventional receiver performed the chain of single executed blocks, which were the MMSE frequency-domain equalizer, deinterleaver and the iterative decoder of turbo code.

The EVA model was used for channel simulation, we performed BER estimation for the static channel (EVA0) and for the maximal Doppler shift equal to 70 Hz (EVA70). The model is described by 9 paths delayed for {0, 30; 150; 310; 370; 710; 1090; 1730; 2510} ns with average gains {0.0; −1.5; −1.4; −3.6; −0.6; −9.1; −7.0; 12.0; −16.9} dB respectively. The multipath fading channel was randomly initialized before each subframe transmission.

During the simulation, we obtained a lot of results in form of BER curves, which show the bit error probability depending on the normalized energy consumptions E_b/N_0. It was shown that for the conventional receiver and for single-path and multipath channels 6 iterations of the turbo code decoder is a good compromise between detection accuracy and computational complexity. Because of this observation, all results shown below for the conventional receiver were obtained for 6 iterations of the decoder. In case of the turbo equalizer, there were simulated 2 outer iterations of the receiver each comprising 3 inner iterations of the decoder. In this simulation technique, the numbers of decoder iterations are equal in the conventional receiver and the turbo equalizer cases. Despite this fact, the iterative receiver remains more complex since a-priori information processing and equalization are performed on each outer iteration. Moreover, we should note that 6 iterations of the turbo code decoder in the turbo equalizer are equal to 5 iterations in the conventional receiver in sense of detection accuracy. That may be true since the first iteration of the iterative decoder actually performs initialization of the extrinsic information exchange process, and the first iteration results are not usually used for making hard decisions about information bits.

We compare the conventional receiver and the MMSE turbo equalizer (the iterative receiver based on the MMSE SISO-equalizer) in fig. 3. The iterative receiver is indicated as TE-MMSE. The conventional receiver is indicated as MMSE. These results were obtained in the 25 resource blocks mode with an ideal channel response estimation.

As we can see from the results, a bit error probability decreases from switching EVA0 to EVA70, and using the signal constellation of higher order leads to larger difference in BER. The possible explanation for this observation is that in the case of EVA0 occurred "bad" realization of the channel response corrupts all SC-FDMA symbols in the subframe, while in the EVA70 case the "bad" realization is likely to be improved during the subframe transmission. The energy gain of the turbo equalizer depends on the signal constellation order and is about 0.6 dB for 16-QAM at 10^{-3} BER. This small gain of the turbo equalizer was also confirmed in the 100 resource blocks mode.

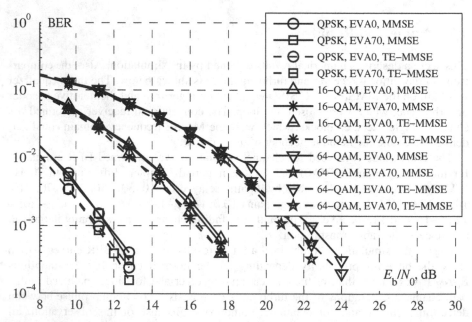

Fig. 3. BER performance of the conventional receiver and the MMSE turbo equalizer in the 25 resource blocks mode, the results were obtained with an ideal channel response estimation

The comparison of turbo equalizer and conventional receiver in case of non-ideal channel response estimation is shown in fig. 4. These results were obtained for 16-QAM signal constellation and allocation of 25 and 100 resource blocks (RB25, RB100). During the simulation, an ideal channel response estimation was corrupted by AWGN; the standard deviation σ_h of the estimation error was chosen in proportion to the standard deviation σ of channel noise.

As we can see from fig. 4, the turbo equalizer also outperforms the conventional receiver, however its energy gain remains rather small. Both of these receivers reduce their energy consumptions by 2 dB at 10^{-3} BER with increase of the number of allocated resource blocks from 25 to 100. The explanation is that in the 100 RB mode each subframe carries several encoded blocks, hence, the bit interleaving inside the subframe leads to better decoding results.

The SIC turbo equalizer (TE-SIC) is compared with the conventional receiver in fig. 5. Because of high complexity of the SIC SISO-equalizer these results were obtained only for the 25 resource blocks allocation and for QPSK, 16-QAM signal constellations. BER curves shown in fig. 5 are actual for the ideal channel response estimation.

As we can see, the SIC turbo equalizer does not demonstrate any energy gain, and in case of the EVA0 channel model it even increases a number of bit errors.

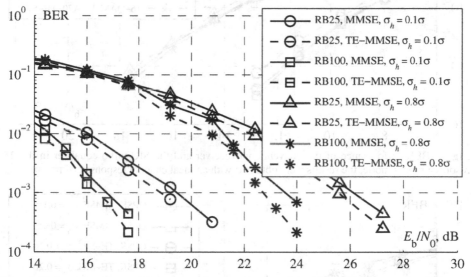

Fig. 4. BER performance of the conventional receiver and the MMSE turbo equalizer, the results correspond to 16-QAM signal constellation, EVA0 channel model and channel response estimation corrupted by white Gaussian noise

The comparison of the conventional receiver and the SIC turbo equalizer in conditions of corrupted channel estimation is shown in fig. 6. As before, the channel response estimation contained additive error which is a realization of white Gaussian noise, the standard deviation of error samples was chosen in proportion to the standard deviation of channel noise.

From the fig. 6 we can see that in case of 16-QAM the use of the SIC turbo equalizer provides an energy gain about 0.8 dB at 10^{-3} BER if the standard deviation of channel response error is equal to 0.2 of the standard deviation of channel noise. Reduction of the error standard deviation leads to reduction of the energy gain provided by the SIC turbo equalizer. Also, we have noticed that in case of QPSK signal constellation the SIC turbo equalizer does not provide any noticeable energy gain.

To summarize the results, we underline the MMSE turbo equalizer as a good compromise between computational complexity and the capability to provide rational energy gain in application to UL LTE signals detection. This marginal contribution of

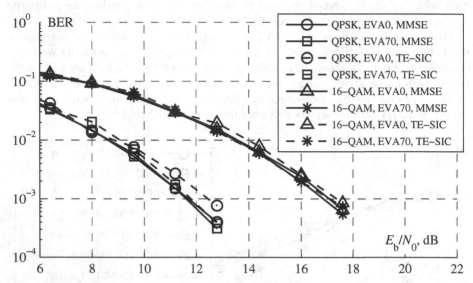

Fig. 5. BER performance of the conventional receiver and the SIC turbo equalizer in the 25 resource blocks mode, the results were obtained with an ideal channel response estimation

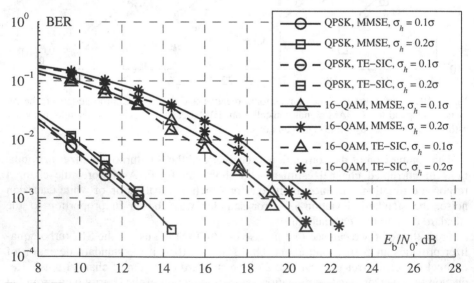

Fig. 6. BER performance of the conventional receiver and the SIC turbo equalizer, the results were obtained in the 25 resource blocks and EVA0 mode with a channel response estimation corrupted by white Gaussian noise

the MMSE turbo equalizer and the fail of the SIC turbo equalizer can be explained by number of reasons. At first, specified in LTE for user data transmission, turbo code, interleaver and rate matching procedure have a good capability of correcting bit errors even in case of multipath fading channels; the use of the turbo equalizer leads to decoding of only small additional number of blocks, which cannot be corrected by the conventional receiver. Moreover, in this paper we consider the EVA model, which is rather hard case of multipath fading channel, while other considered works on turbo equalizers usually examine a data transmission using convolutional codes and "artificial" channel models, which are far from real-life conditions. Obviously, in such conditions the conventional receiver chain containing single-executed independent blocks yields to the iterative receiver, which implies the extrinsic information exchange between equalizer and decoder blocks and their multiple execution.

5   Conclusion

In this paper we have represented the review on MAP symbol decoding, turbo codes and application of a turbo coding concept to the iterative receiver (turbo equalizer). For implementation of the turbo equalizer, two SISO-equalization algorithms were considered in detail. These algorithms perform the signal processing in frequency domain; the first is built on the minimum mean squared error (MMSE) criterion, and the second algorithm embodies the soft interference cancellation (SIC) approach.

In order to estimate efficiency of the turbo equalizer applied to UL LTE signals detection, we implemented the simulation model, which performed SC-FDMA signal generation, passing of the signal through the fading channel simulation filter, addition of white Gaussian noise and signal detection using the conventional receiver or the turbo equalizer. Also, several UL LTE modes were considered with QPSK, 16-QAM and 64-QAM signal constellations and allocation of 25 and 100 resource blocks. The extended vehicular A model (EVA) was used for the fading channel simulation.

We have shown that the iterative receiver based on the MMSE SISO-equalizer provides an energy gain about 0.6 dB for 16-QAM signal constellation at 10^{-3} BER in case of ideal channel response estimation and in presence of channel estimation error (hence, for 10^{-5}–10^{-6} BER one could expect an energy gain about 1-2 dB). The turbo equalizer based on the SIC SISO-equalization algorithm does not demonstrate any energy gain. Moreover, because of high computational complexity of the SIC algorithm its application to UL LTE signals detection is actually impossible in case of allocation of 100 resource blocks.

References

1. Proakis, J., Salehi, M.: Digital Communications, 5th edn. McGraw-Hill, New York (2008)
2. Myung, H.G., Goodman, D.: Single Carrier FDMA: A new air interface for long term evolution. John Wiley & Sons Ltd., Chichester (2008)
3. Douillard, C., Jezequel, M., Berrou, C.: Iterative correction of intersymbol interference: Turbo equalization. Eur. Trans. Telecommun. 6(5), 507–511 (1995)
4. Shannon, C.E.: A mathematical theory of communication. The Bell System Technical Journal 27, 379–423, 623–656 (1948)
5. Viterbi, A.J.: Error bounds for convolutional codes and an asymptotically optimum decoding algorithm. IEEE Trans. Inform. Theory IT-13, 260–269 (1967)

6. Bahl, L., Cocke, J., Jelinek, F., Raviv, J.: Optimal decoding of linear codes for minimizing symbol error rate. IEEE Trans. Inf. Theory **IT-20**, 284–287 (1974)
7. Berrou, C., Glavieux, A., Thitimajshima, P.: Near shannon limit error-correcting coding: turbo codes. In: Proc. IEEE Int. Conf. Commun., Geneva, Switzerland, pp. 1064–1070 (1993)
8. Benedetto, S., Montorsi, G., Divsalar, D., Pollara, F.: Serial concatenation of intereleaved codes: Performance analysis, design and iterative decoding. In: TDA Progr. Rep. 42–126, Jet Propulsion Lab., Pasadena, CA, pp. 1–26 (1996)
9. Benedetto, S., Divsalar, D., Montorsi, G., Pollara, F.: Serial concatenation of intereleaved codes: Performance analysis, design and iterative decoding. IEEE Trans. Inform. Theory **44**, 909–926 (1998)
10. Hagenauer, J., Hoeher, P.: A Viterbi algorithm with soft-decision outputs and its applications. In: Proc. IEEE GLOBECOM, Dallas, TX, pp. 47.1.1–47.1.7 (1989)
11. Tüchler, M., Koetter, R., Singer, A.: Turbo equalization: principles and new results. IEEE Trans. Commun. **50**(5), 754–767 (2002)
12. Tüchler, M., Singer, A., Koetter, R.: Minimum mean squared error equalization using a priori information. IEEE Trans. Signal Process. **50**(3), 673–683 (2002)
13. Glavieux, A., Laot, C., Labat, J.: Turbo equalization over a frequency selective channel. In: Proc. Int. Symp. Turbo Codes Related Topics, Brest, France, pp. 96–102, September 1997
14. Raphaeli, D., Saguy, A.: Linear equalizers for turbo equalization: A new optimization criterion for determining the equalizer taps. In: Proc. Int. Symp. Turbo Codes Related Topics, Brest, France, pp. 371–374, September 2000
15. Trajkovic, V.D.: Novel exact low complexity MMSE turbo equalization. In: IEEE 19th Int. Symp. Personal, Indoor and Mobile Radio Commun., pp. 1–5, September 15–18, 2008
16. Ampeliotis, D., Berberidis, K.: A linear complexity turbo equalizer based on a modified soft interference canceller. In: IEEE 7th WS Signal Proc. Advances in Wireless Commun., pp. 1–5, July 2–5, 2006
17. Benvenuto, N., Tomasin, S.: On the comparison between OFDM and single-carrier modulation with a DFE using a frequency-domain feed forward filter. IEEE Trans. Commun. **50**(6), 947–955 (2002)
18. Huang, G., Nix, A., Armour, S.: Decision feedback equalization in SC-FDMA. In: IEEE 19th Int. Symp. Personal, Indoor and Mobile Radio Commun., pp. 1–5, September 15–18, 2008
19. Wang, Q., Yuan, C., Zhang, J., Li, Y.: A robust low complexity frequency domain iterative block DFE for SC-FDMA system. In: IEEE Int. Conf. on Commun. pp. 5042–5046, June 9–13, 2013
20. Tuchler, M., Singer, A.C.: Turbo equalization: an overview. IEEE Trans. Inf. Theory **57**(2), 920–952 (2011)
21. Wu, B., Niu, K., Gong, P., Sun, S.: An improved MMSE turbo equalization algorithm in frequency domain. In: IEEE 14th Int. Conf. Commun. Technology (ICCT), pp. 444–448, November 9–11, 2012
22. Jar, M., Bouton, E., Schlegel, C.: Frequency domain iterative equalization for single-carrier FDMA. In: IEEE 12th Int. WS Signal Proc. Advances in Wireless Commun., pp. 301–305, June 26–29, 2011
23. 3GPP TS 36.211 3rd Generation Partnership Project; Technical Specification Group Radio Access Network; Evolved Universal Terrestrial Radio Access (E-UTRA). Physical Channels and Modulation
24. 3GPP TS 36.212 3rd Generation Partnership Project; Technical Specification Group Radio Access Network; Evolved Universal Terrestrial Radio Access (E-UTRA). Multiplexing and channel coding
25. Sherman, J., Morrison, W.: Adjustment of an Inverse Matrix Corresponding to a Change in One Element of a Given Matrix. Annals of Mathematical Statistics **21**(1), 124–127 (1950)

The Automated System for Collection, Processing and Transmission of Data for Training and Competitive Process in Ski Jumping

Dmitry Kiesewetter[1(✉)], Konstantin Korotkov[2], and Victor Malyugin[1]

[1] Saint-Petersburg Polytechnic University, Saint-Petersburg, Russia
info@adk-electronics.spb.ru, vim@imop.spbstu.ru
[2] Saint-Petersburg Scientific-Research Institute for Physical Culture, Saint-Petersburg, Russia

Abstract. It is described the system of collection, transmission and processing of data necessary for the organization of competitions and training process on jumps on skis from a springboard that contains the wireless data collection network with judges panels, wireless system remote control start, wired network data collection from sensors of landing and as well as an automated system for determining the distance of a jump on skis from a springboard. Typical signals obtained from the sensors of landing are shown. It is noted that the main problems in automatic detection range jumps are: different speed of sound propagation in the coating springboard, non-point sound source, and other factors. The experience of designing and operation of this system are represented.

Keywords: Data acquisition system · Wireless communication · Wired connection · Ski jumping · Springboard

1 Introduction

Ski jumping is a complex technical sport, equipped with the special springboard facilities, appropriate equipment and infrastructure. The modern ski jump is a complex set of engineering, operating in winter and summer, and designed for viewing jumping by a large number of spectators.

Evaluation of a skier jump by the judging panel is determined by the sum of points for jump length and style of performance, taking into account wind strength and direction [1, 2]. The major technical challenge in conducting competitions in ski jumping is to determine the length of jumps. Until recently, the distance of the jump was determined visually by a panel of judges, who were located along the landing hill. Currently, the major international competitions are using television technology, which allows the operator to view the video with the slowdown and determine the distance of the jump in accordance with the rules of the International Federation of Ski (FIS - Federation Internationale de Ski Jumping). However, as in the first and in the second case, the accuracy of determining the distance may be influenced by the so-called human factor.

© Springer International Publishing Switzerland 2015
S. Balandin et al. (Eds.): NEW2AN/ruSMART 2015, LNCS 9247, pp. 473–480, 2015.
DOI: 10.1007/978-3-319-23126-6_41

During the training process the use of the operator or the judging panel is economically inexpedient and not always possible due to organizational reasons. Therefore, the actual technical problem is to develop an automated system for evaluation of ski jumping length, capable of operating in an automatic mode without human intervention. Attempts to create such a system are known. The principle of operation of such systems is based on the registration and subsequent processing of acoustic signals caused by a skier landing based on several sensors located at different points of the landing hill [3]. The magnitude of the delay pulse detected by various sensors determines the touchdown point of the skier. However, it was not possible to achieve the required accuracy of measurement distances - 0.5 m, which was due to the following reasons:

- On the springboard a multilayer coating is being used. The speed of sound in different materials used in the surface structure of springboard is different. The highest speed of sound is reached in the concrete structural elements along the surface of the landing; the lowest - in the rubber springboard coating. In winter, the lowest rate of acoustic wave propagation takes place in a snowy covering a springboard, and the rate of spread depends on the density and humidity of snow and the geometrical form of snowflakes.
- The source of the acoustic wave – skis of an athlete, - is not a point source of sound. The length of the skis significantly exceeds the required precision of the jump evaluation.
- The point corresponding to the most powerful acoustic signal is not always corresponds to the position of the foot of an athlete and can be located both in front of and behind the foot, depending on the technology of jump performance.
- Long before the landing skis tips may touch the coating of the springboard several times.

Based on the foregoing, it can be concluded that to obtain the required measurement accuracy is necessary to optimize the layout of sensors on the landing hill, to use special techniques for processing complex acoustic signals and to develop devices for determining the velocity of acoustic waves in various structural elements of the springboard [4].

2   Block Diagram of the System

Consider the examples of the hardware implementation of the system for ski jumping ranging using accelerometers or acoustic sensors. Sensors are located along the mountain landing ramp in one or more rows, depending on the width of the hill and the desired landing system reliability. As an example, Fig. 1 shows the location of the sensors on the trampoline in the Molodegnoe settlement of the Leningrad region during testing of one of the working models of the system. Regardless of the layout of sensors, their signals are being received by the multi-channel analog-to-digital converter (ADC) by the electrical screened cable. This cable supplies power to the sensors, accelerometers and preamplifiers.

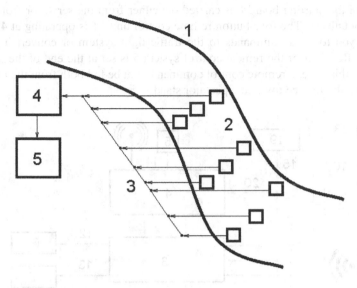

Fig. 1. Arrangement of sensors and data acquisition system on the trampoline in the Molodeg-noe settlement of the Leningrad region: 1 - landing table, 2 - sensors, 3 - connecting cables, 4 - multi-channel ADC, 5 – PC.

Example of the systems for determining ski jumping distance for K-65 and K-40 springboards in the village Toksovo of the Leningrad region is shown in Fig. 2. At the K-65 springboard the analog accelerometers 1, a shielded two-wire connection cable 10, the two 32-channel analog-to-digital converters integrated into one unit 2 (LTR-11) was used. From ADC the digital signal indulges via a cable 11 (UTP-5) to the server 3 by the TCP/IP protocol. The device address was programmed during setting. Processing of the incoming signals was performed by the server, located in the technical room located near the ski jump.

For synchronization the launch of the start we use signals arising in passing by a skier of the optical barriers 19, which at the same time are used to determine the rate of acceleration of the skier on the launch pad. The sensors 19 are connected with the microprocessor 4, the server 3, the ADC module and the transmission line 2 by an electrical cable 15, 20 (UTP-5). The speed of the skier on the acceleration table is calculated by the microprocessor 4 and stored in the server to be displayed in the protocol and used in the determination of the outcome.

Estimates for the jumping technique are made by the judges and come from their personal computers 9 to the server via wireless Wi-Fi. The server is connected to a router or a wireless (Wi-Fi) wire line 16. During training the jumping distance is calculated automatically using the data coming from the sensors. During the competition the distance of the jump is determined by the judges. In this case, the data is sent to the server over the Wi-Fi from judicial PCs. Sporting results are displayed on the display 7. Type of server communication with the display depends on the type of display. In particular, on the trampoline complex in Toksovo settlement communication is provided by RS232 protocol via an electrical cable.

Control of the starting board 6 is carried out either from the server or from the remote control unit 21. The four-button remote control unit 21 is operating at 433 MHz and allows you to send commands to the traffic light system or convert it into the sleep mode. Receiver of the remote control system 5 is set at the end of the acceleration table. In this case, a remote control commands can be fed both from the top of the hill, and from the referee tower or spectator stands.

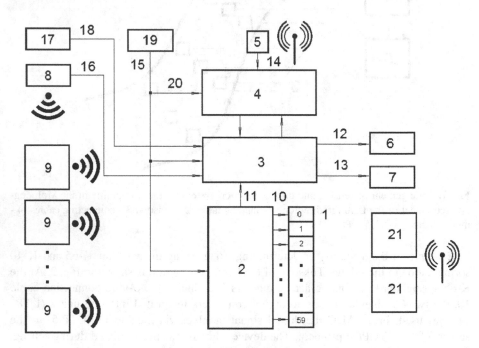

Fig. 2. Block diagram of the system for determining the distance of ski jumping: 1 - sensors located on the landing mountain; 2 - a multi-channel ADC and data transmission system; 3 – Server; 4 - microprocessor control system; 5 - start remote control sensor; 6 - start display, 7 - coaching board; 8 - Wi-Fi router or access point; 9 – judges' laptops or judicial panels; 10 - cables connecting sensors and multi-channel ADC; 11 - cable between the ADC and the server; 12, 13 - data cables to the scoreboard; 14 - communications cable of the remote control sensor and microprocessor system; 15 - synchronization signals cable; 16 - communication cable to the router (not in use in case of Wi-Fi connection to the server); 17 - an ultrasonic sensor of the wind direction and speed; 18 - connection to the server cable; 19 – sensors of a skier speed acceleration, 20 – cables of start synchronization.

Springboard K-65 in the Toksovo settlement is equipped with the three coordinate ultrasonic sensors 17 for wind speed and direction detection. The data from the sensor (LOMO-IVPU) also come into the server and are used in the calculation of points, as well as used as a warning system if the wind speed exceeds the allowable limit. The data exchange is carried out through an electric cable 18 by a TCP/IP or RS232 protocols.

3   Typical Sensors Signals and Data Processing

Parameters of electrical signals coming from the sensors depend on the sensor type, the type of springboard coverage, season, location of the sensor on the landing table and other factors. Typical signals, received from sensors installed on the K-65 springboard in the Toksovo settlement in summer are shown in Fig. 3. Waveforms at the top of Fig. 3 correspond to the sensors located at the top of the landing table. The lower waveforms in Fig. 3 correspond to the sensors positioned lower regarding the separation point. On the springboard K-65 sensors are installed starting from the top mark of 1.5 meters and then every 1 meter down the landing hill. Total amount of sensors is 60, which in Fig. 2 are designated by numbers 0…59. The low-frequency component of the signal is due to fluctuations in the structure of springboard, the high-frequency component is due to the deformation of the outer coating of the landing surface. Before the moment of landing, skier touches coating by skis tips. As we see from the waveforms to determine visually the delay time of the signal and a time corresponding to the landing is almost impossible. In this case, we apply a special signal processing algorithm.

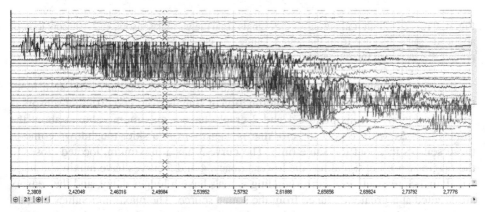

Fig. 3. Signal waveforms from the accelerometers placed at intervals of 1 m along the landing hill; along the axis Ox - time in seconds. On the axis Oy waveform of the channels sequentially shifted with increasing distance from the separation table.

The most universal, but requiring a relatively long computation, method for determining the signal delay is a correlation method. If there are two signals U1(t) and U2(t), the cross-correlation function (CCF) can be calculated by the formula:

$$CCF(\Delta\tau) = \frac{1}{T} \int_{t1}^{t1+T} (U1(t) - \overline{U}1)(U2(t - \Delta\tau) - \overline{U}2)dt, \qquad (1)$$

where $\overline{U}1$ and $\overline{U}2$ are averages of the U1(t) and U2(t) in the time interval T from t1 to t1+T, respectively.

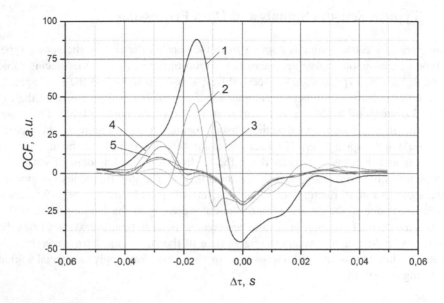

Fig. 4. Cross-correlation functions of the signals from the various sensors: 1, 2 - sensors located about the same distance from the sound source; 3 - closest sensor; 4, 5 - sensors located at a greater distance than the sensors 1 and 2.

The cross-correlation function of the signal with itself is the autocorrelation function. Example of the cross-correlation function of signals from two different sensors is shown in Fig. 4. At this picture 1 and 2 graphs have a maximum value at about the same offset $\Delta\tau$ with respect to zero, that is, signals from a skier touchdown come to these sensors in about the same time. The curve 3 has a maximum, located closer to the origin $\Delta\tau=0$; i.e. signal has the lower latency. 4 and 5 curves have maxima at large $\Delta\tau$, than 1-3, respectively, corresponding signals have a longer delay, and the sensors are located at a greater distance from the sound source. The maximum values of CCF of these signals are also smaller, since with a larger distance between the sensors the useful signal experiences the greater attenuation.

The following typical patterns should be noted. Low-frequency electrical oscillations caused by the mechanical vibrations of springboard construction can complicate finding the required CCF maximum. The cross-correlation function of the two periodic functions for a limited period of time in general has a lot of maximums against which it is sometimes difficult to identify a local maximum, generated by the useful signal. High-frequency components of the signal have significantly higher damping factor in the propagation along the springboard coatings, so we can assume that in this part of the spectrum signals of the sensors are not cross-correlated. Accordingly, in the process of the calculation CCF make an extra noise.

It should also be noted that the pulses from the speed sensors used to determine the rate of acceleration of the skier and to synchronize the launch of the system could be "double" or "triple". This should be considered when developing the algorithm for calculating the speed and overall control of the system.

4   Protection Against Interferences and Unauthorized Access

The most heavily affected by electromagnetic interference is the transmission system and the process of collection of analog signals from the sensors to a multi-channel ADC. The main sources of noise are the light power wires (commercial frequency), as well as signals of cellular phones (high frequency noise). Reducing the amount of interference of industrial frequency is achieved by twisting electrical lighting wires and placing them into a metal corrugated pipe, as well as the use of shielded cables for connection of sensors with the ADC.

As an example Fig. 5 presents a waveform of the input voltage of the ADC and the noise spectrum in the absence of the useful signal in the channel, the most heavily exposed to electromagnetic interferences from the lighting network. At the Fig. 5 the maximum of spectral density at the frequency of about 50 Hz and peaks corresponding to the 2nd, 3rd and 4th harmonic of the industrial frequency are seen. The worst signal to noise ratio in the channels is approximately 20 dB.

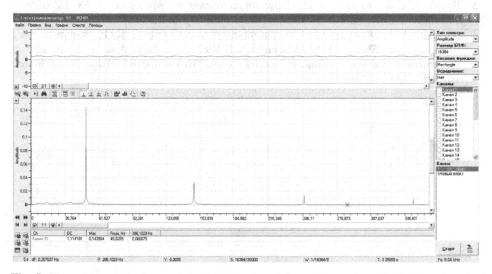

Fig. 5. The voltage at the input of the ADC (top) and the noise spectrum (bottom) in the absence of the signal in the channel, the most heavily exposed to electromagnetic interference from the lighting network.

To prevent the influence of high-frequency interference from mobile devices the following principles were used: shielding the signal wires and power wires, using RC filters in the power supply of sensors and LC filters in the input circuits of power supplies. During operation of the system, the impact of radiation from mobile phones, radio stations, as well as from the start control panels to the signals of the sensors, and the measurement accuracy, respectively, was not revealed.

To organize the exchange of data between the server and the PCs of judges (or judges' panels) a secure wireless computer network protected by a password is used. All the equipment in the Toksovo settlement is equipped with a burglar alarm system with wireless transmission of the alarm signal to the control unit.

5   Conclusion

Application of the presented above hardware and software solutions allowed to obtain the required accuracy (0.5 m) of determining ski jumping range in the automatic mode, as well as manages the start, provide safety and informational support during the competition and training of athletes. The developed system was successfully operated for over 2 years on the complex of ski jumps in the Toksovo settlement of the Leningrad region.

References

1. Regulations for "ski jumping" Sport. Order number 1071 of the Ministry of Sport of Russia, December 25, 2014 (in Russian)
2. FIS: The international ski competition rules (ICR). Book III. Ski jumping 69p. (2012)
3. Bychkov, G.G., Aronov, V.I.: Copyright certificate № 84400, cl. A 63 23/00.1981. The device for determining the length of the ski jumping (in Russian)
4. Slavskii, V.F., Kuramshin, A.M., Kalinin, Y.V., Korotkov, K.G., Sinister, A.A., Zakharov, G.G., Shelkov, O.M., Malyugin, V.I., Kiesewetter, D.V.: A device for measuring the length of the ski jumping. Patent RF N 102789 U8, G01B7/00 (published March 10, 2011) (in Russian)

A Possible Development of Marine Internet: A Large Scale Cooperative Heterogeneous Wireless Network

Shengming Jiang$^{(\boxtimes)}$

College of Information Engineering, Shanghai Maritime University,
1550 Haigang Avenue, Pudong New District, Shanghai, China
smjiang@shmtu.edu.cn

Abstract. Today, terrestrial Internet can be easily accessed with various types of terminals almost anytime and anywhere on the land. But this is not yet the case in the ocean mainly due to huge differences between terrestrial and marine environments. Although satellite Internet services are available in marine environments, at the time-being, they are neither cost-effective nor popular due to their inherent weaknesses in construction, launching and operation. Ever-increasing human activities in the ocean require marine Internet to provide handy, reliable and cost-effective high-speed Internet access not only on surface but also underwater in marine environments. This is due to the fact that a huge number of sensors and things have been deployed underwater, and this number is still increasing. How to interconnect them becomes an important issue that is necessarily addressed in order to form large and sophisticated underwater systems. This paper discusses the major available network technologies and new networking approaches that can be used to develop marine Internet, particularly a large scale cooperative heterogeneous wireless network, along with some further research issues.

Keywords: Marine internet · Cooperative heterogeneous wireless networks · Underwater networking

1 Introduction

Marine Internet aims to provide Internet services in marine environments for users and applications on water surface and under water [17]. One type of user comes from civil sectors, including seamen, fishermen, cruise/yacht passengers, island habitants etc. These users have to be provided with seamless Internet access and kept connected with the rest of the world during their sea voyages. Another kind of user is from industry sectors, such as maritime transportation and offshore oil industry. Marine Internet can be used for real-time control of

---

This work is financially supported by The National Natural Science Foundation of China (NSFC) under Grant No. 61472237.

S. Balandin et al. (Eds.): NEW2AN/ruSMART 2015, LNCS 9247, pp. 481–495, 2015.
DOI: 10.1007/978-3-319-23126-6_42

product line and remote surveillance. It can also be used to transport large volume of data to terrestrial high-performance computing centers for fast analysis and rapid result feedback. This is typically useful to undersea oil exploration, ecosystem monitoring and scientific research in marine environments.

On the other hand, a large number of underwater sensors and things have been deployed in the ocean, and this number is still increasing fast. But due to many challenges for realizing efficient underwater networking [1, 3, 23, 24, 26], a single underwater wireless network can only cover a small area and is often isolated from each other. This is due to the fact that the currently popular underwater communication medium is acoustic wave, which can only provide low transmission rates with large propagation delay [2, 4, 27, 29]. Another underwater communication medium is blue/green laser, which can provide much higher transmission rates with shorter propagation delay but can only propagate well over shorter distance [22]. In this case, marine Internet can be used to connect underwater networks in different locations to form a large and sophisticated underwater system, such as the Internet of Underwater Thing (IoUI) [7].

For the time-being, marine Internet is still one of less developed areas due to huge differences between the terrestrial and marine environments. These differences, mainly residing in geographical environments, climate conditions and user distribution, cause that marine Internet cannot be a simple extension of terrestrial Internet. The ocean is a huge water body containing about 1.3 billion cubic kilometer saltwater with a 3,682-meter average depth, and covers about 71% of the earth's surface, which makes it extremely difficult and costly to deploy network infrastructures therein. Special ocean climate conditions of high humidity, various forms of precipitation and frequent extreme weather also cause problems to communication and networking. For example, humidity and precipitation may severely degrade the performance of satellite communication at super high frequencies, and extreme weather may severely destroy network infrastructure to paralyze the network. On the other hand, most marine Internet users are transient and unevenly distributed at a very low density on average. Marine Internet users are much less than terrestrial ones but the former are distributed over a surface about 2.5 times the terrestrial surface. Most areas in the ocean are unpopulated, while a large number of users may just crowd in a very small place such as a cruise ship, whose capacity ranges from hundreds to thousands with an average of 3000 for ocean liners. These ships often move from one place to another.

This paper discusses the major network technologies available and under development as well as a large scale cooperative heterogeneous wireless network structure for marine Internet in Sections 2, 3 and 4, respectively. The underwater inter-networking in the context of marine Internet is discussed in Section 5. Some issues to be further addressed are discussed in Section 6.

2   Available Network Technologies

This section briefly reviews the major currently available network technologies able to provide marine Internet services, which include maritime radio systems, cellular networks and satellite Internet. The former two are collectively called coastline networks (CLNs), henceforth.

2.1 Maritime Radio Systems

The very high frequency (VHF) refers to frequency bands between 30 and 300 MHz. Particularly, the maritime VHF radio operates in frequency bands between 156 and 162.025 MHz, with a typical channel spacing of 50 and 25 kHz, respectively. Traditionally, it only provides analog voice communication with a maximum communication range of up to about 111 km. Now it has a digital selective calling capability at a rate of up to 1.2 kbps, which can allow a distress signal to be sent by pressing a single button. Some VHF channels have been used to develop VHF data links (VDLs) to provide data communication for automatic identification systems (AISs) at a maximum rate of 9.6 kbps. An AIS is an automatic tracking system used to identify and locate vessels for navigation and vessel collision avoidance. To this end, the position information of a ship is continuously transmitted on AIS VDLs to ensure that all its closest vessels will receive its position report. To improve the safety of maritime navigation and operations especially in adverse conditions, more messages have to be exchanged in real-time, such as weather, ice charts, status of aids to navigation, water level and rapid changes of port status, voyage information, passenger manifest and pre-arrival report etc [10].

Due to the popularity of AIS applications and increasing demands of ship-to-shore and ship-to-ship data exchange, the capacity of the current AIS VDLs becomes inadequate to satisfy these increasing demands. To handle this problem, two additional 25-kHz channels have been proposed to support application specific message (ASM) communication, and six original VHF channels have been enhanced for the VHF Data Enhanced System (VDES) [10]. The VDES aims to provide a maximum rate of 302.2 kbps by merging four of the six VHF channels into a 100-khz channel and the other two channels to a 50-kHz channel [5]. Existing wireless communication technologies such as the Orthogonal Frequency Division Multiplexing (OFDM) modulation and distributed antenna technologies can be used to enable VDESs. Some results of a VDES channel sounding campaign are reported in [28]. Similarly, in the ultra high frequency (UHF), which ranges from 300 to 3000 MHz, six frequencies between 450 and 470 MHz with a 25-kHz channel spacing are also used for on-board communication for maritime operations. To improve spectrum utilization, narrower channel spacings such as 12.5 or 6.25 kHz are suggested so that additional channels can be introduced [11,12].

Both VHF and UHF are very important for maritime radio communication to support maritime operations, especially for safety and rescue. However, due to the limited VHF/UHF bandwidth allocated for marine communications and ever increasing data communication necessary for improving the safety of maritime navigation and operations, there is no adequate channel capacity available for popular Internet applications in marine environments.

2.2   Cellular Networks

Advanced cellular network technologies such as WiMAX and the Long-Term Evolution (LTE) have been used in harbor areas and busy water channels to provide Internet accesses for residents and ships therein. It was reported by the Wall Street Journal in June 24, 2013 that Verizon Wireless had enhanced its 4G LTE coverage in and around Boothbay Harbor in Maine, USA. These technologies can provide transmission rates up to several hundreds of Mbps with a maximum coverage radius of a hundred kilometer. The Huawei eWBB LTE solution can cover a circle of 100 km radius with a downlink and uplink data rates of up to 100 Mbps and 50 Mbps, respectively [13].

Although the ocean is big, most human activities therein take place in water areas near coastlines. For example, many domestic maritime shipping routes are set close to the coastline, say about $2 \sim 20$ nautical miles away from the coastline as suggested in the literature. Therefore, it is necessary to consider how to provide handy, reliable and cost-effective high-speed Internet access for marine Internet users in these water areas. Thus, deploying cellular network like systems along coastlines makes sense. This system is mainly composed of certain terrestrial infrastructures like base stations installed near or along coastlines or on islands. This type of infrastructure can act as a bridge between water areas and terrestrial networks.

The benefit of using such system is twofold. First, some existing technologies such as WiMAX and LTE can be used without need of a long R&D period. Internet users covered by this system can have seamless and direct Internet access with their handsets without paying extra cost of specific devices. Second, newly developed technologies for cellular networks can be continuously used to increase the system capacity and reduce system construction cost. For example, the combination of radio over fiber (RoF) and distributed antenna technologies has been considered as a promising technology to increase the capacity of future wireless networks, with which, the base station can be simplified into an antenna system mainly equipped with a tx/rx module that simply relays analog radio signal. Multiple antennas can be linked through optical fibers to a processing center, which conducts further processing for signaling, communication and networking.

Since this system requires deployment of terrestrial infrastructure, which limits its coverage, it is very difficult for the system to cover deep water areas. With the global coastline of about $L = 356,000$ km, given the maximum radius of a base station's coverage $R = 50$ km (e.g., with WiMax), the overall coverage of the system can be roughly estimated by $L \times R = 1.78 \times 10^7$ km^2, which is only 4.92% of the overall ocean's surface of 3.62×10^8 km^2; if $R = 100$ km (e.g., with LTE), the coverage ratio is doubled to 9.83%.

2.3   Satellites

A geostationary earth orbit (GEO) satellite runs in an orbit of 35,786 km with an orbit period of 24 hours and seems to be fixed in the sky to a ground observer. Theoretically, three GEO satellites can cover most of the earth's surface except the polar areas. They have been used to provide communication services in marine environments for long time. For example, Inmarsat uses GEO satellites to provide voice, IP data services and access to the global maritime distress and safety system (GMDSS) in sea. The major problem of using GEO satellites for Internet applications is the long round-trip propagation delay between two ground stations via a GEO satellite, which is about 250 ms and almost a double of the end-to-end delay bound of voice application. A much larger latency further contributed by queueing delay and processing time for communication and networking will severely affect the performance of network protocols (e.g., TCP) and quality of service (QoS) provisioning for delay-sensitive applications.

Table 1. Some available marine Internet services provided by satellites [9]

Service package	Data rate (kbps)	Cost / MB (USD)	Cost / PM (USD)	Equipment's cost (USD)
FleetBroadband G[†]	≤ 432	$0.40 \sim 20.85$	$0.42 \sim 1.15$	$4,700 \sim 16,914$
MCD-4800-BGAN[†]	448	$4.70 \sim 6.99$	0.98	13,733
FleetPhone Global[†]	2.4	$15 \sim 50$	$0.80 \sim 0.95$	$1,899 \sim 2,349$
Iridium Pilot Global	134	$7.41 \sim 10.90$	$0.65 \sim 1.22$	4,595

[†]which is provided by Inmarsat, MB=Megabyte, PM=Phone Minute

This long propagation delay can be significantly reduced with a medium earth orbit (MEO) satellite (whose orbit ranges from 3000 to 35,786 km), especially with a low earth orbit (LEO) satellite (whose orbit ranges between 200 and 3000 km), such as Iridium and Globalstar systems. However, LEO/MEO satellites fly rapidly around the earth while vessels may also move and rock with water. In this case, it is difficult to track a satellite to keep a line of sight between the satellite and directional antennas to maintain high communication quality. If an omni-directional antenna is used instead, wireless communication quality will be degraded severely. On the other hand, satellite communication at the L-band (1~2 GHz), C-band (4~8 GHz), Ku-band (19 GHz) and Ka-band (29 GHz) is also affected by moisture and various forms of precipitation frequently present in marine environments, to which communication at these bands is susceptible.

Another weakness of satellite systems is the high cost of construction, launching, operation and maintenance of satellites. Hardware upgrading or repairing of a satellite already in the orbit almost means a replacement. Furthermore, for high altitude satellites, radio communication suffers high attenuation and large path loss so that some specific bulky terminals with large transmission power are needed, which further financially burdens the user. With such high cost and very low user density, satellite Internet is not cost-effective for marine Internet users. Table 1 lists some recently available marine Internet services provided by satellites, which are still luxury to the ordinary user.

3   Developing Network Technologies

Due to the weaknesses of the existing network technologies discussed above, several new approaches have been studied in the literature, namely, wireless ad hoc networks (WANETs) and high altitude platforms (HAPs).

3.1   Wireless Ad Hoc Networks (WANETs)

Cellular networks and wireless local area networks need infrastructures like base stations or access points to coordinate communication between terminals, which cannot communicate each other directly. This is just opposite in WANETs, where no infrastructure is required and terminals can communicate each other directly. The ability of WANETs in terms of self-organizing and self-curing makes them suitable for dynamic and unstable networking environments. That is, any vessels and facilities on water surface (e.g., buoys) equipped with wireless communication devices can be used to construct WANETs to enable vessel-to-vessel and vessel-to-shore communication. Reference [8] is among the earliest discussing such idea, in which a WiMAX mesh network is used to provide onboard Internet broadband access for vessels in the Mediterranean without using satellites. A systematical study has been carried out by the project TRITON [25], which tries to set up a WiMax-based mesh network for ship-to-ship and ship-to-shore communication at high rates. This kind of WANET consisting of vessels and facilities on water surface is often called nautical WANET (NANET) [19,21].

A WANET can be used as a complementary technology to the available networking technologies to improve their service coverage and performance with the following extensions. The first one is that any vehicles above water surface such as balloons, airships, helicopters and airplanes can be used to form an aeronautic WANET (AANET) [18] to provide opportunistic networking. Obviously, NANETs plus AANETs can cover much larger water areas and provide more networking opportunities. The second extension is underwater WANETs (UANETs) [20], which can avoid underwater construction of network infrastructure and be used to connect underwater things [7].

The major weaknesses of this WANET-based approach include small network coverage, unstable network connectivity and unreliable network performance due to its dynamic network topology. The end-to-end network performance degrades quickly as the number of hops along a network route increases. It can provide high-speed network connection occasionally at low cost, but the performance largely depends on the density and distribution of network nodes, both of which are highly dynamic in marine environments. Hence, WANETs cannot provide always-on connectivity or service guarantee.

3.2   High Altitude Platforms (HAPs)

A HAP is a quasi-stationary aerial platform in the stratosphere located at an altitude of 17∼22 km above the earth's surface. The radius of a HAP's footprint can be up to 100 kilometers, depending on its altitude and elevation angle. With a 50-MHz bandwidth at 28-GHz frequency, a HAP at a height of 10 km above the ground can provide downlink date rates up to 320 Mbps [30]. HAPs are particularly suitable for large water areas near a coastline such as 200-nautical-mile exclusive economic zones, and can be used in the following scenarios to provide marine Internet services:

- An instant demand for a short time period: For example, when the number of users goes beyond the normal situation due to some occasional events such as a gathering, a HAP can be set up shortly and removed afterwards.
- A short-term solution where neither CLNs nor NANETs are available. A HAP can be in the place for duration of several months and even longer with solar energy supply, and are particularly useful for some events with special networking requirements for months, such as scientific exploration.
- A fundamental part of marine Internet networks in the places where it is difficult to deploy CLNs. This HAP can consist of solar-powered unmanned airships, which can stay in the stratosphere for a very long period.

Actually, there are several projects of HAPs under going. For example, the vulture program of USA's DAPRA aims to develop a single high-altitude unmanned airplane to operate continuously on-station for five years [6]. Google's Loon project (see http://www.google.com/loon/) tries to use balloons to provide Internet access for everyone in the world, and was tested in 2013.

The major advantages of HAPs over satellites include easy and fast deployment, low cost and large capacity with shorter propagation delay [30]. In comparison with a WANET, a HAP can cover much larger areas with more reliable network performance; but their deployment and power supply are more difficult to handle. As discussed in [30], one major challenge is the overall long-term power balance. Relatively mild wind and turbulence in the stratosphere need power for propulsion and station-keeping. Particularly for a balloon platform, the HAP requires wind compensation to stay still in the sky, and for an aircraft platform, the HAP has to fly on a circle to maintain services to some area. Power is also required for the payload, communication and networking. Unlike satellites which can be re-charged by solar power frequently, for HAPs, enough power has to be stored in cells during the day in order to maintain the normal operation throughout the whole night. Thus, a large capacity cell is required, resulting in more payloads and more power consumption. Hence, the ageing of cells is a major factor determining the achievable mission duration of a HAP. A possible solution to this problem is to bring a HAP back to the ground for service, which however will cause service disruption and increase deployment cost.

4   A Hybrid Networking Structure

Table 2 summarizes the suitability for marine Internet (the upper part) and characteristics (the lower part) of the major network technologies discussed above, where strengths are spelt in the italic font. This table shows that none of them alone can provide a cost-effective solution for marine Internet, hence that a hybrid networking structure able to make use of their good features is necessary [17]. With such a structure as depicted in Fig. 1, CLNs are deployed along coastlines to cover water areas in which the majority of marine Internet users will present. Typical CLNs include the maritime radio systems and cellular systems discussed in Section 2.2. To reduce interference to terrestrial wireless networks, directional antennas should be used to focus the signal coverage to the corresponding water areas. These types of antennas can also allow larger transmission power to be used to expand transmission distance and signal coverage when necessary. A CLN also functions as a gateway between water areas and terrestrial networks. It is expected that marine Internet users covered by a CLN can enjoy terrestrial Internet access with acceptable cost and QoS.

To improve QoS in a water area and expand the coverage of a CLN, WANETs can be used jointly to provide low cost and short-term network connections. For example, vessels close to each other can form a NANET, which can further involve buoys and small boats nearby. This kind of NANET can link nodes therein to a CLN for terrestrial Internet access, or be used to support intra-NANET communication. For delay-tolerant applications, an AANET can also be exploited if any to provide opportunistic network connectivity with a store-and-forward transfer mode. When a node is out of the service coverage of the CLNs or NANETs, this kind of AANET may become the only chance for nodes therein to communicate with the outside. Another important application of WANETs is to construct underwater networks at lower cost with more flexibility in comparison with wired underwater networks, which will be discussed more in Section 5.

Table 2. Characteristics of networking technologies for marine Internet

	Satellite	HAP	Cellular	WANET	VHF/UHF
Wide water surface	*Suitable*	Unsuitable			
Underwater network	Unsuitable			*Suitable*	Unsuitable
Channel quality	Vulnerable	*Invulnerable to humidity & precipitation*			
Infrastructure	*Safe*		Unsafe in extreme weathers		
Power supply	Limited		*Unlimited*	(except UANETs)	
Service capacity	Small		*Large*		Too small
Service guarantee	*Yes*			No	*Yes*
Cost-effectiveness	Low	*High*			
Direct user access	Difficult	*Easy*			

A HAP can be developed on-demand for occasional marine Internet users, especially when neither CLNs nor WANETs are available. If the problems of HAPs mentioned in Section 3.2 can be effectively resolved for dense deployments, HAPs can even become a replacement of terrestrial base stations to cover large water areas, with the satellite being used as a backhaul to terrestrial networks. Different from the current situation in which the satellite usually is the only option available for marine Internet, the hybrid networking structure tries to make the satellite to be the last option. For example, the satellite is just used as

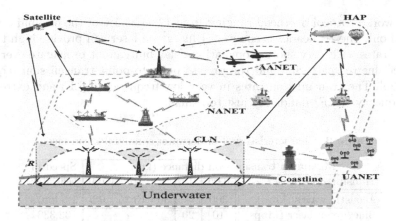

Fig. 1. A hybrid networking structure for marine Internet

a backup solution for emergency situations when none of the above-mentioned networks are available.

It is expected that the hybrid structure can provide a cost-effective network service following application requirements by exploiting any networking opportunity as much as possible with the following preference: CLNs are always selected first if any, then WANETs and/or HAPs can be exploited with the backup of satellites. Actually, if each network in this hybrid structure is treated as a special networking node, this structure itself can be regarded as a super-scale hybrid WANET. One challenge is how this super-scale network can be smart enough to select the most cost-effective network service to satisfy application requirements without requiring user's involvement in real-time.

5   Underwater Inter-networking

Terrestrial Internet usually need not consider underground networking. However underwater networking has to be taken into account by marine Internet to support the Internet of Underwater Thing (IoUI) [7], for example. This is due to the fact that a huge number of sensors, vehicles and other underwater things have been deployed, and this number is still increasing. The network technologies and approaches except WANETs discussed in above sections are not suitable for underwater networks due to difficulties in deploying the required network infrastructure.

Table 3 summarizes the transmission rates, distance and propagation speeds of the major media when they are used for underwater communication. It shows that electromagnetic wave cannot propagate well in seawater, while acoustic wave can propagate over a long distance but at a very slow speed, which will cause long end-to-end latency. The kbps-level capacity of acoustic media is insufficient to support many Internet applications. For example, the audio codec bandwidth is 11.8~128 kbps and 0.25~4 Mbps for video, both of which exclude the network protocol overhead ranging from 12.5~55.5% of the original one for audio if only the IP header is counted. Blue/green laser can provide high transmission rates with very short propagation delay but cannot propagate over long distance due to scattering and the precise point-to-point transmission requirement [26]. The transmission rates in very low frequency (VLF) and extremely low frequency (ELF) bands are just too small.

Table 3. Characteristics of underwater communication media

	Maximum transmission distance (km)					Speed in
Transmission rates	0.05	0.1	1	10	100	seawater
Acoustic waves (kbps)	300	-	30	15.36	0.5	1.5 km/s
Blue/green laser (Mbps)	10	20	-			33,333
VLF/ELF (bps)	300	-	3	Negligible		km/s

To maintain acceptable transmission rates with the current underwater communication technologies, the size of a underwater wireless network should be constrained. The WANET is a reasonable option to avoid underwater infrastructure construction, with which underwater nodes close to each other can automatically form a underwater ad hoc network (UANET). To cover a large underwater area in this case, underwater inter-networking technologies should be used to link different underwater networks.

Basically, there are two underwater inter-networking methods. The first one is to use cables deployed on the sea floor to link underwater things and provide energy to them. This method has been used in some underwater observation systems such as the Monterey Accelerated Research System (MARS). Its major advantages include high-speed reliable network connections and sustainable power supply. The major challenge arises from the difficulty in and the cost of dense infrastructure deployment for covering wide underwater areas.

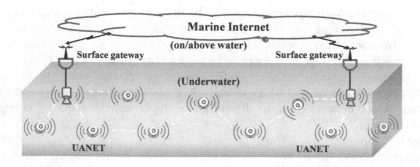

Fig. 2. Underwater inter-networking with marine Internet

The other method is to use surface gateways floating on water to link UANETs especially underwater sensor networks [14] through acoustic wave or green/blue lasers. The marine Internet above water surface can be used to inter-network different UANETs by connecting their surface gateways as illustrated in Fig. 2. Particularly with a star-topology, which is more common than other topologies for underwater networks [24], each UANET has one surface gateway, which collects data from the UANET and sends it to marine Internet. This gateway can also broadcast the data coming from marine Internet to the UANET under its control. The underwater nodes covered by a surface gateway can communicate with it directly, while those out of its coverage needs the relay of other nodes to connect the gateway. A set of surface gateways close to each other and equipped with the same type of air interface can also form a wireless mesh network, but those with different types of air interfaces or far away from each other have to use marine Internet for inter-networking. This method can avoid severe performance degradation problem in WANETs as the number of hops along a route increases, and allow many small-sized UANETs to be connected to cover a large underwater area. This method is more flexible, easier and cheaper than the first one but cannot provide power supply to underwater things.

6   Further Research Issues

This section briefly discusses some major research issues necessarily to be further addressed in order to foster the development of marine Internet.

6.1   CLNs and Satellite Internet

Although CLNs are relatively mature, there are still some issues necessarily to be further addressed to overcome their weaknesses discussed earlier. Particularly for the VHF/UHF maritime radio systems, it is important to improve the utilization of the allocated spectrum bands to support ever-increasing data communication dedicated for safe and rescue as well as maritime operation, while using available bandwidth for general data communications. This may be achieved by using advanced modulation and distributed antenna technologies such as OFDM, Multi-Input-Multi-Output (MIMO) and even their combination. OFDM can reduce channel spacing, and MIMO allows the same carrier to be reused multiple times simultaneously at the same site. The efficiency of the spectrum reuse is subject to the number of transmission and receive antenna-pairs available. In addition, some UHF frequencies allocated to maritime radio overlap with those allocated to International Mobile Telecommunication (IMT) or satellite systems in some countries (e.g., 420-470 MHz). Thus a cognition ability of communication systems on vessels is necessary for a global inter-operability. To this end, cognitive communication and networking need to be addressed. Although 3G and beyond-3G cellular networks are very popular in terrestrial environments, their applications in marine environments still need to be investigated since radio signals can be easily absorbed by seawater and also due to special user distribution in marine environments mentioned earlier.

To increase communication capacity at low cost for satellite Internet, deploying more small satellites into low orbits seems to be promising since the cost for both construction and launching of satellites can be reduced significantly in this case. However, keeping the communication quality of the wireless link between a flying satellite and a moving terminal in progress of communication and dealing with rapid handoff between a satellite's coverage and terminals on the vessels are two challenging issues to be addressed. On the other hand, to enable satellite Internet to attract more terrestrial users is also important to improve investment efficiency so as to further reduce the cost for users in marine environments.

6.2 WANETs, HAPs and Hybrid Networking Structure

Regarding the newly developing network approaches, more research is still needed to improve the performance of WANETs, which can be achieved by exploiting some favorable features uniquely present in NANETs. For example, one feature of NANETs is the availability of information on positions and speeds of vessels, which can be provided by AIS or GPS. This information can be used to improve routing performance especially for opportunistic networking. Research is also required to reduce the cost of HAP deployment in marine environments. To this end, the optimization of power allocation should be used to allow a HAP to stay in the sky for longer time period and to support more communication loads, while guaranteeing successful return of HAPs to the ground for recycle usage.

An open networking structure that enables cooperative communication and networking is necessary to support the hybrid networking discussed above. It has to accommodate various types of networks and make them to cooperatively provide cost-effective service to marine Internet users. One challenge is to handle the heterogeneity of air interfaces, communication capacities, networking capabilities and various requirements for quality of service (QoS) as well as network security. Most of Internet users in marine environments are transient and come from different places through vessels, which may have various communication facilities and radio interfaces following different technical standards. It is also important to enable an efficient collaboration between CLNs, various types of WANETs, HAPs and satellites to maximize network connectivity at low cost. Another challenge is to provide networking services on-demand since it is costly and even impossible to maintain an always-on connection between any nodes in marine environments. For opportunistic connection, all kinds of nodes should be considered to maximize connectivity especially in emergency situations when no other options are available. In this case, how to ensure end-to-end networking security is an important and difficult issue.

6.3 Underwater Inter-networking

If there is no breakthrough for underwater communication technologies to enable large underwater coverage at high transmission rates, a major performance improvement for a single underwater wireless network is the optimization of the data link layer and network layer, especially from medium access control

(MAC) and routing protocols. Particularly with acoustic underwater communication, it is necessary to minimize the use of handshake to avoid long latencies caused by slow signal propagation speed and low transmission rates as discussed below.

The ratio of the time used to transmit a packet (T) to the time used to obtain transmission opportunity (t, the interval between when a packet arrives at a node and when the node starts the transmission), $\frac{t}{T}$, is often used to evaluate a protocol's efficiency. It can be estimated below for once packet transmission:

$$\frac{t}{T} \approx \frac{x \times \frac{\tau}{r} + y \times \frac{d}{v}}{\frac{l}{r}} = x \times \frac{\tau}{l} + y \times \frac{r}{l} \times \frac{d}{v}, \tag{1}$$

where, r is transmission rate, τ is protocol overhead, l is packet length, d is the distance between the sender and the receiver, v is signal propagation speed. Both x and y are a positive coefficient, and their settings depend on protocol design. For a terrestrial wireless network, $v = 300,000 \ km/s$ in the air, so that $\Delta = y \times \frac{r}{l} \times \frac{d}{v}$ in (1) can be very small and even negligible so that $\frac{\tau}{l}$ is a dominant factor of $\frac{t}{T}$. However, for acoustic wave in seawater, $v = 1.5 \ km/s$, Δ becomes too large to be negligible, resulting in a significant increase in $\frac{t}{T}$. When handshaking schemes are used, such as the RTS/CTS adopted by IEEE 802.11 and the RREQ/RREP used by the Ad hoc On-demand Distance Vector (AODV) routing protocol, both x and y will be increased, causing increase in $\frac{t}{T}$ too.

Particularly for a star-topology underwater wireless network as illustrated in Fig. 2, the Code Division Multiple Access (CDMA) can avoid using a handshaking scheme in MAC protocols. With CDMA, each node can send simultaneously in the same frequency band if an orthogonal code is pre-assigned to each node with proper power control. This objective can also be achieved by using the logical MIMO approach [15,16], which can allow multiple nodes to share the same uplink simultaneously, and the number of such nodes depends on the number of receive antennas installed in the surface gateway. Regarding the underwater network routing with this topology, actually the major work can be carried out by the surface gateways rather than by underwater nodes, which can minimize transmission of hand-shaking message between underwater nodes.

7   Conclusions

It is believed that marine Internet will become more and more important in the future for people to expand their activities in marine environments, while there is not yet a cost-effective solution ready for it. This paper carries out an evaluation of the major state-of-the-art network technologies available and under development that can be used to develop marine Internet, particularly a possible solution using a large scale cooperative heterogeneous wireless network consisting of various types of wireless networks. The challenging issues discussed in the paper also show that marine Internet is still in its enfant stage, and more research is required to foster its development.

References

1. Akyildiz, I.F., Pompili, D., Melodia, T.: Challenges for efficient communication in underwater acoustic sensor networks. ACM SIGBED Review **1**(2), 3–8 (2004)
2. Basagni, S., Conti, M., Giordano, S., Stojmenovic, I.: Advances in underwater acoustic networking. In: Melodia, T., Kulhandjian, H., Kuo, L.C., Demirors, E. (eds.) Mobile Ad Hoc Networking: Cutting Edge Directions, 2nd edn., chap. 23. John Wiley & Sons Inc., Mar 2013
3. Chen, K., Ma, M., Cheng, E., Yuan, F., Su, W.: A survey on mac protocols for underwater wireless sensor networks. IEEE Commun. Surveys & Tutorials **16**(3), 1433–1447 (2014)
4. Chitre, M., Shahabudeen, S., Stojanovic, M.: Underwater acoustic communications and networking: Recent advances and future challenges. Marine Tech. Society J. **42**(1), 103–116 (2008)
5. Electronic communication committee (EEC): Information paper on vhf data exchange system (vdes). In: The 3rd Meeting CPG PTC, CPGPTC(2013)INFO 16, Bucharest, Hungary, October 2013
6. DARPA: Vulture program enters phase ii. DAPAR's New Release, September 2010
7. Domingo, M.C.: An overview of the internet of underwater things. J. Network & Computer Applications **35**(1), 1879–1890 (2012)
8. Friderikos, V., Papadaki, K., Dohler, M., Gkelias, A., Agvhami, H.: Linked water. In: IEEE Communications Engineer, pp. 23–27 (April/May 2005)
9. Ground Control Company: Marine Satellite Internet & Phone Solutions, December 2013. http://www.groundcontrol.com/maritime_satellite_internet.htm
10. ITU radiocommunication study groups: aeronautical, maritime and radiolocation issues. In: Agenda Item 1.16, Document 5B/TEMP/281-E, May 2014
11. ITU radiocommunication study groups: aeronautical, maritime and radiolocation issues. In: Agenda Item 1.15, Document 5B/TEMP/283-E, May 2014
12. ITU radiocommunication study groups: technical characteristics of equipment used for on-board vessel communications in the bands between 450 and 470 mhz. In: Document 5B/TEMP/284-E, May 2014
13. HUAWEI TECHNOLOGIES CO. LTD.: Huawei eWBB LTE Solution (2013). http://www.huawei.com/ilink/enenterprise/download/HW_203747
14. Ibrahim, S., Al-Bzoor, M., Liu, J., Ammar, R., Rajasekaran, S., Cui, J.H.: General optimization framework for surface gateway deployment problem in underwater sensor networks. EURASIP Journal on Wireless Communications and Networking (2013). http://jwcn.eurasipjournals.com/content/2013/1/128
15. Jiang, S.M.: A logical MIMO MAC approach for uplink access control in centralized wireless. In: Proc. IEEE Int. Conf. Comm. Systems (ICCS), Guangzhou, China, November 2008
16. Jiang, S.M.: Future Wireless and Optical Networks. Networking Modes and Cross-Layer Design. Springer (2012)
17. Jiang, S.M.: On marine internet and its potential applications for underwater inter-networking (extended abstract). In: Proc. ACM Int. Conf. Underwater Networks (WUWNet), Kaohsiung, Taiwan, pp. 57–58, November 2013
18. Karras, K., Kyritsis, T., Amirfeiz, M., Baiotti, S.: Aeronautical mobile ad hoc networks. In: Proc. European Wireless Conference (EW), Prague, Hungary, June 2008
19. Kim, Y.B., Kim, J.H., Wang, Y.P., Chang, K.H.: Application scenarios of nautical ad-hoc network for maritime communications. In: Proc. MTS/IEEE OCEANS. Biloxi, MS, October 2009

20. Kong, J.J., Cui, J.H., Wu, D.P., Gerla, M.: Building underwater ad-hoc networks and sensor networks for large scale real-time aquatic applications. In: Proc. IEEE Military Comm. Conf. (MILCOM), Atlantic City, USA, vol. 3, October 2005
21. Lambrinos, L., Djouvas, C.: Creating a maritime wireless mesh infrastructure for real-time applications. In: Proc. IEEE GLOBECOM Workshops (GC Wkshps), pp. 529–532, December 2011
22. Lanzagorta, M.: Underwater Communications. MORGAN & CLAYPOOL (2012)
23. Melodia, T., Kulhandjian, H., Kuo, L.C., Demirors, E.: Advances in underwater acoustic networking. In: Basagni, S., Conti, M., Giordano, S., Stojmenovic, I. (eds.) Mobile Ad Hoc Networking: The Cutting Edge Directions, chap. 23, pp. 804–852. Wiley-IEEE Press (2013)
24. Partan, J., Kurose, J., Levine, B.N.: A survey of practical issues in underwater networks. In: Proc. The ACM Int. Conf. on Underwater Networks & Systems (WUWNet), Los Angeles, California, USA, September 2006
25. Pathmasuntharam, J.S., Kong, P.Y., Zhou, M.T., Ge, Y., Wang, H.G., Ang, C.W., Sui, W., Harada, H.: Triton: high speed maritime mesh networks. In: Proc. IEEE Symp. Personal, Indoor & Mobile Radio Commun. (PIMRC), September 2008
26. Pompili, D., Akyildiz, I.: Overview of networking protocols for underwater wireless communications. IEEE Commun. Mag., 97–102, January 2009
27. Preisig, J.: Acoustic propagation considerations for underwater acoustic communications network development. In: Proc. ACM Int. WS. Underwater Networks (WUWNet), Los Angeles, USA, September 2006
28. Safar, J.: Vdes channel sounding campaign - trial report. In: RPT-09-JSa-14, Report Version 1.1, April 2014
29. Stojanovic, M.: Underwater acoustic communications: design considerations on the physical layer. In: Proc. Annual Conf. Wireless on Demand Net. Sys. & Services (WONS), Garmisch-Partenkirchen, January 2008
30. Tozer, T.C., Grace, D.: High-altitude platforms for wireless communications. Electronics & Communication Engineering Journal 13(3), 127–137 (2001)

On Interoperability in Distributed Geoinformational Systems

Elena Velichko[1,2](✉), Aleksey Grishentsev[2],
Constantine Korikov[1], and Anatoly Korobeynikov[3]

[1] Saint Petersburg State Polytechnic University, St. Petersburg, Russia
velichko-spbstu@yandex.ru, korikov.constantine@spbstu.ru
[2] Mechanics and Optics, Saint Petersburg National Research University of Information
Technologies, St. Petersburg, Russia
tigerpost@yandex.ru
[3] Ionosphere and Radio Wave Propagation of the RAS,
St. Petersburg Institute of Terrestrial Magnetism, St. Petersburg, Russia
Korobeynikov_A_G@mail.ru

Abstract. Means for providing in a centralized manner a technical interoperability of distributed geographic information systems are suggested. A "format conversion module" which is optimal from the point of view of security and scalability of geoinformational systems and a low cost of software upgrade is developed. The solution is based on the file-server-embedded software that performs the format conversion hidden from the user. The authors have developed algorithms and implemented computer programs for converting data formats. The format description *.ion was analyzed and restored in the course of modernization of Russian vertical ionosphere sounding stations. The problems of embedding the format conversion module into the geoinformational system structure and also technical problems of integration of Russian geoinformational systems into the structure of World Data Centers are considered. A model for building a common information space of geographic informational systems is suggested.

Keywords: Computer-aided design system · Geographic informational system · Interoperability · Format conversion

1 Introduction

Advances in science and technology are accompanied by development of information technologies, in particular, geoinformational systems (GISs). A group of facilities that ensure information transmission is designed for the satellites that perform Earth remote sensing [1]. The problems of interoperability and the absence of the conceptual unity in the space of data access and processing are vital for modern GISs.

Interoperability is the ability of two or more information systems or components to exchange information and to use the information obtained in the exchange. There are

S. Balandin et al. (Eds.): NEW2AN/ruSMART 2015, LNCS 9247, pp. 496–504, 2015.
DOI: 10.1007/978-3-319-23126-6_43

technical (physical, syntactic, semantic) and organizational (consolidational and coordinational) interoperabilities in distributed GISs.

Modern problems of interoperability in GISs often pass from the technical (hardware) level to the field of software development and the use of specialized protocols of transmission, storage formats, and data processing techniques [2–4]. Note that the "data" mean not only the data obtained in scientific observations but also commands, system status reports, etc. In most cases each GIS element (satellite, air monitoring apparatus, ground-based station) receives and transmits data sets. For example, the system of Earth remote sensing satellites EOS MODIS [1] uses 36 spectral channels to obtain images of the Earth's surface and near-Earth space in different spectral ranges. In addition, each satellite must exchange telemetric data of monitoring the satellite state and controlling the satellite equipment. A vertical ionospheric sounding (VIS) station [4–6] typically records not only ionograms, but also the magnetic field strength and weather data at the site where the VIS station is located.

Analysis of operation of Russian VIS stations revealed considerable problems at the level of technical interoperability. For example, ionograms are typically presented as a graphic in *.png format, which is convenient for visual perception but rather unsuitable for computer processing due to a partial loss of data. Many ionosondes have specific formats, which are incompatible with each other in most cases [6,7]. In recent years, Canadian ionosondes CADI [6] with ionogram formats *.md1, *.md2, *.md3, and *.md4 have been widely used. These formats allow one to save vertical sounding data. A significant advantage of the formats is a compact form of information storage, which contributes to unloading of memory resources and data transmission channels.

The achievement of a higher technical level of interoperability can create the common information space at the scale of one GIS and contribute to integration into the World Data Center structure [2,4–6]. However, to perform the international data exchange, results of observation and preliminary analysis should be presented in specific well-documented formats, which requires the use of general principles of data processing and analysis [2,5,6]. At the same time, there is a need to control access to information because of geopolitical interests, copyright protection, corporate interests, etc., at the levels of international cooperation and the organization responsible for the GIS operation.

The goal of our study was to develop the technical interoperability tools that meet the requirements of hidden (from the user or program) and secure conversion of ionogram data formats, to implement them at vertical ionospheric sounding stations, and also to increase the efficiency of GISs via the use of interoperability principles.

2   Centralized Structure for Providing Interoperability

The equipment constituting distributed GISs [4,6] (Fig. 1) is often deeply integrated with the data formats. For example, the data formation can be realized at the hardware level, so the need for format upgrading in individual GIS components can result in a substantial hardware debugging, which is not always possible and appropriate. Our approach involves data exchange between a unit for coordination and control (UCC) and GIS components on the basis of formats and protocols which prove most suitable among those available at the moment.

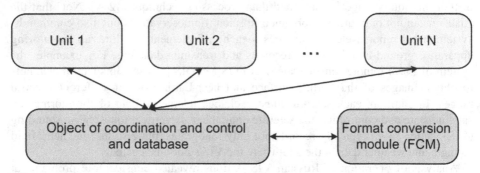

Fig. 1. GIS structure with format conversion module.

To introduce the means of technical interoperability into the GIS architecture under consideration, we suggest that a special resource (Fig. 1) called "format conversion module" (FCM) be used. The physical embodiment of FCM can be a server, virtual computer or a database with special software. In fact, FCM is a version of control system.

The main tasks of FCM are

- to provide code and protocol conversion formats in response to UCC requests,
- to provide a data processing code in a predetermined format in response to UCC requests,
- to support the compatibility and use of the most recent (appointed by an administrator or automatically selected) program versions for data conversion and processing,
- to hide, if possible, conversion procedures from the user,
- to fulfill the security principles. i.e., to hide the format conversion fact from the user and to control rights to access data and programs in accordance with the administrator's requirements.

Because of a considerable variety of hardware and operating systems in the GIS structure, it is reasonable to use the FCM for conversion of protocols and formats in order to reduce the number of required codes. In this case the load on the communication channel between the UCC and FCM can significantly grow [8]. Therefore, it is appropriate to integrate the UCC and FCM into a system connected by a high-speed data exchange channel.

3   Methods for Designing and Integrating the Format Conversion Module

If "compatible" formats are converted, it is possible, as a rule, to save basic information, a part of service information being lost. The loss is acceptable if service information is not necessary to the user and its loss does not affect the quality of data processing and analysis. Service information is often important for the data obtained in research works, so the data should be saved in the original formats (archive) to have a possibility to restore such information. Format conversion with a subsequent data storage in several formats (caching) can lead to a significant increase in the required disk space. On the other hand, if the data in the converted format are not saved, format conversion should be performed at each request, which greatly increases the required computer time. The optimum solution of this problem depends on the available computing resources (disk space, processing power, characteristics of data transmission channels). This is the task of optimizing the use of computing resources of GIS [8].

A possible solution of the format conversion problem is to employ proxy software to hide the conversion from the user. Let us consider program M which is able to accept input data $F1$ and produce result R, so that $R = M(F1)$. Let format $F1$ be compatible (we perform mutual transformation with storage of sufficient data to interpret program M) with format $F2$. The conversion is carried out by program $T: F2 \rightarrow F1$. In order to obtain the result R from the data in the original format $F2$, we perform conversion $R = M(T(F2))$. It is desirable to hide the conversion procedure $T(F2)$ from program M and, hence, from the user (or program). The hidden proxy conversion $T(F2)$ should look for the user as a direct processing of data in format $F2$ by program M.

Let us consider several possible solutions for an operating system which do not contradict the security policy.

1. Program M is requested from program T. After the request of $T(F2)$ a temporary file of format $F1$ is formed, then program $M(F1)$ is requested. This approach is suitable for the applications in which the file name is passed as a parameter of the command and is not applicable when it is necessary to read $F2$ directly from program M.

2. A "hidden" conversion from format $F2$ to format $F1$ is performed by the procedures built in the file server operating in accordance with the FTP protocol (SFTP). Program M (client) and the data $F2$ (server) are at different logical network nodes (perhaps in one computer). On the client side, the $F2$ formats are displayed as $F1$, i.e., all transformations are hidden from the client. This approach (Fig. 2) is well scalable, it also meets the security conditions and does not increase the number of data file copies. A significant advantage offered by this approach is that it is independent of the operating systems and software available on the client side. It also allows a simple organization of remote access and, hence, cloud computing.

3. A direct interception of requests of program M to files $F2$ with a subsequent substitution of $F2$ by $F1 = T(F2)$. For example, it can be performed in Windows by intercepting system calls, which is unsecure.

In the authors' opinion, the most suitable solution for the implementation is solution 2 (Fig. 2) which allows one

  - to fulfill the security conditions,
  - to scale the system without upgrading the software,
  - to arrange a remote access and implement cloud computing methods,
  - to avoid overloading the database with file copies in different formats.

An important feature of this solution is a centralization of the facilities providing technical interoperability.

The solution we suggest makes it possible to implement the technical interoperability methods at the local scale (a distributed GIS) and global scale (international cooperation in the framework of the World Data Center) [2,5,9].

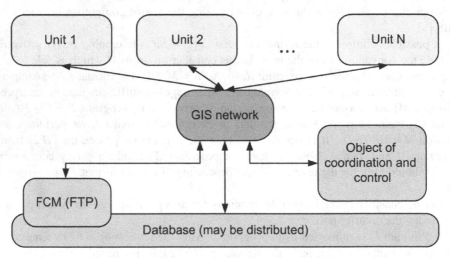

Fig. 2. GIS structure with FCM in the form of a FTP server

4   Formation of the Common Information Space of Distributed GISs

Design of distributed GISs is a continuous dynamic process. Modern distributed GISs with a set of high-tech methods and facilities for data acquisition and processing are the implementation of some models (mathematical, technological, engineering) created by teams of scientists and engineers.

The process of data treatment by the distributed GISs may be divided into

  - routine processes of data acquisition (hardware and software facilities);

- primary data processing (including allocation in a database, systematization, documentation);
- further processing and analysis (simulation, research, prediction, documentation)

The software for primary and subsequent processing and analysis in most cases is updated more often than routine data acquisition processes. In turn, the secondary processing and analysis is updated more often than the primary data processing. There is a tendency of a more frequent updating of software than that of hardware. The facilities at the periphery of a distributed GIS, i.e., the facilities which are interfaces of a distributed GIS, are also updated more often.

The architecture of any GIS includes a variety of mathematical models based on the ideas about the physical phenomena studied and the methods used for their investigation.

Thus, a continuous development of GISs based on advances in scientific knowledge and available technologies allows us to consider GIS as an implementation of mathematical models - from hardware (physical) to software (virtual) (Fig. 3). The set of implementations of these models makes the GIS. In order to provide the GIS operation, it is necessary to "teach" the models (mathematical and physical implementations) that form the GIS how to be interoperable. The basis of the interaction between the models is the exchange of information and the use of information obtained as a result of the exchange, i.e., interoperability. Therefore, the common information space should be formed.

Because of the presence of the common information space of GISs, the distributed GIS can be regarded as an integral structure having all levels of interoperability. Due to the unity of individual components, the GIS

1. facilitates scaling of software and hardware solutions;
2. ensures the same approaches to building integrated facilities for control of design works;
3. contributes to unification of application protocols of information support;
4. provides internal and external consolidation of GISs;
5. enhances the coordination of control processes and, hence, increases the efficiency of control.

Fig. 4 shows the structure of the common information space of GISs. The space of data is formed by a set of data obtained in observations, analysis, predictions, and documentation of a distributed GIS. The space of mathematical models and engineering solutions includes a set of conceptual and partial solutions at the hardware and software levels. The space of users is formed by multiple users of the distributed GIS. The interoperability is a link that allows a "common language" at the technical and organizational levels to be found.

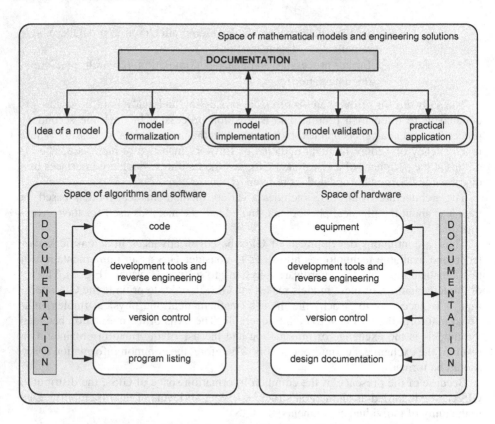

Fig. 3. Space of mathematical models with spaces of hardware and software

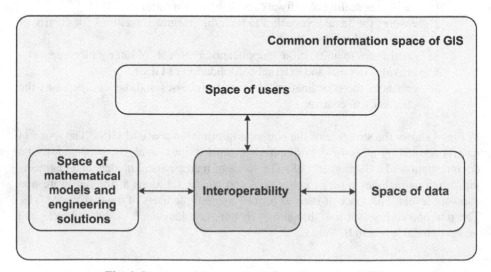

Fig. 4. Structure of the common information space of GISs

The method we suggest was implemented in the framework of modernization of the VIS system of St. Petersburg Institute of Terrestrial Magnetism, Ionosphere and Radio Wave Propagation of the RAS (IZMIRAN). Software libraries for reading files in the *.md1, *.md2, *.md3, *.md4 formats were realized. The libraries may be integrated in Windows or *nux (Linux, UNIX) operating systems [10,11]. In addition, the task of restoration of the *.ion format * (because of a partial loss of the documentation) was fulfilled during modernization of VIS stations of IZMIRAN.

5   Conclusion

Causes and consequences of an insufficient use of the principles of interoperability in distributed GISs have been analyzed. Means for implementation of technical interoperability in the form of the format conversion module have been developed.

The information model of GIS that provides the GIS interoperability at the technical and organizational levels has been developed.

As a result, the authors have solved the problem of ensuring technical interoperability. No changes at GIS nodes are required. Libraries for mutual conversion of files in the *.ion, *.csv, *.md1, *.md2, *.md3, and *.md4 formats have been developed and implemented in the framework of modernization of VIS stations of IZMIRAN and implementation of technical interoperability.

Another important result is the development of the concept of the common information space of a distributed GIS that allows one to regard GIS as an integral system.

References

1. Schowengerdt, R.A.: Remote sensing models and methods for image processing. Elsevier (2007)
2. Laurini, R., Yetongnon, K., Benslimane, D.: GIS Interoperability, from problems to solutions. Advanced Geographic Information Systems. http://www.eolss.net/Sample-Chapters/C01/E6-72-05.pdf
3. Velichko, E.N., Grishentsev, A., Korikov, K., Korobeynikov, A.: Improvement of finite difference method convergence for increasing the efficiency of modeling in communications. In: Balandin, S., Andreev, S., Koucheryavy, Y. (eds.) NEW2AN/ruSMART 2014. LNCS, vol. 8638, pp. 591–597. Springer, Heidelberg (2014)
4. Grishentsev, A., Korobeynikov, A.G.: Design and engineering background for station networks of vertical ionosphere sounding. Scientific and Technical Journal of Information Technologies, Mechanics and Optics 3(85), 61–67 (2013)
5. Piggott, W.R., Rawer, K.: URSI handbook of Ionogram Interpretation and Reduction. INAG (Ionospheric Network Advisory Group) World Data Center A. National Academy of Sciences (1972)

6. Canadian advanced digital ionosonde. System manuals, Canada (2009)
7. Field, P.R., Rishbeth, H.: The response of the ionospheric F2-layer to geomagnetic activity: an analisys of wordwide data. J. Atm. Sol.-Terr. Phys. **59**(2), 163–180 (1997)
8. Grishentsev, A.U., Korobeynikov, A.G.: Solution model of inverse problem of ionosphere vertical sounding. Scientific and Technical Journal of Information Technologies, Mechanics and Optics. **2**(72), 109–113 (2011)
9. NOAA's National Geophysical Data Center (NGDC). http://www.ngdc.noaa.gov
10. Microsoft. Development Network. http://msdn.microsoft.com
11. Stivens, P., Paro, C.: UNIX. Advanced Programming in the UNIX Environment. Addison-Wesley (2005)

Cooperative Spectrum Sensing in Cognitive Radio Networks with QoS Requirements

Jerzy Martyna[✉]

Institute of Computer Science, Faculty of Mathematics and Computer Science,
Jagiellonian University, Ul. Prof. S. Łojasiewicza 6, 30-348 Cracow, Poland
martyna@ii.uj.edu.pl

Abstract. In this paper, we study cooperative multi-channel spectrum sensing in cognitive radio networks (CRNs) with quality of service (QoS) requirements. We first formulate the statistical QoS guarantees in wireless communication in CRNs. Next, the theoretical results on the effective bandwidth to perform the evaluation of the transmission are presented. This paper also proposes a method for multimedia streaming with the required QoS constraints in CRNs. Finally, we consider the cooperative spectrum sensing method among secondary users (SUs) in the CRN with QoS requirements. The simulation results show that the use of cooperative spectrum sensing can improve the performance of SUs with multiuser diversity.

1 Introduction

A study conducted by the Federal Communications Commission (FCC) showed that certain bands of frequencies are partially occupied in specific locations at specific times [6]. This motivates the concept of cognitive radio (CR) [12], which allows unlicensed users (secondary users, SUs) to dynamically and opportunistically access licensed bands allocated to legacy spectrum holders (licensed users, primary users, PUs) temporarily when the spectrum is not being utilised. In other words, the principle of CR requires an alternative handling of the spectrum: a SU can access, at any moment, free frequency bands, which are not occupied by the PU who has a license for those bands. The SU should stop using it once the service is finished or when the PU tries to connect.

A cognitive radio network (CRN) has a hierarchical network architecture, as illustrated in Fig. 1. The CRN can be deployed in network-centric, distributed, ad hoc and mesh architectures, and can serve the needs of both licensed and unlicensed users. Additionally, CRN users can communicate with each other in a multi-hop manner or access the base station. CRN users can operate on both licensed and unlicensed bands. However, we have in CRNs three possible scenarios: a) the CRN system can operate on a licensed band, b) the CRN system can operate on an unlicensed band, c) the CRN system can simultaneously operate on both licensed and unlicensed bands. The third manner provides good example of the cognitive radio concept. In this approach CRN users must constantly monitor the spectrum and then adopt medium access schemes to use the spectrum holes for secondary transmissions, with minimum interference to PUs.

© Springer International Publishing Switzerland 2015
S. Balandin et al. (Eds.): NEW2AN/ruSMART 2015, LNCS 9247, pp. 505–517, 2015.
DOI: 10.1007/978-3-319-23126-6_44

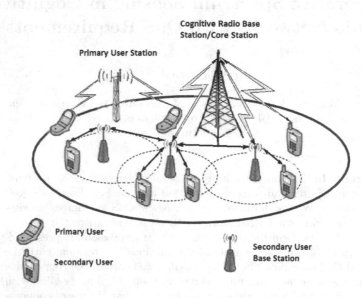

Fig. 1. Basic architecture of downlink/uplink cognitive radio network.

The performance of spectrum sensing can be affected by the degradation of the PU signal due to path loss or shadowing (hidden terminal). These bad sides of spectrum sensing in CRNs can be reduced through collaborative spectrum sensing [7], [15]. Cooperative spectrum sensing with transmit and relay diversity in cognitive radio networks was presented by W. Zhang, and K.B. Lataief [18]. In [7] collaborative spectrum sensing was obtained through a certalised fusion centre combining the SUs' sensing bits using the OR rule for data fusion. Wang et al. [15] propose an evolutionary game model to control the strategies of the SUs and their contributions to the sensing. In general, the use of a centralised spectrum sensing approach reduces path loss and shadowing, but yields a large complexity of computation.

Another approach is based on the multi-channel spectrum sensing problem. The multi-channel spectrum sensing problem was formulated as a coalition game in the paper by X. Hao et al. [9]. An algorithm to find the appropriate channels was proposed as a solution to this problem. Currently, the same authors have formulated a hedonic coalition formation game for cooperative spectrum sensing [10]. However, the above mentioned papers have dealt with the QoS requirements in multimedia traffic flows.

The main contributions of this paper are as follows:

a) We propose an analytical model for multimedia transmission in CRNs. According to this model the cognitive channels are tested by the SUs. Transmission in these channels is possible when the channels are busy and detected as busy or when the channels are idle and detected as busy. The statistical QoS guarantees allow us to determine the probability that the packet delay violates the delay requirement in the cognitive channel.

b) We provide cooperative multi-channel spectrum sensing in CRNs. We assumed that each SU chooses the cognitive channels with the QoS requirements. Then, all the SUs that choose the cognitive channels with the QoS requirements, do cooperative spectrum sensing to determine all of those that are useful in multimedia transmission.

The rest of the paper is as follows. In Section 2 statistical QoS guarantees for wireless communication in CRNs are presented. Section 3 formulates cooperative spectrum sensing among the SUs in CRNs. In Section 4, we provide the results of the performance analysis of our method for various system parameters. The conclusion is provided in Section 5.

2   Statistical QoS Guarantees for Wireless Communication in CRNs

2.1   Cognitive Channel Model for Transmission with QoS Requirements

We assume that a cognitive channel is tested by the SUs. If the secondary transmitter (ST) selects its transmission when the channel is busy, the average power is equal to $\overline{P}_1$ and the rate is equal to r_1. When the channel is idle, the average power is equal to $\overline{P}_2$ and the rate is equal to r_2. In our model the average power $\overline{P}_1 = 0$ denotes that the secondary transmission must be checked in the presence of an active primary unit (PU). Both values of transmission rates, r_1 and r_2, can be fixed or timed depending on whether or not the transmitter has channel side information.

We assume that $\overline{P}_1 < \overline{P}_2$. In the cognitive channel model in the absence of PUs the discrete-time channel input-output relation is given by

$$y(i) = h(i)x(i) + n(i), \quad i = 1, 2, \ldots \tag{1}$$

where $h(i)$ is the channel coefficient, i is the symbol duration. If the PUs are present in the channel, the discrete-time channel input-output relation is given by

$$y(i) = h(i)x(i) + s_p(i) + n(i), \quad i = 1, 2, \ldots \tag{2}$$

where $s_p(i)$ represents the sum of the active PUs faded signals arriving at the secondary receiver (SR), $n(i)$ is the additive thermal noise of the SR and is zero-mean, circularly symmetric, complex Gaussian random variables with variance $E\{| n(i) |^2\} = \sigma_n^2$ for all i.

We assume that the receiver knows the instantaneous value $\{h(i)\}$, while the transmitter has no such knowledge. We have constructed a state-transition model for cognitive transmission by considering cases in which fixed transmission rates are lesser or greater than the instantaneous channel capacity values. In particular, the ON state is achieved if the fixed rate is smaller than the instantaneous channel capacity. Otherwise, the OFF state occurs.

We assume that the maximum throughput can be obtained in the state-throughput model [1], which is given in Fig. 2. Four possible scenarios are associated with the model, namely:

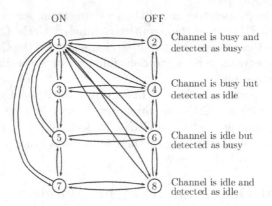

Fig. 2. State transition model with 8 states for the cognitive radio channels.

1) channel is busy, detected as busy (correct detection),
2) channel is busy, detected as idle (miss-detection),
3) channel is idle, detected as busy (false alarm),
4) channel is idle, detected as idle (correct detection).

If the channel is detected as busy, the secondary transmitter sends with power $\overline{P}_1$. Otherwise, it transmits with a larger power, $\overline{P}_2$. In the above four scenarios, we have the instantaneous channel capacity, namely

$$C_1 = B \log_2(1 + SNR_1 \cdot z(i)) \text{channel is busy, detected as busy} \quad (3)$$

$$C_2 = B \log_2(1 + SNR_2 \cdot z(i)) \text{channel is busy, detected as idle} \quad (4)$$

$$C_3 = B \log_2(1 + SNR_3 \cdot z(i)) \text{channal is idle, detected as busy} \quad (5)$$

$$C_4 = B \log_2(1 + SNR_4 \cdot z(i)) \text{channel is idle, detected as idle} \quad (6)$$

where B is the bandwidth available in the system, $z(i) = [h(i)]^2$, SNR_i for $i = 1, \ldots, 4$ denotes the average signal-to-noise ratio (SNR) values in each possible scenario.

The cognitive transmission is associated with the ON state in scenarios 1 and 3, when the fixed rates are below the instantaneous capacity values ($r_1 < C_1$ or $r_2 < C_2$). Otherwise, reliable communication is not obtained when the transmission is in the OFF state in scenarios 2 and 4. Thus, the fixed rates above are the instantaneous capacity values ($r_1 \geq C_1$ or $r_2 \geq C_2$). The above channel model has 8 states and is depicted in Fig. 2. In states 1, 3, 5 and 7, the transmission is in the ON state and is successfully realised. In the states 2, 4, 6 and 8 the transmission is in the OFF state and fails.

2.2   Statistical QoS Guarantees

Real-time multimedia services such as video and audio require bounded delays, or the guaranteed bandwidth. If a received real-time packet violates its delay, it will be discarded. The concept of effective capacity was developed to provide the statistical QoS guarantee in general real-time communication. Among others, in the paper by [3], it was shown that for a queuing system with a stationary ergodic arrival and service process, the queue length process $Q(t)$ converges to a random variable $Q(\infty)$ such that

$$- \lim_{x \to \infty} \frac{\log(Pr\{Q(\infty) > x\})}{x} = \theta \tag{7}$$

Note that the probability of the queue length exceeding a certain value x decays exponentially fast as x increases. The parameter θ ($\theta > 0$) gives the exponential decade rate of the probability of QoS violation.

A framework of statistical QoS guarantees [4] was developed in the context of the wireless communication [16]. In accordance with the effective bandwidth theory, effective capacity can be defined as

$$E_{cap}(\theta) \triangleq - \lim_{t \to \infty} \frac{1}{\theta t} \log \left(E \left[e^{-\theta S[t]} \right] \right) \tag{8}$$

where $S[t] \triangleq \sum_{i=1}^{t} R[i]$ is the partial sum of the discrete time stationary and ergodic service process $\{R[i], i = 1, 2, \ldots\}$.

The probability that the packet delay violates the delay requirement is given by

$$Pr\{Delay > d_{max}\} \approx e^{-\theta \delta d_{max}} \tag{9}$$

where d_{max} is the delay requirement, δ is a constant jointly determined by the arrival process and theirs service process, θ is a positive constant referred to QoS exponent.

3   Cooperative Spectrum Sensing in CRN with QoS Requirements

In this section, we present the problem of cooperative spectrum sensing in cognitive radio networks for QoS provisioning.

We assume that the number of SUs collaborating together is equal to N. All of them are independent and identically distributed (iid) fading/shadowing with the same average SNR. In our approach $\mathcal{H}_0$ and $\mathcal{H}_1$ denote the absence and the presence of the PU, respectively.

The goal of cooperative spectrum sensing is to formulate two fundamental probabilities making up represent the sensing reliability of SUs: probabilities of detection and false alarm for sensing the existence of primary signals at any moment. We define these probabilities as follows:

$$Q_D = 1 - (1 - P_D)^N \tag{10}$$

$$Q_F = 1 - (1 - P_F)^N \tag{11}$$

where P_D and P_F are respectively the individual probabilities of detection and false alarm. It is evident that this cooperative scheme increases the probability of detection as well as the probability of false alarm.

The probabilities of detection and false alarm are given by [5], [7], [8]

$$P_D = P\{Y_j > \lambda \mid \mathcal{H}_1\} = \frac{\Gamma(\frac{M}{2}, \frac{\lambda}{2(1+\gamma_j)})}{\Gamma(\frac{M}{2})}, \ 1 \le j \le N \tag{12}$$

$$P_F = P\{Y_j > \lambda \mid \mathcal{H}_0\} = \frac{\Gamma(\frac{M}{2}, \frac{\lambda}{2})}{\Gamma(\frac{M}{2})}, \ 1 \le j \le N \tag{13}$$

where N is the number of cooperative users, M is the number of samples, λ is a threshold value to decide whether signal is present or not, $\Gamma(.)$, $\Gamma(.,.)$ are complete and incomplete gamma functions, respectively.

Assuming that in a multipath fading environment all the CR users experience Nakagami fading channels with the same average SNR, given by $\overline{\gamma}$, the instaneous power has gamma pdf given by

$$f_\gamma(\gamma, m) = \frac{m^m \gamma^{m-1}}{\overline{\gamma}^m \Gamma(m)} e^{-\frac{m\gamma}{\overline{\gamma}}}, \ \gamma \ge 0 \tag{14}$$

where m is the integer shape factor defined as the Nakagami parameter.

Throughput this paper, we assume that energy detection [14] is used at each SU. Additionally, we suppose that cooperative sensing process consists of three steps [2], namely:

1) Cognitive radio network uses the fusion centre (FC) which selects a channel or frequency of interest for sensing or requests all cooperating SUs to individually perform local sensing;
2) All cooperating SUs report their sensing results over the signalling channel;
3) The FC fuses receive the local sensing information to decide on the acceptation or turn down of the channel with multimedia transmission.

In our approach a role of FC in CRN can fulfil the base station of secondary network which can make the decision by combining them appropriately. We admit here only a soft combination scheme that optimises the detection performance. A soft combination provides better performance than other methods, but it requires a larger bandwidth for the control channel [19].

Let us first consider a problem of conflicting probabilities involved in binary hypothesis testing, namely the detection probability and false probability. In spectrum sensing, statistical hypothesis testing is typically performed to test the sensing results for the binary decision on the presence of PUs. We use here the Neyman-Pearson criterion, equivalent to the likelihood ratio test (LRT) [17], to detect probability for a given false alarm. The following likelihood ratio test is given by

$$LR(\mathbf{Y}) \underset{\mathcal{H}_0}{\overset{\mathcal{H}_1}{\gtrless}} \eta \tag{15}$$

where η is the threshold value determined by the given false alarm probability. $LR(\mathbf{Y})$ can be decomposed as

$$LR(\mathbf{Y}) = \frac{Pr(\mathbf{Y} \mid \mathcal{H}_1)}{Pr(\mathbf{Y} \mid \mathcal{H}_0)} = \prod_{j=1}^{N} \frac{Pr(Y_j \mid \mathcal{H}_1)}{Pr(Y_j \mid \mathcal{H}_0)} \tag{16}$$

where $Pr(Y_j \mid \mathcal{H}_0)$ and $Pr(Y_j \mid \mathcal{H}_1)$ can be determined as follows

$$Pr(Y_j \mid \mathcal{H}_0) = \frac{(\frac{1}{2})^{\frac{M}{2}}}{\Gamma(\frac{M}{2})} Y_j^{\frac{M}{2}-1} e^{-\frac{1}{2}Y_j} \tag{17}$$

and

$$Pr(Y_j \mid \mathcal{H}_1) = \frac{1}{1+\gamma_j} \frac{(\frac{1}{2})^{\frac{M}{2}}}{\Gamma(\frac{M}{2})} \left(\frac{Y_j}{1+\gamma_j}\right)^{\frac{M}{2}-1} e^{\frac{1}{2}\frac{Y_j}{1+\gamma_j}} \tag{18}$$

where M is the number of samples. Then, $LR(\mathbf{Y})$ can be calculated as

$$LR(\mathbf{Y}) - \left(\prod_{j=1}^{N} \frac{1}{1+\gamma_j}\right)^{\frac{M}{2}} e^{\frac{1}{2}\sum_{j=1}^{N} \frac{\gamma_j}{1+\gamma_j}Y_j} = \sum_{j=1}^{N} \frac{\gamma_j}{1+\gamma_j} Y_j \underset{\mathcal{H}_0}{\overset{\mathcal{H}_1}{\gtrless}} \mu \tag{19}$$

where $\mu = 2\ln \eta + M \sum_{j=1}^{N} \ln(1+\gamma_j)$ is the new decision threshold value determined by the given false alarm probability.

In the soft combination scheme with weights w_j, $1 \le j \le N$, the weighted summation of the observed energies can be obtained [11] as

$$Y = \sum_{j=1}^{N} w_j Y_j = \begin{cases} \sum_{j=1}^{N} w_j b_{j0} & \mathcal{H}_0 \\ \sum_{j=1}^{N} w_j(1+\gamma_j)b_{j1} & \mathcal{H}_1 \end{cases} \tag{20}$$

where b_{j0} (or b_{j1}) follow an i.i.d. central chi-square distribution with M degree of freedom for a given hypothesis.

It can be seen that the detection performance of the optimal soft combination is obtained by using only one threshold dividing the whole range of the observed energy into two regions and applying the one-bit combination. In order to reduce the error probability, we divide the observed energy into four parts and use softened the two-bit hard combination. According to the principle of the two-bit combination scheme, given in [11], the decision criterion is equivalent to allocating the 4 regions different weights, $w_0 = 0, w_1 = 1, w_2 = L, w_3 = L^2$. Thus, the weighted summation is given by

$$N_c = \sum_{i=0}^{3} w_i N_i \tag{21}$$

where N_i denotes the number of observed energies falling in region i. If $N_c > L^2$, the signal of PU is declared present, otherwise, it is declared absent.

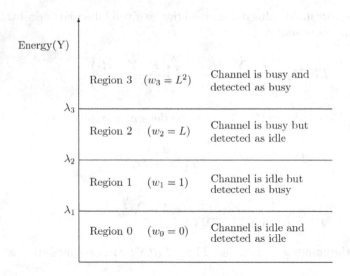

Fig. 3. Principle of two-bit combination scheme for the cognitive radio channel with QoS requirements.

For a softened combination scheme with two-bits, all threshold values, namely $\lambda_1, \lambda_2, \lambda_3$ must be determined to indicate the target overall false alarm probability. All three thresholds in the two-bit scheme, $\lambda_1, \lambda_2, \lambda_3$, are here associated with the state model for the cognitive radio channelswith QoS requirements given in Fig. 2. Then, we obtain the whole range of the observed energy divided into 4 regions with 8 states (see Fig. 3).

The overall false alarm probability, Q_F, of the N SUs in CRN can be determined as follows. Assuming that in order to avoid a false alarm there must be no SU in the region 3, and j users in region 2, $j - i$ users in region 1, and all of the rest $N - i$ users in region 0.

Thus, $N_3 = 0$, $N_2 = j$, $N_1 = i - j$, and $N_0 = N - i$. The probability of the successful detection of $\mathcal{H}_0$, expressed as $1 - Q_f$, can be obtained by summing all of the probabilities of i and j that avoid false alarm and is given by [11]

$$
1 - Q_f = \sum_{i=0}^{I} \sum_{j=1}^{J_1} Pr(N_0 = N - i, N_1 = i - j, N_2 = j, N_3 = 0 \mid \mathcal{H}_0)
$$

$$
= \sum_{i=0}^{I} \binom{N}{i} (1 - P_{F_i})^{N-i} \left\{ \sum_{j=0}^{J_1} \binom{i}{j} (P_{F_1} - P_{F_2})^{i-j} (P_{F_2} - P_{F_3})^{J} \right\} \quad (22)
$$

where $I = L^2 - 1$, $J_i = \min \left\{ \lfloor \frac{L^2 - l - iw_1}{w_2 - w_1} \rfloor, i \right\}$ and $P_{F_l} = P(\mathbf{Y} > \lambda_l \mid \mathcal{H}_0)$ denotes the false alarm probability at each SU corresponding to threshold value $\lambda_l, 1 \leq l \leq 3$, $\lfloor . \rfloor$ indicates the largest integer no greater than the argument.

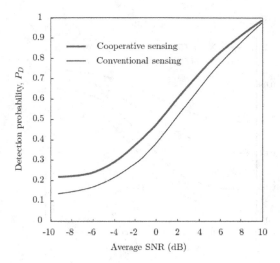

Fig. 4. The detection probability versus the average SNR for the system with cooperative sensing based on two-bit combination scheme and conventional sensing.

In similar manner it is possible to obtain Q_d - the average overall detection probability at each SU. Assuming that I and J_i are defined as earlier, the value of $\overline{Q}_D$ can be expressed by [11]

$$\overline{Q}_D = 1 - \sum_{i=0}^{I} \sum_{j=0}^{J_i} Pr(N_0 = N - i, N_1 = i - j, N_2 = j, N_3 = 0 \mid \mathcal{H}_1)$$

$$= 1 - \sum_{i=0}^{I} \binom{N}{i} (1 - \overline{P}_{D_1})^{N-i} \left\{ \sum_{j=0}^{J_i} \binom{i}{j} (\overline{P}_{D_1} - \overline{P}_{D_2})^{i-j} (\overline{P}_{D_2} - \overline{P}_{D_3})^j \right\}$$

$$(23)$$

where $\overline{P}_{D_l} = P(Y > \lambda_l \mid \mathcal{H}_1)$ indicates the average local detection probability at each SU corresponding to threshold value $\lambda_l, 1 \leq l \leq 3$.

4   Simulation Results

In this section the simulation results are reported in order to demonstrate the compliance of cooperative sensing in CRN with QoS requirements. We assume that there are 2 PUs and 40 SUs composed of CR-Tx CR-Rx pairs under Rayleigh fading channels. We assume that $L = 2$ and the streaming service rate requirements of all the pairs can be increased up to 1 Mbps. The average SNR between the PU and any SU is equal to 10 dB.

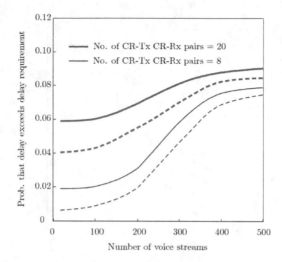

Fig. 5. The probability that delay exceeds the delay requirement versus the number of voice streams. Dash lines indicate the cooperative sensing.

First, we investigated the overall detection probability in the system without cooperative sensing and with cooperative sensing, based on the two-bit combination scheme and without cooperative sensing. Fig. 4 shows that the two-bit combination scheme achieves better performance than the system without cooperative sensing.

Further, we investigated the probability of delay-bound violations of voice stream versus the number of voice streams with and without cooperative spectrum sensing among the CR-Tx CR-Rx pairs (see Fig. 5). We observed that cooperative spectrum sensing reduces the probability of delay-bound violations for all streaming flows.

Fig. 6 shows the mean time delay versus the mean distance between PUs for cooperative spectrum sensing among the CR-Tx CR-Rx pairs with QoS requirement and without QoS requirement. When the observed mean distance between PUs is increased, mean time delay increases - just as the previous cooperative spectrum sensing reduces the mean time delay. It is apparet that sensing cooperation between two or more SUs reduces the probability of a false alarm. Thus, spectrum holes are utilised more efficiently.

Fig. 7 shows the normalised capacity of the QoS capacity of the secondary system versus the mean distance between PU station for all three possible sensing (cooperative sensing with QoS requirements, cooperative sensing without QoS requirements, conventional sensing). The normalised capacity C, in the secondary system can be derived as

$$\frac{C}{B} = \log_2(1 + SNR) \tag{24}$$

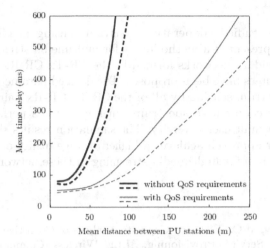

Fig. 6. Mean time delay versus mean distance between PUs for cooperative spectrum sensing among the CR-Tx CR-Rx pairs with QoS requirement and without QoS requirements. Dash lines indicate the cooperative sensing.

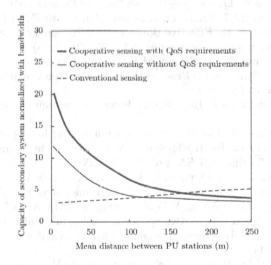

Fig. 7. Normalised capacity of the secondary system versus the mean distance between PU station for cooperative sensing with QoS requirements, and also cooperative sensing without QoS requirements and conventional sensing.

In Fig. 7 we can observe that a significant improvement of the capacity is achieved in cooperative sensing with QoS requirements. This is because the cooperative sensing with QoS requirements senses the channels with QoS requirements than considering all channels including the worse case.

5   Conclusion

In this paper, we studied cooperative spectrum sensing in CRNs with QoS requirements. We presented a methodology for multimedia streaming with QoS constraints in secondary networks formed by the CR-Tx CR-Rx pairs. The statistical QoS guarantees have been proposed. The derived parameters can be used in cooperative spectrum sensing by all of the CR-Tx CR-Rx pairs in the CRNs. We show that the cooperative spectrum can improve the performance of the CR-Tx CR-Rx via multiuser diversity. The simulation results showed that the performance of our approach achieves a higher average utility of the SUs for all of the QoS parameters for multimedia streaming in these networks.

References

1. Akin, S., Gursoy, M.C.: Effective Capacity Analysis of Cognitive Radio Channels for Quality of Service Provisioning. IEEE Wireless Communications **9**(11), 3354–3364 (2010)
2. Cabric, D., Mishra, S., Brodersen, R.: Implementation issues in spectrum sensing for cognitive radios. In: Proc. of Asilomar Conf. on Signals, Systems, and Computers, vol. 1, pp. 772–776 (2004)
3. Chang, C.-S.: Stability, Queue Length, and Delay of Deterministic and Stochastic Queueing Networks. IEEE Trans. on Automat. Control **39**(5), 913–931 (1994)
4. Courcobetis, C., Weber, R.: Effective Bandwidth for Stationary Souces. Probability in Engineering and Information Science **9**(2), 285–294 (1995)
5. Digham, E.F., Alouini, M.-S., Simon, M.K.: On the energy detection of unknown signals over fading channels. In: Proc. of IEEE Int. Conf. on Communications (ICC 2003), pp. 3575–3579 (2009)
6. FCC2002 Spectrum Policy Task Force, Federal Communications Commission. Tech. Rep. (2002)
7. Ghasemi, A., Sousa, E.S.: Collaborative spectrum sensing for opportunistic access in fading environments. In: IEEE Symp. New Frontiers in Dynamic Spectrum Access Networks, Baltimore, USA, pp. 131–136 (2005)
8. Ghasemi, A., Sousa, E.S.: Opportunistic Spectrum Access in Fading Channels Through Collaborative Sensing. Journal of Communications **2**(2), 71–82 (2007)
9. Hao, X., Cheung, M.H., Wong, V.W.S., Leung, V.C.M.: A coalition formation game for energy-efficient cooperative spectrum sensing in cognitive radio networks with multiple channels. In: Global Telecommunications Conference (GLOBECOM 2011), pp. 1–6 (2011)
10. Hao, X., Cheung, M.H., Wong, V.W.S., Leung, V.C.M.: Hedonic Coalition Formulation Game for Cooperative Spectrum Sensing and Channel Access in Cognitive Radio Networks. IEEE Trans. on Wireless Communications **11**(11), 3968–3979 (2012)
11. Ma, J., Zhao, G., Li, Y.: Soft Combination and Detection for Cooperative Spectrum Sensing in Cognitive Radio Networks. IEEE Trans. on Wireless Comm. **7**(11), 4502–4507 (2008)
12. Mitola III, J., Maguire Jr, G.Q.: Cognitive Radio: Making Software Radios More Personal. IEEE Trans. on Personal. Communications **6**(4), 13–18 (1999)
13. Nuttall, A.H.: Some Integrals Involving the Q_M Function. IEEE Trans. on Information Theory **21**(1), 95–96 (1975)

14. Urkowitz, H.: Energy Detection of Unknown Deterministic Signals. Proc. IEEE **55**, 523–531 (1967)
15. Wang, B., Liu, K.J., Clancy, T.: Evolutionary Cooperative Spectrum Sensing Game: How to Collaborate? IEEE Trans. on Communications **58**(3), 890–900 (2010)
16. Wu, D., Negi, R.: Effective Capacity: A Wireless Link Model for Support Quality of Service. IEEE Trans. on Wireless Comm. **2**(4), 630–643 (2003)
17. Varshney, P.K.: Distributed Detection and Data Fusion. Springer-Verlag, New York (1997)
18. Zhang, W., Lataief, K.B.: Cooperative spectrum sensing with transmit and relay diversity in cognitive radio networks - [transaction letters]. IEEE Trans. on Wireless Communications **7**(12), 4761–4766 (2008)
19. Quan, Z., Cut, S., Pour, H.V., Sayed, A.H.: Collaborative Wideband Sensing for Cognitive Radios. IEEE Signal Processing Magazine **25**(6), 63–70 (2008)

The Using of Bluetooth 4.0 Technologies
for Communication with Territorial-Distributed Devices

Pavel Mal'kov[1], Sergei Elyagin[2], Vitalii Dement'ev[2], and Nikita Andriyanov[2]($\boxtimes$)

[1] LLC AIS Gorod, Dimitrovgrad, Russia
[2] Ul'yanovsk State Technical University, Ul'yanovsk, Russia
nik.andrianov@ulstu.ru

Abstract. This paper describes a number of technologies of remote data collection from various kinds of objects. Particular attention is paid to the collection of data from meters of household energy (heat, water, electricity, gas). The main advantages and disadvantages of the most often used technologies of automatic data collection from residential meters are presented in the text. In addition, the paper analyzes the types of available counters in Ul'yanovsk Region. It is proposed the using of Bluetooth 4.0 BLE technology when you develop services of information exchange with territorially-distributed devices, including energy meters. Besides, we propose the settings and configuration of the Bluetooth module for solving the tasks of information exchange. It is shown, that the using Bluetooth 4.0 BLE technology is more efficiency comparing with other technologies.

Keywords: Data collection · Bluetooth 4.0 BLE · Bluegiga · Wireless technologies · Smart house

1   Introduction

In accordance with Federal Law №261 from 21.11.2009 (in the district of 10.06.2012) "About the Energy Saving" [1] the owners of dwellings and buildings in Russia will be required to install the necessary control devices. The meters accounting for consumed energy (water, electricity, gas, heat) will become such devices. So, owners will be pay for utility services according to these meters' indications. But this raises a number of questions about the monitoring of the indications of these counters [2]. First, how is carried out verification of indicators? Second, in what timeframe is carried out verification of indicators? Last, what penalties can expect the owners, which change testimony of counters, etc? At present these problems are solved by the participation of specialized personnel. They periodically bypass the places of the installation of the meters and check the indications. It is clear, that the control issues will become much more acute in the case of total installation of meters. The main reason of such changing is the need to check periodically all households. In addition, ordinary meters installation does not solve the issues related to the analysis of their indications, leaks, over expenditure of resources, reporting to higher authorities and companies and so forth. In this regard, it is urgent to develop of technologies that can reduce these costs by automating the collection and analysis of resource consumption

© Springer International Publishing Switzerland 2015
S. Balandin et al. (Eds.): NEW2AN/ruSMART 2015, LNCS 9247, pp. 518–528, 2015.
DOI: 10.1007/978-3-319-23126-6_45

data. At the same time it is important to ensure conditions of integrity and confidentiality of the transmitted and measured information, as well as the minimum cost of acquisition and maintenance of the devices of end-user.

Currently there are a wide variety of the automated control systems of energy accounting (AMR) [3]. However, the vast majority of them is based either on the use of a wiring (PLC, LAN), and it is not very applicable in the real world because of the difficulty with the installation or expensive wireless (Wi-Fi, GSM) systems [4], which require both the high energy and a power cord. In recent years the new solutions are appeared. They are based on standards ZigBee and M-BUS and their task is to create a wireless data reading from counters. In addition, new devices theoretically must be able to work on independent power supply (conventional battery) [5-6]. However, practice shows that the use of standard solutions to reduce energy consumption is insufficient for the real battery life. For example, household meters can't work for at least several months.

The main advantages and disadvantages common to date technologies for collecting meter data are presented in Table 1.

Table 1. Solutions for data collection from meters of household energy

Title	Physical medium	Advantages	Disadvantages
Wired connections	Cables of various kinds	Speed Reliability	Complexity of the installation, Cost
GSM/GPRS/3G	Radio channel	Ease of installation	High cost of the traffic
PLC	Electrical wiring	Speed	Transfer of the power networks
Sensor networks	Radio channel	Ease of Installation	Battery life

In addition, number of questions concerning the organization of media gateways to the global network appears. These tasks must be solved to create affordable, reliable and extremely simple both to install and operate the modules of the data collection. Besides, the emergence of wireless infrastructure for the collection and processing of measured data in private apartments and homes will allow launching new commercially viable services, focused both on obtaining information about the state of some object and on its control. For example, currently an urgent task is to monitor the state of the premises (temperature, humidity, gas concentration levels and indicators of the electromagnetic field) in the state and municipal institutions (kindergartens, schools, hospitals, etc.). But existing solutions are either expensive and difficult to install and use, or they are based on elementary human control, which doesn't allow to reliably and efficiently managing the situation. Meeting the challenges of monitoring the state of the objects, in turn, opens up new possibilities in the management of these facilities. For example, the automatic maintenance of the temperature in certain buildings and rooms may be such feature. Thus mentioned facts allow making conclusion about the relevance of the development of new technologies to organize both the receipt of information from remote measurement devices and of the necessary control.

In the present work we consider specialized devices based on the use of the protocol Bluetooth 4.0 BLE [7] as tools that solve the problem of the "last mile" in the various monitoring systems. This protocol is the result of Bluetooth technology development. At the same time the submitted modules are open for configuration and setup. A key advantage of Bluetooth 4.0 BLE is the ultra low power consumption of end devices both in standby mode and in peak load. This allows getting the autonomous operation time for endpoint devices from standard batteries about few years. It is also important, that the device providing data to end-user devices is the gadgets available in every modern family. Really, it is smart phones and PDAs. This allows you to create dynamic diagrams for collecting and processing of measurement information. It should be noted, that the use of programmable devices such as smart phones running Android or iOS family allows you to organize storage and transfer the accumulated information to arbitrary treatment centers. This synchronization ensures delivery of the necessary information at all levels of public utilities.

Thus, a very promising direction is use of Bluetooth technology in the development of exchange information services with territorial-distributed devices.

2   The Development of the Interconnection Interfaces with the Measuring Devices

A brief overview of the Russian market of measuring devices shows, that the current resource consumption metering devices are supplied with the following standard interfaces: RS-232, RS-485, CAN and pulse output. In this connection, it is planned to connect the control devices to consumption devices via both RS-232, RS-485 interfaces and interfaces which use a pulsed output. Failure of using the CAN interface is explained by two reasons:

1. It requires a dedicated microcontroller for connect to metering devices, which complicates the monitoring device and makes it more expensive;
2. The relatively short length of the connecting wires (up to 40 meters), which does not allow to build an extended network.

The RS-232 interface is the most common in the modern embedded systems, and it has the only drawback of the method of building a network, that supports only the "point to point" topology. This restriction requires the direct connection of the control device to the metering devices. Then all control devices join at the network.

RS-485 interface allows you to use a single control device to serve multiple (up to 30) meters, connected in the network using RS-485 technology with a length of trunks up to 500 meters. This reduces both the number of control devices and the cost of the newly introduced equipment.

RS-232 and RS-485 interfaces are presented in the Calculator type VCT and electric meters Energomera and Mercury with TTL level. This type of metering devices installed in the Ulyanovsk region. The connection of developing control devices by

RS-232 interface does not cause too much trouble, because the used modems BLE-112 (BlueGiga) and microcontrollers AT91SAM7S256 (Atmel) contain UART port working with TTL level. The only limitation is the need to decrease the voltage on the Rx line to 3 volt for ETRX2 and BLE-112. Voltage decrease performed easily by means of two resistors included by attenuator scheme.

To use the RS-485 interface it is necessary to connect additional UART interface chip (e.g. MAX3430) converting the signal of UART port into a signal of RS-485 interface. The technology uses a RS-485 bi-directional two-wire line, therefore the control outputs for enable/disable transmission/reception are needed for distinguishing between the processes of reception and transmission in interface chips. External control signal should be generated by a control device. Such signal can be generated by the microcontroller AT91SAM7S256, because it contains the hardware driver of RS-485 interface. Using ETRX2 and BLE-112 having no hardware driver leads to the appearance of the echo signal at the receiving side, which requires software filtering of received messages. You should take into account this feature when you are writing the program of processing of the received messages.

Thus, the developed monitoring devices provide backhaul sending of the control messages from the central unit to the metering devices and the back way.

The use of meters with pulse output (water meters and electric meters) requires the pulses counting followed by transfer the amount to the central unit. The circuits with open collector or mechanical contact, working on switching, are used as pulse output metering devices. In both cases, we need the external power supply and a current limiting resistor. The input of developed control devices connects to pulse output, and control devices are adjusted to the mode of termination. Finally, the presence of the voltage pulse leads to adding the "one" to general sum of the pulses.

The technical schemes describing circuit connecting of the BlueGiga modules to the respective interfaces are shown at Fig. 1, 2, 3 and 4.

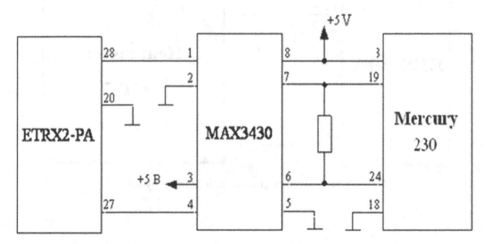

Fig. 1. Scheme of connection to the electricity meter by RS-485

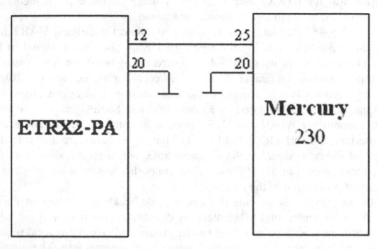

Fig. 2. Scheme of connection to the electricity meter by the pulse output (taking into account the active power).

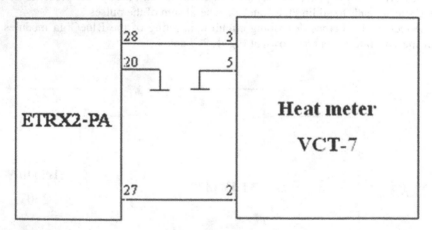

Fig. 3. Scheme of connection to the heat meter by RS-232.

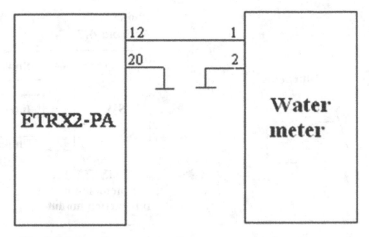

Fig. 4. Scheme of connection to the water meter by the pulse output.

Thus, the analysis showed that it is not only permissible but also desirable to use remote data collection technologies for a number of counters. And the decision based on the technology of Bluetooth 4.0 is able to get an important place among these technologies.

3   Setup and Configuration of the Bluetooth Modules

The following technique was suggested to research the issues of energy efficiency and the range of work of developed modules. It was decided to use an alternative to Zig-Bee technology, namely Bluetooth (configuration BLE). The hardware component of the technologies is identical, and the declared power consumption characteristics are also the same. But Bluetooth can significantly easier be integrated with various devices at least only because of the wide spread of the latter. Furthermore, the final version of developed research project will allow in the future entering the commercial tool also solving the problem of the "last mile" in the monitoring of resource utilities. And it will be possible to exploit specialized devices based on the use of the Bluetooth 4.0 BLE protocol. As previously mentioned, the devices providing data to end-user devices are available in each of the modern family. They are smart phones and PDAs. It allows to generate the following dynamic scheme for the collection and processing of the measuring data (Figure 5).

It should be noted that the use of programmable devices such as a smart phone running Android or iOS family, allows you to organize the transfer of the accumulated information to arbitrary processing centers. Such synchronization performed either manually or automatically allows providing the exchange of the necessary information at all levels of public utilities with minimal cost compared with the known systems. Let us briefly consider the specific implementation of the system, developed by us in accordance with the described technology.

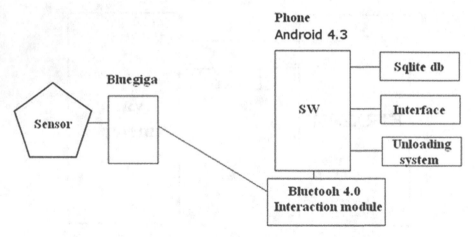

Fig. 5. The structure of the system based on Bluetooth 4.0 technologies

The system is a combination of the following devices:

1. The Bluegiga 112 module as part of a compact transceiver Bluetooth 4.0 BLE (Bluetooth low energy), the integrated modules of memory (up to 1MB), input/output ports (2 analog inputs and RS-485) and the managing microcontroller.
2. The mobile phone running at the operating system Android 4.3 and having a hardware implementation of Bluetooth 4.0 BLE.
3. The mobile application providing the interaction of phone with a remote Bluegiga device.
4. The server software, which allows you to synchronize with the ultimate mobile applications for collecting and updating data.

The Bluegiga module is capable to store the measurement data, correlating them with the measurement time. The transfer of accumulated information is running when the mobile phone with the installed program of analysis and storage of data appears in range of BLE module. Herewith the data acquisition from BLE device is a request of values of so-called characteristics. This is a request for a hexadecimal, unique identifier to a particular device with a specific MAC address. Thus, each device may have several such characteristics and therefore may transmit multiple types of data. These data are in particular the temperature, humidity, light intensity, number of pulses received from the counter of the resources, battery life, signal strength, etc.

The data obtained from the characteristics are stored in the local database of the phone (SQLite). This database provides the storage of the information obtained from the tables, with the further possibility of their processing, editing, and also the selection and sending. The data is sorted by MAC address of the device and by the type of the data.

Thus, nowadays it is possible to create a compact and low-cost devices continuously collecting information from the arbitrary sensors (e.g., temperature sensors, metering devices hot / cold water, electricity meters, etc.) and transmitting this information to devices for collecting and publishing information.

Table 2. Bluegiga module's work at different load levels

Working hours	Load level								
	100%	70%	40%	20%	10%	5%	3%	2%	1%
6 h	95%	97%	98%	99%	99%	99%	99%	99%	99%
12 h	91%	94%	97%	98%	99%	99%	99%	99%	99%
24 h	84%	88%	93%	96%	98%	99%	99%	99%	99%
2 d	70%	76%	83%	91%	95%	98%	99%	99%	99%
3 d	56%	64%	74%	87%	93%	96%	98%	99%	99%
4 d	45%	53%	66%	82%	91%	95%	97%	98%	99%
5 d	32%	42%	53%	76%	83%	91%	96%	98%	99%
7 d	15%	29%	40%	50%	76%	87%	94%	97%	98%
10 d	1%	6%	13%	21%	61%	80%	90%	95%	98%
15 d	0%	1%	2%	16%	56%	78%	87%	93%	95%
20 d	0%	0%	0%	12%	51%	76%	85%	91%	95%
25 d	0%	0%	0%	8%	48%	73%	81%	90%	94%
30 d	0%	0%	0%	4%	24%	61%	78%	86%	91%

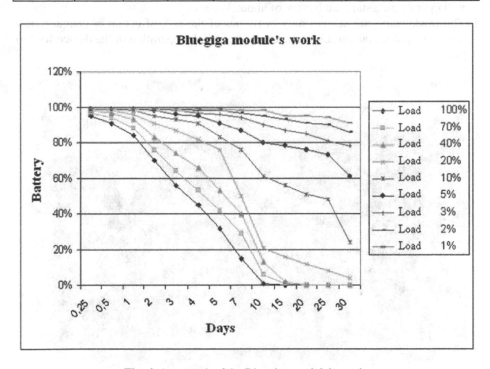

Fig. 6. A research of the Bluegiga module's work

From a research perspective, this product is a handy tool that allows you to receive data in real time on the phone holder and generate relevant reports. The research program was written directly to experiment the Bluegiga 112 module. The program allows you to simulate Bluegiga activities at different levels of load. Table 2 summarizes the results of measurements. Graphs for this table are represented in Fig. 6.

Herewith the time under load is defined here as the relative time of the Bluegiga 112 module's work in the sleeping and the working modes. Telephone receiver located at a distance of 2 m from the module. The battery is a standard battery Samsung EV 3.6V, 900 mA. The module is shown in Fig. 7.

According to the analysis of the data presented in table 2 the conclusions are as follows:

1. Technology of the permanent hibernation used both in ZigBee modules and BlueGiga devices does not lead to premature capacity loss of the battery. It is evident from the comparison of two adjacent columns in the table. Losses associated with a part of the transition from sleep mode operation, are imperceptible.
2. At base load 1% (40 seconds per hour of work) loss of battery capacity is about 10% per month.
3. Analysis of the results presented in the table leads to the conclusion about the possibility of extrapolating the results to the case of less frequent operations. Thus, in the case of using the device 1 times (4 minutes permanent connection) per day, the capacity loss will be less than 3% per month. In other words, in this mode, the unit will operate on a standard battery of about 3 years.
4. The mode using the devices only in frames of the task of obtain of data accumulated to a pulse counter allows you to guarantee the operation of the device for 5-6 years.

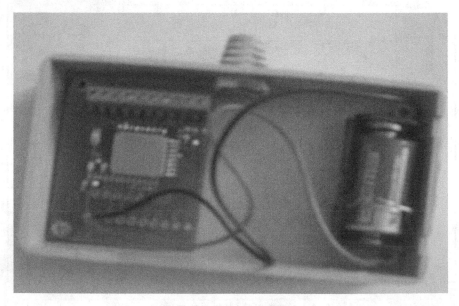

Fig. 7. Bluetooth module

Let's consider some of the technical characteristics of the module [8-10]. Now then, the voltage supply of the module may vary within the range 2-3,6 V.

The transmitter power can be programmed for three different types of data in addressing the specific problems. For example, if you do not need the maximum ranges of the module, the module can be switched to a lower power and greatly increase battery life. The range of depends on the mode. So the range will be change when the mode changes. We note three ranges of the power:

1. In the range of +4 dBm to -93 dBm the range of the module is 150 m (open area).
2. In the range of 0 dBm to -88 dBm the range of the module is 40 m.
3. In the range of +23 dBm to -88 dBm the range of the module is 10 m.

Receiver sensitivity can range from -87 to -93 dBm. The current consumption is 27 mA at an average transmission power of 0 dBm. Operating temperature is -40 ... + 85 ° C.

Indeed, the Bluetooth modules opened to the setup and configuration will take its place in the nearest future, thanks to major advantages, chief among which are:

1. Low energy consumption.
2. High data rate to 24 Mbit / s.
3. Increased range, more than 100 meters.

4 Conclusion

Thus, at present, there are several technologies to exchange information with the distributed devices. The most promising ones are wireless technologies. Their distribution is related to actual tasks of the organization of automatic data collection. However, existing solutions have several drawbacks, the overcoming of which could be achieved using the Bluetooth 4.0 technology. Developed modules can be used not only in conjunction with household energy resources counters, but also in various problems of interaction with a person, for example, smart home systems or in the different kinds of devices, exercising the communication with territorial-distributed devices.

References

1. Federal Law № 261 dated 23.11.2009 On energy saving and energy efficiency improvements and on Amendments to Certain Legislative Acts of the Russian Federation
2. Tarasyuk, V.: Problems of realization of the Federal Law № 261-FZ Energopolis, № 10 - 2012
3. Turutin, A.: Overview circuit solutions for data transmission from the meters of gas in the domestic sector (electronic resource - http://www.gaselectro.ru/stati/obzor_shemnyh_reshenij_pri_peredache_dannyh_so_schetchikov_gaza_v_kommunalnobytovom_sektore/) (Published: January 23, 2015)
4. Dyk, C.: Choosing GSM/GPRS-modem for automatic meter reading Wireless technology, №3 - 2013
5. Pushkarev, O.: Bricks to build a network ZigBee: 802.15.4 transceivers, transceiver, micro-modules and software stack implementation Wireless technology, №1 - 2006
6. Yamanov, A.D.: Remote metering data collection in low-rise residential complexes for ZigBee-network Technology, №1 - 2012

7. Gomez, C., Oller, J., Paradells, J.: Overview and Evaluation of Bluetooth Low Energy: An Emerging Low-Power Wireless Technology Sensors, №12, - 2012, pp. 11734–11753
8. Alexeev, V.: New modules Bluetooth 4.0 BLE series production Bluegiga Wireless technology, №2 – 2011, pp. 16–22
9. Bluegiga, BLE112 Preliminary Data Sheet. Version 0.91 (January 28, 2011)
10. Nordman, T.: BLE112 Bluetooth® low energy module. Bluegiga Partner Briefing and Product Presentation, Finland (2011)
11. Nechay, O.: Economic Bluetooth: Version 4.0 (electronic resource - http://old.computerra.ru/terralab/multimedia/526102/) (Published: April 23, 2010)

Configurable CRC Error Detection Model for Performance Analysis of Polynomial: Case Study for the 32-Bits Ethernet Protocol

Vinaya R. Gad, Rajendra S. Gad$^{(\boxtimes)}$, and Gourish M. Naik

Altera System on Chip Laboratory, Department of Electronics, Goa University,
Taleigão 403206, Goa, India
rsgad@unigoa.ac.in

Abstract. Almost every form of digital information exchange can introduce communication errors. In order to overcome the inherent inaccuracy of information transmission, a few methods for error detection and correction have been developed. Cyclic Redundancy check Codes (CRCs) are used in embedded networks for effective error detection. Paper discusses the development of the simulation model for the configurable CRC-polynomials performance analysis over Binary Symmetric Channels (BSCs). This paper presents a novel model which can be used to investigate several classes of CRC polynomials codes with 'n' parity bits varying from '1' to '64'. It also discuss the hardware implementation on Altera's FPGA Stratix II GX device 'EP2SGX90FF1508C3' for CRC-32 'IEEE-802' and suggest the indirect methodology of CRC-performance using Packet Error Rate (PER) parameter using Altera TSE MegaCore function that supports 10/100/1000 Mbps Ethernet over 1000Base-LX physical media. It is proposed to search good CRC codes of 32, 40 and 64 bit and studying performance for maximum payload over Ethernet protocol.

Keywords: CRC · Gigabit Ethernet · BER · PER · FPGA

1 Introduction

Cyclic Redundancy Check Codes (CRC) is widely used in data communications and storage devices as a powerful method for dealing with data errors. It is also applied to many other fields such as the testing of integrated circuits and the detection of logical faults [1]. CRC-12 is used for transmission of 6-bit character streams, and the others are for 8-bit characters, or 8-bit bytes of arbitrary data. CRC-16 is used in IBM's BISYNCH communication standard. The CRC-CCITT polynomial, also known as ITU-TSS, is used in communication protocols such as XMODEM, X.25, IBM's SDLC, and ISO's HDLC [2]. CRC-32 is also known as AUTODIN-II and ITU-TSS (ITU-TSS has defined both 16- and a 32-bit polynomials). It is used in PKZip, Ethernet, AAL5 (ATM Adaptation Layer 5), FDDI (Fiber Distributed Data Interface), the IEEE-802 LAN/MAN standard, and in some Department of Defense applications.

© Springer International Publishing Switzerland 2015
S. Balandin et al. (Eds.): NEW2AN/ruSMART 2015, LNCS 9247, pp. 529–542, 2015.
DOI: 10.1007/978-3-319-23126-6_46

The challenge to finding ideal CRCs is that the effectiveness of any particular code is computationally expensive to determine, and finding the best code for any particular message length among all possible codes has in the past proven to be computationally intractable [3]. This is particularly true of message lengths beyond 8K bits, which are commonly found on general-purpose computer networks and data storage devices. Philip Koopman team explore the design space of CRC size, message length, and attainable Hamming Distance (HD) and explored the performance analysis in polynomials that primarily maintained high HD values to the longest data word lengths possible, achieved good performance at shorter lengths, achieved good performance at longer lengths than the stated maximum usage length, burst error detection capability up to the size of the CRC width, unidirectional bit error detection and high-noise detection. The study indicates that there are significant opportunities for improving CRC effectiveness because some commonly used CRCs have poor performance [4].

Let us discuss the HD property of interest having application in embedded networks, HD is the minimum possible number of bit inversions that must be injected into a message to create an error that is undetectable by that message's CRC-based Frame Check Sequence. For example, if a CRC polynomial has HD=6 for a given network, that means there are no possible combinations of 1-, 2-, 3-, 4-, nor 5-bit errors (where a bit error is an inversion of a bit value) that can result in an undetected error, but there is at least one combination of 6 bits that, when corrupted as a set within a message, is undetectable by that CRC [3]. Common ubiquitous CCITT-16 polynomial having '0x8810' as hexadecimal value used in software-based CRC implementations, with x16 as the highest bit and an implicit +1 term has three feedback bits set in polynomial fails to detect 84 of all possible 4-bit errors. In comparison, the 16-bit polynomial 0xC86C [5] attains HD=6 at this length. CAN 15-bit polynomial 0x62CC, which is optimized for data word sizes of up to 112 bits, provides HD=6 at this length, missing only 4,314 of all possible 6-bit errors while using one less bit for its 15-bit CRC. Perhaps a surprise, though, is that 12-bit polynomial 0x8F8 can achieve HD=5 at this length, while the best published 12-bit CRC, 0xC07, achieves only HD=4. One may find method developed by T. Fujiwara et al. (1985) for efficiently computing the minimum distance of shortened Hamming codes using the weight distribution of their dual codes is extended to treat arbitrary shortened cyclic codes. Using this method implemented on a high-speed special-purpose processor, several classes of cyclic redundancy-check (CRC) codes with 24 and 32 parity bits are investigated [6]. Also Philip Koopman et al has identified a set of 35 new polynomials in addition to 13 previously published polynomials that provides good performance for 3- to 16-bit CRCs for data word lengths up to 2048 bits [3]. Baicheva, T. et al have investigate the performance of CRC codes generated by polynomials of degree 16 over GF(2) which can be used for pure error detection in communication systems. Also one can find similar work over HD minimum distance, properness and undetected error probability for BSCs for CRC polynomials [7]. Baicheva explored all binary polynomials of degree up to 10 and evaluation of the error control performance and suggested a procedure, based on the computed data, for choosing the best CRC code [8].

Unfortunately, standardized CRC polynomials such as the CRC-32 polynomial used in the IEEE 802.3 (Ethernet) network standard [IEEE85] are known to be grossly suboptimal for important applications. Koopman etal did exhaustive search of the 32-bit CRC design space. Results from previous research are validated and extended to include identifying all polynomials achieving a better HD than the IEEE 802.3 CRC-32 polynomial [9]. For example, the 802.3 CRC can detect up to three independent bit errors (Hamming Distance HD=4) in an Ethernet Maximum Transmission Unit (MTU) having a 1500 byte payload. But, the theoretical maximum is detection of five independent bit errors (HD=6) using identical error detection techniques with a better CRC polynomial.

In this paper, we present a simple but systematic software implementation method using modeling approach for CRC-32 using MATLAB Simulink modeling toolbox (Ver 7.0) with proper modulation schemes over the BSCs. Here, with the help of either random generator or incremental generator we generate the required data and computes the CRC for fixed payload of bytes using Ethernet Frame Generator and append the CRC checksum. Further we modulated the payload using the required scheme and send the data over equivalent noisy channel like BSC or additive white Gaussian noise channel (AWGN) having programmable density distribution and noise intensity. At the receiver end the Ethernet frame receiver detect the error in checksum and calculate the performance parameters like Bit Error Rate (BER) and Packet Error Rate (PER). Compared with the previous works, our approach is very flexible in terms of design methodology and applications, and is suitable for modular design over varied size payload and CRC polynomial, modulation schemes and noisy channels. Also one can upgrade the design of Ethernet frame generator over 10, 100 and 1000 MbPS Ethernet channels to incorporate the integration of detections code and error corrections codes using LDPC coding. Such integration may find application in performance analysis of 10, 40, 100 and 400, 1000Gb/s Ethernet networks in higher generation communication networks having applications in real time. Also this model will have application in automatic identification system modulated using a trellis-coded modulation having receiver which uses the CRC present in the automatic identification system signals for error correction as well as error detection [10]. Recent rate-adaptive optical transceivers exploiting the optimization of available resources in dynamic optical networks, in which different links yield different signal qualities. Such systems use rate-adaptive joint coding and modulation, often called coded modulation (CM) see potential use of our proposed modeling to exploit recent advances in channel modeling [11].

The paper is organized as follows. In section 2, we briefly review the theory and implementation methodologies of CRC polynomial code. Then, section 3, discusses software simulation configuration model for CRC error detection platform for Ethernet Protocols and its equivalent hardware implementation. Further section 4, discusses the ways toward accomplishing two types of performance parameters i.e. BET and PER for the tabulated CRC-32 polynomial (Table 1) for Ethernet protocol over the frame lengths of 64, 128, 256, 512, 1024, 1518 bytes with varied error probabilities over BSCs.

2   Error Detection using CRC

CRC have a long history of use for error detection in computing [12]. Error correction codes provide a means to detect and correct errors introduced by a transmission channel. Two main categories of code exist: block codes and convolutional codes. They both introduce redundancy by adding parity symbols to the message data. Cyclic codes are an important class of linear (n,k) block codes 'C' which are said to be cyclic if for each of its code vectors c=(C0, C1, C2, ... Cn-1) in 'C' the ith right cyclic shift c'=(Cn-1, C0................,Cn-2) is a code vector also in 'C'. CRC codes are a subset of cyclic codes that are also a subset of linear block codes. CRC is a way of providing error control coding in order to protect data by introducing some redundancy in the data in a controlled fashion. It is a commonly used and very effective way of detecting transmission errors during transmissions in various net-works. Common CRC polynomials can detect following types of errors: i. All single bit error, ii. All double bit errors, iii. All odd number of errors, provided the constraint length is sufficient, iv. Any burst error for which the burst length is less than the polynomial length, v. Most large burst errors [13]. CRC use a binary alphabets , 0 and 1. Arithmetic is based on GF(2) i.e. modulo-2 addition (logical XOR and multiplication(logical AND). In the typical coding scheme, systematic codes are used. Suppose $M(x)$ to be the message polynomial, $c(x)$ the code word polynomial and $g(x)$ the generator polynomial. We have $c(x)= m(x)g(x)$ which is also written using the systematic form $c(x)= m(x)x^{n-k} + r(x)$, where $r(x)$ is the remainder of the division of $m(x)x^{n-k}$ by $g(x)$ and $r(x)$ represent the CRC bits. The transmitted message $c(x)$ contain k-information bits followed by 'n-k' CRC bits , for example $c(x)= m_{k-1}x^{n-1}+......+ m_0x^{n-1}+ r_{n-k-1}x^{n-k-1}+......+ r_0$. So encoding is straightforward: multiply $m(x)$ by x^{n-k}, that is , append 'n-k' bits to the message, calculate the CRC bits by dividing $m(x)x^{n-k}$ by $g(x)$, and append the resulting 'n-k' CRC bits to the message. For decoding part, the same algorithm can be used. If $c'(x)$ is the received message , then no error or undetectable errors have occurred if $c'(x)$ is multiple of $g(x)$, which is equivalent to determining that if $c'(x)x^{n-k}$ is a multiple of $g(x)$, that is , if the reminder of the division from $c'(x)x^{n-k}$ by $g(x)$ is 0 [14]. An error is declared to have occurred if any discrepancy between these two sets of parity bits then indicates the presence of transmission errors in the received data. However, as with all digital signature schemes, there is a small, but finite, probability that a data corruption that inverts a sufficient number of bits in just the right pattern will occur and lead to an undetectable error [4]. The minimum number of bit inversions required to achieve such undetected errors (i.e., the HD value) is a central issue in the design of CRC polynomials. It is possible, however, that transmission errors reside in the received frame that cannot be detected by this procedure. Whenever channel noise affects both the information block i and the block r of parity bits in such a way that the received frame ĉ=[ŕ, î] satisfies $ŕ(x)= (x^p . î (x)) \bmod g(x)$, the errors cannot be detected by parity checking. Authors have propose solution for the said problem using double CRC in the payload of the Ethernet frame [15].

2.1   Implementation Methodologies

One finds four major CRC implementation solutions. Many researcher have done comparative studies over CRC implementation [16]. CRC implementation can use either hardware or software methods. In the traditional hardware implementation, a simple shift register circuit performs the computations by handling the data one bit at a time and parallel implementation by handling data in word form [17,18]. Software implementations of CRC encoding/ decoding do not resort to dedicated hardware requirements; however, their applicability is limited to lower encoding rates. In traditional hardware implementations, a simple shift register associated with a specific exclusive-OR (XOR) circuit, which is also known as a linear feedback shift register (LFSR), performs the computations by handling the data one bit at a time. As the demands of high transmission rate requirement in various applications, the need of a simple and systematic method to rapidly calculate the CRC has resulted. The most commonly used CRC digest computation algorithm in iSCSI and Ethernet, the Simultaneous Multiply and Divide (SMD), is derived from the long division algorithm. Sample implementations are provided of both algorithms [19].

2.1.1   Software(SW) Solution

The CRC algorithm can always be implemented as an software algorithm on a standard CPU, with all the flexibility reprogramming then offers. Since there in most communication network terminals exists a CPU, the SW-solution will be cheap or free in terms of hardware cost. The drawback is obviously the computational speed since no general purpose CPU can achieve the same throughput as dedicated hardware. Kounavis and et al implemented novel lookup table based algorithm based on 'slicing-by-x' algorithm which doubles the performance of existing software-based, table driven CRC implementations [20]. Software implementation, which relies on many steps of the polynomial division, is typically slower than other codes. Nguyen et al implemented fast CRCs as well as an effective technique to implement them without using lookup table, to eliminate or to greatly reduce many steps of the polynomial division [21].The CRC algorithm is straightforward to implement in software. However, it requires considerable CPU bandwidth to implement the basic requirements, such as shift, bit test and XOR. Moreover, CRC calculation is an iterative process and additional software overhead for data transfer instructions puts enormous burden on the MIPS requirement of a microcontroller. The CRC engine in 'dsPIC33E/PIC24E' devices calculates the CRC checksum without CPU intervention; moreover, it is much faster than the software implementation. The CRC engine consumes only half of an instruction cycle per bit for its calculation as the frequency of the CRC shift clock is twice that of the dsPIC33E/PIC 24E instruction clock cycle. For example, the CRC hardware engine takes only 64 instruction cycles to calculate a CRC checksum on a message that is 128 bits (16x8) long. If the same calculation is implemented in software, it will consume more than a thousand instruction cycles, even for an optimized piece of code.

2.1.2   Traditional Hardware Solution

Linear Shift Register (LSR) with serial data feed has been used since the sixties to implement the CRC algorithm. As all hardware implementations, these methods simply perform a division and then the reminder which is the resulting CRC checksum, is stored in the registers (delay-elements) after each clock cycle. The registers can then be read by use of enabling signals. Simplicity and low power dissipation are the main advantages. This method gives much higher throughput than the SW solution but still this implementation cannot fulfill all the speed requirements of today network nodes. Since fixed logic is used there is no possibility of reconfigure the architecture and change the generator polynomial using this implementation [22].

2.1.3   Parallel Solution

In order to improve the computational speed in CRC generating hardware, parallelism has been introduced. The speed-up factor is between 4 and 6 when using a parallelism of 8 bits. By using fixed logic, implemented as parallelized hardware, this method can supply for CRC generation at wire speed and therefore it is the pre-dominant method used in computer networks. Like any other combinatorial circuit, parallel CRC hardware could be synthetized with only two levels of gates. This is defined by laws governing digital logic. Unfortunately, this implies a huge number of gates. Furthermore, the minimization of the number of gates is an NP-hard optimization problem. Therefore, when complex circuits must be realized, one generally uses heuristics or seeks customized solutions. Campobella et al derive a recursive formula that can be used to deduce the parallel CRC circuits as in modern synthesis tools [23]. Also, Albertengo and his team designed the parallel CRC encoders based on digital system theory and z-transforms, which allows designers to derive the logic equations of the parallel encoder circuit for any generator polynomial [24]. Glaise [10] used the theory of Galois Fields GF(2m) and employed a GF multiplier for optimized parallel CRC calculation[25]. By inspecting the operations of the single-input LFSR, Pei [26], derived the state transition equation for the parallel CRC circuit by merging a number of the shift and modulo-2 (XOR) operations together within a single clock cycle [27]. Also some attempts of fast CRC computation, which is based on the observation that only a small portion residing in the beginning of an Ethernet frame is changed during the hopes for updating the relevant source and destination address over network. Here, the fast CRC update only calculates the changed portion of a frame and performs a single step update afterwards to obtain the new CRC code. This method dramatically improves the throughput of CRC recalculation which can support a throughput of about 56 Gbps [14]. In above designs if the CRC polynomial is changed or a new protocol is added, new changed hardware must be installed in the network terminal. The lack of flexibility makes this architectures non suitable for use in a protocol processor. Hence, very often new hardware schemes which compute the transition and control matrix of a parallel cyclic redundancy checksum are proposed. This opens possibilities for parallel high-speed cyclic redundancy checksum circuits that reconfigure very rapidly to new polynomials [28].

2.1.4   Configurable Hardware

Configurable hardware are implemented using Look-Up-Tables (LUT). This implementation can be modified by using a larger or smaller LUT. If the size of the LUT is reduced the hardware-cost in terms of power consumption and area will be reduced but in the same time the Combinational Network will be increased so the effect will be cancelled. The optimal solution has not been derived. Another, novel implementation method is the Radix-16 Configurable CRC Unit. Toal and his team proposed architecture using field reprogrammable chips so that it is fully flexible in terms of the polynomial deployed and the input port width. They synthesized circuit and mapped it to 130-nm UMC standard cell [application-specific integrated circuit (ASIC)] technology which is capable of supporting line speeds of 5 Gb/s [29].

3   Configurable CRC Error Detection Platform for Ethernet Protocol

CRC calculation is a core part of the Ethernet processing. CRC is used to generate a checksum over the whole frame at transmission and to check that the frame is still correct at reception. Since, 10 Gigabit Ethernet (GE) uses full duplex bidirectional links each side of the link must be capable of computing two CRC checksums at the same time, one for the received frame and one for the transmitted frame. There are some differences in the requirements on these two CRC calculation units. The reception unit receives the data via the 32 bit wide XGMII (eXtended Gigabit Media Independent Interface) and since the checksum is included at the end of the frame the output is just a one bit flag that tells if the frame can be accepted or must be discarded. The transmission unit, on the other hand, receives data from the MAC (medium access control) as it is transmitted and has to generate the CRC checksum in-line with short latency and then append it to the end of the frame. The interface from the MAC is 64 bits of data, since that is the preferred data width in 10 GE equipment. The output on the other hand is 32 bit wide XGMII. While in fast Ethernet (100 Mb/s) a technique was used that could append the CRC in-line. For Fast Ethernet, MII (Media Independent Interface) was used, which is 4 bit wide. Also, the da ta was handled in 8 bit symbols in the MAC, which is the smallest data unit in Ethernet. Although the interface widths have increased with newer versions of Ethernet, the minimum data unit has not changed. This introduces the problem that the size of a frame must not be divisible with 32 bits and thus the last word on the interface may contain 8, 16, 24, or 32 bits of data. This is solved by having 4 control signals together with the 32 bits of data. Each control bit announces the presence of valid data on the corresponding data lane (8 bits wide). Moreover, the first two words on the XGMII interface contain only start of packet symbol, preamble symbols and a start frame delimiter symbol, so they must not be included in the CRC calculation. After the frame, including the appended CRC, an end-of-packet symbol must be inserted on the next free lane. Whenever no data is transmitted on the XGMII all lanes carry idle symbols and if an error occurs during transmission an error symbol is inserted by the physical layer on the receiver side.

3.1 Simulation Model

Figure 1, illustrates the Matlab model for CRC Error detection. The transmitter is fed with a sequence of Ethernet frames (random bits) of length varying from 64 bytes to 1500 bytes. The frames include four bytes FCS (Frame Check Sequence), which is computed using Matlab code. The transmitter modulates the frame bits using BPSK (Binary Phase Shift Keying) schemes, which is sent through a simulated channel. A controlled amount of noise is added to the transmitted signal. One can generate AWGN (additive white Gaussian noise), by using one of the standard built-in functions in Matlab. This noisy signal then becomes the input to the receiver. The receiver demodulates the signal, producing a sequence of recovered bits. The FCS is recomputed at the receiver to find whether there was an error in the transmitted packet. Finally, we compare the received bits to the transmitted bits, and tally up the errors. Bit-error-rate (BER) performance is depicted by plotting BER at a series of different SNRs (Signal to Noise Ratio).

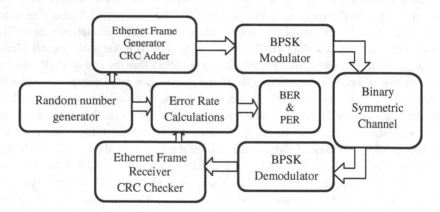

Fig. 1. Generic Matlab model for CRC Error Detection for BSC

The model is user friendly and has a capability to vary the CRC polynomial, size of frames and number of frames .One can also randomly assign the number of packet errors and determine the packet error rate (PER). Table 1 below summarizes the list of CRC-32 codes for which the performance could be studied over model. Here, we have studied IEEE-802 CRC polynomial 'x32 + x26 + x23 + x22 + x16 + x12 + x11 + x10 + x8 + x7 + x5 + x4 + x2 + x + 1' having hexadecimal value of '104C11DB7'.

Table 1. List of the possible CRC codes and their generator polynomials [2].

Sr. No.	CRC Code	Generator Polynomial g(x)
1	CRC-32/8	1F1922815
2	MP-CRC-32/8.1	1404098E2
3	MP-CRC-32/8.2	10884C512
4	CRC-32/6	1F6ACFB13
5	CRC-32/5.1	1A833982B
6	CRC-32/5.2	1572D7285
7	CRC-32/4	11EDC6F41
8	**IEEE-802**	**104C11DB7**
9	CRC-40/GSM	0004820009
10	CRC-64	42F0E1EBA9EA3693

3.2   Equivalent FPGA Based Configurable Hardware

The Altera provides Triple Speed Ethernet (TSE) data path reference design using SOPC Builder system with two serial transceivers. The design enables you to evaluate the TSE MegaCore function for integration into Altera FPGA designs. The reference design has the following features: a. Uses very few pieces of hardware for a complete test: a Stratix® II GX PCI Express development kit, two small form pluggable (SFP) modules, an Ethernet cable assembly, a PC, and a USB-Blaster™ or ByteBlaster™ cable assembly; b. Implements two instances of the Altera TSE MegaCore function and supports 10/100/1000 Mbps Ethernet operations in SGMII mode with auto-negotiation; c. Supports programmable settings such as the number of packets, packet length, payload data type, and source and destination MAC addresses; d. Demonstrates sending and receiving Ethernet packets up to the maximum theoretical data rates without errors; e. Supports external loopback via SFP modules with CAT5 Ethernet cable assembly and optionally through an Ethernet switch; f. Provides a command line interface (CLI) to control the design and monitor TX and RX packet statistics results; f. Provides serial interfaces to configure Copper SFP module. The design is demonstrated in detailed by authors [30] for performance of line rate and throughput on Altera's Stratix II GX device 'EP2SGX90FF1508C3' on PCI Express development kit. The performance of above design was studied for Gigabit Ethernet standards 1000Base-LX, 1000Base-SX and 1000Base-T using various physical media such as Single Mode Fibre, Multimode Fibre and Copper cables (Cat5e) and corresponding SFP Transceivers via a 1.25 Gbps serial transceiver that enables all 10, 100, and 1000 Mbps Ethernet operations. The system was tested by varying frame length and number of frames .

4   Performance Results Conclusion

Fig.4.1(a) and (b) shows the BER and PER performance of model for a BSC .Both the results are plotted for a frame length of 64 bytes to 1518 bytes for 100 packets. The bit error probability of BSC is varied from 0.001 to 0.00001. From Fig.4.1 (a), it can be seen that the BER decreases exponentially with decrease in Error probability. Similarly Fig. 4.1(b), shows that PER approaches zero for .00001 error probability for all frame lengths.

Altera TSE MegaCore function does not support the dropped packets due to error in checksums and hence it's very difficult to implement direct methods of monitoring the packets loss for the performance study of the Ethernet   for the particular CRC polynomial. We have adopted indirect methods of quantifying the performance of the protocol. We use the performance analysis of the said design for the Throughput and the Line Rate for 1000Base-LX physical media for IEEE-802 CRCs

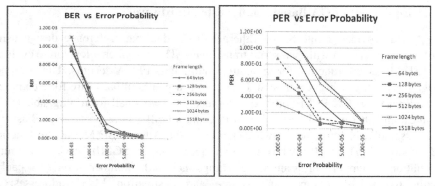

Fig. 2. (a)BER for different block lengths (for **IEEE-802** CRCs) for BSC; (b) PER for different block lengths (for tabulated CRCs).

This study is performed by keeping the total number of bytes (N) sent constant; where 'N' = message length x number of packets. The message length was varied from 64 bytes to 9600 bytes and correspondingly number of packets was changed to keep 'N' constant. The experiment was repeated for two different values of N i.e. 64×105 and 64×107.

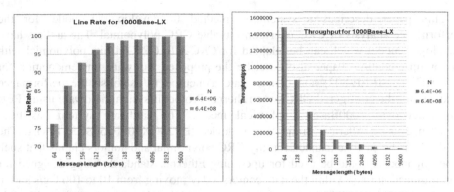

Fig. 3. (a)Line Rate for 1000Base-LX, (b) Throughput for 1000Base-LX

It is found that as the frame length is increased from 64 bytes to 9600 bytes(Fig. 4.2(a) & (b) , the line rate increases and achieves 99.79% for 9600 bytes. The throughput measured is 0 .76 Gbps for 64 bytes frame length and approaches 1 Gbps for 9600 bytes. The results are found to be nearly the same for the different Gigabit Ethernet standards. The Throughput analysis indicates that as the message length increases for higher bytes payload the throughput decreases. This is due to computation time for the CRC over higher bytes payload and also handling time for the payload, while the line rate i.e. bandwidth is fully utilized for higher payload because line remains busy for higher payloads. The study shows similar results for the two different values of N i.e. 64×10^5 and 64×10^7.

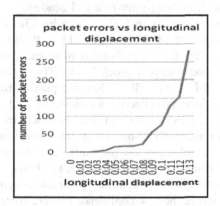

Fig. 4. Packet errors vs longitudinal

Hence with the help of said analysis one can established a method for PER over equivalent AWGN channel by setting the standard deviation of noise over channel. So by setting the channel for the zero PER and one of the Line Rate and Throughput as given in Fig. 4.1 (a) & (b) one can generate the performance of the Ethernet for particular configured CRC polynomial.

This paper describes a systematic method to develop the model for the performance analysis of CRC over configurable CRC polynomial. The authors have developed model which can be adapted for CRC-4 to CRC-64 bits polynomial with minor up gradation. Paper also describe the proposed hardware implementation for 32-CRC (i.e. IEEE-802) and demonstrate a method to establish the performance analysis for the configured CRC over Ethernet Protocols system hardware using PER parameter over 1000Base-LX physical media. The hardware system has been developed using Altera's Stratix II GX device 'EP2SGX90FF1508C3' FPGA. The method can be easily expanded for any CRC polynomial and authors claims that such types of models could be used for upcoming Ethernet standards being designed for communication rates within the data centers today moving from 10 to 100 Gg/s and in future on to 400 or 1000Gb/s. Cisco predicts that global data centers traffic will triple between 2012 to 2017, reaching 7.7 billion terabytes annually. Configurable hardware has also discuss method of performance studies for various payloads, hence this model can also help in exploring the design space of a CRC size and attainable HD in polynomials that primarily maintained high HD values to the longest data word lengths of payload possible. The proposed model uses worldwide algorithm to compute CRC and can assist the system engineers to choose the right CRC code for embedded software error control. One can specify the bounds of the CRC algorithm and make claims about the error probabilities. You can also study the performance for random errors and burst error patterns. Exhaustive exploration CRC polynomials reveals that most previously published CRC polynomials are either inferior to alternatives or are only good choices for particular message lengths. Since the proposed model results have been verified through both software simulations our development is valuable from both industrial and academic points of view. One can find employability of such models in LTE.   Jung-Fu Cheng group has established error detection performance of CRC coding in LTE with turbo decoding. They have established analytical models for the probability of block error and undetected block error at the code block and transport block levels. The analytical models show the setting of 24-bit CRCs in LTE that allows low-complexity, robust and reliable early stopping algorithms to reduce turbo decoding complexity [31].

One can incorporate LDPC coding in the proposed model with the suitable modulation schemes for using CRC not as an error detection tool but rather as a correction method. McDaniel team has used difference between the received CRC and the CRC computed from the received data as a syndrome and developed the basis of the error correction strategy [32].   A strategy suitable for any bit error number has been proposed in [33], where the low confidence bits are corrected in priority. Zhang et al [34] , algorithm for the error correction is based on the BER by modifying the high error probability bits until the received and re-computed CRCs are equal. Finally, a strategy using a convolutional code with the CRC has been proposed by Wang and his team [35]. However, these techniques are not appropriate for data containing bit stuffing. New strategies must therefore be investigated [36].   Hence there is great potential for such configurable models to study the performance analysis of CRC polynomial and LDPC coding for error detection and correction.

Acknowledgements. This work was supported by the University Grant Commission (UGC), New Delhi under Faculty Improvement Program to first author and ALTERA Inc. USA by establishing SoC Laboratory at the Department under University Program.

References

1. Smith, M.J.S.: Application-Specific Integrated Circuits. Addison-Wesley Longman (January 1998)
2. Tanenbaum, A.S.: Computer Networks, 2nd edn. Prentice Hall (1988)
3. Koopman, P.: 32-Bit cyclic redundancy codes for internet applications. In: The International Conference on Dependable Systems and Networks (DSN) (2002)
4. Koopman, P., Chakravarty, T.: Cyclic Redundancy Code (CRC) polynomial selection for embedded networks. In: The International Conference on Dependable Systems and Networks, DSN 2004 (2004)
5. Baicheva, T., Dodunekov, S., Kazokov, P.: Undetected error probability performance of cyclic redundancy-check codes of 16-bit reducndancy. IEEE Proc. Comms. **147**(5) (2000) 253–256
6. Castagnoli, G., Brauer, S., Herrmann, M.: Optimization of cyclic redundancy-check codes with 24 and 32 parity bits. IEEE Transactions on Communications **41**(6), 883–892 (1993)
7. Baicheva, T., Dodunekov, S., Kazakov, P.: Undetected error probability performance of cyclic redundancy-check codes of 16-bit redundancy. IEE Proc. Commun. **147**(5) (2000)
8. Baicheva, T.S.: Determination of the Best CRC Codes with up to 10-Bit Redundancy. IEEE Transactions on Communications **56**(8), 1214–1220 (2008)
9. Koopman, P.: 32-bit cyclic redundancy codes for internet applications. In: Intl. Conf. Dependable Systems and Networks (DSN), pp. 459–468 (2002)
10. Prévost, R., et al.: Cyclic redundancy check-based detection algorithms for automatic identification system signals received by satellite. Int. J. Satell. Commun. Network **31**, 157–176 (2013)
11. Beygi, L., et al.: Rate-Adaptive Coded Modulation for Fiber-Optic Communications. Journal of Lightwave Technology **32**, 2 (2014)
12. Peterson72, Peterson, W., Weldon, E.: Error-Correcting Codes, 2nd edn. MIT Press (1972)
13. Ulf Nordqvist, Thesis:Protocol Processing in Network Terminals, Department of Electrical Engineering, Linkopings Universitet, SE-581 83 Linkoping, Sweden (2004)
14. Lu, W., Wong, S.: A Fast CRC Update Implementation, Computer Engineering Laboratory, Electrical Engineering Department. Delft University of Technology, Delft, Netherlands
15. Gad, V.R., Gad, R.S., Naik, G.M.: Gigabit Ethernet Implementation of CRC-32 in noisy channels. International Journal of VLSI Design, Serial Publications **1**, 22–32 (2011)
16. Nordqvist, U., Henriksson, T., Liu, D.: CRC generation for protocol processing. In: Norchip Turku, Finland, pp. 288–293(2000)
17. Shieh, M.-D., Sheu, M.-H., Chen, C.-H., Lo, H.-F.: A Systematic Approach for Parallel CRC Computations. Journal of Information Science And Engineering **17**, 445–461 (2001)
18. Campobello, G., Patane, G., Russo, M.: Parallel CRC Realization. IEEE Transactions on Computers **52**, 10 (2003)
19. The iSCSI CRC32C Digest and the Simultaneous Multiply and Divide Algorithm Luben Tuikovplentec Ltd. Richmond Hill, Ontario, Canada Vicente Cavannay Agilent Technologies Roseville, California, USA (January 30, 2002)

20. Kounavis, M.E., Berry, F.L.: Novel Table Lookup-Based Algorithms for High-Performance CRC Generation. IEEE Transactions on Computers **57**(11), 1550–1560 (2008)
21. Nguyen, G.D.: Fast CRCs. IEEE Transactions on Computers **58**(10), 1321–1331 (2009)
22. Ulf Nordqvist, Thesis: Protocol Processing in Network Terminals by, Department of Electrical Engineering, Linkopings universitet, SE-581 83 Linkoping, Sweden (2004)
23. Campobello, G., Patane, G., Russo, M.: Parallel CRC Realization. IEEE Transactions on Computers **52**(10) (2003)
24. Albertango, G., Sisto, R.: Parallel CRC Generation. IEEE Micro **10**(5), 63–71 (1990)
25. Shieh, M.-D., Sheu, M.-H., Chen, C.-H., Lo, H.-F.: A Systematic Approach for Parallel CRC Computations. Journal of Information Science And Engineering **17**, 445–461 (2001)
26. Pei, T.-B., Zukowski, C.: High-speed parallel CRC circuits in VLSI. IEEE Transactions on Communications **40**, 653–657 (1992)
27. Ji, H.M., Killian, E.: Fast parallel CRC algorithm and implementation on a configurable processor. In: IEEE International Conference on Communications, ICC 2002, vol. 3, pp. 1813–1817 (2002)
28. Grymel, M., Furber, S.B.: A Novel Programmable Parallel CRC Circuit. IEEE Transactions on Very Large Scale Integration (VLSI) Systems **19**(10), 1898–1902 (2011)
29. Toal, C., McLaughlin, K., Sezer, S., Yang, X.: Design and Implementation of a Field Programmable CRC Circuit Architecture. IEEE Transactions on Very Large Scale Integration (VLSI) Systems **17**(8), 1142–1147 (2009)
30. Gad, V.R., Gad, R.S., Naik, G.M.: Implementation of Gigabit Ethernet Standard Using FPGA. International Journal of Mobile Network Communications & Telematics (IJMNCT) **2**(4) (2012)
31. Jung-Fu, C., Koorapaty, H.: Error Detection Reliability of LTE CRC Coding, Ericsson Research, RTP, NC, USA
32. McDaniel, B.: An algorithm for error correcting cyclic redundancy checks. C/C++ Users Journal (2003)
33. Shi-Yi, C., Yu-Bai, L.: Error correcting cyclic redundancy checks based on confidence declaration. Proc. ITS Telecommunications **6**, 511–514 (2006)
34. Zhang, Y., Yuan, Q.: A multiple bits error correction method based on cyclic redundancy check codes. ICSP Signal Processing **9**, 1808–1810 (2008)
35. Wang, R., Zhao, W., Giannakis, G.B.: CRC-assisted error correction in a convolutionally coded system. IEEE Trans. Comm. **56**(11), 1807–1815 (2008)
36. Prévost, R., Coulon, M., Bonacci, D., LeMaitre, J., Jean-Pierre, M., Jean-Yves, T.: Cyclic redundancy check-based detection algorithms for automatic identification system signals received by satellite. International Journal of Satellite Communications and Networking **31**,157–176 (2013)

Project Management Team Structure
for Internet Providing Companies

Vladimir V. Glukhov, Igor V. Ilin$^{(\boxtimes)}$, and Anastasia I. Levina

Saint Petersburg Polytechnic University, Polytechnicheskaya str. 29,
195251 Saint Petersburg, Russia
vicerector.me@spbstu.ru, ilyin@fem.spbstu.ru, alyovina@gmail.com

Abstract. Services of Internet providing companies are in-demand nowadays when each commercial or residential site needs to have broadband Internet access. Every new customer for the Internet providing company is a new project which requires certain techniques to be managed properly. Plugging a new customer into broadband is a cross-functional challenge and the more effective the project management team is, the more effective would be the whole project. The paper describes the approach to project management team structure (according to PRINCE2 project management standard) for Internet providing companies. The project management team structure proposed in the paper takes into account the traditional organizational structure of telecommunication companies, follows its hierarchy and helps to implement the project management standard into existing business-processes.

Keywords: Broadband Internet · Internet provider · Project management · Project management team

1 Introduction

Until quite recently, the Internet was mostly used for web browsing and sending e-mails. For these purposes a couple of dozen kilobits per second, provided by analog modems, was enough to cover the common needs of ordinary Internet users. Now the opportunities provided by the Internet, have dramatically expanded – users can receive multimedia content, play online games, share large files, chat via VoIP-services etc. All these ways of using Internet require as high-speed as possible Internet access channel.

As broadband Internet is in-demand, no wonder that there are a lot of providing companies willing to capture its part of this market – from small regional companies to state mobile operators. During last 2-3 years in Russia we can see strong centralization trend in internet providing which has a doubtful effect on the efficiency of this market [1]. But anyway a lot of small Internet providing companies, especially in non-metropolitan areas, keep running their business. Under conditions of high competition it is important to provide both high quality internet access and quality of subscribers connection process. The latter depends on the way the job is organized at the provider side. Effective structure of the connection team can result in faster and

© Springer International Publishing Switzerland 2015
S. Balandin et al. (Eds.): NEW2AN/ruSMART 2015, LNCS 9247, pp. 543–553, 2015.
DOI: 10.1007/978-3-319-23126-6_47

cheaper servicing, increasing the number of loyal customers and more effective resource distribution of the Internet providing company.

Plugging into broadband of each object is a unique project. To be run effectively projects need specific project management techniques to be used. In the paper the typical process of broadband Internet providing is analyzed and the possible organizational structure of the connecting team is proposed. The solution proposed is in line with the telecommunication business process frame eTOM [2].

2   Broadband Internet Providing

Broadband is the common name for the technologies which ensure a constant (not by sessions) connection to the Internet. Broadband technology often uses a telephone line (ADSL). Until recently one of the main ways to connect to the Internet was dial-up within the telephone line, but the latter was fully occupied for the period of the Internet session. Broadband Internet provides much higher speed of data transfer than the dial-up (called "high-speed Internet") and does not engage the telephone line. In addition to high-speed information transmission, broadband technology provides a stable continuous connection to the network, and provides a so-called "two-way" communication – the ability to receive and upload data at the same high speeds. [3]

Thanks to broadband Internet, the user can receive digital television services over the Internet, voice data services (IP telephony) at any distance cheaper or even for free, as well as the possibility of remote data large volume storage.

Broadband includes several high-speed transmission technologies such as:

- Digital Subscriber Line (DSL)
- Cable Modem
- Fiber
- Wireless
- Satellite
- Broadband over Powerlines (BPL)

The choice of broadband technology depends on a number of factors: urban or rural area, package of Internet access services (such as voice telephone and home entertainment), price, and availability. [4]

As more and more people are eventually comes to the need for high-speed access to the network, broadband Internet in Russia has great prospects for expansion. ADSL (via the regular telephone network) is a promising way to expand broadband Internet technology in the Russian market. Using this kind of technology, user gets access to the Internet and the telephone at the same time remains available for voice communication.

Another common scheme of the Internet connection of this type is so-called home networks ETTH (Ethernet To The Home), which has a large market share. A fiber-optic backbone is installed directly to the consumer (home, office) and Ethernet commutator is set. Further, individual users are connected by a standard twisted pair. Such connection method consumes more time and costs more than ADSL because of wiring inside the building. But in comparison with the ADSL or using TV cable it provides the best connection speed.

Traditional customers of Internet providers are offices in business centers, large companies with an extensive network, residential estates and residential complexes.

One thing is clear, broadband Internet significantly expands the possibilities of consumers and providers, which means the battle of technology and the struggle for the users.

To understand the way Russian telecommunication companies operate their Internet providing activity the authors of the paper surveyed some representatives of such companies and open-source articles were analyzed. The analysis of business both regional companies [5] and well-known Russian mobile operators [6, 7], has resulted in a business process scheme which is rather common for connecting customers to Internet (*Fig. 1*).

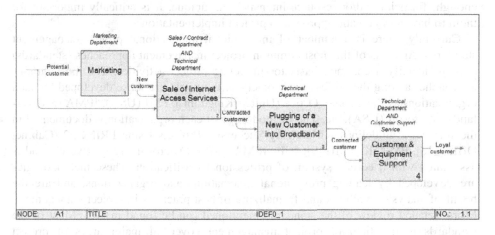

Fig. 1. Business Process of Broadband Internet Providing

By the term "broadband Internet providing" hereinafter we mean connecting the customer's site (office building, residential house etc.) to the Internet, i.e. to install and maintain equipment necessary for providing broadband Internet access for all potential users of the site. The name of business processes and departments can vary from those on Fig.1 for each telecommunication company but the general idea is more or less the same.

After the potential customer has interested in broadband Internet connection, the Sales Department is looking for the ways to make contract with him. To calculate the contract cost the technical inspection is required. It involves checking the possibility of organizing an Internet access point, defining the type and the scope of equipment needed, set necessary hardware configuration, the technical parameters of the connection. Besides, technical details of future project it helps to calculate the cost of services for the contract. After the contract is signed the technical team installs the equipment, checks it and then each single user within the site can be plugged into broadband. Each telecommunication company takes care of its customers via customer support – by phone and technical inspection or repair operations when needed.

3   Project Management Issues of Organizational Structure

A *project* is traditionally defined as "a temporary organization that is created for the purpose of delivering one or more business products" [8].

The project approach is used in different fields: business, social, political, cultural, etc. – in those fields where there is a need to introduce changes, to address unique challenges. In many enterprises, each customer order is considered as a separate project. This type of companies is called project-oriented. Their projects are large enough, financial and/or resource-intensive and unique. It is critically important for them to have a systematic approach to project implementation.

Currently, there is a number of international and national project management standards. Analysis of the most common project management approaches (standards) aims to identify a common basis for modeling project activities. The most famous approaches among the professional society of the world are those developed by such organizations as the Cabinet Office (United Kingdom), PMI (USA), IPMA (Switzerland), Microsoft (USA), etc. The methodology of each organization is documented in the form of a guidelines – Managing Successful Projects Using PRINCE2 (Cabinet Office) [8], PMBoK (PMI) [9], ICB (IPMA), MSF (Microsoft) respectively – and is associated with a certain system of professional certification. These methodologies are developed by leading professional associations and organizations, and are the result of analysis, synthesis and formalizing of best practices in project management. More detailed review of the standards mentioned can be found in [10]. Most of the standards in the field of project management cover all major areas of project management, including cost, risk, quality, personnel management. Each standard addresses to this subsystems from different points of view.

For the purpose of arranging work of telecommunication company based on the revision of management processes and organizational structure, PRINCE2 was chosen as the most clear approach to the organizational structure of the project team (according to authors' opinion). For the purposes of the current analysis the PRINCE2 principle of "defined roles and responsibilities" is particularly important. PRINCE2 allows to create a system of hierarchy and interaction of the participants of the project which forms a well-functioning structure that takes into account the interests of the three main bodies in the project - business, future users and suppliers. This type of structure provides certain subordination levels of project management, each of which controls the interests of different levels, ultimately subordinate to the strategic goals of the business (Fig.2). Responsibility delegation to the higher level is performed according to the «management by exception» principle. This provides the lower levels with more management freedom and the higher ones are not involved into the routine processes of lower level processes. [11]

Upper management level	**Corporate and program management – strategic interests of corporation/project program**		
Project management team	**Directing** – business interests of the project		
	Managing – project interests		
	Delivering – project performing management		

Fig. 2. Project management levels (according to PRINCE2)

In addition to the distinction between project management levels, the organization of the project team according to PRINCE2 implies specific roles and responsibilities, which allow to avoid function duplication, to provide a clear procedure of project control at all stages, as well as ongoing expert and administrative support to the project management team.

The core of the project management team according to PRINCE2 is the vertical management system: Executive – Project manager – Team manager. All roles mentioned are the key figures of the corresponding management level on Fig.2. The PRINCE2 guidance [8] describes these roles as follows.

The Executive role is defined to look after the business interests and to ensure that the project gives value for money. The Executive is ultimately a key decision maker and is accountable to the project's success and. The Executive will then be responsible for designing and appointing the rest of the project management team. He (or she) can consult with the Senior User and Senior Supplier (who form a Project Board). The first one is the representative of the party which will use the results of the project and is responsible for defining the requirements for the project products. The second one provides resources for the project and can act as a consultant of feasibility and availability of the project products. The members of the Project Board can appoint experts in different areas for a Project Assurance role to delegate their responsibility to monitor Project Manager performance.

The Project Manager deals with day-to-day management of a project and factually is responsible for all managing work scope of the project. In projects with no separate person allocated to a Team Manager role, the Project Manager will be responsible for managing work directly with the team members involved. He (or she) can also perform a Project Support role or appoint another person responsible for administrative support of the project.

The Team Manager's primary responsibility is to ensure production of specific products. He (or she) is a leader of specialists of particular area who delivers project products. The number of Team Managers in the project depends on a quantity of professional areas required to complete the project product.

In order to avoid conflicts within a project management team, it is recommended by [8] to appoint company's employees for the project management roles following their natural hierarchy in a company.

4   Project Management Issues for Internet Providing Companies

The telecommunication industry, as many others, has its own standards of running business effectively. Such standards are usually supported by best practices collected as a proven experience of real companies. For telecommunication business such framework, widely accepted as a standard, is developed by the global non-profit industry association TeleManagement Forum (TM Forum) and is documented as the enhanced business process map for communications service providers (eTOM). The standardization of eTOM in the frame of International Telecommunication Union is realized in [2]. This document serves as a reference model or architecture of business processes for service providers and their partners working in the telecommunication industry. Although project management processes are often mentioned throughout eTOM, project management approach is not included in the process frame, while telecommunication industry has its specific projects (the detailed analysis of the latter can be found in [12, 13]). The approach of forming a project management team for the new customer connecting projects, proposed in the paper, can be realized within Operations block of the eTOM process frame (the business process scheme on Fig.1 is in accordance with it).

Projects are traditionally considered to be in conflict with business processes as they both intend to use the scares resources of the company (resource management of telecommunication in analyzed in [14, 15]).

As projects are temporary structures established for delivering unique results, each new customer contract for broadband Internet access can be considered as a project. It meets all the characteristics of a project (see [8]):

- Change: each Internet access contract brings new connected site to the telecommunication company;
- Temporary: as soon as a new customer is plugged into broadband (i.e. as soon as the equipment is installed on the customer's site) the Internet connecting project finishes its existence and a customer is "passed" to the customer support department;
- Cross-functional: Internet connecting project involves at least three parties – Sales Department, Technical Department and customer representatives; depending on a processes of a particular company and a particular customer case there can be more participants involved (other departments, regional authorities, higher level network provider etc.);
- Unique: each site is unique by its geography, facilities, interior lines of communications, Internet requirements etc.;
- Uncertainty: as each connection project is unique, definitely it has a higher risk level then business as usual.

This paper focuses on arranging a "cross-functional" part of the Internet connecting project, i.e. how to organize a cross-functional team in order to perform in the most effective way. Before suggesting a project management team solution, the scope of the Internet connecting project should be clearly defined (Fig.3). The project is open when the interested customer is detected (by Marketing Department or without it) and the Sales Department starts working with the customer. The project is closed when all the necessary equipment is installed and checked and all the acceptance documents are signed by all parties. Marketing activity is not included in the project because it has different object, goal and result. Marketing in a company can be performed either by specific marketing projects or routine processes (which is less effective under conditions of a high competition). Customer support activity is also beyond the scope of the Internet connecting project, because once the customer is plugged into the broadband his servicing is performed as business as usual.

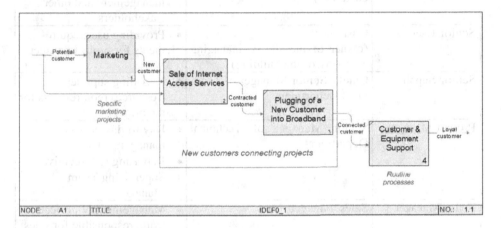

Fig. 3. The scope of the Internet connecting project

The possible organizational structure of the project management team is presented in Table 1. There are only key project management team roles, the rest ones are rather optional or can not be recommended outside of the particular project context. The comments on the appointment choice are given below.

Traditionally all sales people are interested in the number of contracts signed, as they are rewarded proportionally (in more or less explicit form) to sales volume. It is important to motivate sales personnel to take care of a customer through all duration of the project – from the moment of signing a contract till the moment it is considered to be complete. Completing of the contract is the moment when the project is considered to be closed and every participant of the project can get his final premium. As sales person is the one who negotiates with the customer before signing a contract, it seems logical to make him responsible for delivering all the promised benefits. And that is the reason the representative of the Sales Department (of high or even the highest level) was chosen for the role of the Executive – he (or she) will be motivated to successfully complete the whole project but not only the sales part of it. This will create organizational preconditions for the success of the whole project.

Senior User and Senior Supplier appointments seem to be rather obvious.

Table 1. Project Management Team for Internet connecting project

Project Management Team Role	Who?	Responsibilities
Executive	Chief / Senior Manager of Sales Department (depending on the telecommunication company and/or on the importance of a particular customer)	• Overall project management • Ensuring project success • Ensuring resources for the project • Contacts with corporate management and other stakeholders
Senior User	Customer (owner of the site or managing company of the building)	• Providing user requirements for project products
Senior Supplier	Chief / Senior Manager of Technical Department	• Providing supplier (resource) requirements for the project
Project Manager	Senior Manager of Technical Department	• Day-to-day project management • Reporting to Executive • Supervising Team Managers
Team Manager 1	Sales manager	• Management of project team, responsible for sales part: negotiations with the customer, calculating the cost, making a contract etc.
Team Manager 2 (or more teams)	Manager/Senior Manager of Technical Department	• Management of project team (teams), responsible for technical part: inspecting of the site, purchasing of the equipment, cabling, equipment installation, testing the equipment installed etc.

As the most part of the project is a technical one (installation of the equipment) so it is reasonable that a Project Manager would be a technical specialist with experience and management skills. In case of a small-scope project such a person could perform a Team Manager of the technical team at the same time.

New customer connecting projects have two basic parts: sales part and technical one. The first one involves negotiations with a potential customer, making commercial offers, signing a contract which are the competences of Sales Department employees. The second one refers to new site connecting to the Internet access. It includes inspecting of the site, purchasing of the equipment, its installation, cabling, testing of the equipment installed which are technical specialists competences. Depending on the scope of the project and specialization of technical specialists of the providing company, there can be more than one technical team. For example, there can be a separate team for equipment purchasing if it is not a standard one. But anyway such issues are within the competence of technical specialists and one of them should be appointed as a Team Manager.

5   Results and Discussions

Nowadays telecommunication industry delivers one of the most marketable products – communication resource. Running business in such fast developing and high-competitive market telecommunication companies should implement effective managing approaches in order to keep their market position. Analysis of the activity of Russian broadband Internet providing companies revealed that they can be considered as project-oriented ones. Consequently it could be reasonable for them to use some project management techniques.

Providing each new customer with broadband access is a project for a telecommunication company. Effective team work organization within a project is one of the important conditions of project success. The solution proposed in the paper (see Table 1) concerning the structure of the project management team according to PRINCE2 management standard brings the following benefits for the company:

- provides certain roles and responsibilities within a project;
- it is in line with the telecommunication business process framework (eTOM);
- allows to avoid the infamous conflict between project- and process-oriented activities as projects do not "steal" resources from processes but project-oriented parts of the whole process are reorganized according to project management role structure;
- does not break the existing hierarchy levels of the company because of the appropriate project management appointments (Fig.4);
- provides organizational preconditions for project success because of the appropriate appointment of the project leader (Executive in PRINCE2 terms): the one who starts a project is the final responsible for its success.

The solution proposed is rather general as it was developed basing on the information from a number of companies and Internet providing business at all. But the solution can easily be tailored to a particular case without losing any of the benefits listed above.

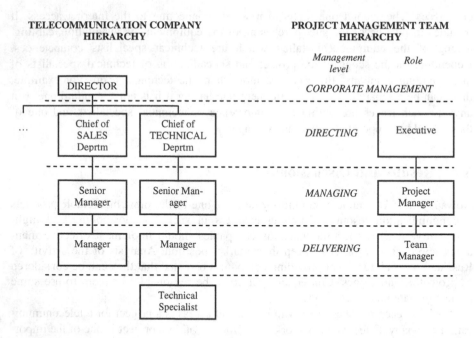

Fig. 4. Correspondence between company's and project management hierarchies

References

1. Prorokov, G.: How Big Providers Capture the Internet (Retrieved February 05, 2015). http://www.lookatme.ru/mag/magazine/russian-internet/207489-decentralization (September 16, 2014), from Look at Me, Internet Journal
2. ITU-T. Enhanced Telecom Operations Map (eTOM) – The business process framework. M.3050.1.International Telecommunication Union (March 2007)
3. Carty, G.: Broadband Networking. McGraw-Hill (2002)
4. FCC. Types of Broadband Connections (Retrieved February 05, 2015). https://www.fcc.gov/encyclopedia/types-broadband-connections (2015), From Federal Communications Commission
5. TELHosting. Virtual Open Day at TELHosting Data Center (Retrieved May 05, 2015). http://habrahabr.ru/company/telhosting/blog/113706/ (February 14, 2011), from Habrahabr, Russian IT Collective Blog
6. BeeLine, V.: Providing Mobile Connection in the Subway - Petersburg Experience (Retrieved May 05, 2015). http://habrahabr.ru/company/beeline/blog/242065/ (October 31, 2014), from Habrahabr, Russian Collective IT Blog
7. BeeLine, V.: 400 Mbit/s via Twisted Pair for Apartments (Retrieved May 05, 2015) (June 18, 2014), from Habrahabr, Russian Collective IT Blog
8. OGC. Managing Successful Projects with PRINCE2. TSO, London (2009)
9. PMI. A Guide to the Project Management Body of Knowledge: PMBOK Guide. 5th edn. Project Management Institute (2013)
10. Ilin, I.V., Antipin, A.R., Lyovina, A.I.: Business architecture modeling for process- and project-oriented companies. Economics and Management, 32–38 (2013)

11. Ilyin, I.V., Levchenko, Y.L., Levina, A.I.: Some issues of the formation of engineering companies' architecture. St Petersburg Polytechnic University Journal, 48–54 (2013)
12. Gluhov, V.V., Ilin, I.V.: Project portfolio structure in a telecommunications company. In: Balandin, S., Andreev, S., Koucheryavy, Y. (eds.) NEW2AN/ruSMART 2014. LNCS, vol. 8638, pp. 509–518. Springer, Heidelberg (2014)
13. Balashova, E.: Projecting resource management of a telecommunications enterprise to ensure business competitive ability. In: Balandin, S., Andreev, S., Koucheryavy, Y. (eds.) NEW2AN/ruSMART 2014. LNCS, vol. 8638, pp. 502–508. Springer, Heidelberg (2014)
14. Balashova, E., Artemenko, E.: Assessment of possible uses of optimization methods and models in resource management of telecommunications enterprises. In: Balandin, S., Andreev, S., Koucheryavy, Y. (eds.) NEW2AN/ruSMART 2014. LNCS, vol. 8638, pp. 526–534. Springer, Heidelberg (2014)
15. Shirokova, S.V., Iliashenko, O.Y.: Decision-Making support tools in data bases to improve the efficiency of inventory management for small businesses. In: Recent Advances in Mathematical Methods in Applied Sciences, Proceedings of the MMAS 2014, Proceedings of the EAS 2014, pp. 204–212 (2014)

Operations Strategies in Info-Communication Companies

Vladimir V. Glukhov and Elena Balashova$^{(\boxtimes)}$

Peter the Great St. Petersburg Polytechnic University, St. Petersburg, Russia
vicerector.me@spbstu.ru, elenabalashova@mail.ru

Abstract. The article deals with the relevance of the managerial focus on the operations strategies in info-communication companies, prioritizes operations strategies. Particular attention is paid to industry specifics of priorities ranking, including analysis of the production costs, quality and reliability of the manufactured product, delivery speed, product delivery process reliability, the company's ability to respond to changes in demand and the rate of new product types implementation.

Keywords: Operations strategy · Competitiveness · Lean production · Costs of production · Quality and reliability · Delivery speed · Product supply

1   Introduction

One of the top targets for the most countries in the world is to ensure a long-term economic growth. Economic growth is accompanied by an increase in production efficiency, reduction of unemployment, prices stability, expansion of foreign economic relations and other positive economic and social processes. These objectives are closely connected with the functioning of the info-communication industry. IT penetration into the economic activity, digitalization of economic processes, extensive use of the Internet opportunities in economics and increase of its impact on the efficiency of enterprises activity led to multiple transformations practically in all markets. The convergence of information and communication technologies has led to the emergence of a new reality - the info-communication services, which differ significantly both from the information services and former telecommunication services. Demand for modern info-communication services, the quantity and quality of which is constantly growing, and that are in demand by many consumers in recent years is maintained at a very high level, which allows to predict their dominance on communication networks in the near future. Not only the total amount of income from the sale of info-communication services is growing, but also the range of these services. However, along with the increasing demand for info-communication services competition is increasing disproportionately in this sector of the economics, which leads to the search for new models of efficient management. Info-communication services markets are in the stage of their formation, researches, there are few studies how to meet the challenges of development, especially in the industrial science [1].

© Springer International Publishing Switzerland 2015
S. Balandin et al. (Eds.): NEW2AN/ruSMART 2015, LNCS 9247, pp. 554–558, 2015.
DOI: 10.1007/978-3-319-23126-6_48

2   Operations Strategy and Competitiveness in the Info-Communication Sector

Modern info-communication – is a global network infrastructure that integrates all available types of communication: Internet communications, telecommunications, mobile, computer, and other networks that can be used for delivery of information products to consumers anywhere in the world. Modern info-communication works integrally in a real time mode. The level of the info-communication development in a state determines a degree of its network readiness to join the emerging information community. Info-communication services markets are characterized by rapid changes in market conditions, methods of market activity shall continually adapt to the realities of the market.

Management of info-communication enterprises for a long enough time treated operations in the industrial sphere without due attention. Only a few companies have considered operating processes as one of the possible sources of competitive advantages. As a rule, the average statistical info-communication company reduced an operations strategy to decrease of the costs and eagerness to maximize the use of labor [2].

In the 1980s practical management was characterized by the fundamentally new trends, the basic idea of which is a concept that the operations should be an integral part of the corporate strategy of a company, which, in turn, shall be capable to respond quickly and accurately to the changing needs of its customers. Due to the high competitiveness of the present management paradigm, it rapidly spreads to all sectors of the economics, covering firstly the entire industrial sector, and then, successively, the sector of services, reaching thus to the info- communication sector. Info-communication companies have seriously embraced the fact that different consumers have different consumer preferences and priorities. The previous idea about the principal role of the maximum production costs reduction was rejected, a new strategy was put into the forefront, which is called the operations strategy. The operations strategy involves a brand new approach to the problems associated with the basic operations and the application of new tools and methods of management. Operations strategy means development of a common policy and plans to use resources of a company, aimed at the most efficient support for its long-term competitive strategy. Operations strategy, in conjunction with the corporate strategy, covers the entire range of the company's activity and involves a long-term process that allows a company to ensure an opportunity to respond quickly to changes in the future.

The present approach is about an overall context, decisions are made within its framework along with the overall strategy of a company, and the needs of customers are taken into account. Companies that have implemented similar transformations have succeeded in introducing the concept of lean production (LP) in their manufacturing process. An important role in this success was played not only by knowledge and skills to apply basic LP tools with industry specifics, not only by huge experience in marketing and general market conditions, but also by a dramatic increase in managerial attention to operational processes. Info-communication operators, who consider themselves world-class companies, recognize that their ability to compete successfully in the market largely depends on whether they comply with the mission of the customer service. Thus, the competitiveness of the info-communication companies is directly dependent on what position they hold relative to their competitors both in the domestic and international market. As a part of this problem, priorities of the operations strategy and criteria to assess their efficiency shall be established.

3   Operations Strategy Priority Ranking

In terms of priority indicators, ranking criteria for customer service shall be specified. Info-communication services markets have no standardized consumer; all info-communication services should be personalized. Info-communication services consumers are prone to innovations; an important task of management is a continuous modernization, development and supply of new services to the market. In early works of S.U. Skinner [3] (Harvard Business School) and later studies of T. Hill [4] (London Business School) the following types of operational priorities are differentiated: the costs of production, quality and reliability of a manufactured product (service), delivery speed, product delivery process reliability, the company's ability to respond to changes in demand and the rate of new product (services) types implementation. Analysis of the literature shows that, for the info-communication sector, this ranking of operations priorities was not used up to nowadays. In our opinion, the above listed priorities surely correspond to the logic of achieving a competitive capacity by the info-communication companies, so the industry manifestation abilities and their ranking in order of importance shall be discussed in more details.

R. Chase [5] notes that in any industry there is a segment of the market, sales volume of which depends entirely on lower production **expenses** of a manufacturing company. In order to compete in such a market niche, an info-communications company shall be a manufacturer with low production expenses, however, it is a required condition, and it is not enough. Its compliance does not always lead to a priori high level of the main activity profitability. As a rule, info-communication product, which sales depend only on the level of expenses, is a daily demand service (e.g., mobile phones). Consumers are often unable to distinguish between similar products from different manufacturers, which can lead to the fact that the price does not act as a main criterion for consumers' decision-making.

Influencing consumer choice, info-communication companies rightly govern the **quality and reliability** of products. As part of the issue two aspects quality management - product quality and process quality shall be considered. To ensure an appropriate level of production quality primarily demands of consumers are to be addressed. An extremely technically complicated info-communication product with an unreasonably high level of quality will be difficult to sell due to an increased price. The reverse situation - poor quality of mobile communication system can lead not only to loss of potential, but also existing customers. Accordingly, LP encourages management while ranking priorities of the operations strategy to pay particular attention to process quality management.

Delivery speed as a priority of the operations strategy has appeared in info-communication recently. This is due to the fact that this performance indicator started to be considered as a competitive advantage because of highly competitive markets emergence. Info-communication industry in Russia for a long time was characterized by the dominance of demand over supply, allowing the main market players to work with a speed comfortable for them. In recent years, the situation has changed, and this operations priority has taken its place among the others.

Product supply process reliability as a priority of the operations strategy, as we see, is of secondary importance for an info-communication operator due to industry-specific core product. This priority is linked, first of all, with the ability of a company to deliver products within declared term or even before that, which is extremely important for the material production of complicated products.

The company's ability to respond to changes in demand and **the rate of new product type's implementation** for the info-communication products is an important determinant of competitiveness. It is well known that if demand for company's products is steadily growing, operations problems are hardly to be faced. In case of lower demand or its long-term stagnation, a company is forced to cut staff and assets. For these reasons, the ability of a company for a long period to respond quickly and adequately market demand dynamics is becoming an essential element of its operations strategy.

4   Results

With the evolution of the Russian info-communications market, a group of companies was established that consider their corporate and marketing strategy on an international scale. Competition in the world market is much tougher, because this market is characterized by a larger number of "players". As a consequence, the Russian info-communication operators should review the concept of the operations strategy. Priorities of its development, dealt with in the present article, in our opinion, should be placed in the following order:

1. the costs of production, quality and reliability of a product (service);
2. the company's ability to respond to changes in demand and the rate of new products (services) types implementation;
3. delivery speed, product supply process reliability.

It should be noted that the prioritization logic could be changed over time. A group of scientists from Boston University [6] conducted a study, the purpose of which was to track changes in the competitive priorities of 212 American companies over the last 10 years. The result showed that with the improvement of development indicators requirements that had to be followed to maintain the level of competitiveness had changed. However, changing of current operational priorities does not negate the idea itself of a managerial focus on the operations strategy. Modern requirements to the quality of business, rapidly changing competitive conditions, general view on management problems suggest that proper operations management is a prerequisite for successful performance and survival of a company.

References

1. Glukhov, V.V., Balashova, E.S.: Economics and Management in Info-Communication: Tutorial. Piter, Saint-Petersburg, 272 (2012)
2. Shirokova, S.V., Iliashenko, O.Y.: Application of Database Technology to Improve the Efficiency of Inventory Management for Small Businesses. WSEAS Transactions on Business and Economics 11, 810–818 (2014)
3. Skinner, W.: The Focused Factory. Harvard Business Review 52(3), 113–121 (1974)
4. Hill, T.J.: Manufacturing Strategy: Text and Cases. 2nd edn. p. 574. Irwin, Burr Ridge (1994)
5. Chase, R.B., Aquilano, N.J., Jacobs, F.R.: Production and Operations Management: Manufacturing and Services. Irwin/McGraw Hill, New York (1998)
6. Kim, J.S.: Search for a New Manufacturing Paradigm. Executive Summary of the 1996 U.S. Manufacturing Futures Survey. Boston University School of Management Manufacturing Roundtable Research (1996)

Cellular Telecommunication Services Cost Formation

Tatyana Nekrasova, Valery Leventsov, and Ekaterina Axionova

Peter the Great St. Petersburg Polytechnic University, Saint Petersburg, Russia
dean@fem.spbstu.ru, vleventsov@spbstu.ru, director@eei.spbstu.ru

Abstract. The paper proposes a calculation method and a costing algorithm of services provided by telecommunications companies. Both the total cost of providing the whole amount of services (operating costs) and prime costs of individual types of services are calculated, with a possibility to estimate each of them at an average for the industry and particularly for a certain telecommunication company. The forecast value of costs on production and sales of cellular telecommunication services is defined. The factors affecting the prime costs of telecommunication services are revealed and the interrelation between them is determined. Some particular cellular telecommunication services are calculated. Based on these calculations of prime costs, prices for cellular telecommunication services are defined.

Keywords: Methods · Algorithms · Cellular services · Telecommunications companies · Operating expenses · Prime costs · Forecast · Factor analysis · Price

1 Introduction

The process of creating and distributing telecommunication services is always related to the use of living and materialized labor, which in its money terms acts as production costs. These costs are expenses on producing and distributing cellular telecommunication services which represent valuation of labor resources, fixed assets and working capital (material, energy, etc.) involved in production during a certain period of time (a year, a quarter). Their estimation is essential for all companies.

2 Main body

In telecommunications prime cost is a monetary value of production and sales costs, which telecommunication companies bear when providing services, maintaining and servicing technical means that are used for information transfer or are made available for customers. There is full prime cost of the total amount of services (operating costs) and prime cost of individual types of services and each of them can be defined at an average for the sector or specifically for a certain telecommunication company.

Prime costs of individual types of services are costs related directly with the transfer of certain types of messages (long-distance phone calls, tele-texts, facsimile messages, etc.) or with servicing technical facilities that are made available for customers

© Springer International Publishing Switzerland 2015
S. Balandin et al. (Eds.): NEW2AN/ruSMART 2015, LNCS 9247, pp. 559–566, 2015.
DOI: 10.1007/978-3-319-23126-6_49

(telephone and telegraph channels, phones, radio outlets). They are defined on the basis of distribution of the full prime cost by types of services and are used to calculate the prices. When estimating the prime cost of cellular communication services, expenses are calculated by each cost item. In order to do this, counting forward methods or enlarged measures are used. The purpose of defining the operating costs is to calculate their desired value so that regular production and commercial activities of the telecommunication company would be possible. According to this, each item is coordinated with other performance indicators, whose levels affect the absolute value of the estimated costs components [1].

The total value of annual costs (C_{total}), related to provision of cellular communication services is defined by formula [6]:

$$C_{total} = \sum_j C_{totalj} = \sum_j c_j * Q_j \tag{1}$$

where j – a type of cellular telecommunication services; c_j – prime cost of a unit of the type j telecommunication service; Q_j – volume of the type j services.

The nomenclature of services for the calculation period which comprises several years can vary.

The basic contents of costs on cellular telecommunication service production and distribution (C_{total}) include main technological costs (C_{tech}) and overheads (C_{overh}).

Main technological costs include the following items:

$$C_{tech} = P + Z_{en} + L + T + A + J_r + Z_{ms} + I \tag{2}$$

where P – expenses on basic and supporting materials purchased on the third hand and used to ensure regular technological process; Z_{en} – energy costs; L – primary production labor costs; T – social costs; A – depreciation costs to recover major industrial production assets; J_r – costs related to repairs of key industrial production assets; Z_{ms} – mutual settlements with communication enterprises; I – communication channel lease.

A specific feature of telecommunication companies is the fact that there is no such element as "raw materials" in the item "cost of materials", necessary for the technological process.

Overheads are related to organizing, managing and planning of production facilities and their maintenance [7].

This category of expenses contains:

$$C_{over} = V + X + B + U \tag{3}$$

where V – general business expenses; U – general production expenses; B – nonproduction expenses; X – other undocumented expenses.

General *production expenses* include expenses related to production process servicing, such as:

- maintenance of general production assets (not key technological ones) in operating condition (technical maintenance, minor and medium repairs and overhaul of equipment, buildings and facilities that are not part of the main technological scheme);
- non-capital expenditure, related to technology improvement, reliability and safety enhancement;
- purchase of spare parts to repair equipment, fittings, facilities, laboratory equipment and other means that do not belong to key assets;
- expenses on equipment certifying;
- outsourcing.

General business expenses include expenses, such as:

- expenses related to enterprise's and its business units' administrative personnel support as well as material, technical and transportation costs associated with their activities (according to the standards established by legislation);
- costs of materials for operation and repairs of general business equipment;
- expenses on service centers.

Non-production expenses are related to service provision and are not directly associated with production sphere (mail and printing expenses, efficiency drive costs, interests on loans, etc.) [8].

Other expenses include costs on marketing research, advertising, R&D, Internet-services, radio frequency subscription fees, subscription fees for the communication building system standard, subscriber servicing, taxes and deductions referred to prime cost and paid under the current legislation, mandatory company's property insurance payments accounted as main production assets, customs tariffs and duties, penalties, fines.

The amount of expenses on production and distribution of cellular telecommunication services is projected from the essential expenses on cellular telecommunication system maintenance at the beginning of the period of time (for the beginning of the calculation period) and additional expenses appearing throughout year t, caused by an increased scale of production. The scale of production grows due to the fact that services are provided to an expanding subscriber market and because new technologies are introduced. Additional expenses expressed in monetary terms in the conditions of the initial period of time (increasing expenses on depreciation and repairs, energy resources, salaries and wages, market research, advertising, service support of subscribers) have to be estimated considering the time factor [5].

The projected value of cellular telecommunication service production and distribution costs for year t (C_t^{pd}) and period T (C_T^{pd}) can be defined as:

$$C_t^{pd} = C_{total}^{init} + C_{total_{t+1}}^{pd} \Big/ (1+r)^t \tag{4}$$

$$C_T^{pd} = C_{total}^{init} + \sum_{t=1}^{T} \frac{C_{total_{t+1}}^{pd}}{(1+r)^t} \tag{5}$$

where C_{gen}^{itit} - cellular telecommunication service production and distribution costs, related to servicing the initial number of subscribers; $C_{total_{t+1}}^{pd}$ - cellular telecommunication service production and distribution costs per year t+1 (for the beginning of the year); r – discount rate.

The projected amount of cellular telecommunication service production and distribution costs considering the period when costs do not grow and the whole subscriber market is served (C_{T1}^{pd}), can be described by the next formula [3]:

$$C_{T1}^{pd} = C_T^{pd}\left(1 + \sum_{t=T+1}^{T1} \frac{1}{(1+r)^t}\right) \tag{6}$$

The algorithm of cellular telecommunication service prime cost formation includes the stages represented in Fig. 1.

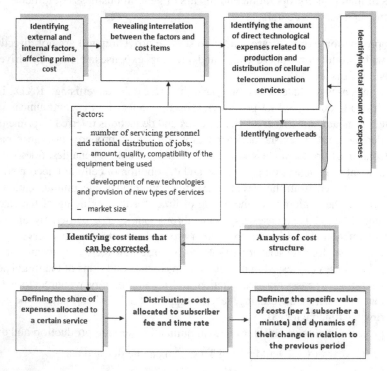

Fig. 1. Prime cost formation scheme of telecommunication services provided by cellular companies

Table 1 includes an example of interrelation between the factors, influencing the prime cost of cellular telecommunication services and the cost items affected [4].

Table 1. Factors affecting the value of prime cost of telecommunication services

Factors affecting	Cost items	Coefficient of elasticity
– number of operative personnel	Labor costs	1.0
– multi-tasking		0.3
– rational assignment of work		0.3
– amount of the equipment being serviced		0.9
– quality of the equipment being used	Depreciation and minor repairs	1.2
– dependence on the equipment deliveries from abroad		0.8
– compatibility of the equipment used		0.4
– cost of the equipment used		0.98
– revaluation of assets from foreign currency to national one		1
– number of operative personnel	General business expenses	0.2
– size of rented space		0.9
– development of new technologies	R&D costs	2.8
– the types of equipment in use that are produced locally or abroad	Training and retraining of personnel	1.2
– new types of services		1.15
– market size	Quality and information protection costs	1.8
	Advertising and market research costs	0.4
	R&D costs	0.1
	Expenses on mutual settlements	0.25

In the prime cost structure of cellular communication services the major share is attributed to the following items: depreciation costs and repairs of buildings, facilities and core process equipment, general business expenses, expenses on mutual settlements, salaries and wages, other expenses (including expenses on information quality and protection, advertising and present-day market research) [2].

Definition of prime cost for a specific cellular communication service is represented in Fig. 2.

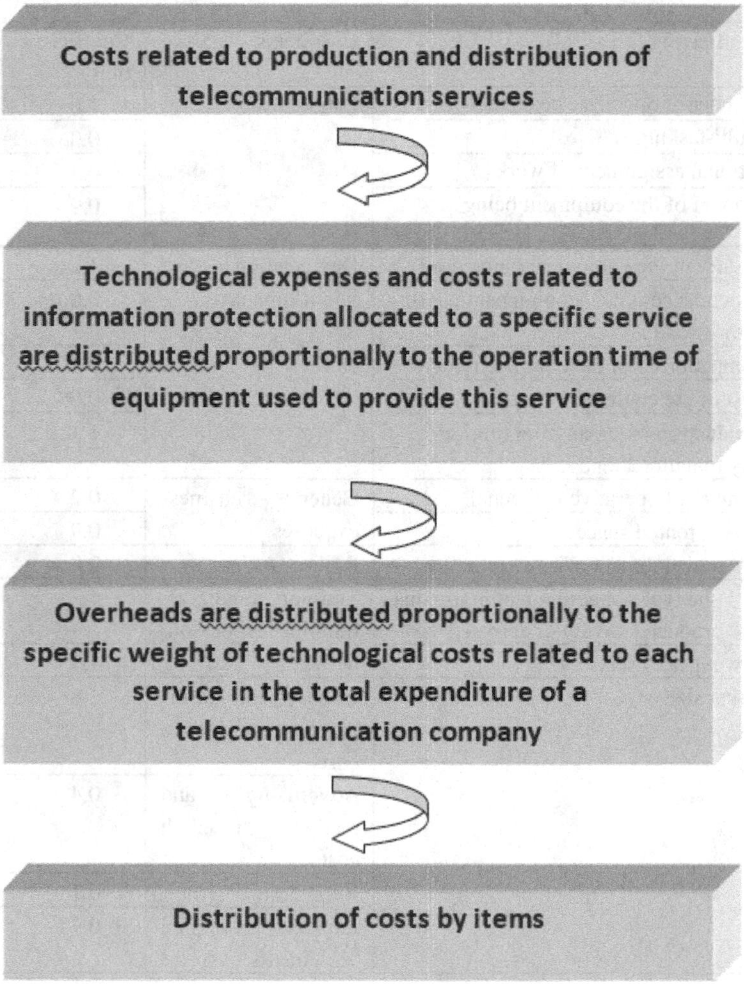

**Costs related to production and distribution of
telecommunication services**

**Technological expenses and costs related to
information protection allocated to a specific service
are distributed proportionally to the operation time of
equipment used to provide this service**

**Overheads are distributed proportionally to the
specific weight of technological costs related to each
service in the total expenditure of a
telecommunication company**

Distribution of costs by items

Fig. 2. Definition of prime cost for a specific cellular communication service

Definition of prime cost for specific types of telecommunication services implies:

- distribution of overheads (to support administrative personnel, service centers) proportionally to the specific weight of each service in the total expenditure of a telecommunication company;
- costs related to technological process of producing telecommunication services can be calculated in proportion to the number of the operative personnel who take part in providing a specific service. If the operative personnel take part in providing different services (maintenance of the equipment) then they are distributed in proportion to the time spent on fulfilling their production functions for each service.

Provision of cellular communication services does not require permanent involvement of personnel, so main technological expenses (labor costs, including wage payments, depreciation and repair costs, energy and auxiliary materials costs, information protection costs) by specific types of services can be distributed in proportion to operation time of the equipment used for each type of services. Expenses in quantification by cost items can be defined based on their monetary terms with a counting forward method.

3   Conclusion

The results of estimating prime cost of cellular telecommunication services are the basis of pricing. When they are calculated for a projected period, the prime cost of services is corrected with due consideration of external and internal factors that reflect various processes (inflation, change of prices on resources, introduction of new technologies, buying power of customers, etc.). A lower prime cost of services is a prerequisite for price reduction, which makes cellular communication services more available for a wide range of consumers. The consequent activation of information exchange results in growing revenues from services provided and increased profits. Thus, prime cost reduction contributes to higher competitiveness of services provided by telecommunication companies. The biggest reserves for prime cost reduction are in the internal factors, which, being actively affected in telecommunication companies, ensure absolute and relative saving of expenses. Moreover, the biggest effect is achieved if measures are taken to reduce the expenses related to those cost items which have the biggest specific weight in the structure of costs.

Efforts of managers in telecommunication companies can be made, for example, to reduce the number of employees or to ensure that the growing scope of work is done by the same number of personnel due to decreased labor intensity and better labor productivity. To solve this problem it is important to make production processes automatic, introduce advanced technologies, replace and modernize outdated equipment, improve labor organization and management.

Since the structure of the prime cost of cellular telecommunication services includes depreciation of the key assets as one of the main elements by its specific weight, then one of the main ways to reduce prime cost is to improve labor tools that are used. Higher return on assets entails bigger operating revenues with the same key assets or advanced rates of revenues growth comparing to the rates of key assets growth. In this case there is relevant decrease in the depreciation costs per prime cost measurement unit.

It is worth noting that improved use of key assets ensures lower prime cost due to decrease in expenses related to current maintenance of communication means, including costs of materials, spare parts and energy, whose amounts depend on the volume of equipment and communication facilities that are maintained.

The next step in decreasing prime cost is to cut down general business, administrative and managerial expenses. It can be done with improved management structure and cheaper administrative personnel expenses which can be achieved if odd management levels are eliminated, performance standards for administrative personnel are introduced, and their functions are made automatic. Saving on utilities and trip expenses positively influences prime cost reduction of telecommunication services.

References

1. Gluhov, V.V., Ilin, I.V.: Project portfolio structure in a telecommunications company. In: Balandin, S., Andreev, S., Koucheryavy, Y. (eds.) NEW2AN/ruSMART 2014. LNCS, vol. 8638, pp. 509–518. Springer, Heidelberg (2014)
2. Huang, T.-L., Liu, F.-H.: Whether service innovativeness has additive effects on mobile banking business from switching costs perspective. International Journal of Mobile Communications 13(2), 204–227
3. Ivakhnenkov, S., Melykh, O.: Financial Controlling: Methods and Information Technologies. Znannia (in Ukr.), Kyiv (2009)
4. Kniazieva, O., Melykh, O.: Functional model for investment projects' evaluation and financial controlling. Economic Annals-XXI 3–4(1), 47–50 (2014)
5. Lechner, A., Klingebiel, K., Wagenitz, A.: Evaluation of product variant-driven complexity costs and performance impacts in the automotive logistics with variety-driven activity-based costing. International MultiConference of Engineers and Computer Scientists 2(2011), 1088–1096 (2011)
6. Liang, H., Payne, J.I., Kim, H.S.: Femto-cells: problem or solution? a network cost analysis. In: GLOBECOM - IEEE Global Telecommunications Conference. Article number 6133791
7. Martin-Escalona, I., Barcelo-Arroyo, F.: Performance evaluation of middleware for provisioning LBS in cellular networks. In: Source of the Document IEEE International Conference on Communications, 4289589, pp. 5537–5544
8. Nekrasova, T., Leventsov, V., Axionova, E.: Forecasting of investments into wireless telecommunication systems. In: Balandin, S., Andreev, S., Koucheryavy, Y. (eds.) NEW2AN/ruSMART 2014. LNCS, vol. 8638, pp. 519–525. Springer, Heidelberg (2014)

Forming a Telecommunication Cluster Based on a Virtual Enterprise

Gueorguy Kleyner and Aleksandr Babkin[✉]

Peter the Great St. Petersburg Polytechnic University, St. Petersburg, Russia
babkin@spbstu.ru

Abstract. The most important role in the development of the national economy is played by telecommunication clusters. According to the research and practice, clusters successfully contribute to solving problems of commercializing innovations. The active involvement of state agencies in forming clusters and supporting their effective performance allows considering clusters as a type of public-private partnership, which is an institutional alliance between the state and business in order to implement particular projects.

The authors have touched upon the issues of creating the telecommunication cluster in the form of a virtual enterprise, which implies the combination of cluster elements, including resources, knowledge, developments, in the virtual environment.

The paper claims that joining enterprises into a virtual cluster will be effective for fulfilling the state order, because it can reduce transactional costs, ensure fulfillment of numerous state orders, ensure involvement in major projects and attract long-term financial resources.

The authors see the trends of further research on creating the telecommunication cluster in the form of a virtual enterprise as an analysis of the virtual enterprise structure in order to make its performance more effective.

Keywords: Telecommunication cluster · Virtual enterprise · Services · State order · Costs

1 Introduction

Telecommunication clusters should play a very important part while reaching the targets of the national economies. Researches and experience have shown that clusters help to solve the most difficult problems of innovation commercialization [4,7,11-14].

As the authorities play an active part in cluster creating and cluster maintenance, clusters can be analyzed as a type of public-private partnership, i.e. an institutional government-business alliance executing certain projects.

As a cluster approach has become popular in developed countries, Michael Porter's ideas have become widespread. According to Porter, "cluster is a geographic concentration of interconnected businesses and associated institutions in a particular field linked by commonalities and complementarities" [3]. Any cluster has the core – its key element usually consisting of the companies producing end-products and services.

© Springer International Publishing Switzerland 2015
S. Balandin et al. (Eds.): NEW2AN/ruSMART 2015, LNCS 9247, pp. 567–572, 2015.
DOI: 10.1007/978-3-319-23126-6_50

Distinctive cluster features include competitive relationships between the cluster companies, which have the same production conditions.

A cluster can work as a sustainable and efficient structure because of the geographical proximity of its members, innovative approach, combination of large-scale producing with small and medium sized businesses, and efficient support of the local authorities.

2   Problem Statement

Cluster policy is any government effort to create and maintain clusters in a particular area. According to Porter, the main goal of cluster policy or concentration policy is to increase the competitiveness in a country, or an area.

At first sight, framework policy, which increases competitiveness, does look more thorough and market-friendly than a conventional approach that includes government help to particular industries, or business. However, cluster policy also involves a selective approach. While dealing with information asymmetry, the government has some limited opportunities to define the best ways to develop an industry, or a business.

The experience has shown that clusters are often formed in certain areas that are considered a priority at some particular moment. They are created "top-down" notwithstanding their being beneficial or useless for the participants. As a result, there emerge some enterprises that are united in name only but, actually, they are independent, and have no common goals or projects. Furthermore, forcing such companies into a cluster may handicap their activity, as they have no cooperation schemes, coordination programs or resource sharing schemes.

At the same time, some alliances between businesses and institutions emerge naturally. They benefit all the participants, and facilitate efficient cooperation, even though such alliances are not officially called clusters. Such alliances do need resource and knowledge integration as well as the government support in order to develop. Cluster policy faces some difficulties finding out such unofficial clusters and rationalizing their support.

To make a cluster efficient we suggest analyzing the formation of a telecommunication cluster as a virtual enterprise. Thus, it is possible to reduce aggregate production costs, and transactional costs, effectively combine resources, increase competitiveness, raise innovativeness, develop new technologies, etc.

3   The Results

Global communication networks, such as the Internet and Intranet, are a new organization kind in the information era. They help create business structures where information shortage is reduced, information efficiency is increased, knowledge is actively accumulated and transferred, trust is highly developed, cooperation between the partner companies is intense, etc. These business structures are more flexible. They can quickly adapt to market changes, and transform into different structures, creating the expertise necessary to offer products and services that fulfil market demands.

These business structures can cooperate and have partner relationships using the Internet and Intranet, while being located anywhere in the world. They are called virtual teams, virtual enterprises, or virtual corporations [8].

A virtual enterprise has the same abilities as a conventional one. However, it has slightly different organization and structure. Virtual enterprise is a temporary willful alliance of several usually independent businesses (institutions, individuals) created in order to optimize their production system and benefit all the participants [6]. The participants come together to share skills or resources to get better results faster, globally, and with minimal costs. From a customer's point of view, the participants act as one united enterprise, using state-of-the-art information and communication technologies.

There is another definition given by J. Hopland, one of the top-managers of DEC. Virtual enterprise is a network computerized structure consisting of inhomogeneous interacting agents that are located in different places. These agents co-develop a project or several projects, while being partners, cooperating, coordinating efforts, etc. Thus, creating a virtual enterprise involves smart modeling of complicated, inhomogeneous, geographically separated agents [6].

The main point of a virtual enterprise is to unite independent business entities (individuals, groups, companies) that are geographically separated but currently cooperate globally and electronically, and share resources and efforts in order to accomplish shared tasks.

Not only have virtual enterprises shared goals and functions, but also roles and responsibilities, ways of communication, benefit distribution rules, etc. There are three types of virtual enterprises, depending on the management system:

— those with centralized administration; participants act upon their organization charge, while one of the participants controls the whole process, i.e. defines a task, instructs others, summarizes the results, and makes the decisions
— those with distributed administrative roles; the knowledge and resources are distributed between the participants, while having a shared command control that is responsible for decision making if there is any conflict
— those with decentralized administration; all the management processes are based on local interactions between the participants.

Virtual enterprises allow sharing unique experience, production capacities, advanced production practices, in order to implement a project that the participants otherwise would not be able to implement individually. Not only a virtual enterprise reduces the companies' transactional costs – it also integrates their interests and functions.

The main advantage of virtual enterprises is the ability to choose the best resources, knowledge, and skills, without spending too much time. Virtual enterprises can:

— accelerate implementation
— decrease total costs
— more thoroughly fulfill the customer's demands
— flexibility in adapting to environmental changes
— remove barriers while entering new markets.

In our opinion, a virtual enterprise can be a functioning telecommunication cluster. To make a cluster efficient, it is useful to unite the cluster elements in the virtual environment, including resources, knowledge, etc. The virtual enterprise will be the core (e.g. based on one of the participating companies). Following processes can be combined in the virtual environment:

— interaction with clients (service consumers), including customer search, marketing, order placement, service providing, etc.
— interaction with suppliers, including material and component delivery
— new technologies, innovations, interaction with research institutes, universities, research and advanced development, etc.
— interaction with the authorities, including state order placement and fulfilment, getting subsidies, licenses, permissions, and guaranties.

The activity of such a virtual enterprise or virtual cluster is represented in fig. 1 (you can see that both market and government can be a customer).

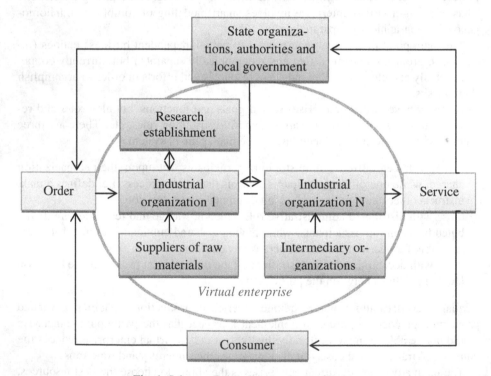

Fig. 1. Scheme of virtual enterprise performance

Advantages of the virtual environment for the telecommunication cluster are as follows:

— effective coordination of the activity in order to achieve common goals and implement joint projects;
— reduction in aggregate costs of production, work fulfillment, service providing;

— reduction in transactional costs (negotiating, searching for information, interacting etc.);
— opportunity to fulfill a great number of state orders and to participate in long-term projects;
— opportunity to obtain state subsidies, grants, benefits;
— flexibility in adapting to environmental changes;
— increase in the innovative activity of a cluster, development of new technologies;
— removing barriers to the entry to new markets, including international ones.

Joining cluster participants in the virtual environment will be effective in fulfilling the state order. It is feasible through the single portal of the state procurement, with state orders placed on web-portals. Hereby the interaction between a virtual cluster and a governmental body will be easier in the virtual environment. Moreover, state services can be provided in the e-form through the system of e-government which is about to be launched.

Thus a cluster can be assigned to fulfill a state order, later the cluster (the core of the cluster in the form of a virtual enterprise) will allocate, run and coordinate the state order fulfillment. Such a collaborative usage of the cluster's opportunities in the form of various resources will help achieve the result along with the reduction in aggregate costs and the creation of the competitive advantage for the cluster in the global scale.

Besides, the formation of a telecommunication cluster based on a virtual enterprise will lead to the rise in the cluster efficiency, resulting into more rational measures of the state policy.

European countries started using the concept of virtual cluster long ago. A lot of research is devoted to this concept. Today the cluster experiences the shift from physical ties within itself to virtual ties which create a competitive advantage in the conditions of the dramatically changing innovation economy.

The effective union of enterprises in the virtual environment can be exemplified by the virtual enterprise «VIRTEC Project», developed by São Carlos School of Engineering at the University of São Paulo, Brazil. The enterprise unites nine small and medium-sized businesses. These businesses worked in the field of electronics, metal - ceramic production, polymeric materials, mechanics, mechatronics, liquid systems, applied software and service. Every business was specialized in manufacturing a single product, e.g. one business manufactures polyurethane hammer, another manufactures polymeric rubber etc. VIRTEC virtual enterprise developed a number of new products, such as biodegradable polyurethane rubber for mechanical assembly, product coating, medical diagnostics. This rubber proved to be cheaper and to have a longer life cycle. VIRTEC enterprise confirmed that a virtual form of the organization allows cutting temporary costs for design and production of a new product, reducing its cost of production and improving quality parameters, compared to similar goods of other manufacturers.

The concept of a virtual cluster is still actually unavailable in Russia, which is caused by the lack of legislation in this area, the lack of necessary economic structure and inefficiency of some institutes. However, the usage of such a type of forming and performing telecommunication clusters is considered to be effective and prospective.

4   Conclusions

This paper considers issues of creating a telecommunication cluster in the form of a virtual enterprise, which implies the combination of cluster elements, including resources, knowledge, developments, in the virtual environment.

Joining enterprises into a virtual cluster will be effective for fulfilling the state order, because it can reduce transactional costs, ensure fulfillment of numerous state orders, ensure involvement in major projects and attract long-term financial resources.

The authors see the trends of further research on creating the telecommunication cluster in the form of a virtual enterprise as an analysis of the virtual enterprise structure in order to make its performance more effective.

References

1. Chang, H.J.: The political economy of industrial policy. St. Martin's, London (1994)
2. Hospers, G., Desrochers, P., Sautct, F.: The Next Silicon Valley? On the relationship between geographical clustering and public policy. International Entrepreneurship Management Journal **5**, 285–299 (2009)
3. Porter, M.: Location, competition, and economic development: Local clusters in a global economy. Economic Development Quarterly **14**, 15–34 (2000)
4. Porter, M.: Location, clusters and company strategy. In: Clark, G.L., Gertler, M.S., Feldman, M.F. (eds.) The Oxford Handbook of Economic Geography. Oxford University Press, New York (2000)
5. Belyakova, G.Y.: Cluster as a form of organizing the industrial production. Journal of SibGTU, Krasnoyarsk (1) (2005)
6. Bugorsky, V.N.: Network economy. M.: Izdatelstvo Finansy and Statistika, p. 237 (2008)
7. Porter, M.: On competition. M.: Izdatelsky dom "Williams" (2003)
8. Serdyuk, V.A., Serdyuk, V.A.: Networking and virtual organizations: state, development, prospects. Management in Russia and Abroad (5) (2012)
9. Sizov, V.V.: Role of clusters in the formation of a regional economic police. TGTU, Tomsk (2008)
10. Is it Possible to Replicate Silicon Valley? Lecture by Pierre Deroche on the international experience in creating economic innovation, April 08, 2011. http://polit.ru/article/2011/04/08/deroche
11. Babkin, A.V., Kudryavtseva, T.J., Utkina, S.A.: Identification and Analysis of Industrial Cluster Structure. World Applied Sciences Journal **28**(10), 1408–1413 (2013)
12. Babkin, A.V., Kudryavtseva, T.J.: Identification and Analysis of Instrument Industry Cluster on the Territory of the Russian Federation. Modern Applied Science **9**(1), 109–118 (2015)
13. Knights, D., Tullberg, M.: Managing masculinity/mismanaging the corporation. Organization **19**, 385–404 (2012)
14. Garud, R., Gehman, J., Kumaraswamy, A.: Complexity Arrangements for Sustained Innovation: Lessons from 3M Corporation. Organization Studies **32**, 737–767 (2011)

Project Controlling in Telecommunication Industry

Sergei Grishunin$^{(\boxtimes)}$ and Svetlana Suloeva

Peter the Great Saint-Petersburg Polytechnic University, St. Petersburg, Russia
sg279sg279@gmail.com, emm@spbstu.ru

Abstract. Telecommunication industry is in transformation stage and character-ised by heavy capital intensity and high-risk environment. Ensuring that capital expenditures projects in the industry meet their promises, i.e. achieve the goals set in strategic plans is a complex task, which cannot be solved by conventional, project management techniques. To solve this task in the environment of high uncertainty we develop project controlling system that identifies the key pro-ject's goals, "tracks" the progress of achieving these goals by using risk-oriented control procedures and suggests remediation measures to reduce the deviations from the project's goals. We developed the reference model of pro-ject controlling in telecommunication industry and described examples of controls used in the system, where and when they should be implemented. We argue that implementation of project controlling can reduce the deviations from goals set by the project's plan by around 50%.

Keywords: Project management · Strategic management · Controlling · Risk management · Internal control

1 Introduction

Global telecommunication industry is currently in transformation stage. Liberalization of regulation, privatization of dominant national carriers along with development of new technologies (such as Internet and wireless networks) has resulted into the reshaping of the industry from a set of state-owned national incumbent carriers, which sell mainly the voice traffic to info-communication space. In this space, telecommunication and information technologies are closely integrated and are used for transmission of information to various distances [1].

The main global industry trends include diminishing role of "traditional" voice services and growing development of high-speed communication technologies and Internet access (including wireless) as well as rendering new services and "rich" content such as entertainment, financial services and e-commerce. Telecommunication companies' ability to create and maintain competitive advantage will rely on their capacity to (1) improve the network by gaining efficient spectrum and investing in fibre to provide greater coverage and capacity; (2) enhance distribution channels; 3) improve the range and affordability of handsets; and (4) support "rich" content [1, 2, 3].

Traditional operators face the growing competition from cable companies, unregulated over-the-top (OTT) content providers, such as Netflix, and alternative internet services

© Springer International Publishing Switzerland 2015
S. Balandin et al. (Eds.): NEW2AN/ruSMART 2015, LNCS 9247, pp. 573–584, 2015.
DOI: 10.1007/978-3-319-23126-6_51

including Skype and WhatsApp. In mobile segment, the growth in competition is driven by liberalization of granting new licenses and frequency spectrums as well as by development of the competing technologies such as 4G/LTE. We expect that higher competition will reduce the flexibility price setting for products and services, which will negatively affect companies' profitability and cash flow generation abilities. Fierce competition results in shortening of products/services lifecycles. This requires significant capital expenditures for frequent and fast innovations.

In Russia, telecommunication industry in the last decade has demonstrated vigorous growth. In 2014, the total capacity of domestic telecommunication market exceeded RUB1.6 trillion (around $30 billion at RUB/$ exchange rate of 55). Since 2000, the market has grown by more than 10 times in rubles or 6 times in US dollars [5]. The key trends in the domestic industry repeat the global trends. They include (1) reduction of revenue from fixed-line telephony and other voice services in favour of mobile and broadband Internet access; and (2) growing incomes from "rich" content and various Internet services. However, the growth of Russian telecommunication market in the next 24-36 months will slow down and can even turn negative. This can be explained by exhaustion of sources of extensive market growth due to attainment of high level of penetration in almost all segments of the market. This deceleration of growth will be further exacerbated by the recession in Russian economy expected in the next 2 years (under the pessimistic scenario Russia's GDP can decrease by 5.5% in 2015 and by 3% in 2016) [7]. In the period, we expect further increase in competition, moderate decrease of companies' revenues and pressure on their profitability.

Telecommunication is a high capital intensive industry. Operators continue to develop new services and products, to modernize their networks to meet the rapidly increasing demands for bandwidth and to engage in mergers and acquisitions to tap new product or geographic markets. Investment project in the industry are exposed to numerous risks, which, if realized will result in significant negative variances of actual projects' outcome vs. goals, for example the lower return on investment capital than that investors anticipated. The range of exposures includes technological, engineering, market, regulatory, operational or financial risks. The risks significantly increase at times of high market uncertainty, such as that expected in the Russian market in the next two years. In such periods, there is a growing demand for innovative and efficient project management systems.

We argue that "conventional" project management techniques, which are applied by Russian telecommunication companies in the last decade, do not suit the new environment. These methods assume that events, which influence the project, can be easily predicted, project management tools and techniques are well known to managers and project tasks can be performed in prearranged and consecutive order. Moreover, "conventional" methods can manage only a single project goal or a very limited set of goals. Conversely, in uncertain environment manager must simultaneously control a balanced set of heterogeneous goals that cover all project's perspectives from technology to finance. These and some other gaps in "traditional" project management can be remediated if the managers apply the new, innovative systems of project management such as investment controlling.

2   Definition and Functions of Investment Controlling

Investment controlling is the project management system ensuring achievement of project goals in the environment of high uncertainty [8]. In telecommunication industry, it is applied to strategically important projects aimed at the development and implementation of new technologies and services, building the strategically important networks, mergers and acquisitions. Investment controlling performs the following main functions (Table 1).

Table 1. Functions of investment controlling

Function	Description of function
Project initiation	• Strategic analysis of environment to search opportunities for new investment projects which can create competitive advantages to the firm
	• Evaluation of investment projects
Project planning	• Identification a set of project goals
	• Decomposition of the project into investment packages
	• Preparation of project master plan and budgets for investment packages
	• Change management: making changes in master plan and budgets
Project administration	• Development of project team
	• Distribution of responsibilities and authority within the project team
	• Creation of motivation system for project team' members
	• Development and maintenance of project's policies and procedures
Information systems	• Identification of requirements to data storage
	• Creation and maintenance of project information system
Risk management	• Identification of project's risks and sources of risks
	• Analysis and evaluation of project's risks
	• Estimation of risks' impact on project's goals and selection of risks for further monitoring and control
	• Revision and re-evaluation of risks in response to the changes
Project control	• Development and implementation of control procedures
	• Execution of control procedures and analysis of variances
	• Adjusting the control procedures in response to changes in project goals, risks, business processes, etc.
Remediation and decision making	• Development and maintenance of project reporting system
	• Development of corrective actions in response to variances
	• Co-ordination and control over decision making

Our own research and analysis of others' research [8] has demonstrated that implementation of investment controlling, despite the cost of its implementation, allows to decrease the deviation of actual time spent and costs over the initial plans by around 50%. It helps to achieve earlier project payback as well as gain sustainable competitive advantage to the firm.

3   Development of Project Controlling System

From the project management perspective, investment controlling can be split into two large parts: "the direction part" responsible for project planning and initial project's setup and "applied part" – project controlling which includes activities directly related to project management (risk management, project control and remediation of variances). Henceforth in this paper, we will discuss project controlling system.

Project controlling is a subsystem of investment controlling which purpose is to ensure that capital expenditures projects meet the goals set in strategic plans. Project controlling's tasks include (1) "tracking" the progress of project's goals achivement by using risk-oriented control procedures; and (2) developing the remediation measures to reduce the negative impact of variances on achievement of project's goals. Project team, created at project initiation stage and consisted of management team and various service groups is ultimately responsible for implementation and maintenance of project controlling system.

Project controlling prevents transformation of the investment project to "zombie project" which, due to cumulated negative variances from the company's strategic plan, fails to fulfil its promises but yet keep being implemented, consuming valuable companies resources without any real hope of having a meaningful impact on the company's revenue, profits and cash flow generation prospects [10].

The reference model of project controlling is presented at Fig 1. The input in the model is the strategic plan which includes: (1) the ultimate strategic goals of the project, (2) interim goals inferred from the strategic goals; (3) detailed budgets (organizational, technical and operational, financial, etc.); (4) analysis of project sensitivity (analysis of project strengths, weaknesses, threats and opportunities (SWOT)); and (5) criteria of materiality of variances from the goals set by the plan.

Strategic goal of the project (SG) is usually defined as achievement of planned investment payoff of future project's cash flows. Such goal can be expressed by the modified profitability index:

$$IP = \frac{(NPI - NPE) + OV}{C} \qquad (1)$$

Where:

IP- profitability index, the ratio of payoff to investment of proposed project;
NPI – present value of cash inflows over the project life;
NPE – present value of cash outflows over the project life;
C – net present value of investments
OV – net present value of real options

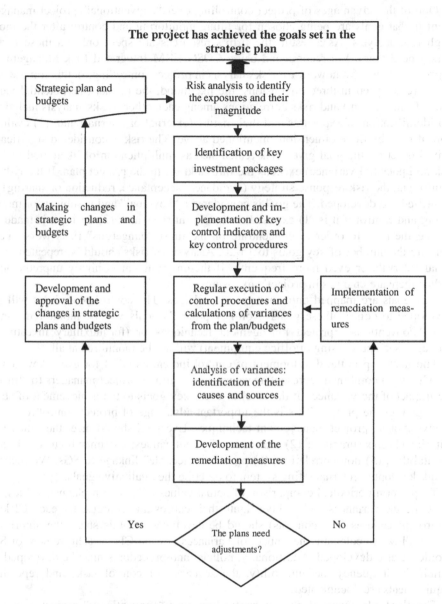

Fig. 1. Reference model of project controlling

The interim project goals are derived from strategic project goal with factor analysis of profitability index by financial, time, technical, macroeconomic and other variables. Interim goals are distributed in investment packages (a set of homogeneous and interrelated jobs).

One of the advantages of project controlling over "conventional" project management is that goals are being chosen for future monitoring and control after the thorough risk analysis. As a result, the firm's resources are spent only on those goals which need them. We recommend using COSO ERM Integrated Risk Management approach [9] as a framework for risk analysis in project controlling. In this framework risks are analysed in three dimensions: the likelihood, the impact (the potential variances from the plan) and risks correlations among each other. Risks analysis includes: (1) identification of exposures and their nature (external or internal) and; (2) evaluation of risks by three dimensions mentioned above. The risk is considered significant if its impact on the goal given risk (calculated as multiplication of likelihood of the risk and potential variance) exceeds the threshold set in the project plan. If the risk is significant, the risk response strategy (avoidance, acceptance, reduction or sharing) is identified and developed. The goal is considered a "key goal" and is subject for monitoring and control if it is (1) exposed to significant risks and; (2) decision is made to reduce the risk. In order to concentrate on the most "dangerous" risks we suggest limiting the number of key goals to 15-20. Analysis of risks should be repeated after equal intervals or even more frequently if the environment is highly unpredictable and/or there are changes in project's exposures.

Key goals are mapped onto investment packages. The control procedures will be developed to control these goals. Non-key project's goals will be monitored either with "conventional" project management techniques or (for auxiliary investment packages such purchasing of office equipment) will not be monitored at all.

The next step is the development of control indicators (CIs) for every key goals. All CIs when combined represent a system, which allows project managers to "trace" the impact of the variances at the level of every key goals to the achievement of strategic goal of the project. This is the important advantage of project controlling over "conventional" project management techniques. Every CI should meet the following criteria: (1) measurability; (2) existence of target values; (3) unambiguity; (4) accountability; (5) non-correlation; (6) provide "traceable" linkage to SGs. We recommend developing internal rating system to describe the qualitative goals.

To perform periodical comparison of actual values of CIs with planned values, to calculate the variances and to investigate their causes and sources, for each CI key control procedures (key controls) should be developed. At this stage, the algorithm which allows to calculate the impact of variances at each CIs on achievement of SG should be also developed. Additionally, policies and procedures must be developed in which the frequency, accountability, the sequence of control tasks and reporting requirements are documented.

The third advantage of project controlling over "conventional" project management systems is that in controlling priority is given to "early warning" controls which can predict variances in achievements of key goals one on more periods ahead and perform remediation measures as early as possible. Let us consider types of key controls in project controlling in more details (Fig. 2).

Project stage	Monitor-ing for key variables	Early warning check	Pre-check	Interim control	Documentary control	Post-factum control	Audit of control system
Allocation of resources from budgets							
Tenders and contracting							
Contracts execution							
Change orders process							
Final tests of equipment, products and services							
Contracts' or project's commissioning							

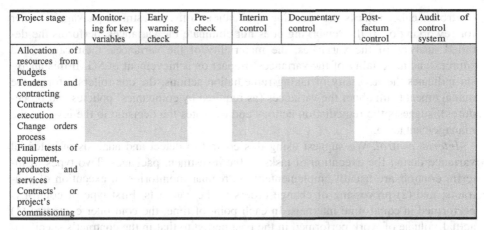

Fig. 2. Types of controls in project controlling

The early warning controls include two types of procedures: *monitoring of key inputs* and *early warnings checks*. The latter is the monitoring the risk zones in the environment with the purpose to receive "weak messages", to "decipher" them into potential threats or opportunities for achievement of project's key goals. The sources for weak messages are the consumers, suppliers and sellers, consultants, competitors, professional associations, media, scholar societies, government, etc. To organize such control procedure, we suggest creating a special monitoring team within the project's organizational structure. The monitoring team will provide the management team with a regular analytical bulletin. It includes the analysis of identified weak messages, forecast of "hidden" variances of key goals for one or more periods ahead and calculation of impact of these variances on achievement of SG.

Monitoring of key inputs is another control tool to identify variances in achievement of project goals at the earliest stage. This control procedure is designed to identify variances in key assumptions/inputs of the project's strategic plan such as supply, demand, prices, availability and cost of capital, competitive environment, viability of project's technology, etc. The same as for early warnings check, this control is organized in form of providing the management team of analytical bulleting on regular basis. The bulletin includes analysis of variances in key assumptions to the strategic plans over time and impact of these variances on achievement of SG.

Pre check. This control is designed to identify the variances from the strategic plan before any task in the investment package starts. The procedure consists of comparison of preliminary estimation of cost and/or duration of the task to those in the budget. The procedure is performed at three checkpoints: (1) allocation of funds from budget; (2) performing choice of supplier/contractor; and (3) conclusion of the contract with the supplier. Let us consider an example. The controller receives from investment package's owner the preliminary valuation of cost and duration of task to be performed. The controller checks if the valuation is valid and reasonable, the correctness of cost allocation to chosen investment package and availability of funds in the budget. If no variances from the plan/budget have been detected then the controller approves the spending. If variances from the plan/budget have been detected, the controller either returns the spending application to the investment package's owner for reworking or, if the reworking does not eliminate the variances, performs the detailed analysis of the variances: the materiality of the variances, their causes and sources and materiality of the variances' impact on achievement of SG. If this analysis indicates the necessity of taking remediation actions, the controllers informs the management team about the variances (as required by companies' policies and procedures), suggests the remediation actions and escalates the decision to the level of the management team.

Interim control. We suggest using this control to detect and analysis of potential variance during the execution of tasks in the investment packages. Two types of interim controls are usually implemented: (1) regular monitoring of execution of contracts; and (2) processing of change orders to the contracts. First type of control is performed in equal time internals. In each point of time, the controller compares (1) actual volume of work performed in the past period to that in the contract's specification; and (2) planned volume of work to be performed until the commissioning of the contract to that in the contract's specification. Control of change orders processing assumes that the controller compares the proposed changes in contract's specification (such as volumes, prices, terms, etc.) to the budget for this particular contract and/or strategic plan. After the variances are identified, the controller performs analysis of these variances as it was described above,

Projects in the telecommunication industry are also exposed to some regulatory risks such as custom, tax, legal risks, etc. To manage these risks we suggest implementing a special type of interim control: a documentary control. The objective of this control is to verify the compliance of legal, accounting, tax and customs paperwork to domestic regulatory rules. This control may also include internal audit of maintenance of large contracts with lengthy terms. For example, the auditor verifies compliance of accounting for these large contracts to local accounting requirements.

Post factum (factual) control. This type of control has inherited by project controlling from conventional project management systems. Its purpose is to uncover accomplished variances from strategic plan/budgets after completion and commissioning of all works either in investment package or in the project as a whole. The first control point is to compare filed-performance characteristics of the technology/product/services or its parts to those in the plan and uncover variances. This control allows project management to demand from suppliers/contractors either to correct deficiencies in equipment/process at the suppliers' expense until equipment's routine use starts or to withhold contingency payments from the suppliers. The second control point is to compare total actual project's costs/amount of time spent to the strategic plan. We consider this control point as of secondary importance in the project controlling. This is because the management team has already known all the variances, which have been identified by the early warning controls described above. Therefore, the main objective of the second control point is the final reconciliation, documentation and final accounting of all identified variances to the strategic plan.

Audit of control system. This control is designed to monitor efficiency and effectiveness of the project control system itself. It is performed by company's internal audit on the basis of annual audit plan. This plan is risk-oriented, i.e. the efficiency and effectiveness of certain control procedures are checked more frequently/ thoroughly than those for the others if the control points where they are located are exposed by critical risks and/or there are some concerns about inefficiencies of that control procedures. The audit also assesses the effectiveness and efficiency of the whole control environment (including the core policies and procedures of project controlling). Internal audit, based on the result of this control, develops recommendations to project team to make changes in certain control procedures and/or to the project control system in general.

To make efficient management decisions aimed at remediation of variances, the project team must receive the appropriate and well-structured information [8]. The reporting system should meet the following criteria: (1) timeliness; (2) reliability; (3) information adequacy; (4) comparability to plans/budgets; (5) verification to source documents and; (6) simplicity.

For projects in telecommunication industry, we recommend using monthly report as a main project report. However, information system should provide the ability to generate less detailed annual and/or quarterly reports, more detailed daily reports or customized ad-hoc report. Monthly report consists of two parts. The first part provides a verbal discussion of tasks performed in the past month as well as the presentation of tasks to be performed in the next 12 months. The second part is in-depth analysis. It contains (1) the network master charts of tasks performed in the past month and tasks to be performed in the next 12 months; and (2) analysis of variances. The latter contains (i) the results of comparison of actual values of each CIs to those in the plan/budgets; (ii) analysis of causes and sources of these variances; (iii) the projections of the future variances at the level of each CIs for the next 12 months and up to the project completion; (iv) calculation and discussion of impact of all these variances on the achievement of project strategic goal.

The last but one of the most important functions in project management is making optimal decisions to remediate variances. In project controlling this process is facilitated by a special subsystem, which provides a set of rules, policies and procedures to ensure that (1) remediation measures are performed timely on the basis of necessity; (2) the cost of resources used for remediation is appropriate given the criticality of the variances and does not exceed the potential gains from the remediation; (3) the process of remediation is monitored and necessary follow-up steps are timely made [8].

For instance, if there is a delay in completion of certain tasks in investment package, management team can speed up the performance of these tasks by reallocation of labour resources and equipment from the other tasks/investment packages. If the variances are more material, for example, there is a delay in supplying necessary technical documentation, an increase in cost of core resources or a violation of contract terms by subcontractors, then the management team will consider making adjustments in the project plan and/or appropriate budgets. These adjustments may be limited by the revision of certain parameters of tasks (costs, terms, etc.) but may result in the complete revision of the plan. Such revision may include development of new network chart and new budgets. The most critical variances in CIs and SG ratio aroused in case of substantial changes in key assumptions/inputs of the project: for instance, the demand for services is lower than originally planned, the technology does not function as it was projected by engineers, financing for the project has become more expensive, equity partners decided not to participate while competitors have done something unanticipated, etc. In this case, management team may switch to the emergency plan and/or analyse project's real options one of which is forced project termination before the project becomes a "zombie."

While developing the decision management subsystem the management team shall (1) define division of authorities and responsibilities among the members of project team; (2) establish the critical levels of variances for all CIs and SG; (3) develop and implement the methodologies, policies and procedures of designing, approving, implementing and monitoring of remediation actions (including criteria of efficiency of those actions); and (4) design the system of incentives for project team members to encourage them both to find/report variances from the goals at the earliest stages and to contribute in developing of remediation measures. The last task is of critical importance as without proper incentives employees will start "hiding" the variances in fear that reporting the variances will ruin their careers. For instance, we recommend establishing a "zombie amnesty" in the company: a period after the launch of the project during which the managers can report a project failure to stakeholders and recommend its termination without being penalized [10]. The best practices also include the development and maintenance of database of efficient remediation measures or even database of failed projects as per previous experience. Additionally, to ease the process of searching for optimal decisions, the management team shall invite experts from other departments of the firm and/or external consultants.

4   Conclusion

In this paper, we discuss the new generation of project management system – project controlling. We argue that in capital intensive and high-risky industries, such as telecommunication, the "conventional" project management methods cannot ensure that investment projects achieve the goals set in strategic plans. This is because these methods assume (1) single or limited number of project's goals under management; (2) the static project environment in which variances from the goals may be prevented by rigorous planning process; and (3) that project tasks are performed in prearranged and consecutive order. In contrast, project controlling does the better job of managing investments in volatile environment as it has the following distinctive advantages over the "conventional" methods:

* Allows to manage a balanced set of heterogeneous (financial and non-financial) goals which cover all project's perspectives from technology to finance
* Uses risk management approach/tools to identify project weaknesses, bottlenecks and external threats and allocates control resources proportionally to potential losses given the exposures
* Builds the system of control indicators, which dynamically links the ultimate project goal with the rest of interim goals. This system allows to "trace" the impact of the variances occurred at the level of interim goals on the achievement of the ultimate goal

- Uses early warning control procedures to predict variances from the plan/budgets in one or more periods before they actually happen
- Performs periodical analysis the real options in the project to "reshape" or even terminate the project before it has been transformed into a "zombie"

We argue that, because of these advantages, implementation of project controlling can reduce the variances from projects goals set in the plan by around 50%.

In the paper, we performed analysis of telecommunication industry in Russia and worldwide which shows that the industry is undergoing dramatic changes. These changes are driven by liberalization of regulations at the majority of the markets, privatization of national carriers and wide dissemination of new technologies such as high-speed wireless communication and Internet applications. Operators suffer from the contraction of revenue from "traditional" voice services and increasing competition from cable companies and Internet start ups.

To survive in such challenging environment, companies in telecommunication industry are increasing investments in research and development, networks expansion and modernisation. Our analysis shows that these investments are exposed to various risks: technological, engineering, market, regulatory, operational and financial. The impact of these risks is reinforced by long payback periods of investments as well as by the significant lags between investments and the moment of receiving cash flows from the project. Therefore, projects in telecommunication require more advanced methods of project management. We argue that project controlling will serve as an efficient project management solution for the industry owning to its advantages over "conventional" methods.

We define project controlling as "applied" project management system which purpose is to ensure that capital expenditures projects meet their goals set in strategic plan. Its main tasks include but not limited to (1) "tracking" the progress of achieving projects goals by using risk-oriented control procedures; and (2) developing the remediation measures to reduce the negative impact of variances from the goals.

In the paper, we developed a reference model of project controlling for companies in telecommunication industry. The model includes description of (1) key inputs to the system; and (2) main steps in development and maintenance of project controlling. We described, in details, the types of controls in the systems, the places and the timing of their application. We also provide several examples of designing of efficient control procedures.

The paper also contains our recommendations how to develop (1) project controlling's reporting system; and (2) the decision management subsystem. In the latter, we emphasized that development of system of incentives for project team is of the critical importance. This is because a proper motivation system helps to avoid employees to hide the variances in fear that reporting of variances will worsen their career in the firm. In this respect, we recommend introducing the period since the launch of the project during which the project managers can recognize the project failure and terminate it without penalties from the senior managers and/or the stakeholders.

Finally, to improve performance of telecommunication companies in the conditions of growing uncertainty, we recommend to expand application of controlling to the all aspects of companies' management. In this case controlling will be aimed at the creation, development and maintenance of companies' competitive advantage owing to co-ordination and integration of all functions and object of management in real time.

References

1. Glukhov, V.V., Balashova, E.S.: Economics and Management in Infocommunnication: Tutorial. Piter, Saint Petersburg (2012)
2. Nekrasova, T.P., Aksenova, E.E.: Economic Evaluation of Investments in Telecommunication Industry. St. Petersburg State Polytechnical University, Saint Petersburg (2011)
3. Moody's Investors Service. https://www.moodys.com/researchdocumentcontentpage.aspx?docid=PBC_177451
4. Puchkova, M.P.: Development and Current State of Russian Telecommunication Market. J. Modern Science: Actual Problems of Theory and Practice. Economics and Law Series, 1–2 (2014). http://www.nauteh-journal.ru/index.php/ru/---ep14-01/1100-a
5. Rosstat. http://www.gks.ru/bgd/reg/b14_11/Main.html
6. Experts: Russian Telecommunication Market Will Slow Growth in 2014-2018. http://www.rbc.ru/rbcfreenews/2013121718428.shtml
7. Moody's Investors Service. https://moodys.com/creditratings/Russia-Government-of-credit-rating-600018921
8. Grishunin, S.V., Suloeva, S.B.: Strategic Controlling and Anti-Crisis Management. In: Strategy and Tactics of Anti-Crisis Management of a Firm. Specialnaya Literatura, Saint-Petersburg (1996)
9. Committee of Sponsoring Organization of The Treadway Commission. Guidance on Enterprise Risk Management. http://www.coso.org/-ERM.htm
10. Anthony, S., Duncan, D., Siren, P.M.A.: Zombie Projects: How to Find Them and Kill Them. J. Harward Business Review (2015). https://hbr.org/2015/03/zombie-projects-how-to-find-them-and-kill-them

Creation of Data Mining Cloud Service on the Actor Model

Ivan Kholod$^{(\boxtimes)}$, Ilya Petuhov, and Nikita Kapustin

Saint Petersburg Electrotechnical University "LETI", ul. Prof. Popova 5,
Saint Petersburg, Russia
iiholod@mail.ru, ioprst@gmail.com, tilamer@yandex.ru

Abstract. This article describes the approach to building data mining cloud service based on actor model. The article describes the mapping of the algorithm decomposed into functional blocks on the set of actors. Also it describes the architecture and implementation of cloud service to perform data mining algorithms for actors. As an example, it describes the implementation and experiments with neural network learning algorithm on the cluster actors.

Keywords: Data mining · Distributed data mining · Cloud computing · Actor model

1 Introduction

Over the past decade as a result of the widespread use of information technology people have accumulated vast amounts of data to analyze for which methods of data mining are widely used. However, their application to large data increases the computational resource requirements. The most effective way to solve this problem is to implement them in computer clusters (the clouds) that can be easily scaled. In this regard, the integration of data mining technologies and cloud computing is very actually. The result of this integration is to create a data mining cloud. This solution has a number of advantages:

- the user always uses the latest version of the algorithm;
- analysis algorithms can use all the available computational resources available in the "cloud";
- algorithms can be applied to the data stored in the "cloud", and outside of it;
- the user can forget about scaling algorithms.

Most of the existing clouds, working in the area close to the data mining are built on MapReduce paradigm [1], and in particular Apache Hadoop. However, not all data mining algorithms can be easily converted to decomposition into two functions map and reduce. In this regard, we propose an alternative approach to building cloud data mining services based on the actor model.

S. Balandin et al. (Eds.): NEW2AN/ruSMART 2015, LNCS 9247, pp. 585–598, 2015.
DOI: 10.1007/978-3-319-23126-6_52

The next section is a review of research in the field building of data mining cloud services. The third section contains the description of the mapping of the algorithm decomposed into functional blocks on the set of actors. The fourth chapter describes the architecture and implementation of cloud service to perform data mining algorithms for actors. The last chapter deals with experiments with the neural network learning algorithm on the cluster actors.

2   Related Work

Few institutions and IT companies are actively working for the creation of Data Mining Cloud. Chinese mobile Institute was one of the first to begin working in this field. In 2007 it began research and development in the field of cloud computing. In 2009, it officially announced a platform for cloud computing BigCloud, which includes tools for parallel execution of algorithms Data Mining Big Cloud-Parallel Data Mining (BC-PDM) [2].

BC-PDM is a SaaS platform that is based on Apache Hadoop. Users can upload data to the repository (hosted in the cloud) from different sources and apply a variety of applications for data management, data analysis and business applications. The application includes analysis of parallel applications to perform: ETL processing, social network analysis, analysis of texts (text mining), data analysis (data mining), and statistical analysis.

In 2009 the company expanded its Amazon cloud services Amazon EC2 and Amazon Simple Storage Service (Amazon S3) with another PaaS service - Amazon Elastic Mapreduce (EMR) [3]. This service is also built on the Apache Hadoop platform and provides a scalable infrastructure to perform user-created (according to certain rules) applications. The service allows you to download Amazon S3 in the desired application and / or data that will later be carried out on job sites of platform Hadoop. The composition of EMR includes examples of applications that can be downloaded to the service, including the crucial tasks of data mining.

In 2012, Google released its new cloud service Google BigQuery [4]. This service allows the processing of large amounts of data stored in the cloud. If you are using a web browser and console applications, the user to process their data must enter SQL-like query. It may include all of the same elements as in the Select request in SQL: FROM, JOIN, WHERE etc. Thus, the user can generate a quite flexible search query according to any data type. It should be noted here that this service does not directly solve the problems of data mining: clustering, classification, and others.

The main disadvantages of the described systems are:

- inability to complete the entire cycle of analysis with their help without the need for complex configuration and refinement;
- using as a platform for distributed computing MapReduce technology (and in particular the Apache Hadoop).

Among other drawbacks, this technology is adapted only for data processing functions, which have the property of list homomorphisms [5]. For example, Apache Mahout [6] is a project of the Apache Software Foundation to produce free implementations of distributed or otherwise scalable machine learning algorithms focused primarily in the areas of collaborative filtering, clustering and classification. Many of the implementations use the Apache Hadoop platform. The project is more than 5 years old. However, last version (0.10 from 11 April 2015) of Apache Mahout includes less than 10 data mining algorithms for MapReduce [7] (algorithms of collaborative filtering, clustering and classification). These algorithms are implemented by splitting on the *"map"* and *"reduce"* functions. Such was created new algorithms for the MapReduce platform. Other example of using the MapReduce conception for data mining algorithms is described in the paper [8]. It is applicable to the algorithms corresponding to the Statistical Query Model (SQM) only. The paper contains only 10 algorithms which have such form. For this purpose it is divided into two phases:

- the first phase is dedicated to the computing of the statistic sufficiency by way of summing up of all the data;
- the second phase is the aggregation of the calculated statistics and obtaining of the final solution.

The listed phases can be well realized through the Map-Reduce distributed computing model.

Thus, if data mining algorithms can not be directly implemented for the MapReduce paradigm then we need substantial modification of the source algorithm or the use of another concept for implementation of distributed computing. An alternative distributed paradigm for building cloud data mining services can be the actor model.

3   Decomposition of Data Mining Algorithms into Actors

3.1   Presentation of Data Mining Algorithms as Functional Expression

The main idea using in proposed approach is an algorithm must be decomposed into thread-safe blocks. The thread safety feature of the blocks allows parallel executing of the algorithm formed of such blocks (block sequences).

Functional languages based on the λ-calculus theory [9] have this feature, because classic functions in the functional languages are pure functions. According to Church-Rosser theorem [10], reduction of functional expression of pure function can be fulfilled in any order, also concurrently.

May make accordance between an algorithm and a functional program:

- algorithm blocks are functions (λ-functions);
- algorithms are functional expressions, presented as a composition of functions;
- execution of algorithm is a functional expression reduction.

A data mining algorithm can be presented as a sequence of function calls:

$dma = fb_n{}^\circ fb_{n-1}{}^\circ...{}^\circ fb_i{}^\circ...{}^\circ fb_1 = fb_n (d, fb_{n-1} (d,.... fb_i (d, ... fb_1(d, nil)...)..)),$ where

fb_i : is a function of the type

$$fb:: D \to M \to M, \text{ where}$$

D : is input data set that is analyzed by the function fb;

M : is mining model that is built by the function fb.

At that each of such functions can also be presented as a composition of functions:

$$fb_i = fb_{i.k}{}^\circ...{}^\circ fb_{i.r}{}^\circ... fb_{i.1} = fb_{i.k} (d, ... fb_{i.r} (d, ... fb_{i.1}(d)...)..) \tag{1}$$

So according to Church-Rosser theorem [10] reduction (execution) of such functional expression (algorithm) can be done concurrently. To make the concurrent execution of a data functional expression, it must be converted into the form in which the function blocks will be invoked as arguments. To this end we should add a function, which will allow data parallelization in the data mining algorithms:

<parallelization_function_name> :: FB
<parallelization_function_name> = join $^\circ fb$ $^\circ split$ where

- *split*: function fulfilling the splitting of the data set D and of the mining model M and returning cortege of two list: separated parts of the data set *[D]* and separated parts of mining model *[M]*:

$$split :: D \to M \to \{[D], [M]\};$$

- *join:* function joining the mining models from the list *[M]* and returning the merged mining model M:

$$join :: [M] \to M.$$

- *fb:* function block executed concurrently.

So, parallel function can present as:

$<parallel\_function\_name>=join([fb_i(split(d, m)[0][0]),..., fb_i(split(d, m)[q][q])])$ (2)

In the join function the elements of the mining model list *[M]* are computed by the fb_i function block, the arguments of which are data sets and mining models from the lists built by the split function.

3.2   Mapping of Functional Blocks to the Actor Model

Actors model is a formalism that allows us to describe parallel and distributed computing. According to the author of the actor model [11, 12] it is more common than formalism of the λ-calculus, allowing also describing parallel computing. Thus in theory of the λ-calculus every call of the λ -function can be regarded as a call of an actor, and the transfer of function parameters as sending messages to the actor.

Thus, with regard to the representation of a data mining algorithm with the expression (1), each actor must perform a separate function block. Execution of an actor will be activated by receiving a message that includes a data set D and a mining model M. After performing the actor will return back the built mining model M. Thus, the work of the actor for the function fb_i can be described by two events (Fig. 1).

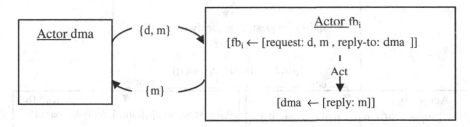

Fig. 1. Presentation of the functional block as actor

The function of parallelization (2) is performed by the fork-join behavior of actors [13]. In accordance with the described display of the functional blocks on the actor model, operation of the parallel algorithm recorded using the functions (1) and (2) can be represented as shown in Figure 2.

In the proposed representation a separate actor is created for each function. This leads to additional overhead, without increasing the efficiency of the algorithm as a whole. The following assumption is proposed in this regard: separate actors are created to perform the functions of the composition forming the whole algorithm to perform parallelization of the function fb_i (called from function (2)). With this assumption, the mapping functions on the actor model can be represented as it is shown in Figure 3.

Thus, the representation of the algorithm in the data mining as a functional expression (as described in the previous section) may be implemented using an actor model. Therefore, the implementation of the functional expression represented in the report [14] in the language of Java, can be integrated with the implementation of the actor model (for example, the tool AKKA [15]). In the next section we will describe the integration of these libraries to perform data mining algorithms on the cluster in the cloud on the actors.

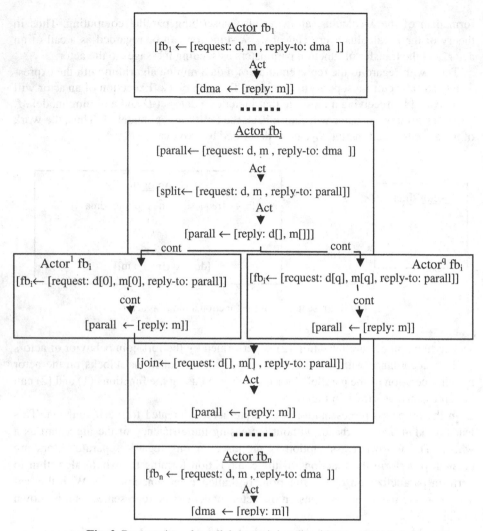

Fig. 2. Presentation of parallel data mining algorithm as actors set

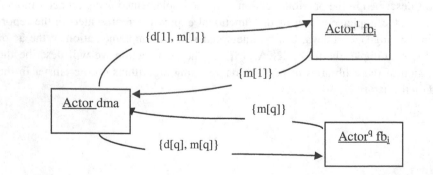

Fig. 3. Optimized implementation of the data mining algorithm by the actor model

4   Integration of Data Mining and Actor Cluster

4.1   Library of Data Mining Algorithms for Distributed Execution

The DXelopes library is extension of Xelopes library [16] (data mining algorithms library) and implement all functions as classes of an object-oriented language Java. Fig. 4 shows the class diagram of these blocks. Here the class Step describes the function of FB type. Since the condition and the cycle are functions of FB type, *DecisionStep* and *CyclicStep* classes corresponding to them are inherited from the class Step. For the implementation of cycle for vectors and cycle for attributes defined *VectorsCycleStep* class and *AttributesCycleStep* class. To form the target algorithm defined the class *MiningAlgorithm*. It contains a sequence of all steps of the algorithm. In the *initSteps* method occurs formation of algorithm structure by creating of the steps which determining of sequence and nesting of their execution. For parallelization of algorithms implemented *ParallelStep* class. As a step of the algorithm, it also inherits from the *Step* class. It contains a sequence of steps that shall be executed in parallel and implements two main split and join methods. For parallelization of algorithms by data added *ParallelByData* class. It inherits from the class *ParallelStep*. The *ExecutionHandler* class is adapter for execution of an algorithm's branch. The library elements were presented in detail at the conference [14].

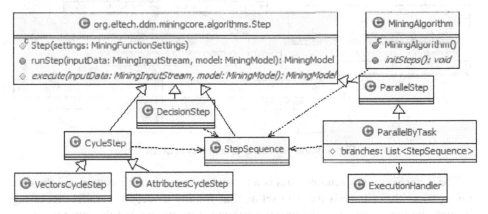

Fig. 4. Class diagram of framework for creating parallel data mining algorithm

To be able to provide the service to implement algorithms for data mining in the cloud on the actor model DXelopes library was expanded in the adapter handler execution.

4.2   Implementation of Adapters for Actors

To implement the handlers that run the functional blocks on actors classes *ActorSettings*, *ActorExecutionHandler*, *ActorExecutionHandlerFactory* were added (see Fig. 5). These classes extend parallel functional blocks of DXelopes library.

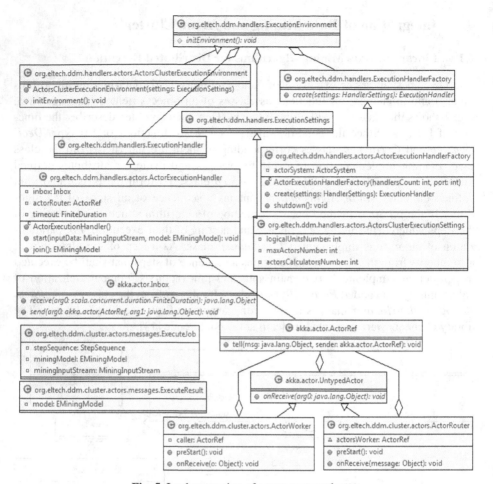

Fig. 5. Implementation of actors system cluster

The *ActorExecutionHandler* class is a proxy between the actor system and the data mining algorithm. It sends the data set and the mining model for parallel processing through the *Inbox*, which is the proxy between actors system and the outer world, to the actor and receives the resulting mining model (see Fig. 6).

To perform the DM algorithms there were implemented two types of actors:

- *ActorWorker* : actor-calculator executes a parallel part of the algorithm.
- *ActorRouter*: actor-router distributes messages from the external environment between actors-calculators.

When placing the algorithm at runtime, actors-calculators are created, each of which is assigned to a parallel branch of the algorithm (sequence of steps implemented by a SequenceStep class). The main sequence of the algorithm (steps of the MiningAlgorithm class) is assigned to a specific actor-calculator.

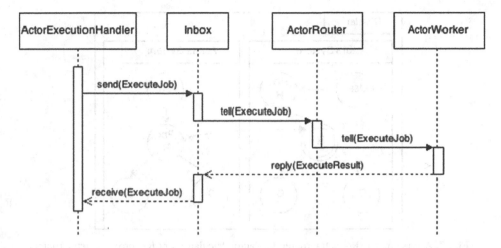

Fig. 6. Sequence of actor system communication

During the execution of the algorithm when the queue reaches the parallel branch, the mining model and the data set are divided between handlers (*ActorExecution-Handler*) and transmitted to them (Fig.6). Then, the main handler starts all parallel handlers and waits for their completion. Then it merges the mining models derived from them in one and continues to calculate the resulting mining model. In turn, each handler sends job assignment to the actor-router (*ActorRouter*), which forwards it to the free actor-calculator (*ActorWorker*). It starts the job, and after its completion sends the result back to handler (*ActorExecutionHandler*), which sent this task.

4.3 Implementation of the System of Actors

To create a system of actors and to perform a data mining algorithm on it an *ActorsClusterExecutionEnvironment* class that inherits from a class *ExecutionEnvironment* was added to the library. This class implements a method *initEnvironment*. This method creates a user-defined number of virtual machines on which the library DXelopes and the system of actors are deployed. The general logical structure of the cluster is shown in the Figure 7.

A cluster consists of machine nodes (individual computers) with logical components - the system of actors with the same names and different network (typically, the pair of an IP address and a port) addresses. Each node contains a number of actors.

An actor-router exists in every actors system in each cluster, but it is considered that the only working actor is the one that is located in the oldest system of all linked into a cluster. The router forwards the message from the external environment (from handler) to actors-calculators. There can be any number of actors-calculators. They are automatically created by actor-router evenly across the cluster.

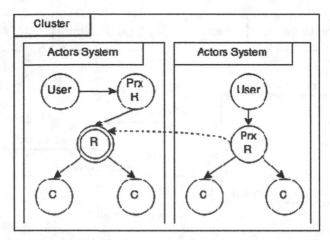

Fig. 7. Actors cluster (R – actor-router, C – actor-calculator, Prx R – proxy of actor-router)

If we imagine an actors cluster, as a hierarchy of objects, we will get the structure shown in the Figure 7. The root of the hierarchy is the cluster itself. Its child nodes are systems of actors. In every system, there is a hierarchy of actors, all actors are created by the user and used for the needs of the user. The descendants of the system actor named *user*. In this case, each system of actors has only one descendant actor *user* - actor intermediary *PrxR* to interact with the lone actor router R. Actors mediators in all systems in the cluster point to the same lone actor, which is situated in only one system across the cluster. The actor router creates all actors-calculators in the system, while it is the parent of calculators only in its system of actors. The router has references to other calculators, and their parents are actually actors intermediary in their systems.

Before starting calculations and creating an environment, the user must configure it. The configuration is done using methods of class *ActorClusterExecutionSettings*:

- specifies the total number of logical units;
- the maximum number of actors;
- a list of root nodes (the nodes through which other nodes will be connected to the cluster);
- the number of actors-calculators for the node.

Thus, to run the data mining algorithm in the environment constructed on the base of the actors, the user needs only to specify the settings of the environment mentioned above, and to initialize it by calling the *initEnvironment* method of the *ActorCluster ExecutionSettings* class.

5   Implementation of Data Mining Algorithms and Experiments

The experiment was conducted on the neural network training algorithm: BackProp. Learning method is back-propagation of errors with parallelization by data.

There are two ways how to train networks (see Fig. 8): stochastic when processing a single vector leads to changes in the mining model (online-learning) and serial when the mining model changes show up only after the processing of all vectors has been finished (batch-learning). The training function has the property of list homomorphism only at the first training stage when the model is formed and the vector set has been sent to actors, since the results are based on the previous steps. Because all vectors are being distributed between actors before calculations and the model is rarely updatable it is acceptable for serial learning. For stochastic training all vectors are being distributed while training in progress and model is frequently updatable. Thus, stochastic training is not homomorphic (i.e. it can not be used on the MapReduce model), but has a higher precision formed models and faster convergence in contrast to serial learning.

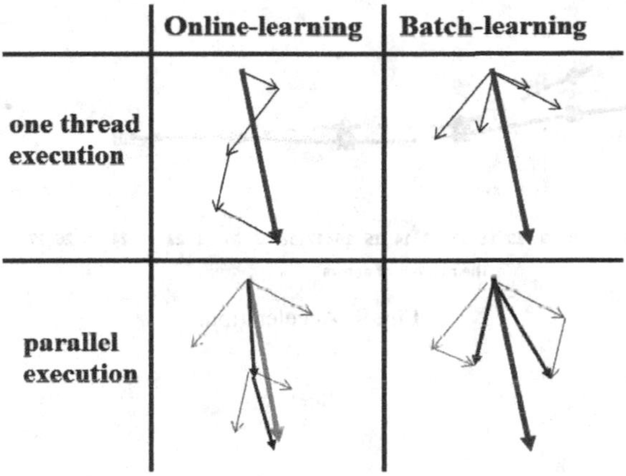

Fig. 8. Online and batch learning

The experiments have been done on a cluster the following configuration: two computers with CPU Intel Xenon (8 cores), 2.90 GHz, 512 Mb. The experimental results are provided in Figures 9 and 10 (horizontal axis indicates number of actors used, and the vertical axis indicates resulting acceleration and efficiency). The experiment was performed on several data sets. The sets of 100, 1k, 10k, 100k vectors were used.

Experimental results show that the execution of the algorithm on a cluster of actors gives a positive result only when count of source vectors is not less than 1k. If cluster of actors is used for calculations with small amount of data the overhead for data transfer and synchronization is more than the cost of the calculations themselves. In addition, the proposed approach of implementing data mining algorithms on actors allows to execute parallel algorithm with synchronization of the results on certain iteration of the vectors cycle. It is main advantage versus the MapReduce paradigm.

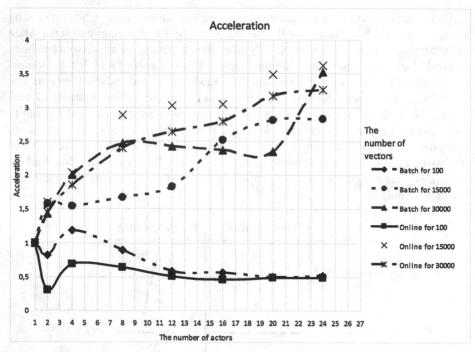

Fig. 9. Acceleration.

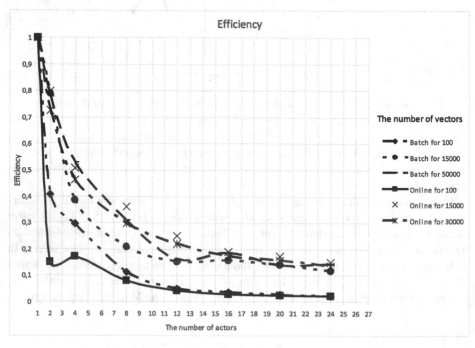

Fig. 10. Efficiency.

This option is more efficient and provides more correct result, but the computational complexity is the same for both versions of the algorithm

Also it may be noted that the acceleration and efficiency obtained with 1k-100k sets of vectors have almost the same value, but it is possible to see the dependence of acceleration decrease with increasing the number of vectors. This is due to the costs of the transfer of information between actors.

Reducing the acceleration by increasing the number of actors in the system caused by the growing time delays in synchronization of all the actors together. When synchronizing all actors should wait each other until they finish their part of the job, so the execution time is reduced to the slowest actor. However, if you doing a training asynchronously, i.e. not waiting for the results of all the actors, you can get good results with a large number of actors.

6   Conclusion

The article describes the implementation of the data mining algorithms executing on cluster systems, implemented on the model of actors. Mapping of algorithm represented as a series of functional blocks in the actor model makes it possible to obtain the advantages of both formalisms:

- the possibility of rebuilding a simple algorithm from serial to parallel;
- representation of parallel and distributed execution of the algorithm by an actor model.

The presence of implementing in a high-level programming language for each of the formalisms allowed integrating them. Integrated solution allows getting practical results in the form of execution of data mining algorithms in the cloud cluster on the actors. In contrast to existing implementations on platforms MapReduce, the proposed implementation allows you to create and run parallel algorithms with a variety of structures. The experiments show the effectiveness of the decision.

Acknowledgments. The work has been performed in Saint Petersburg Electrotechnical University "LETI" within the scope of the contract Board of Education of Russia and science of the Russian Federation under the contract № 02.G25.31.0058 from 12.02.2013. This paper is also supported by the federal project "Organization of scientific research" of the main part of the state plan of the Board of Education of Russia and project part of the state plan of the Board of Education of Russia (task # 2.136.2014/K).

References

1. Dean, J., Ghemawat, S.: MapReduce: Simplified data processing on large clusters. In: Proceedings of Operating Systems Design and Implementation, San Francisco, CA (December 2004)
2. Yu, L., Zheng, J., Shen, W.C., Wu, B., Wang, B., Qian, L., Zhang, B.R.: BC-PDM: Data mining, social network analysis and text mining system based on cloud computing. In: Proceedings of the 18th ACM SIGKDD International Conference on Knowledge Discovery and Data Mining, New York, pp. 1496–1499 (2012)

3. Amazon Elastic MapReduce. Developer Guide. http://s3.amazonaws.com/awsdocs/ElasticMapReduce/latest/emr-dg.pdf
4. Google Developers. Google BigQuery. https://developers.google.com/bigquery/what-is-bigquery?hl=ru
5. Gorlatch, S.: Extracting and implementing list homomorphisms in parallel program development. Science of Computer Programming 33(1), 1–27
6. Grant Ingersoll, "Introducing Apache Mahout". http://www.ibm.com/developerworks/java/library/j-mahout/
7. Mahout 0.10.1 Features by Engine. http://mahout.apache.org/users/basics/algorithms.html
8. Ng, A.Y., Bradski, G., Chu, C.-T., Olukotun, K., Kim, S.K., Lin, Y.-A., Yu, Y.Y.: Map-Reduce for machine learning on multicore. In: Proceedings of the Twentieth Annual Conference on Neural Information Processing Systems, Vancouver, Canada, pp. 281–288 (2006)
9. Church, A., Barkley Rosser, J.: Some properties of conversion. Trans. AMS **39**, 472–482 (1936)
10. Barendregt, H.P.: The Lambda Calculus: Its Syntax and Semantics. Studies in Logic and the Foundations of Mathematics, vol. 103. North-Holland (1981)
11. Hewitt, C., Bishop, P., Steiger, R.: A universal modular ACTOR formalism for artificial intelligence. In: IJCAI, pp. 235–245 (1973)
12. Clinger, W.: Foundations of Actor Semantics, p. 178. Massachusetts Institute of Technology, Cambridge (1981)
13. Hewitt, C., Baker, H.: Actors and continuous functionals, p. 29. Massachusetts Institute of Technology, Cambridge (1977)
14. Kholod, I.: Framework for multi threads execution of data mining algorithms. In: 2015 IEEE NW Russia Young Researchers in Electrical and Electronic Engineering Conference, February 2–4, pp. 74–80 (2015)
15. Akka Documentation. http://akka.io/docs/
16. Barsegian, A., Kupriyanov, M., Kholod, I., Thess, M.: Analysis of Data and Processes: From Standard to Realtime Data Mining, p. 300. Re Di Roma-Verlag (2014)

Efficiency of Coherent Detection Algorithms Nonorthogonal Multifrequency Signals Based on Modified Decision Diagram

Sergey V. Zavjalov$^{(\boxtimes)}$, Sergey B. Makarov, Sergey V. Volvenko, and Anastasia A. Balashova

Peter the Great St. Petersburg Polytechnic University, St. Petersburg, Russia
volk@cee.spbstu.ru

Abstract. Presented an improved detection algorithm for random sequences of Spectrally Efficient Frequency Division Multiplexing (SEFDM) signals based on decision diagram. Application of the modified algorithm makes it possible to reduce the resulting bit error probability. The simulation results presented BER performance and BER depending on the number of subcarriers confirms this. It has been shown that it is possible to reduce the frequency spacing between subcarriers by 10% without an increase of energy losses by applying the modified algorithm based on decision diagram. This allows to increase the transmission rate in proportion by arranging the additional subcarriers in the same frequency band.

Keywords: SEFDM · Spectrally efficient signals · Detection algorithm · Decision diagram · BER performance

1 Introduction

The use of nonorthogonal SEFDM signals can significantly increase spectral efficiency of a data transmission system [1, 2]. Unlike signals with OFDM, which for signal duration equal T have a frequency separation between the subcarrier frequencies $\Delta f = 1/T$ [3], the SEFDM signals is supposed to preserve the values of T with decreasing values of Δf [1]. However, with a decrease of Δf less than a certain value begins to increase sharply energy losses [1, 4, 5], especially when using the algorithm of coherent detection [6]. For example, in the case of $\Delta f = 0.825/T$ and the number of subcarriers $N = 5$ additional energy losses (relative to BER performance of OFDM signals) is about 8 dB for the error probability of 10^{-4}. With increasing N additional energy losses will continue to rise. This fact agrees with the results presented in [5].

Degradation of BER performance associated primarily with the fact that begins to affect inter-frequency interference significantly [7]. The spectra of the signals used on subcarrier frequencies overlap and start to influence each other as interference [2, 5, 8]. Obviously, the greatest influence will have spectra of signals on neighboring subcarrier frequencies [7].

S. Balandin et al. (Eds.): NEW2AN/ruSMART 2015, LNCS 9247, pp. 599–604, 2015.
DOI: 10.1007/978-3-319-23126-6_53

The idea of the detection algorithm, based on the decision diagram [6] is to compare the received noisy signal at a particular subcarrier with signals that are based on the symbols already received in previous subcarrier and all possible symbols in the observed subcarrier. Enumeration of subcarrier frequencies implemented sequentially. As a criterion for selecting the symbols on each subcarrier frequency the objective function acts. Received sequence of symbols by substituting in the objective function must provide a value smaller than the value of the objective function for each different set of symbols. In this case, if the initial characters have been taken incorrectly, the conditions for detection of symbols on subsequent subcarrier frequencies deteriorates.

There are several directions of improvement of detection algorithm based on the decision diagram. The first is to increase the number of passes through the diagram. In addition, the initial subcarrier frequency of pass should be different for different iterations. The result is stored each pass and then used to form received data. However, this approach has several drawbacks. Firstly, the increased complexity and processing time. Secondly, if the initial subcarrier is not an outermost, then deteriorate conditions of symbols detection on a given subcarrier. This will lead to a deterioration of the overall error probability. The second direction improving the algorithm is to change the order of the passage of subcarriers. Initially used subcarriers on the edges of the frequency range, and then passage is from the edges to the center of the occupied bandwidth. This will increase as the accuracy of the detection symbols on subcarriers at the outermost frequencies and to improve the overall error probability.

2   The Detection Algorithm Based on Decision Diagram and Its Improvement

As shown in [6] the essence of the detection algorithm based on decision diagram is to find a sequence of symbols that will minimize the objective function. As the objective function is selected RMS distance:

$$Q = \int_{-T/2}^{T/2} \left| x(t) - s_j(t) \right|^2 dt, \tag{1}$$

where $x(t)$ – received signal (signal is passed through the channel with constant parameters and AWGN), $s_j(t)$ – transmitted signal. Transmitted signal can be written as follows:

$$s_j(t) = A_0 \sum_{n=-(N-1)/2}^{(N-1)/2} a(t) \left[d_{in} \cos\left((\omega_0 + n\Delta\omega)t\right) + \tag{2} \right.$$

$$\left. + d_{qn} \sin\left((\omega_0 + n\Delta\omega)t\right) \right]; (-T/2 \le t \le T/2),$$

where $j = 1,2,3,\ldots 4^N$, values of symbols d_{in} depend on value of index $i = 1, 2$ and index $n = -(N-1)/2,\ldots,(N-1)/2$. In particular, $d_{1n} = 1$; $d_{2n} = -1$.

This expression may be represented as follows:

$$s_j(t) = \sum_{n=-(N-1)/2}^{(N-1)/2} s(t, d_{in}, d_{qn}) = s\left(t; d_{i(-(N-1)/2)}, d_{q(-(N-1)/2)}; ...; d_{i((N-1)/2)}, d_{q((N-1)/2)}\right) \qquad (3)$$

where $s(t, d_{in}, d_{qn}) = s(t, d_{in}) + s(t, d_{qn}) =$

$$= A_0 a(t) \left[d_{in} \cos\left((\omega_0 + n\Delta\omega)t\right) + d_{qn} \sin\left((\omega_0 + n\Delta\omega)t\right) \right].$$

When using N subcarriers requires enumerate m^N combinations of symbols on subcarrier frequencies, where m – the base of the channel alphabet. Thus apparently excessive complexity of the algorithm. Consequently, the procedure requires sequential of the objective function calculations. Then, the objective function can be written as follows:

$$Q(r) = \int_{-T/2}^{T/2} \left| x(t) - \sum_{n=-(N-1)/2}^{r} s(t, d_{in}, d_{qn}) \right|^2 dt, \; r = -(N-1)/2, ..., (N-1)/2. \qquad (4)$$

This expression contains a sequence of enumerating subcarriers. Enumeration subcarriers is performed "one-way" from the leftmost to the rightmost subcarrier. The decision diagram in such a case has the following form shown in Fig. 1.

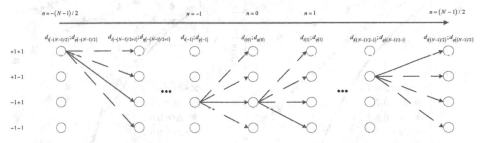

Fig. 1. Decision diagram for the case of sequential computation of the objective function.

The proposed improvement of the algorithm is to change the sequence of pass along the subcarrier frequencies (and construction of the decision diagram). This proposal is because the interference of signals are located on the outermost subcarriers of the occupied frequency band is significantly smaller compared to the mutual influence of the signals in the center frequency band [7]. This proposal is illustrated on Fig. 2.

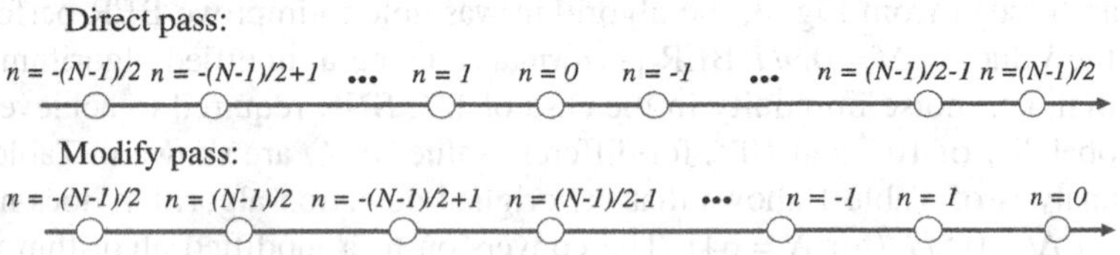

Direct pass:

$n = -(N-1)/2 \; n = -(N-1)/2+1 \; \cdots \; n = 1 \quad n = 0 \quad n = -1 \quad \cdots \quad n = (N-1)/2-1 \; n = (N-1)/2$

Modify pass:

$n = -(N-1)/2 \quad n = (N-1)/2 \; n = -(N-1)/2+1 \quad n = (N-1)/2-1 \quad \cdots \quad n = -1 \quad n = 1 \quad n = 0$

Fig. 2. Sequential and modified enumeration subcarriers.

As seen from Fig. 2, are first processed outermost subcarriers with the numbers $-(N-1)/2$ and $(N-1)/2$. On the next step are processed subcarriers with the numbers $(-(N-1)/2 + 1)$ and $((N-1)/2 + 1)$. At the last step of the algorithm is processed the zero subcarrier frequency.

3   Simulation Results

To investigate the BER performance of the proposed algorithm was built simulation model using Matlab. The number of used subcarriers is $N = 64$. The frequency spacing between subcarriers changed from $0.6/T$ to $1/T$. In all subcarriers are used rectangular envelopes.

Fig. 3 shows the BER performance of detection algorithm based on a decision diagram and BER performance of modified algorithm. The abscissa is the signal to noise ratio (SNR) $h_0^2 = E_b/N_0$, where E_b – bit energy, N_0 – power spectral density of AWGN.

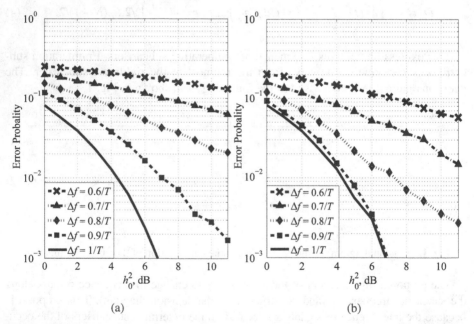

Fig. 3. BER performance of detection algorithm based on decision diagram (a) and modified diagram (b).

As can be seen from Fig. 3, the algorithm was able to improve BER performance. So, for the value of $\Delta f = 0.9/T$ BER performance using a modified algorithm is close to potential, i.e., noise immunity in the case of $1/T$. SNR required to achieve a given error probability of 10^{-2} and 10^{-3}, for different values of Δf are shown in Table 1.

The analysis of Table 1 shows that the original detection algorithm does not allow a value of $\Delta f < 0.8/T$ (for $N = 64$). The conversion to a modified algorithm achieves error probability of 10^{-3} or less. However, additional energy losses are increased relative to the case $\Delta f = 1/T$. For example, for the value of $\Delta f = 0.7/T$, and the required error probability 10^{-3} extra energy losses are 9 dB.

Table 1. SNR for $p_{err} = 10^{-2}$ and $p_{err} = 10^{-3}$.

h_0^2, dB	Detection algorithm based on decision diagram		Detection algorithm based on modified decision diagram	
	$p_{err} = 10^{-2}$	$p_{err} = 10^{-3}$	$p_{err} = 10^{-2}$	$p_{err} = 10^{-3}$
$\Delta f = 0.6/T$	–	–	16.5	> 20
$\Delta f = 0.7/T$	–	–	12.1	15.8
$\Delta f = 0.8/T$	15	–	7.3	12.8
$\Delta f = 0.9/T$	7	12	4.34	6.8
$\Delta f = 1/T$	4.32	6.8	4.32	6.8

Fig. 4 shows the dependence of the BER of the subcarriers number N for a fixed SNR = 5 dB. It can be seen that for $\Delta f = 1/T$ with increasing N from 4 to 256 error probability almost constant. With decreasing Δf starts to affect significantly the mutual influence of adjacent subcarriers. However, an increase of N more than 100 the error probability almost constant. For $N = 100$ detection algorithm based on decision diagram provides error probability $p = 4.5 \cdot 10^{-2}$ when $\Delta f = 0.8/T$. Under the same conditions, a modified algorithm provides error probability $p = 2.2 \cdot 10^{-2}$. Note that the advantage of using a modified algorithm decreases with decreasing values of Δf.

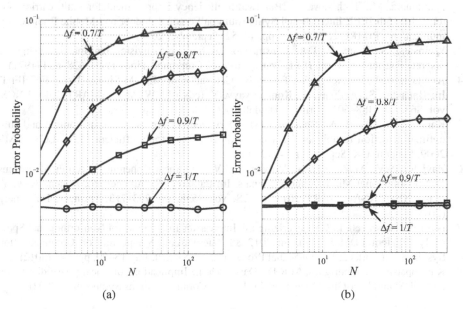

(a)   (b)

Fig. 4. The dependence of BER from N for $h_0^2 = 5$ dB for the algorithm based on decision diagram (a) and for the modified algorithm (b).

4   Conclusions

The technique of improving the BER of SEFDM signals by using a detection algorithm based on decision diagram is presented. Improving BER based on the change of the order of subcarriers passage for the construction of the decision diagram. Initially used subcarriers on the edges of the frequency range, and then passage is from the edges to the center of the occupied bandwidth.

The modified detection algorithm remains effective for values of $\Delta f = 0.6/T$ ($N = 64$), i.e. allows to obtain the error probability 10^{-3} or less (Table 1). However, additional energy losses are increased relative to the case of $\Delta f = 1/T$. For example, for the values of $\Delta f = 0.7/T$, and the required error probability 10^{-3} extra energy losses are 9 dB.

For the value of $\Delta f = 0.9/T$ BER performance is equal to BER performance for $\Delta f = 1/T$ and an arbitrary value N. That is, may increase the data rate by 10% due to additional subcarriers in the same frequency band without deterioration of the error probability.

References

1. Xu, T., Darwazeh, I.: Spectrally Efficient FDM: Spectrum Saving Technique for 5G? In: 2014 1st International Conference on 5G for Ubiquitous Connectivity (5GU), November 26–28, pp. 273–278 (2014)
2. Hamamura, M., Tachikawa, S.: Bandwidth efficiency improvement for multi-carrier systems. In: 15th IEEE International Symposium on Personal, Indoor and Mobile Radio Communications, PIMRC 2004, vol. 1, pp. 48–52 (September 2004)
3. Weinstein, S., Ebert, P.: Data transmission by frequency-division multiplexing using the discrete Fourier transform. IEEE Transactions on Communications 19(5), 628–634 (1971)
4. Kislitsyn, A.B., Rashich, A.V., Tan, N.N.: Generation of SEFDM-Signals Using FFT/IFFT. In: Balandin, S., Andreev, S., Koucheryavy, Y. (eds.) NEW2AN/ruSMART 2014. LNCS, vol. 8638, pp. 488–501. Springer, Heidelberg (2014)
5. Kanaras, Y., Chorti, A., Rodrigues, M., Darwazeh, I.: An overview of optimal and suboptimal detection techniques for a non orthogonal spectrally efficient FDM. LCS/NEMS 2009, London, UK, September 3–4 (2009)
6. Zavjalov, S.V., Makarov, S.B., Volvenko, S.V.: Nonlinear Coherent Detection Algorithms of Nonorthogonal Multifrequency Signals. In: Balandin, S., Andreev, S., Koucheryavy, Y. (eds.) NEW2AN/ruSMART 2014. LNCS, vol. 8638, pp. 703–713. Springer, Heidelberg (2014)
7. Isam, S., Darwazeh. I.: Characterizing the Intercarrier Interference of Non-orthogonal Spectrally Efficient FDM System. In: 2012 8th International Symposium on Communication Systems, Networks & Digital Signal Processing (CSNDSP), July 18–20, pp. 1–5 (2012)
8. Karampatsis, D., Rodrigues, M.R.D., Darwazeh, I.: Implications of linear phase dispersion on OFDM and Fast-OFDM systems. In: London Communications Symposium (2002)

Instantaneous Frequency Measurement Receiver Performance Analysis for AM, FM Signals

Dmitrii Kondakov[✉] and Alexander P. Lavrov

Peter the Great St. Petersburg Polytechnic University, St. Petersburg, Russia
dmitrii.kondakov@spbstu.ru, lavrov@cef.spbstu.ru

Abstract. Measurement of the Instantaneous Frequency (IF) is significant issue. Nowadays this topic has become more attractive due to new broadband electronic components such as power dividers, quadrature hybrids and delay lines. This paper consists of Instantaneous Frequency Measurement (IFM) Receiver simulation results. Particularly responses for continuous wave, amplitude-modulated, frequency-modulated signals in time and frequency domain are presented.

Keywords: Instantaneous frequency receiver · IFM · Frequency discriminator · AM · FM

1 Introduction

Frequency is related to the group of most accurately measured physical quantities. Nature is full of periodic processes. Essentially they surround us everywhere from atoms and molecules oscillations to the movement of the planets in their orbits. However frequency as the number of occurrences of a repeating per unit time is rightfully only for stationary processes when the spectral characteristics are constant with time [1]. The term of instantaneous frequency was first mentioned in 1930s in communication theory [2]. Measurement of the instantaneous frequency can find its application in electronic warfare, seismology, medicine. Moreover this topic is attractive for communication systems diagnostics [3]. Instantaneous frequency (IF) is a basic characteristic which is used in nonstationary processes description. It is defined by the following equation [2]:

$$IF = \frac{1}{2\pi}\frac{d}{dt}\phi(t) \tag{1}$$

where $\{\phi(t)\}$ is a phase of analytical signal.

To the authors knowledge many scientific groups have worked out for the development of the IFM receivers. Prototypes of the IFM based on considered frequency discriminator was implemented via microstrip lines [4],[5] and coplanar waveguides [6].

In [7] the IFM receiver was developed by using photonic methods. Generally speaking the instantaneous frequency measurement can be produced staticstically

© Springer International Publishing Switzerland 2015
S. Balandin et al. (Eds.): NEW2AN/ruSMART 2015, LNCS 9247, pp. 605–611, 2015.
DOI: 10.1007/978-3-319-23126-6_54

and via integral transforms [8]. Moreover IFM receiver can be fully-digital and be designed on modern FPGA [9],[10],[11],[12]. In work [13] the performance comparison of analog and digital IFM receiver is presented. The aim of current article is an analysis of frequency discriminator performance for continuous wave, amplitude-modulated and frequency-modulated signal.

2   Principle of Operation

Typical frequency discriminator is shown in Figure 1. The main idea is concluded in fact that the signal phase is changed linearly with time. Considered frequency discriminator consists of delay line (τ), power divider, three 90-degree and one 180-degree hybrid, low-pass filters, subtractors and DSP block. The input RF signal passes through power divider and separates into two paths.

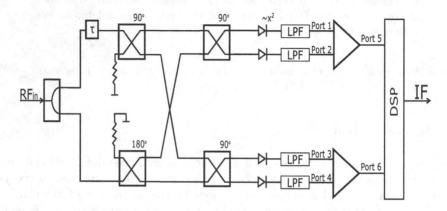

Fig. 1. Frequency discriminator

The signal from first branch goes to the delay line and comes to 90-degree hybrid. Undelayed replica passes through 180-degree hybrid. After two 90-bybrids signals go through detectors. Used in schematic detectors are square-law. Low-pass filters (LPF) leave DC and suppress high-order frequency components. On the next step signals are subtracted in pairs by equations (2),(3) and DSP block performs the operations according to equations (4),(5).

$$Port5 = Port2 - Port1; \tag{2}$$

$$Port6 = Port4 - Port3; \tag{3}$$

$$\theta = atan(\frac{Port5}{Port6}); \tag{4}$$

$$IF = \frac{\theta}{2\pi\tau}. \tag{5}$$

3   Simulation Results

Simulation was performed with MATLAB. Frequency discriminator was tested with 3 type of signals: continuous wave, amplitude-modulated signal (AM) and frequency-modulated (FM) signal. The output of the frequency discriminator was analyzed in time and frequency domain.

3.1   Continuous Wave

Consider the analytical expression for the continuous wave Eq. (6)

$$s_{CW}(t) = A\,cos(2\pi ft) \tag{6}$$

where A is amplitude, f is frequency, t is time

In order to suppress high-ordered components 6-order Butterworth filter with cutoff frequency of 1 was used. For the delay line value of $\tau = 0.01$ was chosen.

It can be seen that positive signal appears after detector (Figure 2). In the next figure 3 filtered signals are presented. It's important to note that response time is defined by order of Butterworth filter.

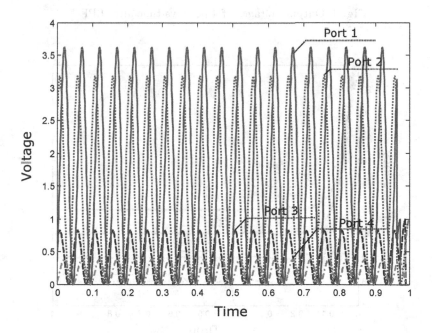

Fig. 2. Output voltages of 4 ports vs time after detector block

In Figure 4 time responses are shown for case when discriminator was excited by continuous wave with constant amplitude $A = 1$ and different frequencies (all the frequency and time values here and below are relative).

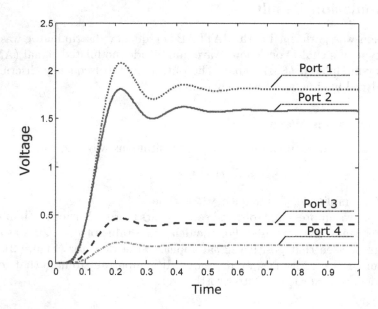

Fig. 3. Output voltages of 4 ports vs time after LPF

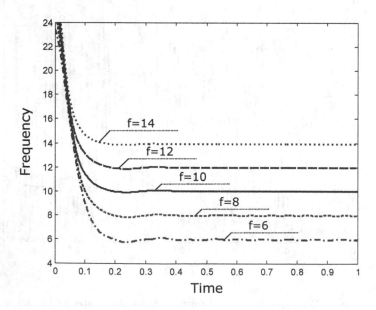

Fig. 4. Time response of continuous wave for different frequency value

Frequency was successfully detected by discriminator for five values of input frequency. For all the cases the detection time was approximately 0.2.

3.2 Amplitude-Modulated Signal

On the next step the frequency discriminator was tested on amplitude-modulated signal, which was defined by eq.(6)

$$s_{AM}(t) = A[1 + m\ cos(2\pi f_1 t)]cos(2\pi f_2 t) \qquad (7)$$

In Figure 5 time response is presented for amplitude-modulated signal for modulation index range from 0.1 to 0.7. For this simulation next values were set $A = 1$; $f_1 = 4$; $f_2 = 10$ respectively.

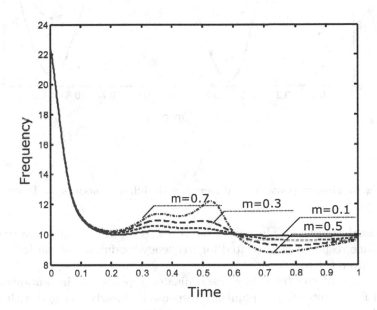

Fig. 5. Time response for AM signals with different modulation index

Simulation of the frequency discriminator with amplitude-modulated signal showed that the measurement error is increased with modulation index.

3.3 Frequency-Modulated Signal

The most interesting case for modeling the frequency discriminator represents signals with variable frequency. Signal with frequency modulation is written by (8):

$$s_{FM}(t) = A\ cos[2\pi f_2 t + m\ cos(2\pi f_1 t)] \qquad (8)$$

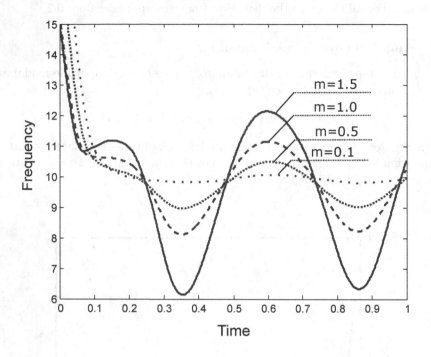

Fig. 6. Time response for FM signals with different modulation index

For this simulation next values were set $A = 1$; $f_1 = 2$; $f_2 = 10$ respectively. In Figure 6 time response is presented for frequency-modulated signal for modulation index from 0.1 to 1.5.

The output of the frequency discriminator repeats the instantaneous frequency of FM input signal. Frequency response is linearly changed with modulation index.

4   Conclusion

The model of the frequency discriminator for IFM receiver was developed. The simulation was performed for continuous wave, amplitude-modulated and frequency-modulated signal. Dependency of the measurement error from modulation index was obtained. Detection time depends on impulse response behavior of used Butterworth filters. In the future optimisation for minimization of the detection time will be done. In addition components losses and phase imbalance will be considered. Analyzed frequency discriminator will be implemented in FPGA and becomes the basis of the IFM receiver prototype.

References

1. Boashash, B.: Estimating and interpreting the instantaneous frequency of a signal. i. fundamentals. Proceedings of the IEEE **80**(4), 520–538 (1992)
2. Boashash, B.: Time-frequency signal analysis-methods and applications. Longman Cheshire (1992)
3. Rutkowski, A.: Analysing of gsm signals by means of ifm receiver. In: International Conference on Microwaves, Radar Wireless Communications, MIKON 2006, pp. 669–672 (May 2006)
4. Tsui, J.: Microwave receivers with electronic warfare applications. Wiley (1986)
5. Tsui, J.: Special Design Topics in Digital Wideband Receivers. Artech House Radar. Artech House, Incorporated (2014)
6. Sinclair, C.: A coplanar waveguide 6–18 ghz instantaneous frequency measurement unit for electronic warfare systems. In: IEEE MTT-S International Microwave Symposium Digest, vol. 3, pp. 1767–1770 (May 1994)
7. Emami, H., Sarkhosh, N., Ashourian, M.: Reconfigurable photonic radar warning receiver based on cascaded grating. Opt. Express **21**(6), 7734–7739 (2013)
8. Blaska, J.: Use of the integral transforms for estimation of instantaneous frequency. Measurement Science Review 1(1) (2001)
9. Herselman, P.L., Cilliers, J.E.: A digital instantaneous frequency measurement technique using high-speed analogue-to-digital converters and field programmable gate arrays: the csir at 60. South African Journal of Science **102**(7&8), 345 (2006)
10. Lee, Y.H., Helton, J., Chen, C.I.: Real-time fpga-based implementation of digital instantaneous frequency measurement receiver. In: IEEE International Symposium on Circuits and Systems, ISCAS 2008, pp. 2494–2497 (May 2008)
11. Helton, J., Chen, C.I., Lin, D., Tsui, J.: Fpga-based 1.2 ghz bandwidth digital instantaneous frequency measurement receiver. In: 9th International Symposium on Quality Electronic Design, ISQED 2008, pp. 568–571 (March 2008)
12. East, P.: Design techniques and performance of digital ifm. Communications, Radar and Signal Processing, IEE Proceedings F **129**(3), 154–163 (1982)
13. Pandolfi, C., Fitini, E., Gabrielli, G., Megna, E., Zaccaron, A.: Comparison of analog ifm and digital frequency measurement receivers for electronic warfare. In: 2010 European Radar Conference (EuRAD), pp. 232–235 (September 2010)

Using the DFT-Based Detection Method for ASK-Manipulated SEFDM Signals

Alexandr B. Kislitsyn and Andrey V. Rashich[✉]

Peter the Great St. Petersburg Polytechnic University, St. Petersburg, Russia
kislitcyn@rambler.ru, rashich@cee.spbstu.ru

Abstract. Spectrally Efficient Frequency Division Multiplexing (SEFDM) signals are considered to be an alternative to OFDM signals in the future telecommunication systems. The state of the art SEFDM DFT receiving techniques can be easily implemented on the basis of OFDM–transceivers. However, such detection algorithms exhibit poor error performance. In this paper it is proved that apart from channel noise the DFT-based detection error performance is determined by both aliasing and intercarrier interference effects. The optimization method is derived, providing intercarrier interference and aliasing impact minimization on DFT-detection BER performance for ASK-manipulated SEFDM signals by determination of appropriate SEFDM parameters. The parameters values was found providing DFT–based SEFDM detection performance approach to ML error performance. The results of the theoretical analysis are confirmed by numerical simulations.

Keywords: OFDM · SEFDM · Bandlimited signals · Antialiasing · Discrete fourier transforms · Frequency division multiplexing · BER performance

1   Introduction

Spectrally efficient frequency division multiplexing (SEFDM) signals exhibit peak-to-average power ratio reduction and spectral efficiency gain of up to 2-3 times compared to orthogonal frequency division multiplexing (OFDM) counterparts ([1], [2], [3]). The main drawbacks of SEFDM signals are poor energy efficiency and significant computational complexity of receiving techniques compared to classic OFDM.

Kanaras et al in [2] investigate the SEFDM maximum likelihood (ML) and minimum mean squared error (MMSE) detection algorithms. In [4] the combination of these two methods is proposed. Nonlinear optimization methods were used by Yang in [1] and Kanaras in [3] to reduce the optimal algorithms complexity. The Sphere Decoder algorithm for SEFDM signals detection is proposed by Kanaras in [3] and further investigated in [5].

However, despite all the efforts the SEFDM detection techniques described in [1], [2], [3], [4] and [5] exhibit either unusably high computational complexity for practical number of subcarriers (128 and more) or low stability and high performance dependence on channel characteristics knowledge.

© Springer International Publishing Switzerland 2015
S. Balandin et al. (Eds.): NEW2AN/ruSMART 2015, LNCS 9247, pp. 612–620, 2015.
DOI: 10.1007/978-3-319-23126-6_55

In [6] and [7] the Discrete Fourier Transform (DFT) is proposed for SEFDM signals generation and receiving, offering easy SEFDM transceivers implementation. Nevertheless, because of SEFDM intercarrier interference and aliasing effects the DFT-detection techniques exhibit poor error performance compared to the methods proposed in [1], [2], [3], [4] and [5]. The motivation behind the research of this paper is derivation of the optimization method, providing intercarrier interference and aliasing impact minimization on DFT-detection performance for ASK-manipulated SEFDM signals.

This paper is organized as follows: section 2 outlines the SEFDM system model. Section 3 introduces the SEFDM DFT-based detection algorithm. The proposed method for aliasing and intercarrier interference impact minimization on DFT-based detection performance is introduced in section 4. Section 5 provides simulation results, whereas section 6 summarizes the paper.

2   System Model

In general, the transmitted analog baseband SEFDM signal $s_{A,\varepsilon}(t)$ can be expressed as:

$$s_{A,\varepsilon}(t) = \sum_{n=-\infty}^{+\infty} \sum_{k=-N/2}^{N/2-1} C_N^{(n)}(k)\, e^{j2\pi k\Delta f(t-nT)}\, \psi_T(t-nT-\varepsilon T),$$

where N is the number of subcarriers; $n \in \mathbb{N}$ is the index of the n-th transmitted SEFDM symbol; $C_N^{(n)}(k)$ represents the manipulation symbol, used for the k-th subcarrier of the n-th SEFDM symbol; Δf is the subcarriers frequency spacing; T is the SEFDM symbol period; $|\varepsilon| \mathrel{<}= 0.5$ is the arbitrary real constant, defining signal offset at the time base, normalized to T ; $\psi_T(t-nT-\varepsilon T)$ is the pulse shape function. In this paper SEFDM signals with the rectangular pulse shape are considered.

The important SEFDM signal parameter is the normalized subcarriers frequency spacing: $\alpha = \Delta f T$. For OFDM-signals $\alpha = 1$, whereas for SEFDM-signals $\alpha < 1$. The n-th SEFDM analog baseband symbol $s_{A,\varepsilon}^{(n)}(t)$ is given by:

$$s_{A,\varepsilon}^{(n)}(t) = \sum_{k=-N/2}^{N/2-1} C_N^{(n)}(k)\, e^{j2\pi k\Delta f t}\, \psi_T(t-\varepsilon T).$$

The spectrum $S_{A,\varepsilon}^{(n)}(f)$ of the signal $s_{A,\varepsilon}^{(n)}(t)$ is equal to:

$$S_{A,\varepsilon}^{(n)}(f) = \sum_{k=-N/2}^{N/2-1} T C_N^{(n)}(k)\, \mathrm{sinc}(\pi T f)\, e^{-j2\pi T f(\varepsilon+1/2)} = \sum_{k=-N/2}^{N/2-1} S_\varepsilon^{(k,n)}(f - k\Delta f), \qquad (1)$$

where $S_\varepsilon^{(k,n)}(f)$ is the k-th subcarrier signals' spectrum, $\mathrm{sinc}(x) = \sin(x)/x$. Fig. 1 illustrates the shape and the relative position of the amplitude spectrums $\left|S_\varepsilon^{(k,n)}(f - k\Delta f)\right|$ over the interval of $fT \in [-5;\, 5]$ for $N = 6$; $\alpha = 1$ (OFDM) and

$\alpha = 1/2$ (SEFDM). Fig. 1 clearly shows the bandwidth reduction as α decreases.

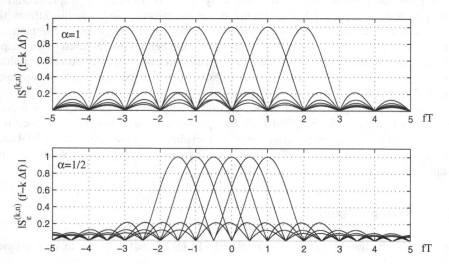

Fig. 1. Amplitude spectrums $\left| S_\varepsilon^{(k,n)}(f - k\Delta f) \right|$ as the components of the spectrum $S_{A,\varepsilon}^{(n)}(f)$, $N = 6$

For the sake of aliasing effect mitigation and transceivers' filters design simplification guard intervals in the frequency domain are used in the digital FDM systems: $C_N^{(n)}(k) = 0$ for $n \in \mathbb{N}$, $k \in [-N/2; -N/2 + N_{GL} - 1]$ and $k \in [N/2 - N_{GR} + 1; N/2]$, where N_{GL} and N_{GR} are the number of guard subcarriers of the left band and right band of the spectrum. Therefore the number of used (manipulated) subcarriers is given by: $N_U = N - N_{GL} - N_{GR}$. Supposed that for the considered SEFDM system $N_{GL} > 1$ and $N_{GR} > 1$, the convenient value of the sampling frequency will be: $F_s = 1/\Delta t = N\Delta f$. As a result: $T/\Delta t = N\alpha = L$. In the remainder of this paper it is assumed that $N\alpha = L \in \mathbb{N}$. Therefore, the discrete representation $s_{D,\varepsilon}^{(n)}(t)$ of the n-th SEFDM symbol, can be expressed as ([8], [9]):

$$s_{D,\varepsilon}^{(n)}(t) = \Delta t \sum_{i=0}^{L-1} \sum_{k=-N/2}^{N/2-1} C_N^{(n)}(k) e^{j2\pi \frac{k(i+0,5+\varepsilon L)}{N}} \delta\left(t - i(\Delta t + 0,5) - \varepsilon T\right).$$

where $\delta(t)$ – is the Dirac delta function, Δt – is the sampling interval, The Δt factor provides dimension consistency for the signals $s_{D,\varepsilon}^{(n)}(t)$ and $s_{A,\varepsilon}^{(n)}(t)$ ([9]). The spectrum $S_{D,\varepsilon}^{(n)}(f)$ of the signal $s_{D,\varepsilon}^{(n)}(t)$ is given by:

$$S_{D,\varepsilon}^{(n)}(f) = \sum_{l=-\infty}^{+\infty} \sum_{k=-N/2}^{N/2-1} S_\varepsilon^{(k,n)}(f - k\Delta f - lF_s) e^{-j2\pi l(0,5+\varepsilon L)}, \qquad (2)$$

where $S_\varepsilon^{(k,n)}(f)$ is defined by (1). Equation (2) shows, that the spectrum $S_{D,\varepsilon}^{(n)}(f)$ of the finite discrete signal $s_{D,\varepsilon}^{(n)}(t)$ is the linear combination of the subcarriers spectrums $S_\varepsilon^{(k,n)}(f)$ shifted in the frequency domain.

3   DFT-Based Detection Algorithm for SEFDM Signals

Let all the manipulation symbols for all the subcarriers used in the analog SEFDM symbol $s_{A,\varepsilon}^{(n)}(t)$ be equal to 0, except for manipulation symbol transmitted on the k'-th subcarrier. Then according to (1) the value of the spectrum $S_{A,\varepsilon}^{(n)}(f)$ at the point $f = k'\Delta f$ is equal to $T C_N^{(n)}(k')$ – the only non-zero manipulation symbol (accurate within a real constant T). Generally, if more than one subcarrier is manipulated, $S_{A,\varepsilon}^{(n)}(k'\Delta f)$ takes the form:

$$S_{A,\varepsilon}^{(n)}(k'\Delta f) = TC_N^{(n)}(k') + T I_\varepsilon^{(n)}(k'),$$

$$I_\varepsilon^{(n)}(k') = \sum_{\substack{k=-N/2 \\ k \neq k'}}^{N/2-1} C_N^{(n)}(k)\operatorname{sinc}\left(\pi\alpha(k'-k)\right)e^{-j2\pi\alpha(k'-k)(\varepsilon+1/2)},$$

where the $I_\varepsilon^{(n)}(k')$ term defines the influence of subcarriers with indexes $k \neq k'$ on the $S_{A,\varepsilon}^{(n)}(k'\Delta f)$ value. This influence is known as *intercarrier interference.*

It can be shown that for discrete SEFDM system the value $S_{D,\varepsilon}^{(n)}(k'\Delta f)$ can be represented as the sum of three terms:

$$S_{D,\varepsilon}^{(n)}(k'\Delta f) = T\sum_{l=-\infty}^{+\infty}\sum_{k=-N/2}^{N/2-1} C_N^{(n)}(k)\operatorname{sinc}\left(\pi\alpha(k'-k-lN)\right)e^{-j2\pi[\alpha(k'-k-lN)(\varepsilon+1/2)+l(0,5+\varepsilon L)]} =$$

$$= TC_N^{(n)}(k') + T I_\varepsilon^{(n)}(k') + TA_\varepsilon^{(n)}(k'),$$

where $A_\varepsilon^{(n)}(k')$ is the term resulting from *aliasing effect*:

$$A_\varepsilon^{(n)}(k') = \sum_{\substack{l=-\infty \\ l \neq 0}}^{+\infty}\sum_{k=-N/2}^{N/2-1} C_N^{(n)}(k)\operatorname{sinc}\left(\pi\alpha(k'-k-lN)\right)e^{-j2\pi[\alpha(k'-k-lN)(\varepsilon+1/2)+l(0,5+\varepsilon L)]}..$$

The values of $\left\{S_{D,\varepsilon}^{(n)}(k')\right\}_{k'=-N/2}^{N/2-1}$ are usually computed using the DFT method ([6], [7]). Consequently *the DFT-based detection algorithm for SEFDM signal $s_{D,\varepsilon}^{(n)}(t)$ can be formalized as follows*: for each received manipulation symbol a detector makes the decision $\hat{C}_N^{(n)}(k)$ from the alphabet set $\mathcal{A}$ if the Euclidian distance between $\hat{C}_N^{(n)}(k)$ and $S_{D,\varepsilon}^{(n)}(k'\Delta f)$ is minimal over all possible values C_i from $\mathcal{A}$:

$$\hat{C}_N^{(n)}(k) = \arg\left\{\min_{C_i \in \mathcal{A}} | TC_i - S_{D,\varepsilon}^{(n)}(k\Delta f) | \right\} =$$

$$= \arg\left\{\min_{C_i \in \mathcal{A}} | \left(C_i - C_N^{(n)}(k)\right) + \left(I_\varepsilon^{(n)}(k) + A_\varepsilon^{(n)}(k)\right) | \right\},$$

(3)

where $\left(I_\varepsilon^{(n)}(k) + A_\varepsilon^{(n)}(k)\right)$ term corresponds to detection systematic error. Therefore in the case of ideal communication channel, the performance of the DFT-based SEFDM detection algorithm depends on the aliasing $A_\varepsilon^{(n)}$ and intercarrier interference $I_\varepsilon^{(n)}$.

4   Minimization of the Interference and Aliasing Impact on DFT-Detection Performance

4.1   Optimization Problem Formalization

According to (3) the impact of the distortions $I_\varepsilon^{(n)}$ and $A_\varepsilon^{(n)}$ on the DFT-based detection performance can be reduced by choosing the appropriate SEFDM parameters values: α^* and ε^*, i.e. solving the following optimization problem:

$$\{\alpha^*, \ \varepsilon^*\} = \text{find } \arg\left\{\min_{\alpha,\varepsilon} E\left(| I_\varepsilon^{(n)}(k') + A_\varepsilon^{(n)}(k') |\right)\right\},$$

$$s.t. \ |\alpha^* - \hat{\alpha}| < \rho, \ C_N^{(n)}(k) \in \mathcal{A},$$

(4)

where $E(\cdot)$ denotes the expected value over all possible SEFDM-symbols and all subcarriers indexes $k' = -N/2...N/2 - 1$, ρ is the maximum deviation of α^* from the desired value $\hat{\alpha}$.

4.2   Solution for Complex Valued Manipulation

Let us consider quadrature phase shift keying (QPSK) manipulation as the example of complex valued subcarriers manipulation method. For QPSK and small N it is possible to compute and depict $g(\alpha, \ \varepsilon) = E\left(| I_\varepsilon^{(n)}(k') + A_\varepsilon^{(n)}(k') |\right)$ values with small axial steps. Fig. 2 shows function $g(\alpha, \ \varepsilon)$ as 3D mesh plot for $N = 16$, $N_U = 10$.

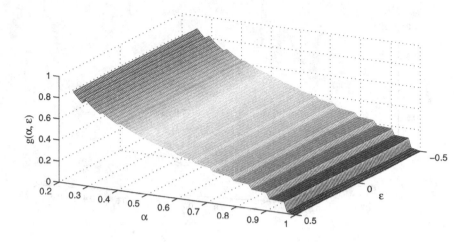

Fig. 2. 3D mesh plot of $g(\alpha, \varepsilon)$ function for $N = 16$, $N_U = 10$

It can be observed in fig. 2 that for any desired $\hat{\alpha}$ function $g(\alpha, \varepsilon)$ has constant value along the ε axis. Furthermore $g(\hat{\alpha}, \varepsilon)$ value grows significantly as $\hat{\alpha}$ decreases. Therefore using SEFDM DFT-based detection for complex valued subcarriers manipulation and small $\hat{\alpha}$ inevitably leads to significant energy loss as compared to ML detection.

4.3   Solution for Real Valued Manipulation

The amplitude shift keying (ASK) is considered as the special case of real valued subcarriers manipulation method. For ASK (3) can be rewritten as:

$$\hat{C}_N^{(n)}(k) = \arg\left\{ \min_{C_i \in \mathcal{A}} |\operatorname{Re}\left[\left(C_i - C_N^{(n)}(k)\right) + \left(I_\varepsilon^{(n)}(k) + A_\varepsilon^{(n)}(k)\right)\right]| \right\},$$

where $\operatorname{Re}(\cdot)$ denotes the real part of complex number. Therefore (4) takes the form:

$$\{\alpha^*, \varepsilon^*\} = \text{find } \arg\left\{ \min_{\alpha, \varepsilon} E\left(\left|\operatorname{Re}[I_\varepsilon^{(n)}(k') + A_\varepsilon^{(n)}(k')]\right|\right) \right\},$$

$$s.t. \ |\alpha^* - \hat{\alpha}| < \rho, \ C_N^{(n)}(k) \in \mathcal{A},$$

(5)

where $\mathcal{A}$ is the ASK-M alphabet set. Fig. 3 shows function $h(\alpha, \varepsilon) = E\left(|\operatorname{Re}[I_\varepsilon^{(n)}(k') + A_\varepsilon^{(n)}(k')]|\right)$ as 3D mesh plot for ASK-4 manipulation and $N = 16$, $N_U = 10$.

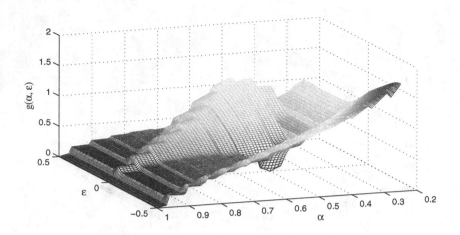

Fig. 3. 3D mesh plot of $h(\alpha,\ \varepsilon)$ function for $N = 16$, $N_U = 10$

It can be observed in fig. 3 that for each desired $\hat{\alpha}$ value there is at least one ε^* value providing $g(\hat{\alpha},\ \varepsilon)$ minimization. Furthermore $h(\alpha,\ \varepsilon)$ function has absolute minimum at the point $\alpha^* = 0.5$, $\varepsilon^* = 0$, achieving zero value. In the neighborhood of this point $h(\alpha,\ \varepsilon)$ reduces significantly providing very effective DFT-based detection use for $0.46 < \hat{\alpha} < 0.54$. The comparative analysis of fig. 3 and fig 4 shows that for any fixed $\hat{\alpha}$, $h(\hat{\alpha},\ \varepsilon_h^*)$ local (or absolute) minimum is always less than $g(\hat{\alpha},\ \varepsilon)$ value. Therefore SEFDM DFT-detection method is potentially much more effective for ASK than for any complex manipulation. For some special cases using proposed (5) method provides DFT-detection error performance approach to ML detection performance.

5   Simulation Results

The efficiency of the proposed optimization method (5) is proved through the analysis of the SEFDM DFT-detection BER performance. BER performance is evaluated in AWGN channel for ASK-2 (BPSK) and ASK-4 – manipulated SEFDM signals with $N = 128$, $N_U = 110$. The ASK-2 BER performance is evaluated for $\hat{\alpha} = 3/4$ and three ε values: $\varepsilon^* = 0$, corresponding to (5) solution for $N = 128$; $\varepsilon = -0.009$ and $\varepsilon = +0.010$, corresponding to optimal solution deviations. Similarly, the ASK-4 BER performance is evaluated for $\hat{\alpha} = 1/2$ and: $\varepsilon^* = 0$, corresponding to (5) solution; $\varepsilon = -0.008$ and $\varepsilon = +0.009$, corresponding to optimal solution deviations. The simulation-derived BER plots are shown in fig. 4.

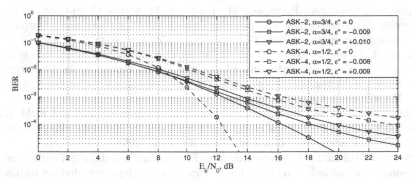

Fig. 4. BER performance of the DFT-based SEFDM detection algorithm for $N = 128$, $N_U = 110$

The analysis of results provided in Fig. 4 shows, that even the small deviations from optimal solution leads to significant energy loss. Considering ASK-2 and $BER = 10^{-3}$, this penalty achieves 0.6 dB for $\varepsilon = -0.009$ and 1.3 dB for $\varepsilon = +0.010$ Considering ASK-4, the loss is equal to 4.6 dB for $\varepsilon = -0.008$ and 5.7 dB for $\varepsilon = +0.009$ Note also that for ASK-4 SEFDM DFT-detection BER performance with optimal parameters $\hat{\alpha} = 1/2$, $\varepsilon^* = 0$, achieves ASK-4 single-carrier ML BER performance.

6   Conclusions

In this paper we proved that apart from channel noise the SEFDM DFT-detection error performance is determined by aliasing and intercarrier interference effects. The optimization method was introduced providing minimization of intercarrier interference and aliasing impact on DFT-detection BER performance for ASK-manipulated SEFDM signals. The minimization is achieved by using specific SEFDM parameters values, which are obtained by solving the optimization problem. Numerical simulation demonstrates that using optimal SEFDM parameters provides significant energy gain as compared to the case of small deviations from optimal values. The SEFDM DFT-detection algorithms can be easily implemented based on the modern OFDM transceivers offering gradual transition from OFDM- to SEFDM-technology.

References

1. Yang, X., Ai, W., Shuai, T., Li, D.: A fast decoding algorithm for non-orthogonal frequency division multiplexing signals. In: International Conference on Communications and Networking in China (CHINACOM), pp. 595–598 (August 2007)
2. Kanaras, I., Chorti, A., Rodrigues, M., Darwazeh, I.: Analysis of Sub-optimum detection techniques for a bandwidth efficient multi-carrier communication system. In: Proceedings of the Cranfield Multi-Strand Conference, Cranfield University, pp. 505–510 (May 2009)

3. Kanaras, I., Chorti, A., Rodrigues, M., Darwazeh, I.: A New Quasi-Optimal Detection Algorithm for a Non Orthogonal Spectrally Efficient FDM. In: International Symposium on Communication and Information Technologies, pp. 460–465 (September 2009)
4. Kanaras, I., Chorti, A., Rodrigues, M., Darwazeh, I.: A combined MMSE-ML detection for a spectrally efficient non orthogonal FDM signal. In: Proceedings of the 5th International Conference on Broadband Communications, Networks and Systems, September 8–11, pp: 421–425. IEEE Xplore Press, London
5. Kanaras, I., Chorti, A., Rodrigues, M., Darwazeh, I.: Spectrally Efficient FDM Signals: Bandwidth Gain at the Expense of Receiver Complexity. In: 2009 Proceedings of the IEEE International Conference on Communications ICC (June 2009)
6. Ahmed, S., Darwazeh, I.: Inverse discrete Fourier transform-discrete Fourier transform techniques for generating and receiving spectrally efficient frequency division multiplexing signals. American Journal of Engineering and Applied Sciences 4, 598–606 (2011)
7. Kislitsyn, A.B., Rashich, A.V., Tan, N.N.: Generation of SEFDM-Signals Using FFT/IFFT. In: Balandin, S., Andreev, S., Koucheryavy, Y. (eds.) NEW2AN/ruSMART 2014. LNCS, vol. 8638, pp. 488–501. Springer, Heidelberg (2014)
8. Sklar, B.: Digital communications, fundamentals and applications, 2nd edn. Prentice Hall PTR, Upper Saddle River (2001)
9. Hentschel, T.: Sample Rate Conversion in Software Configurable Radios. Artech House, Norwood (2002)

Combined Adaptive Spatial-Temporal Signal Processing System Based on Sequential Circuit with Dependent Component Adaptation

Vladimir Grigoryev[✉] and Igor Khvorov[✉]

Department of Wireless Telecommunications, ITMO University,
Saint Petersburg, Russian Federation
{vgrig,khvorov}@labics.ru

Abstract. The state-of-the-art adaptive combined (space-time) signal processing methods provide either optimum results under unrealistic computational times and efforts, or affordable computations under severe service degradation. This paper first analyses the existing methods and derives the motivation for the novel approach to creation of combined systems of the space-time-adaptive processing with *dependent* component adaptation. Then the generic derivations of the weighting factors calculation is shown step by step for a two-component (space-time) combined processing system with dependent components adaptation. Finally, numerical simulations show that through the proposed method it is possible to suppress wideband and narrowband interference, even if the number of interfering signals is higher than the number of antenna array elements.

Keywords: Combined signal processing (CSP) systems · Space-Time-Adaptive processing (STAP) · Minimum mean square error (MMSE) criterion · Weighting factors vector · Space filter · Time filter · Adaptive antenna array · Adaptive filtering

1 Introduction

Nowadays a highly demanded increase in interference immunity is achieved by the following means:

- Improving the selectivity properties of electronic receiving devices
- Other organizational methods of separating frequency-locational resources.

The research emphasis in methods of separating frequency-locational resources lies on dynamic spectrum access and cognitive radio [1], [2], [3]. However, these methods are not in the industrial state-of-the-art. This is why the possibility of radically increasing the interference immunity of electronic receiving devices is now expected from perfecting the means and methods of signal forming and processing.

The modern radio communication systems are characterized with subscriber mobility, frequent change of base station sites. This leads to a dynamically changing

© Springer International Publishing Switzerland 2015
S. Balandin et al. (Eds.): NEW2AN/ruSMART 2015, LNCS 9247, pp. 621–635, 2015.
DOI: 10.1007/978-3-319-23126-6_56

Signal-Interference Situation (SIS) and unpredictable interference parameters. It is therefore necessary to adjust the signal processing entities' parameters for every specific SIS case. Such an adjustment is possible as a part of adaptive signal processing realization.

An interest in adaptive signal processing is increased significantly by its ability to suppress interfering signals without prior information on their parameters [4], [5], [6]. Modern and next generation communication systems, i.e., 4G and 5G, are inseparably linked to the concept of adaptation at all levels: starting from the physical level, up to the user interface and flexible selection of an available communication that is the most suitable for the required communication services.

In order to facilitate this flexibility and adaptation, research looks into *Combined Signal Processing* (CSP) systems. The first attempts towards defining and evaluating the CSP systems as a separate class were done in [7], [8], [9]. CSP meets practically all processing-device structural requirements due to incorporating in the general case the means for polarization, spatial and time signal processing, as well as for correlation-based (coherent) signal processing.

First, in Section 2 the paper analyses the existing approaches to combined signal processing, defining the main methods, analyzing their computational complexity, performance and assumptions. After the conclusions were made, the system with dependent components adaptation is introduced and its design is shown in Section 3. Finally, Section 4 shows the numerical simulation results for the proposed method. In the paper the terms "space" and "spatial" as well as "time" and "temporal" are used as synonyms.

2   Analysis of Existing Realization Options of Adaptive Combined Signal Processing (CSP)

Traditional design principles of adaptive Combined Signal Processing (CSP) systems are:

- Full circuit with parallel processing [10],
- Sequential circuit with independent component adaptation [11], [12], [13], [14], [15].

In this section we analyse and describe the aforementioned principles, pointing out their drawbacks and motivating the proposed approach.

2.1   Full Circuit with Parallel Processing

The full circuit option [10] means that a processing system is designed as a multidimensional matched filter. The dimension of such a system grows sharply with an increase in the number of components. For the CSP system it results in significant implementation costs and long calculation time of the weighting factors. Therefore real-time calculation becomes only possible for small dimensioned filters.

Fig. 1 shows an example of a CPS system, consisting of two filters (Space-Time Adaptive Processing, STAP), according to the full circuit design principle [4]. In the figure X represents the input signal, i.e., signal + noise, and Z is the output signal.

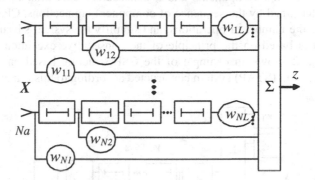

Fig. 1. Combined Signal Processing (CSP) system example designed according to full circuit filter with parallel processing: consists of two Space-Time Adaptive Processing (STAP) filters

Traditionally, an integral equation solution is obtained in discrete form by realizing pulse characteristics based on multi-tapped delay lines with L taps and (L+1) weighting factors. Under the Minimum Mean Square Error (MMSE) criterion, the optimum weights of the STAP are given by the well-known Wiener-Hopf solution in matrix form [10], [16]:

$$\mathbf{R}_{\text{intn}} \mathbf{W} = \mathbf{R}_{Xd} \tag{1}$$

where $\mathbf{R}_{\text{intn}}$ is a Correlation Matrix (CM) with dimensions $N_a(L+1) \times N_a(L+1)$; $\mathbf{W}$ is a Weighting Factors (WF) vector, with the length of $N_a(L+1)$: composed from $(L+1)$ vectors $\mathbf{W}_i$ ($i = 0, L$), with the length of N_a; $\mathbf{R}_{Xd}$ is a correlation vector of input signals and the reference signal, with the length of $N_a(L+1)$.

For optimal signal processing realization, it is necessary to have time delays between the taps, meeting the following inequality in accordance with the sampling theorem:

$$\tau_d < 1/2\Delta F \tag{2}$$

where ΔF is the signal and interference bandwidth.

Despite the fact that realization of the full processing principle provides the highest efficiency (quality) of signal processing under the conditions of stationarity, it is not commonly applied. First of all, it is associated with the high computing and material costs for solving the (1). The required volume of computational operations for calculating WF vector W is proportional to $(N_a L)^3$. This together with the requirement of high calculation accuracy, and final processing time interval imposes a restriction on the maximum number of adaptation channels.

2.2   Sequential Circuit with Independent Component Adaptation

The second design option [11], [12], [13], [14], [15] is associated with independent adaptation of system's components and a consecutive execution of the processing operations, determined by the sequence of component connections. CPS is applied in order to reduce the number of operations, and simplify the system's beam synthesizer. Its application is based on the principle of the consecutive execution of processing operations. Fig. 2 shows an example of the CSP structure based on a Space-Time Adaptive Processing (STAP) system constructed according to this principle.

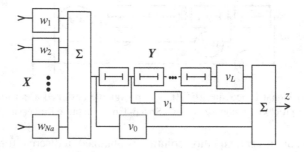

Fig. 2. A Combined Signal Processing (CSP) structure example based on a Space-Time Adaptive Processing (STAP): Sequential circuit with independent component adaptation

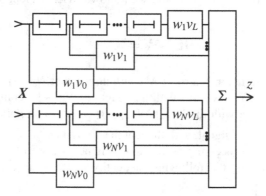

Fig. 3. Equivalent representation of the CSP structure with the weighting factors as in (3), it is equivalent to the optimal solution from the Fig. 1

One can easily see that the presented structure is equivalent to the structure in Fig. 1, if weighting factors, w_{ij} in each of the taps (Fig. 3) are presented as:

$$w'_{ij} = w_i v_j \qquad (3)$$

where w_i $(i = \overline{1, N_a})$ is the weighting factor in i-channel of the spatial processing device; v_j $(j = \overline{0, L})$ the weighting factor in j-tap of the time processing component.

The WF synthesis of an entire signal processing system, Fig. 2, is often reduced to WF synthesis of the individual devices of that system, and optimized by individual quality parameters. This approach is called *independent adaptation*. It assumes factorability of the processing operator into individual component operators. For example, a STAP system could be factorized into space and time components: $L_{STAP}=L_{SP}L_{TP}$; where L_{SP} is the space processing operator and L_{TP} is the time processing operator.

The solution for CSPs, constructed according to the principle of independent optimization, is well-known and can be written in the form of two matrix equations:

$$\mathbf{R}_{XX}\mathbf{W}_{indep} = \mathbf{R}_{Xd} , \quad \mathbf{R}_{YY}\mathbf{V}_{indep} = \mathbf{R}_{Yd} \tag{4}$$

where $\mathbf{R}_{XX}$ is a Correlation Matrix (CM) of signals at the inputs of the spatial component; $\mathbf{R}_{YY}$ is a CM of signals at the inputs of the time component; W - the WF of the space filter (SF); V - the WF of the time filter (TF); $\mathbf{R}_{Xd}$, $\mathbf{R}_{Yd}$ the signal correlation vectors at SF, TF inputs and the reference signal.

Within this design option, it is possible to sharply reduce the number of controlled parameters by representing the full processing operator in a factorized form. Therefore, the system's dimensions are defined by the sum of its component dimensions.

Independent component representation makes it possible to reduce the weighting factor calculation time. However, this solution has its drawbacks related to the limited quality of signal processing, since each component tends to suppress all the interferences active at the input, i.e., including the signals of other components.

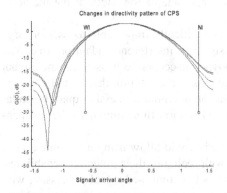

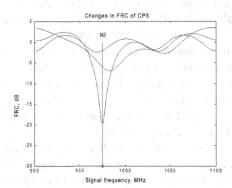

Fig. 4. Resulting antenna array directivity pattern with wideband and narrowband interferences, WI and NI respectively. Both interferences are not suppressed by the filter.

Fig. 5. Resulting time filter frequency response curve, the NI is not fully suppressed.

The following example (Fig. 4, Fig. 5) shows that if at the system input there are more interference signals as N_a -1, the system fails to suppress them completely. The results are obtained for a CSP system, containing two-element antenna array (N_a = 2) and three-element time filter (L=3). It is affected by one wideband interference (WI) signal and one narrowband interference (NI) signal. Thus, WI+NI= Na, and the interference cannot be suppressed completely.

Fig. 4 illustrates resulting antenna array directivity patterns. It can be observed that in the spatial filter, the WI is not fully suppressed (no zero in the resulting directivity diagram), and the rest of the interference signal passes into the temporal filter. Fig. 5 shows time filter frequency response curve. Here also no suppression for WI takes place.

For a signal processing system it means that the dimensions of an antenna array (N_a) or spatial filter are costly and complex to create, while a multitap time filter in comparison to spatial is a cheaper and less complex solution. At the same time, a narrowband interference signals are more common compared to wideband, and there could be a significant amount of NI. Thus, ideally, the lower number of WI shall be suppressed at the spatial filter, whereas a higher number of NI in the temporal filter.

Moreover, it has to be noted that the assumption of processing components as independent from each other operators holds only in some special cases. Therefore it would be necessary to find a solution, considering mutual influence of the components of processing, i.e., using the principle of joint adaptation (optimization).

2.3   Conclusions on the State-of-the-Art

The two analyzed classical design options were: full circuit with parallel processing and sequential circuit with independent component adaptation. Both of the methods, although well-known, have their limitations.

The full circuit with parallel processing is practically impossible to use for large dimensioned filters to realize a reasonable signal to interference adaption time. The independent processing, in its turn, imposes unacceptable losses in the signal processing quality. Therefore, it is natural to look for a solution that would combine signal processing methods in order to provide the potential processing quality similar to the full processing system, while keeping adaptation time similar to the independent processing systems.

Development of such a CSP design approach would allow using a combination of the signal processing methods in real-time with realizing their potential capabilities. Further in the paper, we refer to this solution as the combined signal processing with *dependent* component adaptation. Here the system's component adaptation is done taking into account the other components. This, for example, can allow suppressing wideband interference signals in the space filter without taking into account the action of interferences suppressed in the time filter. Accordingly, in the time filter suppressing interference signals without taking into account interference signals that are suppressed in the space filter.

3   Calculation of Weight Factors in System with Dependent Components Adaptation

Architecture of a two-component combined processing system (STAP) with dependent components adaptation is depicted in Fig. 6. It consists of the following modules: Space Filter (SF), Time Filter (TF) and an Adaptive Processor (AP). Signal X is the signal at the space filter entry, i.e., signal + noise; signal Y is the signal *after* the space filter – at the entry of the time filter, and d is a reference signal. Comparison under Minimum Mean Square Error (MMSE) criterion is done with the signal d.

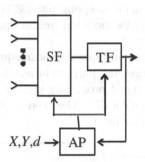

Fig. 6. Two component (Space-Time Adaptive Processing) Combined Signal Processing system with dependent component adaptation

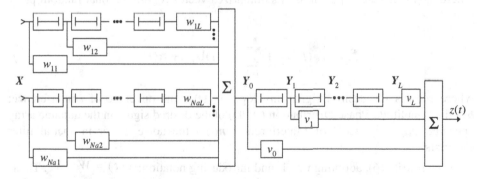

Fig. 7. The structure with the weighting factors for a two component (Space-Time Adaptive Processing) Combined Signal Processing system with dependent component adaptation

The structure with the weighting factors is shown in Fig. 7. The SF consists of N_a channels, each of which contains a weighting device, in general, containing a multi-tap delay line for providing wideband weighting. Weighted signals at the outputs of N_a channels are summed up in the adder of the spatial filter. The output signal of the spatial filter adder is fed to the TF. The TF is realized based on the multi-tap delay line with $(L+1)$-taps, in each of which, weighting devices with weighting factors v_i, $i = \overline{0, L}$ are available. The combined system output signal of spatial-temporal processing is the signal $z(t)$ at the output of the temporal filter adder. Thus, the output signal is the result of weightings with weighting factors W and V in SF and TF:

$$z(t) = \sum_{j=0}^{L} v_j^* \sum_{i=1}^{N_a} w_i^* x_i (t - j\tau_K') = \operatorname{tr}(V^+ \otimes W^+ X), \tag{5}$$

where w_i is the weighting factor in i-channel of SF; v_j is the WF in j-channel of TF; τ_d is the delay element duration in the temporal filter, $\otimes$ is the notation for the operation of the exterior product, "tr" is the notation for the operation for finding the matrix trace.

To find the optimal WFs, the minimum of mean square error (MMSE) criterion is used. According to this criterion, the purpose of adaptation consists in minimization of the error between the combined processing system output signal $z(t)$ and the desired (standard or reference) signal $d(t)$. The expression for the mean of squared deviation is written in the following form:

$$J_{\text{MSD}} = \overline{(z(t) - d(t))^2} = \overline{\left(\sum_{j=0}^{L} v_j^* \sum_{i=1}^{N_a} w_i^* x_i(t - j\tau_K') - d(t) \right)^2} \tag{6}$$

where $x_i(t)$ is the i-component of a summarized vector N_a-dimensional random process:

$$X = u_S(t)\mathbf{b}_S + \sum_{i=1}^{K_.} u_{\text{п}i}(t)\mathbf{b}_{\text{п}i} + \mathbf{n}(t) \tag{7}$$

where $u_S(t)$ is the desired signal envelope; $u_{\text{int}i}(t)$ - the envelope of i-interference; b_S - the Amplitude-Phase Distribution (APD) of the desired signal in the antenna array aperture; $b_{\text{int}i}$ - the APD of i-interference; $n(t)$ - the noise vector in spatial filter channels.

If we rewrite (6), adopting $v_0 = 1$, and introducing notation: $y(t) = W^+ X$. Then:

$$J_{\text{MSD}} = \overline{\left(W^+ X + Y^+ V - d(t) \right)^2} \tag{8}$$

where $Y^{\text{T}} = [y(t), y(t-\tau), ..., y(t-L\tau)]$.

The weighting factors W and V optimality condition is the equality of the derivatives to zero: $\dfrac{\partial J}{\partial \mathbf{W}} = 0$, $\dfrac{\partial J}{\partial V} = 0$.

Differentiation of the (8) is made taking into account the equality $y(t) = = \boldsymbol{W}^+\boldsymbol{X}$, then having removed the brackets, and having executed averaging, we obtain:

$$\begin{cases} \dfrac{\partial J}{\partial W} = 2\left[\boldsymbol{R}_{XX}\boldsymbol{W} + \boldsymbol{R}_{XY}\boldsymbol{V} - \boldsymbol{R}_{Xd} + \sum_{i=1}^{L} v_i^*\left(\boldsymbol{R}_{X_iX}\boldsymbol{W} + \boldsymbol{R}_{X_iY}\boldsymbol{V} - \boldsymbol{R}_{X_id} \right) \right], \\ \dfrac{\partial J}{\partial V} = 2\left[\boldsymbol{R}_{YX}\boldsymbol{W} + \boldsymbol{R}_{YY}\boldsymbol{V} - \boldsymbol{R}_{Yd} \right], \end{cases}$$

where $\boldsymbol{R}_{XX}$ is a correlation matrix with dimensions $N_a \times N_a$ signals at the SF inputs; $\boldsymbol{R}_{XY}$ is a correlation matrix of signals X and Y; $\boldsymbol{R}_{X_iX}$ is the correlation matrix of signal X at the SF output and signal $X_i = X(t - i\tau_d)$ is delayed for $i\tau_d$; $\boldsymbol{R}_{X_iY}$ is the correlation matrix of signals X_i ($i = 1...L$) and signals $Y^\tau = [y_a,...,y_{i-1}, y_{i+1},...,y_L]$; $\boldsymbol{R}_{Xd}$, $\boldsymbol{R}_{X_id}$ and $\boldsymbol{R}_{Yd}$ are the respective cross correlated vectors.

The equations for the vectors WF V and W are summarized as:

$$\begin{cases} \boldsymbol{R}_{XX}\boldsymbol{W} + \boldsymbol{R}_{XY}\boldsymbol{V} - \boldsymbol{R}_{Xd} + \sum_{i=1}^{L} v_i^*\left(\boldsymbol{R}_{X_iX}\boldsymbol{W} + \boldsymbol{R}_{X_iY}\boldsymbol{V} - \boldsymbol{R}_{X_id} \right) = 0, \\ \boldsymbol{R}_{YY}\boldsymbol{V} + \boldsymbol{R}_{YX}\boldsymbol{W} - \boldsymbol{R}_{Yd} = 0. \end{cases}$$

So we obtain:

$$\begin{cases} \boldsymbol{V} = \boldsymbol{R}_{YY}^{-1}\left(\boldsymbol{R}_{Yd} - \boldsymbol{R}_{YX}\boldsymbol{W} \right), \\ \left(\boldsymbol{R}_{XX} - \boldsymbol{R}_{XY}\boldsymbol{R}_{YY}^{-1}\boldsymbol{R}_{YX} \right)\boldsymbol{W} - \boldsymbol{R}_{Xd} + \boldsymbol{R}_{Yd}^+\left[\left(\boldsymbol{R}_{X_iX} - \boldsymbol{R}_{X_iY}\boldsymbol{R}_{YY}^{-1}\boldsymbol{R}_{YX} \right)\boldsymbol{W} \right] - \\ - \boldsymbol{R}_{Yd}^+\left[\boldsymbol{R}_{X_id} - \boldsymbol{R}_{X_iY}\boldsymbol{R}_{YY}^{-1}\boldsymbol{R}_{Yd} \right] - \boldsymbol{W}^+\boldsymbol{R}_{XX}\boldsymbol{R}_{YY}^{-1}\left[\left(\boldsymbol{R}_{X_iX} - \boldsymbol{R}_{X_iY}\boldsymbol{R}_{YY}^{-1}\boldsymbol{R}_{YX} \right)\boldsymbol{W} \right] + \\ + \boldsymbol{W}^+\boldsymbol{R}_{XY}\boldsymbol{R}_{YY}^{-1}\left[\boldsymbol{R}_{X_id} - \boldsymbol{R}_{X_iY}\boldsymbol{R}_{YY}^{-1}\boldsymbol{R}_{Yd} \right] = 0, \end{cases}$$

$$(9)$$

Here, the square brackets designate block matrixes, where the blocks are L matrixes $\boldsymbol{R}_{X_iX}$, $\boldsymbol{R}_{X_iY}$ or vectors $\boldsymbol{R}_{X_id}$. The second equation in (9) is quadratic w.r.t. WF $W;$ its solution is possible, if one of the known numerical methods is used [17].

An analysis of (9) led us to the following conclusions. Firstly, for calculating the spatial filter WF, correlation matrixes of signals $X_i = X(t - i\tau_3)$, are used, which requires the application of L-element delay lines in each antenna element. This significantly complicates the signal processing circuit compared to independent processing. In comparison with the optimal full circuit, dimensionality of matrixes is reduced by a factor of (L+1), and, in fact, one line of the full correlation matrix $\boldsymbol{R} = \{ \boldsymbol{R}_{X_iX_j} \}$ ($i = \overline{0, L}$; $j = \overline{0, L}$) is used.

Secondly, the optimal values of vectors V and W are achieved as a result of the iterative process because of the dependency of the correlation matrixes R_{X_iY}, R_{YY} and vector R_{Yd} on the weighting factors vector W. Such a dependency causes changes in the correlation matrix values, and accordingly, in the WF values.

Thirdly, it should be noted, that the second equation is broken into two parts. One of which $\left(R_{XX} - R_{XY}R_{YY}^{-1}R_{YX}\right)W - R_{Xd}$ is obtained, if by differentiation of (8), the dependency of signal Y on W is not taken into account. The second part of the equation takes into account dependency of matrixes R_{XY} and R_{YY} on W. This makes it possible to obtain more exact approximations for the values V and W.

Further, we consider interferences of two classes: wideband and narrowband. Then, the input random process can be presented as:

$$X_\Sigma = u_S(t)b_S + \sum_{i=1}^{L_{WI}} u_{WI_i}(t)b_{WI_i} + \sum_{j=1}^{L_{NI}} u_{NI_i}(t)b_{NI_i} + n(t),$$

where L_{WI}, L_{NI} are the numbers of wideband and narrowband interferences, respectively. For convenience, we choose the temporal filter unit delay τ_3 to be equal to the correlation interval of the wideband interference:

$$\tau_d = \left(\Delta F_{WI}\right)^{-1},$$

then, correlation of wideband interference and signal at the delay line outputs can be considered as approximately equal to zero. Therefore, the values of vectors $R_{Yd} \equiv 0$ and $R_{X_id} \equiv 0$, and the (9) is rewritten in a simpler form:

$$\begin{cases} V = -R_{YY}^{-1}R_{YX}W, \\ \left(R_{XX} - R_{XY}R_{YY}^{-1}R_{YX}\right)W - R_{Xd} - W^+ R_{XY}R_{YY}^{-1}\left[R_{X_iX} - R_{X_iY}R_{YY}^{-1}R_{YX}\right]W = 0. \end{cases}$$

From where:

$$\begin{cases} V = -R_{YY}^{-1}R_{YX}W, \\ W = \left(R_{XX} - R_{XY}R_{YY}^{-1}R_{YX} - W^+ R_{XY}R_{YY}^{-1}\left[R_{X_iX} - R_{XY}R_{YY}^{-1}R_{YX}\right]\right)^{-1} R_{Xd}. \end{cases} \tag{10}$$

Use of orthogonality principle conditions allows us to define the following system of equations relative to spatial and temporal filter weighting factors:

$$\begin{cases} \overline{X\varepsilon} = 2R_{XX}W + 2R_{XY}V - 2R_{Xd} = 0, \\ \overline{Y\varepsilon} = 2R_{YY}V + 2W^+ R_{XY} - 2R_{Yd} = 0. \end{cases} \tag{11}$$

In matrix form, the system of equations (11) can be defined as follows

$$\begin{bmatrix} R_{XX} & R_{XY} \\ R_{YX} & R_{YY} \end{bmatrix}\begin{bmatrix} W \\ V \end{bmatrix} = \begin{bmatrix} R_{Xd} \\ R_{Yd} \end{bmatrix}$$

Let us obtain optimal values of WF W and V:

$$\begin{cases} W = \left(R_{XX} - R_{XY}R_{YY}^{-1}R_{YX}\right)^{-1}\left(R_{Xd} - R_{XY}R_{YY}^{-1}R_{YX}\right), \\ V = R_{YY}^{-1}\left(R_{Yd} - R_{YX}W\right). \end{cases} \tag{12}$$

The solution to (12) only determines the optimal, according to the orthogonality principle, combined signal processing system's weighting factors based on MMSE criterion, for the *current state* of weighting factors W. Let us explain the above remark. The values of WFs W and V in the equations (12) are obtained with the fixed vector W of spatial filter's weighting factors for the particular matrixes values R_{XX}, R_{XY}, R_{YY}. The matrixes R_{XY} and R_{YY} are formed by the signals depending on the weighting factors W. Therefore, the form of Correlation Matrix (CM) also changes with each new value of W. Hence, the solution the (12) has to be considered as one of the steps in the general iterative process for obtaining the WF optimal values. The conclusion's validity on the iterative nature of the process for obtaining the optimal WF follows from the essential nonlinearity of the objective function, which in turn, is explained by the dependence of CM on the weighting factors.

Analysis of the expressions (12) allows us to draw the following conclusions. In order to calculate WF optimal values of the spatial filter, it is necessary to correct input signal correlation matrixes, having excluded the influence of interferences that are suppressed in the temporal filter. Such a correction can be done with using the matrix $R_{XY}R_{YY}^{-1}R_{YX}$. In addition, the reference vector R_{Xd} is also subjected to displacement through the vector $R_{Xd}R_{YY}^{-1}R_{Yd}$. Optimal current value of temporal filter weighting factors is formed based on CM R_{YY}, where, due to dependence on W, the influence of interferences suppressed in the spatial filter, has already been eliminated (completely or partly). The solutions to (10) and (12) shall therefore be considered as basic, used for constructing various iterative adaptation algorithms.

It can also be noted, that the expression (10) was obtained with the assumption, that WFs of the spatial filter W are fixed for the estimation interval of correlation matrixes R_{XY}, R_{YY}. This restriction made it possible to obtain all specified relationships, since otherwise it would be necessary to take into account the influence of the vector $W(t)$ dynamic changes.

4   Simulation Results

This section presents results of numerical simulations for the proposed model. The analysis follows the same steps as for the state-of-the-art algorithms. The following three figures: Fig. 8 - 10; introduce the results of simulation modelling of the CPS adaptation process, containing seven-element linear spatial filter ($N_a = 7$) and temporal filter with $L = 25$ elements of delay, according to the system of equations (10). The system has to deal with four wideband and four narrowband interferences: WI+NI > N_a (8 > 7).

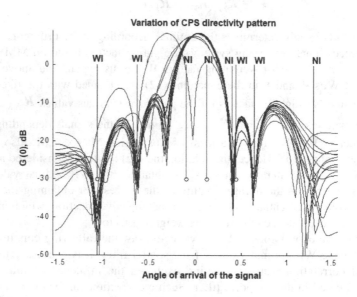

Fig. 8. Resulting directivity pattern of the CSP filter with Space and Time filters: wideband interference is mitigated in the spatial filter through zeros in characteristic after the adaptation. The narrowband interference is ignored as it is filtered in the frequency domain of the time filter, which is shown in the Fig. 9

Fig. 8 shows the resulting antenna directivity pattern after the adaptation process. It can be clearly seen that during the adaptation process, all four wideband interferences are suppressed completely in the spatial filter. The narrowband interferences have no impact on the directivity pattern of the space filter. In its turn the wideband interference does not propagate to the time filter, where it could not be suppressed anymore.

Fig. 9 illustrates the time filter Frequency Response Curve (FRC). It can be seen that the rest of the interference that was left after the space filter, i.e., narrowband interference, is suppressed in the temporal filter due to the dips in its FRC. Thus the proposed method with interdependent component adaptation is able to suppress WI+NI > N_a due to iterative distribution of the interference signals between the filters. This redistribution was impossible under independency assumption.

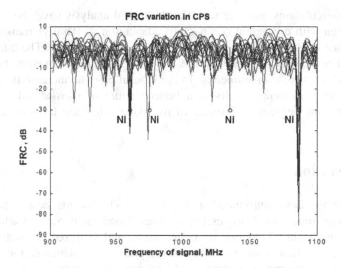

Fig. 9. Resulting time filter frequency response curve: the narrowband interference is filtered out, wideband interference was filtered by the space filter, which is shown in Fig. 8.

The gain of such signal processing method is shown in Fig. 10 that depicts an increase in the signal/ (interference+noise) ratio at the output of CPS due to interference suppression. The SNR value in Fig. 10 is normalized with respect to Na. Thus, if SNR would have reached Na (seven), we could say that all the interference is completely suppressed. The peaks and flops show iterative redistribution of the interference signals between the filters. At about 12th iteration the system stabilizes and further only small fluctuations take place.

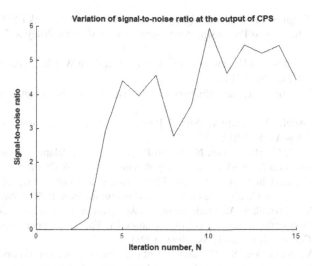

Fig. 10. Growth of the SNR due to interference suppression

Computer simulations together with the theoretical analysis have shown that with CPS realization with dependent components adaptation, an iterative redistribution of the interference signals between the filter components takes place. The quality of the output signal is defined not only by the combined system parameters, but also by the signal and interference parameters. In the general case, an increase in the number of temporal filter elements results in a better capture of narrowband interference, which liberates the degrees of freedom of the spatial filter and leads to reduction in losses.

5   Conclusion

In this paper we have introduced a novel approach towards design of adaptive combined space-time signal processing systems, based on dependent adaptation of processing components. First, we have introduced and analyzed the state-of-the-art methods for CSP system's design. It was shown that the optimum solution is costly to implement as its computation times and accuracy requirements are too high for current practical implementations. A simplification under an assumption of component independency, while solving complexity problems, results in severe signal quality degradation. Thus dependent signal adaptation aimed at preserving high signal processing quality. It was shown in Section 4 that due to iterative interference redistribution it is possible to suppress more interference signals as there are elements in the antenna array (WI+NI > Na). Thus CPS with dependent component adaptation outperforms the independent one.

References

1. Fette, B.A.: Cognitive Radio Technology. Academic Press, New York (2009)
2. Jayaweera, S.K.: Signal Processing for Cognitive Radios. John Wiley & Sons, New York (2014)
3. Matin, M.: Handbook of Research on Progressive Trends in Wireless Communications and Networking. IGI Globa (2014)
4. Monzingo, R., Miller, T.: Introduction to Adaptive Arrays. Wiley and Sons, New York (1980)
5. Haykin, S., Adali, T.: Adaptive Signal Processing: Next Generation Solutions. Wiley-IEEE Press, New York (2010)
6. Christodoulou, C.G., Blaunstein, N.: Radio Propagation and Adaptive Antennas for Wireless Communication Networks. John Wiley & Sons, New York (2014)
7. Grigoryev, V., Kuzichkin, A., Khvorov, I.: Algorithms of combined signal processing. In: The 2nd International Conference on Satellite Communications, ICSC 1996 (1996)
8. Grigoryev, V., Kuzichkin, A., Stadinchuk, A.: Application of methods of combined signal processing in systems of mobile communication. In: The Third International Conference on Satellite Communications, ICSC 1998 (1998)
9. Grigoryev, V., Kurichkin, S.: The analysis of construction variants of onboard systems of combined signals processing. In: The Third International Conference on Satellite Communications, ICSC 1998 (1998)

10. Klemm, R.: Principles of Space-Time Adaptive Processing. IEEPress, Bodmin (2002)
11. Godara, L.: Application of Antenna Arrays to Mobile Communications, Part II: Beam-Forming and Direction-of-Arrival Considerations. Proceedings of the IEEE **85**(8), 1195–1245 (1997)
12. Wu, D., Zhu, D., Shen, M., Zhu, Z.: Time-varying space-time autoregressive filtering algorithm for space-time adaptive processing. Radar, Sonar & Navigation, IET **6**(4), 213–221 (2012)
13. Xiong, Z.-Y., Xu, Z.-H., Zhang, L., Xiao, S.-P.: A recursive algorithm for the design of array antenna in STAP application. In: ET International Radar Conference (2013)
14. Lai, P., Lu, R., Liu, Y.: Space-time interference suppression technology based on sub-band blind adaptive array processing. In: 4th IEEE International Conference on Information Science and Technology (ICIST) (2014)
15. El Khatib, A., Assaleh, K., Mir, H.: Space-Time Adaptive Processing Using Pattern Classification. IEEE Transactions on Signal Processing **63**(3), 766–779 (2015)
16. Yang, K., Zhang, Y., Mizuguchi, Y.: Spatio-temporal signal subspace-based subband space-time adaptive antennas. In: International Symposium on Antennas and Propagation, Fukuoka, Japan (2000)
17. Ortega, J.M., Rheinboldt, W.C.: Iterative solution of nonlinear equations in several variables, Computer science and applied mathematics. Academic Press (1970)

Waveform Optimization of SEFDM Signals with Constraints on Bandwidth and an Out-of-Band Emission Level

Sergey V. Zavjalov[1], Sergey B. Makarov[1], Sergey V. Volvenko[1(✉)], and Wei Xue[2]

[1] Peter the Great St. Petersburg Polytechnic University, St. Petersburg, Russia
zavyalov_sv@spbstu.ru, {makarov,volk}@cee.spbstu.ru
[2] College of Information and Communication Engineering,
Harbin Engineering University, Harbin, China
weihe@mail.ru

Abstract. A problem of reducing frequency bandwidth in case of multi-frequency signals is considered. The main conditions are the minimization of additional energy losses while maintaining the information rate. Frequency bandwidth reduction is carried out by means of doubling the duration of the signals used for each subcarrier frequency. The envelope signals are used as a numerical solution of the optimization problem for a given out-of-band emission reduction rate, energy of the signal and cross-correlation coefficient. The simulation showed that, for the cross-correlation coefficient values equal to 10^{-2}, frequency bandwidth decrease of proposed SEFDM signal is 38%, as compared to the OFDM. At the same time, this energy loss, with respect to the OFDM signals, is no more than 1 dB.

Keywords: SEFDM · OOBE reduction rate · Intersymbol interference · Optimization problem · Envelope

1 Introduction

Multi-frequency signals are used in broadband Wi-Fi, LTE, and digital TV standards DVB-T/T2. Such signals of duration T_c with orthogonal frequency division multiplexing (OFDM) along with advantages have several significant drawbacks. These include a large occupied bandwidth ΔF and low rate out-of-band emissions (OOBE) [1,2]. One way to reduce ΔF multifrequency signals is to reduce the frequency spacing Δf between the subcarrier frequencies. As follows from [3], the OFDM signals have a value of $\Delta f = 1/T_c$. When the frequency spacing $\Delta f < 1/T_c$, such signals are nonorthogonal spectrally efficient frequency division multiplexing (SEFDM) signals and have been proposed for use in [4]. Small values of $\Delta f < 1/T_c$ significantly reduce BER performance [5,6,7]. Degradation of BER performance is caused by an inter-carrier interference ICI [8]. Thus the signals spectra, located on adjacent subcarriers frequencies, overlap and begin to influence each other as random noise [7], [9,10]. In addition, a transition to a reduced frequency spacing Δf, while maintaining an unchanged rectangular shape of the signal envelope $a(t)$ on the subcarrier frequencies, does not increase the OOBE reduction rate at the edges of the occupied bandwidth ΔF [1,2].

© Springer International Publishing Switzerland 2015
S. Balandin et al. (Eds.): NEW2AN/ruSMART 2015, LNCS 9247, pp. 636–646, 2015.
DOI: 10.1007/978-3-319-23126-6_57

To reduce OOBE level [11] it is proposed to use optimal functions, obtained as a solution of the functional equation in the form of spheroidal functions, as envelope signals with QPSK on each subcarrier frequency. The equation itself ([11], Eq. 7) corresponds to the criterion of the maximum energy concentration in the occupied bandwidth. As follows from the results of this paper [11], obtained forms of the envelope signals ensure uniform distribution of low OOBE level. However, the OOBE reduction rate of such signals is small, which does not allow them to increase the number of channels by means of their frequency division multiplexing. In [12] it is suggested to use a narrowband filter to generate signals, resulting in an envelope in the form of a Nyquist pulse. However, at a higher order narrowband filter [12], a significant uncontrolled inter-symbol interference (ISI) appears, which leads to BER performance deterioration. Reduction of the OOBE is considered in [13], where it is proposed to use the Gabor window for this. The signals have a smoothed envelope shape ([13], Fig. 3) on each quadrature channel on the subcarrier frequency. However, uncontrollable inter-symbol interference, appearing in the process of signal generation, leads to BER performance decrease on each subcarrier frequency. Appearance of an uncontrollable ISI is associated with increase of the transmitted signals duration Tc and overlay of adjacent signals, while maintaining a constant bit rate channel symbols. The BER performance reduces due to the inability to include information about this uncontrollable interference into a detection algorithm.

Papers [14] and [15] present techniques for the OOBE reduction: the subcarrier weighting (SW) and the castellation expansion (CE). These techniques can reduce the OOBE level, but the reduction rate of the OOBE level remains low and equal to that of the signals with the OFDM. In addition, use of these techniques leads to a significant degradation of the BER performance and an increase of the peak-to-average power ratio (PAPR) [1], [14].

One of the promising methods that will allow to reduce ΔF is a solution of the corresponding optimization problem. The article [16] deals with the problem of the OOBE reduction rate minimization. Formulation of the optimization problem, a functional form, the constraints on the OOBE reduction rate, and the envelope power are presented. It should be noted that this article is devoted to the problem of optimizing the shape of a single frequency signal. The decision is made by the use of Legendre polynomial. A possibility of obtaining a given OOBE reduction rate is given, the shapes of the envelopes are shown [16, Fig. 2]. However, as one can see, with an increase in the reduction rate of the OOBE level, the frequency bandwidth expands on the level of −30 dB [16, Fig. 3]. Also, it is demonstrated that in this method BER performance degrades relative to potential BER performance, i.e. BER performance of BPSK-signals. So, for the error probability of 10^{-2}, extra energy losses come to more than 3 dB. The question of applicability of these signals for multifrequency systems remains open.

Paper [17] proposes a similar approach. The difference lies in the application of the Fourier series. Formulation of the optimization problem is carried out with constraints on the reduction rate of the OOBE level and the power of the envelope signals. It is also shown that the frequency band occupied by the level of −40 dB expands [17, Table 1]. It should be noted that the BER performance for this case is not shown.

In [18] an optimization problem for minimization of the OOBE reduction rate is presented and solved. However, additional restrictions when solving the optimization problem are not imposed. The subcarriers location is orthogonal. Time and spectral characteristics of the obtained signals are examined. A possibility of reducing the occupied bandwidth with the transition to the smoothed envelope can be seen. One should note that in this case the BER performance is remarkably deteriorated. Smoothed envelopes, when used on each subcarrier frequency, lead to additional energy losses. These come to about 10 dB for the BER 10^{-3}, in the case of coherent bit-by-bit detection algorithm usage, and more than 3 dB in the case of using an algorithm based on decision diagram.

In this paper, we propose an integrated approach to reduce frequency bandwidth, while minimizing additional energy losses, by introducing a controlled ISI in the signals used on each subcarrier.

A natural way to reduce the bandwidth of the multifrequency signals is to increase the duration $T_c > T$ of signals on each subcarrier frequency. While maintaining the transmission rate, interference of signals in time appears, which leads to a decreased BER performance. However, in contrast to [12, 13], when solving an optimization problem, the interference can be controlled. The BER performance can be numerically determined through cross-correlation coefficient K_0 between the adjacent time signals, transmitted under ISI. Also, in this case the frequency plan of subcarriers location needs to be determined. It is proposed to choose the value of Δf as the minimum value for the solution of the equation $|S(\omega)|=0$, where $|S(\omega)|$ – a single frequency spectrum. In this case, mutual influence of adjacent subcarriers will be minimized. Nevertheless, the second and subsequent adjacent subcarriers will remain influential [8]. The degree of this influence will be small (at the level of -30 dB).

The solution of the optimization problem is searched for a given OOBE reduction rate with constraints for signal energy. The duration of the signal is selected Tc>T to reduce the occupied frequency band. Frequency interference is minimized through the use of the proposed frequency plan, and introduced controlled ISI is limited when solving the optimization problem.

2 Location of Subcarriers Frequency

Figure 1 shows the frequency plan for the number of subcarriers $N=3$ when using BPSK signal duration $T_c=2T$, transmitted at rate $R=1/T$. The signal duration and the waveform selection is taken as an example. Of course, the duration of the signal can be arbitrary, and the shape of the signal envelope $a(t)$ is determined by solving the optimization problem. Figure 1, along the axis of ordinate, indicates the value of the normalized module of the spectrum $|S(\omega)|^2/|S(0)|^2$ signals, and the abscissa represents the normalized frequency. The subcarrier frequencies are located on the horizontal axis at the points $(-0.75, 0, 0.75)$, corresponding to the minimum value of the equation solution $|S(\omega)|^2=0$.

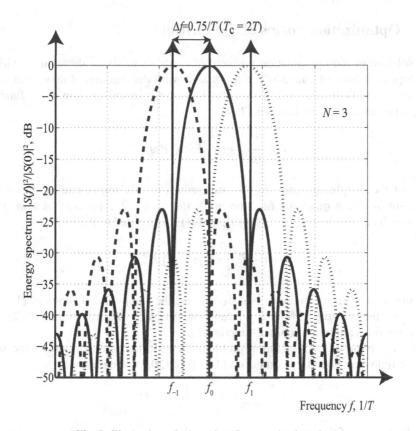

Fig. 1. Illustration of subcarriers frequencies location.

Mutual influence of signals on the same frequency increases with increasing T_c. One can show that the cross correlation coefficient K_0 can be calculated by the following formula:

$$K_0 = \max_{k=1\dots(L-1)} \left\{ \int_0^{(L-k)T} a(t)a(t-kT)dt \right\},$$

where $a(t)$ – the envelope of the signal duration LT. For example, K_0 is equal to 0.5 for the value $T_c=2T$, and $a(t) = \cos(\pi t/2T)$. To minimize the energy loss, value K_0 should be chosen as small as possible, at least not more than 10^{-2}.

Let us consider the solution of the optimization problem to minimize the occupied bandwidth, at which the value of K_0 is as small as possible. Thus, the BER performance will be close to the BER performance of the OFDM signals.

3   Optimization Forms of the Envelope

The solution of the optimization problem of synthesis of signal duration T_c with the envelope $a(t)$ and spectrum $S(\omega)$, in accordance with the criterion of maximum reduction rate of the OOBE level, is associated with the numerical solution of the functional type minimization problem [16,17,18]:

$$J = \frac{1}{2\pi} \int_{-\infty}^{+\infty} g(\omega)|S(\omega)|^2 \, d\omega. \tag{1}$$

Type of the weighting function $g(\omega)$ determines the reduction rate of the signal's spectrum $|S(\omega)|^2$. A quadratic function $g(\omega)=1/\omega^{2n}$ ($n=1, 2, ...$) is used as a weighting function. It can be shown [16, 17], [19], that (1) is converted to the form:

$$J = (-1)^n \int_{-\infty}^{+\infty} a(t)a^{(2n)}(t) \, dt, \tag{2}$$

wherein $a^{2n}(t)$ – $2n$-th derivative of $a(t)$. Thus, the optimization problem (1) transforms into the problem of finding the function $a(t)$ on the interval of time $[T_c/2; T_c/2]$, which provides a minimum of the functional (2).

While solving this optimization problem, a constraint on the energy of the signal and the following boundary conditions [16], [18], [20,21] are imposed:

$$a^{(k)}\Big|_{t=\pm T_c/2} = 0, \, k = 1...(n-1) \tag{3}$$

(All derivatives of the envelope $a(t)$ until the derivative $(n-1)$-th order of $a^{(n-1)}(t)$ does not have discontinuity, and $a^{(n)}(t)$ is finite on the whole time interval).

The function $a(t)$ is even on the interval $[-T_c/2; T_c/2]$, and it can be represented by the following expression [17,18]:

$$a(t) = \frac{a_0}{2} + \sum_{k=1}^{m} a_k \cos\left(\frac{2\pi}{T_c}kt\right), \tag{4}$$

where $a_0 = \dfrac{1}{T_c} \displaystyle\int_{-T_c/2}^{T_c/2} a(t)dt$, $a_k = \dfrac{1}{T_c} \displaystyle\int_{-T_c/2}^{T_c/2} a(t)\cos\left(\dfrac{2\pi}{T_c}kt\right)dt$, m – number of coefficient of the Fourier series.

Restriction on the signal energy with the use of (4) can be written as [18]:

$$E_c = \frac{T_c}{2}\left(\frac{a_0^2}{2} + \sum_{k=1}^{m} a_k^2\right) = 1. \tag{5}$$

Let us present the 2n-th derivative of the function $a(t)$ in the form:

$$a^{(2n)}(t) = (-1)^n \sum_{k=1}^{m} \left(\frac{2\pi}{T_c}k\right)^{2n} a_k \cos\left(\frac{2\pi}{T_c}kt\right). \tag{6}$$

Then, converting the functional (2) to the function of many variables, taking into account the existence of a signal on the interval $[-T_c/2; T_c/2]$, we have (Appendix A):

$$J\left(\{a_k\}_{k=1}^{m}\right) = \frac{T_c}{2} \sum_{k=1}^{m} \left(\frac{2\pi}{T_c}h\right)^{2n} a_k^2. \tag{7}$$

It should be noted that, while implementing this conversion, boundary conditions (3) are converted into additional constraint equations [9,10], [18]:

$$a(t)\big|_{t=\pm T_c/2} = \frac{a_0}{2} + \sum_{k=1}^{m}(-1)^k a_k = 0, \quad a^{(2k)}(t)\big|_{t=\pm T_c/2} = (-1)^k \sum_{k=1}^{m} a_k \left(\frac{2\pi}{T_c}k\right)^{2k} = 0;$$

$$a^{(2k-1)}(t)\big|_{t=\pm T_c/2} \equiv 0; \, k = 1,...,n. \tag{8}$$

Restriction on the cross-correlation coefficient K_0 can be presented as follows:

$$\max_{k=1...(L-1)} \left\{ \int_0^{(L-k)T} a(t)a(t-kT)dt \right\} < K_0 \quad (\text{for } T_c = LT). \tag{9}$$

A rectangular form of the envelope $a(t)$ is used as an initial approximation for the numerical solution of the optimization problem. Initially, restrictions on the reduction rate of the $|S(\omega)|^2$ level are imposed. After that, restrictions on the cross-correlation coefficient K_0 are imposed. Thus, if the target value of $K_0=0.1$, it is necessary to make a transition from the values $K_0=1$ to the value of $K_0=0.1$ with steps of 0.1. At each step of the initial approximation the result of the previous optimization step should be used.

4 The Results of Solving the Optimization Problem

Minimization of the values K_0 will minimize energy losses on the BER performance of the OFDM signals. For distinctness we consider the case $n=2$. For other values of n results can be approximated.

For the duration of the signal $T_c=2T$, the optimization results are shown in Table B.1 (Appendix B). Fig. 2 shows the forms of the envelope $a(t)$ (Fig. 2 (a)), the forms of the normalized spectrum $|S(\omega)|^2/|S(0)|^2$ (Fig. 2 (b)) and the used values of Δf. It is seen that for values $K_0<5\cdot10^{-3}$ the envelope $a(t)$ becomes symmetrical bipolar function, which leads to an increase of the occupied bandwidth. Accordingly, for the purpose of minimizing the occupied bandwidth optimally, the signals for which $K_0>5\cdot10^{-3}$ should be used. The values of the frequency band for this case are shown in Table 1.

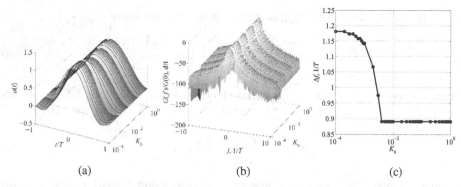

(a)   (b)   (c)

Fig. 2. Forms of the envelopes $a(t)$ (a), the corresponding spectra for $n=2$ (b) and the used values of Δf (c) at $T_c=2T$.

Table 1. Values of the frequency bandwidth on level −30 dB.

	$\Delta F_{-30\,dB}$ $1/T$			
OFDM	$K_0 = 10^{-2}$	$K_0 = 5 \cdot 10^{-3}$	$K_0 = 10^{-3}$	$K_0 = 5 \cdot 10^{-4}$
193	119	119	156	169
$1 - \dfrac{\Delta F_{-30dB}}{\Delta F_{-30dB}(\text{OFDM})}$	38%	38%	19%	12%

To investigate the BER performance of proposed signals, a simulation model was built in the Matlab. The algorithm of coherent bit-by-bit detection is used [22]. The number of the subcarriers equals $N=32$. The BER performance for $T_c=2T$ at different values of K_0 is shown in Fig. 3 as well. The abscissa is the signal-to-noise ratio $h_0^2=E_b/N_0$, where E_b – bit's energy, N_0 – one-sided power spectral density of AWGN.

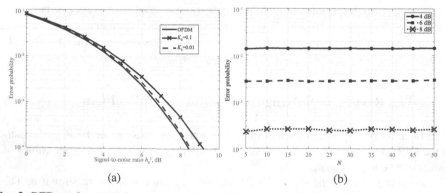

(a)   (b)

Fig. 3. BER performance for the case of using the optimal $T_c=2T$ signals for $N=32$ (a), the dependence of the error probability by N for different signal-to-noise ratio and $K_0=10^{-2}$ (b).

As seen from the comparison of the curves in Fig. 3 (a), for the values of $K_0<0.1$, the required value of h_0^2 is not higher than 9 dB at the error probability of $p=10^{-4}$. Difference at this point h_0^2, relative to the error probability of the OFDM signals, is not higher than 1 dB. Fig. 3 (a) shows dependence of the error probability by N for $K_0=10^{-2}$ and values of $h_0^2=4$, 6, 8 dB. As it can be seen from the comparison of these curves, dependence of the errors probability on N is not revealed. This is explained by the fact that the level of mutual interference of the energy spectrum components of signals, transmitted on adjacent subcarriers, is less than -30 dB.

5   Conclusions

Usage of the signals with the optimum envelope form and introduction of controlled intersymbol interference can reduce the occupied bandwidth. Thus, for the duration of the signal $T_c=2T$, and value of cross correlation coefficient $K_0=10^{-2}$, gain on an occupied bandwidth with respect to OFDM signals is equal to 38%. This makes it possible to proportionally increase the amount of transmitted information by increasing the number of subcarriers in the same frequency band.

Usage of the optimization procedure provides a higher OOBE reduction rate, compared to OFDM signals. The reduction rate is determined by terms of the optimization problem and can be selected within a wide range. Increase of the reduction rate allows to increase a number of group data channels in systems Wi-Fi, LTE by reducing protective frequency intervals between these channels.

Imposing of restrictions on a cross-correlation coefficient, while solving the optimization problem, allows to get signals that provide BER performance almost equal to the BER performance of the OFDM signals. Additional energy losses are less than 1 dB for error probability 10^{-4}. This result can not be obtained by conventional methods of signal processing in a presence of controlled interference.

6   Appendix A

Transition from functional J to multivariable function $J\left(\{a_k\}_{k=1}^m\right)$.

Applying expansion of $a(t)$ in a finite Fourier series functional J can be represented as follows:

$$J\left(\{a_k\}_{k=1}^m\right) = (-1)^n \int_{-T_c/2}^{T_c/2} \left[\frac{a_0}{2} + \sum_{k=1}^m a_k \cos\left(\frac{2\pi}{T_c}kt\right) \right] \times$$

$$\left[(-1)^n \sum_{h=1}^m \left(\frac{2\pi}{T_c}h\right)^{2n} a_h \cos\left(\frac{2\pi}{T_c}ht\right) \right] dt .$$

Let us open the brackets and divide the expression into two terms:

$$J\left(\{a_k\}_{k=1}^m\right) = I_1 + I_2 , \text{ where}$$

$$I_1 = \int_{-T_c/2}^{T_c/2} \frac{a_0}{2} \sum_{h=1}^{m} \left(\frac{2\pi}{T_c} h\right)^{2n} a_h \cos\left(\frac{2\pi}{T_c} ht\right) dt \, ,$$

$$I_2 = \int_{-T_c/2}^{T_c/2} \sum_{k=1}^{m} a_k \cos\left(\frac{2\pi}{T_c} kt\right) \sum_{h=1}^{m} \left(\frac{2\pi}{T_c} h\right)^{2n} a_h \cos\left(\frac{2\pi}{T_c} ht\right) dt \, .$$

In this case, the order of summation and integration can be changed. Then the first term:

$$I_1 = \int_{-T_c/2}^{T_c/2} \frac{a_0}{2} \sum_{h=1}^{m} \left(\frac{2\pi}{T_c} h\right)^{2n} a_h \cos\left(\frac{2\pi}{T_c} ht\right) dt = \frac{a_0}{2} \sum_{h=1}^{m} \left(\frac{2\pi}{T_c} h\right)^{2n} a_h \int_{-T_c/2}^{T_c/2} \cos\left(\frac{2\pi}{T_c} ht\right) dt = 0 \, .$$

The second term can be rewritten as follows:

$$I_2 = \frac{1}{2} \sum_{k=1}^{m} \sum_{h=1}^{m} \left(\frac{2\pi}{T_c} h\right)^{2n} a_k a_h \int_{-T_c/2}^{T_c/2} \cos\left(\frac{2\pi}{T_c} kt\right) \cos\left(\frac{2\pi}{T_c} ht\right) dt \, .$$

Using the known trigonometric formulas can be shown:

$$I_2 = \frac{1}{2} \sum_{k=1}^{m} \sum_{h=1}^{m} \left(\frac{2\pi}{T_c} h\right)^{2n} a_k a_h \int_{-T_c/2}^{T_c/2} \left[\cos\left(\frac{2\pi}{T_c}(k-h)t\right) + \cos\left(\frac{2\pi}{T_c}(k+h)t\right)\right] dt =$$

$$= \frac{1}{2} \sum_{k=1}^{m} \sum_{h=k}^{m} \left(\frac{2\pi}{T_c} h\right)^{2n} a_k^2 \int_{-T_c/2}^{T_c/2} \left[1 + \cos\left(\frac{2\pi}{T_c}(2k)t\right)\right] dt = \frac{T_c}{2} \sum_{k=1}^{m} \left(\frac{2\pi}{T_c} h\right)^{2n} a_k^2 \, .$$

Thus, the transition is made to the problem of minimizing a function of several variables:

$$J\left(\{a_k\}_{k=1}^{m}\right) = \frac{T_c}{2} \sum_{k=1}^{m} \left(\frac{2\pi}{T_c} h\right)^{2n} a_k^2 \, .$$

7   Appendix B

The Expansion Coefficients of the Envelope $a(t)$.
It should be noted that the accuracy of the representation of the function $a(t)$ depends on a number of terms of the finite Fourier series. The number of terms in the Fourier series is defined by means of the standard deviation between the calculation of the function $a_m(t)$ and $a_{m-1}(t)$, which have in the expansion of m and $m-1$ terms of the Fourier series:

$$\varepsilon(m) = \sqrt{\int_{-T/2}^{T/2} \left(a_m(t) - a_{m-1}(t)\right)^2 dt} \, .$$

If we restrict the number of terms of the Fourier series to $m=18$, standard deviation value ε is less than $2 \cdot 10^{-3}$.

Solution of the problem of finding an optimal shape of the envelope $a(t)$, in the form of expansion coefficients of the Fourier series, are presented in Table B.1.

Table B.1. The expansion coefficients of the envelope $a(t)$ for different values of K_0 and $n=2$, $T_c=2T$.

K_0	0,01	0,005	0,001
a_0	1,175001	1,175001	1,036883
a_1	0,555952	0,555952	0,670313
a_2	−0,02407	−0,02407	0,111241
a_3	0,004678	0,004678	−0,02588
a_4	−0,00148	−0,00148	0,00773
a_5	0,000604	0,000604	−0,00321
a_6	−0,00029	−0,00029	0,001536
a_7	0,000157	0,000157	−0,00083
a_8	−9,21E-05	−9,22E-05	0,000487
a_9	5,75E-05	5,75E-05	−0,0003
a_{10}	−3,77E-05	−3,77E-05	0,000199
a_{11}	2,58E-05	2,57E-05	−0,00014
a_{12}	−1,82E-05	−1,82E-05	9,61E-05
a_{13}	1,32E-05	1,32E-05	−6,98E-05
a_{14}	−9,82E-06	−9,85E-06	5,19E-05
a_{15}	7,43E-06	7,42E-06	−3,94E-05
a_{16}	−5,78E-06	−5,73E-06	3,04E-05
a_{17}	4,50E-06	4,53E-06	−2,39E-05

References

1. Selim, A., Doyle, L.: Improved out-of-band emissions reduction for OFDM systems. In: Military Communications Conference, MILCOM 2013, pp. 107–111. IEEE, November 18–20, 2013
2. Selim, A., Doyle, L.: A method for reducing the out-of-band emissions for OFDM systems. In: Wireless Communications and Networking Conference (WCNC), pp. 730–734. IEEE (2014)
3. Weinstein, S., Ebert, P.: Data transmission by frequency-division multiplexing using the discrete Fourier transform. IEEE Transactions on communications **19**(5), 628–634 (1971)
4. Rodrigues, M., Darwazeh, I.: A spectrally efficient frequency division multiplexing based communications system. In: Proc. 8th Int. OFDM Workshop, Hamburg, pp. 48–49 (2003)
5. Xu, T., Darwazeh, I.: Spectrally efficient FDM: spectrum saving technique for 5G? In: 2014 1st International Conference on 5G for Ubiquitous Connectivity (5GU), pp. 273–278, November 26–28, 2014
6. Kislitsyn, A.B., Rashich, A.V., Tan, N.N.: Generation of SEFDM-signals using FFT/IFFT. In: Balandin, S., Andreev, S., Koucheryavy, Y. (eds.) NEW2AN/ruSMART 2014. LNCS, vol. 8638, pp. 488–501. Springer, Heidelberg (2014)
7. Kanaras, Y., Chorti, A., Rodrigues, M., Darwazeh, I.: An overview of optimal and sub-optimal detection techniques for a non orthogonal spectrally efficient FDM. In: LCS/NEMS 2009, London, UK, September 3–4, (2009)

8. Isam, S., Darwazeh, I.: Characterizing the intercarrier interference of non-orthogonal spectrally efficient FDM system. In: 2012 8th International Symposium on Communication Systems, Networks & Digital Signal Processing (CSNDSP), pp. 1–5, July 18–20, 2012

9. Hamamura, M., Tachikawa, S.: Bandwidth efficiency improvement for multi-carrier systems. In: 15th IEEE International Symposium on Personal, Indoor and Mobile Radio Communications, PIMRC 2004, vol. 1, pp. 48–52, September 2004

10. Karampatsis, D., Rodrigues, M.R.D., Darwazeh, I.: Implications of linear phase dispersion on OFDM and Fast-OFDM systems. In: London Communications Symposium, pp. 117–120 (2002)

11. Vahlin, A., Holte, N.: Optimal Finite Duration Pulses for OFDM. IEEE Trans. on Comm. **44**(1), 10–14 (1996)

12. Baas, N.J., Taylor, D.P.: Pulse shaping for wireless communication over time- or frequency-selective channels. IEEE Trans. on Comm. **52**(9), 1477–1479 (2004)

13. Sriram, S., Vijajakumar, N., Aditya Kumar, P., Shetty, A.S., Prasshayth, V.P., Narayanankutty, K.A.: Spectrally efficient multy-carrier modulation using Gabor transform. Wireless Engineering and Technology **4**, 112–116 (2013)

14. Selim, A., Doyle, L.: Real-time sidelobe suppression for OFDM systems using advanced subcarrier weighting. In: Wireless Communications and Networking Conference (WCNC), pp. 4043–4047. IEEE (2013)

15. Selim, A., et al.: Efficient sidelobe suppression for OFDM systems with peak-to-average power ratio reduction. In: IEEE International Symposium on Dynamic Spectrum Access Networks (DYSPAN), pp. 510–516 (2012)

16. Xue, W., Ma, W., Chen, B.: Research on a realization method of the optimized efficient spectrum signals using Legendre series. In: 2010 IEEE International Conference on Wireless Communications, Networking and Information Security (WCNIS), pp. 155–159, June 25–27, 2010

17. Xue, W., Ma, W., Chen, B.: A realization method of the optimized efficient spectrum signals using Fourier series. In: 6th International Conference on Wireless Communications Networking and Mobile Computing (WiCOM), pp. 1–4, September 23–25, 2010

18. Zavjalov, S.V., Makarov, S.B., Volvenko, S.V.: Application of optimal spectrally efficient signals in systems with frequency division multiplexing. In: Balandin, S., Andreev, S., Koucheryavy, Y. (eds.) NEW2AN/ruSMART 2014. LNCS, vol. 8638, pp. 676–685. Springer, Heidelberg (2014)

19. Shasha, Y., Shengli, L., Guomin, L.: FQPSK modulation characteristic and performance simulation. Radio Engineering of China **38**(2), 35–37 (2008)

20. Kabal, P., Pasupathy, S.: Partial-response signaling. IEEE Transactions on communications **com-23**(9), 921–934 (1975)

21. Bennett, W.R.: Introduction to signal transmission. New (Lett.) **57**, 701–702 (1969)

22. Zavjalov, S.V., Makarov, S.B., Volvenko, S.V.: Nonlinear coherent detection algorithms of nonorthogonal multifrequency signals. In: Balandin, S., Andreev, S., Koucheryavy, Y. (eds.) NEW2AN/ruSMART 2014. LNCS, vol. 8638, pp. 703–713. Springer, Heidelberg (2014)

Analyze of Quantum Fourier Transform Circuit Implementation

Ivan Murashko and Constantine Korikov[✉]

Peter the Great St. Petersburg Polytechnic University, St. Petersburg 195251, Russia
ivan.murashko@gmail.com, korikov.constantine@spbstu.ru

Abstract. Quantum Fourier transform circuit is the key element of quantum computation. Originally it was introduced in the Shor's paper containing a classical proof of correctness. We propose another corroboration that shows the role of each element of the circuit and can be used for the scheme analyze.

Keywords: Quantum fourier transform · Shor's algorithm · Quantum computing

1 Introduction

Discrete Fourier transform (DFT) can be used for periodic sequence (functions) analysis. The transformation is defined by the following equation:

$$\tilde{X}_k = \sum_{m=0}^{M-1} x_m e^{-i\frac{2\pi}{M}k \cdot m}, \tag{1}$$

where the input sequence $\{x_m\}$ has M members.

Shor [1] suggested a quantum circuit (see Fig. 1d) that can be used for DFT in quantum algorithms [1–4] especially in the well known Shor's algorithm for factorization problem solving [1].

Quantum Fourier transform [1] operates with the following states

$$|x\rangle = \sum_{k=0}^{M-1} x_k |k\rangle, \tag{2}$$

where the sequence of amplitudes $\{x_k\}$ forms initial sequence for the Fourier transform (1). The basis vector $|k\rangle$ keeps the number of a sequence's element.

The state (2) should be normalized:

$$\sum_k |x_k|^2 = 1.$$

© Springer International Publishing Switzerland 2015
S. Balandin et al. (Eds.): NEW2AN/ruSMART 2015, LNCS 9247, pp. 647–654, 2015.
DOI: 10.1007/978-3-319-23126-6_58

Lets an operator $\hat{F}^M$ (quantum Fourier operator) transforms basis vector $|k\rangle$ by means of a rule defined by equation (1):

$$\hat{F}^M |k\rangle = \frac{1}{\sqrt{M}} \sum_{j=0}^{M-1} e^{-i\omega kj} |j\rangle_{inv},\tag{3}$$

where $\omega = \frac{2\pi}{M}$. Basis vectors $\{|k\rangle\}$ and $\{|k\rangle_{inv}\}$ are the same set of vectors which are enumerated in a different order (see Fig. 1a and Example 1).

From (2) and (3) one can get

$$\hat{F}^M |x\rangle = \sum_{j=0}^{M-1} x_k \hat{F}^M |k\rangle =$$

$$= \frac{1}{\sqrt{M}} \sum_{k=0}^{M-1} \sum_{j=0}^{M-1} e^{-i\omega kj} x_k |j\rangle_{inv} =$$

$$= \sum_{j=0}^{M-1} \left\{ \frac{1}{\sqrt{M}} \left(\sum_{k=0}^{M-1} e^{-i\omega kj} x_k \right) \right\} |j\rangle_{inv} =$$

$$= \sum_{j=0}^{M-1} \tilde{X}_j |j\rangle_{inv} = \left| \tilde{X} \right\rangle_{inv},\tag{4}$$

where

$$\tilde{X}_j = \tilde{X}_j^M = \frac{1}{\sqrt{M}} \sum_{k=0}^{M-1} e^{-i\omega kj} x_k.\tag{5}$$

The equation (5) conforms the classical ones that is used for discrete Fourier transform (1), i.e. one can write

$$|x\rangle \longleftrightarrow \left| \tilde{X} \right\rangle_{inv}.$$

We are going to show that the circuit shown on Fig. 1d does the transformation defined by (5). The subject of analyze consists of several elements: quantum Fourier transform gate for higher bits ($\hat{F}^{\frac{M}{2}}$), phase shift and Hadamard gate (see Fig. 1). If the elements are added to the circuit step by step then the different transformations of initial data are performed. The transformations are described below in details.

2 Input Data

Lets the input data for the system is a quantum state (2) that is a superposition of M basis states $\{|k\rangle\}$ (see Fig. 1a). Lets the number of basis states is a power of 2, i.e. a basis state can be present as a tensor product of $n = \log_2 M$ qbits:

$$|k\rangle = \left| a_0^{(k)} \right\rangle \otimes \left| a_1^{(k)} \right\rangle \otimes \cdots \otimes \left| a_{n-1}^{(k)} \right\rangle,\tag{6}$$

where

$$k = a_0^{(k)} + 2^1 a_1^{(k)} + \cdots + 2^{n-1} a_{n-1}^{(k)},$$
$$a_i^{(k)} \in \{0, 1\}.$$

There is a superposition of M basis states $\{|j\rangle_{inv}\}$ on the input (see Fig. 1a). For the state $|j\rangle_{inv}$ one can get

$$|j\rangle_{inv} = \left| b_{n-1}^{(j)} \right\rangle \otimes \left| b_{n-2}^{(j)} \right\rangle \otimes \cdots \otimes \left| b_0^{(j)} \right\rangle, \tag{7}$$

where

$$j = b_0^{(j)} + 2^1 b_1^{(j)} + \cdots + 2^{n-1} b_{n-1}^{(j)},$$
$$b_i^{(j)} \in \{0, 1\}.$$

Example 1. Input data] Let $M = 8$ i.e. $n = 3$. In the case the number 3 is represented as follows

$$|3\rangle = |1\rangle \otimes |1\rangle \otimes |0\rangle.$$

and

$$|3\rangle_{inv} = |0\rangle \otimes |1\rangle \otimes |1\rangle.$$

Using well known CooleyTukey fast Fourier transform algorithm [5] the (1) can be rewritten in the following form

$$\tilde{X}_k - \sum_{m=0}^{\frac{M}{2}-1} \Gamma_{k,m}^{\frac{M}{2}} x_{2m} + \exp\left(-2\pi i \frac{k}{M}\right) \sum_{m=0}^{\frac{M}{2}-1} \Gamma_{k,m}^{\frac{M}{2}} x_{2m+1}. \tag{8}$$

We can see (8) that if an input signal x consists of $n = \log_2 M$ bits when bit $a_0^{(k)}$ can be used to choose even (the first part of sum (8)) or odd (the second part of sum (8)) members.

Thus the state (2) can be presented in the form of sum of even and odd components:

$$|x\rangle = \sum_{k=0}^{M-1} x_k |k\rangle = \sum_{k=0}^{M-1} x_k \left| a_0^{(k)} \right\rangle \otimes \left| a_1^{(k)} \right\rangle \otimes \cdots \otimes \left| a_{n-1}^{(k)} \right\rangle =$$

$$= \sum_{m=0}^{\frac{M}{2}-1} x_{k=2m} |0\rangle \otimes \left| a_1^{(k)} \right\rangle \otimes \cdots \otimes \left| a_{n-1}^{(k)} \right\rangle +$$

$$+ \sum_{m=0}^{\frac{M}{2}-1} x_{k=2m+1} |1\rangle \otimes \left| a_1^{(k)} \right\rangle \otimes \cdots \otimes \left| a_{n-1}^{(k)} \right\rangle =$$

$$= \sum_{m=0}^{\frac{M}{2}-1} x_{k=2m} |0\rangle \otimes |m\rangle + \sum_{m=0}^{\frac{M}{2}-1} x_{k=2m+1} |1\rangle \otimes |m\rangle =$$

$$= \sum_{m=0}^{\frac{M}{2}-1} x_{2m} |2m\rangle + \sum_{m=0}^{\frac{M}{2}-1} x_{2m+1} |2m+1\rangle,$$

where

$$m = a_1^{(k)} + 2^1 a_2^{(k)} + \cdots + 2^{n-2} a_{n-1}^{(k)}. \tag{9}$$

3   Fourier Transform for High Bits

With Fourier transform for high bits only in $\hat{F}^{\frac{M}{2}}$, i.e. with excluding $a_0^{(k)}$ we can obtain (see Fig. 1b):

$$|x\rangle \rightarrow \hat{F}^{\frac{M}{2}} \sum_{m=0}^{\frac{M}{2}-1} x_{2m} |2m\rangle + \hat{F}^{\frac{M}{2}} \sum_{m=0}^{\frac{M}{2}-1} x_{2m+1} |2m+1\rangle =$$

$$= \hat{F}^{\frac{M}{2}} \sum_{m=0}^{\frac{M}{2}-1} x_{2m} |0\rangle \otimes |m\rangle + \hat{F}^{\frac{M}{2}} \sum_{m=0}^{\frac{M}{2}-1} x_{2m+1} |1\rangle \otimes |m\rangle =$$

$$= \sum_{m=0}^{\frac{M}{2}-1} x_{2m} |0\rangle \otimes \hat{F}^{\frac{M}{2}} |m\rangle + \sum_{m=0}^{\frac{M}{2}-1} x_{2m+1} |1\rangle \otimes \hat{F}^{\frac{M}{2}} |m\rangle. \tag{10}$$

Using (3) it can be got

$$\hat{F}^{\frac{M}{2}} |m\rangle = \sqrt{\frac{2}{M}} \sum_{j=0}^{\frac{M}{2}-1} e^{-i\frac{4\pi}{M} mj} |j\rangle_{inv}. \tag{11}$$

Thus for (10) it holds

$$|x\rangle \rightarrow \sum_{m=0}^{\frac{M}{2}-1} x_{2m} |0\rangle \otimes \hat{F}^{\frac{M}{2}} |m\rangle + \sum_{m=0}^{\frac{M}{2}-1} x_{2m+1} |1\rangle \otimes \hat{F}^{\frac{M}{2}} |m\rangle =$$

$$= \sqrt{\frac{2}{M}} \sum_{j=0}^{\frac{M}{2}-1} e^{-i\frac{4\pi}{M} mj} \sum_{m=0}^{\frac{M}{2}-1} x_{2m} |0\rangle \otimes |j\rangle_{inv} +$$

$$+ \sqrt{\frac{2}{M}} \sum_{j=0}^{\frac{M}{2}-1} e^{-i\frac{4\pi}{M} mj} \sum_{m=0}^{\frac{M}{2}-1} x_{2m+1} |1\rangle \otimes |j\rangle_{inv} =$$

$$= \sum_{j=0}^{\frac{M}{2}-1} \left(\sqrt{\frac{2}{M}} \sum_{m=0}^{\frac{M}{2}-1} e^{-i\frac{4\pi}{M} mj} x_{2m} \right) |j\rangle_{inv} +$$

$$+ \sum_{j=0}^{\frac{M}{2}-1} \left(\sqrt{\frac{2}{M}} \sum_{m=0}^{\frac{M}{2}-1} e^{-i\frac{4\pi}{M} mj} x_{2m+1} \right) \left| \frac{M}{2} + j \right\rangle_{inv} =$$

$$= \sum_{j=0}^{\frac{M}{2}-1} \tilde{A}_j |j\rangle_{inv} + \sum_{j=0}^{\frac{M}{2}-1} \tilde{B}_j \left| \frac{M}{2} + j \right\rangle_{inv}, \tag{12}$$

where

$$\tilde{A}_j = \sqrt{\frac{2}{M}} \sum_{m=0}^{\frac{M}{2}-1} e^{-i\frac{4\pi}{M}mj} x_{2m}$$

$$\tilde{B}_j = \sqrt{\frac{2}{M}} \sum_{m=0}^{\frac{M}{2}-1} e^{-i\frac{4\pi}{M}mj} x_{2m+1} \tag{13}$$

4   Phase Shift

As we can see the equations (12, 13) do not correspond to the required Fourier transform (4). One of the problem is that components x_{2m} and x_{2m+1} in (13) have the same phase but it should differ. The problem can be fixed if a phase shift is added for odd components i.e. for elements with $a_0^k = 1$. As result the circuit shown on Fig. 1c is got:

$$|x\rangle \rightarrow \hat{F}^{\frac{M}{2}} \sum_{m=0}^{\frac{M}{2}-1} x_{2m} |2m\rangle + \hat{R}\hat{F}^{\frac{M}{2}} \sum_{m=0}^{\frac{M}{2}-1} x_{2m+1} |2m+1\rangle =$$

$$= \sum_{j=0}^{\frac{M}{2}-1} \tilde{A}_j |j\rangle_{inv} + \sum_{j=0}^{\frac{M}{2}-1} \tilde{B}_j \hat{R} \left| \frac{M}{2} + j \right\rangle_{inv},$$

$$= \sum_{j=0}^{\frac{M}{2}-1} \tilde{A}_j |j\rangle_{inv} + \sum_{j=0}^{\frac{M}{2}-1} \tilde{C}_j \left| \frac{M}{2} + j \right\rangle_{inv}. \tag{14}$$

With equation

$$\hat{R}_l \left| b_l^{(j)} \right\rangle = cxp\left(2\pi i \frac{b_l^{(j)}}{2^{n-l}} \right) \left| b_l^{(j)} \right\rangle$$

we can get that operator $\hat{R}$ applied to state $\left| \frac{M}{2} + j \right\rangle_{inv}$ will produce the following result:

$$\hat{R} \left| \frac{M}{2} + j \right\rangle_{inv} = \hat{R} |1\rangle \otimes |j\rangle_{inv} =$$

$$= |1\rangle \otimes \hat{R}_0 \left| b_0^{(j)} \right\rangle \otimes \cdots \otimes \hat{R}_{n-2} \left| b_{n-2}^{(j)} \right\rangle =$$

$$= \prod_{l=0}^{n-2} exp\left(-2\pi i \frac{2^l b_l^{(j)}}{2^n} \right) |1\rangle \otimes |j\rangle_{inv} =$$

$$= exp\left(-2\pi i \frac{j}{M} \right) \left| \frac{M}{2} + j \right\rangle_{inv} \tag{15}$$

We used the following fact

$$j = b_0^{(j)} + 2^1 b_1^{(j)} + \cdots + 2^{n-2} b_{n-2}^{(j)}$$

for (15) derivation.

Thus for $\tilde{C}_j$ in (14) it can be got

$$\tilde{C}_j = \sqrt{\frac{2}{M}} \sum_{m=0}^{\frac{M}{2}-1} e^{-2\pi i \frac{j}{M}} e^{-i\frac{4\pi}{M}mj} x_{2m+1} =$$

$$= \sqrt{\frac{2}{M}} \sum_{m=0}^{\frac{M}{2}-1} e^{-i\frac{2\pi}{M}(2m+1)j} x_{2m+1} \qquad (16)$$

5   Hadamard Transformation for the Lowest Bit

The result after phase shift still does not produce the required (Fourier) transform and additional manipulations are required. Especially the result of sum for even components is that required. Another situation with odd ones where sum should be done with additional phase shift. The required transformation is provided by Hadamard gate.

If the Hadamard transformation is applied for the qbit $|a_0\rangle$ then the circuit shown on Fig. 1d is got. The initial state is transformed by the following law:

$$|x\rangle \rightarrow \hat{H}\hat{F}^{\frac{M}{2}} \sum_{m=0}^{\frac{M}{2}-1} x_{2m} |2m\rangle + \hat{H}\hat{R}\hat{F}^{\frac{M}{2}} \sum_{m=0}^{\frac{M}{2}-1} x_{2m+1} =$$

$$= \sum_{j=0}^{\frac{M}{2}-1} \tilde{A}_j \hat{H} |0\rangle \otimes |j\rangle_{inv} + \sum_{j=0}^{\frac{M}{2}-1} \tilde{C}_j \hat{H} |1\rangle \otimes |j\rangle_{inv} =$$

$$= \frac{1}{\sqrt{2}} \sum_{j=0}^{\frac{M}{2}-1} \tilde{A}_j (|0\rangle + |1\rangle) \otimes |j\rangle_{inv} + \frac{1}{\sqrt{2}} \sum_{j=0}^{\frac{M}{2}-1} \tilde{C}_j (|0\rangle - |1\rangle) \otimes |j\rangle_{inv} =$$

$$= \sum_{j=0}^{\frac{M}{2}-1} \frac{\tilde{A}_j + \tilde{C}_j}{\sqrt{2}} |0\rangle \otimes |j\rangle_{inv} + \sum_{j=0}^{\frac{M}{2}-1} \frac{\tilde{A}_j - \tilde{C}_j}{\sqrt{2}} |1\rangle \otimes |j\rangle_{inv} =$$

$$= \sum_{j=0}^{\frac{M}{2}-1} \frac{\tilde{A}_j + \tilde{C}_j}{\sqrt{2}} |j\rangle_{inv} + \sum_{j=0}^{\frac{M}{2}-1} \frac{\tilde{A}_j - \tilde{C}_j}{\sqrt{2}} \left| \frac{M}{2} + j \right\rangle_{inv}. \qquad (17)$$

For (17) members using (13) and (16) one can conclude:

$$\frac{\tilde{A}_j + \tilde{C}_j}{\sqrt{2}} = \sqrt{\frac{1}{M}} \sum_{m=0}^{\frac{M}{2}-1} e^{-i\frac{4\pi}{M}mj} x_{2m} + \sqrt{\frac{1}{M}} \sum_{m=0}^{\frac{M}{2}-1} e^{-i\frac{2\pi}{M}(2m+1)j} x_{2m+1} =$$

$$= \sqrt{\frac{1}{M}} \sum_{m=0}^{M-1} e^{-i\frac{2\pi}{M}mj} x_m \qquad (18)$$

and

$$\frac{\tilde{A}_j - \tilde{C}_j}{\sqrt{2}} = \sqrt{\frac{1}{M}} \sum_{m=0}^{\frac{M}{2}-1} e^{-i\frac{4\pi}{M}mj} x_{2m} - \sqrt{\frac{1}{M}} \sum_{m=0}^{\frac{M}{2}-1} e^{-i\frac{2\pi}{M}(2m+1)j} x_{2m+1} =$$

$$= \sqrt{\frac{1}{M}} \sum_{m=0}^{M-1} e^{-i\frac{2\pi}{M}mj} x_m \frac{1+e^{-i\pi m}}{2} - \sqrt{\frac{1}{M}} \sum_{m=0}^{M-1} e^{-i\frac{2\pi}{M}mj} x_m \frac{1-e^{-i\pi m}}{2} =$$

$$= \sqrt{\frac{1}{M}} \sum_{m=0}^{M-1} e^{-i\frac{2\pi}{M}mj} e^{-i\pi m} x_m = \sqrt{\frac{1}{M}} \sum_{m=0}^{M-1} e^{-i\frac{2\pi}{M}mj} e^{-i\frac{2\pi}{M}m\frac{M}{2}} x_m =$$

$$= \sqrt{\frac{1}{M}} \sum_{m=0}^{M-1} e^{-i\frac{2\pi}{M}m(\frac{M}{2}+j)} x_m \quad (19)$$

Combine (17), (18) and (19) finally it can be got

$$|x\rangle \rightarrow \sum_{j=0}^{\frac{M}{2}-1} \sqrt{\frac{1}{M}} \sum_{m=0}^{M-1} e^{-i\frac{2\pi}{M}mj} x_m |j\rangle_{inv} +$$

$$+ \sum_{j=0}^{\frac{M}{2}-1} \sqrt{\frac{1}{M}} \sum_{m=0}^{M-1} e^{-i\frac{2\pi}{M}m(\frac{M}{2}+j)} x_m \left|\frac{M}{2}+j\right\rangle_{inv} =$$

$$= \sum_{j=0}^{M-1} \tilde{X}_j^M |j\rangle_{inv}$$

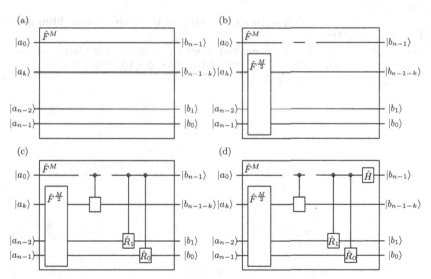

Fig. 1. The quantum Fourier transform circuit based on the fast Fourier transform algorithm. Initial data on the input (a). Step 1 (b): $|x\rangle \rightarrow \hat{F}^{\frac{M}{2}} \sum_{m=0}^{\frac{M}{2}-1} x_{2m} |2m\rangle +$ $\hat{F}^{\frac{M}{2}} \sum_{m=0}^{\frac{M}{2}-1} x_{2m+1} |2m+1\rangle$. Step 2 (c): $|x\rangle \rightarrow \hat{F}^{\frac{M}{2}} \sum_{m=0}^{\frac{M}{2}-1} x_{2m} |2m\rangle + \hat{R}\hat{F}^{\frac{M}{2}}$ $\sum_{m=0}^{\frac{M}{2}-1} x_{2m+1}$. Final circuit (d).

Thus the final scheme (Fig. 1d) provides us the required Fourier transformation (4).

6   Conclusion

As a result of the scheme analyze the proof of correctness was got. The scheme analyze was divided into several steps. The first one included Fourier transform for highest bits only. Unfortunately the result state did not conform the requested one. Especially a phase is not correct. The issue can be fixed with phase shift correction elements that were added in the second step. As result the correct Fourier transform was got for even components but not for odd ones. The problem can be resolved by means of the Hadamard gate that was added on the last step. It was shown that the final circuit performs the requested Fourier transform.

References

1. Shor, P.: Algorithms for quantum computation: discrete logarithms and factoring. In: FOCS, pp. 124–134 (1994)
2. Yu, K.A.: Quantum measurements and the Abelian Stabilizer Problem, arXiv:quant-ph/9511026
3. Lanyon, B.P., Weinhold, T.J., Langford, N.K., Barbieri, M., James, D.F.V., Gilchrist, A., White, A.G.: Experimental Demonstration of a Compiled Version of Shor's Algorithm with Quantum Entanglement. Physical Review Letters **99**(25), 250505 (2007)
4. Ettinger, M., Hoyer, P.: A quantum observable for the graph isomorphism problem, arXiv:quant-ph/9901029
5. Cooley, J.W., Tukey, J.W.: An algorithm for the machine calculation of complex Fourier series. Math. Comput. **19**, 297–301 (1965). doi:10.2307/2003354

On the Synthesis of Optimal Finite Pulses for Bandwidth and Energy Efficient Single-Carrier Modulation

Aleksandr Gelgor[(✉)], Anton Gorlov, and Evgenii Popov

Peter the Great St. Petersburg Polytechnic University, St. Petersburg, Russia
a_gelgor@mail.ru, anton.gorlov@yandex.ru, eugapop@gmail.com

Abstract. Proposed in this paper, multicomponent signals represent a new versatile tool for description of single- and multicarrier signals with amplitude-phase shift keying (APSK). Each component carries a sequence of symbols, which have increased symbol transmission time. Neighboring pulses do not overlap inside the component; hence, each component has zero inter symbol interference (ISI). The value of peak to average power ratio (PAPR) and the ISI measure are accurately calculated for the assembled multicomponent signal, which is represented as the sum of components. We demonstrated the pulse shape optimization to provide a narrow spectrum of single-carrier multicomponent signal with the PAPR value and the ISI measure included as additional constraints. By minimization of the occupied bandwidth for the fixed energy concentration, we achieved a comparable efficiency for the multicomponent signals with QPSK as for conventional orthogonal signals with QPSK, 16-QAM or 64-QAM. Moreover, proposed optimal multicomponent signals clearly outperform in terms of modified energy consumptions, which include a signal PAPR value into account.

Keywords: Partial response signaling · Multicomponent signals · Bandwidth efficiency · Energy efficiency

Introduction

Development of new advanced modulation techniques for physical layers of data transmission systems requires a variety of parameters to be taken into account. On the one hand, the increase of data transmission rate up to the channel capacity is required. On the other hand, the system should meet the set of practical requirements. They are a spectral mask, an allowed PAPR and computational complexity of generation and detection algorithms, which all define the cost and the physical size of devices.

Development of digital modulation techniques originated from frequency shift keying providing the minimal PAPR, then turned to APSK, which provided higher bandwidth efficiency, and finally turned to spectrum spreading techniques such as CDMA and OFDM allowing effective data transmission in multipath fading channels.

Up to the present, the majority of data transmission systems with APSK, at first, use orthogonal signals thus simplifying a detection; at second, they meet constraints on the spectral mask and PAPR value by filtering and clipping of the output signal.

© Springer International Publishing Switzerland 2015
S. Balandin et al. (Eds.): NEW2AN/ruSMART 2015, LNCS 9247, pp. 655–668, 2015.
DOI: 10.1007/978-3-319-23126-6_59

At the same time, many papers propose using partial response (non-orthogonal) signals, allowing a reduction of out-of-band emissions and PAPR.

The pioneer work dedicated to partial response signaling was written by Mazo [1]. A proposed technique called "Faster than Nyquist Signaling" introduces the ISI by increasing the symbol rate with respect to the standard value, providing orthogonality of neighboring pulses. The "Faster than Nyquist" approach was actively examined for many applications including OFDM-signals [2].

A lot of papers propose reduction of out-of-band emissions of single-carrier APSK schemes by using special pulses, or PAPR reduction by using modified schemes such as OQPSK and π/4-QPSK [3] or F-QPSK [4]. These papers generally propose particular solutions. Nevertheless, the best characteristics can be provided by using of optimal pulses, obtained as solutions of the optimization problem, which takes all required characteristics into account. Formulation of such optimization problem and description of the solving technique represent the subject of our research.

For instance, optimal bandwidth-efficient pulses for single-carrier APSK modulation with ISI were represented in [5]. The considered approach uses the criterion of maximal free Euclidean distance between partial response signals; and one of the additional constraints provides the needed bandwidth energy concentration (99% or 99.9%) in a chosen bandwidth. The key advantage of this approach is that the problem of pulse search is solved in terms of linear program. However, including the PAPR value as linear constraint is impossible. Also, the use of another measure of signal bandwidth is rather difficult.

This paper proposes a new approach to the synthesis of optimal pulses for APSK signals. The approach implies the pulse optimization using the criterion of minimal bandwidth; the ISI measure and the PAPR value are included as additional constraints. We also propose a new concept called "multicomponent signals" for APSK signals representation. The weakness of our approach is that it implies dealing with a nonlinear optimization problem.

1   Multicomponent Signals

A baseband APSK signal carrying N symbols can be written as follows:

$$y_N(t) = \sum_{k=0}^{N-1} C_r^{(k)} a(t - kT)$$ (1)

where $C_r^{(k)}$ is a complex-valued symbol of kth time interval; symbols are uniformly chosen from any signal constellation of size M; the index $r = 1, 2, ..., M$ indicates a symbol number inside the constellation; $a(t)$ is some pulse; and T is a symbol time interval. For APSK signaling, the conventional solution is using of so called Nyquist pulses (or root-raised cosine (RRC) pulses):

$$a(t) = \frac{1}{T^{1/2}} \frac{\sin\{(\pi t / T)(1 - \alpha)\} + (4\alpha t / T)\cos\{(\pi t / T)(1 + \alpha)\}}{(\pi t / T)\{1 - (4\alpha t / T)^2\}}$$ (2)

where $0 \leq \alpha \leq 1$ is a roll-off factor. This formula defines infinite pulses, but in practice finite truncated pulses are actually used. In addition, a filtering of output signal or using of windowed pulses can take place in order to meet practical requirements.

In this work, we propose the use of finite pulses thus avoiding additional losses caused by implementation:

$$a(t) = 0 \text{ if } |t| > T/2. \tag{3}$$

Extended pulses can be obtained by scaling of the initial one:

$$a(t, L) = L^{-1/2} a(t / L), \tag{4}$$

where L is a pulse duration as a number of time intervals, the $L^{-1/2}$ factor makes pulse energy independent from the L value, that will be convenient in a further stage of optimization problem formulation. Notice, if the pulse duration is equal to the symbol time interval ($L = 1$), then neighboring pulses do not overlap and the signal has zero ISI. Nevertheless, extended smooth pulses are employed to obtain a narrow bandwidth in practice. Therefore, L is greater than one and transmitted signal has non-zero ISI. Despite this fact, for pulses (2) the matching filter gives the signal with zero ISI at synchronized time instants, and the symbol-by-symbol receiver can be used.

Let a baseband APSK signal be represented as the sum of components; each component carries non-overlapping modulated pulses and has zero ISI. Such representation of the APSK signal is particularly convenient for our goals, including a formulation of optimization problem and definition of the ISI measure. We provide zero ISI in each component by extension of the symbol time interval in L times with respect to the initial value. Thus, APSK signal based on finite pulses can be written as the sum of its L components:

$$y_{L,N}(t) = \sum_{p=1}^{L} y_{L,N}^{(p)}(t) = \sum_{p=1}^{L} \sum_{k=0}^{N-1} C_{r_p}^{(k)} a(t - \Delta t_p - kLT, L) \tag{5}$$

where p is a component number; each component carries N symbols; $\Delta t_p = (p - 1)T$ is a time delay of pth component measured from the time instance $t = -LT/2$ (that is the initial time of the first component).

It should be pointed out that each component can use its own type of M_p-order signal constellation. Also, each component is described by its own numbering order of LT-duration symbol intervals. For instance, the kth time interval of pth component is:

$$T_p^{(k)} = [kT + \Delta t_p - LT/2, \ kT + \Delta t_p + LT/2]. \tag{6}$$

To simplify some further expressions, we agree that $\left| C_{r_p} \right| \leq 1$. Also, we will use the brief notation for the pth component pulse from the kth time interval:

$$a_p^{(k)}(t, L) = L^{-1/2} a((t - \Delta t_p - kLT) / L). \tag{7}$$

For completeness, we suppose that each component can be transmitted on its own subcarrier frequency:

$$y_{L,N}(t) = \sum_{p=1}^{L} \sum_{k=0}^{N-1} C_{r_p}^{(k)} a_p^{(k)}(t, L) \exp(j 2\pi \Delta f_p t), \tag{8}$$

where Δf_p is a frequency shift of pth component measured from the central frequency of transmitted signal. We will call (8) as multicomponent (or L-component) signal.

Some particular combinations of multicomponent signal parameters are shown below to demonstrate universalism of the proposed representation for APSK signals. In the left part of fig. 1, we have shown pulse sequences of four separate components of the 4-component single-carrier signal, each component uses the following pulse:

$$a(t) = \cos^2(\pi t/T), \; |t| \le T/2. \tag{9}$$

As we can see, pulses inside the same component do not overlap and, hence, do not interfere. If the pulse is equal to the truncated Nyquist pulse (2)

$$a(t) = \left(\frac{L}{T}\right)^{1/2} \frac{\sin\{(\pi Lt/T)(1-\alpha)\}+(4\alpha Lt/T)\cos\{(\pi Lt/T)(1+\alpha)\}}{(\pi Lt/T)\{1-(4\alpha Lt/T)^2\}}, \; |t| \le T/2, \tag{10}$$

then at sufficient number of components L the multicomponent signal represents conventional APSK signal with Nyquist pulse. Hereinafter, let the number of components L be an even number. If we use BPSK signal constellation $\{+1; -1\}$ in even-numbered components and rotated BPSK constellation $\{+1j; -1j\}$ in odd-numbered components then our multicomponent signal corresponds to offset QPSK modulation. In addition, if the signal uses the pulse (9) and $L = 2$ then its PAPR is equal to one.

Pulses of three separate components of multicarrier multicomponent signal are shown in the right part of fig. 1. This signal is based on rectangular pulses: $a(t) = 1$ for $|t| \le T/2$. Component frequency shifts are chosen in inverse ratio to the symbol duration $\Delta f_p = (p - L/2)/(LT)$; component time shifts are equal to zero $\Delta t_p = 0$. These conditions make our multicomponent signal equal to OFDM-signal without cyclic prefix.

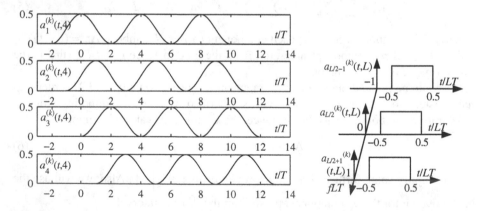

Fig. 1. Left part: pulse (9) sequences of four separate components of the 4-component signal; right part: single pulses of three separate components of multicarrier L-component signal with the rectangular pulse, $\Delta f_p = (p - L/2)/(LT)$ and $\Delta t_p = 0$

In this work we analyze single-carrier multicomponent signals with the component time shifts $\Delta t_p = (p - 1)T$ and zero component frequency shifts $\Delta f_p = 0$.

2   Characteristics of Multicomponent Signals

2.1   Power Spectrum Density

We can represent the power spectrum of multicomponent signal as follows

$$G(f) = \lim_{N \to \infty} \frac{1}{NLT + (L-1)T} \mathrm{E}\left\{\left|\sum_{p=1}^{L} S_{L,N}^{(p)}(f)\right|^2\right\} \tag{11}$$

where $S_{L,N}^{(p)}(f) = \int_{-\infty}^{+\infty} y_{L,N}^{(p)}(t)\exp(-j2\pi ft)dt$ is a spectrum of N-symbol length realization of pth component of the L-component signal, and the averaging is done over all possible realizations.

It can be shown that the spectrum of single-carrier multicomponent signal with symmetrical signal constellations in components is defined as follows:

$$G(f) \equiv G(f, L, a(t)) = (LZ/T)\left|F_a(Lf)\right|^2, \tag{12}$$

where $F_a(f) = \int_{-T/2}^{T/2} a(t)\exp(-j2\pi ft)dt$ is the pulse spectrum, and the factor Z is defined by signal constellation:

$$Z = L^{-1}\sum_{p=1}^{L} M_p^{-1}\sum_{r_p=1}^{M_p}\left|C_{r_p}\right|^2. \tag{13}$$

We omit the Z as argument of G because spectrums are usually normalized respectively to their maximal absolute values. The symmetry condition is valid for the majority of constellations used in practice and can be written as:

$$\sum_{r_p=1}^{M_p} C_{r_p} = 0,\ p = 1,\ 2,\ ...,\ L. \tag{14}$$

According to the Parseval's theorem and taking (12) into account, we can see that the average power of multicomponent signal is

$$P_{\text{av}} = \int_{-\infty}^{\infty} G(f)df = ZE_a/T, \tag{15}$$

where $E_a = \int_{-T/2}^{T/2} a^2(t)dt$ is the pulse energy.

2.2   Peak-to-Average Power Ratio

As for all other signal types, the PAPR value of multicomponent signals is:

$$PAPR = P_{\text{max}}/P_{\text{av}}. \tag{16}$$

The average power P_{av} is defined by (15). Due to inherent periodicity of multicomponent signals, the peak power P_{max} can be calculated at any T-duration time interval. For simplicity of notation, we chose the beginning of the first time interval of Lth component $T_L^{(0)}$: $T_{\text{PAPR}} = [LT/2 - T, LT/2]$. This interval carries the first symbols

$(k = 0)$ of components. Then, throwing for simplicity the k-superscript of constellation point, we write the maximal power as:

$$P_{max} = \max\{|y_{L,N}(t)|^2\} = \max\left\{\sum_{p=1}^{L}\sum_{d=1}^{L}C_{r_p}C_{q_d}^{*}a_p^{(0)}(t,L)a_d^{(0)}(t,L)\right\}, \qquad (17)$$

where the maximum is computed for all values of $t \in T_{PAPR}$ and over all possible combinations of r_p and q_d; the symbol " $*$ " indicates complex conjugation.

It can be shown, that in case of equal and non-rotated constellations used in components, the peak power can be computed as follows

$$P_{max} = \max_{t \in T_{PAPR}}\left\{\left(\sum_{p=1}^{L}\left|a_p^{(0)}(t,L)\right|\right)^2\right\}\max_{r_p}\left\{\left|C_{r_p}\right|^2\right\}. \qquad (18)$$

Finally, the PAPR value for the square M-QAM constellation is following

$$PAPR = \frac{3(\sqrt{M}-1)}{\sqrt{M}+1}\frac{1}{E_a/T}\max_{t \in T_{PAPR}}\left\{\left(\sum_{p=1}^{L}\left|a_p^{(0)}(t,L)\right|\right)^2\right\} = \max_{t \in T_{PAPR}}\left\{P(a(t),L,M)\right\}. \qquad (19)$$

2.3   Correlation Properties

As it was noticed above, each component has zero ISI since modulated pulses do not overlap. However, components interfere in the assembled multicomponent signal. The pulse $a_1^{(1)}(t,4)$ of the first component and parts of pulses $a_2^{(0)}(t,4)$, $a_2^{(1)}(t,4)$ of the second component from the left part of fig. 1 are shown in the left part of fig. 2. It is evident, that any time interval of duration LT chosen from some component partially contains signals of $2(L-1)$ time intervals of other components. In other words, each other component places parts of two neighboring signals on the considered time interval. Although this conclusion is violated on the first and the last time intervals of multicomponent signals, it is fair for general case. Obviously, the measure of ISI must be defined with all $2(L-1)$ interfering signals taken into account. But, for convenience, we first define the partial correlation coefficient as the ISI measure for any pair of signals, and then the group correlation coefficient as the "cumulative" ISI measure.

The partial correlation $PC_{p,d}^{(k,l)}$ is a value of normalized correlation coefficient between the signal from the kth time interval of pth component and the signal from the lth time interval of dth component:

$$PC_{p,d}^{(k,l)} = (1/E_a)\int_{-\infty}^{\infty}C_{r_p}^{(k)}a_p^{(k)}(t,L)C_{q_d}^{*(l)}a_d^{(l)}(t,L)dt. \qquad (20)$$

Notice, that the partial correlation is normalized: $\left|PC_{p,d}^{(k,l)}\right| \le 1$.

The group correlation $GC_p^{(k)}$ is a value of the "composed" correlation coefficient between the signal from the kth time interval of pth component and all other signals:

$$GC_p^{(k)} = \sum_{\substack{d=1\\d\ne p}}^{L}\sum_{l=0}^{N-1}PC_{p,d}^{(k,l)}. \qquad (21)$$

Actually, the group correlation coefficient comprises $2(L-1)$ values of partial correlations, hence $\left|GC_p^{(k)}\right| \leq 2(L-1)$. A normalization procedure for the group correlation is missed intentionally since the analysis of its non-normalized values makes it possible to observe general trends in the behavior of solutions characteristics during the variation of the number of components L.

For signals based on the pulse (10) all partial correlations and the group correlation coefficient are equal to zero since these pulses satisfy the following condition:

$$\lim_{L\to\infty} \int_{-\infty}^{\infty} a_p^{(k)}(t,L)a_d^{(l)}(t,L)dt \overset{k\neq l}{=} 0. \tag{22}$$

Due to this condition, a conventional correlation receiver can be applied for detection of these signals. However, in general case, the matched filter gives symbols corrupted by ISI-noise. The value of ISI-noise is proportional to the group correlation coefficient. Hence, received symbols form a "clouded" constellation (the right part of fig. 2). According to the fact, that bit error rate (BER) in AWGN channel is mainly defined by the value of minimal Euclidean distance between signal constellation points we will use the maximal absolute value of group correlation as the measure of ISI (Maximal Group Correlation, MGC). In absence of a signal constellation rotation, the value of MGC comprises maximal absolute values of partial correlations and can be computed as follows

$$\max_{p,k}\left|GC_p^{(k)}\right| = MGC = \sum_{d=2}^{L}\left|\int_{-T/2+\Delta t_d/L}^{T/2} a(t)a(t-\Delta t_d/L)dt\right|. \tag{23}$$

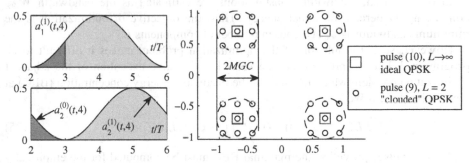

Fig. 2. Left part: the dark gray indicates signal parts involved in the calculation of $PC_{1,2}^{(1,0)}$, and the light gray – of $PC_{1,2}^{(1,1)}$; right part: output symbols from the matched filter

The signal constellation rotation occurs, for instance, in OQPSK modulation. In such case expressions (18), (19) and (23) must be modified, but the above discourse remains relevant and has the same sense.

In [5] the criterion of maximal free Euclidean distance d_{free} is used for the pulse optimization at a chosen bandwidth efficiency, pulse length and signal constellation type. Of course, the analysis of free Euclidean distance is fair way to improve BER performance. But this approach requires a great number of constraints to be included in the optimization, because actually all combinations of symbol errors must be considered. Actually, the value of free Euclidean distance is in inverse monotonic dependence from the MGC.

Thus, we use the MGC constraint instead of d_{free} in order to simplify the optimization complexity and to avoid dependence from any particular type of signal constellation.

3   Formulation of Optimization Problem

A particular mode of multicomponent signaling is defined by the pulse $a(t)$, the number of components L and the signal constellation type. In this work, we consider only QPSK constellation used in each component. However, because of presence of the parameter $a(t)$, any optimization will lead to the inconvenient, particularly for numerical methods, problem of searching functional extremum. The pulse decomposition in some basis is usually employed for such cases. The decomposition in truncated Fourier series is used in this work. It is expected that the truncation of high harmonics will not really deteriorate the optimal solution:

$$a(t) = c_0 / 2 + \sum_{k=1}^{K-1} \left(c_k \cos(2\pi kt / T) + s_k \sin(2\pi kt / T) \right), \tag{24}$$

where the number of sought parameters defining $a(t)$ is equal to $2K - 1$. In such representation of the pulse $a(t)$ any optimization can be transformed to the extremum search problem for the multivariable function which has $2K$ arguments (including L). There are several algorithms, which can compute numerical solutions of such nonlinear optimization problems. We used the "active-set" algorithm built in the fmincon( ) function in MATLAB.

In our research, bandwidth consumptions are estimated as the bandwidth $W_{99\%}$ comprising 99 percent of signal energy. Thus, the objective of optimization is the minimum bandwidth $W_{99\%}$ at a chosen number of components L.

However, such formulation of the optimization problem makes it difficult to be programmed, but the problem can be solved by iterative execution of the adjacent optimization problem which maximizes the bandwidth energy concentration (BEC) at a given bandwidth W:

$$BEC(W, L, a(t)) = (1 / P_{av}) \int_{-W/2}^{W/2} G(f, L, a(t)) df. \tag{25}$$

At first, pulses providing the maximal BEC must be computed for the empirically chosen set of bandwidth W values. After, from the obtained set of BEC values the closest to the 0.99 bounding pair is chosen; the corresponding pair of bandwidth values forms the range of search for the next iteration. Every next iteration executes the multiple optimization for the updated set of bandwidth values. We used the dichotomy method choosing the central value of W from the range obtained at the previous iteration. The iterative optimization procedure was stopped when the obtained BEC was equal to 0.99 with a $5 \cdot 10^{-6}$ tolerance. Such high precision in the BEC was used in order to obtain the bandwidth W value with three decimal digits. Formally, the optimization can be written as follows:

$$(a_{\text{opt}}(t), W_{99\%}) = \underset{W}{\text{argmin}} \{ |\max_{a(t)} \{ BEC(W, a(t)) \} - 0.99| \},$$

$$\left| BEC(W_{99\%}, a_{\text{opt}}(t)) \} - 0.99 \right| \le 5 \cdot 10^{-6}. \tag{26}$$

In order to fix the pulse energy in all optimization modes, the equality constraint $E_a = 1$ was applied to sought pulses. Also, we used two types of additional constraints. First, the measure of ISI (MGC) was bounded by the value of $\kappa \in [0, 2(L-1)]$:

$$MGC = \sum_{d=2}^{L} \left| \int_{-T/2+\Delta t_d/L}^{T/2} a(t)a(t-\Delta t_d / L)dt \right| \leq \kappa. \tag{27}$$

Another additional constraint was applied to solutions in order to obtain multicomponent signals with reduced PAPR values. For multicomponent signals with M-QAM signal constellations used in each component the constraint was performed by the set of N_{PAPR} inequalities:

$$P(a(t_n), L, M) \leq \rho, \ n=1, \ 2, \ \ldots, N_{PAPR}, \tag{28}$$

where the signal power was actually bounded for the N_{PAPR} uniformly chosen time instants $t_n = LT/2 - T + (n-1)\Delta t$, $\Delta t = T / (N_{PAPR} - 1)$; and $\rho \geq 1$. On the one hand, the replacing of continuous time interval T_{PAPR} with the set of discrete time instants leads to an error of PAPR limitation. On the other hand, the constraint accuracy can be improved by increasing the number of tested time instants N_{PAPR}.

Thus, our optimization is transformed to the problem of multiple search of multivariable function maximum with the additional non-linear equality constraint $E_a = 1$ and the ($N_{PAR} + 1$) non-linear inequality constraints (27) and (28).

For the numerical optimization, the number of active harmonics K must be defined. Let the minimal number K' of harmonics is considered to be sufficient if the following inequalities are satisfied:

$$\sigma^2 = (1/T) \int_{-T/2}^{T/2} (a_{K'}(t) - a_{K'+i}(t))^2 \, dt \leq 0.001,$$
$$2 \, | MGC_{K'} - MGC_{K'+i} | / (MGC_{K'} + MGC_{K'+i}) \leq 0.01, \tag{29}$$
$$2 \, | PAPR_{K'} - PAPR_{K'+i} | / (PAPR_{K'} + PAPR_{K'+i}) \leq 0.01,$$
$$i = 1, \ 2, \ 3,$$

where $a_K(t)$ is a pulse defined by K harmonics, MGC_K and PAR_K are values of its MGC and PAPR respectively. Notice that this approach has no accurate mathematical justification, but it was successfully used for the solving of our optimization problem.

Finally, due to possible errors in numerical solutions, we consider the minimal values of MGC and PAPR as

$$\min\{MGC\} = 0.01, \min\{PAPR\} = 1.01. \tag{30}$$

4   Optimal Solutions

In this work we have found optimal pulses for multicomponent signals with $L = 2, 4$ and 8 components with QPSK signal constellation used in each component. For each particular number of components L two basic groups of optimal solutions were obtained: with the only PAPR constraint and with the only MGC constraint. Additional third-kind group of optimal solutions included pulses obtained at fixed bandwidth values for various combinations of MGC and PAPR. Moreover, each time we found pulses in the form of $2K-1$ Fourier coefficients (24), and pulses in the form of even functions, when $s_k = 0$ for $k = 1, 2, ..., K-1$ and thus number of Fourier coefficients was reduced to K. In simplification purposes, we bounded values of K and N_{PAPR}:

$$K < 20 \text{ and } N_{PAPR} < 512. \tag{31}$$

The obtained optimal pulses are represented in fig. 3–4 in various ways, where the PAPR and MGC values are indicated by symbols ρ and κ respectively. We discuss some features of these solutions below.

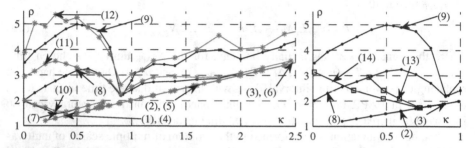

Fig. 3. PAPR values versus MGC values for optimal solutions: (1)–(6) for the only PAPR constraint; (7)–(12) for the only MGC constraint; (1), (4), (7), (10) for $L = 2$; (2), (5), (8), (11) for $L = 4$; (3), (6), (9), (12) for $L = 8$; (13), (14) for $W_{99\%}T = 1.1$ and $L = 4$, $L = 8$ respectively; (1)–(3), (7)–(9) and (13)–(14) for even pulses

First, optimal pulses obtained under the only PAPR constraint are non-negative, i.e. $a(t) \geq 0$ for $|t| < T/2$. Stronger PAPR constraints (smaller ρ) leads to pulses with "longer" edges, which are almost equal to zero. The use of these pulses actually leads to a reduction of number of components. For instance, we can conclude from the right part of fig. 4 that the pulses obtained with $\rho \leq 2.4$ are equal to zero for more than half of their intervals, i.e. $a(t) = 0$ for $|t| \geq T/4$. Hence, we can combine components 1 and 5, 2 and 6 etc. and consider the 4-component signal ($L = 4$) instead of the mode of eight components ($L = 8$). At the same time, this transformation does not violate orthogonality of neighboring pulses carried on the same component.

Also, it must be pointed out, that in case of optimal solutions with zero-valued edges optimization problem converges to several formally different pulses, but actually all of them are cyclically shifted versions of each other. These formally different solutions have the same values of MGC and PAPR. In order to specify the sole solution in this optimization mode we chose the even version of the pulse.

The most interesting observation regarding the pulse optimization under the only PAPR constraint is that the PAPR and the MGC values of obtained optimal pulses accurately fit the linear function: $\rho = \kappa + 1$. Therefore, the lines (1–6) merge in fig. 3. These lines do not originate at the point (0, 1) because of the fact, that for strong PAPR constraints the optimal solution converges to a rectangular pulse, which Fourier decomposition faces well-known problems (of Gibbs effect).

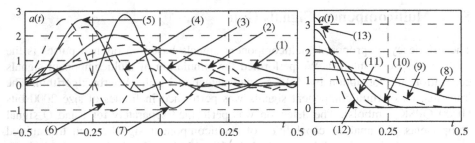

Fig. 4. Optimal pulse's shapes for $L = 8$ and MGC constraint only (1)–(7) and PAPR constraint only (8)–(13); (1)–(7) correspond to $\kappa = 5.9, 4.0, 2.2, 1.2, 0.9, 0.4, 0.01$ respectively; (8)–(13) correspond to $\rho = 6.9, 4.8, 3.2, 2.4, 1.9, 1.4$ respectively; (8)–(13) correspond to even pulses.

During the pulse optimization under the only MGC constraint, the problem also converges to several solutions. The ambiguity occurs if the assumption of solution evenness (44) is not applied. Moreover, in this case the alternative solutions have different PAPR values. In order to specify the sole solution in this optimization mode we chose the minimal PAPR solution. Such optimal pulses of general form (neither even nor odd) always outperform even optimal pulses in sense of bandwidth at fixed MGC.

The analysis of third-kind groups of pulses shows that larger number of components leads to better pulse characteristics. In other words, the use of increased number of components provides the same bandwidth but with smaller MGC and/or PAPR values (the right part of fig. 3).

Probably the most interesting observation regarding characteristics of optimal solutions is the local minimum of the curves (8)–(9) and (11)–(12) in fig. 3. The BER performance curves of the multicomponent signals with $\kappa \leq 0.9$ (with characteristics located to the left from the local minimum) detected by the Viterbi algorithm are the same as for orthogonal signals. Another multicomponent signals with $\kappa > 0.9$ yield to orthogonal signals in sense of BER.

In addition, we analyzed the possibility to obtain Nyquist pulses by searching of optimal pulses for quasi-orthogonal multicomponent signals. We performed the optimization under the assumption of even pulse with the MGC constraint $\kappa = 0.01$. The comparison of Nyquist pulses and obtained optimal solutions for 8-component signals is given in table 1. The PAPR values are represented in the first column, the second column contains roll-off factors of Nyquist pulses, which give corresponding PAPR values. Values of mean-square errors between obtained optimal pulses and corresponding Nyquist pulses are shown in the third column.

Table 1. Comparison of optimal and Nyquist pulses

ρ	α	σ^2
6,00	0,262	0,00008
5,00	0,341	0,00020
4,00	0,446	0,00070

From these results we can see that some of our optimal pulses are actually equal to Nyquist pulses with a mean-square error less than chosen error in (29). This fact proves versatility of the proposed multicomponent signals concept.

5   Multicomponent Signals Gain

The efficiency of proposed multicomponent signals based on optimal pulses is the subject of special interest. We performed the simulation of multicomponent signals generation, its corruption by AWGN, and detection by the Viterbi algorithm. The transmission of multicomponent signals was performed in frames of size 2000 bits (1000 QPSK symbols). The detection was performed separately for I and Q signal components. To analyze capabilities of multicomponent signals, each transmitted frame was terminated by $2L$ zeros, and the Viterbi algorithm had a traceback depth equal to $(1000 + L)$, thus, implementing the optimal receiver.

For estimation of multicomponent signals gain in energy, we used the unit energy consumptions

$$\beta_E = E_b / N_0 = P_{av} / (RN_0), \tag{32}$$

where R is the bit rate; and for bandwidth efficiency comparison, the unit bandwidth consumptions were used:

$$\beta_F = W_{99\%} / R. \tag{33}$$

In the left part of fig. 5, we compare conventional signals based on Nyquist pulses and Gray mapped QPSK, 16-QAM, 64-QAM with the 8-component QPSK signals based on optimal pulses. In addition, we put in fig. 5 the efficiency curve corresponding to the 8-component signals based on pulses obtained under the only free Euclidean distance d_{free} constraint, as it was done in [5]. All experimental points shown in fig. 5 correspond to a 10^{-4} BER. The curve (1) corresponds to the Shannon limit $(\beta_E = (2^{1/\beta_F} - 1)\beta_F)$, which defines the upper bound for reliable data transmission in AWGN channel. It is shown only to clarify the allowed (β_F, β_E) range. The marked points on curves (6)–(8) were computed for roll-off factor values chosen from $\alpha = 0$ to $\alpha = 1$ with a 0.1 step. The maximal values of β_F on curves (6)–(8) correspond to $\alpha = 1$.

For the signals based on Nyquist pulses (10), we chose such minimal number of components L which provided a concentration of no less than 99.99 percent of the pulse energy in its truncated version. This empirical approach provides almost equal falling rate of out-of-band emissions in frequency domain for truncated Nyquist pulses and for obtained optimal pulses.

We give below the analysis of results shown in the left part of fig. 5. At first, optimal pulses obtained under the MGC constraint and under the d_{free} constraint provide almost equal efficiencies. To be more precise, pulses obtained under the d_{free} constraint demonstrate the best efficiency, pulses of general form obtained under the MGC constraint yield, and even pulses obtained under the MGC constraint demonstrate the worst efficiency. Optimal pulses obtained under the only PAPR constraint yield to all of them. At second, the MGC constraint gives the wider range of efficiency, covering quasi-orthogonal signals. At third, although there is no apparent gain of the optimal multicomponent signals with respect to conventional signals with Nyquist pulses, the gain occurs with respect to the Nyquist pulses with roll-off factors $\alpha = 0.15$–0.35, which actually take place in practice (DVB-S2, UMTS etc.).

From the left part of fig. 5 we can see that $\alpha = 0$ provides the best efficiency for Nyquist pulses, but it is not usually used in practical applications due to several reasons. One of them is its PAPR value which is the maximal for $\alpha = 0$. Therefore, the conventional comparison in the (β_F, β_E) plane is not really convenient; a more equitable comparison must include PAPR values of competing signals. In this connection, we propose the modified energy consumptions with a PAPR value taken into account:

$$\beta_E^* = P_{max} / (RN_0) = PAPR \; \beta_E. \tag{34}$$

Signals efficiency curves in the (β_F, β^*_E) plane are shown in the right part of fig. 5. At first, in the modified efficiency plane, the optimal multicomponent signals always demonstrate the best efficiency. In detail, the pulses obtained under the only PAPR constraint outperform the even MGC pulses; the MGC pulses of general form yield to the even MGC pulses. The multicomponent signals based on optimal pulses obtained under the d_{free} constraint yield to all of them, and conventional orthogonal signals with Nyquist pulses demonstrate the worst efficiency. At second, we can see that the choice of roll-off factors from the range $\alpha = 0.15$–0.35 provides a real compromise between bandwidth and modified energy consumptions.

To specify the particular gain of multicomponent signals, we chose the conventional signals with Nyquist pulses, $\alpha = 0.2$ and QPSK signal constellation as reference position, corresponding to the point (0.54, 14.1) in the modified efficiency plane. From this position, the 8-component signals using optimal pulses obtained under the PAPR constraint provide a 3.1 dB energy gain or about 30% of bandwidth gain. Moreover, if we allow a 1.2 dB energy loss for the multicomponent signal then it saves about 37% of bandwidth.

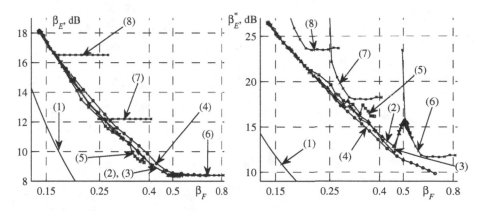

Fig. 5. Consumptions curves in the (β_F, β_E) plane on the left and in the (β_F, β^*_E) plane on the right; (1) – Shannon limit, (2)–(5) – 8-component signals with QPSK; (2)–(3) – the only MGC constraint for general-form and even pulses respectively; (4) – the only PAPR constraint; (5) – the only d_{free} constraint; (6)–(8) – signals with Nyquist pulses and Gray mapped QPSK, 16-QAM, 64-QAM respectively.

Conclusion

The proposed multicomponent signals is a real versatile tool for APSK signals description, including conventional orthogonal signals and signals with ISI. Through the proposed concept, we demonstrated the possibility of pulse optimization under the criterion of maximal bandwidth efficiency and additional constraints on the PAPR value and the ISI measure. Therefore, it gives the opportunity to use the pulse shape which exactly meets the system practical requirements. Our further research includes efficiency estimation of the multicomponent signals combined with error-correcting codes.

Acknowledgements. This work was supported by Russian Federation President's grant 14.124.13.4724-MK.

References

1. Mazo, J.E.: Faster-than-Nyquist signaling. Bell Syst. Tech. J. **54**, 1451–1462 (1975)
2. Anderson, J.B., Rusek, F., Owall, V.: Faster than Nyquist signaling. Proc, IEEE (2013)
3. Proakis, J., Salehi, M.: Digital Communications, 5th edn. McGraw-Hill, New York (2008)
4. Leung, P.S., Feher, K.: F-QPSK–a superior modulation technique for mobile and personal communications. IEEE Transactions on Broad-casting **39**, 288–294 (1993)
5. Said, A., Anderson, J.B.: Bandwidth-efficient coded modulation with optimized linear partial-response signals. IEEE Trans. Inform. Theory **44**(2), 701–713 (1998)

Optimal Input Power Backoff of a Nonlinear Power Amplifier for SEFDM System

Dmitrii K. Fadeev and Andrey V. Rashich[✉]

Peter the Great St. Petersburg Polytechnical University, St. Petersburg, Russia
fadeev_dk@spbstu.ru, andrey.rashich@gmail.com

Abstract. In this paper the impact of nonlinear distortions caused by nonlinear power amplifier (PA) is evaluated for BPSK and QPSK Spectrally Efficient Frequency Division Multiplexing (SEFDM) signals. System performance depends on input power back-off (IBO) and SEFDM parameters, such as modulation scheme, frequency spacing between subcarriers and total number of subcarriers. Performance is quantified with total degradation (TD), defined as the sum (in decibels) of the IBO and the increment Δ in the ratio Eb/No required to maintain a given bit error rate (BER) with respect to the situation of perfectly linear PA. IBO corresponding to minimum TD is obtained.

Keywords: OFDM · SEFDM · BER performance · Input back-off · IBO · Total degradation

1 Introduction

Orthogonal Frequency Division Multiplexing (OFDM) has been widely used in wireless communications due to high bit error rate (BER) performance in channels with inter-symbol interference (ISI). OFDM has been adopted by digital broadband telecommunication systems such as WiFi, WiMAX, DVB-T, DAB, DRM and 4G LTE downlink.

At the present time research throughout the world has started to define modulation schemes for the 5th generation wireless systems (5G). One of potential candidates is the technique termed Spectrally Efficient FDM (SEFDM). SEFDM systems can significantly enhance spectral utilization and save up to 40% of bandwidth with 1.1 dB degradation compared to an equivalent OFDM system [1, 2] (for special modulation schemes).

One of the major drawbacks in OFDM and SEFDM is its high peak-to-average power ratio (PAPR). It forces the PA to work at a reduced efficiency level and results in energy consumption increase.

The reason of a reduced efficiency of a PA is a nonlinearity of amplitude/ amplitude (AM/AM) conversion curve. The highest efficiency of a PA is achieved when operating point for a PA is at the saturation level. In this case a PA introduces some distortion into signals with high PAPR such as OFDM and SEFDM. These distortion cause interference between subcarriers and bandwidth increase. Finally, it degrades BER performance of a system [3].

© Springer International Publishing Switzerland 2015
S. Balandin et al. (Eds.): NEW2AN/ruSMART 2015, LNCS 9247, pp. 669–678, 2015.
DOI: 10.1007/978-3-319-23126-6_60

Distortions can be reduced by shifting the operating point, but reducing the average power of signal degrades BER.

Therefore, there are two factors depending on input back off which are introduced to shift the operating point: the first is the average power of the signal and second is signal distortion caused by the nonlinearity of a PA. The impact of one or another factor depends on number of used subcarriers, subcarrier modulation and subcarrier spacing.

It is shown for OFDM that there is the optimum back-off value, which corresponds to minimum signal-to-noise (SNR) value to achieve target BER [4, 5]. This value depends on subcarrier modulation type.

There is an interference between SEFDM subcarrier frequencies, therefore additional distortions caused by PA are expected to degrade BER performance of SEFDM system. In [6] the effect of clipping in SEFDM system on the spectrum of the signal and the BER performance is evaluated but overall performance of system is not estimated and the presence of the PA is not taken into account.

It is shown in [4] that clipping degrades overall BER performance of the OFDM system.

SEFDM is a promising technique to be used in 5G standard therefore it is useful to evaluate the impact of distortions caused by PA to BER performance and estimate overall performance of SEFDM system taking into account the presence of the PA.

The aim of this work is to determine the operating point for a PA to obtain the best BER performance of SEFDM system for specified parameters: number of subcarriers, subcarrier spacing and subcarrier modulation (for BPSK and QPSK).

2   SEFDM Model Description

The SEFDM signal is the sum of modulated subcarriers and can be written as:

$$x(t) = \frac{1}{\sqrt{T}} \sum_{l=-\infty}^{\infty} \sum_{n=0}^{N-1} s_{l,n} g(t-lT) e^{j2\pi n \Delta f t} \qquad (1)$$

where N is the number of subcarriers, $s_{l,n}$ is the complex manipulation function for n-th subcarrier of l-th symbol, T is the symbol duration, $g(t)$ is the pulse-shaping function (in this work the rectangular pulse-shaping function is used), $\Delta f = \alpha / T$ is the frequency separation between adjacent subcarriers, α is the bandwidth compression factor.

The peak-to-average power ratio (PAPR) P_F is defined as

$$P_F = P_{peak} / P_{avg}, \qquad (2)$$

where $P_{peak} = \max\{|x(t)|^2\}_{t\in[0,T]}$ is peak signal power and $P_{avg} = \frac{1}{T} \int_0^T |x(t)|^2 \, dt$ is the average power of the signal.

3   The Power Amplifier Model

The PA output signal is

$$y(t) = G\big[|\,x(t)\,|\big]e^{j(\phi(t)+\Phi[|x(t)|])} \tag{3}$$

where $x(t) = |\,x(t)\,|\,e^{j\phi(t)}$ is the PA input signal, $G[\cdot]$ and $\Phi[\cdot]$ are the AM/AM and AM/PM conversions, respectively.

In this work solid-state power amplifier (SSPA) model is used. $G[\cdot]$ and $\Phi[\cdot]$ are expressed as [7]:

$$G\big[|\,x(t)\,|\big] = \frac{g_0\,|\,x(t)\,|}{\left[1+\left(\dfrac{|\,x(t)\,|}{x_{sat}}\right)^{2p}\right]^{\frac{1}{2p}}} \tag{4}$$

$$\Phi\big[|\,x(t)\,|\big] = 0$$

where g_0 is the amplifier gain, x_{sat} – is the saturation level, p is the parameter to control the AM/AM sharpness of the saturation region.

AM/AM conversion curve for SSPA model with $p = 2$ in dB scale is shown in figure 1. The value of the input power 0 dB corresponds to saturation level x_{sat}, the value of the output power 0 dB corresponds maximum output power $g_0 \cdot x_{sat}$.

Maximum PA power efficiency is obtained when operating point is at saturation level. As mentioned above, SEFDM signals exhibit high PAPR, therefore input power back-off (IBO) is required to shift the operating point to ensure that the PA operates within linear region, as shown in figure 1.

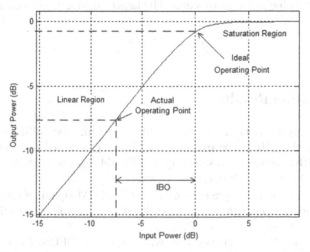

Fig. 1. AM/AM conversion curve for SSPA model with $p = 2$ in dB scale

IBO shows the value of the average input power decrease compared with saturation level, in dB, and can be written as

$$IBO = 10\log_{10}\left(P_{sat} / P_{avg}\right) \tag{5}$$

On the one hand, higher IBO corresponds to less output power and less BER performance. On the other hand, higher IBO corresponds to lower distortions caused by the PA.

For overall performance estimation, Total Degardation (TD) metric is used [8]

$$TD = SNR_{PA}(IBO) - SNR_{AWGN} + IBO \quad [in \ dB] \tag{6}$$

where SNR_{AWGN} is the SNR which is required to achieve a target BER in AWGN channel and $SNR_{PA}(IBO)$ is the SNR which is required to achieve a target BER taking into account distortions caused by the PA at a given IBO.

4   MatLab Model Description

The MatLab simulation model was developed to evaluate TD. The model includes generation of SEFDM symbols, AWGN channel and the receiver of SEFDM-signals. This model is used for SNR_{AWGN} estimation. After that SSPA model unit is added to the model and SNR_{PA} is estimated for the set of IBO values.

The algorithm for SEFDM generation based on Inverse Fast Fourier Transform (IFFT) [9] is used. First, the OFDM symbol for dedicated data vector is generated. FFT size is Q times as great as the number of subcarriers N and data vector is enhanced by zeros. The output sequence of IFFT with length NQ is reduced to length $NQ\alpha$ to produce SEFDM symbol.

Detection algorithm based on decision diagram [10] is used to receive SEFDM signal. The algorithm consists in determining the sequence of symbols, which provides minimum value of target function. The target function is defined as

$$T = \int_{-T/2}^{T/2} \left|x(t) - s_j(t)\right|^2 dt, \tag{7}$$

5   Simulation Results

Simulation results are obtained for two modulation schemes: BPSK and QPSK. Target BER is equal to 10^{-4}. The oversampling factor Q is equal to 8. Figures 2 and 3 show TD vs IBO curves corresponding to SEFDM signals with $\alpha = 0,825; 0,9; 1$. Note that $\alpha = 1$ corresponds to OFDM signals.

Minimum value of TD is greater for less α. SEFDM signals with less α has significant interference between subcarriers, therefore even small distortions can affect BER performance, especially for QPSK.

For BPSK and $\alpha = 0,9$ energy loss with respect to OFDM ($\alpha = 1$) is 0,75 dB, for $\alpha = 0,8$ energy loss is 1,5 dB. For QPSK energy loss is 2,2 dB and 3,5 dB, respectively.

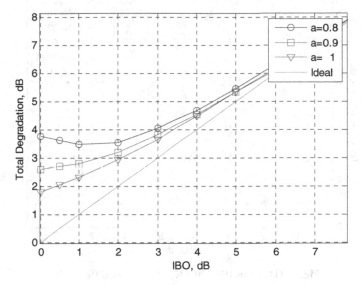

Fig. 2. TD versus IBO, BPSK, $N = 64$

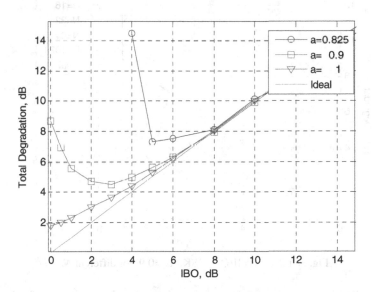

Fig. 3. TD versus IBO, QPSK, $N = 64$

Figures 4 and 5 show TD vs IBO curves for SEFDM signals with different number of subcarriers.

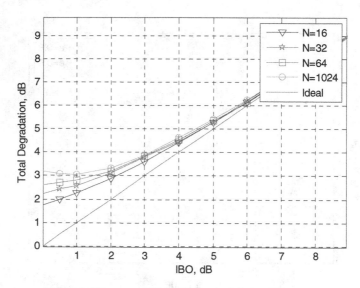

Fig. 4. TD versus IBO, BPSK, α =0.9, for different N.

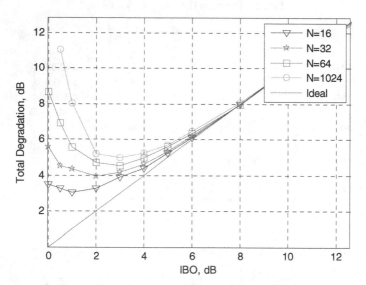

Fig. 5. TD versus IBO, QPSK, α =0.9, for different N.

PAPR of SEFDM signal increases with the increase of the number of sub-carriers therefore greater value of *IBO* is required to keep the level of distortions. Optimal value of input back-off, corresponding to minimum *TD* value, is equal to 1 dB for 16 subcarrier SEFDM signal and 3 dB for 1024 subcarrier SEFDM signal modulated with QPSK.

Figures 2-5 can be used to determine optimal value of IBO which corresponds to minimum total degradation.

The spectrum of distorted SEFDM signals is evaluated to estimate the impact of distortions caused by the PA to the out of band emission. The signal bandwidth is estimated for three threshold values: –25 dB, –20 dB and –15 dB.

Figure 6 shows that curves are similar in shape for different α and constant number of subcarriers. Non-linear distortions affect the spectrum of SEFDM signals the same way as for OFDM. Figures 7-9 shows that curves have different shape depending on number of subcarriers.

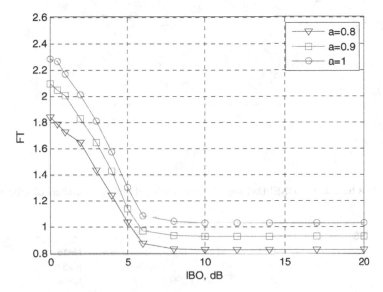

Fig. 6. –25 dB bandwidth of 1024 subcarrier SEFDM signal

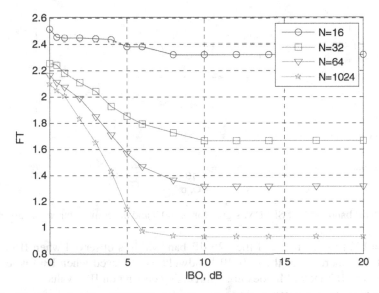

Fig. 7. –25 dB bandwidth of SEFDM signal for α =0,9 and different numbers of subcarriers

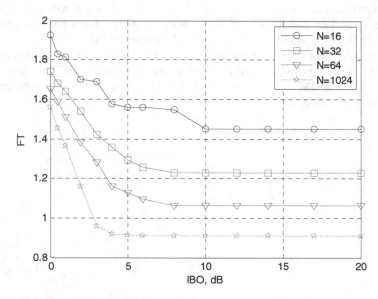

Fig. 8. -20 dB bandwidth of SEFDM signal for α =0,9 and different numbers of subcarriers.

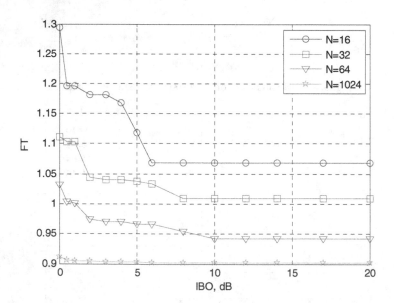

Fig. 9. –15 dB bandwidth of SEFDM signal for α =0,9 and different numbers of subcarriers

For $N = 1024$ the increase of the –25 dB bandwidth is observed when IBO is less than 6 dB. The increase of the –20 dB bandwidth is observed when IBO is less than 3 dB. The –15 dB bandwidth does not appreciably depend on IBO value.

Operating point for a PA should be chosen so that bandwidth requirements are met and minimum TD is obtained. Minimum TD is the major factor when −15 dB and −20 dB bandwidth is required to remain constant. In other cases, minimum value of IBO should be chosen from values, which met bandwidth requirements.

6   Conclusion

In this paper, the impact of nonlinear distortions caused by nonlinear power amplifier on SEFDM signals is evaluated. The SSPA model with typical parameters is utilized. System performance depends on input power backoff (IBO) and SEFDM parameters, such as modulation scheme, frequency spacing between subcarriers and total number of subcarriers. Performance is quantified with total degradation (TD), defined as the sum (in decibels) of the IBO and of the increment Δ in the ratio Eb/No required to maintain a given bit error rate (BER) with respect to the situation of perfectly linear PA.

The MatLab model was developed to evaluate TD. The set of TD is obtained for SEFDM signals with different parameters values (number of subcarriers, subcarrier spacing and subcarrier modulation). The bandwidth of distorted SEFDM signals is estimated.

For BPSK and $\alpha = 0,9$ energy loss with respect to OFDM ($\alpha = 1$) is 0,75 dB, for $\alpha = 0,8$ energy loss is 1,5 dB. For QPSK energy loss is 2,2 dB and 3,5 dB, respectively.

Obtained results can be used to determine the operating point for a PA to obtain the best BER performance of SEFDM system with specified parameters and to meet bandwidth requirements.

References

1. Xu, T., Darwazeh, I.: Spectrally efficient FDM: spectrum saving technique for 5G?. In: 2014 first International Conference on 5G for Ubiquitous Connectivity (5GU), pp. 273–278, November 26–28, 2014
2. Hamamura, M., Tachikawa, S.: Bandwidth efficiency improvement for multi-carrier systems. In: 15th IEEE International Symposium on Personal, Indoor and Mobile Radio Communications, PIMRC 2004, vol. 1, pp. 48–52, September 2004
3. Antognetti, P. (ed.): Power Integrated Circuits: Physics, Design, and Applications, p. 544. McGraw-Hill, New York (1986)
4. Thompson, S.C., Proakis, J.G., Zeidler, J.R.: The effectiveness of signal clipping for PAPR and total degradation reduction in OFDM systems. In: Global Telecommunications Conference, GLOBECOM 2005, vol. 5, p. 2811. IEEE (2005)
5. Guel, D., Palicot, J.: Analysis and comparison of clipping techniques for OFDM peak-to-average power ratio reduction. In: 16th International Conference on Digital Signal Processing, pp. 1–6 (2009)
6. Ahmed, S.I.A.: Spectrally Efficient FDM Communication Signals and Transceivers: Design, Mathematical Modelling and System Optimization. A thesis submitted for the degree of Doctor of Philosophy, pp 182–187 (2011)

7. Rapp, C.: Effects of the HPA-nonlinearity on a 4-DPSK/OFDM signal for a digital sound broadcasting system. In: Proc. 2nd European Conference on Satellite Communications (ECSC), Liège, Belgium, pp. 179–184, October 1991
8. D'Andrea, A.N., et al.: RF Power Amplifier Linearization Through Amplitude and Phase Distortion. IEEE Trans. Commun. 44(11), 1477–1484 (1996)
9. Kislitsyn, A.B., Rashich, A.V., Tan, N.N.: Generation of SEFDM-signals using FFT/IFFT. In: Balandin, S., Andreev, S., Koucheryavy, Y. (eds.) NEW2AN/ruSMART 2014. LNCS, vol. 8638, pp. 488–501. Springer, Heidelberg (2014)
10. Zavjalov, S.V., Makarov, S.B., Volvenko, S.V.: Nonlinear coherent detection algorithms of nonorthogonal multifrequency signals. In: Balandin, S., Andreev, S., Koucheryavy, Y. (eds.) NEW2AN/ruSMART 2014. LNCS, vol. 8638, pp. 703–713. Springer, Heidelberg (2014)

Investigation of Key Components of Photonic Beamforming System for Receiving Antenna Array

Sergey I. Ivanov, Alexander P. Lavrov[⊠], and Igor I. Saenko

St-Petersburg State Polytechnical University, St-Petersburg, Russia
{lavrov,s.ivanov,saenko}@cef.spbstu.ru

Abstract. The performance characteristics of the key components of photonic beamformer for an ultrawideband phased array antenna in the receive mode are investigated. Here we consider beamforming arrangement that can currently provide the required true time delay (TTD) capabilities by using the units and elements available at the market of modern components of fiber-optical telecommunication systems. The essential parameters of accessible analog microwave photonic link's main components as well as performance characteristics of the complete photonic link assembly have been measured. The links under consideration operate in "intensity modulation direct detection" mode with radio frequency range up to 18 GHz and include a set of transmitters with wavelengths in accordance with 100 GHz ITU grid and external electro-optical Mach-Zehnder modulators, 8-channel DWDM multiplexor, switchable time delay unit based on single mode fiber dispersion usage and common photoreceiver unit. The results of key components' parameters measuring are given and discussed.

Keywords: Phased array antenna · Photonic beamforming · True time delay · Fiber optic link · Fiber dispersion · DWDM

1 Introduction

For ultrawideband (up to a few decades) phased array antennas (PAA) the photonic schemes using true-time delay (TTD) beamforming provide a far reaching solution that permit to form scanning PAA beam or a set of switched beams free of squint effect. During the last years TTD beamforming based on photonic technologies has been extensively investigated due to its advanced features: extremely wide bandwidths, fast switching, light weight and immunity to electromagnetic interference [1,2,3]. Most of the proposed architectures using fiber optic link components to provide control of one- or multibeam antenna array pattern require specially developed components, such as chirped fiber Bragg gratings, fast tunable lasers with narrow spectral linewidths, reflection fiber segments, filter arrays and so on. For the last years the implementation of widely developed analog fiber-optic link components to compose the antenna array beamforming arrangement has been extensively investigated [4,5,6].

© Springer International Publishing Switzerland 2015
S. Balandin et al. (Eds.): NEW2AN/ruSMART 2015, LNCS 9247, pp. 679–688, 2015.
DOI: 10.1007/978-3-319-23126-6_61

For receiving antenna array a key problem is whether photonic beamformer can satisfy the noise figure and dynamic range demands. So the fiber link components that comprise the specific beamforming scheme need to be analyzed from this point of view as well as to be tested and adjusted to provide the correct delays the microwave signals from antenna elements. The goal of this paper is to analyze the essential parameters of accessible analog fiber optic link main components and to estimate the main performance characteristics of a chosen beamforming scheme (BFS).

Section 2 briefly discusses the performance principle of beamforming scheme based on fiber chromatic dispersion and dense wavelength division multiplexing (DWDM). Section 3 summarizes the investigation technique and some results of measuring key characteristics of fiber-link components under studying. Section 4 presents conclusions.

2   Principle of Photonic Beamforming Based on Fiber Chromatic Dispersion and DWDM

The approach we have chosen for investigating is based on optical fiber chromatic dispersion and WDM technique and permits to provide stepped true time delay control of a phased array antenna pattern delays by optoelectronic switching of fiber segment's lengths [4,5], [7]. In contrast with most of the configurations for photonic BFS with fiber optic components that we have examined [1,2], [8,9,10,11] this approach can be realized using the units and elements available at the market of components of modern fiber-optical telecommunication systems. The general schematic architecture of optical switched TTD beamformer with fiber chromatic dispersion based delays we have discussed earlier [7], [12], and here we briefly consider the principle of this TTD beamforming. Fig. 1 illustrates the schematic diagram of the necessary set of delays that are formed by using the laser's wavelength comb and fiber dispersion. The emission from a set of lasers whose wavelengths are arranged following a uniformly spaced comb according to DWDM ITU (International Telecommunication Union) grid which is intensity modulated by the received microwave signals and combined into a single fiber by a multiplexer enters a time delay unit (TDU) composed from the switchable optic fiber lengths (L_1 and L_2 in Fig. 1). As all the optical carriers share with the same light path, the time delay differences between adjacent channels are produced by the fiber chromatic dispersion D measured in ps/(nm·km).

The time delay of the optical carrier in the neighboring channels with difference between two lasers wavelength $\Delta\lambda$ is expressed as $\Delta\tau = \Delta\lambda\, L\, D$, giving for different fiber lengths L_1 and L_2 delays $\Delta\tau_1$ and $\Delta\tau_2$ respectively. Hence each fiber length introduces a different time delay law necessary for given angle of PAA beam. Maximal dispersion delays amount the value $(M-1)\,\Delta\lambda\,L\,D$ as it is shown in Fig. 1.

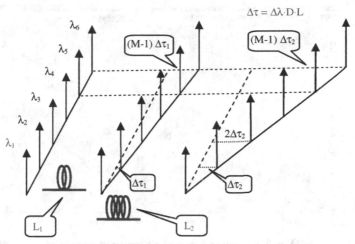

Fig. 1. Principle of photonic beamforming based on a DWDM comb of equidistant laser wavelengths and fiber chromatic dispersion

3 Investigation of Beamformer Key Components Characteristics

3.1 Microwave Photonic Link Components Selection

The initial experimental investigation of performance characteristics of BFS key components were carried out with 'off-the-shelf' ones available in the laboratory. The first analog fiber-optic link was combined of the following units: the laser LDI-DFB-1550-20/80 (Lasercom); the electro-optical Mach-Zehnder modulator IM-1550-20-a (Optilab); the photoreceiver module from SCML-100M18G fiber link (Miteq). Some results of this analog link investigation are published in [7]. For this link some problems were found, for example Mach-Zehnder modulator operating point time and temperature drift. Those problems are typical [13,14]. So for our experimental BFS model we chose new analog fiber-optic links – OTS-2-18 fiber links (Emcore) each consisting of transmitter and receiver modules which can operate at 0.1..18 GHz bandwidth [15]. In-built microprocessor-based unit for control transmitter laser bias and temperature, and Mach-Zehnder modulator bias provides better consistent performance operation and allows for appreciably reduce the measurement results variations. Actual receiver module has an additional microwave amplifier with 15 dB gain after photodetector.

3.2 Transmitters Characteristics

We have measured the performance parameters of some links containing the transmitters with wavelengths from 1555.75 nm to 1551.72 nm corresponding to ITU 100 GHz grid channels from 27 to 32 and the same wideband receiver. Fig. 2 shows optical spectrum of six transmitter modules emission combined by DWDM multiplexer and measured by optical spectrum analyser OSA-20 (Yenista) with resolution 20 pm.

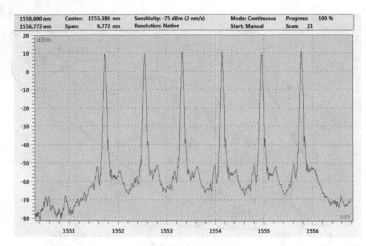

Fig. 2. Optical spectrum of six transmitter modules

One can see fair alignment the laser wavelengths in accordance with 100 GHz ITU grid as well as lasers' emission power passed multiplexer (near 10 dBm).

Half-wave voltage of electro-optical Mach-Zehnder modulator has a great importance for BFS performance calculation [12], [16], so we fulfilled special experiments based on transmitter optical spectrum measurements when modulated by microwave harmonic signal with different power P_{IN}. Fig. 3 shows optical spectrum $P(\lambda)$ of Ch 32 transmitter when modulated by 15 GHz harmonic signal with power $P_{IN} = 16$ dBm (line 1) and unmodulated one (line 2). The intensity modulation gives except for the optical carrier the upper and lower sideband components corresponding to the 1-st, 2-nd and 3-rd harmonics of the microwave oscillation. These components are clearly observed at the figure; their levels and total number depend on microwave signal power. Results of this spectrum analysis are given in Fig. 4: the levels of the carrier $P0$ ("0" label) and of the side components $P1$, $P2$ ("1" and "2" labels) depending on the input microwave signal power P_{IN}.

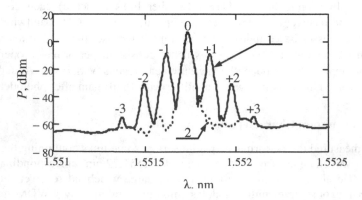

Fig. 3. Optical spectrum of Ch 32 transmitter modulated by microwave 15 GHz harmonic signal (1) and unmodulated one (2)

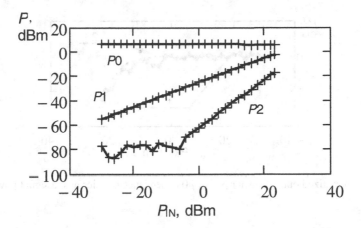

Fig. 4. The optical spectrum components' levels $P0$, $P1$ and $P2$ versus input microwave signal power P_{IN}

The modulation characteristic of electro-optical Mach-Zehnder modulator is approximated by classical interference formula: intensity $I = I_{MAX}0.5[1+\cos(\varphi)]$, where $\varphi = \varphi_0 + \pi V/V_\pi$, φ_0 – operating point location, V_π – the modulator half-wave voltage at microwave input, $V(t) = V_M\cos(2\pi ft)$ – microwave signal voltage. Modulator output optical field $E(t)$ in normalized form is expressed through the Bessel functions of the first kind, k order [17]

$$E(t) = J_0(m)\cos(\varphi_0)\cos(\omega_0 t) + J_1(m)\sin(\varphi_0)\cos[(\omega_0 \pm \omega)t] +$$

$$J_2(m)\cos(\varphi_0)\cos[(\omega_0 \pm 2\omega)t] + J_3(m)\sin(\varphi_0)\cos[(\omega_0 \pm 3\omega)t], \qquad (1)$$

where ω_0 – optical carrier frequency, $\omega = 2\pi f$ – microwave modulation frequency, $m = \pi V_M/V_\pi$ – modulation index. The expression (1) indicates significant difference in spectrum of optical radiation at modulator output and spectrum of photodetector output current. The spectrum of optical radiation contains harmonics of all orders of the microwave modulation frequency, but spectrum of photodetector current for quadrature modulator operating point ($\varphi_0 = \pi/2$) contains only odd harmonics of the microwave modulation frequency. This is necessary to keep in mind.

In Fig. 5 there are given in more detail experimental $p0 = P0/P_{MAX}$ (line 1) and the approximating theoretical $J_0^2(P_{IN})$ (line 2) normalized dependences of an optical carrier power on microwave signal power $P_{IN} = (V_M)^2/2R_0$, where $R_0 = 50$ Ohm – transmitter input characteristic impedance, P_{MAX} – unmodulated optical carrier power. The values of modulation index $m = m(P_{IN})$ and half-wave voltage were found by the approximation, at that V_π coming to 8.7 V. The found V_π value and $m(P_{IN})$ relation allow to calculate theoretical dependences for other optical spectrum side components in a normalized form as $J_1^2(m)/J_0^2(m)$ and $J_2^2(m)/J_0^2(m)$. Fig. 6 shows comparison of these theoretical and experimental dependences. Here experimental dependences are given as $p1 = P1/P_{MAX}$ and $p2 = P2/P_{MAX}$. Good compliance of dependences each other allows to draw a conclusion on accuracy of the found value for the half-wave voltage $V_\pi = 8.7$ V.

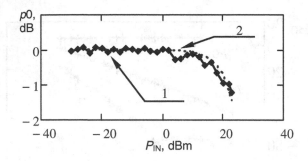

Fig. 5. Normalized optical carrier power $p0$ dependence on microwave signal power P_{IN}

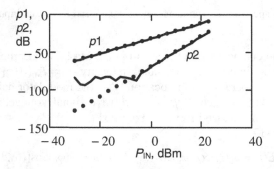

Fig. 6. Comparison of calculated (points) and measured (solid lines) dependences of normalised optical spectrum components $p1$ and $p2$ on microwave input signal power P_{IN}

In BFS and in most other cases links frequency characteristics are highly significant. Shown in figures 7 and 8 frequency responses of two links for frequency band up to 20 GHz are: $|S21(f)|$ in Fig. 7, where line '1' is for link 'Ch 28' and line '2' – for 'Ch 32' one and phase-frequency characteristic as argument of $S21(f) = \Delta\varphi(f)$ – in Fig. 8. One can see the same links frequency behavior for $S21$ and for other S-matrix components also. The traceable curve ripples we connect with the features of receiver module optical entrance. The respective measured links gain is $G_{FOL} \approx -6$ dB for frequency 10 GHz.

Intrinsic time delays brought by OTS-2-18 link modules were measured separately and amounted to 15.6 ns for the receiver and 17.4 ns for the transmitter (Ch 32) modules. Note that time delays in all transmitter modules including multiplexer path are of primarily importance for BFS time delay unit correct operation and they have to be equaled before signals enter TDU.

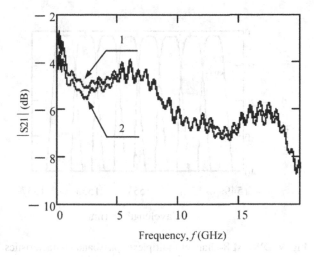

Fig. 7. Frequency responses for 'Ch 28' and 'Ch 32' links

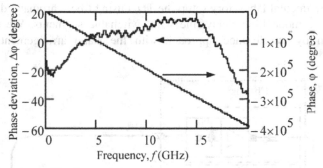

Fig. 8. Phase-frequency response for 'Ch 28' link

3.3   Multiplexer

8-channel DWDM multiplexer (AFW Technology) has 100 GHz channel step for ITU channels from 27 up to 34, insertion loss approx. 1.5 dB, −1 dB channel bandwidth approx. 65 GHz, which is sufficient for throughput intensity modulated optical carrier with RF frequencies up to 20 GHz and more. Nevertheless it should be mentioned that all of the transmitters and corresponding multiplexer channels midmost wavelengths have to be accurately alignment according with ITU channel plan and avoid mismatching during operation. As it was shown in [4] such mismatching could lead to augmentation of the noise power due to DWDM channel slope. We have measured amplitude transmission versus optical wavelength for all of the multiplexer channels with optical spectrum analyzer. The measured multiplexer channels passband characteristics are shown in Fig. 9. They demonstrate allowable ITU grid setting accuracy and flat-shape channel curve behavior.

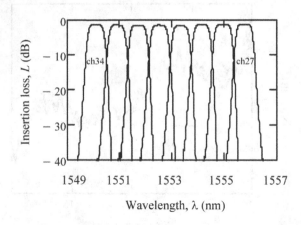

Fig. 9. DWDM 8-channel multiplexer passband characteristics

3.4   Time Delay Unit

In our BFS experimental laboratory setup the TDU using fiber chromatic dispersion is realised as single mode fiber delay line with switchable segments. We tested fiber TDU operation by interference approach with the stand arrangement shown in Fig. 10.

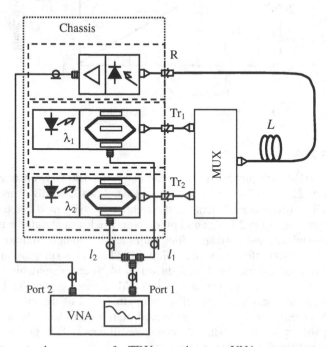

Fig. 10. Laboratory stand arrangement for TDU operation tests, VNA – vector network analyser

Electric length of two channels (Tr transmitters, including the multiplexer) was equaled by a fiber segment to a residual difference Δl = 6.6 cm. Channels Ch 27 and Ch 29 had an optical carriers wavelength difference $\Delta \lambda$ = 1.6 nm. The single harmonious signal from the vector network analyzer ZVA 40 (Rohde & Schwartz) after 50 : 50 division entered the microwave inputs of both channels. During the vector analyzer frequency sweeping the recorded output signal of the receiver ($|S21(f)|$) represented an interference picture, see Fig. 11. Its period $f_{\Delta l}$ and the first minimum localization $f_{MIN} = f_{\Delta l}/2$ are defined unambiguously by a difference: $f_{\Delta l} = c/\Delta l$, where c – velocity of light. In Fig. 11 $f_{\Delta l}$ = 4.5 GHz for Δl = 6.6 cm – line 1, and line 2 shows changed interference picture due to inclusion of a single-mode SMF-28 fiber coil with a length L_{TDU} = 3.1 km (as TDU) after the multiplexer. The dispersion delay in this case is changed by $\tau = D\Delta\lambda L_{TDU}$ = 17•1.6•3.1 = 84.3 ps, that leads to the interference period $f_{\Delta l}$ and the first minimum localization change: $f_{\Delta l+L}$ = 7.3 GHz, which is confirmed by direct measuring.

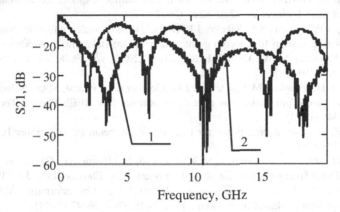

Fig. 11. Interference picture at the receiver output; 1 – for 6.6 cm residual difference of electric lengths of two channels, 2 – with the addition of 3.1 km fiber segment length

Received experimental results of our fiber-optic links basic characteristics investigation showed satisfactory agreement with the estimation values.

The results of measuring the link transfer characteristics will be used to calculate the basic performance of the BFS for receiving PAA in a manner we have reported in [12]. For further studies of the optical BFS and its experimental setup development the components of radiophotonic fiber links with DWDM technology can be used.

4   Conclusions

We have investigated the essential characteristics of new analog fiber optic links main components with the view of their application in true time delay based broadband beamforming system for receiving microwave phased array antenna. We have started development the experimental setup of BFS based on fiber optic dispersion and DWDM technology usage. The obtained results permit to determine the parameters and operation modes of microwave fiber link's components limiting the photonic BFS performance characteristics.

References

1. Riza, N.A.: Selected Papers on Photonic Control Systems for Phased Array Antennas, ser. SPIE Milestone, vol. MS136. SPIE Press, New York (1997)
2. Chazelas, J., Dolfi, D., Tonda, S.: Optical Beamforming Networks for Radars & Electronic Warfare Applications. RTO-EN-028 **9**, 1–14 (2003)
3. Yao, J.P.: A Tutorial on Microwave Photonics – Part II. IEEE Photonics Society Newsletter **26**(3), 5–12 (2012)
4. Blanc, S., Alouini, M., Garenaux, R., Queguiner, M., Merlet, T.: Optical multibeamforming network based on WDM and dispersion fiber in receive mode. IEEE MTT Trans. **54**(1), 402–411 (2006)
5. Yang, Y., Dong, Y., Liu, D., He, H., Jin, Y., Hu, W.: A 7-bit photonic true-time-delay system based on an 8×8 MOEMS optical switch. Chinese Optics Letters **7**, 118–120 (2009)
6. Yilmaz, O.F., Yaron, L., Khaleghi, S., Chitgarha, M.R., Tur, M., Willner, A.: True time delays using conversion/dispersion with flat magnitude response for wideband analog RF signals. Optics Express **20**, 8219–8227 (2012)
7. Lavrov, A.P., Ivanov, S.I., Saenko, I.I.: Investigation of analog photonics based broadband beamforming system for receiving antenna array. In: Balandin, S., Andreev, S., Koucheryavy, Y. (eds.) NEW2AN/ruSMART 2014. LNCS, vol. 8638, pp. 647–655. Springer, Heidelberg (2014)
8. Esman, R.D., Frankel, M.Y., Dexter, J.L., Goldberg, L., Parent, M.G., Stilwell, D., Cooper, D.G.: Fiber-optic prism true time-delay antenna feed. IEEE Photon. Technol. Lett. **5**, 1347–1349 (1993)
9. Soref, R.A.: Fiber grating prism for true time delay beam steering. Fiber Integr. Opt. **15**, 325–333 (1996)
10. Cruz, J.L., Ortega, B., Andrés, M.V., Gimeno, B., Pastor, D., Capmany, J., Dong, L.: Chirped fiber Bragg gratings for phased array antennas. Electron. Lett. **33**, 545–546 (1997)
11. Mitchella, M., Howarda, R., Tarran, C.: Adaptive digital beamforming (ADBF) architecture for wideband phased array radars. Proc. SPIE **3704**, 36–47 (1999)
12. Ivanov, S.I., Lavrov, A.P., Saenko, I.I.: Optical-fiber system for forming the directional diagram of a broad-band phased-array receiving antenna, using wave-multiplexing technology and the chromatic dispersion of the fiber. J. Opt. Technol. **82**(3), 139–146 (2015)
13. Ponomarev, R.S., Zhuravlev, A.A., Khrychikov, A.A., Shevtsov, D.I.: Short-term DC-drift in integrated optical Mach-Zehnder interferometer. Proc. SPIE **8410**(1–6), 841008 (2011)
14. Wooten, E.L., Kissa, K.M., Yi-Yan, A., Murphy, E.J., Lafaw, D.A., Hallemeier, P.F., Maack, D., Attanasio, D.V., Fritz, D.J., McBrien, G.J., Bossi, D.E.: A Review of Lithium Niobate Modulators for Fiber-Optic Communications Systems. IEEE J. of Selected Topics in Quantum Electronics **6**, 69–82 (2000)
15. 50 MHz to 18 GHz Unamplified Microwave Transport System. http://products.emcore.com/avcat/images/documents/dataSheet/Optiva-OTS-2-18GHz-Unamplified2.pdf
16. Froberg, N.M., Ackerman, E.I., Cox, C.: Analysis of signal to noise ratio in photonic beamformers. In: IEEE Aerospace Conference, Paper 1067 (2006)
17. Hilt, A.: Microwave harmonic generation in fiber-optical links. J. of Telecommunications and Information Technology **1**, 22–28 (2002)

All-Optical 4-Channel Demultiplexer with an Arbitrary Access for Full C+L Spectral Range

Viktor M. Petrov[1(✉)], Vladimir V. Lebedev[2],
Nicolai V. Toguzov[2], and Sergey Zukov[3]

[1] Saint Petersburg State Polytechnical University, Polytechnicheskaya Str, 29,
St. Petersburg 195251, Russia
vikpetroff@mail.ru
[2] A.F. Ioffe FTI Institute, Polytechnicheskaya Str, 26, St-Petersburg 194021, Russia
{vladimir_l,ntv_05}@mail.ru
[3] Darmstadt Technical University, Darmstadt, Germany
zukov.tud@gmail.com

Abstract. Development of all-optical devices, i.e., the devices without an intermediate conversion into an electrical signal, for optical (D)WDM systems is a vital problem. Below we present the results of research and development of an optical demultiplexer that allows one to independently select spectral channels in the C and L spectral bands (1530 – 1625 nm). The device we consider here allows operation with information flows of 10 - 64 Gbit/s.

Keywords: Optical demultiplexers · Tunable Fabry-Perot filters

1 Introduction

An efficient way to create all-optical demultiplexers for the optical communication lines operating according to the DWDM standards is to use different tunable filters [1, 2]. When used in such a device, each filter can be tuned independently to the desired wavelength, and thus demultiplexing is carried out. At present different optical filters which can be used for a wide range of applications are available. The most commonly used filters are Fabry-Perot filters [3], filters based on Bragg gratings written in a fiber [4, 5], and Bragg filters recorded in photorefractive crystals [6, 7]. One of the most challenging tasks is to create demultiplexers operating in the C + L spectral bands, i.e., in the wavelength range from 1530 to 1625 nanometers. According to modern standards, a channel bandwidth of no more than 0.2 nm should be provided. Therefore, the second significant requirement is that a constant spectral selectivity should be maintained at a level of 0.2 nm in the entire huge spectral range. As our analysis of the literature has shown, only tunable Fabry-Perot filters can satisfy these requirements. Below we present the results of investigation of the demultiplexer based on Fabry-Perot filters we have developed.

© Springer International Publishing Switzerland 2015
S. Balandin et al. (Eds.): NEW2AN/ruSMART 2015, LNCS 9247, pp. 689–697, 2015.
DOI: 10.1007/978-3-319-23126-6_62

2   Scheme of the Demultiplexer Based on Tunable Fabry-Perot Filters

The demultiplexer has one optical input for N channels in the (D)WDM standard. At each of its four outputs the demultiplexer selects any of the input channels. Let us consider the implementation of a four-channel demultiplexer based on Fabry-Perot filters. The demultiplexer comprises a passive 1-to-4 splitter and four independent tunable wavelength filters, each of which is connected to its splitter output. Thus, an optical signal containing all input channels is fed to the input of each filter. The input power is divided between all four channels in approximately equal fractions.

To select the desired frequency channel, independently tunable optical Fabry-Perot filters are used. Such filters are fabricated on the basis of optical fibers.

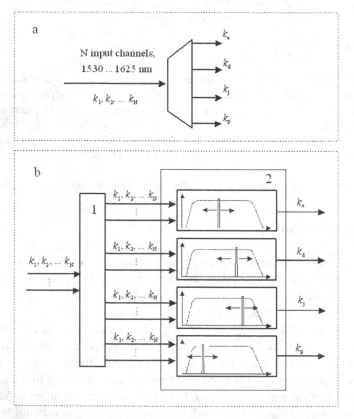

Fig. 1. Full-optical demultiplexer based on Fabry-Perot filters. a: structural scheme. 1 – demultiplexer. The demultiplexer has 1 input and 4 arbitrary tunable outputs. b: functional scheme of the demultiplexer based on tunable filters. 1 – 1-to-4 splitter, 2 – set of tunable Fabry-Perot filters. $k_1, k_2, \ldots k_N$ – N are input channels, k_a, k_d, k_j, k_l – are selected channels (output).

In order to perform the wavelength tuning of the filter, it is necessary to change the distance between the mirrors. This was achieved in our device by means of a piezoelectric actuator by applying an electric voltage. The control voltage ensuring the filter tuning in the entire tuning range was about 18 V.

3   Basic Technical Parameters and Construction of the Filters

To build the demultiplexer, 4 Fabry-Perot fiber-optical, electrically-tunable filters were used. The ranges of the most important parameters are as follows: finesse $F = 1281 \ldots 1331$; free spectral range (tuning range) $\Delta\lambda = 93.1 \ldots 98.8$ nm; spectral selectivity $\Delta\lambda = 0.19 \ldots 0.22$ nm. Table 1 shows as an example a set of parameters for one of the filters.

Table 1.

Finesse F.	1286
Free spectral range (tuning range).	98,8 nm
Spectral selectivity $\Delta\lambda$, C - band.	0,20 nm
Insertion loss, C - band.	1,56 ... 1,72 dB
Spectral selectivity $\Delta\lambda$, L - band.	0,22 nm
Insertion loss, L - band.	1,35 dB / 1,55 dB
Working temperature's range	- 80 ... + 20 C^0

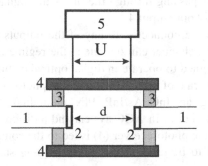

Fig. 2. Electrically tunable Fabry-Perot filter (cross-section). 1 – optical fiber, 2 – interference mirrors, 3 – sokets, 4 – cylindrical piezodriver, 5 – control voltage unit.

Fig. 2 shows the cross-section of the filter. A pair of cleaved fibers (1) and covered by an interference mirrors (2) provides a Fabry-Perot interferometer. The fibers are fixed into the sockets (3). The sockets are connected to the piezoceramic cylinder (4). The distance d between the mirrors can be tuned by applying the electric field from unit 5. The control voltage is about 10 volts, tuning time was approximately 0.1 seconds.

4   Demonstrator

The demonstrator has one optical input and four optical outputs. FC/APC connectors are used. The channel tuning speed in the range of approximately 90 nm is about 0.1 s. It is important to note that this time is determined solely by the USB interface used. Potentially, the tuning time in this range can be 3 ms. The channel spectral resolution is not poorer than 0.09 nm. The optical loss between the channel input and output is about 10 dB. The maximum optical power applied to the input is not more than 40 mW in the operating band. The demultiplexer is connected to a computer through a USB port. The demonstration model of the demultiplexer is in the box 410 x 200 x 130 mm^3 in size. The weight is 3 kg, the power consumption is about 50 W.

5   Optical Scheme of the Demonstrator

The optical scheme of the demultiplexer is shown in Figure 3. The 1-to-4 splitter (1) consists of three directional couplers with a coupling ratio of 1/1. They are followed by 1-to-2 couplers (2) with a coupling ratio of about 1/10. The major fraction of the power (approximately 90%) is fed to the filters (3), the remaining power is fed to photoreceivers (6). After passing through the filters the signals are applied to optical outputs "opt.output 1" ... "opt.output 4".

Photoreceiver (4) can be connected to any of the outputs by means of an optical cable (5). In this case the channel can operate in the regime of spectral analysis. The photoreceiver (4) is designed to operate in digital optical communication lines with an information transmission rate of up to 10 Gbit/s. The photoreceiver is fabricated as an integral unit containing an InGaAs/InP PIN photodiode and a transimpedance low-noise amplifier with a built-in AGC system and a differential output. A digitized electric signal from the the photoreceiver (4) is fed to the computer to be displayed on the monitor screen and to be used in the unit providing stabilization of a selected spectral band.

When experimental studies of the model were carried out, an optical insulator (7) which provided isolation of possible reflected waves from the outputs of the light sources used was placed in front of the device input.

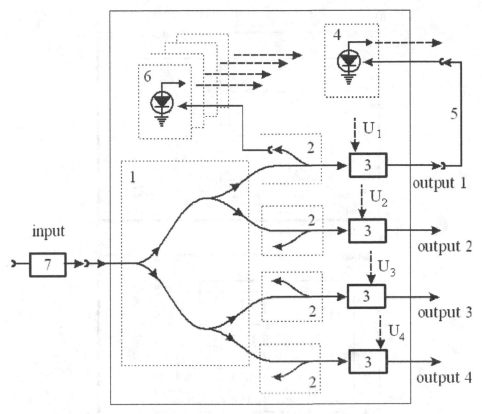

Fig. 3. Optical sheme of the four-channel demultiplexer with an arbitrary access and optical spectrum analyzer. 1 – 1-to-4 splitter, 2 – 1-to-2 splitters, 3 - electrically controlled Fabry-Perot filters, 4 - photoreceiver, 5 - additional optical cable which can be connected to any of the optical outputs 1 ... 4, 6 - set of four control photoreceivers (installed additionally), 7 - optical insulator, "opt. input" - optical input, "opt output 1 ... 4" - optical outputs of corresponding optical channels. $U_1...U_4$ - control voltages. The solid lines show optical links, the dot-and-dash lines show electric connections.

6   Results of experimental investigations

Fig. 4 present experimental results for the model using a signal from a broadband source as an input. The broadband spectrum applied to the input was examined at the output by a spectrum analyzer. Fig. 4, b shows the transfer function of a single channel. The spectral selectivity of the channel estimated from it is about 0.2 - 0.22 nm. Note that it is necessary to take into account the effect of the transfer function of the spectrum analyzer on the measurements.

a)

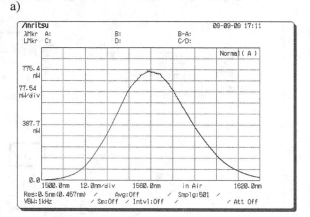

b)

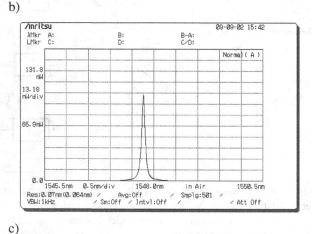

c)

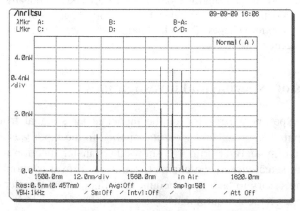

Fig. 4. a – broadband source spectrum (approximately 1520 - 1620 nm) applied to the input, b – demonstration of spectral selectivity of a single spectral channel, c – demonstration of allocation of four channels of the broadband spectrum (the difference in channel amplitudes is caused by the input spectrum shape).

Figs. 5 and 6 presents some examples of experimental results obtained in the studies of the model using two lasers the wavelengths of which differ by 0.2 and 0.4 nm, or by 25 and 50 GHz, respectively. The measurements were performed with the frequency modulation amplitude in the stabilization unit of 0.05 nm.

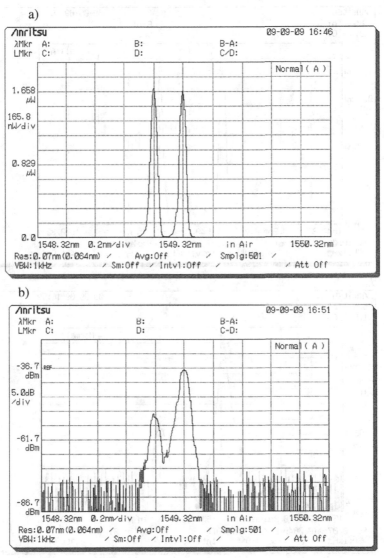

Fig. 5. a – spectra of two input narrow-band signals applied to the input: 1549.12 nm and 1549.32 nm, $\Delta\lambda = 0.2$ nm, b – allocation of channel at 1549.32 nm, crosstalk is approximately 15 dB.

In some experiments the filter was tuned to a lower frequency, in others it was tuned to a higher frequency. This method allows estimation of the crosstalk that arises when closely spaced signals are used.

a)

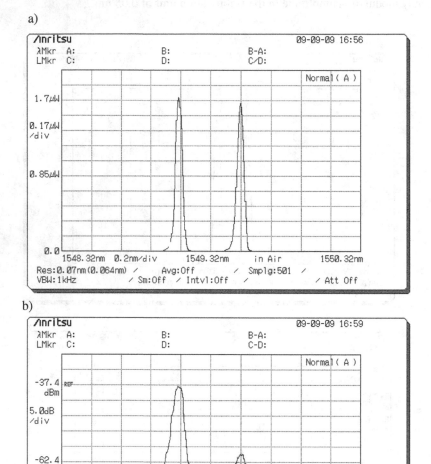

Fig. 6. a – spectra of two input narrow-band signals at 1549.12 nm and 1549.52 nm, Δλ = 0.4 nm, b – allocation of channel at 1549.32 nm, crosstalk is approximately 22 dB.

As the data presented above show, the crosstalk was approximately -15 and -20 dB for the frequency channel spacing of 25 and 50 GHz, respectively.

7   Conclusions and Results

Owing to the use of tunable Fabry-Perot filters, an all-optical four-channel demultiplexer has been developed. This device can operate in the 1530 - 1625 nm band and select four independent spectral channels in the 0.2 or 0.4 nm band. Direct measurements of informational throughput using proper telecommunication equipment have confirmed the possibility to operate with the traffic as large as 64 Gbit / s. The device provides up to 512 separated WDM channels.

Several features of our device should be noted.

1. Separation of the input power into four parallel channels at the device input leads to a considerable optical loss. To compensate for the loss, it is reasonable to use optical amplifiers at each of four outputs.

2. As additional studies have shown, special attention should be paid to the stability of the filter tuning to a desired wavelength. Temperature variations and uncontrollable charge accumulation by the mechanical construction exert the most pronounced influence on the filter stability. The piezoceramic driver used for wavelength tuning is very sensitive to the presence of parasitic (uncontrollable) charges. Such charges cause small changes in the distance between the mirrors and, hence, in the selected wavelength. To eliminate this phenomenon, particular attention should be paid to a reliable grounding and shielding of the filter.

References

1. ITU-T Recommendation G.671, Transmission Characteristics of Passive Optical Components (1996)
2. ITU-T Recommendation G.692, Optical Interfaces for Multichannel Systems with Optical Amplifiers (1998)
3. Born, M., Wolf, E.: Principles of Optics, 3rd edn, p. 808. Pergamon Press, Oxford (1964)
4. Teixeira, A., Beleffi, G. (eds.): Optical Transmission, p. 625. Springer, New York (2011)
5. Almeida, P., Silva, H.A.: Radio-over-Fiber Systems with Support for Wired and Wireless Services, p. 460. Scholars Press, Saarbrücken (2015)
6. Arora, P., Petrov, V.M., Petter, J., Tschudi, T.: Integrated optical Bragg filter with fast electrically controllable transfer function. Opt. Comm. **281**(8), 2067–2072 (2008)
7. Shamray, A.V., Ilichev, I.V., Kozlov, A.S., Petrov, V.M.: Controllable holographic optical filters in photorefractive crystals. Journal of Holography and Speckle **5**, 243–253 (2009)

Implementation of Digital Demodulation for Fiber Optic Interferometer Sensors

Andrei Medvedev[✉], Andrei Berezhnoi,
Aleksei Kudryashov, and Leonid Liokumovich

Peter the Great St. Petersburg Polytechnic University, Saint Petersburg, Russia
medvedev@rphf.spbstu.ru, andrey-berezhnoy@yandex.ru,
kudrjashov_av@svetlanajsc.ru, leonid@spbstu.ru

Abstract. The experimental validation of the fiber interferometer signal demodulation method based on the additional phase modulation and digital signal processing is presented. The results obtained for various target signal shapes and different number of samples per modulation period demonstrated the effectiveness of the processing algorithms applied to the interference systems.

Keywords: Fiber optic sensor · Interferometer · Polarization · Phase modulation · Digital signal processing

1 Introduction

Fiber-optic interferometers are widely used nowadays along with other devices that utilize the modulation of optic wave's parameters by some physical value. This occurs due to the development of laser and fiber-optic technologies as measuring devices. The photodetector output of the interference system is

$$u(t) = U_0 + U_m \cos[\varphi_{S}(t)], \tag{1}$$

where $\varphi_{S(t)}$ is a target phase difference carrying information about the measured value, U_0 is a constant component, U_m is an interferometer signal amplitude. However, such a signal does not provide means to determine $\varphi_{S(t)}$ uniquely. Moreover, the constant component and the interferometer signal amplitude can fluctuate in time. Because of these factors determining of the target phase difference $\varphi_{S(t)}$ becomes a complicated task.

The problem of phase demodulation can be solved by applying additional phase modulation and analog processing of the interferometer signal. Many of these methods use harmonic additional modulation. In this case the interference signal acquires the form

$$u(t) = U_0 + U_m \cos[\varphi_S(t) + \varphi_m \sin(\omega_m t)]. \tag{2}$$

For the signal (2) analog devices are able to determine the target phase difference by synchronous detecting, summation, differentiating, integrating, etc.

© Springer International Publishing Switzerland 2015
S. Balandin et al. (Eds.): NEW2AN/ruSMART 2015, LNCS 9247, pp. 698–704, 2015.
DOI: 10.1007/978-3-319-23126-6_63

A so-called pseudo heterodyne method is another well-known algorithm based on the additional phase modulation. There the signal formed at the photo receiver output becomes

$$u(t) = U_0 + U_m \cos[\varphi_s(t) + \omega_m t]. \tag{3}$$

To demodulate this signal traditional analog phase metric devices can be utilized.

Digital technologies provide new opportunities to demodulate interferometer signals [1]. Of course, digital methods are more effective when they are not just a digital implementation of the analog algorithms [2], but utilize, for example, calculation of inverse trigonometric functions [3], Hilbert transform [4] and even more complicated procedures [5], that cannot be implemented using analog signal processing.

2   Demodulation Method

Here we present the results of implementation of previously described digital demodulation algorithm of the fiber interferometer signal using additional phase modulation [6]. The main concept of this method is that digital processing enables implementation of various mathematical procedures and there is no need to obtain the relatively simple signal like (3). So interferometer output signal used for processing can be of more complex shape but easier to be obtained in optical scheme.

This method enables implementation of demodulation with various parameters. The essential requirement for this approach is using of signal with shape described by the equation (2). In this case, φ_s, U_0, U_m can vary. Oscillation of φ_s, bears useful information. Fluctuations of U_0, U_m are hindering, but considered slow enough in comparison with the period of the modulating signal.

The basis of the demodulation method is calculating of a single value of the target phase φ_s, using three interference signal samples per every modulation period. These samples are taken under different values of the additional phase difference caused by the modulation applied. The necessity of using of three samples can be explained as follows. The function (2) describes the modulated interference signal. It includes three unknown parameters that are supposed to be constant while taking samples: target phase φ_s, U_0, and U_m. Let us denote these three samples $\{u^{(0)}, u^{(1)}, u^{(2)}\}$, and values of the modulating phase signal $\{\varphi^{(0)}, \varphi^{(1)}, \varphi^{(2)}\}$, respectively. Then, the system of equations with these unknown parameters can be written as

$$\begin{cases} u^{(0)} = U_0 + U_m \cdot \cos(\varphi_s + \varphi^{(0)}) \\ u^{(1)} = U_0 + U_m \cdot \cos(\varphi_s + \varphi^{(1)}). \\ u^{(2)} = U_0 + U_m \cdot \cos(\varphi_s + \varphi^{(2)}) \end{cases} \tag{4}$$

Using trigonometric formulas and denoting $cos[\varphi^{(0)}]=C_0$, $cos[\varphi^{(1)}]=C_1$, $cos[\varphi^{(2)}]=C_2$ and $sin[\varphi^{(0)}]=S_0$, $sin[\varphi^{(1)}]=S_1$, $sin[\varphi^{(2)}]=S_2$, U_0 and U_m can be excluded. Hence, the formula for a tangent of the target phase is

$$tg\varphi_S = \frac{(u^{(0)}-u^{(1)})\cdot(C_1-C_2)-(u^{(1)}-u^{(2)})\cdot(C_0-C_1)}{(u^{(1)}-u^{(2)})\cdot(S_1-S_0)-(u^{(0)}-u^{(1)})\cdot(S_2-S_1)} = \frac{a}{b},\tag{5}$$

where

$$\begin{cases} a = \left(u^{(0)} - u^{(1)}\right) \cdot (C_1 - C_2) - \left(u^{(1)} - u^{(2)}\right) \cdot (C_0 - C_1) \\ b = \left(u^{(1)} - u^{(2)}\right) \cdot (S_1 - S_0) - \left(u^{(0)} - u^{(1)}\right) \cdot (S_2 - S_1) \end{cases}.\tag{6}$$

In view of the formula for a tangent, the target phase φ_S in the range $[-\pi, \pi]$ can be found using the following equations:

$$\begin{cases} \varphi_S = arctg\left[\frac{a}{b}\right], \text{for } b > 0; \\ \varphi_S = arctg\left[\frac{a}{b}\right] + \pi \cdot sign(a), \text{for } b < 0. \end{cases}\tag{7}$$

3   Implementation of the Demodulation Method

We have carried out a series of experiments to test the demodulation method, using an experimental model of a fiber-optic polarization sensor designed for electric field measurement [7]. It contained an additional polarization modulator that provided the sine modulation to recover the target signal $\varphi_{S(t)}$ (Fig. 1). The output signal of this system can be described by the equation (2), which allows application of the considered demodulation method.

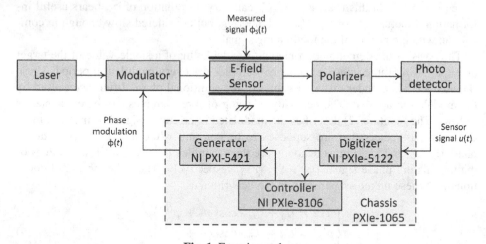

Fig. 1. Experimental setup

The LabVIEW program controlled the hardware, displayed the results and wrote them to file. An additional modulation signal was formed using 16-bit arbitrary waveform generator NI PXI-5421. Signal parameters were chosen according to the demodulation method protocol [6]. The additional modulation frequency was 10 kHz.

The digitizer NI PXIe-5122 registered the interference signal from the photodetector output. The signal sampling frequency was set in accordance with the number of samples processed per every modulation period. Cases with samples per period $N=3$, 4 and 5 were considered.

Samples of the additional modulation signal φ_n and of the interference signal u_n can be written as follows

$$\begin{cases} \varphi_n = \varphi(t_n) = \delta\varphi_m \sin[2\pi f_M(t_n + \Delta t)] = \delta\varphi_m \sin\left[\left(\frac{2\pi n}{N}\right) + \theta_0\right] \\ u_n = u(t_n) = U_0 + U_m \cos[\varphi_S + \varphi_n] = U_0 + U_m \cos\left[\varphi_S + \delta\varphi_m \sin\{\left(\frac{2\pi n}{N}\right) + \theta_0\}\right] \end{cases}, \quad (8)$$

where n is the sample number starting from the beginning of the first modulation period ($n=0, 1, 2, \ldots$), t_n is the time of the n-th sample, Δt and $\theta_0=2\pi f_M\Delta t$ are time delay and a corresponding phase shift of the first sample relatively to the beginning of the first modulation period. In our experiments the sampling sequence was synchronized with the modulating signal without delay ($\Delta t=0$) regardless of the number of samples N chosen.

Fig. 2 shows the acquisition points for different values of N marked on the experimental modulation and interference signals charts.

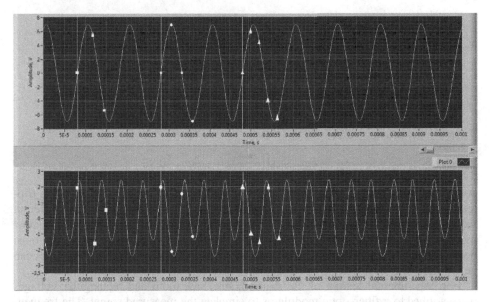

Fig. 2. Acquisition points for the modulation signal (upper trace) and interference signal (lower trace) for different sampling rates

Points for the case of $N=3$ are marked with squares. According to the method modulation phase values are

$$\varphi_0 = \delta\varphi_m \sin\left[\left(\frac{2\pi \cdot 0}{3}\right) + \theta_0\right] = 0,$$

$$\varphi_1 = \delta\varphi_m \sin\left[\left(\frac{2\pi \cdot 1}{3}\right)\right] = \delta\varphi_m \cdot 0{,}866,$$

$$\varphi_2 = \delta\varphi_m \sin\left[\left(\frac{2\pi \cdot 2}{3}\right)\right] = -\delta\varphi_m \cdot 0{,}866.$$

Acquisition points for $N=4$ are marked with circles. The corresponding modulation phase values are

$$\varphi_0 = \delta\varphi_m \sin\left[\left(\frac{2\pi \cdot 0}{4}\right)\right] = 0,$$

$$\varphi_1 = \delta\varphi_m \sin\left[\left(\frac{2\pi \cdot 1}{4}\right)\right] = \delta\varphi_m,$$

$$\varphi_2 = \delta\varphi_m \sin\left[\left(\frac{2\pi \cdot 2}{4}\right)\right] = 0,$$

$$\varphi_3 = \delta\varphi_m \sin\left[\left(\frac{2\pi \cdot 3}{4}\right)\right] = -\delta\varphi_m.$$

Acquisition points for $N=5$ are marked with triangles. The corresponding modulation phase values are

$$\varphi_0 = \delta\varphi_m \sin\left[\left(\frac{2\pi \cdot 0}{5}\right)\right] = 0,$$

$$\varphi_1 = \delta\varphi_m \sin\left[\left(\frac{2\pi \cdot 1}{5}\right)\right] = \delta\varphi_m \cdot 0{,}95,$$

$$\varphi_2 = \delta\varphi_m \sin\left[\left(\frac{2\pi \cdot 2}{5}\right)\right] = \delta\varphi_m \cdot 0{,}59,$$

$$\varphi_3 = \delta\varphi_m \sin\left[\left(\frac{2\pi \cdot 3}{5}\right)\right] = -\delta\varphi_m \cdot 0{,}59,$$

$$\varphi_4 = \delta\varphi_m \sin\left[\left(\frac{2\pi \cdot 4}{5}\right)\right] = -\delta\varphi_m \cdot 0{,}95.$$

As it was demonstrated in [6], for the proper calculation of target phase φ_S using triplet of samples $\{u^{(0)}, u^{(1)}, u^{(2)}\}$, three phase samples $\{\varphi^{(0)}, \varphi^{(1)}, \varphi^{(2)}\}$ must have different values caused by the additional phase modulation. For example, for the case of $N=4$ modulation triplet $\{\varphi_0, \varphi_1, \varphi_3\}$ or $\{\varphi_1, \varphi_2, \varphi_3\}$ with corresponding signal triplet must be picked out for calculation. In the same way, the rule of choosing phase samples can be formulated for other N.

For experimental testing of the method, sine, saw tooth and rectangular voltages were applied to a fiber-optic modulator to simulate the measured signal. The frequency of the test voltage was 50 Hz.

The demodulation method was implemented for the number of samples $N=3, 4, 5$. As the results of the interference signals demodulation, the target signals $\varphi_{S(t)}$ were extracted. The sensor output demodulated signals for $N=3$ are shown in Fig. 3.

The spectrum of the sine-shape demodulated signal is shown in Fig. 4. This spectrum helps to evaluate such parameters as noise level and harmonic distortion. The noise level appears to be -90 dB, which corresponds to the minimal registered phase shifts of $3 \cdot 10^{-5}$ rad/Hz$^{1/2}$. The harmonic distortion for the sine signal for $N=3$ was 0.5%.

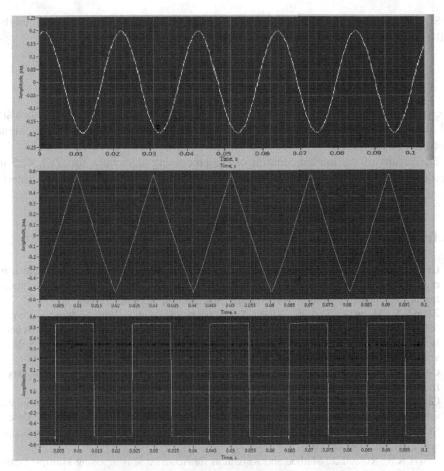

Fig. 3. Demodulated signals of different shapes for $N=3$

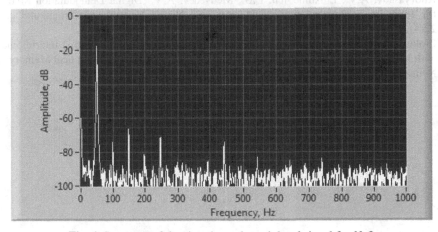

Fig. 4. Spectrum of the sine-shape demodulated signal for $N=3$

Similar results were obtained for noise levels and harmonic distortions for other values of N and corresponding optimal modulation amplitudes.

4   Conclusion

The paper presents the experimental validation of the method of the interference signal demodulation with additional harmonic phase modulation and digital signal processing in LabVIEW.

The results of the implementation of the processing algorithms for N=3, 4, 5 for various signal shapes demonstrated their effectiveness in application to the interference systems with signals described by the equation (2).

References

1. Griffin, B., Connelly, M.J.: Interferometric fiber optic sensor interrogation system using digital signal processing and synthetic-heterodyne detection. In: Proc. of SPIE 17th International Conference on Optical Fibre Sensors, vol. 5855, pp. 619–622 (2005)
2. Zhang, N., Meng, Z., Rao, W., Xiong, S.: Investigation on upper limit of dynamic range of fiber optic interferometric sensors base on the digital heterodyne demodulation scheme. In: Proc. of SPIE 22nd International Conference on Optical Fiber Sensors (OFS 2012), vol. 8421, pp. 8421BE-2–8421-BE4 (2012)
3. Cekorich, A.C., Davis, J.G.: Demodulator and Method for Interferometric Outputs of Increased Accuracy. US Patent 6,556,509 B1 (2003)
4. Giulianelli, L.C., Buckman, A.B., Walser, R.M., Becker, M.F.: Digital demodulation scheme for wide-dynamic-range measurements with a fiber optic interferometer. In: Proc. of SPIE Smart Structures and Materials 1994: Smart Sensing, Processing, and Instrumentation, vol. 2191, pp. 314–323 (1994)
5. Lewis A.B., Russell S.: Method and Apparatus for Acoustic Sensing Using Multiple Optical Pulses. US Patent 2015/0160092A1 (2015)
6. Kudryashov, A.V., Liokumovich, L.B., Medvedev, A.V.: Digital Demodulation Methods for Fiber Interferometers. Optical Memory and Neural Networks (Information Optics) 22(4), 236–243 (2013)
7. Liokumovich, L.B., Medvedev, A.V., Petrov, V.M.: Fiber-Optic Polarization Interferometer with an Additional Phase Modulation for Electric Field Measurements. Optical Memory and Neural Networks (Information Optics) 22(1), 21–27 (2013)

Optical Coder with A Synthesized Transfer Function for Optical Communication Lines

Viktor M. Petrov[1(✉)] and Roman V. Kiyan[2]

[1] St.-Petersburg State Polytechnical University, Polytechnicheskaya Str, 29,
St.-Petersburg 195251, Russia
`vikpetroff@mail.ru`
[2] Laser Zentrum Hannover e.V., Hollerithallee 8 30419, Hannover, Germany
`r.kiyan@lzh.de`

Abstract. Almost all modern optical communication lines, both local and trunk ones, use the wavelength-division multiplexing principle (WDM or DWDM). We suggest that frequency encoded OCDMA algorithms be used as most suitable tools for local and metropolitan area networks. Coding is performed in this case in the frequency domain, i.e., code sequences can be mapped inside the optical source spectrum. A broadband light source is used in this approach. The spectral amplitude of the light source is modulated with the code that specifies certain components of the spectrum to be on or off. In such a network, different transmitters use different codes which are orthogonal to one another. The receiver can then select the data from the desired transmitter by correlating the spectrally modulated signal with an appropriate code. Furthermore, the channels are independent in the sense of transmitted data bit rates.

The coding algorithm we suggest is performed in a narrow spectral band, within one (D)WDM channel, i.e., within a spectral range of 0.4 to 0.8 nm. In this case the available communication lines can be used. However, due to the fact that such lines use amplitude modulation for information encoding, the signal with a spectral encoding will be nearly inaccessible for an unauthorized user.

Keywords: Optical coder · Tunable bragg gratings

1 Coding Principle

We implemented the coding based on electro-optical synthesis of the transfer function of an electrically-controllable optical Bragg filter [1, 2]. The optical filter with the synthesized transfer function can be regarded in this case as a controllable optical information coder. In our approach the coder is the device based on an electrooptic crystal in which a reflection Bragg grating in the form of a periodic refractive index variation is formed in the light propagation direction [3]

$$n = n_0 + n_1 \cos 2\pi x/\Lambda \qquad (1)$$

© Springer International Publishing Switzerland 2015
S. Balandin et al. (Eds.): NEW2AN/ruSMART 2015, LNCS 9247, pp. 705–711, 2015.
DOI: 10.1007/978-3-319-23126-6_64

Here, n_0 is the average refractive index of the substrate, n_1 is the refractive index variation amplitude, and Λ is the grating period. Let us explain the principle of spectral coding by using the following example (see Fig. 1).

Fig. 1(a) (left) shows a device of length L and grating period Λ. The readout light propagates from the left to the right. In accordance with the Bragg diffraction law, such a grating will reflect the light at wavelength λ_B.

$$\lambda_B = 2n\Lambda \tag{2}$$

The width of reflected spectral band $\delta\lambda$ can be estimated as

$$\delta\lambda = \lambda_B \, \Lambda/L \tag{3}$$

Fig. 1 (a) (right) shows the transfer function of such a coder. Here, the transfer function means the dependence of the reflected light power I on the wavelength λ. It can be seen that it has only one reflected spectral line $\delta\lambda$ at the central wavelength λ_B.

Fig. 1 (b) (left) shows the coder with the same length and the same grating period, but the phase shift $\varphi_1 - \varphi_2 = \Delta\varphi = 180°$ is introduced in the middle of the grating. To put it otherwise, the periodic grating consists of two identical sections the phase difference between which is 180°. The transfer function in this case has the form shown in the right part of Fig.1(b). It should be noted that at wavelength λ_B the reflected light power is zero.

Fig.1(c) presents a grating consisting of five identical sections. There are phase shifts $\Delta\varphi = 180°$ in the second and fourth sections. In this case the transfer function has two reflected spectral lines at λ_{+1} and λ_{-1}, the distance between the peaks (maxima) of which is equal to the spectral selectivity of the filter $\delta\lambda$.

Fig. 1 (d) shows the filter that also consists of five identical sections, but the phase shift between the second and fourth section is $\Delta\varphi = 118°$. In this case the transfer function has three reflected spectral lines λ_B, λ_{+2} and λ_{-2}, the distance between the peaks (maxima) of which is also equal to the spectral selectivity of the filter.

Fig. 1 (e) shows the same grating, but the phase shift between the second, third and fourth sections is 180°, 115° and 180°, respectively. For such a phase distribution the transfer function of the coder has four reflected spectral lines at λ_{+1}, λ_{-1}, λ_{+3}, λ_{-3}.

Fig. 1 (f) and (g) show the cases of the phase distribution along the grating formed by using a piecewise linear approximation. For the examples shown, the transfer function had one reflected spectral line shifted either to the right or left with respect to the wavelength λ_B. The shift is equal to the filter spectral selectivity.

No doubt, all the transfer function types presented here are statistically independent .i.e., they differ from each other), and hence they may be used to code information. The number of statistically independent transfer functions depends on the number of sections determining the phase distribution along the filter and on the phase set for a given number of sections that can be obtained.

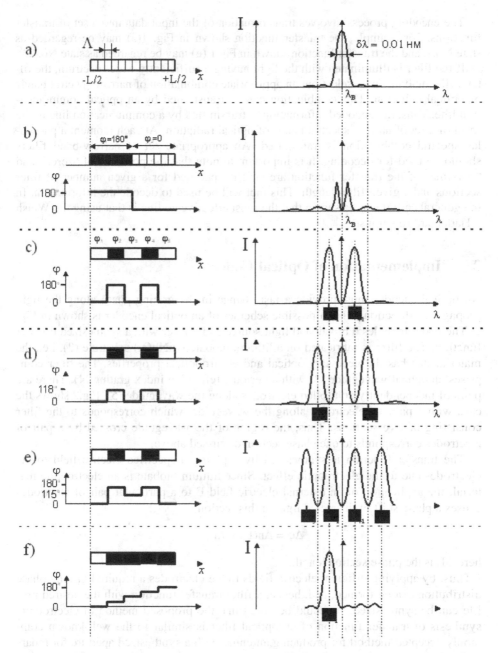

Fig. 1. Explanation of spectral coding principles. Left – examples of phase distribution along the grating. Right - the corresponding filter transfer function. L – grating length, Λ – grating period, φ_i - number of phase sections.

The encoding process involves transformation of the input data into a set of transfer functions. For example, the transfer function shown in Fig. 1(a) may be regarded as state No.1, and the transfer function shown in Fig.1 (e) may be regarded as state No.4.

If the filter is illuminated with the light having a sufficiently wide spectrum, the filter will simultaneously select out an appropriate combination of narrow spectral bands of optical radiation. This combination can be transmitted by an optical communication line. Thus, the encoded information is transmitted by a communication line in the form of a set of narrow spectral bands of optical radiation. At each moment a particular spectral combination is transmitted. An appropriate set of narrow-band filters should be used for decoding. It is important to note that the positions of "zeros" and "maxima" of the transfer function are strictly specified for a given number of filter sections and a given filter length. This fact will be used to decode the information. In the general case it can be shown that the most effective coding is that using the Walsh – Hadamard functions.

2   Implementation of Optical Coder

An optical encoder must provide a fast change in the grating phase along the light propagation direction. One of possible schemes of an optical encoder is shown in Fig. 2. The coder includes a broadband light source and a filter with a synthesized transfer function. The filter is fabricated on a lithium niobate ($LiNbO_3$) substrate (2), i.e., the material that has the necessary optical and electrooptical properties. The filter comprises an optical waveguide (3) with a periodic refractive index grating (4). There are pairs of independently controlled electrodes along the waveguide (5). Fig.2 shows the case with 5 pairs of electrodes along the waveguide, which corresponds to the filter consisting of five sections. Thus, the region of the waveguide crossed by a pair of electrodes corresponds to one phase section discussed above.

The transfer function is synthesized by applying a specified electric field to the electrodes due to the electrooptic effect. Since lithium niobate is an electrooptic material, the application of an external electric field E to a particular pair of electrodes causes a phase shift of diffracted light in this section

$$\Delta\varphi = \Delta n(E)\Delta L/n \tag{2}$$

here ΔL is the phase section length.

Thus, by applying different electric fields to the electrodes a required grating phase distribution can be formed, and, hence, a filter transfer function with the desired profile can be synthesized. It should be noted that the proposed method of electrooptic synthesis of transfer function of an optical filter is similar to the well-known commonly accepted method for producing antennas with a synthesized aperture for radars and sonars.

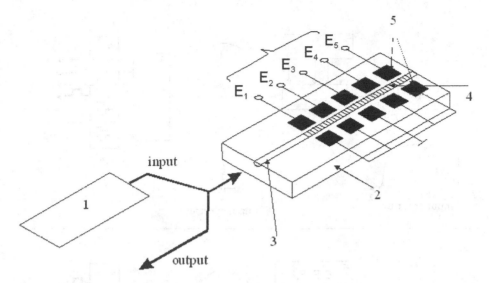

Fig. 2. Simplified scheme of optical encoder. 1 - broadband light source (for example, LED), 2 - lithium niobate substrate, 3 - optical waveguide, 4 - refractive index grating, 5 - electrode pairs. The power supply and controller are not shown. The scheme operated in the "reflection" mode, i.e., a reflected spectrum is launched into the communication line. To isolate the input and output radiation, an element equivalent to the 3-dB circulator is used.

3   Operation of Communication Line

Fig.3 shows one of possible realizations of a communication line using the coder - decoder system we suggest. The communication line consists of a broadband light source (1), a filter (2), an electronic controller (5), an optical fiber (6), and a set of output filters (3) and photodetectors (4), each of which is connected to its own filter.

The transmitted signal is fed to the input of the electronic controller which transforms it into the desired combination of control voltages which are applied to the electrodes. The light from the broadband source is fed to the filter input. The filter selects a set of spectral components which are transmitted through the optical fiber. There is a parallel set of filters at the output. Each filter is tuned to the wavelength, which is either "zero" or "maximum" of the transfer function of the filter-encoder. In the case the eigenfrequency of the filter-decoder coincides with the maximum transmitted light power, a "high" electric signal arises at the photodetector output. In such a way the frequency modulation of optical radiation is converted into the amplitude modulation of electric voltage.

It can be shown that for this example the number of statistically independent states is $N = 8$. In this case, according to Shannon, this system contains $Log_2 N = 3$ bits of information. The time of synthesis of a unit transfer function is determined by the interelectrode capacitance and is equal to about 400 ns at present, which corresponds to a clock frequency of 2.5 MHz. Thus, informational throughput of the encoder comprising 5 pairs of electrodes is 2.5 MHz x 3 bits = 7.5 Mbit/s. Potentially, the number of electrode pairs may be 10 ... 20, and the synthesis time can be reduced to 100 ns. In this case, the informational throughput can be 1.2 ... 24 Gbit/s.

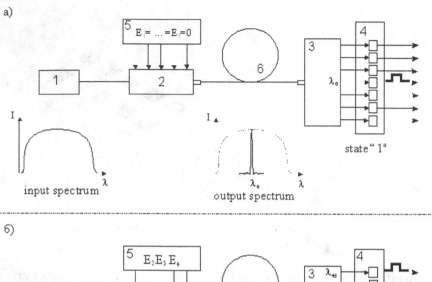

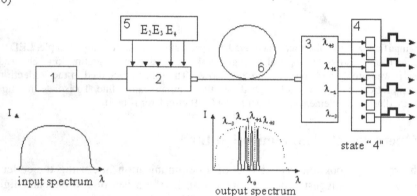

Fig. 3. Example of communication line implementation. 1 - broadband light source, 2 - filter with a synthesized transfer function, 3 – set of output filters, 4 - set of photoreceivers, 5 - electronic controller, 6 - optical fiber. a) transmission of state "1". All control electric fields are zero. The filter transfer function corresponds to the characteristic shown in Fig.1(a). In this state the filter selects only one spectral band at wavelength λ_B. The output is the signal from the corresponding photoreceiver. b) transmission of state "4". The electric fields that provide the transfer function shown in Fig.1(e) are applied to the electrodes. The filter selects four spectral lines. The output is the signals from four respective photoreceivers that arise simultaneously.

4   Discussion

The major advantage of the coding technique we suggest is its novelty. To our knowledge, this technique is not used at present. First of all, this is caused by the lack of necessary controlled narrow-band filters. Other advantages offered by this coding technique are:

There is no problem of amplitude signal fluctuations and no problem of chromatic signal dispersion. One more important fact should be noted (see Fig.4). The spectral selectivity of the filters under consideration is approximately 0.01 nm. It can be seen from the example given above that the entire coding bandwidth is about 0.1 nm. The bandwidth of a WDM channels is about 0.8 nm, the DWDM channel bandwidth is about 0.4 nm. Thus, the spectral coding technique we suggest can be used within one standard (D)WDM channel, which is very important for the use of the available standard optical communication lines. On the one hand, all the control elements of the communication line (switches, amplifiers, routers, de/multiplexers, etc.) will be "transparent" for the coding technique. On the other, the available amplitude detection systems are unable to receive a signal with a frequency modulation.

References

1. Heinisch, C., Lichtenberg, S., Petrov, V.M., Petter, J., Tschudi, T.: Phase – Shift Keying of an Optical Bragg Cell Filter. Opt. Comm. **253**, 320–331 (2005)
2. Arora, P., Petrov, V.M., Petter, J., Tschudi, T.: Integrated optical Bragg filter with fast electrically controllable transfer function. Opt. Comm. **281**(8), 2067–2072 (2008)
3. Petrov, V.M., Lichtenberg, S., Petter, J., Tschudi, T., Chamrai, A.V., Bryksin, V.V., Petrov, M.P.: Optical on-line controllable filters based on photorefractive crystals. J. Opt. A.: Pure and Appl. Opt. **5**, 471–476 (2003)

Fiber-Optics System for the Radar Station Work Control

Vadim V. Davydov[1], Natalya V. Sharova[1(✉)], Elena V. Fedorova[1],
Evgenia P. Gilshteyn[1], Kirill Yu Malanin[2], Igor V. Fedotov[2],
Vasiliy A. Vologdin[1], and Anton Yu Karseev[1]

[1] Peter the Great St. Petersburg Polytechnic University, St. Petersburg, Russia
Davydov_vadim66@mail.ru, sharova.natalia0510@yandex.ru,
e.v.fedorova@Onegroup.ru, evgenia.gilshteyn@skolkovotech.ru,
Joy21@rambler.ru, Antonkarseev@mail.com
[2] SRC Leninetz Plant Inc., St. Petersburg, Russia
kymalanin@gmail.com, i.v.fedotov@Onegroup.ru

Abstract. The purpose of this paper was to develop fiber-optics system for the multifunctional radar stations control in the frequency range of 0.1-12 GHz. Power losses and total attenuation of the optical signal was calculated for developed fiber-optics systems. Different methods for fiber-optics system characterization were developed, and the results of input signal frequency, temperature and vibration influence was experimentally investigated. Methods for fiber-optics system improvement were identified.

Keywords: Fiber-optic communication line · Line delay · Target simulator · Optical fiber · Optical transmitter · Optical receiver · Loss · Dispersion

1 Introduction

Fiber-optics data transmission (FODT) systems have a large number of applications on airplanes, ships and submarines [1-4]. One of the actual directions in the area of modern fiber-optics data transmission systems is focused on the development of super-high-frequency (SHF) range devices for radar systems. Such systems can be efficiently used in different channels of multicomponent phased antenna arrays [2, 3, 5-7].

Optical signal modulated by a SHF signal can be transmitted by the optical fiber on a distances of tens of kilometers almost without attenuation (less than 0.5 dB per km) and without the necessity to retransmit the signal. The analog fiber-optics communication line can be used to transmit ultra-broadband signals due to the fact, that its frequency range starts from units of megahertz and lasts to 12–14 GHz (in prospect – to 30 GHz). Neither of the previously used SHF signal transmission lines (coaxial, waveguide, open-wire) is capable to carry it out. This fact open up huge opportunities for the fiber-optics system development for spaced radar systems, antennas, remote from the control and information processing points, and systems for routing heterodyne, control, and synchronizing signals in phased antenna arrays.

© Springer International Publishing Switzerland 2015
S. Balandin et al. (Eds.): NEW2AN/ruSMART 2015, LNCS 9247, pp. 712–721, 2015.
DOI: 10.1007/978-3-319-23126-6_65

Delay lines based on optical fibers exhibit unique properties because its analog signal delay in a fiber is frequency -independent, which allows to create false radar targets even for the use of ultra-broadband probing signals. Delay lines which working mechanisms is based on other physical principles do not exhibit these advantages.

The use of fiber-optics systems on board of air and sea vehicles reduces energy consumption, radically decreases sensitivity to any electromagnetic interference, appreciably increases the reliability of information transmission, and considerably reduces weight. For example, the most modern vehicle data transmission cables based on copper replacement by vehicle fiber-optic cables reduces weight of cable interconnections in more than 10 times.

One of the biggest challenges facing the developers of radar stations (RS) is antenna complex set-up and verification, as well as its individual components and units in the enterprise conditions. In some cases, the antenna system should be checked after the assembly and installation (e.g., on the ship). Simulators of false target can be used to test aircrafts and ships radar, due to various restrictions in distance to the possible target, as well as in the possible use.

Since for the radar transmission of microwave signals fiber-optics communication line is used (FOCAL), to simulate the false target it should be developed a fiber-optics system (FOS) with different delay times t_d, which can be used both at the enterprise and polygon, and on board of ship [2-5].

The advantage of fiber-optics line is that it does not distort transmitted signal in the presence of huge electromagnetic radiation, noises, cross talks, which are particularly present at the polygon and warship [2-5]. Furthermore, it has less power losses than standard delay lines used for microwave signals.

2   Fiber-Optics System for the Microwave Radar Signal Delay

Fiber-optics systems for signal delay constructively can be divided into two types: passive – fiber-optics delay line (FODL) without optoelectronic transmitter-receiver modules and active – with transmitter-receiver modules. First type has disadvantages due to the complexity of their connection while radar testing [5, 7]. FOS connected by standard connectors of transmitter-receiver module eliminates these complexities and makes device more versatile for using.

Fig. 1 shows schematic diagram of radar checking process in an enterprise or polygon conditions without using of aircrafts. This technique is more widely used for radar testing.

Fiber-optical system (FOS) is located in radar and provides time delay of detected microwave signal before its further processing. While testing in the enterprise distance between the transceiver antenna and system of corner reflectors does not exceed 150 m. In reality, the target is located on a distance of 5 km and more. Therefore, for complex radar testing FOS with different delay times of 25 to 200 μs (t_d) should be used, located in one unit with common power supply.

714   V.V. Davydov et al.

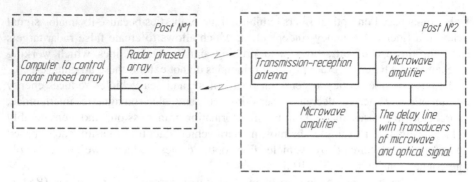

Fig. 1. Schematic diagram of radar testing.

Single-mode fiber can be used for transmission of analog signal in a microwave radar. Multimode fiber has larger losses and dispersion, while the most important disadvantage is low coefficient of the broadband of 1200 MHz*km. For transmission in 1310 nm range fiber complying with the standard ITU-T G.652 must be used. Due to wide availability of the fibers provided by the Corning company on domestic market, fiber Corning® SMF-28e+® complying with this standard was selected for this FOS.

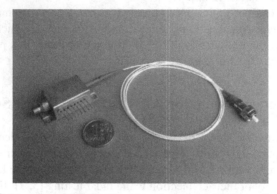

Fig. 2. Module DMPO131-23 with optical fiber of 0.9 mm diameter.

Transmitters with direct modulation in the 1310 nm window are used in FOS. Currently, radar should comply with the requirement - maximum use of domestically produced elements. Therefore, after preliminary studies high frequency Dilaz laser module DMPO131-23 (Figure 2) and the receiving optical module DFDMH40-16 (Figure 3) are used in FOS.

These devices have following parameters: frequency range of 0.1 to 16.2 GHz; transmission coefficient – 32 dB; uneven gain over the entire bandwidth of ± 2 dB; compression point of 12 dBm; maximum RF input level 10 dBm; noise factor of 25 dB; input and output impedance of 50 ohms; laser power 4 - 9 mW; receiver operating wavelength 980-1650 nm; operating photodetector bandwidth ΔF = 15.99 GHz; operating temperature range –40 to +60°C.

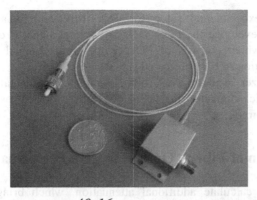

Fig. 3. Module DFDMH*40-16* with optical fiber of 0.9 mm diameter.

Power driver SL.5210.00.000RE must be connected to the transmission module. Standard driver mode is mode of constant temperature and medium power laser diode in transmitting optical module (TOM) provided by maintaining of constant thermistor resistance and photocurrent of built into the TOM photo monitor.

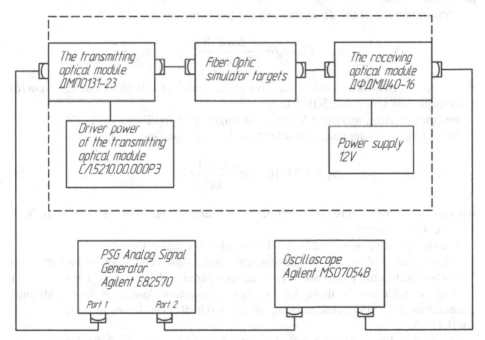

Fig. 4. Experimental setup for imitators delay time measurements.

Main factors while choosing of connectors are their return losses, and strength of the connector itself for the developed FOS. Best return losses values demonstrate connectors with APC (Angled Physical Contact) polish, while one of the most reliable and convenient are FC connectors. Therefore FC-APC connectors will be used in FOS.

Fig. 4 shows experimental setup for delay time and dynamic characteristics measurements of developed the FOS. Delay of optical signal is carried out in fiber-optics simulator, which consists of FOLS with predetermined length and optical isolator for elimination of back-reflected signals.

Spectrum analyzer Agilent PXA N9030A for experimental measurements of simulators dynamic characteristics (Fig. 4) must replace oscilloscope Agilent MSO7954B in the circuit.

3   Evaluation of Fiber - Optical System Performance

It is necessary to calculate additional attenuation, which brings developed FOS connected to the fiber-optics radar system. All A_Σ losses in the fiber can be defined as follows [5, 7]:

$$A_\Sigma = \alpha_a + \alpha_d + \alpha_i + \alpha_{ex}, \tag{1}$$

where α_a – absorption losses, α_d – scattering losses, α_i – losses due to impurities, α_{ex} – additional losses.

Absorption losses are determined by the following equation [2, 7]:

$$\alpha_a = 8.69 \frac{\pi \cdot \tan \delta \cdot n_1}{\lambda}, \tag{2}$$

where n_1 – core refractive index, we have chosen value of 1.4676 for our fiber, tan (δ) – dielectric loss tangent ($2.5 \cdot 10^{-12}$).

For the operating wavelength ($\lambda = 1310$ nm) $\alpha_a = 0.076$ dB/km.

Scattering losses are determined by the following equation [2, 7]:

$$\alpha_d = 4.34 \cdot 10^3 \cdot 8\pi^3 \frac{n_1^2 - 1}{3\lambda^4} k \beta T, \tag{3}$$

where T – temperature of fiber core manufacturing, $\beta = 8.1 \cdot 10^{-11}$ m^2/N – compressibility factor.

For the operating wavelength ($\lambda = 1310$ nm) $\alpha_d = 0.234$ dB/km.

Fiber Corning SMF-28e$^+$ is manufactured using high technology, so its impurities losses and additional losses are very small compared with α_a and α_d. Then $A_\Sigma = 0.31$ dB/km. In addition to these losses, there are return losses in four additional connections (FC-APC connectors) $\alpha_c = 4 \cdot 0.2 = 0.8$ dB. The losses in optical isolator are 0.41 dB.

In addition, losses of laser radiation entering the fiber are $\alpha_l = 0$ dB. Losses while docking of detector with fiber are $\alpha_s = 0$. Admission to temperature changes under given conditions is $\alpha_T = 1$ dB. Admission to the parameters degradation in time for the following combination of elements: Laser + pin – photodiode $\alpha_{lf} = 4$ dB. Then, additional admission to these losses is as follows:

$$\alpha_D = \alpha_{lf} + \alpha_T = 5 \text{ dB}.$$

Then all losses associated with the inclusion of fiber-optics system are $\alpha_{im} = (\alpha_a + \alpha_d) \cdot L_{op} +$ 6.21, where L_{op} – length of the fiber in delay line.

For system energy reserve of 2 dB maximum possible length l_{max} of FOS delay line piece of regeneration [2, 7]:

$$l_{max} = \frac{A_{max} - M - n_{pc}\alpha_{pc}}{\alpha + \dfrac{\alpha_{nc}}{l_{ct}}},$$

where n_{pc} – number of detachable joints, α_{pc} – losses in detachable joints, l_{ct} – construction cable length ($l_{ct} = 63000$ m), α_{nc} – loss in permanent joints (for welding of construction cable lengths $\alpha_{nc} = 0.05$ dB), $\alpha = \alpha_{im}$ for calculation $L_{op} = l_{ct}$, M – system reserve on damping in the area of regeneration ($M = 2$ dB) – maximum value of the equipment damping.

As the result $l_{max} = 124900$ m. These results show presence of great opportunities for the use of this fiber type for FOS creation.

4   Experimental Study of Fiber - Optical System

The research work was carried out by means of FOS experimental setup (Fig. 4). As an example, the results of experiments with FOS described by $t_d = 39.58$ µs is represented (fiber length $L_{op} = 8$ km). The waveforms of rectangular pulses at frequency of 5 GHz is shown on Fig. 5. The bottom signal is obtained without the inclusion of an optical simulator, top signal – with the simulator.

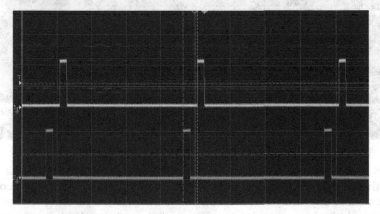

Fig. 5. Transit signal time through the simulator with a delay of 39.58 µs. Horizontal scale is 100 µs.

One of the FOS parameters for testing the radio locating stations (radar) is resulting in minimization of distortion in the shape of the transmitted signals. Fig. 6-8 shows the results of rise time and fall time comparison of the delayed signal in the optical signal simulator compared with signal obtained without simulator at frequency of 10 GHz.

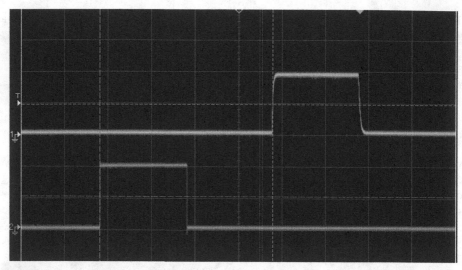

Fig. 6. Transit signal time through the simulator with a delay of 39.58 μs. Horizontal scale is 10 μs.

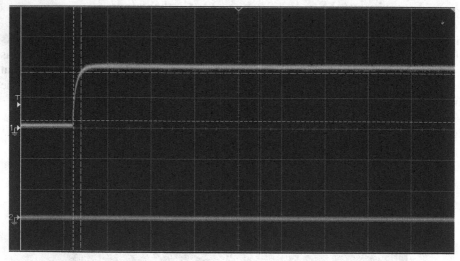

Fig. 7. Distortion of the delayed pulse rise time. Horizontal scale is 2 μs.

Rise time of delayed pulse increased by 361 ns and fall time – by 746 ns, compared to pulse without any delays. This result shows that introduced distortions are of the order 2% with the lowest used delay time t_d of about 40 μs. With increase of t_d the percentage of distortion will decrease.

Moreover, it was obtained experimentally that for different transmission frequencies used at FOS it is necessary to limit the input signal at the optical transmitting module, to avoid distortion. Fig. 9 shows the distorted view of delayed signal (signal below on the waveform) at the input of transmitting optical module with the input signal capacity of 9 dB. Due to the distortion of delayed pulse rise time t_d increased to 40.71 μs, which does not allow to test the radar.

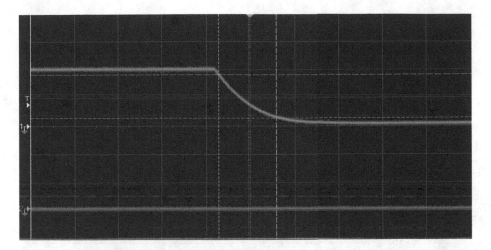

Fig. 8. Distortion of delayed pulse fall time. Horizontal scale is 2 µs.

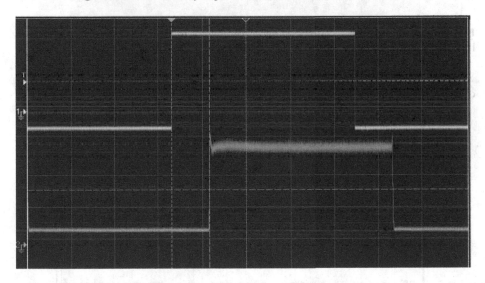

Fig. 9. Transit signal time through the optical simulator with delay. Horizontal scale is 50 µs.

Fig. 10 shows the dynamic characteristics example of the developed FOS – dependence of the transmitted through the FOS signal power on the input signal power with $t_d = 39.58$ µs.

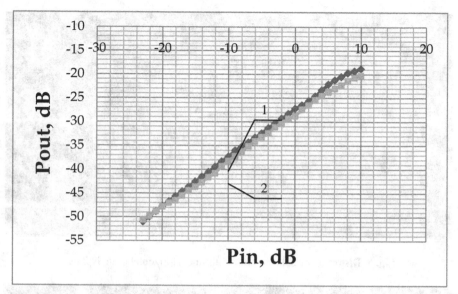

Fig. 10. FOS dynamic characteristics. Line 1 corresponds to the input frequency of F = 9 GHz, line 2 – to the F = 11 GHz.

Figure 11 shows, as an example, the experimental dependence of the loss factor Kp on frequency F of the input signal in FOS.

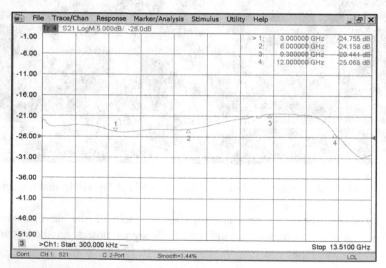

Fig. 11. Dependence of the losses in FOS with t_d = 39.58 μs on the input RF signal frequency.

On the base on experimental data shown on Figure 11 one can make following conclusions: the average value of losses with the delay t_d = 39.58 μs is – 26 dB and unevenness of frequency response is ± 4 dB. This open up opportunity to test radar by the developed FOS without any additional signal amplification after its switching.

5   Conclusion

The experimental and theoretical results show that developed fiber-optical system for delay of different output microwave radar signals is satisfied with all parameters requirements and is ready to be uses for radar testing. The only drawback of developed FOS is lack of the delay time t_d adjustment option. To eliminate this deficiency FOS consists of 8 blocks, each of which realize a predetermined delay t_d, with common supply. It made FOS design more cumbersome compared to a traditional designs of microwave delay lines.

References

1. Agrawal, G.P.: Light wave technology: telecommunication systems, 480 p. Wiley – Inter Science, NJ (2005)
2. Friman, R.K.: Fiber – optic communication systems, 496 p. Tekhnosfera, Moskow (2012)
3. Davydov, V.V., Ermak, S.V., Karseev, A.U., Nepomnyashchaya, E.K., Petrov, A.A., Velichko, E.N.: Fiber-optic super-high-frequency signal transmission system for sea-based radar station. In: Balandin, S., Andreev, S., Koucheryavy, Y. (eds.) NEW2AN/ruSMART 2014. LNCS, vol. 8638, pp. 694–702. Springer, Heidelberg (2014)
4. Davydov, V.V., Dudkin, V.I., Karseev, A.U.: Nuclear Magnetic Flowmeter – Spectrometer with Fiber – Optical Communication Line in Cooling Systems of Atomic Energy Plants. Optical Memory & Neural Networks (Information Optics) 22(2), 112–117 (2013)
5. Davydov, V.V., Dudkin, V.I., Yu, K.A.: Fiber – Optic Imitator of Accident Situation for Verification of Work of Control Systems of Atomic Energy Plants on Ships. Optical Memory & Neural Networks (Information Optics) 23(3), 170–176 (2014)
6. O'Mahony, M.J.: Future optical networks. IEEE OSA Journal of Light wave Technology 24(12), 4684–4696 (2006)
7. Slepov, N.N.: Fiber-optic communication system, 512 p. Technosfera, Moskow (2011)

Optimization of Angle-of-Arrival GPS Integrity Monitoring

Igor A. Tsikin and Antonina P. Melikhova[✉]

St. Petersburg Polytechnic University, 29 Politechnicheskaya St., St. Petersburg 195251, Russia
tsikin@mail.spbstu.ru, antonina_92@list.ru

Abstract. The paper considers the navigation satellite systems integrity monitoring method which is based on the differences between measured and calculated azimuth and elevation of the navigation signal source respectively. Probability-based integrity monitoring characteristics such as probability of false detection (false alarm probability) and the probability of missing the violation of integrity (missing probability) were obtained. Different kinds of decision-making procedures were analyzed, and for the most important methods logical "AND" and logical "OR" optimal choice of decision thresholds was found. It was concluded that the logic "OR" method had significant advantage over the logic "AND" method in the most practical interest area of missing probability values for fixed value of false alarm probability.

Keywords: Interference mitigation · Global navigation satellite systems · GNSS · GPS · Angle of arrival · GPS integrity monitoring · Violation of integrity · Direction finding

1   Introduction

Due to intentional or inadvertent interference to global navigation satellite systems (GNSS), the measured positions of the users can significantly deviate from their true values. A case when the error value exceeds an admissible limit referred to as GNSS integrity failure, and a procedure to detect such a failure is known as GNSS integrity monitoring [1]. Widely known integrity monitoring methods [2,3,4,5] suffer from severe shortcomings such as a lack of detectable integrity failure kinds, low noise immunity, high mass and dimensions, and long time required for decision.

On the other hand, the receiver-autonomous Angle-of-Arrival (AOA) integrity monitoring method is expected to have significant advantages arising from the use of the spatial separation of structural interference sources [6,7]. However, despite the number of publications about this method [6,7,8,9], its analysis in terms of probability-based characteristics cannot be considered fully completed. In particular, successful AOA method implementation needs the optimal decision threshold setting, which allows the minimum of missed integrity failure detection probability (P_{MD}) to be achieved when the false integrity failure alarm probability (P_{FA}) is fixed.

S. Balandin et al. (Eds.): NEW2AN/ruSMART 2015, LNCS 9247, pp. 722–728, 2015.
DOI: 10.1007/978-3-319-23126-6_66

2   Angle-of-Arrival Integrity Monitoring Method

The AOA method uses both measured and calculated direction finding (DF) parameters of navigation signal sources [10]. Calculated DF parameters estimations μ_c (azimuth) and η_c (elevation) take into account the user's and the satellite's coordinates obtained from the navigation message [11,12]. The results of the calculation procedure simulation shows that μ_c and η_c can be considered as unbiased and normally distributed [8].

On the other hand, DF parameters of navigation signal source can be measured by the direct two-dimensional source bearing-finding procedure. Parameters μ_m and η_m obtained as a result of the procedure are called measured. Under large signal to noise ratio, these parameters can also be considered as unbiased and normally distributed [10]. In case of normal GNSS integrity the differences $\Delta\mu = \mu_m - \mu_c$ and $\Delta\eta = \eta_m - \eta_c$ between measured and calculated parameters are zero-mean Gaussian with a known standard deviations (STD) $\sigma_{\Delta\mu}$ and $\sigma_{\Delta\eta}$, which take a value in the range 0.2...1.0 degrees in the working area of signal-to-noise ratio [6]. But in case of GNSS integrity failure differences $\Delta\mu$ and $\Delta\eta$ are nonzero-mean, and their exceeding of the thresholds $\Lambda_{\Delta\mu}$ and $\Lambda_{\Delta\eta}$ is interpreted as integrity failure.

There are three kinds of decision-making procedures to detect GNSS integrity failure. The first one involves using only one of the differences $\Delta\mu$ (or $\Delta\eta$) ("single" method) whereas the other procedures use both of them. In the latter case it is possible to use two ways of decision-making – when both differences exceed the thresholds $\Lambda_{\Delta\mu}$ and $\Lambda_{\Delta\eta}$ simultaneously (logic "AND"), and at least one of them exceed the threshold (logic "OR"). The probability-based characteristics for each method are analytically given by:

"single" method:

$$P_{FA}^{\eta} = 1 - erf\left(\Lambda_{\Delta\eta}/\sqrt{2}\sigma_{\Delta\eta}\right)$$

$$P_{MD}^{\mu} = \frac{1}{2}erf\left(\Lambda_{\Delta\mu} + \beta_{\Delta\mu}/\sqrt{2}\sigma_{\Delta\mu}\right) + \frac{1}{2}erf\left(\Lambda_{\Delta\mu} - \beta_{\Delta\mu}/\sqrt{2}\sigma_{\Delta\mu}\right)$$

$$P_{FA}^{\eta} = 1 - erf\left(\Lambda_{\Delta\mu}/\sqrt{2}\sigma_{\Delta\mu}\right),$$

$$P_{MD}^{\eta} = \frac{1}{2}erf\left(\Lambda_{\Delta\eta} + \beta_{\Delta\eta}/\sqrt{2}\sigma_{\Delta\eta}\right) + \frac{1}{2}erf\left(\Lambda_{\Delta\eta} - \beta_{\Delta\eta}/\sqrt{2}\sigma_{\Delta\eta}\right),$$

where $\beta_{\Delta\mu} = m_1\{\Delta\mu\}$ and $\beta_{\Delta\eta} = m_1\{\Delta\eta\}$ – mean values of $\Delta\mu$ and $\Delta\eta$ respectively; $erf(x) = \dfrac{2}{\sqrt{\pi}}\int\limits_0^x e^{-t^2}\,dt$;

logic "AND" method:

$$P_{FA} = P^{\mu}_{FA} \cdot P^{\eta}_{FA} \quad \text{and} \quad P_{MD} = P^{\mu}_{MD} + P^{\eta}_{MD} - P^{\mu}_{MD} \cdot P^{\eta}_{MD} \; ;$$

logic "OR" method:

$$P_{FA} = P^{\mu}_{FA} + P^{\eta}_{FA} - P^{\mu}_{FA} P^{\eta}_{FA} \quad \text{and} \quad P_{MD} = P^{\mu}_{MD} \cdot P^{\eta}_{MD} \; .$$

It is obvious that the most informative methods are logic "AND" and logic "OR" using both direction-finding parameters simultaneously.

False alarm probability P_{FA} surfaces are shown in Fig.1a ("AND" method) and Fig.1b ("OR" method). On the basis of these curves the combinations of relative thresholds $\Lambda_{\Delta\mu}/\sigma_{\Delta\mu}$ and $\Lambda_{\Delta\eta}/\sigma_{\Delta\eta}$ providing the predetermined probability P_{FA} values were obtained (Fig.2a and Fig.2b for each of the two-dimensional methods respectively).

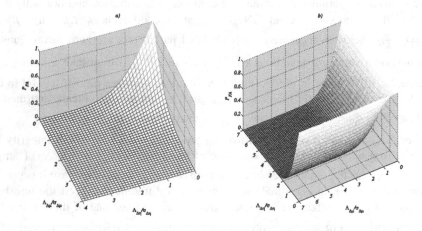

Fig. 1. False alarm   probability P_{FA} for two-dimensional methods "AND" (a) and "OR" (b)

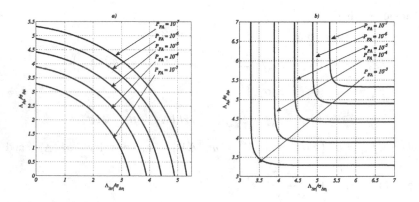

Fig. 2. Relative thresholds combinations for two-dimensional methods "AND" (a) and "OR" (b)

3   Optimal Thresholds Setting

The view of P_{MD} surfaces is shown in Fig.3 for typical $P_{FA} = 10^{-5}$ and the relative thresholds pairs with $\Lambda_{\Delta\mu}/\sigma_{\Delta\mu} - \Lambda_{\Delta\eta}/\sigma_{\Delta\eta} = 1$. Unfortunately, the optimal values of the relative thresholds depend on $\beta_{\Delta\mu}$ and $\beta_{\Delta\eta}$, which are not known in advance, so there exists a problem of choosing thresholds, providing the lowest losses in the value of P_{MD} when $\beta_{\Delta\mu}$ and $\beta_{\Delta\eta}$ change within a given range.

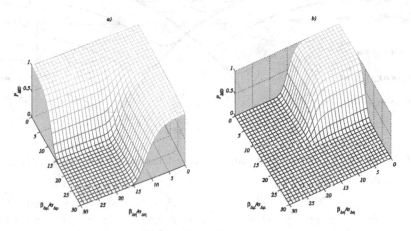

Fig. 3. Probability P_{MD} surfaces for logic "AND" (a) and "OR" (b) when $P_{FA} = 10^{-5}$ and $\Lambda_{\Delta\mu}/\sigma_{\Delta\mu} - \Lambda_{\Delta\eta}/\sigma_{\Delta\eta} = 1$

For logic "AND" method and $P_{FA} = 10^{-5}$ Fig.4 illustrates dependences P_{MD} on differences $\Lambda_{\Delta\mu}/\sigma_{\Delta\mu} - \Lambda_{\Delta\eta}/\sigma_{\Delta\eta}$ for different pairs of relative deviations $\beta_{\Delta\mu}/\sigma_{\Delta\mu}$ and $\beta_{\Delta\eta}/\sigma_{\Delta\eta}$. The presented curves show that each pair $\beta_{\Delta\mu}/\sigma_{\Delta\mu}$ and $\beta_{\Delta\eta}/\sigma_{\Delta\eta}$ has their own optimal relative thresholds combination. An attempt to select a combination of thresholds based on the "local" (for concrete pair of relative deviations) probability minimization immediately leads to a significant (dozens of times) increase in the probability P_{MD} when a pair of relative deviations would, in fact, be the other. Besides, rather strong dependence P_{MD} on differences $\Lambda_{\Delta\mu}/\sigma_{\Delta\mu} - \Lambda_{\Delta\eta}/\sigma_{\Delta\eta}$ may be noted for a given pair of relative deviations $\beta_{\Delta\mu}/\sigma_{\Delta\mu}$ and $\beta_{\Delta\eta}/\sigma_{\Delta\eta}$. On the other hand, the choice of equal relative thresholds ($\Lambda_{\Delta\mu}/\sigma_{\Delta\mu} - \Lambda_{\Delta\eta}/\sigma_{\Delta\eta} = 0$) provides the smallest change of P_{MD} at least when the values $\beta_{\Delta\mu}/\sigma_{\Delta\mu}$ and $\beta_{\Delta\eta}/\sigma_{\Delta\eta}$ vary within the most interesting range (for $P_{MD} = 10^{-3}...10^{-5}$).

Similarly, the probability P_{MD} was analyzed for logic "OR" method (Fig.5). In this case for given pair of relative deviations $\beta_{\Delta\mu}/\sigma_{\Delta\mu}$ and $\beta_{\Delta\eta}/\sigma_{\Delta\eta}$ dependence P_{MD} on differences $\Lambda_{\Delta\mu}/\sigma_{\Delta\mu} - \Lambda_{\Delta\eta}/\sigma_{\Delta\eta}$ is not as strong as in Fig.4. Moreover, optimal values of $\Lambda_{\Delta\mu}/\sigma_{\Delta\mu} - \Lambda_{\Delta\eta}/\sigma_{\Delta\eta}$ differ very little for different relative deviations, and these optimal values are very close to zero. As a result it is reasonable again to choose equal relative thresholds ($\Lambda_{\Delta\mu}/\sigma_{\Delta\mu} - \Lambda_{\Delta\eta}/\sigma_{\Delta\eta} = 0$).

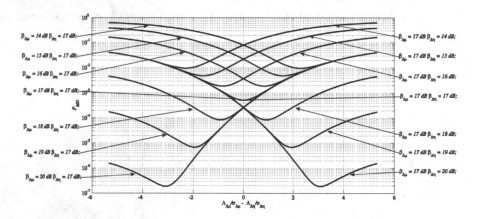

Fig. 4. Probability P_{MD} for different pairs of relative deviations $\beta_{\Delta\mu}/\sigma_{\Delta\mu}$ and $\beta_{\Delta\eta}/\sigma_{\Delta\eta}$ (logic "AND" method)

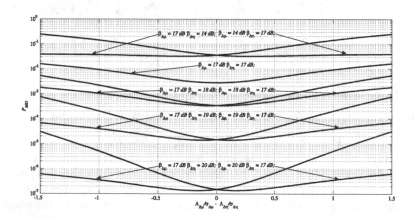

Fig. 5. Probability P_{MD} for different pairs of relative deviations $\beta_{\Delta\mu}/\sigma_{\Delta\mu}$ and $\beta_{\Delta\eta}/\sigma_{\Delta\eta}$ (logic "OR" method)

It is interesting to compare the areas of $\beta_{\Delta\mu}/\sigma_{\Delta\mu}$ and $\beta_{\Delta\eta}/\sigma_{\Delta\eta}$ where the probability P_{MD} does not exceed a given value. The solid lines in Fig.6 illustrate the borders of areas (to their right) where P_{MD} is less than 10^{-3}, 10^{-5}, 10^{-7} for logic "OR" method ($P_{FA} = 10^{-5}$). The dotted lines illustrate the same for logic "AND" method. In both cases $\Lambda_{\Delta\mu}/\sigma_{\Delta\mu} - \Lambda_{\Delta\eta}/\sigma_{\Delta\eta} = 0$.

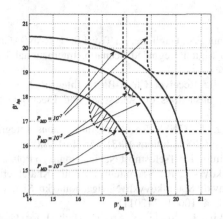

Fig. 6. The areas of relative deviations $\beta_{\Delta\mu}/\sigma_{\Delta\mu}$ and $\beta_{\Delta\eta}/\sigma_{\Delta\eta}$ for P_{MD} less than given values ($\Lambda_{\Delta\mu}/\sigma_{\Delta\mu} - \Lambda_{\Delta\eta}/\sigma_{\Delta\eta} = 0$, $P_{FA} = 10^{-5}$)

As it can be seen from these curves, the logic "OR" method has the advantage over the logic "AND" method when $P_{MD} < 10^{-6}$. The same holds in the region $10^{-3} < P_{MD} < 10^{-6}$ except in very small areas of this region (shaded). Working on logic "OR" method, the integrity monitoring system is capable of fixing direction-finding parameters deviations in azimuth and elevation starting with values 2...10 degrees (depending on the values of $\sigma_{\Delta\mu}$ and $\sigma_{\Delta\eta}$ in the range 0.2...1.0 degrees for working area of signal-to-noise ratio) for integrity failure detection probability P_{MD} no more than 10^{-6} when the false integrity failure alarm probability P_{FA} is fixed on value $P_{FA} = 10^{-5}$.

4   Conclusions

In solving the problem of parametric optimization of the Angle-Of-Arrival GPS Integrity Monitoring method there is a reasonable solution to minimize the maximum of missed integrity failure detection probability P_{MD} when the false integrity failure alarm probability P_{FA} is fixed. The solution is to choose equal values of the relative decision thresholds for each of the direction-finding parameters. In this case in the most practical interest area $10^{-3} < P_{MD} < 10^{-6}$ the logic "OR" method has significant advantage over the logic "AND" method except very small areas with inconsiderable disadvantage (no more than 1 dB in relative deviations $\beta_{\Delta\mu}/\sigma_{\Delta\mu}$, $\beta_{\Delta\eta}/\sigma_{\Delta\eta}$).

References

1. Castaldo, G., Angrisano, A., Gaglione, S., Troisi S.: P-RANSAC: An Integrity Monitoring Approach for GNSS Signal Degraded Scenario. International Journal of Navigation and Observation (2014)
2. Bednarz, S., Misra, P.: Receiver Clock-Based Integrity Monitoring for GPS Precision Approaches. IEEE Transactions on Aerospace and Electronic Systems $42(2)$, 636–643 (2006)
3. Hewitson, S., Wang, J.: GNSS Receiver Autonomous Integrity Monitoring (RAIM) Performance Analysis. GPS Solutions $10(3)$, 155–170 (2006)
4. Veremeyenko, K., Zimin, R.: Tselostnost Navigatsionnogo Polya. ISNS 4, 38–42 (2009)
5. Senatorov, M., Syatkovskiy, R.: Sravnitelnyy Analiz Kharakteristik Metodov Kontrolya Tselostnosti Globalnykh Sputnikovykh Navigatsionnykh Sistem. Bezopasnost Informatsionnykh Tekhnologiy 4, 106–108 (2011)
6. Melikhova, A., Tsikin, I.: Angle of Arrival Method for Global Navigation Satellite Systems Integrity Monitoring. St.Petersburg State Polytechnical University Journal. Computer Science. Telecommunications and Control Systems $212(1)$, 37–49 (2015)
7. Jafarnia-Jahromi, A., Broumandan, A., Nielsen, J., Lachapelle, G.: GPS Vulnerability to Spoofing Threats and a Review of Antispoofing Techniques. International Journal of Navigation and Observation (2012)
8. Montgomery, P.Y., Humphreys, T.E., Ledvina, B.M.: Receiver-autonomous spoofing detection: experimental results of a multi-antenna receiver defense against a portable civil GPS spoofer. In: Proceedings of the ION International Technical Meeting, pp. 124–130 (2009)
9. Sathyamoorthy, D.: Global Navigation Satellite System (Gnss) Spoofing: a Review of Growing Risks and Mitigation Steps. Defence S&T Technical Bulletin $6(1)$, 42–61 (2013)
10. Denisov, V., Dubinin, D.: Fazovyye radiopelengatory: Monografiya. Tomskiy gosudarstvennyy universitet sistem upravleniya i radioelektroniki, Tomsk (2002)
11. Kaplan, E., Hegarty, C. (eds.): Understanding GPS: Principles and Applications. Artech house (2005)
12. Navstar GPS Space Segment/Navigation User Segment Interfaces, IS-GPS-200. http://www.gps.gov

Ultra-Wideband Feed for Radio Telescope of a New-Generation Radio Interferometric Network

Vitaliy K. Chernov[1], Alexander V. Ipatov[1], Vyacheslav V. Mardyshkin[1],
Sergey I. Ivanov[2(✉)], and Artem A. Roev[2]

[1] Institute of Applied Astronomy of Russian Academy of Sciences,
St-Petersburg, Russian Federation
mardyshkin@rambler.ru
[2] Peter the Great St. Petersburg Polytechnic University, St-Petersburg, Russian Federation
serg.i.ivanov@mail.ru, artr@nxt.ru

Abstract. Experimental model of a dual-polarized ultra-wideband feed operating from 3 to 18 GHz was developed. The feed supports two orthogonal linear polarizations. The feed is designed for the antenna of a new-generation global radio interferometric network. Adjustment and tuning of the ultra-wideband feed experimental model were performed. Technical solutions to optimize the feed performance were proposed and their validity was confirmed by simulation and performance measurements. Measurements of the main feed parameters: phase and amplitude radiation pattern in E-, H-, and D- planes, gain, voltage standing-wave ratio, and isolation between the channels were performed. The experimental results are in good agreement with the simulation results.

Keywords: Radio telescope · Ultra-wideband feed · Quadruple-ridged waveguides · Reflector antenna feeds · VGOS

1 Introduction

The two-element radio interferometer is being designed according to the modern concept of Very Long Baseline Interferometry (VLBI). VLBI [1] is able to solve the problems of coordinate-time support in the interests of different areas of science and technology, including the operation of the global navigation satellite system (GLONASS). The technical realization of the project is based on the use of a radio interferometric network of small antennas RT-13 with the diameter of the primary reflector of 13 meters. A new radio telescope RT-13 will be equipped with two types of receivers: a tri-band S / X / Ka and ultra-wideband. Institute of applied astronomy of Russian Academy of Sciences has decided to use the frequency band of 3-18 GHz in the international program of global radio astronomy observations (VGOS) [2]. This decision can be explained by the high level of interference caused by mobile phones and advanced wireless technologies at the low frequency part of the operating band. One of the most complex elements of such a radio telescope antenna system is a broadband feed providing 6:1 frequency ratio.

© Springer International Publishing Switzerland 2015
S. Balandin et al. (Eds.): NEW2AN/ruSMART 2015, LNCS 9247, pp. 729–738, 2015.
DOI: 10.1007/978-3-319-23126-6_67

The goals of this project are the design and development of an ultra-wideband feed with the frequency band of 3-18 GHz for the RT-13 radio telescope, which is able to work as a part of the international VGOS receiving complex. Section 2 briefly discusses the specifics of the ultra-wideband feed design, the general requirements, and the possible prototypes. Section 3 summarizes electrodynamic design and calculations of the main performance parameters of the selected prototype – the voltage standing-wave ratio (VSWR) and radiation pattern. Section 4 briefly discusses some results of the prototype feed construction and presents measured data – E-, H- plane radiation pattern, cross-polarization level, VSWR, and isolation between channels. Finally, conclusions are presented in Section 5.

2   General Requirements and a Prototype Choice

The radio telescopes use reflector antennas manufactured by Vertex Antennen technik (Germany) that are based on the geometry proposed by J. K. Lee (axisymmetric two-mirror antenna with off-axis parabolic generator). The geometry of the radio telescope antenna complex determines the requirements to the feed: radiation pattern should have circularly symmetrical shape close to the Gaussian and its beam width should be $\pm 65^{\circ}$ measured at the -16 dB level. The feed should provide the simultaneous reception of two orthogonal linear polarizations.

There are several types of ultra-wideband feeds that are currently used in radio astronomy: a log-periodic structure of the cascade-connected loop antenna (called Eleven Feed [3]), a feed based on the self-complementary structure [4], the log-periodic antenna, and Quadruple-Ridged Flared Horn. A Quadruple-Ridged Flared Horn [5], which most closely meets the requirements of the project, was chosen as a prototype. It has the following advantages over other types:

• Frequency ratio of more than 6: 1 with the relatively frequency independent radiation pattern in the E- and H-planes;
• Two separate RF outputs for the received signals with two orthogonal linear polarizations;
• Output impedance of 50 Ohms providing easy matching with modern low-noise amplifiers;
• The nominal beam width is $\pm 65^{\circ}$ measured at the -16 dB level;
• The changes in the position of the phase center with frequency and the cross-polarization levels are acceptable;
• Relatively simple technical realization, good repeatability, and low cost.

3   Results of the Theoretical Research and Computer Simulation

Geometry of the horn, and its internal longitudinal ridges were optimized for the operating band of 2.3 - 14 GHz by authors of [6, 7]. The desired feed geometry for the operating band of 3 - 18 GHz was based on scaling the previous design down by a factor of 0.767, assuming that the horn and ridges are perfectly conductive. A lot of technical problems caused by the scaling of the excitation region of the horn were successfully solved [8]. Scaling down by a factor of 0.767 leads to unachievable geometrical dimensions of the feed excitation region. Optimization of the geometry of the excitation region over the operating frequency band required numerical and analytical calculations and computer simulations. Actual task is the development of analytical methods for calculating the electrodynamic characteristics of complex shape ultrawideband feed. An elliptical waveguide with longitudinal ridges was considered as an electrodynamic model of the feed's excitation region. Elliptical coordinate system (u, v) was used to solve the equations of the electromagnetic field in the elliptical waveguide. The curves $u = const$ form a family of confocal ellipses, the distance between the foci is equal to $2C_0$. The curve $u=u_0$ corresponds to the inner surface of the elliptical waveguide, curve $v=v_0$ corresponds to the ridge surface. Variation of parameters u and v allows varying the distance d between the ridges and their shape. The wave equation for the longitudinal component of the magnetic field H_z in the elliptical coordinate system is:

$$\frac{\partial^2 H_z}{\partial u^2} + \frac{\partial^2 H_z}{\partial v^2} + 2h^2(\text{ch}2u - \cos2v)H_z = 0 .\tag{1}$$

Equation (1) can be solved by the method of separation of variables with constant separation χ. Resulting differential equations are Mathieu equations. The continuity of the normal magnetic field component on the conductive surfaces gives the following boundary conditions: $H_u(u_0)=0$, $H_v(v_0)=0$. These boundary conditions are satisfied for certain separating parameters χ and h^2 - eigenvalues of Mathieu equation. Parameter h determines the critical wavelength for a waveguide with a given shape of the ridges. Fig. 1 shows the dependence of the critical wavelength λ_{cr} on distance d between the ridges for three ridge widths. The graph shows that if the ridges are thinner and closer to each other in the cross section of the waveguide, then the critical wavelength is longer and the frequency band of the waveguide is wider.

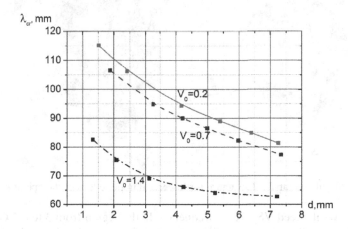

Fig. 1. Critical wavelength λ_{cr} vs. distance d for the ridged elliptical waveguide

The studies based on the proposed waveguide eigenfunctions calculation method show that the use of ridges inside the horn is an effective method of expanding the feed's operating band that leads to critical frequency decrease of more than 1.5 times. The proposed method also allows simplifying and speeding up the calculation process at the same time providing required accuracy.

For checking the correctness of the selected technical solutions the main characteristics of the feed were calculated using ANSYS HFSS (High Frequency Structure Simulator) software that uses the finite element method to solve the Maxwell's equations.

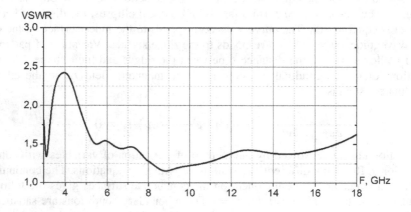

Fig. 2. Simulated VSWR vs. frequency

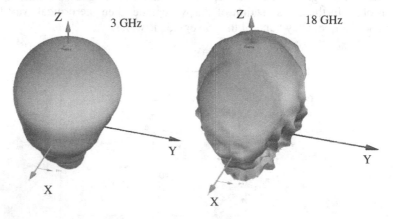

Fig. 3. Simulated far-field structure for lower and upper edges of the operating range

Fig. 2 shows the feed VSWR vs. frequency. In the region from 3 to 4.5 GHz VSWR reaches 2.4, but in the other parts of the operating frequency band it is about 1.5, which meets the technical requirements. Fig. 3 and 4 show the simulated 3-D radiation patterns of the feed produced by HFSS software. Fig. 3 shows the results of simulation in the lower (3GHz) and upper (18 GHz) parts of the operating frequency range. The field distribution is the same for all frequencies which gives the stability of basic parameters of the feed and ensures its effective operation in the RT-13 antenna system.

Fig. 4 and 5 show the feed's gain G (normalized to the maximum value of G) as a function of the angle α in E- and H- planes, respectively. The graphs confirm the small change in the shape of the mainlobe of the radiation pattern in the E and H planes. The width of mainlobe is about ± 65° at a -16 dB level. Sidelobe level is about -20 dB, which determines the low antenna noise. The gain of the feed is about 10 dB.

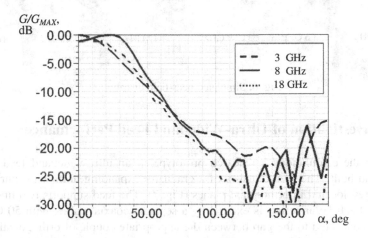

Fig. 4. Simulated E-plane radiation pattern G (normalized to G_{MAX}) for 3, 8, 18 GHz

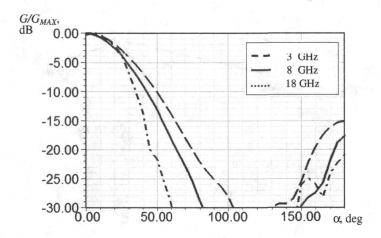

Fig. 5. Simulated H-plane radiation pattern G (normalized to G_{MAX}) for 3, 8, 18 GHz

Fig. 6 shows the structure of the near fields. Beam width is determined by the ratio of the effective aperture size to the wavelength. The separation of the field from the feed's walls restricts the effective aperture, and hence the radiation pattern at various frequencies remains unchanged. In general, the structure of the field is the same at all of the frequencies, which gives the stability of the basic parameters of the feed.

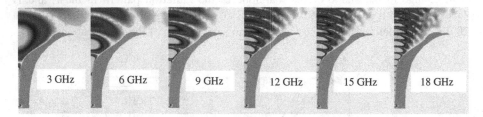

Fig. 6. Simulated near-field distribution

4   Investigation of Ultra-Wideband Feed Performance

Based on the calculations a research prototype of an ultra-wideband feed was designed and built. The feed consists of a gradually expanding horn with four longitudinal ridges, located in orthogonal planes (Fig. 7). The feed supports two linear polarizations. Each polarization is excited by a separate coaxial cable, with 50 Ohm impedance, connected to the gap between the appropriate couple of orthogonally placed ridges. The ridges with complex profile and horn were made using high-precision equipment with computer numerical control. All parts were made by enterprises located in St. Petersburg.

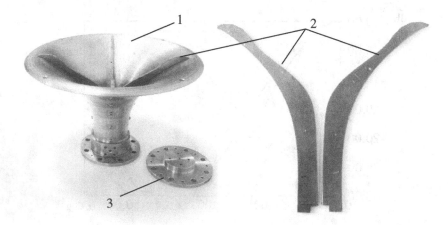

Fig. 7. Parts of the ultra-wideband feed experimental model: 1 - horn; 2 - ridges, 3 - cap

The tuning device for tuning of the matching and decoupling consist of a movable contact, clamp and nut. This arrangement allows to keep the symmetry of the excitation region during tuning and reduce the impact of technological errors.

After the tuning by the movable contact the following results were obtained (Fig. 8, 9): channels isolation I across the operating range is no worse than 16 dB, VSWR in the range 5-14 GHz is less than 1.8: 1, in the range of 3-16 GHz is less than 2.5: 1. Results of the VSWR measurement near the low frequency boundary of the operating band agree well with the model of the waveguide with ridges, the cross section of which is the same as the cross section of the feed in the excitation region.

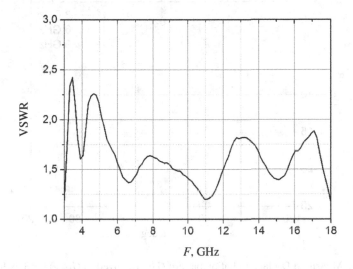

Fig. 8. Measured VSWR vs. frequency

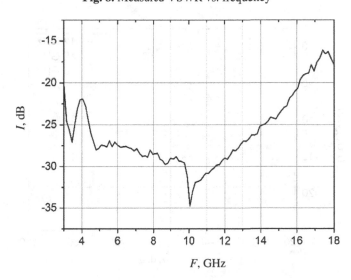

Fig. 9. Measured isolation I between channels of linear polarization versus frequency

Measurement of the complex radiation pattern was done using a broadband horn, adapted to operation with the linear polarization, as a transmitting antenna. Vector network analyzer was used for plotting amplitude and phase radiation patterns. Figures 10 and 11 show the radiation pattern G (normalized to the maximum value of G) in the D- planes (diagonal) respectively for the six frequencies from the operating range. Preliminary sidelobe level measurements show that the side lobe level is less than -25 dB and an anechoic chamber is required for precise measurements.

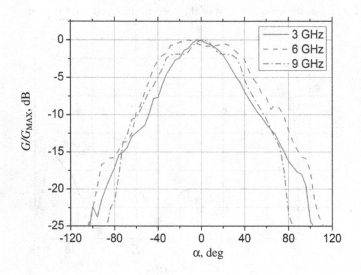

Fig. 10. Measured D-plane radiation pattern G (normalized to G_{MAX}) for 3, 6, 9 GHz

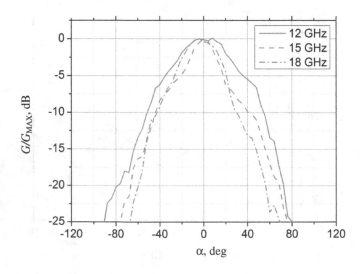

Fig. 11. Measured D-plane radiation pattern G (normalized to G_{MAX}) for 12, 15, 18 GHz

The measured radiation patterns in E- and H-planes satisfy specifications and are close to the corresponding calculation and simulation results. The H- plane main lobe is narrower, due to the ridges' presence. Potential aperture efficiency of RT-13 is not less than 60%. An important goal in the development of an ultra-wideband feed is the independence of the phase center on frequency and its alignment in the E- and H- planes. Figures 12, 13 show the phase diagram φ versus angle α in the diagonal direction (-D) plane. The axis of rotation is aligned with the position of the calculated phase center. If the feed phase center is displaced, there is a phase change depending on the feed's rotation angle. Figures 12, 13 show that ultra-wideband feed has a stable phase center, so it can be used in the antenna system of RT-13 radio telescope.

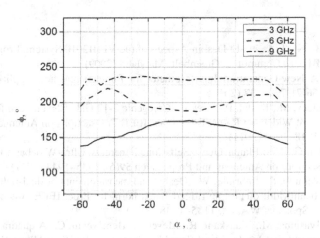

Fig. 12. Measured D-plane phase radiation pattern φ for 3, 6, 9 GHz

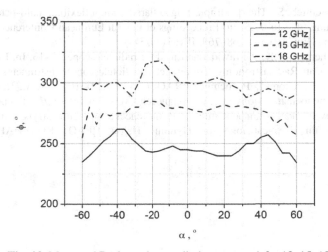

Fig. 13. Measured D-plane phase radiation pattern φ for 12, 15, 18 GHz

5   Conclusions

Results of development of a dual-polarized ultra-wideband feed with the operating frequency range of 3-18 GHz are presented. The measurements of main performance parameters of the ultra-wideband feed confirm that it can be effectively used in RT-13 antenna systems for new-generation global radio interferometric network [2]. The proposed design methods and construction solutions also can be used for development of the ultra-wideband feed for modern wireless telecommunication systems, which provide a wide frequency range.

References

1. Petrachenko, B., Niell, A., et al.: Design Aspects of the VLBI2010 System. Progress Report of the IVS VLBI2010 Committee. Greenbelt, Maryland (2009)
2. Ipatov, A.V.: A New-Generation Interferometer for Fundamental and Applied Research. J. PHYS-USP 56(7), 729–737 (2013)
3. Yang, J., Pantaleev, M., Kildal, P.-S., et al.: Cryogenic 2-13 GHz Eleven Feed for Reflector Antennas in Future Wideband Radio Telescopes. IEEE Transactions on Antennas and Propagation 59(6), 1918–1934 (2011)
4. Cortes-Medellin, G.: Non-Planar Quasi-Self-Complementary Ultra-Wideband Feed Antenna. IEEE Transactions on Antennas and Propagation 59(6), 1935–1944 (2011)
5. Akgiray, A., Weinreb, S., Imbriale, W.: Design and measurements of dual-polarized wideband constant-beamwidth quadruple-ridged flared horn. In: Proc. IEEE Antennas Propag. Soc. Int. Symp., Spokane, WA, pp. 1135–1138 (2011)
6. Beukman, T., Ivashina, M., Maaskant, R., Meyer, P., Bencivenni, C.: A quadraxial feed for ultra-wide bandwidth quadruple-ridged flared horn antennas. In: Proceedings of the 8th European Conference Antennas and Propagation (EuCAP). IEEE, Hague, pp. 3312–3316 (2014)
7. Akgiray, A., Weinreb, S.: The quadruple-ridged flared horn: a flexible, multi-octave reflector feed spanning f/0,3 to f/2.5. In: Proceedings of the 7th European Conference Antennas and Propagation (EuCAP), pp. 768–769. IEEE (2013)
8. Roev, A.A., Chernov, V.K.: Ultrawideband feed for radiotelescope RT-13. In: Proceedings of the All-Russian Radioastronomy Conference "Radiotelescopes, instruments and techniques of radio astronomy". Pushchino, PRAO, p. 67 (2014). (Роев А.А., Чернов В.К. Сверхширокополосный облучатель радиотелескопа РТ-13 // Всероссийская радиоастрономическая конференция "Радиотелескопы, аппаратура и методы радиоастрономии". Тезисы докладов. Пущино: Изд-во ПРАО АКЦ ФИАН, 2014. – С.67)

Improvement Frequency Stability of Caesium Atomic Clock for Satellite Communication System

Alexander A. Petrov[✉] and Vadim V. Davydov

Peter the Great Saint-Petersburg Polytechnical University, St. Petersburg, Russia
alexandrpetrov.spb@yandex.ru, davydov_vadim66@mail.ru

Abstract. One of the directions of modernization of the caesium atomic clocks is considered. A new implementation of a digital frequency converter for atomic clocks is presented. The new design of frequency converter is based on method of direct digital synthesis. The theoretical calculations and experimental research showed decrease step frequency tuning by several orders and improvement the spectral characteristics of the output signal of frequency converter. A range of generated output frequencies is expanded, and the possibility of detuning the frequency of the neighboring resonance of spectral line that makes it possible to adjust the C-field in quantum frequency standard is implemented. Experimental research of the metrological characteristics of the quantum frequency standard on the atoms of caesium - 133 with a new functional unit showed an improvement in the daily frequency stability.

Keywords: Atomic clocks · Quantum frequency standards · Frequency converter · A digital frequency synthesis · Frequency stability · Allan variance

1 Introduction

Most of modern technologies based on the use of precision measuring instruments. Operation of navigation services, measuring time systems, various electronic equipment, as well as, communication and information data transmission devices is impossible without reliable operation of precision devices.

A necessary condition for reliable operation of info communication systems is coordinated work of the primary generator and receiver, so that the receiving unit can correctly interpret the digital signal. The difference in the synchronization of various units in a network, may lead to missing or to reread the information by receiving unit.

In order to solve this problem atomic clocks also called quantum frequency standards are used. Quantum frequency standards - devices which generate signals different types and frequencies. Thus, the frequency standards can synchronize the work of many sophisticated devices and instruments.

Standard frequencies such as rubidium or caesium, used as a clock generator in the communications equipment and in a data transmission device, applied in the satellite navigation systems GLONASS and GPS as a clock generator and in the various metrological services. Also these standards perform a role of the reference signals with high precision and stability in radio equipment [1].

© Springer International Publishing Switzerland 2015
S. Balandin et al. (Eds.): NEW2AN/ruSMART 2015, LNCS 9247, pp. 739–744, 2015.
DOI: 10.1007/978-3-319-23126-6_68

With the development of scientific - technical progress operating conditions of caesium atomic clocks are constantly changing. Therefore new requirements for measurement accuracy, reliability and weight - dimensional characteristics are produced to frequency standard. This leads constantly upgrade existing and develop new models of caesium atomic clocks.

Development and commissioning of new atomic clock are very long and costly process, which in most cases is not enough funds and time. Therefore, in most cases, for specific tasks related to the operating conditions of frequency standards modernization is occur. Therefore, applied research in this field, in contrast to the fundamental, aimed at solving problems of improving and modernization existing designs of frequency standards [2 - 5].

For this aims it is reasonable to carry out research, develop new methods and find a new design solutions based on the latest appearing electronic components and discovered physical phenomena.

The process of modernization of frequency standards includes various directions: change the weight and dimensions, reduced energy consumption, improved metrological characteristics. And for frequency standards characterized by the fact that modernization may not be for the whole construction and may be for individual units or blocks.

In present work one of the directions of modernization of the caesium atomic clock is considered. A new implementation of a digital frequency converter for atomic clocks is presented. Improved output signal parameters frequency converter leads to improvement of characteristics of quantum frequency standard in particular the frequency stability.

2   Frequency Converter of the Caesium Atomic Clock

The work of a caesium atomic clock is based on the principle of adjustment of the frequency of a highly stable crystal oscillator to quantum transition frequency of atoms of caesium -133 [6]. Fig. 1 shows a block diagram of a caesium atomic clock.

The output signal frequency of 5MHz of the crystal oscillator 8 is supplied to the frequency converter 7. Frequency converter consists of the frequency synthesizer, the mixer signals and the multiplier signals. In the frequency synthesizer input signal frequency of 5MHz is converted to the signal frequency of 12,631772 MHz and supplied to the input of mixer signals. In the multiplier signals input signal frequency of 5 MHz is multiplied to the frequency of 270 MHz and then to frequency of 9180 MHz. This signal frequency of 9180 MHz is also supplied to the input of mixer signals. As a result, the output signal of the frequency converter is the signal of ultrahigh frequency of 9192,631772 MHz This signal is supplied in the waveguide 6 and then on the input of atomic beam tube.

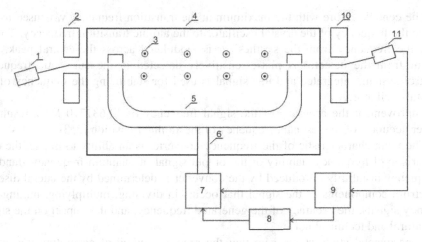

Fig. 1. Block diagram of a caesium atomic clock. 1 - a source of caesium atoms, 2 - magnet polarizer 3 - magnetic field 4 - magnetic shield 5 - Ramsey resonator 6 - the waveguide 7 - frequency converter 8 - crystal oscillator 9 - automatic frequency control system 10 - magnet analyzer, 11 – detector

In caesium atomic clock with the help of magnet polarizer 2 the atoms are prepared such that they are either in the F=4, m_F=0 or in F=3, m_F=0 state. Afterwards the atoms interact with an electromagnetic field that induces transitions into the former unoccupied state. The atoms in this state are detected and allow one to determine the frequency of the interrogating field where the transition probability has a maximum. The observed transition frequency is corrected for all known frequency offsets that would shift the transition frequency from the unperturbed transition and is used to produce a standard frequency or pulse per second every 9192631772 cycles [6, 7, 8].

Scanning the frequency ν of the atomic resonance leads to a detector current like the one shown on the Fig. 2. The signal shows the Ramsey resonance structure on a broader, so-called, Rabi pedestal.

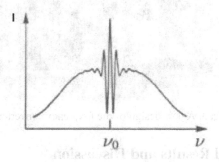

Fig. 2. Ramsey resonance structure on the Rabi pedestal

The central feature with the maximum at the transition frequency v_0 is used to stabilize the frequency of the crystal oscillator to the atomic transition frequency. To this end, the frequency from the synthesizer is modulated across the central peak. The signal from the detector is phase-sensitively detected in the automatic frequency control system, integrated and this signal is used for stabilizing the frequency of the crystal oscillator.

Improvement the accuracy of the signal frequency of 12,631770 MHz results in better accuracy of the resonant frequency of the atomic transition [2, 3].

The main characteristic of the frequency converter is an ability to impact the characteristic of frequency stability of the output signal of quantum frequency standard. Frequency instability introduced by the converter is determined by the lateral discrete spectrum components of the signal that occurs in dividing, multiplying, mixing frequency signals, the accuracy of the generated frequency, and the impact on the signal of natural and technical noise.

Experimental study showed up that the present method of generating the output signal of the frequency converter needs to increase the accuracy. The large resolution of step frequency is necessary. New scheme of the frequency converter is designed using direct digital synthesis (DDS - Direct Digital Synthesis) [9]. This method allows to generate the output signal of the converter with accuracy more than 10^{-6} Hz.

The implementation of our proposed method enables us to control the frequency of the output signal frequency converter in real time. This ensures a high rate of frequency tuning, which makes it possible to more efficiently adjust the crystal oscillator frequency in contrast the previously used scheme.

The application of direct digital synthesis gave the possibility of obtaining the generated frequencies in a wide range (0-3MHz), in contrast to previous schemes, where this feature was absent [2].

In Fig. 3 a new design of the frequency converter is presented.

Fig. 3. A new design of the frequency converter

3   Experimental Results and Discussion

Experimental study of the output signal parameters of a new scheme frequency converter confirmed the simulation results and showed great advantages in comparison with previously used scheme.

In Fig. 4 and Fig. 5 as an example, oscillograms measured in the band of 1 kHz and 600 kHz of the output signal of a new scheme (a) and a previously used (b) of the frequency converter are presented.

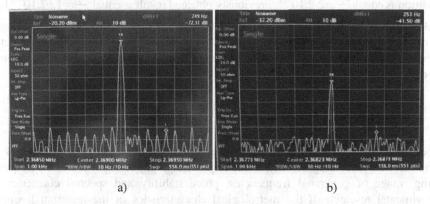

a)   b)

Fig. 4. Suppression of the amplitude of the lateral components in the band of 1 kHz.

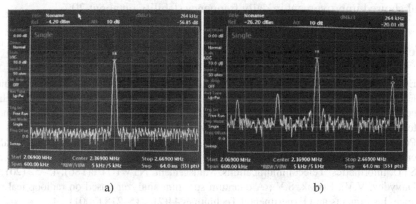

a)   b)

Fig. 5. Suppression of the amplitude of the lateral components in the band of 600 kHz.

The experimental results show that the suppression of lateral discrete components in the spectrum of microwave-excitation signal in the band of 1 kHz is improved on 75% and in the band of 600 kHz is improved about two times.

The shift of the peak of the line resonance atomic transition Δf is arising due to there are lateral discrete components in the microwave-excitation signal spectrum. The shift is defined by the following expression [2]:

$$\frac{\Delta f}{f} = \frac{A}{I} \frac{(\delta f)^2}{f(f-f_s)},$$ (1)

where A is the amplitude of the lateral components, I is the amplitude of the carrier, δf is the width of the spectral line, $(f - f_s)$ is the detuning of the lateral components relative to peaks of the spectral line.

In the developed scheme of the frequency converter the suppression of lateral components is not less than -72 dB, spectral line width is equal 500 Hz and the band of the lateral components is equal 1 kHz. Then relative frequency shift of the atomic transition is equal $\frac{\Delta f}{f}$ =2.2.10^{-14}. Previously obtained result is equal $\frac{\Delta f}{f}$ =1.62.10^{-13} Hz.

By reducing the frequency shift of the atomic transition Δf the more fine-tuning on the center of the resonance line is occur. This leads to a more accurate determination of the value of the nominal output frequency of frequency standard and, consequently, improves the long-term frequency stability of a caesium atomic clock.

4   Conclusion

Experimental study of frequency converter showed improvement parameters of the microwave-excitation signal, such as the step of frequency tuning, time of the frequency tuning, range of generated frequencies, phase stability and spectral characteristics. Experimental research of the metrological characteristics of the quantum frequency standard on the atoms of caesium - 133 with new scheme of the microwave – excitation signal showed improvement in daily frequency stability on about 20%.

References

1. Ermak, S.V., Semenov, V.V., Pyatyshev, E.N., Kazakin, A.N., Komartsev, I.M., Velichko, E.N., Davydov, V.V., Petrenko, M.V.: Production and research of integrated cells for compact frequency standard on the effect of coherent population trapping. Scientific - technical statements STU. Physical - mathematical sciences 1(213), 61–68 (2015)
2. Petrov, A.A., Davydov, V.V., Shabanov, V.E., Zaletov, D.V.: Digital frequency synthesizer for a quantum frequency standard caesium atoms – 133. Scientific - technical statements STU. Informatics. Telecommunications. Management, NTV IUT 6(186), 45–52 (2013)
3. Davydov, V.V., Ermak, S.V.: A quantum spectrum analyzer based on radiooptical resonance. Instruments and Experimental Techniques 44(2), 215–218 (2001)
4. Davydov, V.V., Semenov, V.V.: Nutation line of nuclear magnetic spectrometer with a flowing sample. Radiotekhnika I Elektronika (Journal of Communications Technology and Electronics) 44(12), 1528–1531 (1999)
5. Semenov, V.V., Nikiforov, N.F., Ermakov, S.V., Davydov, V.V.: Calculation of stationary magnetic resonance signal in optically oriented atoms induced by a sequence of radio pulses. Radio Engineering and Electronics 35(10), 2179–2183 (1990)
6. Riehle, F.: Frequency standards. Basics and applications, p. 496 (2004)
7. Oduan, K, Gino, B.: Chronometry and basics of GPS, p. 400 (2002)
8. Dudkin, V.I., Pachomov, L.N.: Quantum electronics. Devices and their applications, p. 422 (2012)
9. Ridiko, L.I.: Components and Technology, 27–39, 83 (2005)
10. Ushakov, V.: Optical devices in radio engineering. Radiotechnika, Moskow (2009)

Cyber-Physical Approach in a Series of Space Experiments "Kontur"

Vladimir Zaborovsky, Vladimir Muliukha, and Alexander Ilyashenko[✉]

Peter the Great St. Petersburg Polytechnic University, St. Petersburg, Russia
vlad@neva.ru, vladimir@mail.neva.ru, ilyashenko.alex@gmail.com

Abstract. The paper analyzes network-centric methods for control cyber-physical objects. Cyber-physical objects interact with each other by transmitting information via computer networks. In the framework of this cyber-physical approach proposed a structure for interactive control system for on-surface robot from International Space Station. This system implements the circuit-torque sensitization algorithms for network delays while transferring data over the computer telecommunication network.

Keywords: Cyber-physics · Robots · Force-torque feedback · Space experiment

1 Cyber-Physical Approach

We live in a world where the results of the development of science and technology make a radical change in the entire human environment. That is why the old aphorism holds true: a fundamental physics today – is a technology of tomorrow. The content of the fundamental physics laws is descriptive in nature and can be viewed as a result of analysis and following mathematical formalization of properties that characterize features of natural objects and processes of matter in motion. Technology is created by man in accordance with the laws of physics, as opposed to the objects of nature; it is the result of specific engineering works, including the assembly process and the formation of the rules of its application.

Process of creative engineering is the process of synthesis of new objects of physical reality. This process is inseparable from the different aspects of management, communication and control that are studied within the framework of cybernetics. Useful to note that the term "cybernetics" was used by Plato to denote the science of management of the objects containing people. Such objects may be, for example, a city populated by people or Persian soldiers in a chariot or a trireme – an ancient battle ship.

According to Plato, built and equipped ship is just a thing, but the ship with the crew and passengers – it is kiberno (κοβερνω) or "cyber-object". This "cyber-object" under the control of a skilled specialist or helmsman gets a fundamentally new properties that are not possessed by its parts if was taken separately. So sailboat cannot itself go against the wind, but a skilled crew of a sailors copes with this task easily. In

S. Balandin et al. (Eds.): NEW2AN/ruSMART 2015, LNCS 9247, pp. 745–758, 2015.
DOI: 10.1007/978-3-319-23126-6_69

Plato's definition, the important word is "skilled", i.e. with experience of successful execution of certain operations.

To implement the new properties, first of all a "part" of the cyber-object must be given special resources – namely, memory, which captures the "correct" sequence of operation leading to the success of the chosen scenario of the action. Unlike simple mechanical manipulations, such as use of the lever of Archimedes, whose actions are defined by clear mathematical formula, this scenario describes complex communication between connected states or parts of cyber-object. The attributes of the physical reality are complex and hard to describe formally, because this concept expresses not only the causal relations, but also the events, i.e., represents stochastic or information relation. In the twentieth century scientists began to treat cybernetics as field of research of the general laws, principles and methods of data processing, communication and control of complex systems.

With the advent of computers – physical systems, all processes in which take place in accordance with the instructions stored in computer memory, begins a new stage in the development of cybernetics. New development phase of cybernetics was coupled with the design and creation of automatic and automated control systems. In these systems, the role of Plato's "helmsman" plays a symbolic program matching coded targets and logical operations with controlled states of computer components. As a result, the problem of complexity transformed into a problem of correct operation of computer programs.

With the development of manycore and multithreaded computer technology and concept of quantum computing the problem of control complexity has been recognized as one of the fundamental problems of modern physics. Admittedly, management principles based on feedback, first formulated in cybernetics in order to stabilize the trajectories of technical objects, acquired all scientific significance.

In recent years, the application of scientific results was increasingly focused on applied problems associated with the use of network-centric technology, remote control, and information exchange. Consequently, the interpretation of the physical laws that reflect the characteristics of non-linear processes, chaotic or quantum phenomena from the standpoint of cybernetic principles, is actively paving the way for the new paradigm of studying the phenomena of nature, which is called cyber physics. Within this paradigm have been successfully solved complex problems of control of chaotic systems and purposeful change of flow bifurcation in the systems with continuous state space.

Although, while these considerations touched upon the problems of complexity generated by the sensitivity of the trajectories of dynamical systems to initial conditions or the probabilistic description of the motion of systems with non-reparable states, they paid no attention to the original problem of kiberno, that was indicated in the works of Plato, namely, the control of objects whose behavior occurs in the context of the state of their memory. Notable theoretical results associated with the use of Markov chains, although they may be considered a special case of kiberno control, never the less do not fulfil the requirements of cyber physical applications including group robotic actions, multiservice operational infrastructure for "Internet-of-things" or organization of MESH networks between dynamic objects.

Our work is dedicated to resolution of stated above various applied problems of cyber physics as a new interdisciplinary science. In subsequent sections of this chapter we focus on the organization of the engineering infrastructure and software architecture that provide means of communication between physical, mechanical and information-computing processes arising during network-centric interaction of intelligent robots and their virtual "avatars". Operating space for virtual objects is formed by specific cloud infrastructure based on robust communication channels and Hadoop data repository. The issues of implementation of the engineering infrastructure, and of the concept of cloud computing as applied to embedded software, regarded as the carrier of the control algorithms. The effectiveness of the proposed solutions is illustrated by the example of a network-centric management system of planetary robots whose operations at the level of calculating trajectories or maintaining information integrity and functional robustness are controlled from the board of ISS.

Complex engineering tasks concerning control for groups of mobile robots are developed poorly. In our work for their formalization we use cyber-physical (CPh) approach, which extends the range of engineering and physical methods for a design of complex technical objects by researching the informational aspects of communication and interaction between objects and with an external environment.

It is appropriate to consider control processes with cyber physical perspective because of the necessity for spatio-temporal adaptation to changing goals and characteristics of the operational environment. Thus the priority task is to organize the reliable and high-performance system of information exchange between all entities involved in the realization of all requirements. Hereinafter, by cyber physical object we mean an open system for the information exchange processes. Data in such system is transmitted through the computer networks, and its content characterizes the target requirements achieved through execution of physical and mechanical operations, energy being supplied by the internal resources of the object (Figure 1).

An example of a cyber-physical object is a mobile robot that does complex spatial movement, controlled by the content of the received information messages that have been generated by a human-operator or other robots that form a multi-purpose operation network. An ontological model of informational open cyber physical object may be represented by different formalisms, such as a set of epistemic logic model operations parameterized by data of local measurements or messages received from other robots via computer connection.

The selection of CPh systems as a special class of designed objects is due to the necessity of integrating various components responsible for computing, communications and control processes («3C» – computation, communication, control). Although in modern science there are different approaches to the use of information aspects of the physical objects, but only within cybernetics, such approaches have had structural engineering applications. The conceptual distinction between closed and open systems in terms of information and computational aspects requires the use of new models, which take into account the characteristics of information processes that are generated during the operation of the physical objects and are available for monitoring, processing and transmission via computer network.

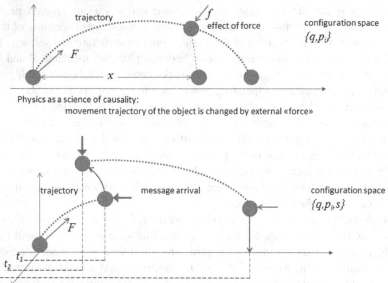

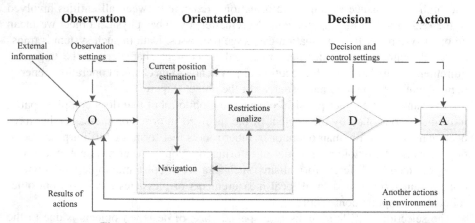

Fig. 1. Physical and cyber physical motion

Fig. 2. Cyber-physical model of proposed control system

According to Fig. 2, CF model of control system can be represented as a set of components, including following units [1]: information about the characteristics of the environment (Observation), analysis of the parameters of the current state for the controlled object (Orientation), decision-making according to the formal purpose of functioning (Decision), organization and implementation of the actions that are required to achieve the goal (Action). The interaction of these blocks using information exchange channels allows us to consider this network structure as a universal platform, which allows using various approaches, including: the use of algorithms and feedback mechanisms or reconfiguration of the object's structure for the goal's re-strictions entropy reduction or the reduction of the internal processes' dissipation.

2   Cyber-physical Approach for Space Robotics

Space Robotics Background

Currently in the development of space activities a major focus is on the use of space robotics. Both during transit and at their destination, astronauts need advanced autonomous and robotic systems to free them from routine tasks and to maximize their valuable scientific exploration time. That is why it is so important to increase autonomous operation time for space robots, which we consider the cyber physical objects.

It is clear that current robotic systems not based on cyber physical approach have serious limitations. Planetary surface rovers have never been designed for real-time interaction with each other or with astronauts. Moreover, controlling the robotic systems currently requires extensive human resources (e.g., a room full of controllers for one Mars rover), far beyond the capabilities of a crew on Mars tasked with real-time operations. We propose modular cyber physical robotic systems that can be fielded more quickly, are more reliably, and are easier to control than current systems. Modular designs will allow multiple robots to interact as peers among themselves and with humans, and improve the flexibility of robots. Because space robotic systems have to operate in highly unstructured environments with limited sensory information, limited actuator capability, limited power sources, limited communication bandwidth for control and coordination, and stochastic time delays, space-based research in autonomy and robotics is an incubator of innovative technologies for terrestrial robotics that can improve the quality of life for the earth's growing and aging population, and which will have utility in manufacturing and service industries, in medicine, and in disaster management.

Adaptation of Autonomous Cyber Physical Objects for Lunar and Planetary Exploration

The research involves development of control methods for tasking cyber physical systems that make it as intuitive and natural to interact with autonomous agents as it is to interact with human teammates. There are four major components in this work:

- Collaborative Human-Robot Plan Diagnosis – Reduction of cognitive load on human supervisors will require human and robot teammates to adapt to complex, changing, and sometimes conflicting mission goals. We propose to address this challenge by developing a capability for human and robot teams to collaboratively negotiate in order to diagnose problems with mission requirements, so that a feasible initial plan can be produced using high performance cloud-centric software. This process will be repeated if circumstances or goals change as the plan is being executed. The communication between humans, actors in cloud environment, and their autonomous cyber physical teammates must be efficient and natural, as if the humans were communicating with other humans.

- Automatic Learning of New Skills and Models – Using cloud-centric software we propose to develop learning and execution monitoring capabilities that allow task execution to be more easily adapted to new robotic hardware, mission tasks, and human teammate behavior. Robots are adapted to specific missions by supplying them with a set of models of hardware, allowed robot actions, and user behavior. We propose to reduce the time of developing new models by having robots learn new behavior directly in cloud environment using hardware and tasks requirements. The ability to recognize actions and task sequences is crucial. In order for a robot to provide assistance, it must understand what the human is trying to accomplish, particularly when operating in an unstructured environment, where new skills may have to be developed on the fly in order to adapt to unforeseen circumstances.

- Reactive Planning – In an unstructured environment, execution of a plan will often be disturbed by unforeseen changes in the environment, in the condition of the robots, or in the mission goals themselves. Dealing with unforeseen disturbances requires three capabilities. First, the autonomous system that is monitoring plan execution must recognize whether a disturbance is severe enough that it will make successful execution infeasible. Second, the autonomous system must be able to quickly change a plan that it estimates will fail due to a disturbance. Third, the autonomous system must understand risk and must be able to predict the likelihood of failure of the current plan. We propose to integrate these capabilities into our cloud environment for motion planning algorithms, as well as our collaborative diagnosis capabilities, so that risk is considered at every decision point.

- Perspective Space Robotic Network – To integrate information, intelligent and robotics technology within robust human-robot infrastructure we need to have solid cyber physics platform that merge together formal models of human knowledge, operation ontology, high performance computing resource, and control goals into space robotic network. For this purpose we will implement perspective cloud computing approach and extend IaaS/PaaS concepts to robot as a service model (RaaS model).

Cloud Based Functional Models of Aggregate-Modular Robots for Space Explorations

The creation of an autonomous robot base on the Moon surface is the first step to verify the technology of Human Exploration of the Moon and other celestial bodies. At the moment researches that are focused on the robotic Lunar exploration, are included in the space programs of many countries, including USA, China and Russia. For example, the well-known Chinese Lunar Exploration Program incorporates different types of robots: lunar orbiters, landers, rovers and sample return spacecraft.

According to the regulatory documents of Russian Space Agency, the main component of the robotic lunar base would be a group of robots, which would be capable of solving all kinds of problems arising during the Moon exploration, namely: manipulation with objects, transportation of cargo, installation and maintenance of different facilities, construction and installation works, mining, etc.

There are two main approaches to form such a group of robots:

- Few multifunctional, complex to manufacture and maintain robots, each of which performs a wide range of tasks;
- Lots of different robots that are easy to manufacture and operate, each of them performs a limited range of tasks.

The main advantage of the second approach is the simplicity of robots' and functional modules' manufacturing, and, as a consequence, a higher reliability of the overall system, that in the context of space exploration is a critical parameter. Also it should be noted that the specialized robots, which have been formed from different functional modules, have less redundancy for the task. This improves the efficiency of the robotics system and help to achieve the optimum of the quality criteria, namely: accuracy rate of the operation cost of operation in the terms of energy and fuel, run time of the task and etc.

One of the most common methods to implement this approach is the aggregate-modular design of robots. This method is based on the universal interfaces for the information and the mechanical interactions between components of a robot. It gives us the opportunity to construct a specialized robot on the base of limited number of standardized components. Such construction meets the requirements of the current task in the best possible way, so this specific solution has minimal redundancy.

The main purpose of our work is to research and develop an algorithm for automatic construction of the optimal configuration for the robot from the set of standardized modules to implement a task. It is assumed that such standardized modules have uniform mechanical and informational interfaces. The main objectives of our work are:

- Classification of robots and their modules;
- Development of a model of aggregate-modular robot;
- Researching the limits of the robot model based on the requirements for the optimal realization of the task;
- Development of method and algorithm for automatic construction of the optimally configured robot.

At the first phase of the research standardized robotics modules have been classified. Following groups have been identified:

- handling systems and control devices;
- manipulators;
- transport platforms and movement systems for mobile robots;
- sensors and sensor systems.

On the basis of the proposed classification, we have developed mathematical model of combined aggregate-modular robot that meet following requirements:

- Each module is presented as a graph vertex with emanating from him "hanging" outgoing edges that represent unified interfaces that may be attached to other nodes. Examples of various unified modules are shown in Figure 3;

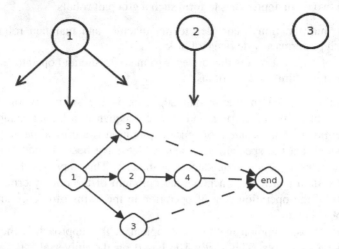

Fig. 3. Various modules and example of representation robot as a network: 1 – transport platform with three unified interfaces; 2 – manipulator with one interface; 3,4 – sensors

- All robotic objects are described as a network graph G (Figure 3). A source of such network is always a transport platform or another movement system for mobile robot. A drain of this network is a dummy node;
- Mathematical model assumes the existence of a vector estimation function of the current solution – network graph G. This function is named $F(G)$, and should estimate the performance of combined robot using characteristics of individual modules;
- Mathematical model assumes the existence of a system of functional limitations $f_i(F(G))$, which is formed according to functional requirements of the robot's task;
- Mathematical model requires an optimization criterion of functioning quality for a particular purpose $J(F(G))$.

The requirements for the problem of finding the optimal configuration of aggregate-modular robots can be formulated as follows: it is necessary to generate a graph G^*, which satisfies the functional limitations $f_i(F(G))$ and gives minimum to the optimization criterion $J(F(G))$.

The proposed mathematical model describes the NP-complete optimization problem. During the research of the following types of methods and algorithms that can be used for solving NP-complete problems were considered:

- exhaustive search algorithms or brute-force search;
- approximate and heuristic methods that involve the use of a priori rates and heuristics to select the elements of the solution;
- branch and bound method, involving discarding obviously suboptimal decisions according to some estimates.

3 Robot Control from ISS

We consider the application of the proposed above principles to control physical objects: on-surface robot, the motion of which is set and controlled from an orbital space station (Fig. 4) [2].

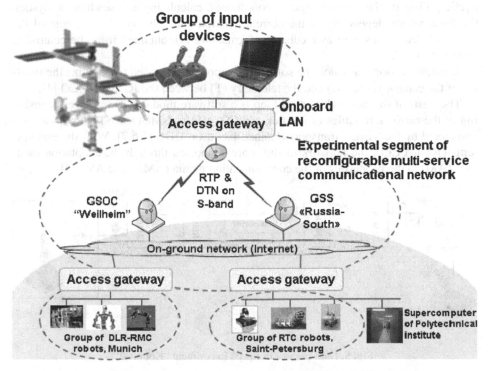

Fig. 4. Scheme of the Space Experiment "Kontur-2"

The feature of this Space Experiment (SE) "Kontur-2", which allows carrying it to the class of cyber-physical experiments, is that it uses the telepresence technology for the operator to simulate the robotic movement while the parameters of the environment (delays in communication channels and obstacles on the planet's surface) may vary in sporadic way. The designed control system allows in real-time to make palpable the results of robotic operations by analyzing the information about the values of the current axes of movement or moments in the joints of the manipulator, which are transmitted via computer communication network with a frequency of 500 packets per second.

The use of the circuit-torque delays sensitization effect allows the operator to adjust the speed and movement direction of the robot and by using force feedback effects on the joystick feel the impact of the network environment and generate an assessment of the environment state in which operates on-surface robot. In the considered control system the processes of the information exchange between joystick and robot can be decomposed into two processes of local commands realization in hard real-time and the commands' delivery process via network infrastructure using the TCP/IP stack.

Physical structure of the data streams in such control system is shown in Fig. 5 and includes:

1. The local loop, in which the software module "Joystick controller" (JC) provides cyclic polling of the current joystick coordinates, calculating and sending in joystick the force vector depending on the current position and velocity of movement of the joystick's handle, as well as feedback information (T'), obtained from the controlled object (CO) [3];

2. Network loop, in which the software components are used to organize the transfer of the control vector (C) and the telemetry (T) between the JC and the CO [4].

The basis of the network control loop is a software module "Transporter" consisting of the network modules of joystick and CO (NMJo, NMCO, see Fig. 5) which are connected by the virtual transport channel, based on UDP [5,6,7]. With the end systems (JC and CO), the network modules are connected through the adaptation modules (AM) to the properties of the communication media (AMJo and AMCO).

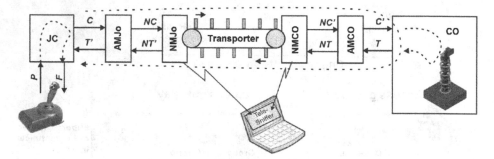

Fig. 5. Data streams in Space Experiment "Kontur-2"

Software module "Transporter" delivers data using UDP, providing isochronous communication for local controllers: vector control samples that are uniformly received from the JC should also be uniformly (but, of course, delayed) transferred to the CO. Similarly, in the opposite direction is transferred the vector data obtained from the telemetry system of the robot. Thus, the "digital" reality of the physical processes is updated with a sampling frequency equal to the frequency of sending packets. Taking into account that the data delivery delay in digital communication channels is not constant, some packets may be lost, and delivery order may be disturbed, adaptation module provides recovery of the missing packets using automata model of data transfer processes, the adequacy of which is verified using the probabilistic model of packet delivery.

Module "Transporter" also ensures the delivery of asynchronous messages about events that are relevant to the remote control. The example of such events may be the pressing the button on the joystick's handle. These buttons may be used to control the operating mode of the CO or for some other action. Asynchronous event occurs sporadically at any given time. However, it should be guaranteed to reach the operator, saving the time reference to the transmitted isochronous packets' stream.

The computing resources of adaptation module allow to implement the methods of predictive modeling and, in the case of insufficient data, to predict the behavior of the control object or operator without any delay in the transmission of control signals, and thereby ensure the smooth control for the robot. Control circuit scheme is presented in Fig. 6.

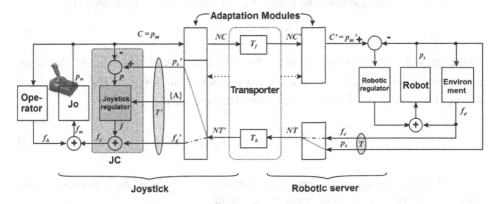

Fig. 6. "Kontur-2" control circuit scheme

In Fig. 6, there is the separation of the control circuit into three main parts:

1. The loop of Joystick (Jo) (the left part of the scheme), whose output is the control vector c that is showing the current position of the Jo handle p_m. Movement of the handle is determined by the force f, which is summable from the operator's impact forces f_h, force f_c, which is formed by Joystick, and force f_e', displaying the impact of the environment f_e on robot. P input of JC is the mismatch between the current Jo position (p_m) and the display of the current position of the robot (p_s). The controller parameters vector $\{A\}$ can be changed dynamically.
2. The means of communication (Transporter), which implements delays in delivery, and adaptation modules that are counteract the effect of these delays. In general, the delays in the direct (T_f) and the reverse channels (T_b) can be different.
3. The loop of the controlled device (the right part of the scheme), generating the impact on the robot using a mismatch between its current position p_s and displayed position of the Jo p_m'. A robot may also be affected by the environment with force f_e. As the feedback, the vector of current robot position p_s can be used, and if there are appropriate sensors, the force vector of environmental impact f_e also can be used.

The remote robot control with force-torque sensitization requires that during the control process, operator has to be able to "feel" the current state of the robot, and communication network. Therefore, to describe the configuration space of the controlled system is proposed to use a model of the virtual spring that is fixed at the one end to the base of the joystick, and the initial position of which coincides with the current position of the controlled robot. In the initial position, while not moving, the robot does not affect the joystick and the operator. However, if the operator starts the control process and move the joystick's handle from the initial position, the virtual string will try to return it to the initial position. Farther the joystick will be deflected from the initial position, the greater force will the operation of stretching virtual springs require from the operator. Virtual elastic force of the spring is calculated based on analysis of the current situation, taking into account current positions of joystick and information from the robot. It is calculated using the formula (1):

$$p = p'_s - p_m,$$ (1)

where p'_s is a vector of robot's coordinates, which is currently available for the adaptation module of joystick , p_m is a current position of the joystick. The impact force felt by the operator is calculated according to the formula (2):

$$f(t) = A_0 \int_0^t p(\tau)d\tau + A_1 p + A_2 \dot{p} + A_3 \ddot{p},$$ (2)

where A_0 is an astatism in system , A_1 is virtual spring stiffness, A_2 is viscosity of the control medium, A_3 is a virtual mass of the handle.

During the control process, the initial position of the virtual spring will change to coincide with the current position of the robot. The proposed model allows taking into account the information about the state of the communication channel by making adjustments during the process of sensitization. In other words, the virtual stiffness of the spring must be increased not only according to the divergence between positions of the robot and joystick, but also using the value of the data delay in the communication channel, thus increasing the inertia of the network control loop.

To implement such interaction mode we will need to handle the additional information about the state of the communication channel: sequence of RTT samples that is ordered by time and the percentage of the lost network packets (PLP) [5,7]. Thus, the stiffness of the virtual spring and a moving speed of the robot will vary proportionally average values of RTT and PLP, which are calculated for the time period according to the formula (3). This period is commensurate with the time constant for the closed-loop control system.

$$f(t) = (A_0 \int_0^t p(\tau)d\tau + A_1 p + A_2 \dot{p} + A_3 \ddot{p}) + A_4 \cdot RTT \cdot PLP \cdot \frac{p}{\|p\|},$$ (3)

where A_4 is a coefficient of influence of the environment on the force-torque feedback. As a result, while increasing the latency, the virtual spring will not allow the operator to change robot's position quickly. This circumstance reduces speed of the robot while moving, but allows the operator to adjust the results of mining operations, analyzing data of control actions, despite the fact that this data will be available to the operator is delayed.

The considered organization of the remote control system was developed for the on-ground tests of algorithms and software debugging of scientific equipment for the space experiment "Kontur-2". On the space oriented stand the parameters of the communications system and the software module Transporter (see Fig. 6) are modified by taking into account the limitations of S-band communication channel, capacity of which is estimated based on the necessity of the telepresence conditions, which requires to transmit IP packets containing 22 bytes of application data every 2 ms. These requirements allow us to modify the format of the messages and take into account the properties of network processes, to reduce the length of packet's numbering field and timestamps by dividing the packet number into even and odd. Even-numbered packets are used to transmit isochronous traffic (5 values in 32-bit floating point format), the odd ones are used for transmission of asynchronous messages and service information that facilitates determination of channel parameters with the implementation of network protocols and the use of "light" version of LwIP stack.

According to the requirements of bilateral control the software for the remote control system is symmetrical, namely the structure of software modules on the side of Joystick and the robot matches. Local controllers like Joystick or robot are implemented as PID controllers and differ from each other only by settings. The developed software for the Joystick is implemented using 32-bit ARM microcontroller.

The active phase of the experiment begins in August, 2015. Thus, the detailed description of the software as well as numerical data obtained during the experiment will be presented later. In our future works, we plan to expand the boundaries of the application for cyber-physical approach to control the groups of on-surface robots. In the next experiment, which we plan to start in 2016 will consider the data exchange processes during the control of on-surface robotics group. Some robotic routine operations will be automated by using high-performance cloud computing system [8,9] in the control loop.

4   Conclusion

The paper describes the structure of a cyber-physical system for remote force-torque control of a robot located at the Earth's surface, from the manned orbital space station. The feature of the proposed solution is the use of dual-circuit system in which the "almost analog" objects (the robot and the joystick with their control loops) are connected by the network channel. This communication link brings in the control loop the variable discrete and significant delays during data delivery. Force-torque sensitization implementation allows the operator to feel not only the state of the controlled device, but also to obtain information about the network channel, which indicates about possible unsafe further control. The control method and software architecture for its implementation were proposed. As the operating system for network nodes was chosen freely distributed real-time system FreeRTOS, which provides a high frequency cycles in local control loops and network interaction between the operator and the robot. Software systems for the interactive remote control are used in bilateral mode for both joystick and robot that allows developing effectively algorithms for circuit-torque feedback to the various options for network infrastructure.

References

1. Fradkov, A.L.: Cybernetics physics: Principles and Examples. Nauka, SPb, 208 p. (2003) (in Russian)
2. Zaborovsky, V.S., Kondratiev, A.S., Silinenko, A.V., Muliukha, V.A., Ilyashenko, A.S., Filippov, M.S.: Remote Control for Robotic Objects in Space Experiments of "Kontur" series. Scientific and Technical Statements of SPbSPU. Computer. Telecommunications. Control **6**(162), 23–32 (2012). (in Russian)
3. Artigas, J., Jee-Hwan, R., Preusche, C., Hirzinger, G.: Network representation and passivity of delayed teleoperation systems. In: 2011 IEEE/RSJ International Conference on Intelligent Robots and Systems (IROS), pp. 177–183, September 25–30, 2011
4. Niemeyer, G., Slotine, J.-J.E.: Telemanipulation with Time Delays. Int. Journal of Robotics Research, **23**(9), 873–890
5. Zaborovsky, V., Zayats, O., Mulukha, V.: Priority queueing with finite buffer size and randomized push-out mechanism. In: Proceedings of The Ninth International Conference on Networks (ICN 2010), Menuires, The Three Valleys, French Alps, April 11–16, 2010. Published by IEEE Computer Society, pp. 316–320 (2010)
6. Zaborovsky, V., Gorodetsky, A., Muljukha, V.: Internet performance: TCP in stochastic network environment. In: 1st International Conference on Evolving Internet, INTERNET 2009, pp. 21–26 (2009). doi:10.1109/INTERNET.2009.36
7. Muliukha, V., Ilyashenko, A., Zayats, O., Zaborovsky, V.: Preemptive Queuing System with Randomized Push-Out Mechanism. Communications in Nonlinear Science and Numerical Simulation **21**(1–3), 147–158 (2015)
8. Zaborovsky, V., Lukashin, A., Kupreenko, S., Mulukha, V.: Dynamic Access Control in Cloud Services. International Transactions on Systems Science and Applications **7**(3/4), 264–277 (2011)
9. Zaborovsky, V., Muliukha, V., Popov, S., Lukashin, A.: Heterogeneous virtual intelligent transport systems and services in cloud environments. In: Proceedings of The Thirteenth International Conference on Networks (ICN 2014), Nice, France, pp. 236–241, February 23–27, 2014

Reflectivity Properties
of Graphene-Coated Silica

Constantine Korikov[✉]

Peter the Great St. Petersburg Polytechnic University,
St. Petersburg 195251, Russia
korikov.constantine@spbstu.ru

Abstract. The reflection coefficients on graphene-coated substrate at real frequencies are calculated. Simple analytic expressions for the reflection coefficients at high frequencies are found. The reflectivities of graphene-coated silica, as a prospective material for nanocommunications, are compared with the reflectivities of uncoated silica.

Keywords: Graphene · Silica · Reflection coefficients · Polarization tensor · Nanocommunications

1   Introduction

Graphene, which is a (quasi) one-atom-thick layer of carbon atoms has many unique mechanical, electrical and optical properties making it one of the most interesting materials in condensed matter physics [1–3]. In fact, graphene allows a wide potential usage of its properties in nanocommunications.

During the last few years, a two-dimensional sheet of carbon atoms attracted much experimental and theoretical attention [1]. In addition, there are theoretical suggestions of other graphene-like nanostructures, for instance based on Si (silicene) [4], Ge (germanene) [5] and also Sn (stanene) [6]. Recent experimental approaches to investigations of silicene [7] and germanene [8] show significant prospectives for nanotechnologies.

The principle feature of all these materials is that the quasi-particles in them are described by the Dirac equation, but move with a Fermi velocity instead of the velocity of light [9].

The purpose of the current work is to calculate the reflectivity properties of a plate coated with graphene at real frequencies. The plate material is characterized by the frequency-dependent dielectric permittivity $\epsilon(\omega)$ and the reflectivity properties of graphene are described by the polarization tensor.

The paper is organized as follows. In Sec. 2 the reflection coefficients on substrate coated with graphene film are derived. For this purpose the results of [10] applied to real frequencies are used. Both the transverse magnetic and transverse electric reflection coefficients on a substrate coated with graphene are found. Sec. 3 contains analytic asymptotic expressions for the reflection coefficients at high frequencies and numerical comparison between reflectivities of the graphene sheet deposited on a silica substrate and a silica plate with no coating. Sec. 4 contains thes conclusions.

© Springer International Publishing Switzerland 2015
S. Balandin et al. (Eds.): NEW2AN/ruSMART 2015, LNCS 9247, pp. 759–764, 2015.
DOI: 10.1007/978-3-319-23126-6_70

2   Reflection Coefficients for Graphene-Coated Substrate

Let us consider the reflection of the electromagnetic waves incident on one-atom-thick film spaced above a thick plate (semispace) parallel to it in vacuum separated by a gap of thickness d.

The amplitude reflection coefficient $r_{1,2}$ and the transmission coefficient $t_{1,2}$ represent the reflective properties of the film and the plate, respectively.

Taking into account the structure of the system (the film above the plate), it has to be considered as a case of multiple reflections [10]. The reflection coefficient from this system is

$$r = r_1 + t_1 r_2 t_1 e^{-2dq} \sum_{n=0}^{\infty} \left(r_1 r_2 e^{-2dq} \right)^n = r_1 + \frac{t_1 r_2 t_1 e^{-2dq}}{1 - r_1 r_2 e^{-2dq}}, \tag{1}$$

where

$$q \equiv \left[\frac{\omega^2}{c^2} - k_\perp^2 \right]^{1/2} = \frac{\omega}{c} \cos\theta_i, \tag{2}$$

because for real photons $k_\perp = \omega \sin\theta_i / c$ and θ_i is the angle of incidence. Equation (1) can be used for the transverse magnetic (TM) and transverse electric (TE) reflection coefficients defined for the two independent polarizations of the electromagnetic field. This equation can be rewritten for the case of deposited film on a thick plate. In this case one should put $d \to 0$ in (1) and arrive at

$$r = r_1 + \frac{t_1 r_2 t_1}{1 - r_1 r_2}. \tag{3}$$

For graphene film (f) in vacuum it was shown that [11]

$$t_{\mathrm{TM}}^{(f)} = 1 - r_{\mathrm{TM}}^{(f)}, \tag{4}$$

$$t_{\mathrm{TE}}^{(f)} = 1 + r_{\mathrm{TE}}^{(f)}. \tag{5}$$

Let us apply (3) to a graphene film (f) deposited on a material substrate (s) characterized by the frequency-dependent dielectric permittivity $\epsilon(\omega)$. For the TM mode one should put $r_1 = r_{\mathrm{TM}}^{(f)}$, $t_1 = t_{\mathrm{TM}}^{(f)} = 1 - r_{\mathrm{TM}}^{(f)}$, $r_2 = r_{\mathrm{TM}}^{(s)}$ in Eq. (3) and obtain

$$r_{\mathrm{TM}}^{(f,s)} = r_{\mathrm{TM}}^{(f)} + \frac{r_{\mathrm{TM}}^{(s)} \left(1 - r_{\mathrm{TM}}^{(f)} \right)^2}{1 - r_{\mathrm{TM}}^{(f)} r_{\mathrm{TM}}^{(s)}} = \frac{r_{\mathrm{TM}}^{(f)} + r_{\mathrm{TM}}^{(s)} \left(1 - 2 r_{\mathrm{TM}}^{(f)} \right)}{1 - r_{\mathrm{TM}}^{(f)} r_{\mathrm{TM}}^{(s)}}. \tag{6}$$

For the TE mode $r_1 = r_{\mathrm{TM}}^{(f)}$, $t_1 = t_{\mathrm{TM}}^{(f)} = 1 + r_{\mathrm{TM}}^{(f)}$, $r_2 = r_{\mathrm{TM}}^{(s)}$ and Eq. (3) results in

$$r_{\mathrm{TE}}^{(f,s)} = r_{\mathrm{TE}}^{(f)} + \frac{r_{\mathrm{TE}}^{(s)} \left(1 + r_{\mathrm{TE}}^{(f)} \right)^2}{1 - r_{\mathrm{TE}}^{(f)} r_{\mathrm{TE}}^{(s)}} = \frac{r_{\mathrm{TE}}^{(f)} + r_{\mathrm{TE}}^{(s)} \left(1 + 2 r_{\mathrm{TE}}^{(f)} \right)}{1 - r_{\mathrm{TE}}^{(f)} r_{\mathrm{TE}}^{(s)}}. \tag{7}$$

The reflection coefficients on the boundary between a vacuum and a semispace are the Fresnel coefficients

$$r_{\mathrm{TM}}^{(s)} = \frac{\epsilon(\omega)\cos\theta_i - \sqrt{\epsilon(\omega) - \sin^2\theta_i}}{\epsilon(\omega)\cos\theta_i + \sqrt{\epsilon(\omega) - \sin^2\theta_i}}, \tag{8}$$

$$r_{\mathrm{TE}}^{(s)} = \frac{\cos\theta_i - \sqrt{\epsilon(\omega) - \sin^2\theta_i}}{\cos\theta_i + \sqrt{\epsilon(\omega) - \sin^2\theta_i}}. \tag{9}$$

For a graphene film in a vacuum the reflection coefficients in terms of the polarization tensor take the form [12,13]

$$r_{\mathrm{TM}}^{(f)} = \frac{c\cos\theta_i\,\Pi_{00}(\omega,\theta_i)}{2i\omega\hbar\sin^2\theta_i + c\cos\theta_i\,\Pi_{00}(\omega,\theta_i)}, \tag{10}$$

$$r_{\mathrm{TE}}^{(f)} = -\frac{c^3\Pi(\omega,\theta_i)}{-2i\hbar\omega^3\sin^2\theta_i\cos\theta_i + c^3\Pi(\omega,\theta_i)}. \tag{11}$$

The explicit expressions for Π_{00} and Π can be found in [12–14].

Let us introduce the new functions:

$$\widetilde{\Pi}_{00}(\omega,\theta_i) \equiv -i\frac{c}{\omega}\frac{\cos\theta_i}{2\hbar\sin^2\theta_i}\Pi_{00}(\omega,\theta_i), \tag{12}$$

$$\widetilde{\Pi}(\omega,\theta_i) \equiv i\frac{c^3}{\omega^3}\frac{1}{2\hbar\cos\theta_i\sin^2\theta_i}\Pi(\omega,\theta_i). \tag{13}$$

Using (12) and (13), the reflection coefficients can be transformed to

$$r_{\mathrm{TM}}^{(f)} = \frac{\widetilde{\Pi}_{00}(\omega,\theta_i)}{1 + \widetilde{\Pi}_{00}(\omega,\theta_i)}, \tag{14}$$

$$r_{\mathrm{TE}}^{(f)} = -\frac{\widetilde{\Pi}(\omega,\theta_i)}{1 + \widetilde{\Pi}(\omega,\theta_i)}. \tag{15}$$

Substituting (8), (9), (14) and (15) in (6) and (7), one obtains the reflection coefficients from the graphene deposited on a material plate

$$r_{\mathrm{TM}}^{(f,s)} = \frac{\epsilon(\omega)\cos\theta_i - \sqrt{\epsilon(\omega) - \sin^2\theta_i} + 2\widetilde{\Pi}_{00}\sqrt{\epsilon(\omega) - \sin^2\theta_i}}{\epsilon(\omega)\cos\theta_i + \sqrt{\epsilon(\omega) - \sin^2\theta_i} + 2\widetilde{\Pi}_{00}\sqrt{\epsilon(\omega) - \sin^2\theta_i}}, \tag{16}$$

$$r_{\mathrm{TE}}^{(f,s)} = \frac{\cos\theta_i - \sqrt{\epsilon(\omega) - \sin^2\theta_i} - 2\widetilde{\Pi}\cos\theta_i}{\cos\theta_i + \sqrt{\epsilon(\omega) - \sin^2\theta_i} + 2\widetilde{\Pi}\cos\theta_i}. \tag{17}$$

3   Reflectivity at High Frequencies

In this section we find analytic asymptotic expressions for the reflection coefficients (16) and (17) at high frequencies satisfying the condition $\omega \gg \omega_T$, where ω_T is the thermal frequency expressed as

$$\omega_T \equiv \frac{k_B T}{\hbar}. \tag{18}$$

As was shown previously [12,13], thermal effects do not significantly contribute to the polarization tensor at high frequencies. Thus, one can use the expressions for the polarization tensor at zero temperature [12–14]

$$\Pi_{00}(\omega, \theta_i) \approx {\Pi_{00}}^{(0)}(\omega, \theta_i) \approx i\hbar\pi\alpha \frac{\omega}{c} \sin^2 \theta_i, \tag{19}$$

$$\Pi(\omega, \theta_i) \approx \Pi^{(0)}(\omega, \theta_i) \approx -i\hbar\pi\alpha \frac{\omega^3}{c^3} \sin^2 \theta_i. \tag{20}$$

Substituting these equations in Eqs. (12) and (13), one obtaines

$$2\widetilde{\Pi_{00}} \approx 2\widetilde{\Pi_{00}}^{(0)} = -i\frac{c}{\omega}\frac{\cos\theta_i}{\hbar\sin^2\theta_i}\left[i\hbar\pi\alpha\frac{\omega}{c}\sin^2\theta_i\right] = \pi\alpha\cos\theta_i, \tag{21}$$

$$2\widetilde{\Pi} \approx 2\widetilde{\Pi}^{(0)} = -i\frac{c^3}{\omega^3}\frac{1}{\hbar\cos\theta_i\sin^2\theta_i}\left[-i\hbar\pi\alpha\frac{\omega^3}{c^3}\sin^2\theta_i\right] = \frac{\pi\alpha}{\cos\theta_i}, \tag{22}$$

As a result, the expressions for the reflective coefficients at high frequencies $\omega \gg \omega_T$ are

$$r_{TM}^{(f,s)}(\omega, \theta_i) = \frac{\epsilon(\omega)\cos\theta_i - \sqrt{\epsilon(\omega) - \sin^2\theta_i}[1 - \pi\alpha\cos\theta_i]}{\epsilon(\omega)\cos\theta_i + \sqrt{\epsilon(\omega) - \sin^2\theta_i}[1 + \pi\alpha\cos\theta_i]}, \tag{23}$$

$$r_{TE}^{(f,s)}(\omega, \theta_i) = \frac{\cos\theta_i - \sqrt{\epsilon(\omega) - \sin^2\theta_i} - \pi\alpha}{\cos\theta_i + \sqrt{\epsilon(\omega) - \sin^2\theta_i} + \pi\alpha}. \tag{24}$$

At the normal incidence ($\theta_i = 0$) the coefficients have the following form

$$r_{TM}^{(f,s)}(\omega, 0) = -r_{TE}^{(f,s)}(\omega, 0) = \frac{\sqrt{\epsilon} - 1 + \pi\alpha}{\sqrt{\epsilon} + 1 + \pi\alpha} = \frac{n - 1 + \pi\alpha + ik}{n + 1 + \pi\alpha + ik}, \tag{25}$$

where $n + ik = \sqrt{\epsilon}$ is a complex refractive index and ϵ, n and k are functions of frequency. Equations (23)–(25) coincide with respective expressions of Ref. [12], where they were written without rigorous derivation.

The thermal frequency at room temperature $T = 300K$ has the value $\omega_T|_{T=300K} = 3.9 \cdot 10^{13} \text{rad/s} \approx 0.026 \text{eV}$. This means that the asymptotic expressions (23) and (24) are applicable in the optical range and all higher frequencies.

The representation (25) is convenient for numerical analysis. As an example, one can choose materials widely used in experiments, such as silicene on Ag [7], germanene on Au [8] or graphene on silica [10]. The reflectivities $R^{(f,s)} = \left|r^{(f,s)}\right|^2$ for the graphene-coated glass and for pure glass were obtained using the measurment data for the complex index of refraction of selica [15]. Comparison of the results is presented in Table 1.

Table 1. Numerical comparison between the reflectivities of the graphene sheet deposited on a silica substrate, $R^{(g,s)}$, and silica plate with no graphene coating, $R^{(s)}$.

ω (eV)	n	k	$R^{(s)}/R^{(g,s)}$
92.53	0.977	$8.7 \cdot 10^{-3}$	8.17
8.0	1.702	$3.2 \cdot 10^{-5}$	0.95
0.35	1.406	0	0.91
0.27	1.365	$2.56 \cdot 10^{-4}$	0.90

It is seen that the influence of graphene coating on the reflectivity properties depends on the frequency. In the range of frequencies where $n < 1$ graphene makes glass more transparent. In the range where absorption is absence graphene improves the reflectivity properties.

4   Conclusions

In this paper, the reflection coefficients on a graphene-coated substrate are calculated at real frequencies using the formalism of the polarization tensor in (2+1)-dimensional space-time and dielectric permittivities of substrate material. Additionally,simple analytic expressions for the coefficients at high frequencies have been found. The numerical comparison was performed between reflectivities of graphene deposited on a silica substrate and a silica plate with no graphene coating at different frequencies of the incident light.

Taking into account that graphene and other graphene-like nanostructures are of potential use in electromagnetic communications, the investigation of their optical properties is topical for future technological applications.

Acknowledgments. The author is grateful to G. L. Klimchitskaya and V. M. Mostepanenko for the formulation of this problem and help in the work.

References

1. Katsnelson, M.I.: Graphene: Carbon in Two Dimensions. Cambridge University Press, Cambridge (2012)
2. Castro Neto, A.H., Guinea, F., Peres, N.M.R., Novoselov, K.S., Geim, A.K.: The Electronic Properties of Graphene. Rev. Mod. Phys. **81**, 109–162 (2009)

3. Bonaccorso, F., Sun, Z., Hasan, T., Ferrari, A.: Graphene Photonics and Optoelec-tronics. Nature Photonics **4**, 611–622 (2010)
4. Takeda, K., Shiraishi, K.: Theoretical Possibility of Stage Corrugation in Si and Ge Analogs of Graphite. Phys. Rev. B **50**, 14916–14922 (1994)
5. Cahangirov, S., Topsakal, M., Aktürk, E., Şahin, H., Ciraci, S.: Two- and One-Dimensional Honeycomb Structures of Silicon and Germanium. Phys. Rev. Lett. **102**, 236804 (2009)
6. Garcia, J.C., de Lima, D.B., Assali, L.V.C., Justo, J.F.: Group IV Graphene- and Graphane-Like Nanosheets. The Journal of Physical Chemistry C **115**, 13242–13246 (2011)
7. Lalmi, B., Oughaddou, H., Enriquez, H., Kara, A., Vizzini, S., Ealet, B., Aufray, B.: Epitaxial Growth of a Silicene Sheet. Applied Physics Letters **97**, 223109 (2010)
8. Dávila, M.E., Xian, L., Cahangirov, S., Rubio, A., Lay, G.L.: Germanene: a novel two-dimensional germanium allotrope akin to graphene and silicene. New Journal of Physics **16**, 095002 (2014)
9. Matthes, L., Pulci, O., Bechstedt, F.: Massive Dirac quasiparticles in the Optical Absorbance of Graphene, Silicene, Germanene, and Tinene. Journal of Physics: Condensed Matter **25**, 395305 (2013)
10. Klimchitskaya, G.L., Mohideen, U., Mostepanenko, V.M.: Theory of the Casimir Interaction from Graphene-Coated Substrates using the polarization tensor and comparison with experiment. Phys. Rev. B **89**, 115419 (2014)
11. Stauber, T., Peres, N.M.R., Geim, A.K.: Optical Conductivity of Graphene in the Visible Region of the Spectrum. Phys. Rev. B **78**, 085432 (2008)
12. Klimchitskaya, G.L., Mostepanenko, V.M., Petrov, V.M.: Reflectivity proper-ties of graphene and graphene-coated substrates. In: Balandin, S., Andreev, S., Koucheryavy, Y. (eds.) NEW2AN/ruSMART 2014. LNCS, vol. 8638, pp. 451–458. Springer, Heidelberg (2014)
13. Bordag, M., Klimchitskaya, G.L., Mostepanenko, V.M., Petrov, V.M.: Quantum Field Theoretical Description for the Reflectivity of Graphene. Phys. Rev. D **91**, 045037 (2015)
14. Fialkovsky, I.V., Marachevsky, V.N., Vassilevich, D.V.: Finite-Temperature Casimir Effect for Graphene. Phys. Rev. B **84**, 035446 (2011)
15. Palik, E.D.: Handbook of Optical Constants of Solids. Academic Press, New York (1997)

Nano Communication Device with Embedded Molecular Films: Effect of Electromagnetic Field and Dipole Moment Dynamics

Elena Velichko[1,2(✉)], Tatyana Zezina[1], Anastasia Cheremiskina[1], and Oleg Tsybin[1]

[1] Peter the Great St. Petersburg Polytechnic University, St. Petersburg, Russia
{velichko-spbstu,zezinat}@yandex.ru, a.cheremiskina@gmail.com,
otsybin@rphf.spbstu.ru
[2] Saint Petersburg National Research University of Information Technologies,
Mechanics and Optics, St. Petersburg, Russia

Abstract. The communication among molecular networks may be specifically realized by nanomechanical, acoustic, and electromagnetic fields and molecular transport. Here, experimental and theoretical studies of peptide and protein films and single molecules in static and radiofrequency electromagnetic fields are reported. Impedance (dielectric) electrochemical spectroscopy revealed nonlinear properties of glycine, alanine and albumen films in the external electromagnetic field in frequency range 0.5–100 MHz. Computer "all atom" simulation allows one to calculate the nanoelectromagnetic field of molecular systems and to evaluate the self-assembled supramolecular architectures. Theoretical studies revealed the dipole moment dynamics of polyalanine peptides. Further, we combine both approaches, thus providing a prediction model of nanoelectromagnetic field generation, and molecular transportation/communication.

Keywords: Molecular communication · Peptide · Protein films · Computer simulations · Electromagnetic field · Dipole moment

1 Introduction

Important part of modern and emerging nanotechnologies is presented by nanodevices of micrometer – sub-micrometer scale containing embedded molecular films [1–3]. This research area requires a board interdisciplinary approach, including physics, chemistry, biology, electronics and biotechnology.

The thin peptide/protein films based on the self-assembly principles are considered to be the advanced materials for modern biomolecular nanodevices. The films are active media which are synthetic molecular systems including those consisting of single molecules and multiparticle systems assembled as metamaterials in solution or on a solid surface by using known techniques. The electric properties study of protein and peptide films, their ordered semi-crystalline or hydrogel supramolecular architectures and conformational changes under the influence of electric field is of great interest for many research areas, such as nano electromagnetic and molecular communications [3–5].

© Springer International Publishing Switzerland 2015
S. Balandin et al. (Eds.): NEW2AN/ruSMART 2015, LNCS 9247, pp. 765–771, 2015.
DOI: 10.1007/978-3-319-23126-6_71

Such investigations are effectively realized by electrochemical or dielectric spectroscopy measurements [6–8], and molecular dipole moment dynamic calculations [9]. Control of dipole motion by external variations is a promising novel direction of research which can result in development of metamaterial-based nanonetworks. The description of the state of a molecular system by using a single integrated characteristic, the dipole moment, is a promising foundation for research and devel-opment of nano-devices with built-molecular films.

Here, we present two approaches:

a)   the experimental study of peptide/protein films by electrochemical imped-ance radiofrequency spectroscopy which is based on the fact that the imped-ance is determined by the dipole moments of film (micro) domain;

b)   theoretical study of the dynamics of dipole moments in electromagnetic fields. This is based first of all on the fact that the dynamics of dipole mo-ments of in-dividual molecules in an electric field can be revealed. It is essen-tial for build-ing dipole (micro) domain of a film. Second, computer simula-tion of dipole moments of peptides have already been recognized in science as trustworthy, accurate and reliable. Control of positions of electric dipoles by an external electric field seems to be most efficient and convenient, which is confirmed by the results given in Section 3.2. Moreover, it is pretty practi-cal to use the electromagnetic field control from the point of view of a typical nano device design

We combine both methods in order to provide a predictive model of nanodevice processing and some examples of probable communication scenario.

2   The Methods

We discuss studies on the dielectric and electrophysical properties of polypeptide films containing L-alanine, glycine and albumin. Radiofrequency spectroscopy and computer simulation studies on peptide films in electromagnetic fields are reported. In our experimental studies we used glycine, alanine and albumin films, formed from distilled water solutions of these biomolecules in concentration 0.01 %.

2.1   Dielectric Spectroscopy

Dielectric spectroscopy also known as electrochemical impedance spectroscopy is considered to be a powerful tool for measuring the dielectric and transport properties of biomolecular systems and their response on applied electrical field [10].

The method is based on the interaction of an external field with the electric dipole moment of the sample, and measures the dielectric properties of a medium as a func-tion of frequency [11]. Measurements of impedance characteristics allow one to study structure, interactions and kinetics of supramolecular systems assembled on a solid or liquid interfaces.

The objects of our studies were protein films in liquid and solid state. The films were formed from alanine, glycine and albumen water solutions and located on Teflon plate between electrodes. We measured the complex impedance and phase shifts of protein films in frequency range 0.5–110 MHz (Radiofrequency Impedance Meter BM538, Tesla).

The experimental setup is presented in figure 1.

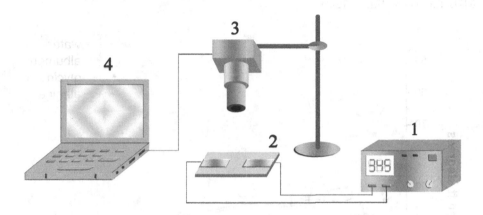

Fig. 1. Experimental setup: 1 – RF impedance meter, 2 – the special object-plate, 3 – microscope for visual control, 4 – computer

The thin films of biomolecules were placed on a teflon substrate within the 4 mm gap between the Cu foil electrodes of the object plate. The position of the formed films was controlled by the microscope 3. The values of signal phase and amplitude were recorded in frequency range 0.5–110 MHz.

2.2   Theoretical Modeling

For interpretation of experimental results the theoretical modelling of supramolecular architectures based on computer simulation of dipole moment vs conformations of polyalanine oligomers was developed. The model allows evaluation of dipole moment distribution, conformations, and translational motion. For computer simulation the following programs were used: Avogadro program [12], VMD [13] and NAMD [14] packages, and CHARMM.

We used the Avogadro program to create the pdb-files containing coordinates of all atoms in the molecule, followed by generation of the water molecules boxes around the peptides if needed by the VMD program. The NAMD package was used to perfom the MD simulation after a 1000 steps energy minimization of the molecular system. Also VMD program was applied to do the time-dependent dipole moments calculations for both watered and single molecules.

3   Results and Discussion

3.1   Experimental Results

The experimental results revealed the specific nonlinear characteristic of phase vs frequency dependences, in the frequency range 5-15 MHz for glycine films and 90-100 MHz for alanine films (fig.2).

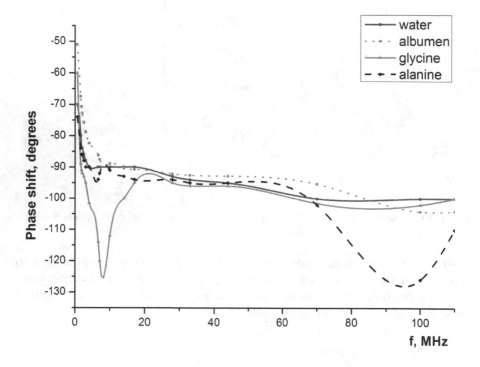

Fig. 2. Phase shift dependences on frequency for peptide films

These nonlinearities were reproducible in in the temperature range 0-30 ^{0}C. Preliminary results showed that nonlinearities might be caused by collective effects in the molecular self-assembled films.

3.2   Theoretical Results

A number of dynamic scenarios were pefomed for model peptides PolyAla (2-24) at 300 K, situated in vacuum and in the water environment with 1 fs time-step and runtime up to 2 ns. For example, one can see the dipole moment transition occurred simultaneously with spontaneous dramatic changing of model peptide Ala12 conformation in two selected points (fig. 3).

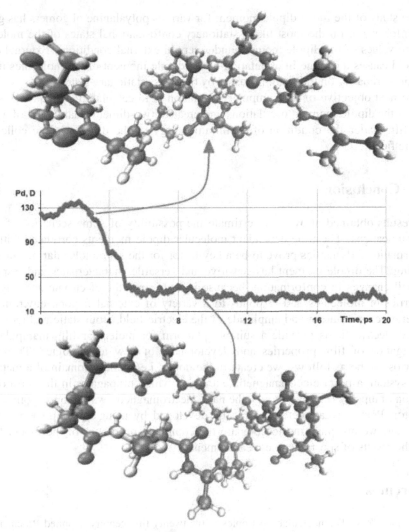

Fig. 3. Computer simulated time-dependence of dipole moment and conformation of peptide Ala 12

In fig. 3 the dipole moment value of the computer modeled alanine-12 peptide in vacuum at a temperature of 300 K is shown, two examples of conformational states in concerted times of 3.84 ps (above) and 10.24 ps (bottom) are represented. Here, we draw attention to the general view of the backbone primary circuit of the molecule spatial structure. It is seen, as expected, that the radical conformational transition in the form of backbone compressing is accompanied by a dramatic dipole moment decreasing. In the next segment of the simulation time up to 2 ns observed were moderate correlated oscillations around an average magnitude of both conformational states and values of the dipole moment.

The study of the mean dipole moment for various polyalanine oligomers has given new information on the most likely stationary conformational states of the molecule and the values of the dipole moment under certain external conditions. External electric field causes a change in orientation of the dipole moment of the molecules in the direction primarily on the field, followed by the quasi-stationary state.

The next objective of the computer simulation is to calculate the frequency spectrum of the dipole moment oscillations and create a two-dimensional model of a peptides film under the condition of minimum energy of the dipoles in the collective electric field.

4   Conclusion

The results obtained allow us to estimate the possibility of some scenarios of nano communications. The time-dependent molecular dipole moments correlated with the conformational dynamics prove to be a key factor for the supramolecular metamatrial building. The dipole moment has sensitive and versatile characteristics and responds to small changes in conformation, because it determines the electronic and physical properties of molecules and responds to a variety of external factors: environment, temperature, orientation and amplitude of the electric field. Both static and radio frequency electric fields provide a suitable platform for molecular film manipulation, investigation of film properties and development of new technologies. Therefore, scenarios can be as follows: we create in the donor, i.e., a local domain of a metamaterial system, a point electromagnetic excitation which propagates in the form of relaxation of dipoles (in other words, the nano electromagnetic wave) to the point of the acceptor. Practical nanonetworks can be developed by using this approach. At the final stage we are planning to compare the computed properties of the model films with the results of spectroscopic measurements.

References

1. Phadke, R.S.: Biomolecular electronics in the twenty-first century. Applied Biochemistry and biotechnology. **96**, 279–286 (2001)
2. Offenhäusser, A., Rinaldi, R.: Nanobioelectronics– for Electronics, Biology, and Medicine. Springer, Canada (2009)
3. Akyildiz, I.F., Jornet, J.M.: Electromagnetic wireless nanosensor networks. Nano Communication Networks **1**, 3–19 (2010)
4. Jiang, C., Chen, Y., Liu, K.J.R.: Nanoscale molecular communication networks: a game-theoretic perspective. EURASIP Journal on Advances in Signal Processing **5**, 1–15 (2015)
5. Akyildiz, I.F., Jornet, J.M., Hana, C.: Terahertz band: Next frontier for wireless communications. Physical Communication **12**, 16–32 (2014)
6. Wua, C.C., Kua, B.C., Koa, C.H., et al.: Electrochemical impedance spectroscopy analysis of A-beta (1-42) peptide using a nanostructured biochip. Electrochimica Acta. **134**, 249–257 (2014)

7. Biela, A., Watkinson, M., Meier, U., et al.: Disposable MMP-9 sensor based on the degradation of peptide cross-linked Hydrogel films using electrochemical impedance spectroscopy. Biosensors and Bioelectronic **1** (2015)

8. Nakanishi, M., Sokolov, A.P.: Protein dynamics in a broad frequency range: Dielectric spectroscopy studies. J. Non-Crystalline Solids **407**, 478–485 (2015)

9. Kelly, C.M., Northey, T., Ryan, K., Brooks, B.R., Kholkin, A.L., Rodrigueza, B.J., Buchetea, N.V.: Conformational dynamics and aggregation behavior of piezoelectric diphenylalanine peptides in an external electric field. Biophysical Chemistry **196**, 16–24 (2015)

10. Macdonald, J.R.: Impedance Spectroscopy Theory, Experiment, and Applications (2005)

11. Kremer, F., Schonhals, A., Luck, W.: Broadband Dielectric Spectroscopy. Springer-Verlag (2002)

12. Hanwell, M.D., Curtis, D.E., et al.: Avogadro: an advanced semantic chemical editor, visualization, and analysis platform. J. Chem. Informatics **4**, 17 (2012)

13. Humphrey, W., Dalke, A., et al.: VMD: visual molecular dynamics. J. Molec. Graphics **14**, 33–38 (1996)

14. Phillips, J.C., Braun, R., et al.: Scalable Molecular Dynamics with NAMD. Journal of Computational Chemistry **26**, 1781–1802 (2005)

Nano-device with an Embedded Molecular Film: Mechanisms of Excitation

Oleg Tsybin[✉]

Peter the Great St. Petersburg Polytechnic University, St. Petersburg, Russia
otsybin@rphf.spbstu.ru

Abstract. Mechanisms of a fast energetic excitation in a nano-device with an embedded molecular film are discussed herein. Excited intermolecular transport of protons, electrons and other quantum carriers could create communication channels inside a molecular network. We suggest that such channels could be analyzed by means of electron capture dissociation reactions of a protonated peptide, protein or a DNA to reveal the intramolecular pathways of mobile carrier motion in an isolated dehydrated molecule. In general, initial processing stage in a nano-device with a molecular film consists of a target direct activation followed by energy transport from target excitations into a surface molecular film which produces an indirect activation. We analyzed selected indirect molecular excitations produced by a metal or a semiconductor target directly activated by means of a radiofrequency electromagnetic field. Further processing stage after the activation consists of molecular relaxation and transportation phenomena providing nano-electromagnetic field, molecular flow, and communication.

Keywords: Nanotechnology · Nanodevice · Nanocommunication · Nano electromagnetic field · Molecular

1 Introduction

In the field of modern nanotechnology, the most promising nano-machines (NMs) are the nano-devices of micrometer – sub-micrometer active operation zone size which contain molecular films built on an electro-conductive target. The base materials for these targets are biosensors, medical biochips, molecular transmitters, and molecular electronic devices [1,2,3]. Thus employed molecular mono- or multi-layer films on solid state target surfaces usually consist of organic molecules or biomolecules, e.g., peptides and proteins.

In the so-called "bottom-up" case, a metal or a semiconductor target has a smooth clean polycrystalline surface as a substrate covered by a molecular layer. Lateral size of a surface operation zone can be as small as ~ 0.1 – 10 nm. In contrast, "top-down" case describes a microelectronic solid-state device in which at least a part of a surface contains integrated elements assembled by molecules. In such "bio-hybrid" nano-device, operation zone size can be larger, up to and exceeding 1000 nm, than in the bottom-up case. Two types of processing can be energetically activated: i) an

© Springer International Publishing Switzerland 2015
S. Balandin et al. (Eds.): NEW2AN/ruSMART 2015, LNCS 9247, pp. 772–777, 2015.
DOI: 10.1007/978-3-319-23126-6_72

intramolecular internal processing, and ii) an external processing of a communication channel among domains of a molecular network, or among separate NMs. The four agents are used for activation: electromagnetic fields (EMF), acoustic phonons, nano-mechanics, and molecular transport. At least two of them, electromagnetic fields and molecular transport, are directly appropriate for wireless nano-communication [2]. A comprehensive description of energy phenomena for NMs is still missing.

Nano electromagnetic field generation takes place when the movement of charges occurs in a molecular unit. Charge migration refers to a charge transfer mechanism occurring on a temporal scale that precedes notable nuclear motion.

In this manuscript, we analyze from a physical point of view some specific models of NMs fast energetic activation by EMF, including both external and internal processing. For that we use our experimental and theoretical results, as well as published data of biomolecular research.

2 The Methods

2.1 Intramolecular Direct Excitation

In spectroscopy and mass spectrometry, a number of methods of intramolecular electronic and atomic subsystems excitation by external electrons and photons have been developed. These excitations result in molecule's internal energy increase and subsequent molecular fragmentation. The ultrafast dynamics of charge transfer processes initiated in biomolecules by photoionization have been investigated in various experimental and theoretical papers. It was suggested that if an electron is suddenly removed from a molecular orbital, the temporal evolution of the electronic wave packet produces charge oscillations. These oscillations have been referred to as charge migration to distinguish them from charge transfer mediated by nuclear motion. Direct experimental access to such processes is mandatory to image electronic dynamics in complex molecules. If an electron is selectively ionized from a chromophore at the C-terminal end of a peptide, the location of the charge can be probed by using the shift in absorption of the chromophore. Charge migration in an amino acid induced by a UV laser pulse is characterized by fast oscillations appearing in the quantum yield of a specific charged fragment. We believe that the reverse reaction of an electron attachment to the protonated molecule also leads to the movement of electrons and holes, hence introduces a novel and radically different method for the communication on the molecular scale by exploiting a well-known, physically-controllable phenomenon, electron capture dissociation (ECD). Upon irradiation of gaseous protonated biomolecules with low-energy (~1 eV) electrons, there is a specific exothermic ECD reaction, which determines the probability of a rupture of certain, mainly N-Cα, peptide backbone bonds of biomolecules [4,5]. Released intramolecular electron-proton neutralization energy could be equal to the energy difference between the electron affinity of the proton and position-dependent potential energy of the proton (hole) on a molecular orbital. Intramolecular transport of charge and energy carriers could create communication channels between parts of the same excited molecule and/or connect a few domains inside the same molecular network.

Detailed mechanisms of energy delivery to the intramolecular remote site of fragmentation have not been studied sufficiently. In isolated proteins, peptides, and DNA some mechanisms of charge – energy transport are known based on electron, hole and proton carriers [6,7,8,9,10,11]. The most adopted are the two basic models of carriers transport: a coherent super exchange tunneling over short distances, probability of which decreases exponentially with the length of the transfer (the damping constant is on the order of 10 nm^{-1}), and series of jumps, responsible for migration to the far distance of about 20 nm. Additionally, neutralization energy can be delivered to the remote sites, where the vibrational energy is released, by vibrons, or quantum excitations of the C = O groups, bound together by electromagnetic interaction into a unique system [12].

For amphipathic peptides and proteins statistically significant spatially inhomogeneous distribution of the ECD MS/MS fragment ion abundances along biomolecular backbone was reported, showing pronounced and periodic amplitude modulation [4,5]. We use this finding as a first analytical evidence, which may reveal the intramolecular pathway of charge - excitation energy carrier transport. More complex picture includes molecular conformation dynamics combined with charge and energy transport [4,5,12].

2.2   Indirect Molecular Excitation Processing on a Target Surface

In a framework of the second method, we analyzed indirect molecular excitation through the target activation by means of a radiofrequency EMF. Other well-known, but less suitable for NMs methods (laser, fast atom, or plasma irradiation of a target) are not considered here. In the considered here indirect, so called "skin-current", method, initial processing stage is based on a target activation by short, 1-10 ns, EMF pulse having wide electromagnetic spectrum with the highest frequency of 1000 MHz. In contrast to the known UV laser photoelectron spectroscopy, direct interaction of the external EMF with molecules cannot remove an electron from molecular orbital due to much lower frequency range. Hot electrons and acoustic phonons are excited simultaneously in the skin-current layer of a few micrometers depth nearby the surface of a solid state target. Further, the next processing stage at the target surface is presented by electron and phonon transport from an activated target into surface molecular excitations, followed by excited molecules relaxation and transportation phenomena, resulting in nano-electromagnetic field and communication scenario.

Schematic representation of the target, excited via EMF, and the molecular layer are shown in figure 1. Unfortunately, there is still no known theory of high-frequency electromagnetic excitation of surface covered with a molecular film. By means of a time-of-flight mass - spectrometry, instantaneous desorption/ionization/fragmentation of molecular particles in vacuum has been observed when a short nanosecond time duration pulse (1-10 ns) with an over-threshold current propagates through a metal or a semiconductor target, irradiated by EMF [13,14]. Upon excitation of the electron subsystem of the target by external short pulse EMF, hot electrons from the conduction band in the skin layer likely interact with the adsorbed molecules and thus stimulate: a) hot electron emission from molecular orbitals; b) desorption, induced by excitation of surface - molecular bonds; c) tunnel neutralization of desorbed ions; d) electron capture/transfer dissociation-type reactions. In parallel, desorption of neutrals is enhanced by an acoustic phonons flow from the surface.

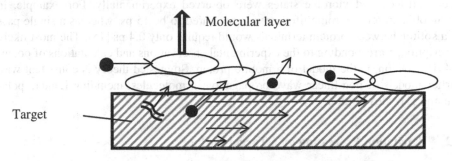

Fig. 1. Schematic representation of the target and the molecular layer, excited via EMF.   Black circles represent electrons, double curve line denotes acoustic phonons, and the arrows indicate the direction of their movement. Curly arrow shows the flow of desorbed molecules and their fragments.

3   Experimental and Theoretical Results

3.1   Intramolecular Excitation

We suggest that fragment ion abundance distributions in electron capture dissociation mass spectrometry (ECD MS/MS) reveal the most probable positions of the knots of mobile proton transfer in an internal structure of the molecule. For example, consider two-dimensional ECD MS/MS maps of a doubly and a triply protonated membrane peptide M2TMP [4]. For a doubly protonated M2TMP the knots of proton localiza- tion are concentrated near the N- and C-terminals, due to a specific location of the two most basic sites in this particular peptide. The proton lifetime is the longest in these two terminal locations.

When molecular ion is charged by three protons, or ion internal energy of a doubly protonated species is increased by infrared activation, the length of a proton route is increased within the same hydrogen spines (H spines). The evident asymmetry of N- and C- terminal branches could be explained by an influence of N –C molecular dipole field on the protons. One also cannot exclude mutual Coulomb repulsion between protons originated at C and N terminals. Indeed, the calculations show that the positions of fragmentation sites mainly correspond to the local minima potential and are limited by potential barriers. To clarify this picture, additional computation studies of the potential distributions are required.

The ECD MS/MS mass spectra contain fragments of two types: (i) even-electron ions (prime ions) and (ii) odd-electron ions (radicals). Same backbone cleavage may produce fragment ions of both types. The ratio of the fragment ion prime to radical components is a measure of an internal molecular energy. This measure may be used as a method of experimental determination of a level of ion internal energy upon molecular activation via infrared activation [4,5].

In the suggested schemes, direct excitation of protonated molecular species by low energy (~1 eV) electrons induces motion of intramolecular mobile protons, electron-hole pairs and vibrons along both molecular backbone and H-bonds. Indeed, long-lived self-localized vibronic states were observed experimentally. For example, in myoglobin protein vibronic lifetime was measured to be 15 ps, whereas a single pass of a soliton between protein terminals would require only 0.4 ps [15]. The most likely description, corresponding to the experimental observations and calculations of potential distribution, is the model of a mobile proton. Suggested theory is consistent with earlier models of non-linear wave propagation in molecules, including Landau polaron and Davydov soliton.

3.2   Excitation Processing on a Target Surface

Pulsed flows of neutral molecules, positive and negative ions and electrons have been obtained as desorbed from the surface of the metal and semiconductor targets under the skin-current activation (see Figure 1) [13,14]. The two carriers in a physical model are: (i) accelerated non-thermalized conduction electrons (hot electrons) in the skin layer, and (ii) acoustic phonons which face the surface levels of the adsorbed molecules. Acoustic phonons are generated by thermo-elastic stresses in an atomic lattice of a substrate.

Due to a high electric conductivity of a substrate, the main component of a current which flows through this substrate is a convection current, whose magnitude is proportional to the field strength. In the molecular layer, the main components are the displacement current and the tunnel current, the values of which are non-linearly related to the field strength. The above-mentioned provisions in this section are included in the computer simulation of processes, shown schematically in Figure 1. When a deposited biomolecular layer co-adsorbs water molecules on the target surface, ion desorption and detection with a mass spectrometer indicates a formation of extremely stable, long-lived molecular ions in the desorption mass spectrum (data not shown). For instance, desorbed molecular flows contain hydronium ion H_3O^+ and its clusters $H_3O^+(H2O)_n$, among which most frequently detected are Zundel ion (n = 1) and Eigen ion (n = 3, first hydrated shell). Due to their very stable configuration and high probability of proton transfer reaction to biomolecules, these ions are promising candidates for a use in nano-communication channels.

4   Conclusion

Molecular film in which molecules are coupled and ordered can be viewed as an emerging meta material, possesses novel features, and capable of transporting particles, charges, excitations like nonlinear waves. Integrated with nano-device, such meta material extends the functionality of these systems. In particular, the communication channels and their networks could be created in the film. To switch on the channel it is necessary to create the required activation, consistent with the properties of the meta material.

In this paper we discuss two phenomena: (i) direct excitation of protonated biomolecules by low-energy electrons which may initiate intramolecular charge and energy transfer that creates electromagnetic nano-communication channels; and (ii) indirect molecular excitation by means of a radiofrequency skin-current activation of a target which could induce both intramolecular communication channels and pulsed molecular flow. This flow consists of selected stable self-organized species, potentially appropriate for molecular communication channels between remote NMs.

References

1. Offenhäusser, A., Rinaldi, R.: Nanobioelectronics - for Electronics, Biology and Medicine, p. 337. Springer–Verlag (2009)
2. Akuildiz, I.F., Jornet, J.M.: Electromagnetic wireless nanosensor networks. Nanocommunication networks **1**, 3–19 (2010)
3. Heath, J.R., Ratner, M.A.: Molecular Electronics. Physics Today, 43–49 (May 2003)
4. Ben Hamidane, H., He, H., Tsybin, O.Y., Emmet, M., Hendrikson, C.L., Marshall, A.G., Tsybin, Y.O.: Periodic Sequence Distribution of Product Ion Abundances in Electron Capture Dissociation of Amphipathic Peptides and Proteins. Journal Am. Soc. Mass. Spectrom. **20**(6), 1182–1192 (2009)
5. Ben Hamidane, H., Chiappe, D., Hartmer, R., Vorobyev, A., Moniatte, M., Tsybin, Y.O.: Electron Capture and Transfer Dissociation: Peptide Structure Analysis at Different Ion Internal Energy Levels. J. Am. Soc. Mass Spectrom. **20**(4), 567–575 (2009)
6. Winkler, J.R.: Electron tunneling pathways in proteins. Current Opinion in Chemical Biology **4**, 192–198 (2000)
7. Symons, M.C.R., Taiwo, F.A., Svistunenko, D.A.: Electron Paramagnetic Resonance Studies of Hole Mobility and Localisation in Haemoglobin. J. Chem. Soc. Faraday Trans. **89**(16), 3071–3073 (1993)
8. MacDonald, B.I., Thachuk, M.: Gas-phase proton-transfer pathways in protonated histidyglycine. Rapid Commun. Mass Spectrom. **22**, 2946–2954 (2008)
9. Lewis, F.D., Liu, X., Liu, J., Miller, S.E., Hayes, R.T., Wasielewski, M.R.: Direct measurement of hole transport dynamics in DNA. Nature **406**, 51–53 (2000)
10. Berlin, Y.A., Burin, A.L., Ratner, M.A.: Charge Hopping in DNA. J. Am. Chem. Soc. **123**, 260–268 (2001)
11. Schlag, E.W., Sheu, S.Y., Yang, D.Y., Selzle, H.L., Lin, S.H.: Distal charge transport in peptides. Angewandte Chemie-International Edition. **46**(18), 3196–3210 (2007)
12. Pouthier, V., Tsybin, Y.O.: Amide-I relaxation-induced hydrogen bond distortion: an intermediate in electron capture dissociation mass-spectrometry of alpha-helical peptides? Journal of Chemical Physics **129**, 095106 (2008)
13. Tsybin, O., Mishin, M.: Ion desorption from skin-current induced metal surface. ZTF Lett. **22**(4), 21–24 (1996). (In Russian)
14. Zamiatin, A.V., Tsybin, O.Y.: Surface skin-current activated emission of electrons and ions. In: 20th International Workshop on Beam Dynamics and Optimization, BDO 2014, art. no. 6890100 (2014)
15. Xie, A., van der Meer, L., Hoff, W., Austin, R.H.: Long-Lived Amide I Vibrational Modes in Myoglobin. Phys. Rev. Lett. **84**(23), 5435–5438 (2000)

Analysis of in-Plane Conductivity of La$_{1-x}$Sr$_x$F$_{3-x}$ Superionic Thin Films

T. Yu Vergentev[1], E. Yu Koroleva[1,2], L. Rissing[3], and A.V. Filimonov[1(✉)]

[1] Peter the Great St. Petersburg Polytechnic University,
195251, Polytechnicheskaya 29, St. Petersburg, Russia
tikhon.v@gmail.com, e.yu.koroleva@mail.ioffe.ru,
filimonov@rphf.spbstu.ru
[2] Ioffe Physico-Technical Institute, 194021, Polytechnicheskaya 26, St. Petersburg, Russia
[3] Institute for Microproduction Technology,
Leibniz Universität Hannover, 30823 Garbsen, Germany
rising@impt.uni-hannover.de

Abstract. Electrical properties of thin films of La$_{1-x}$Sr$_x$F$_{3-x}$ solid solutions with x = 0 ÷ 0.24 were measured in temperature range from RT to 300°C and wide frequency range from 10^{-1} to 10^6 Hz by impedance spectroscopy method. The spectrums of impedance were analyzed with equivalent circuits contain RC and Warburg parts. DC-conductivities were calculated from RC-circuit of impedance and activation energies determined from Arrhenius-Frenkel equation

$$\sigma_{DC}T = \sigma_0 e^{\left(-\frac{E_{\sigma T}}{kT}\right)}.$$ Diffusion coefficients and their temperature dependencies were determined from Warburg part of impedance for different SrF$_2$ content.

Keywords: Electrical properties · Ionic conductivity · Activation energy · Hodograph · Impedance

1 Introduction

Ionic conductor fluorite lanthanum (LaF$_3$) has been studied since 1960s [1,2], but until now this material has been continued to be interesting, due to its unique properties such as high value of ionic conductivity [1], non-typical for solid state materials. LaF$_3$ has a tysonite structure and belongs to $P\bar{3}c1$ space group [3,4]. Doping of tysonite phase RF$_3$ (R – rare-earth elements) with fluoride phase MF$_2$ (M – alkaline) of low concentrations saves the structure type [5], but leads to further increase the conductivity. The increase of ionic conductivity of $R_{1-x}M_x$F$_{3-x}$ solid solutions is caused by heterovalent substitutions R^{3+} for M^{2+}, that generates defects V_F^+ in ionic sublattice of LaF$_3$, and these defects play the roles of mobile carriers [5,6]. Investigations of solid solutions $R_{1-x}M_x$F$_{3-x}$ have shown an increase of conductivity compared to pure LaF$_3$ by two orders [7,8] at concentrations ∼3-7% of MF$_2$. Comparison bulk and thin films electric properties of solid solutions La$_{1-x}$Sr$_x$F$_{3-x}$ has been done by us in [9], however

© Springer International Publishing Switzerland 2015
S. Balandin et al. (Eds.): NEW2AN/ruSMART 2015, LNCS 9247, pp. 778–785, 2015.
DOI: 10.1007/978-3-319-23126-6_73

the accurate analysis of in-plane conductivity of films has not been carried out. Herein, based on AC-measurement of films in wide frequency and temperature ranges we'll find the corresponding equivalent circuit and determine its parameters and their temperature dependencies. Also we'll analyze the behavior of DC-conductivity and near-electrode processes.

2   Samples Preparation and Measurement Procedure

Thin films of La$_{1-x}$Sr$_x$F$_{3-x}$ with x=0÷0.24 on glass-ceramic substrates were grown by molecular-beam epitaxial method in ultra-high vacuum chamber with vacuum less than 10^{-8} Pa. Chamber equipped a heater allows varying temperature of a sample up to 1200^{0}C. Glass-ceramic substrates with 8x10mm^2 and 0.4mm thickness were one-side polished. The chemical preparation of substrates consisted of few steps: degreasing boiling in 3:3:1 solution H$_2$O$_2$:H$_2$O:NH$_4$OH, purification of the reaction products in boiling steel water, and dehydration in isopropyl pairs. After that the substrate was mounted in special sample holder and moved to heater in vacuum chamber. Before growing process the dehydration at 200^{0}C took one hour in high vacuum. The temperature of substrates was 400^{0}C for all samples during evaporation process. Average rate of growing process was 2nm per minute, which was measured by quartz crystal microbalance (QCM). The concentration of doped component (SrF$_2$) was evaluated with $\frac{v_{SrF_2}}{v_{LaF_3}+v_{SrF_2}}$ ratio, where v_{LaF_3}, v_{SrF_2} – average rates of growing. After short exposure at the air the gold electrodes were deposited up to the film.

Fig. 1. Photography of a sample.

Electrical properties were measured in frequency range from 10^{-1} to 10^6Hz with Alpha-A analyzer of Novocontrol BDS80 dielectric spectrometer equipped a cryosystem. Measurements were carried on with standard two-contact method with special designed inter-digital electrodes (IDE) (a gap 0.2mm and a step 0.6mm) (figure 1). That allows to measure planar properties of films, and to increase the measuring signal. To separate a substrate and a film contributions into measured signal correctly, the special theoretical calculations were used [10,11].

3   Results and Discussions

The principles of frequency domain spectroscopy consist in measuring of current response I of a sample at the fixed frequency ω to the applied voltage with the same frequency and phase shift between current and voltage φ. For a sample with a liner electromagnetic response, the measured impedance of the sample capacitor $Z^* = \frac{I^*}{U^*}$ is connected with the dielectric function of the sample material by

$$\varepsilon^*(f) = \varepsilon' - j\varepsilon'' = \frac{j}{2\pi f Z^*(f)}\frac{1}{C_0},$$

where C_0 – is the capacity of the empty sample capacitor. The specific conductivity σ^* is related to the dielectric function by $\sigma^* = \sigma' - \sigma'' = j2\pi f \varepsilon_0 (\varepsilon^* - 1)$. Diagrams $Z'' = f(Z')|_{T=const}$ are called hodographs of impedance. Form of impedance hodograph depends on the test material. For solid electrolytes, it consists of a semicircle and a sloped line [7] in most bulk materials. For ceramics or heterogeneous materials consisted of core and shell having different conductivities, the spectrum of

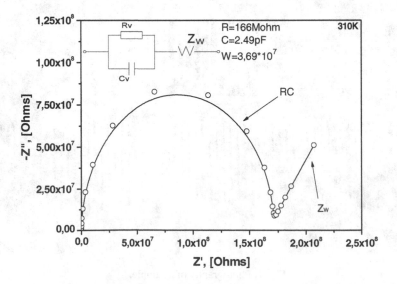

Fig. 2. Hodograph of impedance for sample #6171 (5%) and the equivalent circuit.

impedance has two semicircles, which corresponds to a response inside of grain and grain boundaries, that was shown in AgI embedded in porous glass case [12]. Impedance hodographs of all the investigated films consisted of a section of one semicircle and a segment of the inclined line (figure 2). Fitting of the curve by an electric circuit with a pattern of R, C elements demonstrates success fit of spectrum by one RC-circuit and Warburg impedance $Z_W = W(1 - j)\omega^{-0.5}$ (figure 2). Capacity part C is ~2÷4pF for all samples and independent of temperature There is no deformation of impedance spectrum up to 23.4% content of SrF$_2$ component, that corresponds there is no coexistence of LaF$_3$ and SrF$_2$ without solid solution formation, as was noticed at 50% of SrF$_2$ in bulk [13].

Composition and thickness of all investigated samples are presented in Table 1. Electrical properties of all samples have been measured in temperature range from 300 to 570K. DC-conductivities of films are calculated according to [11] as

$$\sigma_{DC}^{film} = \frac{1}{F}\left[\frac{1}{R\frac{K}{\varepsilon_0}} - \sigma_{DC}^{sub}\right] + \sigma_{DC}^{sub},$$

where F,K – parameters of IDE structure analogous C_0, and R – fit parameter of equivalent circuit. In this temperature range all conductivities are linearized in Arrhenius-Frenkel's plot $\ln(\sigma T)$ vs $\frac{1000}{T}$ (figure 3), that characterized typical for ionic conductors thermally activated behavior of La$_{1-x}$Sr$_x$F$_{3-x}$ films. Activation energies and pre-exponential factor σ_0 are estimated from law $\sigma T = \sigma_0 e^{-\frac{E_{\sigma T}}{kT}}$ are presented in Table 1. Influence of thickness of films in the conductivity was not revealed. The highest value of conductivity is noticed for film 95%LaF$_3$ +5%SrF$_2$ that qualitatively correlates with result for bulk solid solutions R_{1-x}Sr$_x$F$_{3-x}$ [6,8], where R – rare-earth elements. Increasing concentration of SrF$_2$ above 5% leads to a decrease in σ_{DC}.

Table 1. Samples numbers, composition and thickness of films, and fitting parameters of DC-conductivities.

Sample	% of SrF$_2$	Thickness, nm	$\ln(\sigma_0$, S*K/cm)	E$_{\sigma T}$, meV
#6186	0	95	10.35±0.14	680±5
#6208	2	100	12.35±0.17	515±4
#6167	3,5	160	11.67±0.12	486±4
#6171	5,2	90	12.36±0.17	496±6
#6209	7,3	85	11.29±0.13	505±3
#6169	8,2	160	12.49±0.06	527±2
#6170	9	90	12.27±0.09	529±3
#6168	10,6	240	13.47±0.08	562±3
#6213	15,6	100	13.28±0.33	643±11
#6214	23,4	100	12.53±0.06	698±2

In the frames of the hopping conductivity model DC-conductivity is $\sigma_{DC} = \frac{q^2 d^2 n}{\gamma kT} v_0 e^{-\frac{\Delta H_v}{kT}}$, where q – charge, d – jump distance, n – density of the charge carriers $n \sim e^{-\frac{\Delta G_f}{2kT}}$, ΔG_f – energy of Frenkel defect formation (the mechanism of formation of "impurity" vacancies $La_{La}^x + 3F_F^x \Rightarrow Sr_{La}^- + V_F^+ + 2F_F^x$), γ – lattice type parameter, ν_0 –the frequency of attempts of carrier to hop in the neighbor position, ΔH_ν – enthalpy of migration, determining temperature dependence of mobility $\mu = \frac{\mu_0}{T} e^{-\frac{\Delta H_\nu}{kT}}$ [5,8,14]. Then $\sigma_{DC}T \sim e^{-\frac{\Delta G_f}{2kT} - \frac{\Delta H_\nu}{kT}}$ and $E_{\sigma T} = \frac{1}{2}\Delta G_f + \Delta H_\nu$. We can determine the migration enthalpy directly from frequency dependencies of AC-conductivity $\sigma_{AC} = \sigma_{DC}\left(1 + \left(\frac{H}{\nu_h}\right)^\alpha\right)$, as $\nu_h \sim e^{-\frac{\text{Д}H_H}{kT}}$, where ν_h- hopping frequency. The activation energy obtained from the temperature dependence of ν_h, within the error of our measurements coincided with the value obtained for the DC-conductivity. The energy $E_{\sigma T} \simeq \Delta H_\nu$ and then the concentration of mobile carriers is temperature independent [8] in the experimental temperature range.

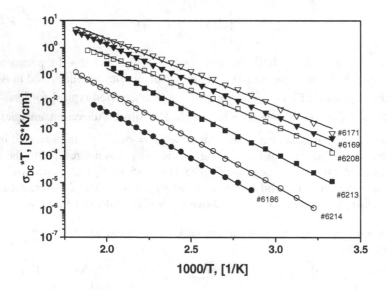

Fig. 3. DC-conductivities of samples #6171 (5%), #6169 (8%), #6208 (2%), #6213 (16%), #6214 (23%), #6186 (LaF₃), and presented as Arrhenius-Frenkel's plot.

Activation energy $E_{\sigma T}$ characterizes a slope of the $\sigma_{DC}T$ dependence physically is the Gibbs free energy barrier $\Delta G_m = \Delta H - T\Delta S$ in the region the defects are not thermally produced [14]. Distribution of activation energies in dependence of SrF₂ content is shown in figure 4, and the values of energies are 475-700meV. The minimum of energy lies in range 5-7% of SrF₂, which coincides with the maximum of DC-conductivity of La₁₋ₓSrₓF₃₋ₓ solid solutions. Physically, it is a barrier necessary for realize ionic conductivity in the materials, and if there is no mechanism of formation of defects then the process of conductivity is a diffusion of carriers into materials under applied electric field.

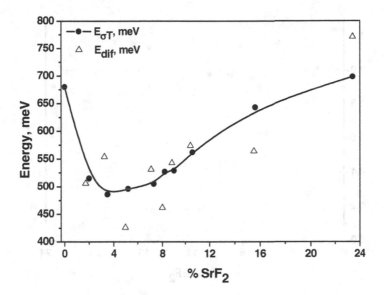

Fig. 4. Activation energies of DC-conductivities (closed dots •) and diffusion (opened triangles △) in dependence of doping SrF$_2$ component.

Migration of carriers under electric field, when the frequency of applied field is enough slow, is formed an accumulation charge zone near the electrodes. This mechanism is observed on hodograph of impedance as a segment of the inclined line. In the approach of linear semi-infinite diffusion the near-electrode processes describe Warburg impedance Z_W on equivalent circuit (figure 2). Warburg impedance is $Z_W = W(1 - j)\omega^{-0.5}$. Hodographs of impedance of all investigated materials are well fitted by the Warburg impedance and we obtained temperature dependencies of W for all samples. Coefficient of diffusion is thermally dependence $D = D_0 e^{-\frac{E_{dif}}{kT}}$, where E_{dif} – activation energy of diffuse mechanism. Definition of diffuse coefficients from spectrum of impedance is determined by equation $W = \left\{ \frac{V_m}{nFA\sqrt{2D}} \right\} \frac{dE}{dx}$, where V_m – molar volume, F – Faraday constant, A – Avogadro constant [15,16,17]. Analysis of Warburg impedance $Z_W = W(1 - j)\omega^{-0.5}$, and where $D \backsim \frac{1}{W^2}$ vs $\frac{1000}{T}$ estimates activation energy of diffusion E_{dif} (figure 4). As we can see, founded energies E_{dif} are consistent with $E_{\sigma T}$ with 15% accuracy (figure 5). The energy barriers are equals. It is indicates that the recombination of mobile carriers in near-electrode region is not important and mechanism of conductivity in diffusion layer and in the volume of the materials are the same.

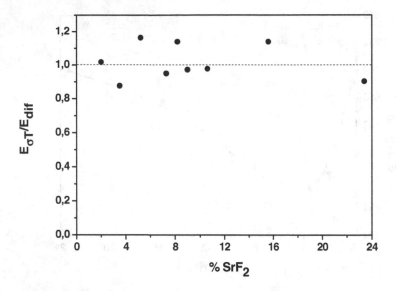

Fig. 5. Ratio of activation energies of DC-conductivities and near-electrode diffusion process.

4   Conclusion

Based on AC-measurements of $La_{1-x}Sr_xF_{3-x}$ solid solutions of films on glass-ceramic substrates the DC-conductivity and its temperature behavior were founded with analysis of hodograph of impedance. The activation energies of DC-conductivities were estimated and compared with energy barriers of mobility and of diffusion. Analysis of these parameters explains some aspects of mechanism of conductivity in $La_{1-x}Sr_xF_{3-x}$ solid solutions with low concentration of SrF_2 component. Close value of activation energy of DC-conductivity and energy barrier of mobility confirms no formation of additional mobile carriers under the temperature gradient in the investigated temperature region. The conductivities are mediated by the mechanism of formation of impurity vacancies through heterovalent replacements. Close values of activation energies of DC-conductivities and energy barriers of diffusion founded with linear semi-infinite approach revealed a lack of noticeable effect processes of recombination and generation of carriers in near-electrode region of solid solutions $La_{1-x}Sr_xF_{3-x}$.

Acknowledgements. We are grateful to A.G. Banschikov for help in growing the films and to N.S. Sokolov in discussions.

The work of T.Yu. Vergentev and A.V. Filimonov was performed under the government order of the Ministry of Education and Science of RF.

References

1. Sher, A., Solomon, R., Lee, K., Muller, M.W.: Phys. Rev. B **144**(2), 594 (1966)
2. Lilly Jr., A.C., LaRoy, B.C., Tiller, C.O., Whiting, B.: Journal of the Electrochemical society **120**(12), 1673 (1973)
3. Cheetham, A.K., et al.: Acta Crystallogr **32**, 94 (1976)
4. Belzner, A., Schulz, H.: Zeitschrift fur Kristallographie **209**, 239 (1994)
5. Sobolev, B.P., Sorokin, N.I.: Crystallography Reports, **59**(6), 807–830 (2014)
6. Sorokin, N.I., Sobolev, B.P.: Crystallography reports **52**(5), 842 (2007)
7. Sorokin, N.I., Fominykh, M.V., Krivandina, E.A., Zhmurova, Z.I., Sobolev, B.P.: Crystallography reports **41**(2), 292 (1996)
8. Sorokin, N.I., Sobolev, B.P.: Physics of the Solid State **50**(3), 416 (2008)
9. Vergentyev, T.Y., Koroleva, E.Y., Banshchikov, A.G., Sokolov, N.S., Chibisov, A.G.: Russian Journal of Electrochemistry **49**(8), 783–787 (2013)
10. Kidner, N.J., Homrighaus, Z.J., Mason, T.O., Garboczi, E.J.: Thin Solid Films (496), 539 (2006)
11. Kidner, N.J., Meier, A., Homrighaus, Z.J., Wessels, B.W., Mason, T.O., Garboczi, E.J.: Thin Solid Films (515), 4588 (2007)
12. Vergent'ev, T.Y., Koroleva, E.Y., Kurdyukov, D.A., Naberezhnov, A.A., Filimonov, A.V.: Physics of the Solid State **55**(1), 175–180 (2013)
13. Ivanov-Schitz, A.K., Sorokin, N.I., Sobolev, B.P., Fedorov, P.P: Int. Symp. on Systems with Fast Ionic Transport. Bratislava, p. 99 (1985)
14. Murin, I.V., Glumov, O.V., Gunber, W.: Ionics **1**, 274 (1995)
15. Drozhzhin, O.A., Vorotyntseva, M.A., Maduara, S.R., Khasanova, N.R., Abakumova, A.M., Antipov, E.V.: Electrochimica Acta **89**, 262–269 (2013)
16. Oberschmidt, J., Lazarus, D.: Phys. Rev. B, **21**(12), 5823 (1980)
17. Wurz, M.C., Shaganov, A., Rissing, L., Filimonov, A., Vakhrushev, S.: ECS Transactions **50**(10), 147 (2012)

Studies of Biomolecular Nanomaterials for Application in Electronics and Communications

Elena Velichko[1,2(✉)], Maxim Baranov[1], Elina Nepomnyashchaya[1],
Anastasia Cheremiskina[1], and Evgeni Aksenov[1]

[1] Peter the Great St. Petersburg Polytechnic University, St. Petersburg, Russia
velichko-spbstu@yandex.ru,
{baranovma1993,elina.nep,a.cheremiskina,et.aksenov}@gmail.com
[2] Saint Petersburg National Research University of Information Technologies,
Mechanics and Optics, St. Petersburg, Russia

Abstract. Experimental studies of biomolecular materials for goals of biomolecular electronics and communications are discussed. The joint use of laser correlation spectroscopy and dielectric spectroscopy methods of investigation of electrophysical properties of biomolecular objects in liquid state is considered. The aspects of complexity of studies of electrophysical properties of liquid samples are discussed. The preliminary results of studies of molecular solutions are presented. These studies are actual for nanoscale molecular technologies.

Keywords: Nanomaterials · Biomolecular studies · Spectroscopy · Molecular structure · Liquid state

1 Introduction

Nanoelectronic devices based on biomolecules are of great interest for relatively new scientific area of biomolecular electronics [1–2]. Scientists are exploring ways of manipulating, assembling and applying biomolecules on integrated circuits and artificial devices. Investigations of nanomaterials for nanoscale molecular communications are widely discussed in modern publications [3–5]. Molecular communication is considered as the most promising approach for nanocommunication networks [4,5].

The use of novel nanomaterials to build new devices is required. Biomolecular materials (usually peptide or protein), for example, films of nanometer sizes are considered to be integrated into artificial electronic devices. Thus, the study of electric properties of biomolecular materials, their supramolecular architectures and conformational changes under the influence of electric field is an actual task.

Biomolecular solutions may become promising materials for the goals of biomolecular electronics because liquid state is natural for the biomaterials, amino acids, proteins, etc. But there are a number of challenges in implementation of integration of liquid biomolecular objects in electronic devices. A lot of works consider molecular films as possible way of application of biomolecular structures. Such films are often

S. Balandin et al. (Eds.): NEW2AN/ruSMART 2015, LNCS 9247, pp. 786–792, 2015.
DOI: 10.1007/978-3-319-23126-6_74

formed from liquid medium and this is one more reason to study electrophysical properties of molecular solutions.

In our work we discuss the challenges of electrophysical characteristics studies and consider the experimental testing of biomolecular objects in liquid state.

2   The Methods

We used radiofrequency and laser correlation spectroscopies for preliminary studies of different biomolecular objects in liquid state. These methods were chosen as rather accessible methods which allow one to evaluate the conformational and impedance properties of biomolecular solutions. We also discuss challenges of traditional volt-ampere characteristics and present the result of amplitude frequency response.

2.1   Impedance Spectroscopy

The data on radiofrequency spectroscopy in different frequency ranges were analyzed [2,6,7]. It is known that individual molecular and sub-molecular rotations and vibrations are intrinsic to microwave and infrared ranges of electro-magnetic fields. It is expected that for intermolecular interaction in case of strong coupling with the medium (in solution) the frequencies should be lower — in radiofrequency range. It is may be illustrated as shown in fig.1.

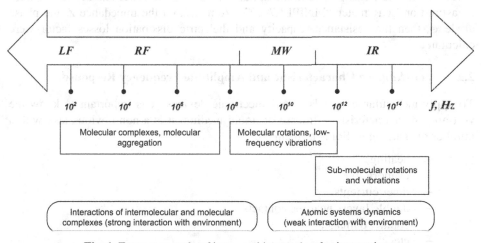

Fig. 1. Frequency scale of intra- and intermolecular interactions

In our previous studies the nonlinearities of phase-frequency characteristics were revealed for glycine liquid films in 5–10 MHz range. Glycine is the smallest of the 20 amino acids commonly found in proteins.

Thus, for the further studies we used solutions of albumen – a globular protein. It was expected that larger-sized molecules may reveal aggregation in lower frequency range.

For the studies an experimental setup was constructed (fig.2).

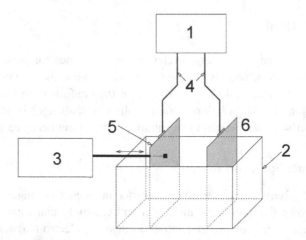

Fig. 2. Experimental setup: 1 — LCR-meter, 2 — optical cell with water solution of biomolecules, 3 — micrometer, 4 — coaxial wires, 5 — moving electrode; 6 — motionless electrode.

Tested solution was placed in the optical cell 2, and dielectric characteristics were measured on LCR meter "MNIPI E7-20". We measured the impedance Z and phase angle φ, then the resistance, capacity and dielectric dissipation losses factor were calculated.

2.2   Volt-Ampere Characteristic and Amplitude Frequency Response

To apply molecular materials to any electronic devices, it is important to know the volt-ampere characteristics. In case of liquid medium it is a non-obvious task with a number of challenges. Some of them are:

- heating,
- evaporation,
- inner currents,
- double layers near the contacts,
- contact damage,
- etc.

In such situation registration of traditional volt-ampere characteristic is problematic. One of the possible solutions of this issue is use of the alternate current, but the appropriate frequency range should be chosen.

We have tested frequency ranges from 20 Hz to 200 MHz, and a number of nonlinearities were revealed. The results of these preliminary studies are discussed in our report.

Amplitude frequency response allows evaluation of molecular conformation response in different frequency ranges. To analyze the amplitude frequency response and volt-ampere characteristics we used the scheme (Fig. 3) which allowed us to measure the parameters of molecular materials in the following configurations:

- a drop of liquid on the surface;
- a thin liquid film on the surface, covered by a glass;
- a solid film after drying of the biomolecular solution.

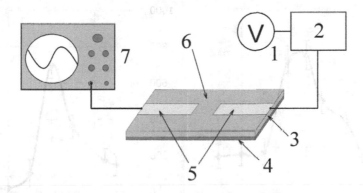

Fig. 3. Experimental setup: 1 — voltmeter, 2 — pulse generator, 3— dielectric surface, 4 — metal layer, 5 — electrodes, 6 — capacitive gap, 7 — oscilloscope

The 3 V signal from high-frequency generator (G4-143) with frequency range to 400 MHz was applied to the capacitive gap with biomolecular solution. The oscilloscope signal was recorded and analyzed.

2.3   Laser Correlation Spectroscopy

To control the sizes of molecules in the test solutions the method of laser correlation spectroscopy was used. This method allows evaluation of molecular sizes and aggregation dynamics in biomolecular solutions [8].

3   Results and Discussion

3.1   The Joint Use of Impedance and Laser Correlation Spectroscopies

The experimental results revealed the similar behaviors of characteristics, obtained by dielectric spectroscopy (the maximal frequency of dielectric losses peak angle tangent was analyzed) and laser correlation spectroscopy (the diameters of the molecular structures were calculated). An example of such characteristics for the parameters in albumin water solutions with different pH is presented in fig.4.

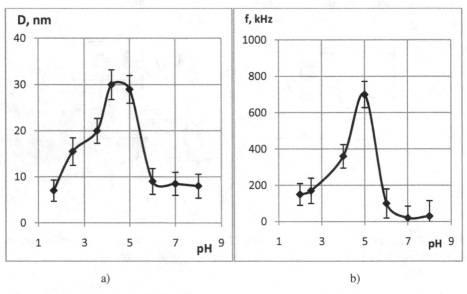

a)   b)

Fig. 4. Dependence of molecular sizes (a) and frequencies of maximal dielectric loss angle tangent (b) on acidity of albumin solutions

The maximal sizes of albumin molecules and maximal frequencies of dielectric loss angle tangent peaks were observed near the isoelectric point of albumin (pH = 4.8). These results testify the efficiency of suggested approach for analysis of biomolecular dynamics under the influence of different factors.

3.2   Amplitude Frequency Response

The volt-ampere characteristic did not reveal any nonlinearity. The amplitude-frequency response had some nonlinearities (fig. 5) which are discussed in the report.

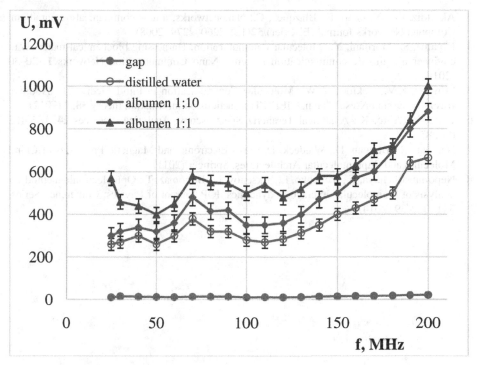

Fig. 5. Amplitude frequency response of different solutions on capacitive gap

4 Conclusion

The preliminary results of nanomaterials studies for the goals of biomolecular electronics are presented. Novel bio nanomaterials, biomolecules and nanoparticles show new properties and behaviors not observed at the microscopic level of traditional solid state materials. Using the suggested methods, biomolecular aggregation dynamics and changes in electrophysical properties on transition from one molecular structure to another in liquid state may be investigated.

These results may be useful for investigation of different biomolecular samples for goals of biomolecular electronics. The observed nonlinearities give evidence of opportunities of usage of biomolecular structures as an electronic components for biomolecular electronics and communications.

Acknowledgments. The authors would like to thank Professor Oleg Yu. Tsybin for the valuable scientific discussions and support.

References

1. Phadke, R.S.: Biomolecular electronics in the twenty-first century. Applied Biochemistry and biotechnology **96**, 279–286 (2001)
2. Velichko, E.N., Tsybin, O.Y.: Biomolecular electronics SPbSTU, Russia (2011)

3. Akyildiz, I.F., Brunetti, F., Blazquez, C.: Nanonetworks: a new communication paradigm. Computer Networks Journal (Elsevier) **52**(12), 2260–2279 (2008)
4. Hiyama, S., Moritani, Y.: Molecular communication: Harnessing biochemical materials to engineer biomimetic communication systems. Nano Communication Networks **1**, 20–30 (2010)
5. Srinivas, K.V., Eckford, A.W.: Molecular Communication in Fluid Media: The Additive Inverse Gaussian Noise Channel. IEEE Transactions on information theory **58**, 7 (2012)
6. Ueno, Y., Ajito, K.: Analytical Terahertz Spectroscopy. Analytical Sciences **24**, 185–192 (2008)
7. Naaman, R., Beratan, D., Waldeck, D. (eds.) Electronic and Magnetic Properties of Chiral Molecules and Supramolecular Architectures. Springer (2011)
8. Nepomnyashchaya, E., Velichko, E., Aksenov, E., Bogomaz, T.: Optoelectronic method for analysis of biomolecular interaction dynamics. IOP Journal of Physics: Conference Series. **541**, 012039 (2014)

Ultrabroadband Dielectric Spectroscopy of Lead-Free Relaxor Ferroelectric Na$_{1/2}$Bi$_{1/2}$TiO$_3$

Alexey V. Filimonov[1($\boxtimes$)], Ekaterina Yu Koroleva[1,2], Alexander A. Naberezhnov[1,2], Sergej B. Vakhrushev[1,2], and Tikhon Yu Vergentiev[1]

[1] Peter the Great St. Petersburg Polytechnic University,
Polytechnicheskaya 29 195251, St.-Petersburg, Russia
filimonov@rphf.spbstu.ru,
{e.yu.koroleva,alex.nabereznov,s.vakhrushev}@mail.ioffe.ru,
tikhon.v@gmail.com
[2] Ioffe Institute, Polytechnicheskaya 26 194021, St.-Petersburg, Russia

Abstract. The results of the detailed study of the dielectric properties of single crystal $Na_{1/2}Bi_{1/2}TiO_3$ (NBT) in wide frequency (0.1 Hz–10 MHz) and temperature (300–1000 K) ranges are reported. Analyses of dispersion curves of the complex dielectric response within existing phenomenological models are done. Relaxation processes are identified and their dynamics are traced. Activation energies of Arrhenius' relaxation processes are obtained by fitting. Besides we have found a relaxation process described by Vogel-Fulcher law, indicating that in the system occurs freezing at 370 K.

Keywords: Relaxor · Dielectric spectroscopy · Ferroelectrics · Phase transition · Sodium bismuth titanate · NBT

1 Introduction

The sodium bismuth titanate $Na_{1/2}Bi_{1/2}TiO_3$ (NBT) belongs to perovskite-type materials with two different isovalent ions Na and Bi at the A-site of the ABO_3 structure. It has recently attracted attention as environmentally friendly lead-free relaxor ferroelectrics, which exhibits very attractive piezoelectric properties [1,2]. NBT undergoes a very peculiar sequence of phase transitions [3-8], and shows unusual dielectric [4,9-12] and ferroelectric [2,4,12] properties. NBT undergoes at least two structural phase transitions: from cubic phase at 813K to tetragonal, and to rhombohedral at ~470K. Also a broad maximum of dielectric permittivity occurs about 593K. Nevertheless, X-ray [13], neutron scattering [4], Raman scattering [14] measurements do not show any structural phase transition around this temperature. Regarding this contradiction, the antiferroelectric order or coexistence of tetragonal and rhombohedral phases have been suggested [15]. In our diffuse neutron scattering measurements [16] we found that the rhombohedral phase is incommensurately modulated along the four-fold axis of the precursor tetragonal phase. It was suggested that the existence of the atomic ordering in the A-sublattice may be responsible for the formation of the modulated

© Springer International Publishing Switzerland 2015
S. Balandin et al. (Eds.): NEW2AN/ruSMART 2015, LNCS 9247, pp. 793–798, 2015.
DOI: 10.1007/978-3-319-23126-6_75

structure in the ordered regions. Although in the rhombohedral phase the thermodynamically stable modulation should be along the rhombohedral three-fold axis, we argued that the observed modulation direction is imposed by the symmetry of the atomic ordering in the precursor phase. The results of most detailed studies of the dielectric properties of NBT in a wide temperature range on medium measuring frequencies were presented in [10]. Large low-frequency permittivity peak in NBT between 900 and 600 K was explained in terms of superparaelectric clusters existing in the tetragonal phase. Below 640K they observed two phase transitions - from a superparaelectric tetragonal to an antiferroelectric or incommensurate trigonal phase, and the lower transition is to a ferroelectric phase of trigonal structure. However until now, the interpretation of these data is discussed and the question of carrying out a thorough study of the NBT dielectric response on the middle and low measuring frequency and analysis of its behavior remains actual. In this paper, the results of the detailed study of the dielectric properties of single crystal NBT in wide frequency and temperature ranges are reported.

2   Samples and Experiment

The measurements were carried out on single crystal $Na_{1/2}Bi_{1/2}TiO_3$ (NBT). Single crystals were grown up by Chochralski' method. Sample had shape of the thin plate of 9x8x1 mm^3. The thin plate has been cut out with the normal to its plane parallel to the (001). The dielectric response was studied at frequency range 0.1 Hz - 10 MHz and temperature region 300-1000 K using ultrabroadband dielectric spectrometer Novocontrol BDS80 with Novotherm-HT cryosystem. Platinum electrodes were deposited on the flat sides of the crystal.

3   Results and Discussion

Temperature dependencies of real (ε') and imaginary (ε") parts of dielectric permittivity of single crystal NBT for several measuring frequencies obtained in the heating mode are presented on Fig.1. One can see three dielectric anomaly regions – dielectric dispersion lower 500K, asymmetric diffuse maximum of ε near 610 K and low-frequency peak at ~850-900 K, position and the value of which depend strongly on measuring frequency. Above 500 K there observed a sharp increase in the dielectric losses at low frequencies.

On Fig.2 temperature dependencies of real (ε') and imaginary (ε") parts of dielectric permittivity of single crystal NBT in cooling run for the same measuring frequencies (as in heating) are presented. You can see noticeable difference between the behavior of the dielectric constant of the sample upon heating and cooling. Asymmetrical peak at 610K significantly broadens and becomes symmetrical, but the position and the amplitude don't change significantly. Also the widening of the high-temperature of the permittivity maximum during cooling compared with heating.

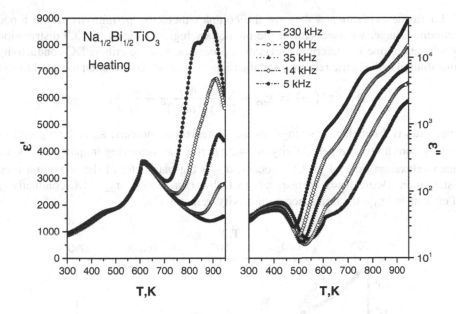

Fig. 1. Temperature dependencies of real (ε') and imaginary (ε'') parts of dielectric permittivity of single crystal NBT for several measuring frequencies upon heating.

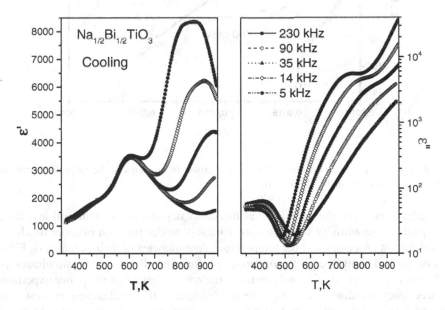

Fig. 2. Temperature dependencies of real (ε') and imaginary (ε'') parts of dielectric permittivity of single crystal NBT for several measuring frequencies upon cooling.

To fit the experimental data for the complex dielectric permittivity in the broad frequency range, we used sum of the phenomenological Cole-Cole (CC) distribution of relaxation time for describing relaxation processes, term describing DC-conductivity contribution to dielectric response and the higher frequency contribution to permittivity:

$$\varepsilon^*(\omega) = \varepsilon_\infty + \sum_j \frac{\Delta\varepsilon_j}{1+(i\omega\tau_j)^\alpha} + \frac{i\sigma_{DC}}{\varepsilon_0\omega},$$

where $\Delta\varepsilon_j$ is the dielectric strength of the jth relaxation process, ε_∞ is the higher frequency contribution to permittivity, $\omega = 2\pi f$, f- is the measuring frequency, τ_j is the mean relaxation time of jth CC process, α – is the parameter of the relaxation time distribution ($0<\alpha<1$) $\alpha=1$ corresponds to Debye distribution, σ_{DC} – DC-conductivity of crystal and ε_0 – the free space permittivity, respectively.

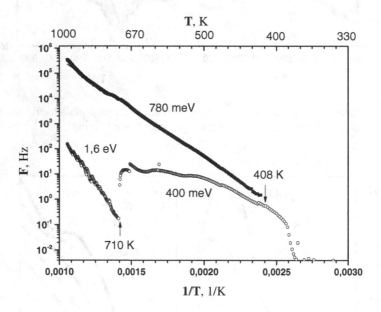

Fig. 3. Temperature dependencies of the characteristic frequencies of the relaxation processes in the Arrhenius coordinates (lgF(1/T)).

We have analyzed the experimental frequency dependencies of the real and imaginary parts of permittivity upon cooling with this model function in order to identify the relaxation processes and the temperature dependencies of their parameters. Fitting was performed using specially developed software that allows treating simultaneously both parts of ε^*, that significantly reduces the error of calculations. In our experimental frequency window, we were able to distinguish three relaxation processes described by CC distribution and estimated the contribution of DC-conductivity. The temperature dependencies of the dielectric strength and the mean relaxation time of CC processes were obtained. The figure 3 shows the temperature dependencies of the characteristic frequencies of the relaxation processes in the Arrhenius coordinates (lgF(1/T)). The high frequency mode in the temperature range 1000-400 is well described by Arrhenius law:

$$f = f_0 \cdot \exp\left(-\frac{E_a}{kT}\right),$$

where E_a is activation energy, k- Boltzmann constant. Obtained from fitting the value of activation energy is 780 meV. In the temperature range 670 - 410 K there is also another relaxation process with an activation energy of 400 meV, which merges with the high-frequency process at 410K. On further decreasing temperature characteristic relaxation frequency decreases sharply, and its behavior is no longer described by the Arrhenius law, but well described by Vogel-Fulcher law:

$$f = f_0 \cdot \exp\left(-\frac{E_a}{k(T-T_f)}\right),$$

where T_f – freezing temperature. Obtained from fitting freezing temperature is 370K. At high temperatures 700-1000K there is observed a low-frequency relaxation process with activation energy of 1.6 eV disappearing at 710K.

Fig. 4 shows temperature dependence of derived DC- conductivity of NBT. Starting at 470 K, the exponential growth of the DC-conductivity of NBT is observed.

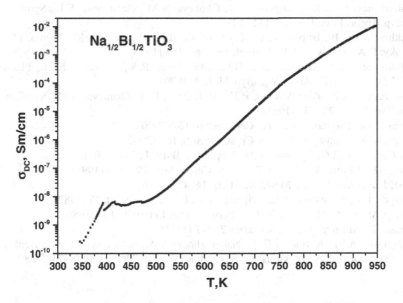

Fig. 4. Temperature dependencies of DC-conductivity of NBT.

Nature of exponential increase of DC-conductivity of NBT in this temperature region still remains unclear. Probably, DC-conductivity of NBT associates with increasing sodium ion mobility.

4   Conclusion

The detailed study of the complex dielectric response of single crystal $Na_{1/2}Bi_{1/2}TiO_3$ in wide frequency (0.1 Hz - 10 MHz) and temperature (300-1000 K) ranges have done. Cole-Cole relaxation processes are identified from analysis of dispersion curves within existing phenomenological models. and their dynamics are traced. From fitting activation energies of Arrhenius' relaxation processes are obtained. It is discovered relaxation process whose behavior is described by Vogel-Fulcher law, indicating that in the system there happening freezing at 370K.

Acknowledgements. This work was performed at SPbPU by a grant of the Russian Scientific Foundation (Project №14-22-00136).

References

1. Emelianov, S.M., Rayevsky, I.P., Smotrakov, V.G., Savenko, F.I.: Fiz. Tverd. Tela **26**, 1897 (1984)
2. Roleder, K., Franke, I., Glazer, A.M., Thomas, P.A., Miga, S., Suchanicz, J.: J. Phys.: Condens. Matter **14**, 5399 (2002)
3. Vakhrushev, S.B., Kvyatkovsky, B.E., Okuneva, N.M., Plachenova, E.L., Syrnikov, P.P.: J. Exp. Theor. Phys. Lett. **35**, 111 (1982)
4. Vakhrushev, S.B., Isupov, V.A., Kvyatkovsky, B.E., Okuneva, N.M., Pronin, I.P., Smolensky, G.A., Syrnikov, P.P.: Ferroelectrics **63**, 153 (1985)
5. Vakhrushev, S.B., Kvyatkovsky, B.E., Malysheva, R.S., Okuneva, N.M., Plachenova, E.L., Syrnikov, P.P.: Kristallografiya **34**, 154 (1989)
6. Vakhrushev, S.B., Kvyatkovsky, B.E., Malysheva, R.S., Okuneva, N.M., Syrnikov, P.P.: Fiz. Tverd. Tela **27**, 737 (1985)
7. Jones, G.O., Thomas, P.A.: Acta Cryst. B **56**, 426 (2000)
8. Jones, G.O., Thomas, P.A.: Acta Cryst. B **58**(2), 168 (2002)
9. Tu, C.-S., Siny, I.G., Schmidt, V.H.: Phys. Rev. B **49**, 11550 (1994)
10. Park, S.E., Chung, S.J., Kim, I.T.: J. Am. Ceram. Soc. **79**, 1290 (1996)
11. East, J., Sinclair, D.C.: J. Mater. Sci. Lett. **16**, 422 (1997)
12. Zvirgds, I.A., Kapostins, P.A., Zvirgzde, I.V.: Ferroelectrics **40**, 75 (1982)
13. Zhang, M., Scott, J.F., Zvirgds, I.A.: Ferroelectrics Letters **6**, 147 (1986)
14. Sakata, K., Masuda, Y.: Ferroelectrics **7**, 347 (1974)
15. Balagurov, A.M., Koroleva, E.Y., Naberezhnov, A.A., Sakhnenko, V.P., Savenko, B.N., Ter-Oganessian, N.V., Vakhrushev, S.B.: Phase Transit. **79**(1–2), 163–173 (2006)

Author Index

Printed in the United States
By Bookmasters

Printed in the United States
By Bookmasters